# INDUSTRIAL ORGANIC CHEMICALS

## Starting Materials and Intermediates

## VOLUME 6

Weinheim · New York · Chichester · Brisbane · Singapore · Toronto

# INDUSTRIAL ORGANIC CHEMICALS

### VOLUME 1
**Acetaldehyde** to **Aniline**

### VOLUME 2
**Anthracene** to **Cellulose Ethers**

### VOLUME 3
**Chlorinated Hydrocarbons** to **Dicarboxylic Acids, Aliphatic**

### VOLUME 4
**Dimethyl Ether** to **Fatty Acids**

### VOLUME 5
**Fatty Alcohols** to **Melamine and Guanamines**

### VOLUME 6
**Mercaptoacetic Acid and Derivatives** to **Phosphorus Compounds, Organic**

### VOLUME 7
**Phthalic Acid and Derivatives** to **Sulfones and Sulfoxides**

### VOLUME 8
**Sulfonic Acids, Aliphatic** to **Xylidines**

**Index**

# INDUSTRIAL ORGANIC CHEMICALS

## Starting Materials and Intermediates

### VOLUME 6
**Mercaptoacetic Acid and Derivatives** to **Phosphorus Compounds, Organic**

Weinheim · New York · Chichester · Brisbane · Singapore · Toronto

This book was carefully produced. Nevertheless, authors and publisher do not warrant the information contained therein to be free of errors. Readers are advised to keep in mind that statements, data, illustrations, procedural details or other items may inadvertently be inaccurate.

Library of Congress Card No.: Applied for.
British Library Cataloguing-in-Publication Data: A catalogue record for this book is available from the British Library.

Die Deutsche Bibliothek – CIP-Einheitsaufnahme
**Industrial organic chemicals** : starting materials and intermediates ;
an Ullmann's encyclopedia. – Weinheim ; New York ;
Chichester ; Brisbane ; Singapore ; Toronto : Wiley-VCH
    ISBN 3-527-29645-X
Vol. 6. Mercaptoacetic Acid and Derivates to Phosphorus Compounds, Organic. – 1. Aufl. – 1999.

© WILEY-VCH Verlag GmbH, D-69469 Weinheim (Federal Republic of Germany), 1999
Printed on acid-free and chlorine-free paper.
All rights reserved (including those of translation in other languages). No part of this book may be reproduced in any form – by photoprinting, microfilm, or any other means – nor transmitted or translated into machine language without written permission from the publishers. Registered names, trademarks, etc. used in this book, even when not specifically marked as such, are not to be considered unprotected by law.

Composition and Printing: Rombach GmbH, Druck- und Verlagshaus, D-79115 Freiburg
Bookbinding: Wilhelm Osswald & Co., D-67433 Neustadt (Weinstraße)
Cover design: mmad, Michel Meyer, D-69469 Weinheim
Printed in the Federal Republic of Germany

# Contents

## 1 Mercaptoacetic Acid and Derivatives

1. Introduction . . . . . . . . . . . . . . . . . 3221
2. Physical Properties . . . . . . . . . . . . 3221
3. Chemical Properties . . . . . . . . . . . 3222
4. Production . . . . . . . . . . . . . . . . . . 3222
5. Analysis . . . . . . . . . . . . . . . . . . . . 3223
6. Transportation and Storage . . . . . . 3223
7. Toxicology . . . . . . . . . . . . . . . . . . 3224
8. Uses . . . . . . . . . . . . . . . . . . . . . . . 3224
9. References . . . . . . . . . . . . . . . . . . 3225

## 2 Metallic Soaps

1. Introduction . . . . . . . . . . . . . . . . . 3227
2. Production . . . . . . . . . . . . . . . . . . 3229
3. Fire and Explosion Protection . . . . . 3233
4. Occupational Health and Environmental Protection . . . . . . . . . . . . . 3234
5. Individual Metallic Soaps . . . . . . . . 3234
6. Quality Specifications . . . . . . . . . . 3247
7. References . . . . . . . . . . . . . . . . . . 3248

## 3 Methacrylic Acid and Derivatives

1. Introduction . . . . . . . . . . . . . . . . . 3249
2. Properties . . . . . . . . . . . . . . . . . . . 3251
3. Production . . . . . . . . . . . . . . . . . . 3254
4. Quality Specifications . . . . . . . . . . 3261
5. Storage and Transportation . . . . . . 3261
6. Uses . . . . . . . . . . . . . . . . . . . . . . . 3262
7. Economic Aspects . . . . . . . . . . . . . 3263
8. Toxicology and Occupational Health 3263
9. References . . . . . . . . . . . . . . . . . . 3265

## 4 Methane

1. Introduction . . . . . . . . . . . . . . . . . 3269
2. Physical and Chemical Properties of Biogas . . . . . . . . . . . . . . . . . . . . . 3271
3. Biological Methanogenesis . . . . . . . 3272
4. Technological Biomethanation . . . . 3273
5. Feedstocks for Biomethanation . . . 3277
6. Process and Environmental Parameters for Stable, Efficient Biomethanation . . . . . . . . . . . . . . 3278
7. Environmental Impact of Biomethanation . . . . . . . . . . . . . . . . . . . . . . 3281
8. Storage, Pretreatment, Upgrading and Uses of Biogas . . . . . . . . . . . . . . . 3282
9. Analysis . . . . . . . . . . . . . . . . . . . . 3283
10. Biogas in Developing Countries . . . 3283
11. Economic Aspects . . . . . . . . . . . . . 3284
12. References . . . . . . . . . . . . . . . . . . 3285

V

## 5 Methanol

1. Introduction .................. 3288
2. Physical Properties ............ 3288
3. Chemical Properties ........... 3290
4. Production .................... 3292
5. Process Technology............ 3299
6. Handling, Storage, and Transportation ..................... 3306
7. Quality Specifications and Analysis . 3308
8. Environmental Protection ....... 3310
9. Uses.......................... 3310
10. Economic Aspects.............. 3315
11. Toxicology and Occupational Health 3317
12. References.................... 3320

## 6 Methyl *Tert*-Butyl Ether

1. Introduction .................. 3337
2. Physical and Chemical Properties .. 3338
3. Resources and Raw Materials..... 3339
4. Production .................... 3340
5. Environmental Protection ....... 3342
6. Quality Specifications .......... 3343
7. Chemical Analysis ............. 3343
8. Storage and Transportation ...... 3343
9. Legal Aspects................. 3344
10. Uses.......................... 3344
11. Toxicology and Occupational Health 3346
12. References.................... 3347

## 7 Methylamines

1. Introduction .................. 3325
2. Physical Properties ............ 3326
3. Chemical Properties ........... 3326
4. Production .................... 3327
5. Environmental Protection ....... 3330
6. Quality Specifications .......... 3331
7. Storage and Transportation ...... 3331
8. Uses.......................... 3332
9. Economic Aspects.............. 3333
10. Toxicology and Occupational Health 3334
11. References.................... 3334

## 8 Naphthalene Derivatives

1. Introduction .................. 3364
2. Naphthalenesulfonic Acids....... 3365
3. Naphthols..................... 3372
4. Hydroxynaphthalenesulfonic Acids . 3384
5. Aminonaphthalenes ............ 3401
6. Aminonaphthalenesulfonic Acids... 3406
7. Aminonaphthols .............. 3425
8. Aminohydroxynaphthalenesulfonic Acids ...................... 3427
9. References.................... 3443

## 9 Naphthalene and Hydronaphthalenes

1. Physical and Chemical Properties .. 3351
2. Production .................... 3352
3. Uses.......................... 3354
4. Alkylnaphthalenes ............. 3355
5. Hydronaphthalenes............. 3358
6. Economic Aspects.............. 3358
7. Toxicology.................... 3359
8. References.................... 3360

## 10  Naphthoquinones

1. Introduction . . . . . . . . . . . . . . . 3447
2. 1,4-Naphthoquinone . . . . . . . . . . 3448
3. Other Naphthoquinones . . . . . . . 3451
4. Naphthodiquinones . . . . . . . . . . . 3452
5. Polynaphthoquinone . . . . . . . . . . 3452
6. References . . . . . . . . . . . . . . . . . 3452

## 11  Nitriles

1. Introduction . . . . . . . . . . . . . . . 3455
2. Aliphatic Nitriles . . . . . . . . . . . . 3456
3. Aromatic and Araliphatic Nitriles . . 3467
4. References . . . . . . . . . . . . . . . . . 3474

## 12  Nitrilotriacetic Acid

1. Introduction . . . . . . . . . . . . . . . 3479
2. Properties . . . . . . . . . . . . . . . . . 3479
3. Production . . . . . . . . . . . . . . . . 3481
4. Analysis . . . . . . . . . . . . . . . . . . 3482
5. Uses . . . . . . . . . . . . . . . . . . . . . 3483
6. Economic Aspects . . . . . . . . . . . . 3484
7. Toxicology and Environmental Aspects . . . . . . . . . . . . . . . . . . . . 3484
8. References . . . . . . . . . . . . . . . . . 3485

## 13  Nitro Compounds, Aliphatic

1. Introduction . . . . . . . . . . . . . . . 3487
2. Properties . . . . . . . . . . . . . . . . . 3488
3. Production . . . . . . . . . . . . . . . . 3488
4. Quality Specifications and Analysis . 3492
5. Storage and Transportation . . . . . . 3492
6. Uses of Nitroalkanes and their Derivatives . . . . . . . . . . . . . . . . . . . 3494
7. Toxicology and Occupational Health 3498
8. References . . . . . . . . . . . . . . . . . 3500

## 14  Nitro Compounds, Aromatic

1. Introduction . . . . . . . . . . . . . . . 3504
2. Nitration . . . . . . . . . . . . . . . . . 3505
3. Nitro Aromatics . . . . . . . . . . . . . 3509
4. Nitrohalo Aromatics . . . . . . . . . . 3526
5. Nitroamino Aromatics . . . . . . . . . 3538
6. Nitroaromatic Sulfonic Acids and Derivatives . . . . . . . . . . . . . . . . . 3548
7. Nitrohydroxy and Alkoxy Aromatics 3560
8. Nitroketones . . . . . . . . . . . . . . . 3567
9. Nitroheterocycles . . . . . . . . . . . . 3568
10. References . . . . . . . . . . . . . . . . 3574

## 15  Oxalic Acid

1. Introduction .................. 3577
2. Physical Properties ............. 3578
3. Chemical Properties ........... 3580
4. Production Processes and Raw Materials.................... 3581
5. Chemical Analysis ............ 3590
6. Uses...................... 3590
7. Economic Aspects.............. 3592
8. Storage, Handling, Transportation, Waste Disposal .............. 3593
9. Derivatives ................. 3593
10. Toxicology................... 3597
11. References.................. 3597

## 16  Oxocarboxylic Acids

1. Introduction .................. 3599
2. Pyruvic Acid ................. 3600
3. Acetoacetic Acid and Derivatives... 3602
4. Levulinic Acid ............... 3606
5. Acetonedicarboxylic Acid........ 3607
6. Other Oxocarboxylic Acids....... 3608
7. References................... 3609

## 17  Pentanols

1. Introduction,................. 3611
2. Physical Properties ............. 3611
3. Chemical Properties ........... 3613
4. Production .................. 3615
5. Uses...................... 3618
6. Quality Specifications and Chemical Analysis .................. 3620
7. Storage and Transportation ...... 3620
8. Toxicology and Occupational Health 3621
9. References................... 3622

## 18  Peroxy Compounds, Organic

1. Introduction .................. 3630
2. Alkyl Hydroperoxides .......... 3632
3. Dialkyl Peroxides ............. 3638
4. Peroxycarboxylic Acids ........ 3641
5. Diacyl Peroxides.............. 3649
6. Peroxycarboxylic Esters ......... 3653
7. α-Oxyperoxides .............. 3662
8. α-Aminoperoxides ............ 3669
9. Safety Measures.............. 3672
10. Toxicology................... 3675
11. References.................. 3678

## 19  Phenol

1. Introduction .................. 3689
2. Physical Properties ............. 3690
3. Chemical Properties ........... 3691
4. Production .................. 3693
5. Environmental Protection ....... 3704
6. Quality Specifications .......... 3705
7. Storage and Transportation ...... 3705
8. Economic Aspects.............. 3706
9. Toxicology and Occupational Health 3709
10. References.................. 3710

## 20  Phenol Derivatives

1. Alkylphenols . . . . . . . . . . . . . . . . .  3714
2. Catechol . . . . . . . . . . . . . . . . . . . .  3758
3. Trihydroxybenzenes . . . . . . . . . . .  3763
4. Bisphenols (Bishydroxyarylalkanes) .  3768
5. Hydroxybiphenyls . . . . . . . . . . . . .  3774
6. Phenol Ethers . . . . . . . . . . . . . . . .  3778
7. Halogen Derivatives of Phenolic Compounds . . . . . . . . . . . . . . . . . .  3782
8. References . . . . . . . . . . . . . . . . . . .  3793

## 21  Phenylacetic Acid

1. Physical Properties . . . . . . . . . . . .  3805
2. Chemical Properties . . . . . . . . . . .  3805
3. Production . . . . . . . . . . . . . . . . . .  3806
4. Uses . . . . . . . . . . . . . . . . . . . . . . .  3806
5. Quality Specifications . . . . . . . . . .  3806
6. Economic Aspects . . . . . . . . . . . . .  3806
7. Toxicology . . . . . . . . . . . . . . . . . . .  3807
8. References . . . . . . . . . . . . . . . . . . .  3807

## 22  Phenylene- and Toluenediamines

1. Introduction . . . . . . . . . . . . . . . . .  3809
2. Physical Properties . . . . . . . . . . . .  3810
3. Chemical Properties . . . . . . . . . . .  3810
4. Production . . . . . . . . . . . . . . . . . .  3811
5. Environmental Protection . . . . . . .  3813
6. Quality Specifications . . . . . . . . . .  3813
7. Storage and Transportation . . . . . .  3813
8. Uses . . . . . . . . . . . . . . . . . . . . . . .  3814
9. Toxicology and Occupational Health  3816
10. References . . . . . . . . . . . . . . . . . .  3817

## 23  Phosphorus Compounds, Organic

1. Introduction . . . . . . . . . . . . . . . . .  3820
2. Phosphines . . . . . . . . . . . . . . . . . .  3820
3. Halophosphines . . . . . . . . . . . . . . .  3825
4. Phosphonium Salts . . . . . . . . . . . . .  3827
5. Phosphine Oxides and Sulfides . . . .  3830
6. Phosphonous Acid Derivatives . . . .  3834
7. Phosphinic Acids and their Derivatives . . . . . . . . . . . . . . . . . . . . . . . .  3835
8. Phosphites and Hydrogenphosphonates . . . . . . . . . . . . . . . . . . . .  3838
9. Phosphonic Acids and their Derivatives . . . . . . . . . . . . . . . . . . . . . . . .  3842
10. Esters of Phosphoric Acid . . . . . . .  3847
11. Esters of Thiophosphoric Acid . . . .  3851
12. Economic Aspects . . . . . . . . . . . . .  3855
13. Toxicology . . . . . . . . . . . . . . . . . . .  3856
14. References . . . . . . . . . . . . . . . . . .  3860

# Mercaptoacetic Acid and Derivatives

ROBERT RIPPEL, Hoechst AG, Frankfurt, Federal Republic of Germany

| | | | | |
|---|---|---|---|---|
| 1. | Introduction ............ 3221 | 6. | Transportation and Storage .. | 3223 |
| 2. | Physical Properties ........ 3221 | 7. | Toxicology .............. | 3224 |
| 3. | Chemical Properties ........ 3222 | | | |
| 4. | Production ............. 3222 | 8. | Uses .................. | 3224 |
| 5. | Analysis ............... 3223 | 9. | References .............. | 3225 |

## 1. Introduction

Mercaptoacetic acid [68-11-1], thioglycolic acid, $HSCH_2COOH$, is the simplest and industrially most important mercaptocarboxylic acid. The compound was first prepared in 1862 by CARIUS from chloroacetic acid and potassium hydrogen sulfide. There are numerous patents dealing with the applications of mercaptoacetic acid in the cosmetic and plastic industries.

## 2. Physical Properties

Mercaptoacetic acid is a clear, colorless liquid with a characteristic odor. Some physical properties are as follows:

| | |
|---|---|
| $M_r$ | 92.11 |
| mp | −16.5 °C |
| bp (2.3 kPa) | 110–112 °C |
| bp (0.1 kPa) | 79–80 °C |
| $n_D^{20}$ | 1.5027 |
| $d_4^{20}$ | 1.325 |
| Heat of combustion | 1450 kJ/mol |
| Vapor pressure (30 °C) | 0.02 kPa |
| Vapor density | 3.18 |
| Flash point | 126 °C |
| Dissociation constants (in $NaClO_4$ solution at 20 °C) | |
| $K_1$ | $(3.82 \pm 0.1)$ $pK_a$ |
| $K_2$ | $(9.30 \pm 0.1)$ $pK_a$ |

Mercaptoacetic acid is miscible with water, mono- and polyalcohols, ethers, ketones, esters, chlorinated hydrocarbons, and aromatic hydrocarbons, but not with aliphatic hydrocarbons. The esters of mercaptoacetic acid are mostly colorless liquids with characteristic odors. The cetyl and stearyl esters are pale yellow waxes [1].

## 3. Chemical Properties

Mercaptoacetic acid has two reactive centers: the mercapto group and the carboxyl group. The compound forms salts of the carboxyl and mercapto groups, esters, amides, anilides, thioethers, etc. Oxidation of mercaptoacetic acid, which occurs even on standing in air, gives dithiodiglycolic acid [505-73-7], $HOOCCH_2SSCH_2COOH$. The reaction is accelerated by traces of metals such as copper, iron, and manganese [2]. In concentrated solutions (80%), linear and cyclic polycondensation products are formed, including a tetracarboxylic acid with a 1,4-dithiane ring structure [3]. Thioethers are formed by reactions of the sodium or potassium salts of the mercapto group with alkyl halides or by addition of mercaptoacetic acid or its derivatives to double or triple bonds. Depending on the catalyst used, either the Markownikow or anti-Markownikow product is obtained [4]. Mercaptoacetic acid and its derivatives react with aldehydes and ketones to form mercaptals, mercaptols, or $\alpha,\beta$-unsaturated thioethers. The addition reactions are catalyzed by acids (e.g., mineral acids and toluenesulfonic acids) [5]. In the formation of oxathiolanes, the mercapto and carboxyl groups react simultaneously with a carbonyl group. For example, the reaction of mercaptoacetic acid with cyclohexanone affords 2,2-pentamethylene-1,3-oxathiolane-5-one [6]. The strongly reducing nature of mercaptoacetic acid is of importance both in investigations of biochemical redox and enzyme systems and in industrial applications such as cold-waving preparations and modification of wool. The use of mercaptoacetic acid for the permanent waving of hair is based on the reductive cleavage of the cystine disulfide bridges of keratin [7].

## 4. Production

Industrially, mercaptoacetic acid is produced from chloroacetic acid or its salts and sodium or potassium hydrogen sulfide. Thiodiglycolic acid, dithiodiglycolic acid, and glycolic acid may be formed as side products. The mercaptoacetic acid is isolated from the acidified reaction mixture by extraction with organic solvents (ethers, alcohols, or chlorinated hydrocarbons) and purified by distillation. Both batch and continuous processes have been described in the patent literature as well as processes that operate under a partial pressure of hydrogen sulfide or carbon dioxide [8]. In one process, sulfur wastes, which arise from the reduction of aromatic nitro compounds, are used instead of alkali-metal hydrogen sulfides [9], while in another process mercaptoacetic acid is obtained by cleavage of imidazolyl mercaptoacetic acid with sodium sulfide [10]. It can also be produced through cleavage of the Bunte salt $NaO_3SSCH_2COOH$, prepared from chloroacetic acid and sodium thiosulfate, with dilute sulfuric acid; by reduction of dithioglycolic acid (prepared from chloroacetic acid and alkali-metal polysulfides); and by decomposition of the xanthate ester obtained from chloroacetic acid and potassium ethylxanthate [11]. Anhydrous mercaptoacetic acid is obtained by distillation of technical mercaptoacetic acid with toluene. During production, it is important to avoid contact of the mercaptoacetic acid with metals that catalyze its dimerization.

Esters of mercaptoacetic acid can be produced either by esterification with alcohols, whereby the alcohol can simply be used as the extractant after the synthesis of the acid, or by reaction of the corresponding alkyl chloroacetate with sodium thiosulfate [12].

# 5. Analysis

Mercaptoacetic acetic acid gives color reactions with numerous metal salts. Iron salts and sodium pentacyanonitrosylferrate(III) give a violet coloration [13]. Mercaptoacetic acid is analyzed quantitatively by titration with iodine [14]. Spectroscopic and chromatographic methods in industrial use include determination by HPTLC [15], analysis of cold-waving preparations by HPLC [16], and detection by gas chromatography [17]. Reference spectra for spectroscopic investigations can be found in [18], [19].

# 6. Transportation and Storage

Commercial forms of mercaptoacetic acid are the 99% acid ($d_4^{20}$ 1.325) and the 80% acid ($d_4^{20}$ 1.27). Standardized mercaptoacetic acid should contain not more than 2% dithiodiglycolic acid and a maximum of 1 ppm iron, lead, copper, and arsenic [20]. Isooctyl mercaptoacetate [25103-09-7] and glycerol monothioglycolate [30618-84-9] are sold water-free. Mercaptoacetic acid must be transported in stainless steel ($V_4A$, $V_2A$) or polyethylene-lined polyester containers. The following transportation codes apply [21]:

| | |
|---|---|
| IMDG (GGVSee) | Class 8 D E–F 8154 UN no. 1940 |
| RID (GGVE) | Class 8 Rn 801 Section 32 b |
| ADR (GGVS) | Class 8 Rn 2801 Section 32 b |
| ADNR | Class 8 Rn 6801 Section 32 b |
| GefStoffV | 607-090-00-6 |
| GB Blue Book | Corrosives and IMDG code E 8154 |
| United States | CFR 49: § 172.101 Corrosive M |
| ICAO/IATA-DGR | Class 8 UN no. 1940 |

3223

# 7. Toxicology

Contact with mercaptoacetic acid leads to irritation of the skin and eye. The vapor causes irritation of the mucous membranes, coughing and shortness of breath after inhalation, badly healing wounds, and blisters [22]. Protective gloves and eyewear are therefore necessary when working with mercaptoacetic acid [22]. The TLV value has been set at 1 ppm (5 mg/m$^3$) [23]. The oral LD$_{50}$ in the rat is 261 mg/kg [24]; more recent tests on mice are reported in [25]. Most derivatives of mercaptoacetic acid and commercial products containing them, for which legal maximum concentrations apply, are considerably less toxic and therefore do not present a health risk in normal use [26].

# 8. Uses

Since ca. 1950 mercaptoacetic acid and its derivatives, in particular the ammonium and ethanolammonium salts and to an increasing extent glycerol monothioglycolate, have been used for the production of cold-waving preparations [27]. The calcium salt is used in depilatories. Mercaptoacetic acid and derivatives are also used for industrial unhairing (e.g., leather production) and for permanent forming of wool (e.g., in hat making). Sulfur-containing organotin compounds of the type dialkyltin bis(isooctyl thioglycolate) and alkyltin tris(isooctyl thioglycolate) are the most important thermostabilizers for PVC [28]. The low toxicity and extremely low migration of, for example, dioctyltin bis(isooctyl thioglycolate) with various amounts of octyltin tris(isooctyl thioglycolate), has led to their use worldwide for the production of food packaging materials and pipes for drinking water [29]. Esters, particularly of higher alcohols such as isooctanol and decanol, and amides of mercaptoacetic acid are used as polymerization catalysts and antioxidants for the production of plastics and rubbers [30]. The sodium salt of mercaptoacetic acid is a component of the Brewer nutritive medium for anaerobic bacteria [31].

Mercaptoacetic acid is used in analytical chemistry for the separation of iron and aluminum and as a reagent for iron [13], nitrite, uranium, vanadium, chromium, copper, molybdenum, and palladium [32]. It is used in fast atom bombardment mass spectrometry as a matrix compound [33]. A mild process for the isolation of lignin involves the use of mercaptoacetic acid with boron trifluoride [34]. Other economically important uses include the production of derivatives of thiophene and dihydrothiophene, which are used as precursors for pharmaceuticals [35], and the synthesis of intermediates for dyes (e.g., S-aryl mercaptoacetic acids). A review of chelate complexes of mercaptoacetic acid is given in [36]. World production of mercaptoacetic is estimated to be $15-20 \times 10^3$ t/a.

# 9. References

[1] G. T. Walker, *Seifen Öle Fette Wachse* **88** (1962) 11.
[2] M. Kharasch, R. R. Legaut, A. B. Wilder, R. W. Gerard, *J. Biol. Chem.* **113** (1936) 537.
[3] A. Schöberl, G. Wiehler, *Justus Liebigs Ann. Chem.* **595** (1955) 101.
[4] *Houben-Weyl*, **9**, 199.
[5] *Houben-Weyl*, **7/2 b**, 1938.
[6] Carlisle Chemical Works Inc., US 3 209 012, 1965 (C. H. Miller et al.). P. D. Klemmensen, O. Z. Mortensen, S. O. Lawesson, *Tetrahedron* **26** (1970) 4641.
[7] T. Bersin, J. Steudel, *Ber. Dtsch. Chem. Ges.* **B 71** (1938) 1015.
[8] G. T. Walker, *Seifen öle Fette Wachse* **88** (1962) 402, 431. Denki Kagaku Kogyo K. K., JP 8 034 794, 1977. Hoechst, DE-OS 2 711 867, 1977 (H. Klug).
[9] Martin-Luther-Universität, Halle-Wittenberg, DD 238 229, 1986 (H. Matschner, H. Stange, E. Mendow).
[10] C. F. Spiess u. Sohn, DE-OS 2 832 977, 1978 (R. Himmelreich).
[11] *Houben-Weyl*, **9**, 3 ff.
[12] VEB Arzneimittelwerk Dresden, DD 270 063, 1989 (H. Stange, E. Mendow, E. Loewe, G. Nauwald).
[13] D. Geffken, K. H. Surborg, *Dtsch. Apoth. Ztg.* **128** (1988) 1235.
[14] *DAB 7* (1968) p. 317.
[15] R. Klaus, *Chromatographia* **20** (1985) 235.
[16] Y. Fukuda, F. Nakamura, Y. Morikawa, *Eisei Kagaku* **31** (1985) 235. J. Rooselar, D. M. Liem, *Int. J. Cosmet. Sci.* **3** (1981) 37.
[17] N. Goetz, *Cosmet. Sci. Technol. Ser.* **4** (1985) 65. N. Goetz, P. Gataud, P. Bore, *Analyst (London)* **104** (1979) 1062. U. Hannestad, B. Soerbo, *J. Chromatogr.* **200** (1980) 171.
[18] *The Aldrich Library of Infrared Spectra*, 3rd ed., Aldrich Chemical Company, Milwaukee 1981, p. 3096.
[19] *The Aldrich Library of NMR-Spectra*, vol. **II**, Aldrich Chemical Company, Milwaukee 1974, p. 176 A.
[20] *Kirk-Othmer* 2nd ed., **20**, 200.
[21] GefStoffV, 26th Aug. 1986, *Product Safety Data, Mercaptoacetic Acid*, E. Merck, Darmstadt.
[22] Kühn-Birett: *Merkblätter Gefährliche Arbeitsstoffe*, 13th Supplement, Ecomed Verlag, Landsberg/Lech 1980.
[23] *Hazards in the Chemical Laboratory*, 3rd ed., The Royal Society of Chemistry, London 1981.
[24] E. Merck, *Product Safety Data, Mercaptoacetic Acid*, Darmstadt 1988.
[25] E. W. Schafer, Jr., W. A. Bowles, Jr., *Arch. Environ. Contam. Toxicol.* **14** (1985) 111.
[26] B. Bach, *Seifen Öle Fette Wachse* **105** (1979) 405, 448.
[27] M. E. Marti, *Seifen Öle Fette Wachse* **116** (1990) 23.
[28] A. Holger in G. Becker, D. Braun (eds.): *Kunststoffhandbuch*, 2nd ed., vol. **2/1**, Hanser Verlag, München 1985, p. 527.
[29] R. Franck, H. Mühlschlegel: *Kunststoffe im Lebensmittelverkehr*, Carl Heymanns Verlag, Köln 1977. United States Food and Drug Administration, *Fed. Regist.* **48** (1983) no. 35, 7169.
[30] E. Ceausescu et al., *J. Macromol. Sci. Chem.* **A 22** (1985) 525. *Comprehensive Polymer Science*, vols. **1 – 7**, Pergamon Press, Oxford – New York – Beijing – Frankfurt – Sao Paulo – Sydney – Tokyo – Toronto.

*Encyclopedia of Polymer Science and Engineering,* 2nd ed., vols. **1–17** (incl. Suppl. Vol.), J. Wiley and Sons, New York – Brisbane – Chichester – Toronto 1985–1989.
[31] *DAB 9,* (1986) p. 423.
[32] E. Merck: *Organische Reagenzien für die Spurenanalyse,* Darmstadt 1975.
[33] J. L. Gower, *Biomed. Mass. Spectrom.* **12** (1985) 191.
[34] I. Mogharab, W. G. Glasser, *Tappi* **59** (1976) 110.
[35] DE 1 055 007, 1959 (H. Fießelmann). DE 1 083 830, 1960 (H. Fießelmann).
[36] J. Malowska, J. Szmich, *Zesz. Nauk. Politech. Lodz. Technol. Chem. Spozyw.* **38** (1984) 5.

# Metallic Soaps

Alfred Szczepanek, Faistenlohestr. 5, München, Federal Republic of Germany

Gunther Koenen, Peter Greven Fettchemie GmbH, Bad Münstereifel, Federal Republic of Germany

| | | | | | |
|---|---|---|---|---|---|
| 1. | Introduction | 3227 | 5.2. | Magnesium Soaps | 3236 |
| 2. | Production | 3229 | 5.3. | Calcium Soaps | 3236 |
| 2.1. | Raw Materials | 3229 | 5.4. | Barium Soaps | 3238 |
| 2.2. | Industrial Processes | 3230 | 5.5. | Aluminum Soaps | 3238 |
| 2.2.1. | Double Decomposition | 3230 | 5.6. | Lead Soaps | 3240 |
| 2.2.2. | Direct Reaction with Metal Compounds | 3231 | 5.7. | Copper Soaps | 3241 |
| 2.2.3. | Direct Reaction with Metals | 3232 | 5.8. | Zinc Soaps | 3242 |
| 2.3. | Plant Equipment | 3232 | 5.9. | Cadmium Soaps | 3243 |
| 2.4. | Storage | 3233 | 5.10. | Manganese Soaps | 3244 |
| 3. | Fire and Explosion Protection | 3233 | 5.11. | Iron Soaps | 3244 |
| 4. | Occupational Health and Environmental Protection | 3234 | 5.12. | Cobalt Soaps | 3246 |
| | | | 5.13. | Nickel Soaps | 3246 |
| 5. | Individual Metallic Soaps | 3234 | 6. | Quality Specifications | 3247 |
| 5.1. | Lithium Soaps | 3235 | 7. | References | 3248 |

## 1. Introduction

Originally the term metallic soaps was only used for metal salts of fatty acids of naturally occurring animal and vegetable fats. Today, metallic soaps are understood to refer to the sparingly soluble or insoluble salts of saturated and unsaturated, straight-chain and branched, aliphatic carboxylic acids with 8–22 carbon atoms. Examples are saturated fatty acids like stearic acid (octadecanoic acid), lauric acid (dodecanoic acid), 12-hydroxystearic acid, and mixtures of acids with 8–22 carbon atoms; unsaturated fatty acids like oleic acid (*cis*-9-octadecenoic acid) and linoleic acid (9,12-octadecadienoic acid); synthetic carboxylic acids like isostearic acid, 2-ethylhexanoic acid, dimethylhexanoic acids, trimethylhexanoic acids; and mixtures of synthetic aliphatic isocarboxylic acids.

The salts of the alicyclic naphthenic acids and resin acids (abietic acid) are also frequently regarded as metallic soaps due to properties related to their manufacture and applications. However, the metal salts of resin acids (resinates) no longer have great significance. Naphthenic acids are a byproduct of oil refining. They are monocarboxylic acids of cyclic alkanes and have the following structure:

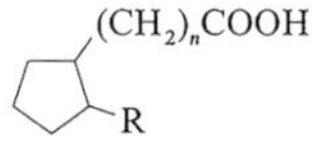

$$\text{cyclopentane}\begin{cases}(CH_2)_n COOH \\ R\end{cases}$$

R = H, CH$_3$, cyclopentyl
n = 60–30

The production and properties of soaps of almost all metals are described in the literature. However, only the metallic soaps of the following metals (arranged according to the periodic table) are industrially important: lithium (group 1); magnesium, calcium, barium (group 2); zirconium (group 4); manganese (group 7); iron (group 8); cobalt (group 9); nickel (group 10); copper (group 11); zinc, cadmium (group 12); aluminum (group 13); lead (group 14); as well as cerium and mixtures of rare earths.

**Uses.** Metallic soaps are used as thickeners for oils and other organic media, as lubricants and separating agents, as well as drying agents (dryers). Dryers are metallic soaps that accelerate drying as, for example, in the film formation of paints based on drying oils and alkyd resins. Lithium soaps are especially important as components of lubricating greases.

**Properties.** The physicochemical properties of metallic soaps are determined by the nature of the metal ion and the organic acid. The metallic soaps of saturated straight-chain carboxylic acids are solids and either have a sharp melting point or melting range, or carbonize on heating without melting. Pastelike products contain water.

Unsaturated fatty acids and branched synthetic acids form metallic soaps with plastic properties. They are, therefore, usually produced and used in solution.

**Types.** Metals (M) can form the following metallic soaps depending on their valence and properties:

| | |
|---|---|
| *Neutral soaps*: | R–COOM(I) |
| | (R–COO)$_2$M(II) |
| | (R–COO)$_3$M(III) |
| *Acid soaps*: | (R–COO)$_2$M(II) · R–COOH |
| | (R–COO)$_3$M(III) · R–COOH |

Acid soaps contain bound free organic acid that can be removed by extraction.
*Basic soaps*: (R–COO)$_2$M(II) · (MO)$_x$, $x = 1-2$
Complexed metal oxides can also be present in the form of metal hydroxides or as mixtures of metal oxides and hydroxides.
*Mixed soaps*: (R–COO) (R$^1$–COO)M(II)
Mixed soaps consist of mixed crystals or mechanically produced mixtures.
*Complex soaps*: Lead can form complex basic salts: (PbO)$_x$ · PbSO$_4$ · [(R–COO)$_2$Pb]$_y$
Other inorganic anions can also be present instead of sulfate.

**Table 1.** Metallic soaps

| Element | Stearate | 12-Hydroxy-stearate | Laurate | Oleate | Naphthenate | Resinate | 2-Ethyl-hexanoate | Reference |
|---|---|---|---|---|---|---|---|---|
| Ag | + | + | + | + | + | + |   | [1], [2] |
| Al | + |   | + | + | + | + | + | [3]–[5] |
| Ba | + | + | + |   | + |   | + | [6] |
| Be | + |   | + |   |   |   |   | [7] |
| Bi | + |   |   | + |   | + | + | [8] |
| Ca | + |   | + |   | + | + | + | [9] |
| Cd | + | + | + | + | + |   | + | [10] |
| Ce | + |   |   |   | + |   |   | [11] |
| Co | + |   |   | + | + | + | + | [12] |
| Cr | + |   |   |   | + |   | + | [13] |
| Cu | + |   |   | + | + |   | + | [14] |
| Fe | + |   |   | + | + |   | + | [13] |
| Hg | + |   | + | + | + |   |   | [7] |
| Li | + | + |   |   |   |   |   | [15] |
| Mg | + |   | + | + | + | + | + | [16] |
| Mn | + |   |   | + | + | + | + | [17] |
| Ni | + |   |   | + | + | + | + | [13] |
| Pb | + |   |   | + | + | + | + | [18] |
| Sb |   |   |   | + |   |   |   | [19] |
| Sn | + |   |   | + | + |   | + | [20] |
| Sr | + |   |   |   | + |   |   | [7] |
| Th | + |   |   |   |   |   |   | [7] |
| Ti | + |   |   |   |   | + |   | [21] |
| U | + |   |   | + |   |   |   | [7] |
| V |   |   |   |   | + |   |   | [22] |
| Zn | + | + | + | + | + | + | + | [23] |
| Zr | + |   |   |   | + | + | + | [21] |

**Overview.** A selection of commercially available metallic soaps and those described in the literature is given in Table 1. The soaps are listed in alphabetical order according to the chemical symbol of the metal. The products selected are representative of other homologous soaps. At least one metallic soap of each type is described in the list of references.

# 2. Production

## 2.1. Raw Materials

The organic acids used for the production of metallic soaps are either solid or liquid. They are supplied in sacks, drums, containers, or tank cars. The use of products packed in paper sacks should be minimized because of high handling costs and danger of impurities. Acids purchased in drums and containers should be stored at the recommended temperature to avoid total or partial crystallization. For large-scale production, molten acids are generally supplied in tank cars. A tank farm is required; the acid should not attack the tank material because it then becomes colored and impure.

Blanketing with an inert gas (e.g., nitrogen or carbon dioxide) may be necessary to prevent darkening caused by oxidation. Acids such as stearic acid, which are solid at room temperature and are supplied as liquids, must be kept above the melting point, preferably by circulation of warm water to avoid overheating at the wall. Stainless steel and aluminum are used as tank materials.

The acids are best dosed in the liquid state; the accuracy of pumps working by volume dosage is sufficient if temperature and density are taken into consideration.

The melting of solid organic acids in the reaction vessel reduces throughput; a separate melting vessel should therefore be used.

The metal salts, which are usually solids, are supplied in sacks, drums, or containers; they are dosed by means of weighing systems. Because the salts may be very toxic, appropriate measures for personal and environmental protection must be taken.

## 2.2. Industrial Processes

Because of the great number of possible combinations of metal and acid groups, only general production processes can be described here. Three types of reaction are possible:

1) Double decomposition (precipitation)
2) Direct reaction with metal oxides, metal hydroxides, or metal salts of volatile acids
3) Direct reaction with metals

### 2.2.1. Double Decomposition

In the first step of the double decomposition (precipitation) process the organic acid is saponified with alkali. In the second step, the resulting alkali soap is precipitated by addition of a solution of a water-soluble metal salt:

$$2\ R\text{–COOH} + 2\ NaOH \longrightarrow 2\ R\text{–COONa} + 2\ H_2O$$
$$2\ R\text{–COONa} + Ca^{2+} \longrightarrow (R\text{–COO})_2Ca + 2\ Na^+$$

The soap solution can also be added to the metal salt solution, or both solutions can be run simultaneously into a separate precipitation vessel. For sparingly soluble alkali soaps, alkaline solutions are used as the reaction medium. Plastic metallic soaps are precipitated in the presence of a water-immiscible solvent which absorbs the metallic soaps formed.

## 2.2.2. Direct Reaction with Metal Compounds

The reaction of organic acids with metal oxides, metal hydroxides, or salts of volatile acids, such as carbonates, acetates, or formates can be carried out by the following types of process:

1) Melt process
2) Reaction in the aqueous phase
3) Reaction in solvents or lubricants

**Melt Process.** In the melt process, metal oxides, metal hydroxides, or metal salts of volatile acids are fed into a stirred melt of the organic acid at elevated temperature. The rate of addition should be adapted to the reaction rate to prevent foaming caused by evaporation of water or volatile acid. With unreactive metal compounds, delays can occur in the initial phase. To initiate the reaction, small quantities of water or, for metal oxides, volatile acids such as acetic acid are often added.

The melt process is suitable for metallic soaps with melting points below 140 °C, which produce relatively nonviscous melts. These are generally metallic soaps with sharp melting ranges such as lead, zinc, or cadmium stearate. Alkaline-earth stearates, for example, are not suited to this process because they have high melting points and exhibit rubber–elastic properties in the melt. Because alkaline-earth metallic soaps are often used in combination with low-melting metallic soaps [e.g., as poly(vinyl chloride) (PVC) stabilizers in the combinations barium – cadmium, barium – zinc, or calcium – lead], a low-melting acidic metallic soap is first produced in the melt; the high-melting soap is then formed in the melt of the low-melting one [24].

**Reaction in the Aqueous Phase.** In the aqueous-phase reaction, metal oxides or hydroxides react in the presence of catalysts such as amines, sparingly soluble alcohols, or other wetting agents. The processes are applicable only to solid metallic soaps (e.g., barium, magnesium, zinc, lead, and calcium stearates) [17].

**Reaction in solvents** is preferred if the direct melt process is impossible or gives low-quality products, or if the metallic soaps are to be used in solution. Dryers and liquid PVC stabilizers based on branched-chain aliphatic carboxylic acids and naphthenic acids are mainly produced by this process. The solvent (e.g., solvent naphtha or white spirits) also acts as an entrainer for removing reaction water by azeotropic distillation [25].

For solid metallic soaps the solvent can be replaced by organic compounds such as alkanes or fatty alcohols which are solid at room temperature and cannot be saponified by metal oxides. The reaction medium cannot be removed and remains in the final product, which may, for example, be used as a lubricant for plastics [26]. Lubricating greases provide a further example of a reaction product that remains in the reaction medium.

**Continuous Process.** Metal oxides are dissolved in aqueous ammonia and converted to basic metal carbonates by passage of gaseous carbon dioxide. The molten fatty acids in aqueous ammonia are converted to saponified emulsions. The two components are continuously mixed in the blade region of a rotor – stator mixer with interlocking radial surfaces. The metallic soaps are produced in pastelike to semidry form; they are removed continuously, dried, and ground [27].

**Dust-Free Granules.** For many uses, metallic soaps do not have to be in powder form. To prevent explosion and promote better handling and dosage, dust-free granules are preferred. The granules are produced from straight-chain fatty acids ($C_8 - C_{22}$) and metal oxides, hydroxides, or carbonates. Granule formation is controlled by the rate of addition of the metal compound, stirring rate, temperature, and water content of the reaction mixture, which is adjusted to ca. 10 wt%. The reactors are mixers equipped for vacuum drying. In the case of unreactive metal oxides (e.g., zinc oxide), chelating acids like citric, tartaric, or glycolic acid are added as catalysts. Because of the small quantity of water involved and the absence of filtration, drying, and pulverization stages, the process runs very economically by using the heat of reaction [28].

### 2.2.3. Direct Reaction with Metals

Organic acids are treated with metal granules, powders, or gauze at elevated temperature, usually with a solvent as the reaction medium. Direct reaction with metals is used when it offers economic advantages or allows the replacement of unreactive metal oxides by metals [29].

## 2.3. Plant Equipment

**Direct Reaction in Aqueous Media.** Reaction vessels, filtration, and drying equipment are necessary for reaction in aqueous media (the precipitation process).

Stainless steel *reaction vessels* are preferred for the production of alkali soaps because fibers from wooden vessels can lead to impurities. The acids are saponified at 60 – 90 °C at a dilution of 1:5 – 1:10 depending on the type.

Metal compounds are dissolved at 40 – 50 °C, depending on their solubility, in plastic-coated iron containers. Tanks for intermediate storage of reaction mixtures are equipped with slow stirrers or recirculation pumps.

In the *filtration* of metallic soaps, the filter cake should be as dry as possible for economic reasons. Nutsche (pan) filters, drum filter presses, and plate filter presses have now been largely replaced by membrane filter presses or continually functioning automatic filters. In the direct reaction in the aqueous phase, the filter water produced is used as circulation water and must be put into intermediate storage.

Belt, drum, and rotary dryers have proved successful as *drying units*. In drying units without integrated pulverization and classifying facilities, pulverization units are necessary.

**Melt Process.** For direct reaction by the melt process or in lubricants, stainless steel or glass-lined reactors fitted with stirrers, heaters (up to 160 °C), and vapor extractors are used. The melts are then processed into a solid, dust-free form by using flaking equipment, cooling belts, or spraying devices. If desired they can be ground.

**Precipitation or Direct Reaction in Solvents.** For precipitation in the presence of solvents or direct reaction in solvents, stainless steel or glass-lined steel reactors are used that are fitted with stirrers, heaters (up to 160 °C), vacuum connections, and reflux condensers with water separators. Suitable filter equipment is necessary for filtration of viscous solutions. Reaction with metals is also possible in this type of plant.

## 2.4. Storage

The liquid to pastelike, solvent-containing metallic soaps are generally stored in packaging forms ready for sale (e.g., cans, drums, or containers). Bags or Big Bags are often used for powdered metallic soaps. For storage in large silos, protective measures against fire and explosion are necessary (see Chap. 3).

## 3. Fire and Explosion Protection

**Fire Protection.** Metallic soaps are flammable substances that burn with intense smoke production. The resulting oxides are carried into the air and may be toxic (heavy metals). This should be taken into consideration during fire fighting. Foam and carbon dioxide extinguishers are suitable fire-fighting agents.

For metallic soap solutions in organic media, local regulations for the handling and storage of flammable liquids must be observed. Plastic metallic soaps are not readily flammable because of their small surface area. Powdered metal soaps have a relatively large surface area and are flammable. With metallic soaps based on unsaturated fatty acids such as oleic and linoleic acids, self-ignition can occur if the packed products are not cooled sufficiently. Self-ignition is promoted by soaps of heavy metals, which are oxidation catalysts.

**Explosion Protection.** When organic solvents are used in the production of metallic soaps, plants must be protected against explosion by suitable equipment in compliance with prevailing regulations. Metallic soaps in powder form can, like other organic dusts, cause dust explosions, which are most likely to occur during drying, pulverizing, and storage. To prevent buildup of static charge (ignition source), good grounding (earthing) should be ensured. Formation of sparks by metal particles is difficult to avoid. Magnetic separators do not offer any safety improvements because the metals are either nonmagnetic or coated by the metallic soaps. To avoid serious damage, relevant parts of the plant must be equipped with pressure-release devices. If this precaution is not taken, a pressure wave can lead to severe secondary dust explosions and destruction of the plant.

# 4. Occupational Health and Environmental Protection

**Occupational Health.** Soaps of heavy metals are toxic substances that can lead to acute or chronic poisoning. Suitable measures to protect the worker are dust extraction, protective clothing, dust masks (when dust sources cannot be avoided), and regular medical checkups. The MAK and TLV values are given for individual metals (Chap. 5).

**Wastewater.** The filter water produced in the direct reaction in the aqueous phase is collected and reused as reaction water; complete circulation eliminates wastewater production. In the double decomposition reaction, considerable quantities of neutral salts such as sodium sulfate, sodium chloride, and sodium nitrate are produced, which are difficult or impossible to remove by water treatment. This fact should be given particular consideration in the choice of plant location. Other sources of water pollution are bleeding of the filter presses and excess metal salts used for precipitation. Before wastewater is sent to the water treatment plant, the heavy metals are therefore precipitated by addition of lime and/or passing through $CO_2$ to give a pH of 8.5 – 9.

**Air Pollution.** Solvent emissions must be reduced in accordance with local regulations by means of a scrubber or absorption tower. Steam- or water-jet vacuum pumps are often used and entrain small quantities of mostly water-immiscible solvents; oil separators are therefore necessary.

The dryers for powdered metallic soaps may release residual dust-laden or water-vapor-containing air into the surroundings. To purify the air, filtration units with large surface areas are necessary. Automatic dust-measuring instruments built into the exhaust stack switch off the plant if the dust concentration becomes too high (e.g., because of a tear in the filter). The residual emission is discharged into the surroundings via a high exhaust stack. Permitted emission values are stipulated in national air purity regulations.

# 5. Individual Metallic Soaps

Metallic soaps based on lithium, magnesium, calcium, barium, aluminum, lead, copper, zinc, cadmium, manganese, iron, cobalt, and nickel are supplied as commercial products and are described in this chapter.

**Table 2.** Technical data for some commercial soaps of lithium and magnesium

| Compound | CAS registry no. | Appearance | Free fatty acid*, wt% | Total ash or metal content, wt% | Moisture*, wt% | Bulk density*, g/L | Melting range, °C |
|---|---|---|---|---|---|---|---|
| Lithium stearate | [4485-12-5] | white powder | 1 | 2.4 Li | 0.5 | 230 | 190 – 210 |
| Lithium 12-hydroxystearate | [7620-77-1] | white powder | 1 | 2.3 Li | 0.5 | 400 | > 200 |
| Magnesium stearate, technical | [557-04-0] | white powder | 2 | 7 – 8 | 4 – 5 | 250 | 115 – 135 |
| Magnesium behenate | [43168-33-8] | white powder | 2 | 6 – 7 | 6.0 | 250 | ca. 140 |

* Approximate values.

## 5.1. Lithium Soaps

**Production.** The lithium soaps of stearic and 12-hydroxystearic acid have the most industrial importance. They are produced by direct reaction in aqueous medium. A very dilute solution of lithium hydroxide hydrate, $LiOH \cdot H_2O$ (concentration ca. 1:40, i.e., 2.5 wt%), is slowly added with intense stirring to the fatty acid dispersed in water in the ratio 1:20 at 90 °C. Lithium soaps are difficult to filter; flocculation is therefore improved by using an excess of lithium hydroxide or by adding neutral salts. The resulting lithium soap dispersion is, however, often not filtered but spray-dried, even though this involves evaporation of a large quantity of water. Under modified reaction conditions, lithium hydroxide hydrate can be replaced by lithium carbonate and ammonia.

**Properties.** See Table 2.

**Uses.** Lithium soaps have particularly good swelling properties. Lithium stearate is used to thicken natural and synthetic oils; to raise the melting point and flexibility of microcrystalline waxes and paraffins; as an additive for coating waxes to increase water repellence; and as a lubricant in the production of injection-molded articles from light metal. Multipurpose lubricating greases have particularly good penetration properties, high oxidation stability, and dropping points up to ca. 200 °C. Lithium 12-hydroxystearate is used primarily for the production of lubricating greases based on synthetic oils (e.g., from silicon and ester oils). These lubricants cannot be produced by the usual saponification in mineral oil because the synthetic oils are attacked under the saponification conditions employed.

## 5.2. Magnesium Soaps

Magnesium soaps are produced as neutral salts by the double decomposition process, using water-soluble magnesium salts (primarily magnesium chloride) and alkali salts of organic acids in an aqueous medium, or by direct reaction with magnesium oxide. Magnesium soaps in powder form based on stearic and behenic acids are particularly important. The degree of purity required, especially with regard to heavy-metal impurities, determines the choice of raw materials in pharmaceutical products. Pharmaceutical quality must fulfill the requirements of the European and U.S. pharmacopoeias.

**Properties.** See Table 2. Magnesium soaps of long-chain fatty acids have a high specific surface area and can absorb considerable quantities of liquid, giving a creamy consistency. They are insoluble in alcohols, esters, ketones, ethers, and hydrocarbons.

**Uses.** Magnesium soaps are used in the plastics industry as lubricants and mold-release agents for thermosetting plastics and thermoplasts; in the pharmaceutical industry as processing aids in the production of dragées and as lubricants for tablet pressing; in the cosmetics industry as powder components and for consistency in water-free ointments; in the wax industry to increase the retention of creams based on semisolid wax products and in the production of lubricating waxes; and as anticaking and waterproofing agents for hygroscopic substances.

## 5.3. Calcium Soaps

Calcium soaps are generally neutral salts. They are obtained by double decomposition or direct reaction. The raw materials consist of saturated straight-chain aliphatic acids ($C_8$–$C_{22}$), montan wax acids, unsaturated straight-chain carboxylic acids (e.g., oleic and linoleic acids), branched-chain aliphatic carboxylic acids (e.g., ethylhexanoic acids), naphthenic acids, and resin acids. Calcium chloride and calcium hydroxide are used as the metal compounds.

Very finely divided calcium soaps of the aliphatic straight-chain carboxylic acids are produced by double decomposition in aqueous medium. The stearates, myristates, and laurates are commercial products.

Direct reaction in an aqueous medium is the most common production method for solid calcium soaps; 2–3% ammonia (based on the fatty acid) is added to the fatty acid solution. Ammonium soaps are first formed, which are effective emulsifiers and react with calcium hydroxide to liberate ammonia. The ammonia reacts with the fatty acid again to form ammonium soaps until conversion of the fatty acid is complete. The products are free of water-soluble salts; their purity depends on the quality of calcium hydroxide used.

**Table 3.** Technical data for some commercial soaps of calcium and barium

| Compound | CAS registry no. | Appearance | Free fatty acid*, wt% | Total ash, wt% | Moisture*, wt% | Bulk density, g/L | Melting range, °C |
|---|---|---|---|---|---|---|---|
| Calcium stearate, precipitated | [1592-23-0] | white powder | 1–2 | 9–10 | 2–3 | 120–130 | 120–160 |
| Calcium stearate, direct reaction | [1592-23-0] | white powder | 0.5 | 10–11 | 2–3 | ca. 250 | 120–160 |
| Calcium arachidate | [22302-43-8] | white powder | 1 | 8–9 | 1.5 | 200 (tapped) | ca. 150 |
| Calcium 12-hydroxystearate | [3159-62-4] | white powder | 1 | 8.5–9.5 | 2–3 | 250 (tapped) | ca. 140 |
| Calcium laurate | [4696-56-4] | white powder | 1 | 13–14 | 2–3 | 300 (tapped) | ca. 140 |
| Calcium oleate | [142-17-6] | yellow mass | 2 | 10–11 | 1 | | ca. 120 |
| Calcium 2-ethyl-hexanoate | [136-51-6] | yellowish powder | 2 | 16–17 | 1 | | ca. 160 |
| Calcium ricinoleate | [6865-33-4] | yellowish powder | 2 | 10–11 | 1.5 | | ca. 130 |
| Calcium naphthenate | [61789-36-4] | solid brown mass | | 11–12 | 1 | | |
| Barium stearate | [6865-35-6] | white powder | 1 | 27–28 | 0.5 | 250 | 200 |
| Barium laurate | [4696-57-5] | white powder | 1 | 35–36 | 0.5 | 275 | 200 |
| Barium 2-ethyl-hexanoate | [2457-01-4] | yellowish paste | | 29–30 | 0.6 | | |
| Barium naphthenate | [61789-67-1] | dark paste | | 24–25 | 2 | | |

* Approximate values.

Aqueous calcium stearate pastes with a solid content of 40–50% are also produced by the direct process and are used as additives for paper coating (slips).

Calcium soaps which are not in powder form and are therefore used mostly as solutions in organic media (e.g., 2-ethylhexanoates and naphthenates) are produced by direct reaction in an appropriate medium. Neutral calcium soaps cannot be produced by the melt process because of their high melting point range and viscous melts. Acid soaps can be obtained by partial reaction in which the excess organic acid serves as the reaction medium.

**Properties.** See Table 3.

**Uses.** Calcium soaps are used as physiologically inert lubricants and secondary stabilizers in the plastics industry to improve the flow properties and prevent the caking of hygroscopic substances; as lubricants and mold-release agents along with magnesium soaps in pharmaceutical tablet pressing; as waterproofing additives in agents used for building protection and for surface treatment of fillers; and in the paper industry in recording paper, diagram paper, and coating slips.

## 5.4. Barium Soaps

The most important organic acids in barium soaps are 12-hydroxystearic, lauric, 2-ethylhexanoic, and naphthenic acids. The metal compounds used are barium chloride and barium hydroxide (mono- or octahydrate). The TLV–TWA value for barium is 0.5 mg/m$^3$.

**Properties.** See Table 3.

**Uses.** Barium soaps are used in the paint industry for wetting pigments and as dispersion agents; in metal forming as a component of wire-drawing and other auxiliaries; and in the plastics industry as lubricants and costabilizers.

## 5.5. Aluminum Soaps

Commercial aluminum soaps are generally produced by reacting alkali soaps with water-soluble aluminum salts. In precipitation in aqueous systems, agglomerates of aluminum hydroxide and carboxylic acids are first produced which then react with each other to form salts during the subsequent drying process.

Because aluminum is trivalent, three types of soap are possible:

1) Aluminum monosoaps Al(OH)$_2$(OOC–R)
2) Aluminum disoaps Al(OH)(OOC–R)$_2$
3) Aluminum trisoaps Al(OOC–R)$_3$

The polymeric structure of aluminum soaps is not conveyed by the formulas [30].

All types of organic acids are used for aluminum soaps, but particularly stearic, oleic, naphthenic, and 2-ethylhexanoic acids. Aluminum sulfate is preferred as the water-soluble aluminum salt, along with potassium aluminum sulfate and aluminum chloride. In industrial production of the three types of aluminum soap, the organic acids are saponified with 1, 2, or 3 mol of alkali, respectively. Because the three types differ greatly in their thickening properties, commercial products are available with intermediate metal contents. The metal, ash, and free fatty acid contents are given to characterize the soaps.

To achieve special thickening properties, above all for the production of lubricating greases based on mineral oil, fatty acids are combined with aromatic acids (e.g., benzoic or phthalic acid) or with dimeric fatty acids. Lubricating grease with high fastness to fulling resistance and a high dropping point is obtained in this way.

Aluminum soaps in powder form are obtained by double decomposition in aqueous medium. Catalyzed reactions between fatty acids and freshly precipitated aluminum oxide hydrate are known but not used industrially. Aluminum oxide hydrate ages too quickly and is not suitable for melting or direct reaction processes in nonaqueous

**Table 4.** Technical data for some commercial aluminum soaps

| Compound | CAS registry no. | Appearance | Free fatty acid*, wt% | Total ash, wt% | Moisture*, wt% | Tapped bulk density, g/L | Melting range, °C |
|---|---|---|---|---|---|---|---|
| Aluminum tristearate | [637-12-7] | white powder | 16 – 22 | 7 – 8 | 1.2 | 300 | 110 – 130 |
| Aluminum distearate | [300-92-5] | fine white powder | 3 – 7 | 10 – 11 | 1.2 | 82 | 150 – 160 |
| Aluminum monostearate | [7047-84-9] | white powder | 1 – 3 | 13 – 14 | 1.2 | 170 | 180 – 250 |
| Aluminum di-12-oxystearate | [7047-84-9] | fine white powder | 4 – 6 | 7.5 – 8.5 | 1.5 | 110 | ca. 140 |
| Aluminum dipalmitate | [14236-50-1] | fine white powder | 5 – 7 | 10.5 – 11.5 | 1 | 300 | ca. 160 |
| Aluminum monopalmitate | [555-35-1] | very fine white powder | 2 – 4 | 15.5 – 16.5 | 1.5 | 280 | ca. 200 |
| Aluminum dimyristate | [56639-51-1] | fine white powder | 6 – 9 | 10.5 – 11.5 | 1 | 300 | ca. 145 |
| Aluminum dilaurate | [817-83-4] | white powder | 4 – 6 | 10.5 – 11.5 | 1 | 280 | ca. 150 |

* Approximate values.

media because of its low reactivity. An exception is the in situ production of lubricating greases in which aluminum alcoholates and freshly precipitated aluminum hydroxide hydrate are used. The TLV – TWA value for aluminum is 10 mg/m$^3$ for metal dust and 5 mg/m$^3$ for pyro powders and welding fumes.

**Properties.** See Table 4. Aluminum soaps of long-chain fatty acids are insoluble in water, lower alcohols, esters, and ketones; they are soluble in hydrocarbons, toluene, xylene, ethylene glycol, chlorinated hydrocarbons, and vegetable and mineral oils. The most striking property of aluminum soaps is their capacity for thickening organic media and for forming solid gels with them. Gel consistency increases in the order trisoaps, disoaps, and monosoaps. The gels are destroyed by polar substances such as primary amines, ammonia, pyridine, and lower alcohols.

**Uses.** The gel-forming property of aluminum soaps is exploited in the mineral oil, cosmetics, and pharmaceutical industries. They are also employed in the paint industry as antisettling and matting agents and for wetting pigments. Aluminum soaps in powder form are used as lubricants in thermosetting plastics and thermoplasts, as waterproofing agents in building protection, and as flow promoters for caking powders.

## 5.6. Lead Soaps

**Neutral lead soaps** are produced by double decomposition, by direct reaction in the melt, or by direct reaction in aqueous media in the presence of catalysts such as triethanolamine and sparingly soluble alcohols [17].

**Basic lead soaps** are produced by direct reaction in the presence of catalysts in aqueous media. The addition salts contain a maximum of two molecules of lead(II) oxide to one molecule of neutral lead soap:

$$3\ PbO + 2\ HOOC-R \rightarrow 2\ PbO \cdot Pb(OOC-R)_2 + H_2O$$

Direct reaction in the melt does not go beyond the addition of one molecule of lead(II) oxide because of an increase in the melting point. In the presence of a fusible organic medium (e.g., $C_{16}-C_{18}$ fatty alcohols, or a combination of fatty alcohols and paraffin or long-chain fatty acid estes), however, a further molecule of lead(II) oxide can be added to give dispersions of dibasic lead soaps [24]. The basicity of basic lead soaps in the direct reaction in solvents depends on their solubility. In general, only one molecule of lead(II) oxide can be added.

**Basic Lead Complex Soaps.** The starting material for basic lead complex soaps is an aqueous dispersion of a basic inorganic lead salt which reacts with fatty acids in the presence of catalysts in an aqueous medium [31]:

$$4\ PbO \cdot PbSO_4 + 2\ HOOC-R \rightarrow 3\ PbO \cdot PbSO_4 \cdot Pb(OOCR)_2 + H_2O$$

**Production.** Technical-grade stearic, oleic, 2-ethylhexanoic, and naphthenic acids are used for the production of the most important lead soaps. The principal metal source is lead oxide. Lead acetate is used as the water-soluble lead salt in the double decomposition process. Direct reaction in aqueous medium is performed with reactive lead(II) oxide (litharge, commercial specification "canary yellow"). The red lead oxide content should be very low because this modification cannot form basic addition salts. The red modification is, however, suitable for the production of neutral salts in the melt or in an aqueous medium. In the melt process, the lead metal content of the lead oxide should be very low to avoid blockage of valves by agglomeration.

Regulations for personal and environmental protection must be observed, particularly with lead soaps in powder form. Even the absorption of small quantities of lead can lead to chronic illness. The MAK value for lead is 0.1 mg/m$^3$ and the TLV–TWA value is 0.15 mg/m$^3$.

**Properties.** See Table 5.

**Table 5.** Technical data for some commercial soaps of lead and copper

| Compound | CAS registry no. | Appearance | Free fatty acid*, wt% | Total ash or metal content, wt% | Moisture*, wt% | Bulk density, g/L | Melting range, °C |
|---|---|---|---|---|---|---|---|
| Neutral lead stearate | [1072-35-1] | white powder | 1 | 27–28 Pb | 0.2 | 500 | 105–110 |
| Monobasic lead stearate | [90459-52-2] | white powder | 1 | 41–42 Pb | 1 | 250–300 | 110–120 |
| Dibasic lead stearate | [56189-09-4] | white powder | 1 | 51–52 Pb | 0.5 | 450 | >200 (decomp.) |
| Lead 2-ethylhexanoate | [16996-40-0] | white, plastic | 1 | ca. 45 | 0.5 | | |
| Lead oleate | [1120-46-3] | brown mass | 1 | 30–31 | 1 | | ca. 65 |
| Lead naphthenate | [61790-14-5] | dark, viscous mass | 1 | 30–31 | 1 | | |
| Copper stearate | [7617-31-4] | blue-green powder | 1 | 14–15 | 1 | | ca. 155 |
| Copper laurate | [19179-44-3] | blue-green powder | 2 | 17–18 | 1 | | ca. 125 |
| Copper 2-ethylhexanoate | [22221-10-9] | blue mass | | 23–24 | 1 | | |
| Copper oleate | [10402-16-1] | blue-green mass | | 13–14 | 1 | | |
| Copper naphthenate | [1338-02-9] | gray-green paste | | 14–15 | 1 | | |

\* Approximate values.

**Uses.** Neutral and basic lead stearates are used as heat stabilizers and lubricants in PVC processing; as lubricants in the pencil industry; and as additives for specialty papers. Lead 2-ethylhexanoate, naphthenates, and resinates are used in the paint industry for wetting pigments and as dryers, and in the mineral oil industry as components of high-pressure lubricants.

## 5.7. Copper Soaps

Copper forms neutral soaps which are produced either by double decomposition or by the direct process in the melt or in organic media. Production from metallic copper is also known: copper powder is dispersed in solvent naphtha; copper soaps that are soluble in solvent naphtha are formed by passage of air in the presence of an organic acid.

Commercial copper soaps are produced from stearic, naphthenic, oleic, and 2-ethylhexanoic acids. Water-soluble copper salts, mainly copper sulfate, are used as the copper source for the double decomposition; copper hydroxide or basic copper carbonate is used for the direct reaction.

Production of copper soaps in powder form by double decomposition requires a separate plant, because cleaning the drying and filtration units is almost impossible. The MAK values for copper are 0.1 mg/m$^3$ (fume) and 1 mg/m$^3$ (dust); the TLV – TWA values are 0.2 mg/m$^3$ (fume) and 1 mg/m$^3$ (dusts and mists).

**Properties.** See Table 5.

**Uses.** Copper soaps are used mainly as fungicides in wood and ship paints, textiles (e.g., tents, awnings, sandbags, camouflage netting, ropes), paper, and cardboard. Compared with inorganic copper compounds, they have the advantage of being difficult to wash out. Naphthenates, oleates, and 2-ethylhexanoate are used as solutions in organic media.

## 5.8. Zinc Soaps

**Neutral zinc soaps** are produced by double decomposition, direct reaction in the melt, direct reaction in aqueous or organic media, and reaction of zinc metal powder with organic acids.

**Basic zinc soaps** produced by double decomposition are obtained in powder form; the basic moiety is usually in the form of zinc hydroxide or carbonate, and they are not therefore regarded as true basic soaps. In organic media, soluble zinc soaps (particularly naphthenates, 2-ethylhexanoate, and oleates) can be rendered basic during the direct reaction by addition of excess zinc hydroxide. Soaps with 1.7 equivalents of zinc hydroxide to 1 equivalent of acid can be obtained.

**Commercial zinc soaps** are produced from stearic, lauric, oleic, 2-ethylhexanoic, and naphthenic acids. Water-soluble salts, preferably zinc sulfate and chloride, are used as the zinc source for double decomposition; zinc oxide, hydroxide, carbonate, and acetate are employed for the direct reaction. When several types of product are prepared in a single plant, attention must be paid to the production order because zinc stearate acts as a gel breaker with gels from aluminum soaps. Normally calcium, magnesium, zinc, and aluminum soaps are produced in a single plant.

**Properties.** See Table 6.

**Uses.** Zinc soaps are used for the following purposes:

1) in the plastics industry as lubricants, release agents, and components of PVC stabilizers;
2) in the rubber industry, as release agents for unvulcanized products (wettable zinc stearate soaps and zinc stearate dispersions in water are also used);
3) in paints, as matting agents and abrasives;
4) in building protection, as waterproofing agents;
5) for textiles, also as waterproofing agents;
6) in the cosmetics and pharmaceutical industries, as additives to body and face powders;

**Table 6.** Technical data for some commercial soaps of cadmium and zinc

| Compound | CAS registry no. | Appearance | Free fatty acid*, wt% | Total ash, wt% | Moisture*, wt% | Bulk density, g/L | Melting range, °C |
|---|---|---|---|---|---|---|---|
| Cadmium stearate | [2223-93-0] | white powder | 2 | 19–20 | 1 | 180 | 105–115 |
| Cadmium 12-hydroxy-stearate | [69121-20-6] | white powder | 2 | 18–19 | 1 | 220 | 110–118 |
| Cadmium laurate | [2605-44-9] | white powder | 2 | 26–27 | 1 | 300 | |
| Cadmium oleate | [10468-30-1] | brown mass | | 19–20 | 2 | | |
| Cadmium 2-ethyl-hexanoate | [2420-98-6] | yellowish paste | | 32–33 | 2 | | |
| Zinc stearate | [557-05-1] | white powder | 1 | 13–14 | 1 | 160 | 118–122 |
| Zinc 12-hydroxy-stearate | [35674-68-1] | white powder | 1 | 12.5–13.5 | 1 | 220 | 145–155 |
| Zinc laurate | [2452-01-9] | white powder | 1 | 17–18 | 1 | 220 | 122–126 |
| Zinc arachidate | [22302-43-8] | white powder | 1.5 | 12–13 | 1 | 180 | 120–124 |
| Zinc oleate | [557-07-3] | bright mass | | 13–14 | 0.3 | | |
| Zinc 2-ethylhexanoate | [136-53-8] | clear, highly viscous liquid | | 23–24 | 1 | | |
| Zinc naphthenate | [12001-85-3] | brown mass | | 14–15 | 0.4 | | |

\* Approximate values.

7) as components of wire-drawing greases (special zinc arachidate is used in tube drawing because of its low ash content); and

8) in paints, as components of dryers and as anticaking agents for powdered products.

## 5.9. Cadmium Soaps

Neutral and basic cadmium soaps are produced by double decomposition, direct reaction in the melt, or direct reaction in organic media. Stearic, lauric, 12-hydroxystearic, oleic, and 2-ethylhexanoic acids are commonly used. Cadmium sulfate, chloride, and nitrate are used in double decomposition; cadmium oxide and hydroxide in the direct reaction. Cadmium soaps are highly toxic. Cadmium accumulates in organs such as the liver and adrenal gland and is released very slowly. Absorption of very small quantities of cadmium over a long period therefore leads to chronic illness. The TLV–TWA value is 0.05 mg/m$^3$.

**Properties.** See Table 6.

**Uses.** The use of cadmium soaps as components of PVC stabilizers is decreasing because of their toxic properties. Solid PVC stabilizers contain cadmium soaps of saturated fatty acids, whereas liquid stabilizers contain soaps of unsaturated or branched, short-chain fatty acids. The soaps act as heat and light stabilizers, and soaps of saturated fatty acids are used as lubricants in the processing of PVC. Unlike other metallic soaps used in the PVC sector, cadmium soaps do not affect transparency and are suitable for hard, transparent PVC products.

## 5.10. Manganese Soaps

**Neutral manganese(II) soaps** are produced by double decomposition using alkali soaps and water-soluble manganese salts in an aqueous medium or in the presence of organic solvents, and by direct reaction in the melt or in organic media with manganese oxides, hydroxides, or salts of volatile acids. The oxides and hydroxides are very unreactive and must be freshly prepared; conversion is generally incomplete. In the reaction with metallic manganese in powder form or as divided gauze, lower carboxylic acids (e.g., acetic acid) are added to the solvent as catalysts.

**Basic manganese(II) soaps** are produced by double decomposition in an aqueous medium; oversaponified alkali soaps (i.e., more than 1 mol of alkali is used per mole of acid) react with water-soluble manganese salts. The excess alkali forms manganese hydroxide or manganese oxide hydrate, which adds to the neutral manganese soap. Whether a real addition salt or a simple mixture is formed is unclear. The basic manganese soaps are insoluble or sparingly soluble in organic solvents. Stearic, oleic, naphthenic, and 2-ethylhexanoic acids are commonly used. Sources of manganese for the double decomposition are water-soluble manganese salts such as manganese chloride and sulfate; for the direct reaction, manganese oxide, hydroxide, and acetate.

Manganese soaps are hazardous to health. A particular danger arises from dust formation by products in powder form. Because of their brown color, separate production plants are necessary for the powdered products. The TLV–TWA values for manganese are 5 mg/m$^3$ (dust) and 1 mg/m$^3$ (fume).

**Properties.** See Table 7.

**Uses.** Manganese soaps are used mainly as dryers in paints, printing inks, and varnishes based on drying oils or alkyd resins. Basic powdered products are also added to printing inks.

Manganese soaps can be used as catalysts (e.g., in paraffin oxidation or the reduction of fatty acids to alcohols).

## 5.11. Iron Soaps

Only neutral iron(III) soaps are produced industrially. Formation of basic soaps by addition of iron oxide or hydroxide is known but is not exploited commercially.

Iron soaps are obtained by double decomposition using alkali soaps with water-soluble iron salts in an aqueous medium with or without organic solvents. Direct reaction in the melt with iron oxides does not go to completion and is very slow; the unreacted free fatty acid is extracted with alcohol.

**Table 7.** Technical data for some commercial soaps of iron, cobalt, nickel, and manganese

| Compound | CAS registry no. | Appearance | Free fatty acid*, wt% | Total ash or metal content, wt% | Moisture*, wt% | Melting range, °C |
|---|---|---|---|---|---|---|
| Iron(III) stearate | [555-36-2] | red-yellow powder | 5 | 14–15 | 1 | ca. 90 |
| Iron oleate | [23335-74-2] | red paste | | 9–10 | 2 | |
| Iron 2-ethylhexanoate | [6535-20-2] | reddish paste | | 16–17 | 1 | |
| Cobalt stearate | [13586-84-0] | blue–violet powder | 1 | 13–14 | 1 | ca. 150 |
| Cobalt 2-ethylhexanoate | [136-52-7] | blue–violet mass | | 22–23 | 1 | |
| Cobalt naphthenate | [61789-51-3] | dark blue mass | | 10–11 Co | | |
| Nickel stearate | [2223-95-2] | bright green powder | 1 | 14.5–15.5 | 1 | 80–86 |
| Nickel oleate | [13001-15-5] | greenish paste | 1 | 22–23 | 1 | |
| Nickel 2-ethylhexanoate | [4995-91-9] | green paste | | 22–23 | 1 | |
| Nickel naphthenate | [61788-71-4] | deep green mass | 1 | 8.5–11.5 Ni | 1 | |
| Manganese stearate | [3353-05-7] | bright brown powder | | 13–14 | 1 | ca. 80 |
| Manganese(II) oleate | [23250-73-9] | brown mass | | 12–13 | 1 | |
| Manganese 2-ethylhexanoate | [15956-58-8] | brown mass | | ca. 16 Mn | | |
| Manganese naphthenate | [1336-93-2] | dark brown mass | | ca. 10 Mn | 1 | |

* Approximate values.

Raw materials used are stearic, oleic, naphthenic, and 2-ethylhexanoic acids. In double decomposition, iron(III) chloride is generally used as the iron source; in the direct reaction, iron oxide, hydroxide, or carbonate are used. Because of the intense color of iron soaps, separate precipitation units are necessary. Although iron soaps do not pose a danger to health, local regulations regarding personal and environmental protection for dust-producing products must be observed during handling.

**Properties.** See Table 7.

**Uses.** Iron soaps are components of dryer mixtures. Because of their intense color, they are used only with dark colors. The function of the iron as a dryer is based on polymerization at elevated temperature rather than catalysis of oxidation.

Further uses are as hydrogenation catalysts, as dry-film lubricants in the form of monomolecular adhesive layers, in the production of fast-copy paper, in waterproofing, and in pharmaceuticals as oil-soluble iron compounds for the treatment of anemia.

## 5.12. Cobalt Soaps

Neutral and weakly basic cobalt(II) soaps are known. They are produced by double decomposition using alkali soaps with water-soluble cobalt salts in an aqueous medium with or without organic solvents, or by direct reaction via the melt process, or in solvents. Reaction of metallic cobalt in powder or gauze form in organic media is also possible; air acts as an oxidizing agent.

Stearic, oleic, naphthenic, and 2-ethylhexanoic acids are used. For double decomposition, cobalt(II) sulfate is used as the water-soluble salt; for the direct reaction, cobalt acetate, cobalt(II) hydroxide, and cobalt(II) oxide are used.

Cobalt soaps have an intense blue-violet color. Soaps in powdered form must be produced in separate plants. Dust from cobalt soaps is hazardous to health. Local regulations concerning personal and environmental protection and the compulsory labeling of packaging must be observed. The TLV–TWA value for cobalt is 0.05 mg/m$^3$.

**Properties.** See Table 7.

**Uses.** Cobalt soaps are very effective dryers for printing and writing inks because they catalyze oxidation. Cobalt stearate is used as an adhesive for rubber–metal bonding.

## 5.13. Nickel Soaps

Divalent nickel forms neutral to weakly basic soaps. They are produced by double decomposition using alkali soaps with water-soluble nickel salts in an aqueous medium in the presence of organic solvents or by direct reaction in the melt or in solvents with nickel oxide, hydroxide, or salts of volatile acids. In the melt process, decomposition occurs above 200 °C with formation of a black powder. The usual acids are stearic, oleic, 2-ethylhexanoic, and naphthenic acids. Nickel sulfate and chloride are used as water-soluble salts for double decomposition, and nickel(II) hydroxide, oxide, acetate, or carbonate for the melt process.

Nickel soaps in powder form must be produced in separate plants because of their intense green color. Dust from nickel soaps is hazardous to health. Local regulations concerning personal and environmental protection and compulsory labeling must be observed. In processing nickel oxide and carbonate, protective measures must be taken because of their carcinogenic properties. The TLV–TWA value for nickel is 1 mg/m$^3$.

**Properties.** See Table 7.

**Uses.** Nickel soaps are used as oxidation catalysts in dryers; as oil-soluble hydrogenation catalysts, which liberate finely divided nickel on heating or decomposition; as detergents in petroleum ether; and as additives in lubricating oils for preventing cyclization and resinification.

# 6.  Quality Specifications

Metallic soaps can be characterized by a series of data. Certain application-orientated properties, such as solubility and swelling, may have to be determined in special tests on test formulations.

**Solid Metallic Soaps.** The following parameters are determined:

*Moisture Content.* The sample is weighed before and after heating at 110 °C for 2 h in a drying oven.

*Metal Content.* Metal content is determined by one of the usual analytical methods after acid decomposition of the metallic soap in an aqueous extract.

*Total Ash.* The substance is preheated carefully in an incinerator and then heated slowly to 800 – 1000 °C in a muffle furnace as soon as all the organic substance has decomposed. After being cooled in a desiccator, the ash is determined by back-weighing. For lead soaps, attention must be paid to reduction to metallic lead.

*Washed Ash.* Residue from the determination of total ash is boiled in distilled water and filtered; the residue and the filter paper are incinerated. The final residue is expressed as a percentage of the weight of substance used for the total ash determination. This method is not applicable to water-soluble metal oxides.

*Water-Soluble Salts.* The difference between total ash and washed ash is described as soluble ash or water-soluble salt. For metallic soaps produced by double decomposition, this parameter gives a good indication of whether the salts have been washed sufficiently.

*Free Fatty Acids.* The substance is refluxed with acetone for 10 min and filtered; the fatty acid content is determined by titration with alcoholic potassium hydroxide using thymol blue and cresol red as a mixed indicator.

*Particle Size.* Particle size is determined as residue on a 71 µm air-stream sieve or by wet sieving of an aqueous dispersion with a surfactant.

*Tapped Bulk Density.* Tapped bulk density is determined in a commercial tamp volumeter by using a fixed number of impacts.

*Bulk Density.* The substance is poured loosely into a measuring cylinder up to the 100-mL mark and weighed.

*Melting Range.* Melting range can be determined in a commercial melting point apparatus.

**Metallic Soap Solutions.** For metallic soap solutions (e.g., used as PVC stabilizers or dryers), metal content, density, viscosity, solid content, and Gardner color index must be determined. Industrial supply conditions are to be observed for metallic soap dryers [32].

# 7. References

[1] I. W. H. Oldham, A. R. Ubbelohde, *J. Chem. Soc.* 1941, 368–375.
[2] P. N. Cheremisinoff, *J. Am. Oil. Chem. Soc.* **28** (1951) 278–279.
[3] Colgate-Palmolive-Peet Co., US 2 447 064, 1948.
[4] G. H. Smith et al., *J. Amer. Chem. Soc.* **70** (1948) 1053–1054.
[5] The Commonwealth Eng. Co., US 2 768 996, 1954.
[6] C. I. Boner, *Ind. Eng. Chem.* **29** (1937) 58.
[7] A. S. C. Laurence, *Trans. Faraday Soc.* **34** (1938) 665–669.
[8] E. A. Nikitina et al., *J. Gen. Chem.* **19** (1949) 1108–1114.
[9] G. H. Smith, S. Ross, *Oil Soap (Chicago)* **23** (1946) 77–80.
[10] H. I. Braun, *Chem. Ztg.* **53** (1929) 913–914.
[11] C. W. Hill, C. W. Stoddart, *Am. Soc.* **33** (1911) 1076.
[12] I. Marwendel OHG, DE 845 800, 1950.
[13] P. N. Cheremisinoff, *Am. Paint. J.* **37** (1953) 90–94.
[14] A. Davidsohn, *Ind. Chem.* **26** (1950) 385–394.
[15] Battenfeld Grease and Oil Corp., US 2 753 364, 1951.
[16] Socony Vacuum Oil, US 2 389 873, 1945.
[17] I.G. Farben, GB 335 863, 1929.
[18] National Lead Co., US 2 650 932, 1953.
[19] I.G. Farben, GB 478 587, 1938.
[20] Socony Vacuum Oil, US 2 311 310, 1943.
[21] L. W. Ryan, W. W. Pletscher, *Ind. Eng. Chem.* **26** (1934) 909–910.
[22] I.G. Farben, US 2 095 508, 1937.
[23] R. C. Pink, *Trans. Faraday Soc.* **37** (1941) 181.
[24] Hoesch-Chemie GmbH, US 3 519 571, 1967.
[25] Tenneco Chemicals Inc., DE-AS 2 159 347, 1971.
[26] US 3 779 962, 1970, (G. Koenen, A. Szczepanek).
[27] Supraton F. I. Zucker GmbH, US 4 376 079, 1981.
[28] Mallinckrodt, Inc., EP 0 086 362, 1983.
[29] Tenneco Chemicals Inc., EP-A 0 058 792, 1981.
[30] Alexander Gray, *J. Phys. Colloid Chem.* **53** (1949) 9–39.
[31] Chemische Fabrik Hoesch KG, DE 1 068 237, 1955.
[32] ISO 4619: "Driers for Paints and Varnishes," Jan. 1982.

# Methacrylic Acid and Derivatives

WILLIAM BAUER, JR., Rohm and Haas Co., Philadelphia, Pennsylvania 19477, United States

| | | | | |
|---|---|---|---|---|
| 1. | Introduction ............. 3249 | 3.3. | Methacrylic Acid from Isobutyric Acid .......... | 3258 |
| 2. | Properties .............. 3251 | 3.4. | Methacrylic Acid From | |
| 2.1. | Physical Properties ....... 3251 | | Ethylene ............... | 3260 |
| 2.2. | Chemical Properties....... 3251 | 4. | Quality Specifications....... | 3261 |
| 3. | Production ............. 3254 | 5. | Storage and Transportation .. | 3261 |
| 3.1. | Methacrylic Acid and Methyl Methacrylate from Acetone Cyanohydrin ............ 3254 | 6. | Uses ................... | 3262 |
| | | 7. | Economic Aspects ......... | 3263 |
| | | 8. | Toxicology and Occupational Health.................. | 3263 |
| 3.2. | Methacrylic Acid from Isobutene ............... 3256 | 9. | References............... | 3265 |

# 1. Introduction

Methacrylic acid [*79-41-4*], α-methylacrylic acid, 2-methylpropenoic Acid, $CH_2=C(CH_3)COOH$, $M_r$ 86.09, is a colorless, moderately volatile, corrosive liquid with a strongly acrid odor. It was first prepared in 1865 from ethyl methacrylate, in turn obtained by dehydration of ethyl α-hydroxyisobutyrate [1].

**Historical Aspects.** OTTO RÖHM's doctoral thesis (1901) on polymerization products of acrylic acid derivatives represented the start of interest in acrylates and methacrylates as materials of potential commercial value [2]. RÖHM reported the preparation of colorless, clear, rubbery materials that he thought could be of practical utility. He and OTTO HAAS formed a company to manufacture leather treatment enzymes in 1907. Although laboratory studies of the preparation of acrylates and methacrylates continued, it was not until the 1920s that significant effort was spent on the development of commercial processes for the monomers and derived polymers.

After World War I, RÖHM developed a route for commercial production of acrylates from ethylene cyanohydrin, in turn derived from ethylene chlorohydrin, an intermediate used in the production of gases for chemical warfare. Later, he and WALTER BAUER developed an analogous synthesis of methyl methacrylate using acetone cyanohydrin. By 1931 BAUER succeeded in producing a methyl methacrylate polymer as a clear, transparent solid that softened above 100 °C. This first "organic glass" could be worked with saws, drills, and other common tools, which suggested significant practical applications. Research on forming cast sheet from methyl methacrylate was carried out in the 1930s by both Rohm and Haas and Du Pont in the United States, by Röhm and Haas AG in Germany, and by ICI in England.

Commercial production of methyl methacrylate monomer began in Darmstadt, Germany in 1933 (two years after the commercial production of acrylates) by the original acetone cyanohydrin route. The cyanohydrin was converted to ethyl α-hydroxyisobutyrate by treatment with ethanol and dilute sulfuric acid. This was then dehydrated using $P_2O_5$ [3]–[7]. In 1934 a patent was issued to ICI for the conversion of acetone cyanohydrin to methacrylamide sulfate using concentrated sulfuric acid [8]. The methacrylamide sulfate could be hydrolyzed to methacrylic acid and esterified to form methyl methacrylate [80-62-6]. Röhm then cross-licensed his technology for manufacture of cast acrylic sheet with ICI so that the more efficient ICI process could be used for producing methyl methacrylate monomer. Both ICI and Röhm and Haas AG began commercial production of methyl methacrylate using the methacrylamide sulfate route in 1937. Rohm and Haas in the United States (a branch of the company that had become independent in 1917) was unable to obtain a similar license until the mid-1940s, so they continued to use the original process for several years [2].

The methacrylamide sulfate route provided the basis for commercial production of methacrylic acid and methyl methacrylate until the 1983 introduction of isobutene oxidation processes, developed independently in Japan by Japan Catalytic Chemical and Mitsubishi Rayon. Although additional isobutene-based methacrylic acid capacity is anticipated in Japan, and a new ethylene-based facility is under construction by BASF in Ludwigshafen, most of the world demand for methyl methacrylate ($1.4 \times 10^6$ t in 1988) continues to be met via acetone cyanohydrin.

Most acetone cyanohydrin-based operations lead directly to methyl methacrylate monomer, which is the most important methacrylate derivative. Methacrylic acid produced by other routes also serves as a key intermediate to methyl methacrylate.

Much smaller quantities of ethyl methacrylate [97-62-3], n-butyl methacrylate [97-88-1], and isobutyl methacrylate [97-86-9], as well as higher methacrylates (e.g., lauryl methacrylate [142-90-5]and stearyl methacrylate [32360-05-7]) are manufactured by direct esterification of methacrylic acid or transesterification of methyl methacrylate. In addition to these, 2-hydroxyethyl methacrylate [868-77-9], 2-hydroxypropyl methacrylate [923-26-2], and aminoalkyl methacrylates such as 2-dimethylaminoethyl methacrylate [2867-47-2] are also commercially available. Methacrylamide [79-39-0] is offered as a specialty monomer; it can be obtained from the intermediate methacrylamide sulfate.

# 2. Properties

## 2.1. Physical Properties

The most important physical properties of methacrylic acid and its derivatives are listed in Tables 1 and 2 [3], [9]–[13]. Liquid–liquid equilibrium data for the methanol–water–methacrylic acid–methyl methacrylate system are available [14]–[16], as are key vapor–liquid equilibrium data for methacrylic acid–water [17] and related methacrylic ester systems [18], [19]. Solid–liquid equilibria for methacrylic acid–water have also been published [20], as have the physical properties of aminoalkyl methacrylates [21].

## 2.2. Chemical Properties

Methacrylic acid and its esters are very reactive, displaying reactions typical of both the vinyl function and the carboxylic acid or ester group. The electron-withdrawing effect of the carboxylic acid or ester group polarizes the double bond and enhances its reactivity, but this reactivity is less pronounced than in the acrylates because of the electron-donating methyl group. Thus, nucleophilic Michael and Michael-like additions to the double bond take place only with electron-rich reagents.

**Addition to the Carbon–Carbon Double Bond.** Addition of hydrogen cyanide, hydrogen halides, hydrogen sulfide, mercaptans, alkyl amines, alcohols, phenols, or phosphines leads to β-substituted α-methyl propionates [3], [5], [22], [23].

$$ZH + CH_2=C(CH_3)-COOR \longrightarrow Z-CH_2-CH(CH_3)-COOR$$

**Diels–Alder Reactions.** Diels–Alder reactions occur with dienes such as butadiene and cyclopentadiene [3], [23], [24].

**Reactions of the Carboxylic Acid Function.** Esterification of methacrylic acid with alcohols in the presence of catalytic quantities of sulfuric or sulfonic acids provides the corresponding esters. Acid-catalyzed addition of olefins also gives esters [25].

**Table 1.** Physical properties of methacrylic acid, methyl methacrylate, and butyl methacrylate

| Property | Methacrylic acid | Methyl methacrylate | Butyl methacrylate |
|---|---|---|---|
| $M_r$ | 86.09 | 100.12 | 142.20 |
| mp, °C | 15.8 | −48 | −50 |
| bp (101.3 kPa), °C | 162 | 101 | 163 |
| Density ($d_4^{25}$), g/cm$^3$ | 1.015 | 0.939 | 0.889 |
| Refractive index, $n_D^{25}$ | 1.4288 | 1.4120 | 1.4220 |
| Vapor pressure, kPa | | | |
|   at 20 °C | 0.09 | 3.87 | 0.20 |
|   at 60 °C | 1.33 | 25.2 | 2.0 |
| Viscosity (24 °C), mPa · s | 1.38 | 0.53 | 0.92 |
| Solubility (20 °C), g/kg | | | |
|   in water | complete | 15.9 | 1.0 |
|   water in | complete | 11.5 | ca. 1 |
| $T_{crit}$, °C | 370 | 291 | |
| $P_{crit}$, Pa | 4.70 | 3.68 | |
| $V_{crit}$, cm$^3$/mol | 270 | 323 | |
| Heat of vaporization (101.3 kPa), kJ/mol | 0.418 | 0.36 | |
| Specific heat capacity, J g$^{-1}$ K$^{-1}$ | 2.1 | 1.9 | 1.9 |
| Flash point, °C | | | |
|   Cleveland open cup | 77 | | 66 |
|   Tag closed cup | 67 | | |
|   Tag open cup | | 13 | |
| Autoignition temperature, °C | 400 | 435 | 294 |

**Table 2.** Physical properties of methacrylic acid derivatives

| Compound | mp, °C | bp (101.3 kPa), °C | Refractive index ($n_D^{25}$) | Density ($d_4^{20}$), g/cm$^3$ |
|---|---|---|---|---|
| Methacrolein | −81 | 68 | 1.4144[b] | 0.837 |
| Methacrylonitrile | −36 | 90 | 1.3989 | 0.800 |
| Methacrylamide | 110 | | | |
| Methacrylic anhydride | | 75[a] | 1.4520 | |
| Methacryloyl chloride | | 96–98[a] | 1.4435 | 1.087[c] |

[a] At 0.67 kPa.
[b] At 20 °C.
[c] At 25 °C.

$$CH_2=\underset{\underset{CH_3}{|}}{C}-COOH + CH_3CH=CH_2 \longrightarrow$$

$$CH_2=\underset{\underset{CH_3}{|}}{C}-COO-\underset{\underset{CH_3}{\backslash}}{\overset{\overset{CH_3}{/}}{CH}}$$

Transesterification of methyl methacrylate is another convenient technique for the manufacture of esters. Catalysts include dialkyl tin oxides [26], sodium phenoxide

[27], and zinc chloride [28]. Methanol formed during the reaction is removed as a binary azeotrope with methyl methacrylate.

Reaction of the methacrylic acid group with epoxides permits the synthesis of hydroxyethyl and hydroxypropyl methacrylates. Appropriate catalysts include anion-exchange resins [29], ferric chloride [30], and lithium salts [31].

$$CH_2=\underset{\underset{CH_3}{|}}{C}-COOH \;+\; \underset{\diagdown\!\diagup}{\overset{O}{CH_2-CH_2}} \longrightarrow CH_2=\underset{\underset{CH_3}{|}}{C}-COOCH_2CH_2OH$$

Methacryloyl chloride [*920-46-7*] is synthesized by reaction of the acid with thionyl chloride or phosphorus trichloride. Methacrylic anhydride [*760-93-0*] is prepared by reaction of the acid with acetic anhydride.

Methacrylamide is most conveniently prepared by reaction of acetone cyanohydrin with sulfuric acid [6], [32], [33]. Crude methacrylamide sulfate prepared as described in Section 3.1 is first neutralized with ammonia, after which water is added. The resulting methacrylamide can be separated from the aqueous ammonium sulfate by solvent extraction, filtration, or as a melt. It can be purified further by recrystallization. Alternatively, methacrylamide can be prepared by reaction of methyl methacrylate with ammonia.

Methacrylonitrile [*126-98-7*] is produced by ammoxidation of isobutene. It may be converted to methyl methacrylate by a process similar to the cyanohydrin route.

**Oxidation.** In the absence of free radical polymerization inhibitors, air oxidation of methyl methacrylate leads to a polymeric peroxide that decomposes to formaldehyde and methyl pyruvate.

**Polymerization.** Methacrylic acid, its esters, and other methacrylate derivatives polymerize readily upon heating in the presence of free radical initiators (e.g., peroxides, UV light, or ionizing radiation). Because the usual phenolic inhibitors function by reaction with peroxy radicals rather than with carbon radicals, these do not interfere with initiated polymerization. However, oxygen must be excluded during the process because it converts alkyl radicals to hydroperoxy radicals.

Poly(methyl methacrylate) is unusual among polymers in that its pyrolysis gives high yields of the monomer. Heating scrap or off-grade polymer above 300 °C gives > 80 % recovery of monomer, which can be purified by fractional distillation.

# 3. Production

This chapter is restricted to production methods for methacrylic acid and methyl methacrylate, since the preparation of methacrylic acid derivatives is summarized in Section 2.2. Disregarding the role of $C_1$ molecules (methanol, carbon monoxide, formaldehyde, and hydrogen cyanide), the commercially practicable routes to methacrylic acid and methyl methacrylate utilize ethylene (C-2 route), propene or acetone (C-3 routes), and isobutene (C-4 route) as carbon sources. Economic considerations for the various processes depend upon regional raw material supplies and costs; none of the newer processes is so superior to the acetone cyanohydrin route as to warrant scrapping existing cyanohydrin-based plants. Further, no single alternative process is clearly superior in an economic sense to the others.

## 3.1. Methacrylic Acid and Methyl Methacrylate from Acetone Cyanohydrin

The most common approach to methacrylic acid synthesis is the hydrolysis of methacrylamide sulfate, obtained from acetone cyanohydrin [6], [7], [23]. Methyl methacrylate may be prepared directly in a similar way by adding methanol in the final reaction step.

Dry acetone and hydrogen cyanide react in the presence of a basic catalyst to give the cyanohydrin, which is then reacted with excess concentrated sulfuric acid (1.4–1.8 mol per mole of cyanohydrin) to form methacrylamide acid sulfate:

$$CH_3-\underset{\underset{OH}{|}}{\overset{\overset{CH_3}{|}}{C}}-C\equiv N \xrightarrow{H_2SO_4} CH_3-\underset{\underset{OSO_2OH}{|}}{\overset{\overset{CH_3}{|}}{C}}-CONH_2 \cdot H_2SO_4$$

The sulfuric acid serves both as a specific reactant and as a solvent for the reaction, which appears to involve an α-sulfatoamide intermediate. If insufficient sulfuric acid is used, the reaction mass becomes a slurry or solid that is difficult or impossible to cool and pump. Both the sulfuric acid and the acetone cyanohydrin must be anhydrous in order to minimize hydrolysis of the sulfato derivative to α-hydroxyisobutyramide. The initial reaction is carried out continuously in a series of stirred tank reactors. Good heat transfer is required to assure removal of the heat of reaction. Thorough mixing is also necessary to avoid decomposition of the cyanohydrin to acetone and hydrogen cyanide, which can react to form byproduct acetone sulfonates and formamide sulfate. After initial reaction is completed at 80–110 °C, the mixture is subjected to brief thermal cracking at ca. 125–160 °C to convert most of the α-hydroxyisobutyramide byproduct to methacrylamide sulfate, along with some acetone, carbon monoxide, and water. Total residence time at this stage of the process is about 1 h.

In a second stage, the methacrylamide sulfate stream is either hydrolyzed with excess water to give methacrylic acid and ammonium acid sulfate, or it is treated with aqueous methanol in a combined hydrolysis–esterification step to produce a mixture of methyl methacrylate and methacrylic acid:

$$CH_2=\underset{CH_3}{C}-CONH_2 \cdot H_2SO_4 \xrightarrow[CH_3OH]{H_2O} \begin{array}{l} CH_2=\underset{CH_3}{C}-COOH \\ +NH_4HSO_4 \\ CH_2=\underset{CH_3}{C}-COOCH_3 \end{array}$$

Several modifications are possible in the hydrolysis–esterification step. For example, the methacrylamide sulfate, water, methanol, and recycle streams can be led through a series of continuous reactors at 80–110 °C with a 2–4 h residence time. The reactor effluent then passes to a stripping column where crude methyl methacrylate, water, and excess methanol are removed overhead. The waste acid ammonium sulfate residue can be either treated with ammonia for conversion to fertilizer or burned to regenerate sulfuric acid. Crude methyl methacrylate is extracted with water to recover methanol, which is concentrated and recycled to the esterification reactor. Washed ester is then purified by further distillation. The overall yield based on acetone cyanohydrin is in the range of 80–90%.

In the manufacture of methacrylic acid, methacrylamide sulfate is reacted with water under conditions similar to those used for formation of the ester. The reactor effluent separates into two phases. The upper organic layer is distilled to provide pure methacrylic acid. The lower layer is steam stripped to recover dilute aqueous methacrylic acid, which is recycled to the hydrolysis reactor. The waste acid stream is treated as in the manufacture of the ester.

In an alternative process, shown in Figure 1, the methanolysis reaction is carried out at a pressure of ca. 800 kPa in one or more reactors (d) operated at 100–150 °C. The reactor effluent is separated into two layers (e) while still under pressure. The lower layer is steam stripped (g) to recover methacrylic acid for recycle to the esterification reactor (d), and acid waste is treated as described previously. The upper layer is passed to a distillation column (f) where light ends (primarily dimethyl ether) are removed. The bottoms from this column are washed with aqueous ammonia (h) to recover methanol and methacrylic acid for recycle to the esterification reactor. The crude, washed methyl methacrylate is then dehydrated in a downstream distillation column (i) and distilled in a product column (j) to provide pure methyl methacrylate.

A sulfuric acid regeneration plant is usually operated in conjunction with the methyl methacrylate plant, because approximately 1.6 kg of sulfuric acid is required to produce each kilogram of methyl methacrylate. The presence of a regeneration facility avoids the

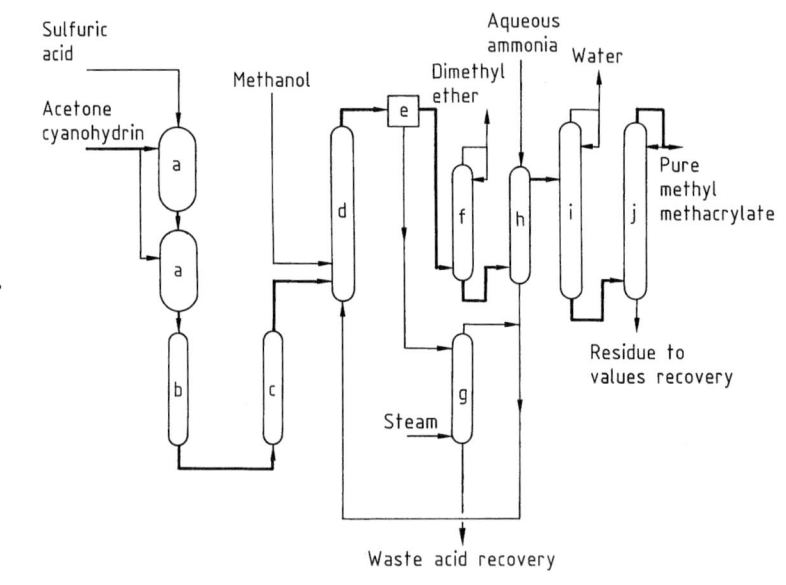

**Figure 1.** Production of methyl methacrylate from acetone cyanohydrin
a) Hydrolysis reactors; b) Cracker; c) Cooler; d) Esterification reactor; e) Separator; f) Flash column; g) Acid stripper; h) Wash column; i) Dehydration distillation column; j) Product distillation column.

need to dispose of large quantities of ammonium sulfate contaminated with organic material.

One of the driving forces for development of the alternative routes discussed in Sections 3.2, 3.3, 3.4 has been the desire to eliminate the need for sulfuric acid regeneration. An additional concern is the hazard associated with transporting hydrogen cyanide, which is not always generated at the methacrylate plant site. On the other hand, both acetone (from phenol manufacture) and hydrogen cyanide (from acrylonitrile manufacture) have the economic advantage of being themselves industrial byproducts.

Asahi Chemical manufactures methyl methacrylate by the sulfuric acid hydrolysis of methacrylonitrile, using a plant originally designed for the acetone cyanohydrin process. The requisite methacrylonitrile is obtained by ammoxidation of isobutene, avoiding the need for hydrogen cyanide.

## 3.2. Methacrylic Acid from Isobutene

In recent years many companies have investigated the manufacture of methacrylic acid by two-stage catalytic oxidation of isobutene or *tert*-butanol. Nihon Methacryl Monomer (a joint venture of Sumitomo and Nippon Shokubai) and Mitsubishi Rayon have both constructed commercial plants using this technology.

In the first stage of the process [4], isobutene is oxidized to methacrolein, and in a second stage the methacrolein is oxidized to methacrylic acid:

$$CH_2=C(CH_3)-CH_3 + O_2 \longrightarrow CH_2=C(CH_3)-CHO + H_2O$$

$$CH_2=C(CH_3)-CHO + 1/2\,O_2 \longrightarrow CH_2=C(CH_3)-COOH$$

A published account of process and catalyst developments [34], [35] contrasts methacrylic acid production from isobutene with a similar process for preparing acrylic acid from propene. Selectivity of the second-stage catalysts is best at modest conversions (65–85%). In the Sumitomo–Nippon Shokubai process shown in Figure 2, the first-stage reactor is operated at high conversion, and its effluent passes directly to the second oxidation reactor [36]. Conversion in the second stage is kept low in order to ensure good catalyst selectivity and increase catalyst life. Unreacted methacrolein from the second-stage reactor effluent is separated and recycled. The overall yield of methacrylic acid from isobutene is about 65–70%. Typical catalysts for the first stage are multicomponent metal oxides containing bismuth, molybdenum, and several other metals to promote activity and modify selectivity [37]–[41]. Second-stage catalysts are based on phosphomolybdic acid, but they usually contain an alkali metal to control the acidity. Other elements such as copper and vanadium may also be present [42]–[46].

Reactor effluent from the second-stage oxidation passes to a quencher (b) where crude aqueous methacrylic acid is obtained. The gaseous effluent from the quencher is passed to an absorber (c) where the unreacted methacrolein is absorbed, usually in aqueous carboxylic acid. Absorber off-gases are sent to a combustion unit (d) before being discharged to the atmosphere. A portion of the incinerated gases may be recycled to the first-stage reactor to provide inert gas diluent for the feed. The methacrolein absorbate is transferred to a methacrolein recovery tower (e) from which methacrolein is recycled to the second-stage oxidation reactor; recovered absorbent solution is returned to the absorber.

The crude, aqueous methacrylic acid is sent to a solvent extraction unit (f) for methacrylic acid recovery. Next, a solvent recovery/dehydration tower (h) affords crude methacrylic acid as a bottoms product. The overhead organic solvent layer is recycled to the extraction step, whereas the overhead aqueous layer is combined with extractor raffinate and sent to a solvent stripping tower (g) before being subjected to wastewater treatment.

Dry, crude methacrylic acid from the extract stripper can be further purified if methacrylic acid is the desired end product, or it can be sent directly to an esterification reactor (i) where catalyst and methanol are added for conversion to methyl methacrylate. Crude ester is extracted with water (j) to recover excess methanol, which is removed by distillation (k) and recycled to the esterification reactor. The washed, crude ester is sent to a light ends stripper (l) for removal of light byproducts (e.g., methyl acetate) and then distilled (m) to provide pure methyl methacrylate. The bottoms from

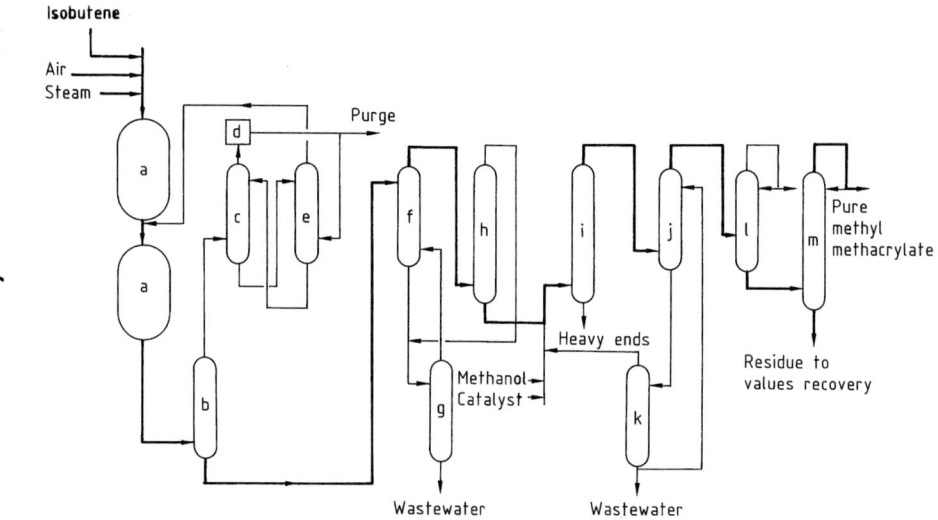

**Figure 2.** Production of methyl methacrylate from isobutene
a) Oxidation reactors; b) Quencher; c) Absorber; d) Combustion unit; e) Methacrolein recovery tower; f) Solvent extraction unit; g) Solvent stripping tower; h) Solvent recovery/dehydration tower; i) Esterification reactor; j) Extractor; k) Distillation unit; l) Light ends stripper; m) Product distillation column.

the final distillation column are recycled to the first extraction step with the exception of a small bleed for removal of inhibitor residues.

Plants based on the C-4 route were introduced in Japan by Nippon Shokubai in 1982 and by Mitsubishi Rayon in 1983 [35]. A joint venture plant operated by Sumitomo and Nippon Shokubai came on stream in 1984. Mitsui Toatsu and Kyowa Gas have also announced plans to construct a plant of this type. Several firms, including Nippon Kayaku, Mitsui Toatsu, Rohm and Haas, and Oxirane (a joint venture of Halcon SD and ARCO) have carried out extensive research on the isobutene process. In 1987 ARCO acquired exclusive worldwide rights to Halcon SD technology using C4-feedstocks [47].

Halcon has proposed a variation of this process that commences with dehydrogenation of isobutane [48], [49].

## 3.3. Methacrylic Acid from Isobutyric Acid

Acid-catalyzed carbonylation of propene to isobutyric acid, followed by oxidative dehydrogenation, presents still another route to methacrylic acid. In this case the starting material is propene itself rather than the oxydized derivative acetone, as in the acetone cyanohydrin route. Although the propene – isobutyric acid – methacrylic acid route is not currently in commercial use, several major methyl methacrylate manufacturers have research efforts aimed at its commercialization.

In the first stage of the process, propene, carbon monoxide, and water are reacted in the presence of a strong acid catalyst to produce isobutyric acid. Sulfuric acid, hydrogen fluoride, and boron fluoride have all been reported to be effective catalysts [50]–[53]. Patents to Röhm [54] and Ashland Oil [55] describe variations that involve preliminary preparation of isobutyroyl fluoride, which is then hydrolyzed to isobutyric acid. Alternatively, isobutyric acid may be synthesized directly by including carefully controlled amounts of water in the carbonylation step:

$$\begin{matrix} CH_3 \\ \diagdown \\ CH \\ \diagup \\ CH_2 \end{matrix} + CO + H_2O \longrightarrow CH_3-\underset{\underset{H}{|}}{\overset{\overset{CH_3}{|}}{C}}-COOH$$

$$CH_3-\underset{\underset{H}{|}}{\overset{\overset{CH_3}{|}}{C}}-COOH + 1/2\,O_2 \rightarrow CH_2=\underset{\underset{CH_3}{|}}{C}-COOH + H_2O$$

Hydrogen fluoride (which acts as both solvent and catalyst), carbon monoxide, and propene (14–40:1.5:1) are reacted in the presence of a slight stoichiometric deficiency of water relative to propene to generate isobutyric acid. Reaction conditions are included in the patent literature; these range from about 30 °C at 20 MPa to 120 °C at 14 MPa. Depending upon the temperature and pressure, residence time during the carbonylation step varies from about 5 to 30 min. Reactor effluent is passed to staged flash tanks, in the first of which excess carbon monoxide can be recovered for recycle to the carbonylation reactor. The second tank permits removal of inert gases, which can be passed to the atmosphere after scrubbing with caustic solution to remove any hydrogen fluoride or isopropyl fluoride. The bulk of the hydrogen fluoride is separated overhead in a distillation tower for recycle to the carbonylation reactor. The bottoms from this tower pass to a hydrolysis stripper where any remaining fluorine-containing materials are reacted with water; hydrogen fluoride is stripped off for recycle. A final distillation step provides isobutyric acid overhead for feed to the second part of the process. Bottoms from this distillation contain small amounts of $C_7$ and $C_{10}$ carboxylic acids resulting from multiple condensations of propene prior to reaction with carbon monoxide. The overall selectivity with respect to propene is reported to be 95–97%. The preceding steps must be carried out in a facility carefully designed to prevent fugitive fluoride emissions.

In the second stage of the process, isobutyric acid, steam, and air are passed over a fixed-bed catalyst [56]–[60] in a multitubular reactor, causing oxidative dehydrogenation to methacrylic acid. The reactor effluent is sent to a quench tower from which an aqueous methacrylic acid stream is obtained. This part of the process is similar to the methacrolein oxidation step in the C-4 process described in Section 3.2. In addition to carbon monoxide and carbon dioxide, which are incinerated with other noncondensable gases, the crude methacrylic acid stream contains acetone and acetic acid as byproducts. If desired, and with proper choice of quench tower conditions, the acetone can be directed to the incinerator along with the noncondensables.

If methacrylic acid is to be isolated, the crude product may be extracted into a solvent and dehydrated in a distillation tower. Isobutyric and acetic acids are then separated by distillation as light ends prior to final distillation of the methacrylic acid. The distillative separation of isobutyric acid from methacrylic acid is very difficult; normal boiling points of the two materials are 155 and 162 °C, respectively.

Catalysts for the oxidative dehydrogenation of isobutyric acid to methacrylic acid are of two general types. The first series, often referred to as Mo–P–V mixed oxide catalysts, were developed by Mitsubishi Chemical Industries [57], Röhm [58], [59], and others, and they are similar in composition to catalysts used in the oxidation of methacrolein. Most are phosphomolybdic acid derivatives, usually with some replacement of molybdenum by vanadium or tungsten. The better catalysts frequently contain at least some copper, and they are partially neutralized by cesium, rubidium, or potassium. Some catalysts are reported to achieve conversions of 99.8% with selectivities above 74% [59].

A second type of catalyst has been intensively studied by Ashland Oil [61]. These include iron phosphate materials and give selectivities of about 84–85% at conversions in the 85–95% range. Such catalysts are used at about 400 °C, in contrast to ca. 300 °C for the phosphomolybdate catalysts. Iron phosphate catalysts require high levels of steam in the reactor feed for optimum selectivity and life.

According to patents issued to Röhm [62], crude reactor product from the oxidative dehydrogenation step can be sent directly to the esterification reactor, which constitutes the starting point for the third section of the process. In this case, the product methyl methacrylate must be separated from methyl acetate and methyl isobutyrate. Separation of methyl isobutyrate from methyl methacrylate is difficult because its boiling point (92 °C) is close to that of methyl methacrylate (101 °C). The esterification step is similar to that described in Section 3.2 for the C-4 process.

Norsolor has an exclusive European license to the Ashland Oil technology, which Norsolor has further developed in a pilot plant at St. Avold (France) [63]. Röhm is developing its own process for this route in a pilot plant at Darmstadt [64].

An alternative process involves the hydroformylation of propene to give isobutyraldehyde, followed by oxidation of the aldehyde to isobutyric acid [65].

## 3.4. Methacrylic Acid From Ethylene

Other routes to methacrylic acid include the condensation of formaldehyde with propionic acid to generate methacrylic acid and the condensation of formaldehyde with propanal to give methacrolein [66]–[69].

BASF has developed such a process based on ethylene, synthesis gas, and formaldehyde. A plant with a production capacity of 36 000 t/a is expected to come on stream at Ludwigshafen in 1989–1990 [70].

Ethylene is first hydroformylated to give propanal, which is then condensed with formaldehyde to produce methacrolein. Catalytic airoxidation of methacrolein to methacrylic acid completes the synthesis, a step that is common to the C-2 and C-4 routes:

$$CH_2=CH_2 + CO + H_2 \longrightarrow CH_3CH_2CHO$$

$$CH_3CH_2CHO + HCHO \longrightarrow CH_2=\underset{\underset{CH_3}{|}}{C}CHO + H_2O$$

An alternative process would entail oxidation of propanal to propionic acid, condensation of which with formaldehyde would give methacrylic acid directly.

# 4. Quality Specifications

Glacial methacrylic acid is typically available in a purity exceeding 99.4% (analysis by GC and HPLC). Water (ca. 0.1–0.2 wt%) is the primary contaminant. Traces of α-hydroxyisobutyrate may also be present. The monomethyl ether of hydroquinone (MEHQ) at 100–250 ppm is used as an inhibitor during storage and transportation. Commercial methyl methacrylate is typically 99.9% pure, with traces of acidity (< 0.003%) and water (< 0.05%). The storage and transportation inhibitor for the methyl ester is generally 10–50 ppm of MEHQ or 25–60 ppm of hydroquinine, though other phenolic inhibitors can also be used. Phenothiazine is an excellent inhibitor in this case because it has both anaerobic and aerobic activity. Nevertheless, it is not commonly used in products intended as polymer intermediates.

# 5. Storage and Transportation

The heat of polymerization of methacrylic acid is −56.5 kJ/mol; corresponding values for the methyl, ethyl, and n-butyl esters are −57.7, −57.7, and −56.5 kJ/mol, respectively. These highly exothermic reactions may occur with violence if the monomers are not properly inhibited against the formation of free radicals. Appropriate inhibitor levels have been established with great care by the manufacturers, resulting in excellent stability to polymerization under normal conditions of storage and transportation. Once stabilized, methacrylic acid and its derivatives can be handled as flammable liquids. The flammability limits of methyl methacrylate in air at 20 °C are 2.12–12.5 vol%.

The effectiveness of typical phenolic inhibitors depends upon the presence of oxygen. For this reason, methacrylic monomers should always be stored in contact with a vapor space whose oxygen concentration is 5–21% (dry air) in order to maintain a sufficient level of dissolved oxygen [11].

Like many other unsaturated organic materials, methacrylic monomers react slowly with oxygen to give peroxides, hydroperoxides, and other compounds. Formation of such materials is greatly retarded by polymerization inhibitors and accelerated by increased temperature, but deterioration of product quality and stability must be anticipated upon extended storage (several months to a year). Storage temperatures should generally be kept below 25 °C.

Methacrylic acid has a relatively high freezing point (15.8 °C), and inhibitors tend to partition strongly into the liquid phase during freezing. Hence, thawing of frozen methacrylic acid can give initial melts that are significantly depleted in inhibitor. If such a liquid is heated, thermal initiation can give rise to violent polymerization. Care should thus be taken to avoid the possibility of freezing. If freezing takes place, thawing should preferably be conducted at 25 °C, and never above 40 °C. Furthermore, material should never be removed until melting is complete and the liquid has been thoroughly mixed to restore uniform inhibitor levels.

The preferred construction material for methacrylic acid storage vessels is type 316 stainless steel. Although less corrosion-resistant materials are suitable for methacrylate esters, stainless steel should be considered for these as well, because rust or iron salts can induce free radical decomposition of the peroxides that accumulate on long storage.

**Transport Regulations.** Methacrylic acid is classified as a flammable and corrosive liquid. Transport is governed by the IMDG-Code no. D 8257, class B; UN-no. 2531; RID/ADR class 8, number 31 c; CFR 49:172.102 corrosive material.

# 6. Uses

Methacrylic acid and methacrylate esters are used to prepare a wide range of polymers. Poly(methyl methacrylate) is the primary polymer in this category, and it provides water-clear, tough plastics that are used in sheet form in glazing, signs, displays, and lighting panels. Automotive lighting lenses and similar products can be prepared from molding pellets. Methyl methacrylate incorporated into copolymers forms the basis for durable coatings and inks. Higher methacrylate polymers are useful in the manufacture of oil additives, solventless inks and coatings, and binders for xerography. Salts of poly(methacrylic acid) can serve as the basis for water-soluble thickeners and detergent additives.

# 7. Economic Aspects

The cost of methyl methacrylate varied widely among the three major economic regions of the world in 1988. Japanese prices in early 1988 approached $ 1.76/kg (80 cents/lb), the price in Europe was ca. $ 1.43 – 1.48/kg (65 – 67 cents/lb), while the United States price was $ 1.17 – 1.23/kg (53 – 56 cents/lb). Current world capacity is of the order of $1.4 \times 10^6$ t/a. The principal manufacturers and their existing capacities are listed in Table 3 [71], [72]. Projected capacities after completion of currently announced expansions and probable retirements of facilities are also given. The normal growth rate in demand is 3 – 3.5 %/a. Announced expansions by BASF in Germany, by Rohm and Haas and Du Pont in the United States, and by Sumitomo – NSKK and Mitsui Toatsu – Kyowa Gas Chemical in Japan are expected to satisfy demand into the 1990s.

# 8. Toxicology and Occupational Health

Acrylic monomers have been the subject of animal toxicity studies covering a variety of exposure routes. Larger esters are found to be less well-absorbed and less toxic than smaller esters. Generally speaking, methacrylates are less toxic than acrylates. Relevant data are provided in Table 4 [11], [73], [74].

With respect to acute toxicity, single exposures to acrylic monomers by the oral or dermal route appear to be slightly to moderately dangerous. This conclusion is based on lethality studies with rats and rabbits. The mucous membranes of the eyes, nose, and throat are particularly sensitive to irritation. Acrylic monomers can produce eye and skin irritation ranging from slight to corrosive depending upon the monomer.

Full eye protection should always be worn during handling of methacrylate monomers. Prolonged exposure to either liquid or vapor can result in permanent eye damage and possibly blindness. Overexposure to vapors causes nose and throat irritation, as well as dizziness or drowsiness (solvent narcosis) and possibly central nervous system depression. Thus, respiratory protection and proper ventilation are required in cases of exposure to high vapor concentrations.

Swallowing methacrylate monomers may lead to severe irritation of the mouth, throat, esophagus, and stomach, causing discomfort, vomiting, diarrhea, dizziness, and possibly collapse. Direct contact with the monomers causes skin redness and irritation ranging from slight to severe, perhaps including corrosion. Methacrylic acid is more corrosive than its esters. Because the monomers irritate and may be absorbed through the skin, gloves and protective clothing are required when handling these materials.

Repeated exposure to methacrylate monomers can produce allergic dermatitis (or skin sensitization), which results in rash, itching, or swelling. After one or more

Table 3. Worldwide methyl methacrylate capacity

| Company | Location | Capacity, $10^3$ t/a | | Process * |
|---|---|---|---|---|
| | | Current | Projected | |
| **United States** | | | | |
| CyRo | Fortier (Louisiana) | 80 | 92 | ACN |
| Du Pont | Memphis (Tennessee) | 134 | 191 | ACN |
| Rohm and Haas | Deer Park (Texas) | 300 | 360 | ACN |
| Total | | 514 | 643 | |
| **Western Europe** | | | | |
| Norsolor | St. Avold (France) | 60 | 60 | ACN |
| Röhm | Worms (FRG) | 115 | 115 | ACN |
| Degussa | Wesseling (FRG) | 45 | 45 | ACN |
| BASF | Ludwigshafen (FRG) | | 36 | C-2 |
| Vedril (Montedison) | Rho (Italy) | 50 | 50 | ACN |
| ICI | Billingham (UK) | 105 | 105 | ACN |
| Paular | Tarragona (Spain) | 30 | 30 | ACN |
| Monacril (Röhm) | Palos-de-la-Fronters (Spain) | 20 | 20 | ACN |
| Total | | 405 | 441 | |
| **Japan** | | | | |
| Mitsubishi Rayon | Ohtake | 75 | 75 | ACN |
| | | 95 | 95 | C-4 |
| Japan Methacryl Monomer (Sumitomo/NSKK) | Niihama | 40 | 80 | C-4 |
| Asahi Chemical | Kawasaki | 60 | 60 | MAN |
| Mitsui Toatsu | Mobara | 14 | 0 | ACN |
| Kyowa Gas Chemical | Nakajo | 42 | 0 | ACN |
| Mitsui Toatsu/Kyowa Gas Chemical | | | 40 | C-4 |
| Total | | 326 | 350 | |
| Other countries | | ca. 150 | ca. 150 | |

* ACN = acetone cyanohydrin; MAN = methacrylonitrile.

exposures to a given monomer, allergic responses sometimes are produced by single exposures to the same or other acrylic monomers.

Repeated exposure of animals to extremely high (near lethal) concentrations of monomer vapor has produced inflammation of the respiratory tract and degenerative changes in the liver, kidneys, and heart muscle. Liver and kidney changes have also been observed during prolonged oral exposure. These effects were the result of concentrations far above the threshold levels known to cause irritation. No such effects have been reported for humans, who presumably would leave a highly contaminated area rather than tolerate the irritation.

Current TLV – TWA and OSHA PEL values appear in Material Safety Data Sheets provided by methacrylate manufacturers upon request. Values for 1989 are also listed in Table 4 [73], [74].

Protection required for the safe handling of methacrylic acid and its esters includes chemical-resistant gloves and clothing, splash-proof goggles, and good ventilation of the workplace. Contact with these materials should be followed by flushing with copious amounts of water. Medical attention should be sought if symptoms appear.

**Table 4.** Toxicity data for methacrylic monomers

|  | Methyl methacrylate | Butyl methacrylates | Methyl acrylic acid |
|---|---|---|---|
| Oral $LD_{50}$ (rat), g/kg | 7.9–9.4 | 20.3 | 1–2 |
| Dermal $LD_{50}$ (rabbit), g/kg | >9.4 | 10.2 | >2 |
| Skin irritation (rabbit) | slight to moderate | moderate | corrosive |
| Eye irritation (rabbit) | slight to moderate | slight | corrosive |
| Inhalation $LC_{50}$ (rat), ppm | 7093 | 1720 | >1300 |
| OSHA PEL, ppm [74] |  |  | 20 |
| TLV–TWA (1989–1990) [73], | | | |
| mg/m³ | 410* |  | 70 |
| ppm | 100* |  | 20 |
| Odor threshold, | | | |
| ppm | 0.083 | 0.016 | 10 |

* These also represent the MAK values.

# 9. References

[1] E. Frankland, B. F. Duppa, *Ann. Chem. Pharm.* **136** (1865) 1–31.
[2] S. Hochheiser: *Rohm and Haas,* University of Pennsylvania Press, Philadelphia 1986.
[3] E. H. Riddle: *Monomeric Acrylic Esters,* Reinhold Publ. Corp., New York 1954.
[4] J. W. Nemec, W. Bauer, Jr.: "Acrylic and Methacrylic Acid Polymers," in *Encyclopedia of Polymer Science and Engineering,* 2nd ed., vol. **1**, J. Wiley and Sons, New York 1985, pp. 211–234.
[5] *Ullmann,* 4th ed., **16,** 609–614.
[6] H. Rauch-Puntigam, T. Volker: *Acryl- und Methacrylverbindungen,* Springer Verlag, Berlin 1967.
[7] M. Salkind, E. H. Riddle, R. W. Keefer, *Ind. Eng. Chem.* **51** (1959) 1232–1238.
[8] ICI, GB 405 699, 1934 (J. W. C. Crawford); *Chem. Abstr.* **28** (1934) 4745–9.
[9] L. S. Luskin: "Acrylic Acid, Methacrylic Acid and Related Esters," in E. C. Leonard (ed.): *High Polymers, Vinyl and Diene Monomers,* vol. 24, part 1, Wiley Interscience, New York 1970.
[10] L. S. Luskin: "Acrylic and Methacrylic Acids and Esters," in F. D. Snell, C. L. Hilton (eds.): *Encyclopedia of Industrial Chemical Analysis,* vol. **4,** J. Wiley and Sons, New York 1967, pp. 181–218.
[11] *Storage and Handling of Acrylic and Methacrylic Esters and Acids,* Bulletin 84C7, Rohm and Haas Co., Philadelphia 1987.
[12] *Specifications and Typical Properties of Acrylic and Methacrylic Monomers,* Bulletin 77S2, Rohm and Haas Co., Philadelphia 1989.
[13] *Glacial Methacrylic Acid and Glacial Acrylic Acid,* Publication CM41A, Rohm and Haas Co., Philadelphia 1975.
[14] J. Kooi, *Rec. Trav. Chim. Pays Bas* **68** (1949) 34–42.
[15] A. F. Frolov, M. A. Longinova, B. F. Ustavshchikov, *Zh. Obshch. Khim.* **36** (1966) 180–184.
[16] A. F. Frolov, *Russ. J. Phys. Chem. (Engl. Transl.)* **39** (1965) 1538–1541.
[17] A. F. Frolov, M. A. Longinova, A. P. Saprykina, A. B. Kundakova, *Zh. Fiz. Khim.* **36** (1962) 3282–3283.
[18] E. A. Frolova, B. F. Ustavshchikov, S. Yu. Pavlov, *Zh. Fiz. Khim.* **48** (1974) 1865.
[19] G. A. Chubarov, S. M. Danov, R. V. Efremov, *Zh. Fiz. Khim.* **48** (1974) 1047–1048.

[20] G. A. Chubarov, S. M. Danov, G. V. Brovkina, *Zh. Prikl. Khim. (Leningrad)* **51** (1978) 1899–1900.
[21] L. S. Luskin: "Basic Monomers: Vinyl Pyridines and Aminoalkyl Acrylates and Methacrylates," in R. H. Yocum, E. B. Nyquist (eds.): *Functional Monomers, their Preparation, Polymerization and Application*, vol. 2, Marcel Dekker, New York 1974, pp. 663–734.
[22] I. N. Azerbaev, B. M. Butin, Y. G. Bosyakov, *Zh. Obshch. Khim.* **45** (1975) 1730–1734.
[23] *Kirk-Othmer*, 3rd ed., **15**, 346–376.
[24] Kohjin, DE-OS 2 217 623, 1972; US 3 813 438, 1974 (A. Oshima, K. Tsuboshima, N. Takahashi).
[25] Rohm and Haas, US 3 087 962, 1963 (N. M. Bortnick).
[26] Nitto Chemical Industry, DE-OS 2 752 109, 1978; US 4 301 297, 1981 (Y. Kametani, Y. Ino).
[27] American Cyanamid, US 4 059 617, 1977 (T. Foster, T. S. Dawson).
[28] Japan Catalytic Chemical Industry, DE-OS 2 311 007, 1977; US 3 872 161, 1975 (S. Fukuchi, N. Shimizu, T. Ohara).
[29] Distillers, GB 1 120 301, 1968 (E. J. Percy, J. A. Wickings); *Chem. Abstr.* **69** (1968) 58 829 t.
[30] Japan Gas-Chemical, DE-OS 2 027 444, 1970 (M. Murayama, K. Abe); *Chem. Abstr.* **74** (1971) 63 938 k.
[31] Japan Oil and Fats, JP 72 51 328, 1972 (Y. Tanizaki, Y. Kubo); *Chem. Abstr.* **79** (1973) 19 377 f.
[32] ICI, US 2 140 469, 1938 (J. W. C. Crawford, J. McGrath).
[33] BASF, US 3 002 023, 1961; GB 873 603, 1961 (H. Fikentscher, H. Wilhelm).
[34] S. Nakamura, H. Ichihashi in T. Seiyama, K. Tanabe (eds.): "Studies in Surface Science and Catalysis," *Proceedings 7th International Congress on Catalysis*, Tokyo, 30 June–4 July 1980, Elsevier, New York 1981, pp. 755–767.
[35] T. Nakamura, T. Kita, *CEER Chem. Econ. Eng. Rev.* **15** (1983) 23–27.
[36] Nippon Shokubai Kagaku Kogyo Co. and Sumitomo Chemical Co., *Hydrocarbon Process.* **64** (1985) 143.
[37] Nippon Kayaku KK, US 4 012 449, 1977 (Y. Shikakura, F. Sakai, H. Shimizu).
[38] Ube Industries, US 4 171 328, 1979 (S. Umemura, K. Ohdan, K. Suzuki, T. Hisayuki).
[39] Mitsubishi Rayon Co., DE-OS 2 427 670, 1975; US 3 972 920, 1976 (H. Ishii, H. Matsuzawa, M. Kobayaashi, K. Yamada).
[40] Rohm and Haas, US 4 306 090, 1981; US 4 280 928, 1981 (L. S. Kirch, W. J. Kennelly).
[41] Rohm and Haas, US 4 358 622, 1982 (J. W. Nemec, M. S. Cholod).
[42] Sohio, US 4 136 110, 1979 (J. F. White, J. R. Rege).
[43] Ube Industries, US 4 364 844, 1982 (S. Umemura et al.).
[44] Sohio, US 4 301 031, 1981 (W. G. Shaw, P. L. Kuch, C. Paparizos).
[45] The Halcon SD Group, US 4 374 757, 4 374 759, 1983 (S. Khoobiar).
[46] Nippon Kayaku, US 4 172 051, 1979; US 4 273 676, 1981; US 4 467 113, 1984 (M. Matsumoto, A. Sudo, H. Sugi).
[47] *Chem. Week* (1987) Aug. 26, 41.
[48] R. V. Porcelli, B. Juran, *Hydrocarbon Process.* **65** (1986) 37–43.
[49] The Halcon SD Group, US 4 413 147, 1983; US 4 532 365, 1985; US 4 535 188, 1985 (S. Khoobiar).
[50] Sinclair Refining, US 2 975 199, 1961 (B. S. Friedman, S. M. Cotton).
[51] Chem. Systems, US 4 256 914, 1981 (J. A. Jung, J. Peress).
[52] Ashland Oil, US 4 303 594, 1981 (R. Norton et al.); US 4 495 110, 1985 (J. E. Corn).
[53] Röhm, DE-OS 3 033 655, 1982; US 4 504 675, 1985 (S. Besecke, G. Schroeder, H. Siegert).
[54] Röhm, DE-OS 3 213 395, 1983; US 4 452 999, 1984 (S. Besecke, G. Schroeder, H. Siegert).
[55] Ashland Oil, US 4 270 983, 1981 (B. C. Trevedi, D. Grote, T. O. Mason).

[56] Ashland Oil, US 4 299 980, 1981 (C. Daniel, P. Brusky).
[57] Mitsubishi Chemical Industries, US 4 061 673, 1977 (T. Onoda, M. Otake).
[58] Röhm, DE-OS 3 019 731, 1981 (H. Siegert).
[59] Röhm, US 4 370 490, 1983 (W. Gruber, G. Schroeder).
[60] M. Akimoto, Y. Tsuchida, K. Sato, E. Echigoya, *J. Catal.* **72** (1981) 83–94; M. Akimoto, K. Shima, H. Ikeda, E. Echigoya, *J. Catal.* **86** (1984) 173–186.
[61] Ashland Oil, US 4 410 728, 1983; US 4 439 621, 1984 (C. Daniel).
[62] Röhm, DE-OS 3 146 191, 1983 (W. Gaenzler, H. Siegert, H. J. Hohage, G. Schroeder); Chem. Abstr. 99 (1983) 38 912 f.
[63] *Chem. Week* (1985) Oct. 23, 10–11.
[64] *Chem. Ind. Int. (Engl. Transl.)* (1986) no. 1, 4.
[65] M. Otake, T. Onoda in T. Seiyama, K. Tanabe (eds.): "Studies in Surface Science and Catalysis," *Proceedings 7th International Congress on Catalysis, Tokyo 1980*, Elsevier, New York 1981, pp. 780–791.
[66] Röhm, US 4 147 718, 1979 (W. Gaenzler, K. Kabs, G. Schroeder).
[67] Rohm and Haas, US 3 933 888, 1976 (F. Schlaefer).
[68] Sohio, US 4 339 598, 1982 (A. T. Guttmann, R. K. Grasselli).
[69] Rohm and Haas, US 4 490 476, 1984 (R. J. Piccolini, M. J. Smith).
[70] S. J. Ainsworth, D. Hunter, S. Ushio, E. Johnson, *Chem. Week* **144** (1988) 64–65.
[71] *Chem. Market. Rep.* **233** (1988) no. 19, 3.
[72] A. S. Bakshi, *Oil Gas J.* **83** (1985) 99–101.
[73] *Threshold Limit Values and Biological Exposure Indices for 1989–1990*, American Conference of Governmental Industrial Hygienists, Cincinnati, Ohio 1989.
[74] *Fed. Regist.* **54** (1989) no. 12, 2631.

# Methane

EDMOND-JACQUES NYNS, Unité de Génie Biologique, Université Catholique de Louvain, Louvain-la-Neuve, Belgium

| | | | | |
|---|---|---|---|---|
| 1. | Introduction ............ 3269 | 7. | Environmental Impact of Biomethanation .......... 3281 |
| 2. | Physical and Chemical Properties of Biogas ....... 3271 | 8. | Storage, Pretreatment, Upgrading and Uses of Biogas 3282 |
| 3. | Biological Methanogenesis ... 3272 | 9. | Analysis ................ 3283 |
| 4. | Technological Biomethanation 3273 | 10. | Biogas in Developing Countries ............... 3283 |
| 5. | Feedstocks for Biomethanation .......... 3277 | 11. | Economic Aspects ........ 3284 |
| 6. | Process and Environmental Parameters for Stable, Efficient Biomethanation ... 3278 | 12. | References.............. 3285 |

## 1. Introduction

Methane [74-82-8], $CH_4$, $M_r$ 16.03, $mp$ −182 °C, $bp$ −162 °C, is a colorless gas that burns with a blue flame. Some physical properties of methane are as follows:

| | |
|---|---|
| Specific heat capacity at −100 °C | 5.34 J g$^{-1}$ K$^{-1}$ |
| Density at −170 °C | 0.4362 g/cm$^3$ |
| Viscosity at −170 °C | 0.142 mPa · s |
| Surface tension at −170 °C | 15.8 mN/m |
| Latent heat of fusion | 0.942 kJ/mol |
| Latent heat of evaporation at $Kp_{101.3 kPa}$ | 8.185 kJ/mol ≈ 0.510 kJ/g |
| Minimum calorific value | 50.0 kJ/g |
| Heat of formation | −74.897 kJ/mol |
| Entropy | 186.31 J K$^{-1}$ mol$^{-1}$ |
| Free energy of formation | 50.83 kJ/mol |
| Critical values: | |
|    Temperature | −82.5 °C |
|    Pressure | 4.67 MPa |
|    Density | 0.162 g/cm$^3$ |
| Minimum auto-ignition temperature (MAIT) (VDE 0173) | 535 °C |
| MAIT varies according to method of determination | between 601 and 748 °C |
| Flame velocity | 43.4 cm/s |

| | |
|---|---|
| Flame velocity varies according to method of determination | between 34 and 44 cm/s |
| Explosion limits in air at 20 °C and 0.1 MPa | |
| lower | 5.0 vol% |
| upper | 15.0 vol% |
| Compressibility factor | 0.299 |
| Solubility in water | |
| at 20 °C, 101.3 kPa $CH_4$ | 33.8 mL/L |
| at 35 °C, 101.3 kPa $CH_4$ | 25.4 mL/L |
| Henry's constant in water | |
| at 20 °C | $1.45 \times 10^{-3}$ mol/L atm |
| at 35 °C | $1.005 \times 10^{-3}$ mol/L atm |

Methane is more soluble in organic solvents than in water. Ethanol dissolves 32 cm$^3$ methane per 100 cm$^3$ at 0 °C. Diethylether dissolves 106.6 cm$^3$ methane per 100 cm$^3$ at 0 °C. Methane forms a hydrate, $CH_4 \cdot 6\,H_2O$, which has a decomposition temperature at 0.1 MPa of −29 °C, a critical decomposition temperature of 21.5 °C and a dissociation pressure at 0 °C of 2.6 MPa.

Naturally occuring methane is essentially of biological origin. Methane has been produced in nature by the dismutation of organic matter since the origin of life on earth.

All organic matter, left to itself, will ultimately be transformed (dismutated) by microorganisms under anaerobiosis into carbon dioxide and methane.

Indeed, nature obtains energy through redox reactions. The ultimate reducing step is the reduction of carbon to methane, and the ultimate oxidizing step, the oxidation of carbon to carbon dioxide [1].

The biological process by which organic matter is transformed into methane and carbon dioxide is called methanogenesis. When this biological process is housed in a reactor to create a technological process, it is known as biomethanation. The term anaerobic digestion is also widely used, but may lead to confusion with other anaerobic digestion processes that do not involve generation of methane. The mixture of carbon dioxide and methane is called biogas, a registered trade name of the Institute of Gas Technology in Chicago, United States, but widely used by the public.

Biomethanation is of interest, firstly because the biogas produced is an energy vector of economic interest, and secondly because the raw material or feedstock is often waste, which can hence be treated by the process [2], [3]. Wastewaters can be treated by biomethanation to yield acceptable effluents for receiving surface waters, and sludges can be treated to yield stabilized end products with useful agricultural properties [4]–[6].

**Table 1.** Composition of biogas

| Component | Content, vol% |
|---|---|
| Methane | 52–95 |
| Carbon dioxide | 9–45 |
| Hydrogen sulfide | 0.001–2 |
| Hydrogen | 0.01–2 |
| Nitrogen | 0.1–4 |
| Oxygen | 0.02–6.5 |
| Argon | 0.001 |
| Carbon monoxide | 0.001–2 |
| Ammonia | trace |
| Organics | trace |

**Table 2.** Densities of methane and biogas of various composition

| Methane content, % | 50 | 60 | 80 | 90 | 100 |
|---|---|---|---|---|---|
| Gas density relative to air | 1.040 | 0.942 | 0.745 | 0.652 | 0.555 |

## 2. Physical and Chemical Properties of Biogas

Biogas contains variable proportions of carbon dioxide and methane, usually somewhat more of the latter than of the former. Biogas is usually saturated with water and also contains minor and trace compounds, the major of which is hydrogen sulfide (see Table 1). The composition of biogas depends firstly upon the composition of the feedstock from which it originates and secondly upon the physicochemical equilibrium reached or maintained in the digesting mixed liquor [7].

Biogas has many of the chemical and physical properties of its parent constituent methane. Of course, the presence of carbon dioxide changes the chemical and physical properties. As an example, Table 2 shows the dependence of the density of biogas upon composition.

Whereas methane has a minimum calorific value of 50.0 MJ/kg (35.9 MJ/m$^3$), upgraded biogas containing 90 vol% methane has a minimum calorific value of 45.0 MJ/kg (32.3 MJ/m$^3$). Mean biogas, with 60 vol% methane, has a minimum calorific value of 30.0 MJ/kg or 21.5 MJ/m$^3$.

The flammability of biogas is shown in Figure 1. The chemical and physical properties of biogas are collected in [8].

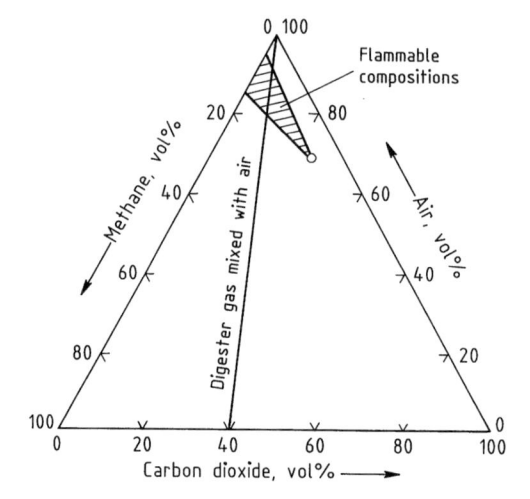

**Figure 1.** Flammability limits of biogas
The flammability limits of biogas are included in the dashed triangle. Point ○ refers to a gas mixture of 24% carbon dioxide, 68% air, and 8% methane
From ref. [8], with permission from Elsevier Applied Science

## 3. Biological Methanogenesis

Microorganisms transform organic matter into methane by the process depicted in Figure 2. The process consists basically of two steps, linked by interspecies hydrogen transfer. In the first step, fermentative bacteria dissolve, depolymerize, and metabolize organic matter into acetate, inorganic carbon (e.g., $CO_2$, $HCO_3^-$, and $CO_3^{2-}$), and hydrogen. In the second step, methanogenic microbial species dismutate acetate into methane and inorganic carbon, and reduce inorganic carbon with hydrogen to produce methane:

$CH_3COO^- + H_2O \rightleftharpoons CH_4 + HCO_3^-$    $\Delta G^{0\prime} = -31.0$ kJ/formular conversion
$4 H_2 + HCO_3^- + H^+ \rightleftharpoons CH_4 + 3 H_2O$    $\Delta G^{0\prime} = -135.6$ kJ/formular conversion

Whenever the first fermentative step produces organic metabolites other than acetate, for example, anions of short-chain carboxylic acids such as propionate and butyrate and alcohols such as ethanol or butanol, a link process, mediated by obligate hydrogen-producing acetogenic bacteria, transforms these metabolites into acetate and hydrogen. For thermodynamic reasons, the latter transformation can only take place at very low partial pressures of hydrogen. In stable, efficient methanogenic processes, the hydrogen partial pressure ranges between 1 and 10 Pa. An abrupt 2–4-fold increase in hydrogen partial pressure or an hydrogen partial pressure above 20 Pa is indicative of misfunctioning of the process. The increased partial pressure of hydrogen results in accumulation of intermediate organic metabolites, and in the vast majority of cases, the pH drops below 6 and methane generation stops.

The methanogenic microorganisms are archaebacteria [9], which are probably the oldest known living organisms. Other microorganisms interfere with hydrogen scaven-

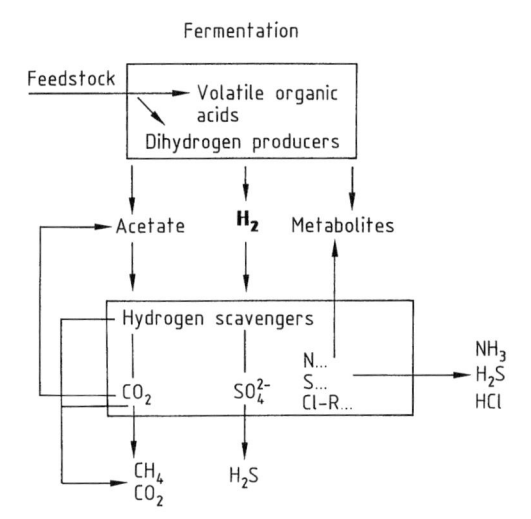

**Figure 2.** Flowsheet for biological methanogenesis

ging. Acetogenic bacteria can reduce inorganic carbon with hydrogen to produce acetate, and sulfate-reducing bacteria reduce sulfate with hydrogen to form hydrogen sulfide:

$$H^+ + SO_4^{2-} + 4\,H_2 \rightleftharpoons HS^- + 4\,H_2O \quad \Delta G^{0\prime} = -152 \text{ kJ/formular conversion}$$

More recently, it has been suspected that some hydrogen-scavenging microorganisms are also able to reduce organic molecules containing nitrogen, sulfur, or halogens. The methanogenic consortium appears in this way to play a fundamental role in detoxication and, hence, in the environmental protection of humans.

# 4. Technological Biomethanation

The biological process of methanogenesis was discovered by VOLTA in 1776, but it was over 100 years later that this biological process was first exploited technologically. MOURAS in France (1881), SCOTT-MONCRIEFF (1890) and CAMERON (1892) in the United Kingdom, HOUSTON (1892) and TALBOT (1894) in the United States, and IMHOFF (1905) in the Federal Republic of Germany, were the first to apply methanogenesis to reduce the organic matter content of sewage sludge and wastewater [10].

There are three generations of reactors for biomethanation [11]. The first generation includes the completely mixed reactor, run either in batch or in continuous mode (Fig. 3 A and B). In the batch system, the feedstock is introduced at once and left in the reactor until the reaction is terminated. In the continuous mode, the feedstock is continuously loaded into the reactor and the effluent continuously removed (e.g., by overflow) at the same flow rate. Batch systems are mainly used for more solid feedstocks, whereas the continuous mode is better adapted to more fluid feedstocks.

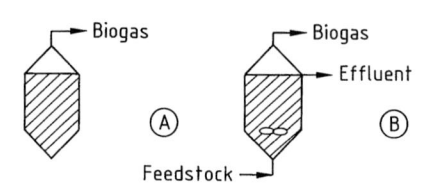

**Figure 3.** Schematics of the main designs of biomethanation reactors

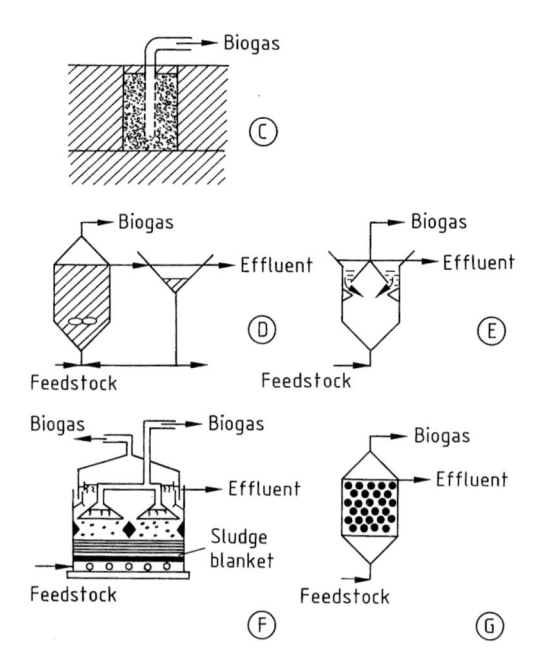

As an example, in the European Community, there were in 1988 about 27 400 aerobic municipal wastewater treatment plants. Sewage sludge produced by these plants can be evaluated at $9 \times 10^6$ t/a of dry matter. The biomethanation of two-thirds of this represents a potential production of biogas equivalent to $1.5 \times 10^6$ t of oil per year [12]. Figure 4 shows a typical egg-shaped anaerobic digester for municipal sewage sludge.

As another example, animal manure has been frequently biomethanized in reactors run in the continuous mode to produce biogas as an energy source and stabilized manure as a fertilizer and soil conditioner. In the European Community in 1989, about 470 biogas plants on farms were in existence. In rural areas of developing countries, reactors of 5–20 m$^3$ capacity, run in the continuous mode on animal waste, mainly cattle dung, exist in very large numbers for producing energy and fertilizer at the family level. Currently, more than three million such digesters are in use in China, more than one million in India, and several thousand in other countries.

An interesting example of biomethanation in the batch mode is given by landfills (Fig. 3C) [13]. Very large amounts of domestic solid waste are disposed of worldwide as huge landfills (several hundreds to several millions of tonnes each). Large amounts of

**Figure 4.** Egg-shaped reactor for the two-stage biomethanation of municipal sludge in the continuous mode at Schijnpoort, Antwerp, Belgium (courtesy of Biotim, Antwerp, Belgium)

biogas (several thousands of cubic meters per hour) are produced by natural fermentation in these landfills over long periods of time (10–20 years). The landfill gas escapes to the atmosphere, presenting a fire and explosion hazard, a source of damage to nearby vegetation, and a substantial contribution to the greenhouse effect. It can be made cost effective to extract and utilize the biogas produced in landfills. This environmental biotechnological industry started in the United States in the early 1970s, mainly in California, and spread in Europe, mainly in the Federal Republic of Germany and the United Kingdom. In the European Community, 107 landfills where biogas is exploited were identified in 1988, and this number is growing.

However, it quickly appeared that the first generation of anaerobic reactors, operating either in the batch mode or in the continuous mode without recycle (known in the United States as once-through mode), present severe limitations of performance for a number of reasons, the most important of which is the lowering in the concentration of active microorganisms (often referred to as active biomass) that arises because active biomass is lost when the effluent is removed.

A second generation of reactors for biomethanation was developed, in which the active biomass is either trapped in the anaerobic reactor or is recycled to it after sedimentation (Fig. 3 D and E). Sedimentation occurs either in a separate decanter or in a built-in decanter. The latter type was developed as early as 20 years ago under the trade name Clarigester. In the Clarigester, the liquid waste enters at the bottom of the reactor and flows upwards through a bed of active biomass. It then overflows into a built-in decanter, in which the active biomass sediments and returns to the reactor, while the effluent escapes from the top of the digester.

It was found that the active biomass, under conditions that are still not well understood, organizes itself as heavy granules. This led to the upflow anaerobic sludge blanket (UASB) system (Fig. 3F), which was developed since 1970 by Lettinga in the Netherlands [14] and is a milestone in the application of biomethanation to the anaerobic digestion of agro-industrial and other wastewaters.

In a variant, the active biomass is either trapped within a polymeric foam, notably polyurethane, or adsorbed on an inert carrier (e.g., baked clay or plastic rings) (Fig. 3 G). The latter system carries the generic name anaerobic filter. The active

**Figure 5.** Tower anaerobic filter for the two-step anaerobic treatment of starchy-containing wastewater
The towers are 22 m high and are filled with 16 layers of polyurethane foam matrices of 1.2 m thickness
By courtesy of Biotim, Antwerp, Belgium

biomass is referred to as a fixed film. When the inert carrier, usually large particles or matrices, is mainly immobile in the reactor, the system is known as a stationary bed. When the inert carrier, usually small particles such as sand or active carbon, expands and becomes mobile in the reactor, the system is termed a fluidized bed. Figure 5 shows a typical full-scale anaerobic filter for the anaerobic treatment of starch-containing wastewaters.

The third generation of biomethanation systems includes combinations of the former designs. For example, because the biological process consists of two steps, it has been found advantageous in some cases to optimize each biological step in a separate reactor. This is referred to as a two-step system and was first introduced 20 years ago [15]. Other two-stage systems have an active biomethanation in the first stage and a finishing biomethanation stage with effluent clarification in the second stage. These are often applied to the biomethanation of sewage sludge and this explains, as can be seen in Figure 4, why egg-shaped digesters usually operate in pairs.

# 5. Feedstocks for Biomethanation
[16]

Feedstocks for biomethanation include energy crops and organic wastes. Energy crops are plants grown mainly for producing energy. They include terrestrial and aquatic plants; the latter have been more intensively investigated. Typical aquatic energy crops are water hyacinths (*Eichhornia*) in soft waters and marine algae such as kelps (*Macrocystis*).

Organic wastes have become classical feedstocks for biomethanation [17]. They can be classified as domestic, industrial, or agricultural, according to their origin [18]. Feedstocks of domestic origin include municipal wastewaters, sewage sludges, and municipal solid wastes. Municipal wastewaters are progressively being investigated for anaerobic treatment. Sewage sludges have been stabilized by biomethanation for over 100 years. Domestic solid wastes are routinely biomethanized in landfills but are also being considered as feedstocks for biomethanation in engineered reactors.

Industrial wastes as feedstocks for biomethanation presently include mainly food wastes, originating from breweries, distilleries, fermentation industries, processing of coffee, fat, meat, milk, shellfish, fruit and vegetables, olive and palm oil mills, slaughterhouses, and starch and sugar refineries. They also include an increasing proportion of other agro-industrial wastes such as those originating from pulp and paper mills and the textile and tanning industries. Even pharmaceutical, (petro)chemical and coal gasification wastes are presently being considered for biomethanation. Industrial wastes range in physical state from wastewaters through sludges to solid material. Agro-industrial wastewaters are now routinely treated anaerobically. More than one hundred mature, economic, and reliable full-scale plants have been built worldwide [19].

Agricultural wastes as feedstocks for biomethanation consist mainly of animal wastes [20]. Whereas all kinds of animal manure are suitable for biomethanation, cattle and pig manure are the main targets, not only for energy and methane production but also for pollution control. Very large scale biogas plants, being developed progressively in Denmark, Italy and The Netherlands, among others, that are intended for a group of farms appear more reliable and profitable than smaller biogas plants on single farms. Agricultural and animal wastes for biomethanation are of increased interest for developing countries.

In the European Community, the potential for biogas from domestic waste is equivalent to $10 \times 10^6$ t of oil; the potential for biogas from agricultural waste is equivalent to $23 \times 10^6$ t of oil and the potential for biogas from energy crops could be $9-57 \times 10^6$ t of oil [19]. In the European Community, there already exists a total biomethanation reactor volume of $6 \times 10^5$ m$^3$ producing $0.49 \times 10^6$ m$^3$ of methane per day (equivalent to $0.15 \times 10^6$ t of oil per year).

# 6. Process and Environmental Parameters for Stable, Efficient Biomethanation

The process parameters characteristic for running a biomethanation system are the space loading rate, the concentration of the feedstock, and the mean residence time. The space loading rate is the mass of feedstock (as total dry matter, organic matter, or COD) per unit digester working volume per day. The ratio of the concentration to the space loading rate gives the mean residence time (excluding any accumulation factor arising from trapping of solid feedstock by sedimentation in imperfectly mixed reactors), usually in days. The state parameters characteristic of the performance of a biomethanation system are the rate of gas production and the residual concentration of dry organic matter in the effluent or its residual COD. The yield is calculated either by the ratio of produced methane to added or consumed feedstock, or from the residual concentration in the effluent. Stability of the methanogenic process is an important aspect of performance. The process parameters related to stability are the concentration of volatile fatty acids (indicative of the equilibrium within the methanogenic microbial consortium), the hydrogen partial pressure (see Chap. 3), and the alkalinity or bicarbonate concentration (linked to the acidification potential of the feedstock).

Degradation of 1 kg of COD produces stoichiometrically 350 L (at STP) or 250 g of methane. One kilogram of average organic matter corresponds to 1–1.4 kg of COD, depending on its mean oxidation state.

In completely mixed systems conducted in the continuous mode without recycle, practical space loading rates remain below 3 kg per cubic meter per day. The mean retention time must remain above 14 d, and this severely limits the concentration of the feedstock. Gas production rates range between 0.5 and 1.5 (sometimes up to 3) volumes of gas per unit digester volume per day.

In systems with trapped active biomass, ca. 10–20 and, in exceptional cases, up to 40 kg of active biomass can be housed in 1 m$^3$ of digester volume. The food to microorganism ratio is the potential amount of feedstock transformed per day per kilogram of active biomass. Acceptable values of food to microorganism ratio range around 1, sometimes higher. Space loading rates in these systems with trapped biomass can thus be increased up to 10–40 kg per cubic meter per day. Mean residence times can be lowered to one day or less with favorable feedstocks, i.e., those for which the fermentative step is fast, such as sugar-containing wastewaters. Gas production rates can increase to 4–10 volumes of gas per unit reactor volume per day.

Yields vary from 30% to over 95%. Animal manure usually gives low yields: 30% for cattle manure and up to 45% for pig manure. Sugar-containing wastewaters can be 95% digestible (at least in terms of BOD). Planning of biogas plants has been reviewed in [21], [22].

Several factors govern the process of methanogenesis in the technology of biomethanation, namely temperature, pH, anaerobiosis, carbon to nitrogen ratio, micronutrients, toxic elements in the feedstock, and inoculum.

There are two temperature ranges for optimum biomethanation, the mesophilic range at ca. 35 °C and the thermophilic range at ca. 60 °C. Thermophilic biomethanation results in faster gas production and higher yields, but the process is often less stable. The requirement for process temperatures above ambient temperature means that methane digesters need to be well insulated and that external process heat is often required. Part of the biogas produced (ca. 30%) is usually used for this purpose. Between 20 and 30 °C, both methane production and methanogenic microbial growth are still possible, but at lower rates and yields. Below 20 °C, methane production remains possible, but microbial growth becomes progressively slower.

Biological methanogenesis occurs at pH 6.8–8.2. A pH below 6.5 inhibits the second, methanogenic step of the biological process but not the first, fermentative step. Under these conditions, the metabolization of organic carbon yields organic anions, for which the only possible pairing cation is the hydrogen ion. As a result of the accumulation of acid anions, the pH of the digesting mixed liquor drops. Bicarbonate alkalinity limits this pH drop by acting as a buffer and through formation of carbon dioxide.

$$H_3O^+ + HCO_3^- \xrightleftharpoons[]{-H_2O} H_2CO_3 \rightleftharpoons H_2O + CO_2 \text{(sol)} \xrightleftharpoons[]{-H_2O} CO_2 \text{(g)}$$

This explains in part why biogas contains carbon dioxide besides methane. The buffering action of nitrogenous compounds is discussed below.

Methanogenesis is a strictly anaerobic process. Some species of the methanogenic consortium are very sensitive to oxygen. However, the methanogenic consortium also includes facultative aerobic–anaerobic microbial species. Therefore, the methanogenic fermentation mixture can easily overcome an occasional limited exposure to air. The fermentative first step of the biological process of methanogenesis occurs at low redox potentials ($E^0$ well below 0 mV). The methanogenic second step occurs at even lower redox potentials ($E^0$ a couple of hundred mV below).

The carbon to nitrogen ratio is an important parameter for efficient biomethanation. First, nitrogen is required for cell growth. Average cell yield is 15 g of dry cells per 100 g of transformed feedstock in anaerobiosis. The nitrogen requirement is 1.7 g per 100 g feedstock. Second, the mineralization of organic nitrogen yields ammonia and ammonium ions. Ammonia produced in this way exerts a buffering capacity complementing that of bicarbonate and is therefore useful in pH control. Optimum mass ratios of carbon to nitrogen given in the literature range between 16 and 19. Shortage of nitrogen in a feedstock can best be overcome by mixing it with another feedstock richer in nitrogen.

Micronutrients are essential for efficient biomethanation. Metal ions seem to be necessary for methanogenic archaebacteria, notably up to 50 mM $Na^+$ and 2–100 mM

Fe(II). Cobalt (50 nM) and nickel (100 nM) are essential trace elements for the reduction of carbon dioxide to methane by hydrogen.

Toxic compounds are often present in feedstocks at concentrations that inhibit biomethanation. For example, ammonia is inhibitory at concentrations of 1.7 g/L (as $NH_4^+/NH_3$). These concentrations are often encountered in pig manure. Heavy metal cations inhibit biological methanogenesis. However, most heavy metal cations form insoluble sulfides. Appropriate amounts of sulfate in the feedstock, a substrate for sulfate-reducing bacteria, normal hydrogen-scavenging members of the methanogenic consortium, help to overcome this potential inhibitory effect. This phenomenon has been conceived as a possible environmental biotechnology for the removal of heavy metals by concentration of insoluble compounds.

Many organic compounds are toxic to methanogenic microorganisms; an example is chloroform. Antibiotics such as monensin are used in animal feed to prevent methanogenesis in the rumen. Hence, animal manure may contain inhibitory concentrations of these antibiotics.

However, the methanogenic microbial consortium exhibits astonishing acclimation to toxic compounds [23]. Furthermore, the methanogenic microbial consortium includes detoxifying microbial species.

The methanogenic microbial consortium is more widely spread in nature than previously thought. However, due to the slow growth rate of several members of the consortium, and the even slower process of stabilization of the equilibrium between species in the consortium, the use of a large inoculum shortens the start-up period. This is especially true for solid feedstocks such as those in landfills. Methanogenic microorganisms do not seem to migrate rapidly over long distances (ca. 1 m). Appropriate and well mixed inocula may well prove to be essential for enhanced landfill gas generation.

When methanogenic microorganisms with adhesion properties are required for 2-d or 3-d generation methane reactors, it has proved more successful to use as inoculum the active flocculent sludge from aerobic wastewater treatment, as harvested at the bottom of the secondary decanter, rather than sludge obtained from the anaerobic stabilization reactor.

The use of specialized and acclimated inocula for given toxic or recalcitrant feedstocks can help reduce the start-up and acclimation periods. Enzymatic preparations have been claimed to be able to enhance biomethanation; however, no full proof for this exists. In the case of septic tanks, the enhancement was found to be due to inorganic contaminants which acted as inert carriers for microbial adhesion.

# 7. Environmental Impact of Biomethanation

Methanogenesis in nature and biomethanation by humans have a double impact on the environment. On the one hand, waste treatment has a positive impact, but biogas may have a negative impact.

Biomethanation is an environmental technology that can be applied at various scales from the family-size in rural tropical areas through farm-size in developed countries to industrial scale. In some cases, this environmental biotechnology can be made cost effective and give rise to an economic market.

As only the carbon of the organic matter is transformed to methane, all other elements remain in the digested effluent. The digested effluent of biomethanation is stabilized; that is, it undergoes further transformation only very slowly. Hence, it can be applied to soil without damage to plants.

Because it contains nitrogen, phosphorous and potassium, digested effluent has a fertilizer value. When the effluent is liquid, it can be used for "fertirrigation". When biomethanation proceeds from agricultural waste, it allows recycling of the valuable mineral elements. The digested effluent also contains lignocellulosic material, which is only poorly degraded because lignin is resistant to anaerobic degradation. This fibrous residue has excellent soil conditioning properties.

Biogas is flammable and explosive when appropriately mixed with air (see Fig. 1) and put in contact with an ignition source. Biogas generated in landfills migrates by diffusion and along pressure gradients. Biogas migrating in the root zones of plants creates asphyxia which can result ultimately in the death of vegetation. Biogas migrating in confined areas, such as the basements of buildings, creates explosion hazards. Several cases have been reported, some with fatal consequences for people. Nevertheless, landfill gas can easily be monitored and its migration is easily controlled. Furthermore, the gas can be profitably extracted, so that in the long term the concept of landfilling is not endangered.

It has been estimated that $1.2 \times 10^9$ t of methane are emitted annually into the atmosphere [24], resulting in a concentration of 1.4 ppm in air. Ten percent of this is fossil methane; 10% originates from animal wastes and human sewage; 5% from the anaerobic fermentation of human domestic wastes; and 5% from biomass burning, notably from forest fires. Natural ecosystems, including notably termites, which account for 10–20%, contribute to the remaining methane in the atmosphere. Methane has been calculated to have a mean residence time in the atmosphere of 3.3 years. Atmospheric methane concentration has doubled within a few centuries. Another doubling is expected within the next 60 years. The atmospheric warming associated with the increase in methane concentration in the atmosphere during the last decade has been calculated to amount to about 38% of the effect of carbon dioxide [25].

# 8. Storage, Pretreatment, Upgrading and Uses of Biogas

Biogas can be stored in low pressure gasholders, in medium-pressure tanks ($1-100$ m$^3$; $1-2$ MPa), in high-pressure gas cylinders ($30-50$ L; 20 MPa), or as liquefied gas below the critical temperature of methane ($-82.5$ °C). Low-pressure gasholders can be divided into wet and dry gasholders. Wet gasholders consist of floating rigid bells, usually of steel, with a capacity of $50-5000$ m$^3$. Dry gasholders are made of rubber or plastics and store from $1-1000$ m$^3$ of biogas.

Biogas must usually be pretreated before enduse. In any case, its 5% water vapor content must be condensed. Condensed water is a major source of corrosion. Hydrogen sulfide whenever present in too large amounts must be removed. Chemical absorption of hydrogen sulfide onto iron oxides is one of the possible treatments. Whenever necessary, trace elements are removed by adsorption on active charcoal.

Biogas sometimes needs to be upgraded to methane, to be used as a substitute or a complement of natural gas, as town gas, or as a motor fuel. Carbon dioxide is removed by absorption in water or by molecular sieves. The pressure swing method, consisting of absorption under high pressure and release under low pressure periods, is often used. Selective membranes for the separation of methane and carbon dioxide are also widely used.

Biogas can be used to generate light using glowing mantles, a common practice in rural developing areas, and for heating and cooking; atmospheric and forced draft burners need to be adapted.

Biogas does not belong to any of the three families of combustible gases [8] and is thus not interchangeable with any gas belonging to these three families. The gas jet of burners must be enlarged and the pressure of the gas supply lowered. The air supply must be slightly increased. In some cases, modifications of the burner design is necessary.

Biogas can be used for mechanical power generation. Spark ignition, internal combustion, Otto engines require the addition of a gas carburettor to provide a stoichiometric air–biogas mixture to the combustion chambers. Compression ignition, internal combustion, diesel engines usually work with biogas on the dual-fuel engine basis. The latter engines use a homogeneous biogas–air mixture compressed below its autoignition conditions and ignited by the injection of pilot diesel fuel. Engine modifications concern air and biogas supply and diesel injection.

Electrical power can be generated using a stationary biogas engine connected to a generator. Combined heat and power generation is very interesting in the case of biogas. The recovery of waste heat, as occurs for example in total energy modules (TOTEM), can provide the process heat necessary for controlling the temperature of the digester.

Whenever high rates of methane production are available, for example from landfills, gas turbines can be used.

# 9. Analysis

Biogas volumetric production rates are easily measured using conventional domestic gas flowmeters. Biogas composition is best measured by gas chromatography using Porapak Q as an inert support in a first column, a molecular sieve in a second column, and a catharometer for detection. Gases appear in the order hydrogen, oxygen, nitrogen, carbon dioxide, and methane. The ratio of carbon dioxide to methane in biogas can be estimated approximatively with a very simple device. A known volume of biogas is pumped into a small portable vessel, and the pressure is adjusted to an arbitrary 100% value. Carbon dioxide is then absorbed in sodium hydroxide solution; the ratio is calculated from the pressure drop.

The COD or the content of organic dry matter are the most common parameters for assessing feedstock and effluent. Individual short-chain carboxylic acids are best determined by gas chromatography with Chromosorb W 100 as inert support, acidified Carbowax as stationary phase and a flame ionisation detector. The total concentration of acids can be determined by steam distillation followed by titration. Alkalinity is another important process parameter; it is usually determined by acid titration, removal of carbon dioxide and back titration [26]. Low partial pressures of hydrogen (1–1000 ppm) can easily be determined by polarography.

# 10. Biogas in Developing Countries
[27], [28]

In rural areas of developing countries, biogas from domestic and agricultural wastes is a potential substitute for fuel wood. Most designs are based on the Chinese and Indian designs. A hybrid family digester designed by the German Technical Cooperation Agency (Gesellschaft für technische Zusammenarbeit, GTZ) is depicted in Figure 6. The optimum volume for a family digester is ca. 8–10 $m^3$. Between 10 and 15 kg of dry organic waste will produce about 3 $m^3$ of biogas containing 55% of methane. This daily feedstock requirement corresponds to 60–80 kg fresh dung originating from six to eight head of cattle, half time in stabulation. The daily produced biogas suffices for 7.5 h use of a cooking burner, or alternatively 6 h of a cooking burner and 3 h of a biogas-fueled glowing-mantle lamp, or 3 h of a cooking burner, 3 h of lighting and 24 h of a 80 L biogas-fueled refrigerator, or 2 h of a 1 kW engine [29]. The replacement of fuel wood by biogas in developing countries will however require one or more generations.

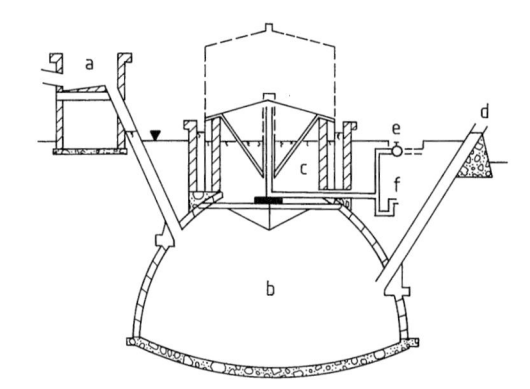

**Figure 6.** German–Chinese methane digester for a rural tropical family
a) Feedstock inlet; b) Digestion chamber; c) Floating dome gasholder; d) Overflow outlet; e) Gas outlet; f) Water trap

## 11. Economic Aspects

Environmental biotechnology, for example wastewater treatment, is usually considered as a necessary cost, without any profit. Biomethanation should not follow this way of thinking.

A thousand methane digesters have been constructed in Europe for an average investment per cubic meter of working volume ranging from ca. DM 600 for biogas plants on farms to ca. DM 800 for industrial biogas plants. Methane digesters produce biogas which has an energy value and can hence be used as a substitute for conventional energy and can be valued for the substituted energy. In Denmark, the public authorities add a financial incentive equivalent to the energy taxes on top of that value. As an average, on the basis of the present energy prices, the value of biogas covers 20% of the financial costs of biomethanation. The digested effluent, mainly when it is solid material has a potential value equivalent to that of compost, that is, ca. 40 ECU per ton. However, when the feedstock is municipal solid waste, marketing of the compost-like end-product is still proving difficult. The digested effluent can also be priced at the value of substituted energy. Preparing one ton of compost requires 6200 MJ. A digested effluent containing nitrogen, phosphorous and potassium can be valued for these elements. In a fertilizer, nitrogen values 18.5 MJ, potassium, 3.35 MJ, and phosphorous, 2.31 MJ.

Whenever the feedstock is a waste, it has a cost to be paid by its producer as municipal taxes for domestic solid waste disposal or as pollution taxes for wastewater treatment. Pollution-tax abbatments for agro-industries that treat their wastewaters by biomethanation contribute 80% of the financial costs of an anaerobic treatment plant. When the biogas produced can be used in the plant (e.g., to generate process heat), this waste treatment technology ends with a no-profit, no-cost balance in these industries. In developing countries, it now appears possible to build family-size digesters for a price acceptable to the buyer, for example the price of a head of cattle in Burundi or the price of a bicycle in China, with a minimum federal financial incentive, corresponding to the macro-economic savings (deforestation, hard currency).

# 12. References

[1] E.-J. Nyns, *Trib. CEBEDEAU* **34** (1981) 351–356.
[2] B. Lagrange: *Biomethane* vol. 1. *Une Alternative Crédible;* vol. **2**. *Principe – Techniques – Utilisation,* Edisud, Aix-en-Provence 1979.
[3] D. A. Stafford, D. L. Hawkes, R. Horton: *Methane Production from Waste Organic Matter,* CRC Press, Boca Raton 1980.
[4] E. C. Price, P. N. Cheremisinoff: *Biogas. Production and Utilization,* Ann Arbor Sci., Ann Arbor 1981.
[5] R. Braun: *Biogas. Methangärung Organischer Abfallstoffe,* Springer Verlag, Wien 1982.
[6] D. L. Wise: *Fuel Gas Systems,* CRC Press, Boca Raton 1983.
[7] A. Pauss, H. Naveau, E.-J. Nyns in D. O. Hall, R. P. Overend (eds.): *Biomass: Regenerable Energy,* Chichester 1987, pp. 273–291.
[8] M. Constant, H. Naveau, G.-L. Ferrero, E.-J. Nyns: *Biogas End-Use in the European Community,* Elsevier Applied Science, London 1989.
[9] W. E. Balch et al., *Microbiol. Rev.* **43** (1979) 260–296.
[10] P. L. McCarty in D. E. Hughes et al. (eds.): *Anaerobic Digestion,* Elsevier, Amsterdam 1981, pp. 3–22.
[11] E.-J. Nyns in H. J. Rehm, G. Reed (eds.): *Biotechnology,* VCH, Weinheim 1986, pp. 207–267.
[12] R. Buvet in M. P. Ferranti, G.-L. Ferrero, P. L'Hermite (eds.): *Anaerobic Digestion. Results of Research and Demonstration Projects,* Elsevier, London 1986, pp. 210–217.
[13] M. M. Schumacker: *Landfill Methane Recovery,* Noyes Data Corp., Park Ridge 1983.
[14] G. Lettinga et al., *Biotechnol. Bioeng.* **22** (1980) 699–734.
[15] F. G. Pohland, S. Gosh, *Environ. Lett.* **1** (1971) 255–266.
[16] D. L. Wise: *Fuel Gas Production from Biomass,* **2** vols., CRC Press, Boca Raton 1981.
[17] K. J. Thomé-Kozmiensky: *Biogas. Anaerobtechnik in der Abfallwirtschaft,* EF Verlag für Energie und Umwelttechnik, Berlin 1989.
[18] E.-J. Nyns in O. Kitani, C. W. Hall (eds.): *Biomass Handbook,* Gordon and Breach, New York 1989.
[19] M. Demuynck, E.-J. Nyns, W. Palz: *Biogas Plants in Europe,* Reidel, Dordrecht 1984.
[20] P. N. Hobson, S. Bousfield, R. Summer: *Methane Production from Agricultural and Domestic Wastes,* Applied Science, London 1981.
[21] R. Kloss: *Planung von Biogasanlagen,* Oldenbourg, München 1986.
[22] P.-J. Meynell: *Methane. Planning a Digester,* Prism Press, Dorchester 1982.
[23] R. E. Speece, *Adv. Solar Energy* **2** (1985) 51–123.
[24] J. C. Sheppard et al., *J. Geophys. Res.* **87** (1982) 1305–1312.
[25] H. Craig, C. C. Chou, *J. Geophys. Res.* **99** (1982) 1221–1224.
[26] R. Buvet, E.-J. Nyns in R. Buvet, M. F. Fox, D. J. Picken (eds.): *Biomethane, Production and Uses,* Bowskell, Exeter 1984, pp. 109–119.
[27] D. House: *Biogas Handbook,* Peace Press, Culver City 1948.
[28] M. M. El-Halwagi (ed.): *Biogas Technology, Transfer and Diffusion,* Elsevier Appl. Sci. London 1984.
[29] E.-J. Nyns, *Enzyme Microbial Technol.* **12** (1990) 151–152.

# Methanol

ECKHARD FIEDLER, BASF Aktiengesellschaft, Ludwigshafen, Federal Republic of Germany
GEORG GROSSMANN, BASF Aktiengesellschaft, Ludwigshafen, Federal Republic of Germany
BURKHARD KERSEBOHM, BASF Aktiengesellschaft, Ludwigshafen, Federal Republic of Germany
GÜNTHER WEISS, BASF Aktiengesellschaft, Ludwigshafen, Federal Republic of Germany
CLAUS WITTE, BASF Aktiengesellschaft, Ludwigshafen, Federal Republic of Germany

| | | |
|---|---|---|
| 1. | Introduction | 3288 |
| 2. | Physical Properties | 3288 |
| 3. | Chemical Properties | 3290 |
| 4. | Production | 3292 |
| 4.1. | Principles | 3292 |
| 4.1.1. | Thermodynamics | 3292 |
| 4.1.2. | Kinetics and Mechanism | 3293 |
| 4.1.3. | Byproducts | 3295 |
| 4.2. | Catalysts | 3296 |
| 4.2.1. | Catalysts for High-Pressure Synthesis | 3296 |
| 4.2.2. | Catalysts for Low-Pressure Synthesis | 3296 |
| 4.2.3. | Production of Low-Pressure Catalysts | 3297 |
| 4.2.4. | Catalyst Deactivation | 3298 |
| 4.2.5. | Other Catalyst Systems | 3299 |
| 5. | Process Technology | 3299 |
| 5.1. | Production of Synthesis Gas | 3300 |
| 5.2. | Synthesis | 3302 |
| 5.3. | Reactor Design | 3303 |
| 5.4. | Distillation of Crude Methanol | 3305 |
| 5.5. | Construction Materials | 3305 |
| 6. | Handling, Storage, and Transportation | 3306 |
| 6.1. | Explosion and Fire Control | 3306 |
| 6.2. | Storage and Transportation | 3307 |
| 7. | Quality Specifications and Analysis | 3308 |
| 8. | Environmental Protection | 3310 |
| 9. | Uses | 3310 |
| 9.1. | Use as Feedstock for Chemical Syntheses | 3310 |
| 9.2. | Use as Energy Source | 3312 |
| 9.3. | Other Uses | 3315 |
| 10. | Economic Aspects | 3315 |
| 11. | Toxicology and Occupational Health | 3317 |
| 11.1. | Toxicology | 3317 |
| 11.2. | Occupational Health | 3319 |
| 12. | References | 3320 |

# 1. Introduction

Methanol [67-56-1], CH$_3$OH, $M_r$ 32.042, also termed methyl alcohol or carbinol, is one of the most important chemical raw materials. Worldwide production capacity in 1989 was ca. $21 \times 10^6$ t/a. About 85% of the methanol produced is used in the chemical industry as a starting material or solvent for synthesis. The remainder is used in the fuel and energy sector; this use is increasing.

**Historical Aspects.** Methanol was first obtained in 1661 by Sir ROBERT BOYLE through the rectification of crude wood vinegar over milk of lime. He named the new compound *adiaphorus spiritus lignorum*. JUSTUS VON LIEBIG (1803–1873) and J. B. A. DUMAS (1800–1884) independently determined the composition of methanol. The term "methyl" was introduced into chemistry in 1835 on the basis of their work.

From ca. 1830–1923, "wood alcohol," obtained by the dry distillation of wood, remained the only important source of methanol. As early as 1913, A. MITTASCH and coworkers at BASF successfully produced organic compounds containing oxygen, including methanol, from carbon monoxide and hydrogen in the presence of iron oxide catalysts during developmental work on the synthesis of ammonia. The decisive step in the large-scale industrial production of methanol was made by M. PIER and coworkers in the early 1920s with the development of a sulfur-resistant zinc oxide–chromium oxide catalyst. By the end of 1923 the process had been converted from the developmental to the production stage at the BASF Leuna Works.

Processes based on the above work were performed at high pressure (25–35 MPa) and 320–450 °C. They dictated the industrial production of methanol for more than 40 years. In the 1960s, however, ICI developed a route for methanol synthesis in which sulfur-free synthesis gas containing a high proportion of carbon dioxide was reacted on highly selective copper oxide catalysts. This and other related low-pressure processes are characterized by fairly mild reaction conditions (5–10 MPa, 200–300 °C). Methanol can now be produced much more economically worldwide by these low-pressure methods.

# 2. Physical Properties

Methanol is a colorless, neutral, polar liquid that is miscible with water, alcohols, esters, and most other organic solvents [1], [2]; it is only slightly soluble in fat and oil. Because of its polarity, methanol dissolves many inorganic substances, particularly salts.

The most important physical data for methanol follow [3], [4]:

| | |
|---|---|
| Density (101.3 kPa), liquid | |
| at 0 °C | 0.8100 g/cm$^3$ |
| at 25 °C | 0.78664 g/cm$^3$ |
| at 50 °C | 0.7637 g/cm$^3$ |
| Critical pressure | 8.097 MPa |
| Critical temperature | 239.49 °C |
| Critical density | 0.2715 g/cm$^3$ |

| | |
|---|---|
| Critical volume | 117.9 cm³/mol |
| Critical compressibility | 0.224 |
| mp | −97.68 °C |
| Heat of fusion (101.3 kPa) | 100.3 kJ/kg |
| Triple-point temperature | −97.56 °C |
| Triple-point pressure | 0.10768 Pa |
| bp (101.3 kPa) | 64.70 °C |
| Heat of vaporization (101.3 kPa) | 1128.8 kJ/kg |
| Standard enthalpy of formation | |
|   at 25 °C (101.3 kPa), gas | −200.94 kJ/mol |
|   at 25 °C (101.3 kPa), liquid | −238.91 kJ/mol |
| Free enthalpy of formation | |
|   at 25 °C (101.3 kPa), gas | −162.24 kJ/mol |
|   at 25 °C (101.3 kPa), liquid | −166.64 kJ/mol |
| Standard entropy | |
|   at 25 °C (101.3 kPa), gas | 239.88 J mol⁻¹ K⁻¹ |
|   at 25 °C (101.3 kPa), liquid | 127.27 J mol⁻¹ K⁻¹ |
| Specific heat, $c_p$ | |
|   at 25 °C (101.3 kPa), gas | 44.06 J mol⁻¹ K⁻¹ |
|   at 25 °C (101.3 kPa), liquid | 81.08 J mol⁻¹ K⁻¹ |
| Viscosity (25 °C) | |
|   Liquid | 0.5513 mPa · s |
|   Vapor | $9.68 \times 10^{-3}$ mPa · s |
| Thermal conductivity (25 °C) | |
|   Liquid | 190.16 mW m⁻¹ K⁻¹ |
|   Vapor | 14.07 mW m⁻¹ K⁻¹ |
| Electrical conductivity (25 °C) | $(2-7) \times 10^{-9}$ Ω⁻¹ cm⁻¹ |
| Dielectric constant (25 °C) | 32.65 |
| Dipole moment | $5.6706 \times 10^{-30}$ C · m |
| Refractive index $n_D^{20}$ | 1.32840 |
|   $n_D^{25}$ | 1.32652 |
| Surface tension in air (25 °C) | 22.10 mN/m |
| Flash point (DIN 51 755) | 6.5 °C |
|   Open vessel | 15.6 °C |
|   Closed vessel | 12.2 °C |
| Explosion limits in air | 5.5 – 44 vol% |
| Ignition temperature (DIN 51 794) | 470 °C |

The temperature dependence of selected physical properties is given in [5]; thermodynamic data can be found in [6] and the heat capacity and enthalpy of the liquid in [7].

The vapor pressure of methanol is determined according to [8] by a Wagner equation of the form

$$\ln p = 8.999 + \frac{512.64}{T} \left( -8.63571\, q + 1.17982\, q^{3/2} - 2.4790\, q^{5/2} - 1.024\, q^5 \right)$$

where $q = 1 - T/512.64$; $T$ is the absolute temperature, and $p$ the pressure in kilopascals. Further vapor pressure correlation data in the temperature range 206 – 512 K are given in [9], and critical data in [10]. A selection of binary azeotropes is shown in Table 1, and a comprehensive summary is given in [11].

**Table 1.** Binary azeotropic mixtures of methanol

| Component | bp of component, °C | bp of azeotrope, °C | Methanol content of azeotrope, wt % |
|---|---|---|---|
| Acetonitrile | 81.6 | 63.45 | 19 |
| Acrylonitrile | 77.3 | 61.4 | 61.3 |
| Acetone | 56.15 | 55.5 | 12 |
| Ethyl formate | 54.15 | 50.95 | 16 |
| Methyl acetate | 57.1 | 53.9 | 17.7 |
| Furan | 31.7 | < 30.5 | < 7 |
| Thiophene | 84 | < 59.55 | < 55 |
| Methyl acrylate | 80 | 62.5 | 54 |
| 2-Butanone | 79.6 | 64.5 | 70 |
| Tetrahydrofuran | 66 | 60.7 | 31.0 |
| Ethyl acetate | 77.1 | 62.25 | 44 |
| Methyl propionate | 79.8 | 62.45 | 47.5 |
| Methyl methacrylate | 99.5 | 64.2 | 82 |
| Cyclopentane | 49.4 | 38.8 | 14 |
| n-Pentane | 36.15 | 30.85 | 7 |
| Benzene | 80.1 | 57.50 | 39.1 |
| Cyclohexane | 80 | 54 | 38 |
| Cyclohexene | 82.75 | 55.9 | 40 |
| Toluene | 110.6 | 63.5 | 72.5 |

Viscosity data of the pure components have been published in [5], [12], [13] for the liquid phase, and in [13] for the vapor. The viscosity and density of aqueous methanol solutions at 25 °C are shown in Table 2. Temperature-dependent densities of the binary mixture are given in [15] and [16]; viscosities are documented in [15] and [17]. The pressure dependence of viscosity has been measured [18], and isothermal compressibilities, coefficients of thermal expansion, partial molar volumes, and excess factors accounting for the difference between real and ideal behavior can be found in [19]. Information on the liquid–solid phase equilibrium in the methanol–water system is given in [20].

Data on the thermal conductivity of liquid methanol appear in [21]; the electrical conductivity of the pure liquid and dielectric properties are given in [22] and [23], respectively. Safety aspects have also been discussed [24], [25].

# 3. Chemical Properties

Methanol is the simplest aliphatic alcohol. As a typical representative of this class of substances, its reactivity is determined by the functional hydroxyl group [26]–[28]. Reactions of methanol take place via cleavage of the C–O or O–H bond and are characterized by substitution of the –H or –OH group (→ Alcohols, Aliphatic) [29]. In contrast to higher aliphatic alcohols, however, β-elimination with the formation of a multiple bond cannot occur.

Important industrial reactions of methanol include the following (Fig. 1):

**Table 2.** Viscosity and density of aqueous methanol solutions at 25 °C

| Mole fraction of methanol | Kinematic viscosity, mm$^2$/s | Density, kg/m$^3$ | Absolute viscosity, mPa · s |
| --- | --- | --- | --- |
| 0.0 | 0.893 | 997.1 | 0.890 |
| 0.0507 | 1.126 | 983.4 | 1.107 |
| 0.1125 | 1.385 | 966.9 | 1.339 |
| 0.1411 | 1.480 | 960.2 | 1.421 |
| 0.2276 | 1.657 | 941.1 | 1.559 |
| 0.2927 | 1.683 | 925.7 | 1.558 |
| 0.4198 | 1.593 | 898.4 | 1.431 |
| 0.4856 | 1.505 | 884.5 | 1.331 |
| 0.5542 | 1.396 | 869.9 | 1.214 |
| 0.7133 | 1.149 | 837.7 | 0.963 |
| 0.8040 | 0.992 | 821.0 | 0.814 |
| 0.8345 | 0.952 | 816.0 | 0.777 |
| 0.9140 | 0.825 | 800.1 | 0.660 |

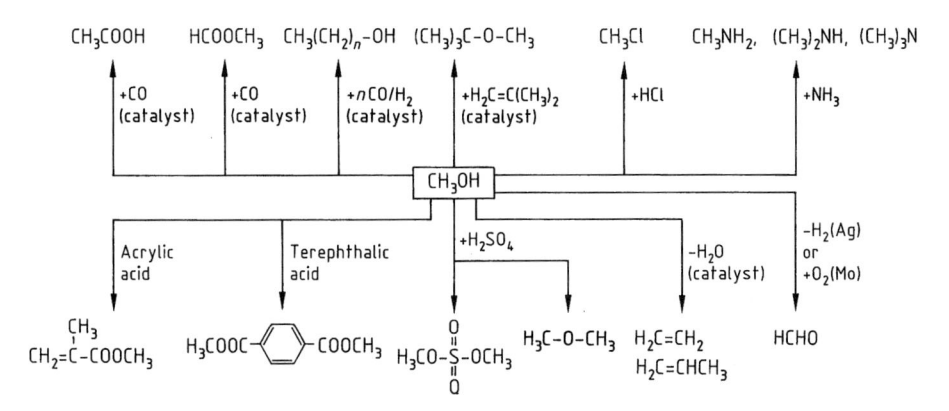

**Figure 1.** Industrially important reactions of methanol

1) Dehydrogenation and oxidative dehydrogenation
2) Carbonylation
3) Esterification with organic or inorganic acids and acid derivatives
4) Etherification
5) Addition to unsaturated bonds
6) Replacement of hydroxyl groups

# 4. Production

## 4.1. Principles

### 4.1.1. Thermodynamics

The formation of methanol from synthesis gas can be described by the following equilibrium reactions:

$$CO + 2\,H_2 \rightleftharpoons CH_3OH \qquad \Delta H_{300\,K} = -90.77\ \text{kJ/mol} \quad (1)$$
$$CO_2 + 3\,H_2 \rightleftharpoons CH_3OH + H_2O \qquad \Delta H_{300\,K} = -49.16\ \text{kJ/mol} \quad (2)$$

Reaction enthalpies are determined from the standard enthalpies of the reactants and products [30]. Both reactions are exothermic and accompanied by a decrease in volume. Methanol formation is thus favored by increasing pressure and decreasing temperature, the maximum conversion being determined by the equilibrium composition.

In addition to the two methanol-forming reactions, the endothermic reaction of carbon dioxide and hydrogen (Eq. 3, the reverse water-gas shift reaction) must also be taken into account:

$$CO_2 + H_2 \rightleftharpoons CO + H_2O \qquad \Delta H_{300\,K} = 41.21\ \text{kJ/mol} \quad (3)$$

For the sake of simplicity, Equations (1) and (3) can be discussed as independent reaction pathways. The conversion of carbon dioxide to methanol (2) is then the overall result of Equations (1) and (3), and the equilibrium constant $K_2$ can be described as $K_2 = K_1 \cdot K_3$. When the nonideal behavior of gases is taken into account, the equilibrium constants are determined as follows:

$$K_1 = \left[\frac{f_{CH_3OH}}{f_{CO} f_{H_2}^2}\right] = \left[\frac{\varphi_{CH_3OH}}{\varphi_{CO}\, \varphi_{H_2}^2}\right]\left[\frac{p_{CH_3OH}}{p_{CO}\, p_{H_2}^2}\right] = K_{\varphi 1} \cdot K_{p1} \qquad (4)$$

$$K_3 = \left[\frac{f_{CO} f_{H_2O}}{f_{CO_2} f_{H_2}}\right] = \left[\frac{\varphi_{CO}\, \varphi_{H_2O}}{\varphi_{CO_2}\, \varphi_{H_2}}\right]\left[\frac{p_{CO}\, p_{H_2O}}{p_{CO_2}\, p_{H_2}}\right] = K_{\varphi 3} \cdot K_{p3} \qquad (5)$$

where $f_i$ is the fugacity, $\varphi_i$ the fugacity coefficient, and $p_i$ the partial pressure of the $i$-th component.

A number of numerical formulations exist for calculating the temperature-dependent equilibrium constants $K_1$ [31]–[38] and $K_3$ [36]–[39]; their results differ widely [40]. The binomial formulations of CHEREDNICHENKO (Eq. 6) [34] and BISSET (Eq. 7) [39] are examples:

$$K_1 = 9.740 \times 10^{-5} \exp\left[21.225 + \frac{9143.6}{T} - 7.492 \ln T + 4.076 \times 10^{-3} T - 7.161 \times 10^{-8} T^2\right] \quad (6)$$

$$K_3 = \exp\left[13.148 - \frac{5639.5}{T} - 1.077 \ln T - 5.44 \times 10^{-4} T + 1.125 \times 10^{-7} T^2 + \frac{49170}{T^2}\right] \quad (7)$$

The fugacity coefficients can be determined according to [41] by assuming ideal solubility for the individual pure components, or they can be calculated from suitable equations of state [42], [43].

The carbon monoxide and carbon dioxide conversions up to attainment of equilibrium are shown as a function of pressure and temperature in Table 3 [35]. A synthesis gas formed by steam reforming was chosen as the starting gas (15 vol% CO, 8 vol% $CO_2$, 74 vol% $H_2$, and 3 vol% $CH_4$). Equations (6) and (7) were used to establish temperature dependence, and the fugacity coefficients were determined according to the Soave–Redlich–Kwong equation. The negative sign for the carbon monoxide conversion denotes carbon monoxide formation by back-conversion [44].

## 4.1.2. Kinetics and Mechanism

The formation of methanol, as a typical heterogeneously catalyzed reaction, can be described by an absorption–desorption mechanism (Langmuir–Hinshelwood or Eley–Rideal). The nature of the active centers in the copper–zinc oxide–alumina catalysts used under industrial conditions is still a subject of discussion (see Chap. 5). The active species in low-pressure methanol synthesis may be a solution of copper(I) ions in the zinc oxide phase [45]. On the other hand, evidence can be found that copper(0) also catalyzes methanol formation. The feed gas composition (particularly the proportions of $CO_2$ and $H_2O$) also plays an important role in determining the activity and selectivity of catalysts in methanol production. Investigations have shown that various routes must exist for the formation of methanol via carbon monoxide or carbon dioxide, and that different active centers in the catalyst are involved [46]–[50].

According to [51], alumina exists in an X-ray amorphous form. The proposed functions of alumina in copper–zinc oxide–alumina catalysts include:

1) prevention of sintering of the fine copper particles by the formation of zinc spinel;
2) stabilization of the highly disperse copper–zinc oxide catalyst system; and
3) formation of surface defects by the incorporation of alumina clusters in the copper lattice [46], [51].

Which effect prevails in methanol synthesis is still not clear. However, alumina has an important function as a structural promoter in copper–zinc oxide catalysts by improving their mechanical stability and long-term activity.

**Table 3.** Temperature and pressure dependence of the carbon monoxide and carbon dioxide equilibrium conversions

| Temperature, °C | CO conversion * | | | $CO_2$ conversion | | |
|---|---|---|---|---|---|---|
| | 5 MPa | 10 MPa | 30 MPa | 5 MPa | 10 MPa | 30 MPa |
| 200 | 96.3 | 99.0 | 99.9 | 28.6 | 83.0 | 99.5 |
| 250 | 73.0 | 90.6 | 99.0 | 14.4 | 45.1 | 92.4 |
| 300 | 25.4 | 60.7 | 92.8 | 14.1 | 22.3 | 71.0 |
| 350 | −2.3 | 16.7 | 71.91 | 9.8 | 23.1 | 50.0 |
| 400 | −12.8 | −7.3 | 34.1 | 27.7 | 29.3 | 41.0 |

\* Negative sign denotes CO formation via Equation (3) [44]: $CO_2 + H_2 \rightleftharpoons CO + H_2O$.

Recent kinetic investigations have concentrated on the role of carbon dioxide in methanol synthesis, which aroused a great deal of controversy during the 1980s [40], [46], [52]–[54]. Until the beginning of the 1980s, mechanistic considerations were based almost exclusively on the hydrogenation of carbon monoxide to methanol (Eq. 1, see 4.1.1) [55]–[59]. The increased yield achieved by adding carbon dioxide was ascribed to the displacement of the reverse water-gas shift equilibrium (Eq. 3). In addition, carbon dioxide was assumed to influence the oxidation state of the active centers in the catalyst [45].

In contrast, KAGAN et al. [60] proposed that methanol was formed solely according to Equation (2) from carbon dioxide. Recent experiments with isotope-labeled reactants show that both reaction pathways (Eqs. 1 and 2) are possible [61], [62]. Similar results were obtained in other studies [63]–[65]. However, according to [62], formation via carbon dioxide predominates under conditions of large-scale industrial methanol synthesis.

## 4.1.3. Byproducts

Commercially available $Cu-ZnO-Al_2O_3$ catalysts for the low-pressure synthesis of methanol permit production of the desired product with high selectivity, typically above 99 % referred to the added $CO_x$.

The following impurities are important for the large-scale industrial process:

1) Higher alcohols formed by catalysis with traces of alkali [66]–[68]

$$n\,CO + 2\,n\,H_2 \rightleftharpoons C_nH_{2n+1}OH + (n-1)\,H_2O$$

2) Hydrocarbons and waxes formed by catalysis with traces of iron, cobalt, and nickel according to the Fischer–Tropsch process [67], [69], [70]

$$CO + 3\,H_2 \rightleftharpoons CH_4 + H_2O$$
$$CO_2 + 4\,H_2 \rightleftharpoons CH_4 + 2\,H_2O$$
$$n\,CO + (2\,n-1)\,H_2 \rightleftharpoons C_nH_{2n+2} + n\,H_2O$$

3) Esters [68], [70], [71]

$$(CH_2O)_{ads} + (RCHO)_{ads} \rightleftharpoons CH_3COOR$$

4) Dimethyl ether [70], [72]

$$2\,CO + 4\,H_2 \xrightleftharpoons{Al_2O_3} CH_3-O-CH_3 + H_2O$$

5) Ketones [73]

$$RCH_2CH_2OH \rightleftharpoons RCH_2CHO + H_2$$
$$2\,RCH_2CHO \rightleftharpoons RCH_2COCHRCH_3 + O_{ads}$$

The formation of most byproducts from synthesis gas, particularly $C_2^+$ species, is thermodynamically favored over methanol synthesis. Because methanol constitutes the main product, however, reactions yielding impurities are controlled kinetically rather than thermodynamically [40]. In addition to catalyst constituents and feed gas composition, the residence time at the catalyst [68], as well as the temperature [69], [70], mainly determine the extent of byproduct formation: an increase in these parameters raises the proportion of byproducts. A detailed discussion of individual byproduct classes is given in [40].

## 4.2. Catalysts

### 4.2.1. Catalysts for High-Pressure Synthesis

The first industrial production of methanol from synthesis gas by the high-pressure process employed a catalyst system consisting of zinc oxide and chromium oxide. This catalyst, which was used at 25–35 MPa and 300–450 °C, was highly stable to the sulfur and chlorine compounds present in synthesis gas [44], [54], [74], [75].

Production of methanol with zinc oxide–chromium oxide catalysts by the high-pressure process is no longer economical. A new generation of copper-containing catalysts with higher activity and better selectivity is now used. The last methanol plant based on the high-pressure process closed in the mid-1980s. For a detailed discussion of high-pressure methanol catalysts, see [74].

### 4.2.2. Catalysts for Low-Pressure Synthesis

Well before the industrial realization of low-pressure methanol synthesis by ICI in the 1960s, copper-containing catalysts were known to be substantially more active and selective than zinc oxide–chromium oxide catalysts. Copper oxide–zinc oxide catalysts and their use in the production of methanol were described by BASF in the early 1920s [76], [77]. These catalysts were employed at 15 MPa and 300 °C.

Their industrial use was prevented, however, by a serious disadvantage: impurities such as hydrogen sulfide and chlorine compounds in synthesis gas rapidly deactivated the catalysts. Nevertheless, the copper-containing catalyst systems proved to be the most promising candidates for producing methanol industrially at lower temperature and pressure. A series of publications on this topic appeared between 1925 and 1955 [74], [78], [79]. Investigations of copper catalysts continue to this day [53].

A low-pressure catalyst for methanol synthesis was first used industrially in the process developed by ICI in 1966. This copper oxide–zinc oxide catalyst was thermally stabilized with alumina. It was used to convert extremely pure (i.e., largely free of sulfur and chlorine compounds, $H_2S$ < 0.1 ppm) synthesis gas to methanol [80]. Because this

**Table 4.** Summary of typical copper-containing catalysts for low-pressure methanol synthesis

| Manufacturer | Component | Content, atom % | Reference |
|---|---|---|---|
| IFP | Cu | 25–80 | [83] |
| | Zn | 10–50 | |
| | Al | 4–25 | |
| Süd Chemie | Cu | 65–75 | [84] |
| | Zn | 18–23 | |
| | Al | 8–12 | |
| Shell | Cu | 71 | [85] |
| | Zn | 24 | |
| | rare-earth oxide | 5 | |
| ICI | Cu | 61 | [86] |
| | Zn | 30 | |
| | Al | 9 | |
| BASF | Cu | 65–75 | [87] |
| | Zn | 20–30 | |
| | Al | 5–10 | |
| Du Pont | Cu | 50 | [88] |
| | Zn | 19 | |
| | Al | 31 | |
| United Catalysts | Cu | 62 | [88] |
| | Zn | 21 | |
| | Al | 17 | |
| Haldor Topsoe | Cu | 37 | [88] |
| | Zn | 15 | |
| | Cr | 48 | |

copper catalyst was extremely active, methanol synthesis could be carried out at 220–230 °C and 5 MPa. Premature aging due to sintering of copper was thereby avoided. The high selectivity of the new catalyst gave a methanol purity > 99.5 %. The formation of byproducts (e.g., dimethyl ether, higher alcohols, carbonyl compounds, and methane) associated with the old high-pressure catalyst, was drastically reduced or, in the case of methane, completely eliminated.

All currently used low-pressure catalysts contain copper oxide and zinc oxide with one or more stabilizing additives (Table 4). Alumina, chromium oxide, or mixed oxides of zinc and aluminum have proved suitable for this purpose [81], [82].

## 4.2.3. Production of Low-Pressure Catalysts

Catalysts now used in low-pressure methanol synthesis plants and based on copper–zinc–aluminum (or chromium) are obtained as metal hydroxycarbonates or nitrates by coprecipitation of aqueous metal salt solutions (e.g., nitrates) with sodium carbonate solution. Precipitation may occur in one or several stages. The quality of the subsequent catalyst is determined by the optimum composition of the metal components, the precipitation temperature, the pH used for precipitation, the sequence of metal salt additions, and the duration of precipitation. The stirring rate, stirring energy, and shape of stirrer also affect catalyst quality.

The precipitated catalyst precursors (largely metal hydroxycarbonates) are filtered off from the mother liquor, washed free of interfering ions (e.g., sodium), and dried at ca. 120 °C. Examples of such hydroxycarbonates are malachite rosasite $(Cu, Zn)_5(CO_3)(OH)_2$, hydrozincite $(Cu, Zn)_5(OH)_6(CO_3)_2$, and aurichalcite $(Cu_{0.3}Zn_{0.7})_5(OH)_6(CO_3)_2$ [40], [47], [82]. Aurichalcite derivatives with the composition $Cu_{2.2}Zn_{2.8}(OH)_6(CO_3)_2$, containing small amounts of alumina for stabilization, are obtained by coprecipitation of metal nitrates [87], [88]. The catalyst precursor is converted to finely divided metal oxide by subsequent calcination at ca. 300–500 °C [80]. The calcined product is then pelleted to commercial catalyst forms. Cylindrical tablets 4–6 mm in diameter and height are common [46], [47], [82], [89].

The catalysts still have a total BET surface area of 60–100 $m^2$/g and have to be activated [47]. They are activated by controlled reduction with 0.5–2% hydrogen in nitrogen at 150–230 °C. Particular care must be taken to avoid hot spots, which lead to premature catalyst aging. In their reduced (i.e., active) form, the synthesis-active copper surfaces of commercial catalysts have a surface area of 20–30 $m^2$/g [81].

Catalysts for the low-pressure synthesis of methanol can also be produced by other methods, e.g., impregnating a carrier with active components, kneading metal compounds together, and leaching Raney alloys [17], [82].

Catalysts must be devoid of interfering impurities. Alkali compounds reduce the useful life and adversely affect the selectivity of catalysts. Even iron or nickel impurities in the parts-per-million range promote the formation of hydrocarbons and waxes. Acidic compounds such as silicon dioxide increase the proportion of dimethyl ether in crude methanol [90].

### 4.2.4. Catalyst Deactivation

As mentioned in Section 4.1, efficient catalysts for low-pressure synthesis of methanol should have a highly disperse distribution of active centers stabilized by structural promotors. The longer a catalyst can retain these properties under industrial conditions, the more valuable it is for industrial operation: downtimes for catalyst replacement are reduced. Catalysts normally have useful lives of 2–5 years. Many factors can drastically reduce catalyst activity and, thus, useful life. Detailed review articles on catalyst deactivation and poisoning can be found in [40], [47], [90].

Even during catalyst production, manufacturing faults can seriously affect the complex structure of the active centers (see Section 4.2.3). Catalyst damage and, consequently, premature deactivation may also occur during reduction. The temperature conditions, hydrogen concentration of the reducing gas, and gas load must be strictly controlled. Deviations from specified reduction procedures may lead to hot spots in the pellets, resulting in sintering of the copper constituents; copper becomes mobile at 190 °C and can agglomerate from its finely divided form into fairly large crystallites.

Reduction must be complete to obtain the entire active mass from the precursor compounds (see Section 4.2.3). Deviations from the specified reduction conditions may permanently decrease the active BET surface area and thus irrevocably damage the catalyst.

Another important point regarding the deactivation of copper catalysts is their high sensitivity to impurities in synthesis gas. Chlorine- and sulfur-containing contaminants long prevented the use of copper-containing catalyst systems in industrial methanol plants. These catalyst poisons must be removed from the feed gas prior to methanol synthesis. A certain degree of protection against deactivation by sulfur is afforded by catalysts containing zinc oxide because the sulfur is bound as zinc sulfide. After deactivation, the catalyst is still able to absorb large quantities of sulfur to protect subsequent catalyst layers against poisoning. Other synthesis gas impurities (e.g., silicon compounds, nickel carbonyls, or iron carbonyls) also cause catalyst damage [90].

The catalyst can also be deactivated by overheating during operation. Thermal damage to the catalyst can occur after use of nonoptimum recycled gas compositions, incorrect temperature control, or overloaded catalyst in the startup phase. The active surface area of the catalyst is decreased and phase transformations occur. The formation of copper spinels as well as malachite rosasite is observed. In effect, this removes active centers for methanol synthesis from the catalyst [40], [45], [91].

### 4.2.5. Other Catalyst Systems

A number of modified copper – zinc oxide – alumina catalysts have been prepared by doping with boron, manganese, cerium, chromium, vanadium, magnesium, or other elements [92] – [100]. Other catalyst systems have also been investigated. Three basic types of catalysts are mentioned in recent literature: Raney copper catalysts, copper alloys with thorium or rare-earth oxides, and supported precious-metal catalysts [101] – [106]. Only copper alloy catalysts are reported to have a higher activity than conventional copper – zinc oxide – alumina catalysts [107]. Until now, however, exclusively copper-containing zinc oxide – alumina catalysts have been used in industrial methanol plants. These catalysts have high activity, very good selectivity, long-term stability, and favorable production costs. They are still the most cost effective catalysts.

## 5. Process Technology

The oldest process for the industrial production of methanol is the dry distillation of wood, but this no longer has practical importance. Other processes, such as the oxidation of hydrocarbons and production as a byproduct of the Fischer – Tropsch synthesis according to the Synthol process, have no importance today.

Methanol is currently produced on an industrial scale exclusively by catalytic conversion of synthesis gas. Processes are classified according to the pressure used:

1) High-pressure process 25 – 30 MPa
2) Medium-pressure process 10 – 25 MPa
3) Low-pressure process 5 – 10 MPa

The main advantages of the low-pressure process are lower investment and production costs, improved operational reliability, and greater flexibility in the choice of plant size.

Industrial methanol production can be subdivided into three main steps:

1) Production of synthesis gas
2) Synthesis of methanol
3) Processing of crude methanol

## 5.1. Production of Synthesis Gas

All carbonaceous materials such as coal, coke, natural gas, petroleum, and fractions obtained from petroleum (asphalt, gasoline, gaseous compounds) can be used as starting materials for synthesis gas production. Economy is of primary importance with regard to choice of raw materials. Long-term availability, energy consumption, and environmental aspects must also be considered.

Natural gas is generally used in the large-scale production of synthesis gas for methanol synthesis. In a few processes (e.g., acetylene production), residual gases are formed which have roughly the composition of the synthesis gas required for methanol synthesis.

Synthesis gases are characterized by the stoichiometry number $S$:

$$S = \frac{[H_2] - [CO_2]}{[CO] + [CO_2]}$$

where the concentrations of relevant components are expressed in volume percent. The stoichiometry number should be at least 2.0 for the synthesis gas mixture. Values above 2.0 indicate an excess of hydrogen, whereas values below 2.0 mean a hydrogen deficiency relative to the stoichiometry of the methanol formation reaction.

**Natural Gas.** Most methanol produced worldwide is derived from natural gas. Natural gas can be cracked by steam reforming and by partial oxidation (Fig. 2).

In *steam reforming* the feedstock is catalytically cracked in the absence of oxygen with the addition of water and possibly carbon dioxide. The reaction heat required is supplied externally. In *partial oxidation*, cracking takes place without a catalyst. Reaction heat is generated by direct oxidation of part of the feedstock with oxygen. In a combination of the two processes, only part of the natural gas stream is subjected to

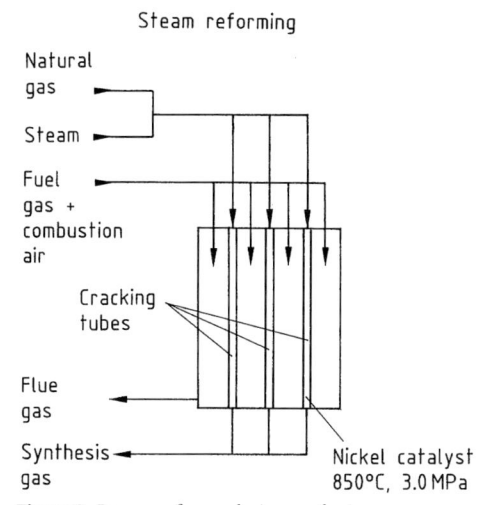

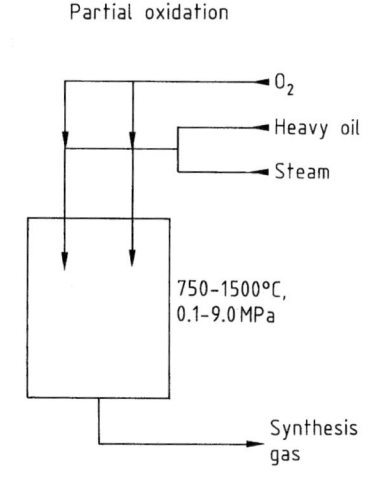

**Figure 2.** Processes for producing synthesis gases

steam reforming [108]. The remainder passes with the reformed gas to an autothermal reformer where the natural gas is partially oxidized by oxygen.

Only the production of synthesis gas by steam reforming is discussed here in some detail.

The catalysts used in steam reforming are extremely sulfur sensitive; sulfur concentrations < 0.5 ppm quickly poison the catalyst. A gas purification stage therefore precedes the reformer stage. If sulfur occurs primarily in the form of higher boiling compounds (e.g., mercaptans), batchwise adsorption on a regenerable activated charcoal bed is recommended. In the case of hydrogen sulfide, zinc oxide is used as adsorbent to remove sulfur as zinc sulfide at 340–370 °C. Hydrogenating desulfurization becomes necessary if organic sulfur compounds (e.g., COS) are present that cannot be removed with charcoal. Hydrogen (e.g., in the form of purge gas from methanol synthesis) is mixed with the gas stream to be desulfurized and passed over a cobalt or nickel–molybdenum catalyst at 290–370 °C. The sulfur compounds are converted into hydrogen sulfide, which can be removed in a subsequent zinc oxide column.

In the reformer, natural gas is catalytically cracked in the presence of steam:

$CH_4 + H_2O \rightleftharpoons CO + 3 H_2$  $\Delta H_{300 K} = 206.3$ kJ/mol
$CO + H_2O \rightleftharpoons CO_2 + H_2$  $\Delta H_{300 K} = -41.2$ kJ/mol

The first of these reactions is endothermic and leads to an increase in volume, whereas the second is exothermic and proceeds without change in volume. The degree of conversion of methane increases with increasing temperature, increasing partial pressure of steam, and decreasing absolute pressure.

The interfering Boudouard equilibrium

$2 CO \rightleftharpoons CO_2 + C$  $\Delta H_{300 K} = -172.6$ kJ/mol

which would lead to carbon deposits on the catalyst or on the walls of reformer tubes, can largely be prevented by using excess steam and avoiding long residence times in the critical temperature range above 700 °C.

To reach the stoichiometry necessary for methanol synthesis, carbon dioxide, if available, is mixed with exit gas from the steam reformer. If carbon dioxide is not available, the conversion must be performed with an excess of hydrogen. Hydrogen accumulates in the synthesis recycle gas and must be removed.

**Other Raw Materials.** Natural gas is not the only raw material for synthesis gas used in methanol production plants. Higher hydrocarbons (e.g., liquefied petroleum gas, refinery off-gases, and particularly naphtha) are also used; they are processed mainly by steam reforming. Crude oil, heavy oil, tar, and asphalt products can also be converted into synthesis gas, but this is more difficult than with natural gas. Their sulfur content is considerably higher (0.7 – 1.5% $H_2S$ and COS) and must be removed. Synthesis gas also contains excess carbon monoxide and must, therefore, be subjected to shift conversion.

Coal can be converted into synthesis gas with steam and oxygen by a variety of processes at different pressures (0.5 – 8 MPa) and temperature (400 – 1500 °C). Synthesis gas must be desulfurized and subjected to shift conversion to obtain the required stoichiometry for methanol synthesis.

## 5.2. Synthesis

Important reactions (Eqs. 1 – 3) for the formation of methanol from synthesis gas are discussed in Section 4.1. In one pass only about 50% of the synthesis gas is converted because thermodynamic equilibrium is reached; therefore, after methanol and water are condensed out and removed, the remaining synthesis gas must be recycled to the reactor. A simplified flow diagram for methanol synthesis is shown in Figure 3. The make-up synthesis gas is brought to the desired pressure (5 – 10 MPa) in a multistage compressor (f). The unreacted recycle is added before the recycle stage. A heat exchanger (b) transfers energy from the hot gas leaving the reactor to the gas entering the reactor. The exothermic formation of methanol takes place in the reactor (a) at 200 – 300 °C. The heat of reaction can be dissipated in one or more stages. The mixture is cooled further (c) after passing through the heat exchanger (b); the heat of condensation of methanol and water can be utilized at another point in the process.

Crude methanol is separated from the gas phase in a separator (d) and flashed before being distilled. Gas from the separator is recycled to the suction side of the recycle compressor (e). The quantity of purge gas from the loop is governed by the concentration and absolute amount of inert substances and the stoichiometry number. If hydrogen is used to adjust the composition of the fresh gas to give the required stoichiometry number it can be recovered from the purge gas by various methods (e.g., pressure swing absorption). The purge gas is normally used for reformer heating.

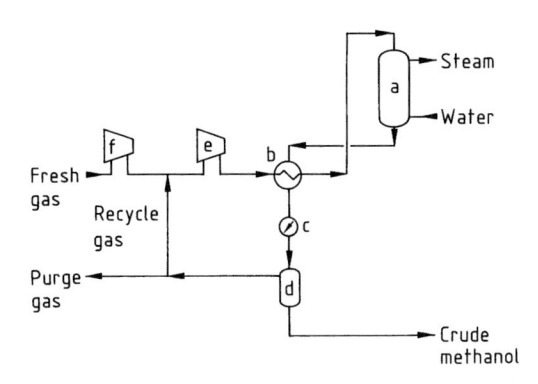

**Figure 3.** Methanol synthesis
a) Reactor; b) Heat exchanger; c) Cooler;
d) Separator; e) Recycle compressor; f) Fresh gas compressor

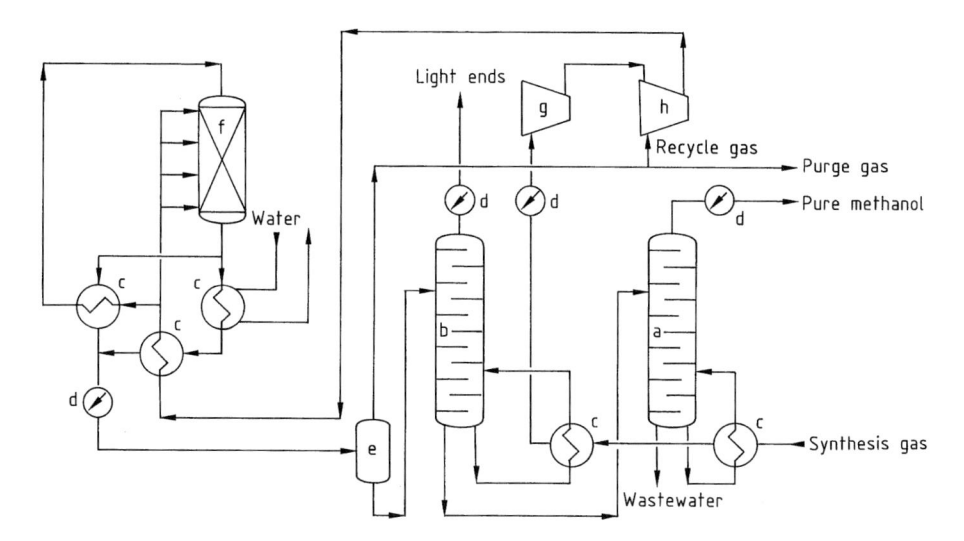

**Figure 4.** The ICI low-pressure methanol process
a) Pure methanol column; b) Light ends column; c) Heat exchanger; d) Cooler; e) Separator; f) Reactor; g) Compressor; h) Compressor recycle stage

## 5.3. Reactor Design

Current industrial processes for producing methanol differ primarily in reactor design. Many different reactors are available [109]; they may be either adiabatic (e.g., ICI) or quasi-isothermal (e.g., Lurgi). The ICI process (Fig. 4) accounts for 60%, and the Lurgi process (Fig. 5) for 30% of worldwide methanol production.

**Adiabatic Reactors.** The ICI process (Fig. 4) uses an adiabatic reactor with a single catalyst bed [110]. The reaction is quenched by adding cold gas at several points. Thus, the temperature profile along the axis of the reactor has a sawtooth shape.

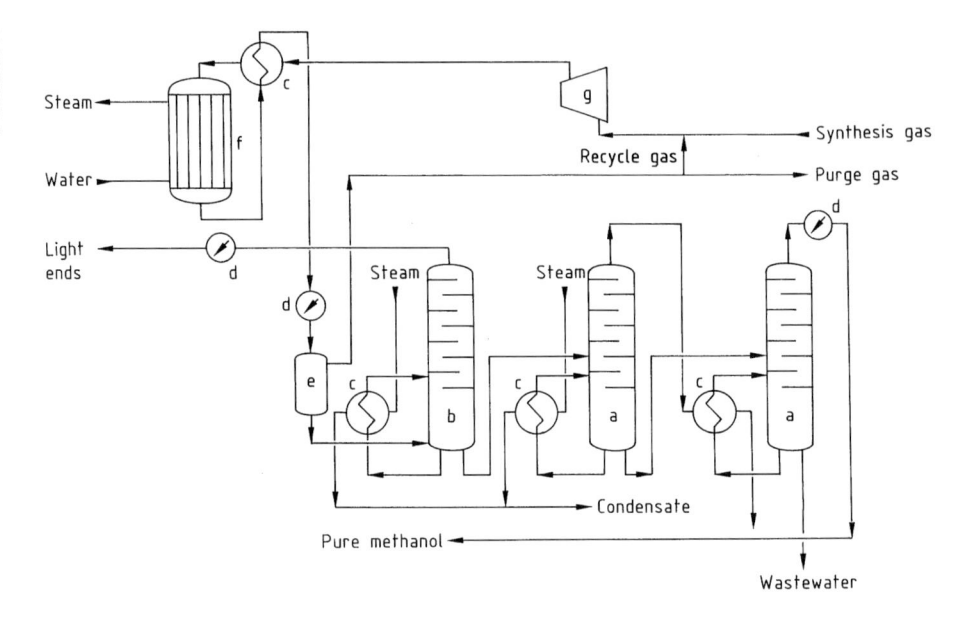

**Figure 5.** Lurgi low-pressure methanol process
a) Pure methanol columns; b) Light ends column; c) Heat exchanger; d) Cooler; e) Separator; f) Reactor; g) Compressor recycle stage

In the Kellogg process, synthesis gas flows through several reactor beds that are arranged axially in series [111]. In contrast to the ICI quench reactor, the heat of reaction is removed by intermediate coolers. The Haldor Topsoe reactor operates on a similar principle, but synthesis gas flows radially through the catalyst beds [112].

Ammonia – Casale S. A. has developed a reactor that employs a combination of axial and radial flow (mixed flow). This type of reactor initially developed for ammonia plants is offered by Davy McKee in ICI license [113].

**Quasi-Isothermal Reactors.** The Lurgi process (Fig. 5) employs a tubular reactor (f) with cooling by boiling water [114]. The catalyst is located in tubes over which water flows. The temperature of the cooling medium is adjusted by a preset pressure.

The Variobar reactor developed by Linde [115] consists of a shell-and-tube reactor coiled in several tiers, whose cooling tubes are embedded in the catalyst packing. The reactor temperature is adjusted by water cooling. As in other processes, the heat of reaction is utilized to produce steam, which can be used, for example, to drive a turbine for the compressor or as an energy source for subsequent methanol distillation.

Whereas synthesis gas flows axially through the two above-mentioned reactors, Toyo offers a reactor through which it flows radially [116]. The advantages, as in the Variobar reactor, lie in a high heat transfer rate with only slight pressure loss.

The Mitsubishi Gas Chemical (MGC) process uses a reactor with double-walled tubes that are filled in the annular space with catalyst [117]. The synthesis gas first flows through the inner tube to heat it up and then, in countercurrent, through the catalyst between the two tubes. The outer tubes are cooled by water, Mitsubishi considers the main advantage of this process to be the high conversion rate (ca. 14% methanol in the reactor outlet).

## 5.4. Distillation of Crude Methanol

Crude methanol leaving the reactor contains water and other impurities (see Section 4.1). The amount and composition of these impurities depend on reaction conditions, feed gas, and type and lifetime of the catalyst. Crude methanol is made slightly alkaline by the addition of small amounts of aqueous caustic soda to neutralize lower carboxylic acids and partially hydrolyze esters.

The methanol contains low-boiling and high-boiling components (light and heavy ends). The light ends include dissolved gases, dimethyl ether, methyl formate, and acetone. The heavy ends include higher alcohols, long-chain hydrocarbons, higher ketones, and esters of lower alcohols with formic, acetic, and propionic acids. Higher waxy hydrocarbons consisting of a mixture of mostly straight-chain $>C_8-C_{40}$ compounds are also formed in small amounts. They have low volatility and thus remain in the distillation bottoms, from which they can easily be removed because of their low solubility in water and low density.

The impurities in crude methanol are generally separated in two stages. First, all components boiling at a lower temperature than methanol are removed in a light ends column (Fig. 4 b, Fig. 5 b). Pure methanol is then distilled overhead in one or more distillation columns (Fig. 4 a, Fig. 5 a). If the columns operate at different pressures, the heat of condensation of the vapors of the column operating at higher pressure can be used to heat the column at lower pressure.

## 5.5. Construction Materials

Low-molybdenum steels are normally used as construction materials in methanol synthesis. Because organic acids are especially likely to be encountered in the methanol condensation stage, stainless steels are generally used then. Damage due to acids can also be prevented by the addition of small amounts of dilute caustic soda.

Stainless steels are normally employed in equipment operating at temperatures in which the formation of iron pentacarbonyl is likely (i.e., 100–150 °C). This applies, for example, to heat exchangers and compressors. Contamination with iron pentacarbonyl should be avoided because it decomposes at the temperatures used for methanol synthesis. Iron deposited on the catalyst poisons it and promotes the formation of higher hydrocarbons (waxy products).

# 6. Handling, Storage, and Transportation

## 6.1. Explosion and Fire Control

The flammability of methanol and its vapors represents a potential safety problem. The flash point is 12.2 °C (closed cup) and the ignition temperature 470 °C; in the Federal Republic of Germany methanol is thus included in ignition group B of the VbF [119].

Methanol vapor is flammable at concentrations of 5.5 – 44 vol%. The saturated vapor pressure at 20 °C is 128 kPa; a saturated methanol–air mixture is thus flammable over a wide temperature range. Methanol is included in ignition group G1, explosion class 1 (ExRL).

In premises and workshops in which the presence of methanol vapor is likely, electrical equipment must be designed in accordance with the relevant regulations:

Guidelines for explosion protection (ExRL)
Regulations governing electrical equipment in explosionhazard areas (ElExV)
DIN VDE 0165
DIN EN 50 014 – 50 020

For international guidelines on the handling of methanol publications of the Manufacturing Chemists' Association should be consulted [118].

Pure, anhydrous methanol has a very low electrical conductivity. Measures to prevent electrostatic charging must therefore be adopted when transferring and handling methanol.

**Fire Prevention.** The VbF restrictions on the amount of methanol that can be stored in laboratory premises should be observed. When large amounts of methanol are stored in enclosed spaces, monitoring by means of lower explosion limit monitors is desirable.

Permanently installed fire-extinguishing equipment should be provided in large storage facilities. Water cannons are generally installed in storage tank farms to cool steel constructions and neighboring tanks in the event of fire. Large tanks should have permanently installed piping systems for alcohol-resistant fire-extinguishing foams.

**Fire Fighting.** Conventional fire-extinguishing agents such as powder, carbon dioxide, or Halon can be used for small fires. Water is unsuitable as an extinguishing agent for fires involving large amounts of methanol because it is miscible with the compound; mixtures containing small amounts of methanol may also burn. Protein-based alcohol-resistant foams are suitable.

A methanol flame is practically invisible in daylight, which complicates fire fighting. The methanol flame does not produce soot, although formaldehyde and carbon monoxide form during combustion when oxygen is lacking. Respirators must therefore be worn when fighting fires in enclosed areas.

## 6.2. Storage and Transportation

**Small-Scale Storage.** Fairly small amounts ($\leq 10$ L) of methanol for laboratory and industrial use are stored in glass bottles or sheet-metal cans; amounts up to 200 L are stored and transported in steel drums. Some plastic bottles and containers cannot be used because of their permeability and the danger of dissolution of plasticizers. High-density polyethylene and polypropylene are suitable, whereas poly(vinyl chloride) and polyamides are unsuitable.

**Large-Scale Storage.** Large amounts of methanol are stored in tanks that correspond in design and construction to those used for petroleum products; cylindrical tanks with capacities from a few hundred cubic meters to more than 100 000 m$^3$ are normally used. With fixed-roof tanks, special measures (e.g., nitrogen blanketing) should be adopted to prevent the formation of an ignitable atmosphere in the space above the liquid surface. Emission of methanol may occur if the level fluctuates. To avoid these problems, large tanks are often equipped with floating roofs; attention should therefore be paid to guard against entry of rainwater.

For anhydrous and carbon dioxide-free methanol tanks, pipelines and pumps can be constructed from normal-grade steel; seals can be made from mineral fiber, graphite, and metal. Styrene–butadiene rubber, chlorine–butadiene rubber, and butyl–chlorobutyl rubber can be used for shaft seals.

**Large-Scale Transportation.** Methanol is traded worldwide. The recent trend toward relocating production to sites that are remote from industrial centers where inexpensive natural gas is available, has meant that ca. 30 % of methanol produced worldwide must be transported by sea to consumer countries (Japan, Europe, United States). Specially built tankers with capacities up to 40 000 t are available for this purpose; ships built to transport petroleum products are also used.

The most important European transshipment point for methanol is Rotterdam. Methanol is distributed to inland industrial regions mainly by inland waterways on vessels with capacities up to 3000 t. Boats specialized for methanol transport are the exception; impurities can therefore be introduced into the methanol due to frequent change of cargo. Analysis prior to delivery is generally essential.

Methanol is also transported by road and rail tank cars. Permanently coupled trains consisting of several large tank cars with common filling, discharge, and ventilation lines are used to supply large customers.

Transportation via pipeline is only of importance for supplying individual users within enclosed, self-contained chemical complexes.

**Safety Regulations Governing Transportation.** The transportation of methanol as less-than-carload freight in appropriate vessels, containers, and bulk, is governed by specific regulations that differ from country to country. An effort is being made, and is already well advanced, to coordinate these regulations within the EC. Relevant legal regulations governing less-than-carload and bulk transportation by sea, on inland waterways, and by rail, road, and air are as follows [120]:

| | |
|---|---|
| IMDG Code (D-GGVSee) | D 3328/E-F 3087, Class 3.2, UN No. 1230 |
| RID (D-GGVE) | Class 3, Rn 301, Item 5 |
| ADR (D-GGVS) | Class 3, Rn 2301, Item 5 |
| ADNR | Class 3, Rn 6301, Item 5, Category Kx |
| European Yellow Book | No. 603-001-00-X |
| EC Guideline/D VgAst | No. 603-001-00-X |
| FRG (Land, VbF) | B |
| Great Britain | Blue Book: flammable liquid and IMDG Code E 3087 |
| United States | CRF 49, Paragraph 172.1.1, flammable liquid |
| IATA | RAR, Art. No. 1121/43, flammable liquid |

# 7. Quality Specifications and Analysis

**Methanol for Laboratory Use.** Methanol is available commercially in various purity grades for fine chemicals:

1) "Synthesis" quality (corresponding to normal commercial methanol)
2) Certified analytical quality
3) Extremely pure qualities for semiconductor manufacture

**Commercial Methanol.** In addition to laboratory grades, commercial methanol is generally classified according to ASTM purity grades A and AA (Table 5). Methanol for chemical use normally corresponds to Grade AA.

In addition to water, typical impurities include acetone (which is very difficult to separate by distillation) and ethanol. When methanol is delivered by ships or tankers used to transport other substances, contamination by the previous cargo must be expected.

Comparative ultraviolet spectroscopy has proved a convenient, quick test method for deciding whether a batch can be accepted and loaded. Traces of all chemicals derived from aromatic parent substances, as well as a large number of other compounds, can be detected.

Further tests for establishing the quality of methanol include measurements of boiling point range, density, permanganate number, turbidity, color index, and acid

**Table 5.** Federal specifications for pure methanol in the United States

| Property | Grade A | Grade AA |
|---|---|---|
| Ethanol content, mg/kg | | < 10 |
| Acetone content, mg/kg | | < 20 |
| Total acetone and aldehyde content, mg/kg | < 30 | < 30 |
| Acid content (as acetic acid), mg/kg | < 30 | < 30 |
| Color index (APHA) | < 5 | < 5 |
| Sulfuric acid test (APHA) | < 30 | < 30 |
| Boiling point range (101.3 kPa), must include 64.6 ± 0.1 °C | < 1 | < 1 |
| Dry residue, mg/L | < 10 | < 10 |
| Density (20 °C), g/cm$^3$ | 0.7928 | 0.7928 |
| Permanganate number | > 30 | > 30 |
| Methanol content, wt% | > 99.85 | > 99.85 |
| Water content, wt% | < 0.15 | < 0.10 |
| Odor | typical, non-persistent | |

number. More comprehensive tests include water determination according to the Karl Fischer method and gas chromatographic determination of byproducts. However, the latter is relatively expensive and time consuming because several injections using different columns and detectors must be made due to the variety of byproducts present.

The most important standardized test methods for methanol are

| | |
|---|---|
| DIN 51 757 | density |
| ASTM D 941 | density |
| ASTM D 1078 | boiling range |
| ASTM D 1209 | color index |
| ASTM D 1353 | dry residue |
| ASTM D 1363 | permanganate number |
| ASTM D 1364 | water content |
| ASTM D 1612 | acetone content |
| ASTM D 1613 | acid content |

Apart from pure methanol, methanol obtained directly from synthesis without any purification, or with only partial purification, is sometimes used. This crude methanol can be used for energy (fuel methanol), for the manufacture of synthetic fuels, and for specific chemical and technical purposes; it is not normally available commercially. Composition varies according to synthesis conditions; principal impurities include, 5 – 20 vol% water, higher alcohols, methyl formate, and higher esters. The presence of water and esters can cause corrosion during storage due to the formation of organic acids (see Section 6.2); remedies include alkaline adjustment with sodium hydroxide and, if necessary, the use of corrosion-resistant materials.

# 8. Environmental Protection

Methanol is readily biodegraded; most microorganisms possess the enzyme alcohol dehydrogenase, which is necessary for methanol oxidation. Therefore, no danger exists of accumulation in the atmosphere, water, or ground; the biological stages of sewage treatment plants break down methanol almost completely. In the Federal Republic of Germany methanol has been classified as a weakly hazardous compound in water hazard Class 1 (WGk I, § 19 Wasserhaushaltsgesetz). In accidents involving transport, large amounts of methanol must be prevented from penetrating into the groundwater or surface waters to avoid contaminating drinking water. Little is known about the behavior of methanol in the atmosphere. Emissions occurring during industrial use are so small that harmful influences can be ignored. That situation could alter, however, if methanol were used on a large scale as an alternative to petroleum-based fuels.

In methanol production, residues that present serious environmental problems are not generally formed. All byproducts are used when possible; for example, the condensate can be processed into boiler feedwater, and residual gases or low-boiling byproducts can be used for energy production. The only regularly occurring waste product that presents some difficulties is the bottoms residue obtained after distillation of pure methanol; it contains water, methanol, ethanol, higher alcohols, other oxygen-containing organic compounds, and variable amounts of paraffins. The water-soluble organic substances readily undergo biological degradation; the insoluble substances can be incinerated safely in a normal waste incineration unit. In some cases this residual water is also subjected to further distillative purification; the resultant mixture of alcohols, esters, ketones, and aliphatics can be added in small amounts to carburetor fuel.

The spent catalysts contain auxiliary agents and supports, as well as copper (synthesis), nickel (gas generation), and cobalt and molybdenum (desulfurization) as active components. These metals are generally recovered or otherwise utilized.

Modern steam reformers can be fired so that emission of nitrogen oxides ($NO_x$) in the flue gas is maintained below 200 mg/m$^3$ without having to use secondary measures.

# 9. Uses

## 9.1. Use as Feedstock for Chemical Syntheses

Approximately 70% of the methanol produced worldwide is used in chemical syntheses: in order of importance formaldehyde, methyl*tert*-butyl ether (MTBE), acetic acid, methyl methacrylate, and dimethyl terephthalate. Only a small proportion is utilized for energy production, although this use has great potential.

**Formaldehyde.** Formaldehyde is the most important product synthesized from methanol (→ Formaldehyde); in 1988, 40% of the methanol produced worldwide was used to synthesize this product. The annual estimated increase in formaldehyde production from methanol is ca. 3%, but because other bulk products have higher growth rates its share as a proportion of methanol use will decrease.

The processes employed are all based on the oxidation of methanol with atmospheric oxygen. They differ mainly with regard to temperature and nature of the catalyst used.

**Methyl *tert*-butyl ether** is produced by reacting methanol with isobutene on acid ion exchangers (→ Methyl tert-Butyl Ether). Increasing amounts of methanol are used in this form in the fuel sector. The ether is an ideal octane booster and has become extremely important due to the introduction of unleaded grades of gasoline and awareness of the possible harmfulness of aromatic high-octane components. In 1988, 20% of worldwide methanol production was used for MTBE synthesis; annual increase rates of up to 12% are expected. The availability of isobutene is becoming an increasing problem in MTBE synthesis, although the situation has recently been improved by the construction of plants for the isomerization of butane and subsequent dehydrogenation of isobutane.

**Acetic Acid.** Another 9% of the methanol produced is used to synthesize acetic acid, and annual growth rates of 6% are estimated. Acetic acid is produced by carbonylation of methanol with carbon monoxide in the liquid phase with cobalt–iodine, rhodium–iodine, or nickel–iodine homogeneous catalysts (→ Acetic Acid). The older BASF process operates at 65 MPa, whereas more modern processes (e.g., the Monsanto process) operate at 5 MPa. By varying operating conditions the synthesis can also be modified to produce acetic anhydride or methyl acetate.

**Other Synthesis Products.** In the intensive search after the oil crisis for routes to alternative fuels, processes were developed that allowed fuels to be produced from synthesis gas with methanol as an intermediate. Mobil in the United States has contributed decisively to the development of such processes, which involve mainly the reaction of methanol on zeolite catalysts. The most important and, up to now, the only industrially implemented process is methanol to gasoline (MTG) synthesis. A plant for producing and converting 4500 t/d of methanol from natural gas into 1700 t/d of gasoline has been built and operated as a joint venture between the New Zealand government and Mobil. Since the prices of petroleum products have not risen as expected, ways are now being sought to process the methanol from this plant into pure methanol and to market it as such.

Further synthesis routes that could become important in the event of a scarcity of petroleum products are the methanol to olefins (MTO) and methanol to aromatic compounds (MTA) processes [121].

A product that received great attention as a result of the discussion of environmental damage caused by chlorofluorocarbons is dimethyl ether (→ Dimethyl Ether). It can be used as an alternative propellant for sprays. Compared to propane–butane mixtures also used as propellants, its most important feature is its higher polarity and, thus, its better solubilizing power for the products used in sprays. Dimethyl ether is also used as a solvent, organic intermediate, and in adhesives.

Methanol is used to synthesize a large number of other organic compounds:

| | |
|---|---|
| Formic acid | preservatives, pickling agents |
| Methyl esters of organic acids | solvents, monomers |
| Methyl esters of inorganic acids | methylation reagents, explosives, insecticides |
| Methylamines | pharmaceutical precursors, auxiliaries, absorption liquids for gas washing and scrubbing |
| Trimethylphosphine | pharmaceuticals, vitamins, fragrances, fine chemicals |
| Sodium methoxide | organic intermediates, catalyst |
| Methyl halides | organic intermediates, solvents, propellants |
| Ethylene | organic intermediates, polymers, auxiliaries (→ Ethylene) |

## 9.2. Use as Energy Source

Methanol is a promising substitute for petroleum products if they become too expensive for use as fuels. As a result of the oil crisis in the early 1970s, a number of projects were started based on the assumption that the use of methanol produced from coal would be more economical in the medium term than the use of petroleum products. The estimates made at the beginning of the 1980s proved to be too optimistic, however, with regard to costs and to overcoming technical or environmental problems involved in producing synthesis gas from coal, and too pessimistic with regard to the price and availability of crude oil (Table 6). Nearly all the large-scale projects for coal utilization have been discontinued. Large-scale operational plants (e.g., Cool Water, United States and Rheinbraun, Wesseling, FRG) are being shut down or modified for use with other feedstocks [122].

**Methanol as a Fuel for Otto Engines.** The use of methanol as a motor fuel has been discussed repeatedly since the 1920s. Use has so far been restricted to high-performance engines for racing cars and aeroplanes. The combustion of methanol in four-stroke engines has been investigated for a long time. Methanol has been found to be an ideal fuel in many respects. Because of its high heat of vaporization and relatively low calorific value, a substantially lower combustion chamber temperature is achieved than with conventional motor fuels. Emissions of nitrogen oxides, hydrocarbons, and carbon monoxide are lower. This is offset, however, by increased emission of formaldehyde.

The important properties of methanol for use as a fuel are compared with those of a conventional fuel (gasoline) in Table 7. Consumption is higher because of the lower calorific values.

Methanol can be used in various mixing ratios with conventional petroleum products:

**Table 6.** Comparison of the efficiencies of natural gas conversion in liquid fuels

| Energy carrier | Higher heating value, Gcal/t | Yield, t/t $CH_4$ | Higher heating value, Gcal/t $CH_4$ | Theoretical efficiency | Stoichiometric factor, [H]−[20]/[C] | Technical efficiency** |
|---|---|---|---|---|---|---|
| Methane | 13.2 | 1 | 13.28 | 100 | 4 | |
| Synthesis gas, partial oxidation | 6.36 | 2 | 12.70 | 95.7 | 2 | 85−90 |
| Synthesis gas, steam reforming | 7.96 | 2.12* | 16.91* | | 6 | |
| Methanol | 5.36 | 2 | 10.72 | 80.7 | 2 | 68−72 |
| Ethanol | 7.14 | 1.43 | 10.25 | 77.2 | 2 | |
| Kerosene | 11.00 | 0.87 | 9.57 | 72.1 | ca. 2.05 | |
| Diesel fuel | 10.70 | 0.87 | 9.34 | 70.4 | ca. 2 | 55−60 FT |
| Gasoline (average) | 10.50 | 0.86 | 9.06 | 68.2 | ca. 1.8 | 58−63 MTG 55−60 FT |
| Benzene | 10.02 | 0.81 | 8.13 | 61.2 | 1 | |

\* With extra heat energy for reformer.
\*\* FT = Fischer−Tropsch; MTG = methanol to gasoline.

**Table 7.** Comparison of methanol and a typical fuel (gasoline) for use in Otto engines

| Property | Gasoline | Methanol |
|---|---|---|
| Density, kg/L | 0.739 | 0.793 |
| Calorific value, kJ/kg | 44 300 | 21 528 |
| Air consumption, kg/kg | 14.55 | 6.5 |
| Research octane number | 97.7 | 108.7 |
| Motor octane number | 89 | 88.6 |
| Mixed research octane number | | 120−130 |
| Mixed motor octane number | | 91−94 |
| Reid vapor pressure, kPa | 64 | 32 |
| Boiling point range, °C | 30−190 | 65 |
| Heat of vaporization, kJ/kg | 335 | 1174 |
| Cooling under vaporization with stoichiometric amount of air, °C | 20 | 122 |

| | |
|---|---|
| M 3 | Mixture of 3% methanol with 2−3% solubilizers (e.g., isopropyl alcohol) in commercially available motor fuel. This system is already widely used because modification of motor vehicles and fuel distribution systems is not required. |
| M 15 | Mixture of 15% methanol and a solubilizer with motor fuel; alterations to the motor vehicles are necessary in this case. The proposed use of M 15 to increase the octane number in unleaded gasoline has been supplanted by the large increase in the use of MTBE. |
| M 85 | Methanol containing 15% $C_4$−$C_5$ hydrocarbons to improve cold-start properties. Modified vehicles and fuel distribution systems are necessary. |
| M 100 | Pure methanol—vehicles must be substantially modified and fully adapted to methanol operation. |

The necessary modifications for methanol operation involve the replacement of plastics used in the fuel system (see Section 6.2). The ignition system and carburetor or fuel injection unit also have to be adapted. With M 85 and M 100 the fuel mixture must be preheated because vaporization of the stoichiometric amount of methanol in the carburetor results in a cooling of 120 K.

In mixtures with a low methanol content (M 3, M 15) phase separation in the presence of traces of water must be avoided. Absolutely dry storage, transportation, and distribution systems must be available for mixed fuels to prevent separation of water–methanol and hydrocarbon phases.

A further restriction on the use of methanol in gasoline is imposed by the increase in gasoline vapor pressure (Reid vapor pressure, RVP). In some warm regions of the United States, legal restrictions on the RVP have already been introduced to reduce hydrocarbon emissions, which are an important factor in the formation of photochemical smog and increased ozone concentration in the lower atmosphere. As a result, methanol can not longer be added to motor fuel because it increases the vapor pressure of the butane used as a cheap octane booster.

Widespread use of methanol as an exclusive fuel for cars is currently being inhibited by its high cost and the lack of a suitable distribution system. Possible solutions to the latter problem include the construction of dual-purpose vehicles (flexible fuel vehicles), which can use either methanol or normal fuel. Another solution is to use methanol for company or government car fleets, which refill their tanks at a few specific filling stations. Trials based on this concept are underway in several countries; the largest is taking place in California [123].

**Methanol as Diesel Fuel.** Exclusive operation with methanol is not possible in diesel engines because methanol has a cetane number of 3 and will therefore not ignite reliably. To ensure ignition the engine must have an additional injector for normal diesel fuel; methanol is injected into the cylinder after ignition of the diesel fuel [124]. Additives are being developed to improve ignition performance.

**Other Uses of Methanol in the Fuel Sector.** In contrast to pure methanol, the use of MTBE in Otto engine fuels is not limited by considerations of miscibility or vapor pressure. The use of methanol for MTBE synthesis could soon quantitatively overtake its conventional uses. Arco, the world's largest producer of MTBE, is also promoting the use of oxinol, a mixture of methanol and *tert*-butanol.

An additional development in the use of methanol is the Lurgi Octamix process. Use of an alkali-doped catalyst and modified conditions (higher temperature, lower $CO_2$ concentration, higher CO concentration) in methanol synthesis yields a mixture of methanol, ethanol, and higher alcohols [125]. This mixture can be used directly in the engine fuel. The presence of higher alcohols is desirable not only because of the increase in octane number, but also because they act as solubilizers for methanol. However, this process is not yet used on an industrial scale.

**Other Energy Uses of Methanol.** A use that has been discussed particularly in the United States and implemented in pilot projects is the firing of peak-load gas turbines in power stations (peak shaving). Benefits include simple storage and environmentally friendly combustion in the gas turbine. The use of methanol as a fuel in conventionally fired boilers obviates the need for costly flue gas treatment plants but is not yet economically viable.

The gasification of methanol to obtain synthesis gas or fuel gas has often been proposed. Apart from exceptions such as the production of town gas in Berlin, here too, economic problems have prevented technical implementation.

## 9.3. Other Uses

Methanol's low freezing point and its miscibility with water allow it to be used in *refrigeration systems*, either in pure form (e.g., in ethylene plants) or mixed with water and glycols. It is also used as an *antifreeze* in heating and cooling circuits; compared to other commonly used antifreezes (ethylene glycol, propylene glycol, and glycerol), it has the advantage of lower viscosity at low temperature. It is, however, no longer used as an engine antifreeze; glycol-based products are employed instead.

Large amounts of methanol are used *to protect natural gas pipelines* against the formation of gas hydrates at low temperature. Methanol is added to natural gas at the pumping station, conveyed in liquid form in the pipeline, and recovered at the end of the pipeline. Methanol can be recycled after removal of water taken up from natural gas by distillation.

Methanol is also used as an *absorption agent* in gas scrubbers. The removal of carbon dioxide and hydrogen sulfide with methanol at low temperature (Rectisol process, Linde and Lurgi) has the advantage that traces of methanol in the purified gas do not generally interfere with further processing [126].

The use of pure methanol as a *solvent* is limited, although it is often included in solvent mixtures.

## 10. Economic Aspects

**Economics of Methanol Production.** The costs of methanol production depend on many factors, the most important being direct feedstock costs, investment costs, and costs involved in logistics and infrastructure.

Natural gas, naphtha, heavy heating oil, coal, and lignite are all used as feedstocks in methanol plants. In heavy oil-based plants and to an increasing extent in coal-based plants the principal cost burden is accounted for by capital costs. This means that, given the currently prevailing low energy prices, such plants have high fixed costs and are, therefore, uneconomical or economical only under special conditions. Under present conditions the balance between investment and operating costs clearly favors natural gas-based plants. All large plants currently being planned are designed for use with natural gas and some plants built for operation with naphtha have been converted. Less than $2 \times 10^6$ t of the currently installed worldwide capacity of ca. $21 \times 10^6$ t is based on raw materials other than natural gas.

**Methanol on the World Market.** After ammonia, methanol is quantitatively the largest product from synthesis gas. Worldwide capacity at the beginning of 1989 was $21 \times 10^6$ t. In 1988, $19 \times 10^6$ t of methanol was produced worldwide. The mean annual production growth rate is ca. 10%. The production curve for methanol since 1965 is illustrated in Figure 6.

The methanol industry underwent radical structural changes during the 1980s. Previously, companies that consumed large quantities of methanol produced the compound themselves from the most readily accessible raw materials at the site of use (i.e., highly industrialized countries with expensive energy sources). Since then the number of plants that produce methanol at remote sites exclusively for sale to processors has risen dramatically.

After the energy crisis of the 1970s, intensive oil prospecting led to the discovery of large natural gas fields in many remote regions. Because little demand for natural gas existed in these regions, the relevant countries in South America, Asia, and the Caribbean were interested in selling natural gas as such or in another form to industrialized countries.

Another, hitherto little-used energy source is the associated gas, which is often flared off. In addition to the transportation of liquefied methane and its use as a starting material for ammonia production, methanol production is often the most suitable alternative for marketing such gases. The technology of methanol production is relatively simple, and transport and storage involve inexpensive technology. On the basis of these considerations, 14 new natural gas-based plants producing methanol for export were built from 1974 to 1985 [128]. The largest single train plant based on this concept is located at Punta Arenas in southern Chile; it came on stream in 1988 and has an output of 750 000 t/a. As a consequence of this development, older methanol plants in industrialized countries such as the United States, Japan, and the Federal Republic of Germany have been shut down. The shift in capacities is illustrated in Figure 7.

Since a close relationship between supply and demand no longer exists, large price fluctuations occur, which are hardly justified by actual market conditions. This makes long-term price forecasts impossible and increases economic risks for new projects.

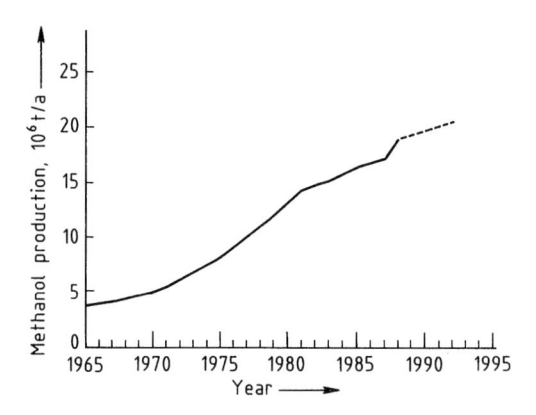

**Figure 6.** Worldwide methanol production
The estimate for 1989–1992 is based on a utilization of 80% capacity [126], [127]

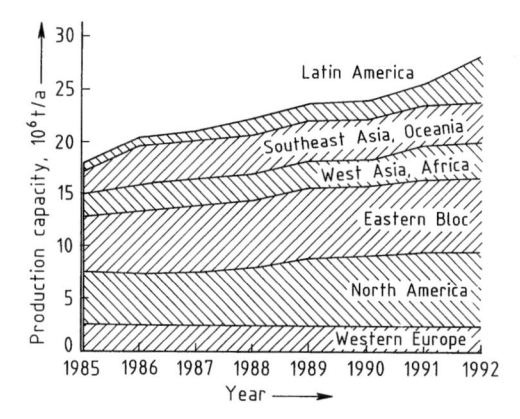

**Figure 7.** Distribution of existing and planned production capacity for methanol according to region [127]

# 11. Toxicology and Occupational Health

## 11.1. Toxicology

**Human Toxicology.** The first accounts of the poisonous action of "methylated spirits" were published in 1855 [129]. However, the number of cases of poisoning increased only after the production of a low-odor methanol. In 1901, DE SCHWEINITZ reported the first cases of industrial poisoning [130].

Liquid methanol is fully absorbed via the gastrointestinal tract [131] and the skin [132] (absorption rate, 0.19 mg cm$^{-2}$ min$^{-1}$). Methanol vapor is taken up in an amount of 70–80% by the lungs [133]. The compound is distributed throughout body fluids and is largely oxidized to formaldehyde and then to formic acid [134]. It is eliminated unchanged through the lungs [132] and in the urine. Elimination half-life is ca. 2–3 h.

The metabolism of methanol to formic acid in humans and primates is catalyzed by the enzyme alcohol dehydrogenase in the liver. This enzyme can be inhibited competitively by ethanol. Formic acid is oxidized to carbon dioxide and water in the presence of folic acid. Because folic acid is not available in sufficient amount in primates, formic acid may accumulate in the body. This leads to hyperacidity of the blood (acidosis), which is ultimately responsible for methanol poisoning [134].

The symptoms of methanol poisoning do not depend on the uptake route (percutaneous, inhalational, oral) and develop in three stages. An initial narcotic effect is followed by a symptom-free interval lasting 10–48 h. The third stage begins with nonspecific symptoms such as abdominal pain, nausea, headache, vomiting, and lassitude, followed by characteristic symptoms such as blurred vision, ophthalmalgia, photophobia, and possibly xanthopsia. Depending on the amount of methanol, individual sensitivity, and the time when treatment is initiated, visual disturbances can either improve or progress within a few days to severe, often irreversible impairment of sight or even to blindness [136]–[139]. The symptoms are accompanied by increasing hyperacidity of the blood due to the accumulation of formic acid, with disturbances in consciousness, possibly deep coma, and in severe cases, death within a few days. The lethal dosage is between 30 and 100 mL per kilogram of body weight. Sensitivity to methanol varies widely, however. Cases have been reported in which no permanent damage occurred after drinking relatively large amounts of methanol (200 or 500 mL) [135], [140]; in another case, however, irreversible blindness resulted after consumption of 4 mL [141].

The treatment of acute oral methanol poisoning [137] should be initiated as quickly as possible with the following measures:

1) Administration of ethanol: In suspected cases of methanol poisoning, 30–40 mL of ethanol (e.g., 90–120 mL of whiskey) is administered immediately as a prophylactic before the patient is referred to a hospital. Because ethanol has a greater affinity for alcohol dehydrogenase than methanol, oxidation of methanol is inhibited; the production of formaldehyde and formic acid from methanol is thus suppressed.
2) Gastric lavage
3) Hemodialysis
4) Treatment with alkali: sodium bicarbonate is infused to control blood hyperacidity.
5) Administration of CNS stimulants (analeptics)
6) Drinking larger volumes of fluid
7) Eye bandage: the eyes should be protected against light
8) The patient should be kept warm

Methanol has a slight irritant action on the eyes, skin, and mucous membranes in humans. Concentrations between 1500 and 5900 ppm are regarded as the threshold value of detectable odor.

Chronic methanol poisoning is characterized by damage to the visual and central nervous systems. Case histories [142], [143] have not been sufficiently documented; whether poisoning is caused by chronic ingestion of low doses or ingestion of intermittently high (subtoxic) amounts is uncertain.

**Animal Toxicology.** Experiments on animals have shown that methanol does not cause acidosis or eye damage in nonprimates (e.g., rats, mice); it generally has a narcotic, possibly lethal, effect. Investigations on laboratory animals cannot, therefore, be extrapolated to humans, at least in the higher dosage range.

In a study on reproductive toxicology, methanol was administered to rats by inhalation during pregnancy. No embryotoxic effects were found after exposure to 5000 ppm [144]. The authors conclude that observance of the recommended concentrations (MAK or TLV values) offers sufficient protection against fetal abnormalities in humans.

In the Ames test, the sex-linked lethal test on *Drosophila melanogaster* and the micronucleus test in mice, methanol was not mutagenic [145], [146].

## 11.2. Occupational Health

No special precautions need be taken when handling methanol because it is not caustic, corrosive, or particularly harmful environmentally. If methanol is released under normal conditions, no danger exists of buildup of acutely toxic concentrations in the atmosphere. (Chronic poisoning via the respiratory tract or oral ingestion is described in Section 11.1.) However, absorption through the skin does constitute a danger, and methanol should be prevented from coming in direct contact with skin.

Appropriate workplace hygiene measures should be adopted if methanol is handled constantly. Rooms in which methanol is stored or handled must be ventilated adequately. The TLV – TWA value (skin) is 200 ppm (262 mg/m$^3$), and the TLV – STEL value is 250 ppm (328 mg/m$^3$). The MAK value is 200 ppm (262 mg/m$^3$). Gas testing tubes can be used to measure the concentration in air. The peak limit should correspond to category II, 1: i.e., the MAK value may be exceeded by a maximum of 100% for 30 min, four times per shift [146]. Respirators must be worn if substantially higher concentrations are present. Filter masks (filter A, identification color brown) can be used only for escape or life-saving purposes because they are exhausted very quickly. Respirators with a self-contained air supply and heavy-duty chemical protective clothing should be used for longer exposures to high methanol concentrations (> 0.5 vol%).

# 12. References

[1] R. C. Weast, D. R. Lide: *CRC Handbook of Chemistry and Physics*, 70th ed., CRC Press, Boca Raton 1989.

[2] H. Stephen, T. Stephen (eds.): *Solubilities of Inorganic and Organic Compounds*, Pergamon Press, Oxford – London – Edinburgh – New York – Paris – Frankfurt 1964.

[3] T. E. Daubert, R. P. Danner: *Data Compilation Tables of Properties of Pure Compounds*, Design Institute for Physical Property Data and American Institute of Chemical Engineers, New York 1985.

[4] J. A. Riddick, W. B. Bunger: "Organic Solvents" in *Techniques of Chemistry*, 3rd ed., vol. **2**, Wiley-Interscience, New York 1986.

[5] K. N. Marsh et al. (eds.): "Non-Hydrocarbons," *TRC Thermodynamic Tables*, Suppl. 61, Thermodynamics Research Center 1989.

[6] J. B. Pedley et al.: *Thermochemical Data of Organic Compounds*, 2nd ed., Chapman and Hall, London – New York 1986.

[7] *Engineering Sciences Data*, vol. 79 028, ESDU International plc, London 1979.

[8] *Engineering Sciences Data*, vol. 88 005, ESDU International plc, London 1988.

[9] B. D. Smith, R. Srivastava: "Thermodynamic Data for Pure Compounds," in *Physical Sciences Data 25*, part B, Elsevier, Amsterdam – Oxford – New York – Tokyo 1986.

[10] K. H. Simmrock et al.: "Critical Data of Pure Compounds," in D. Behrens, R. Eckermann (eds.): *Chemistry Data Series*, vol. **2**, part 1, Dechema, Frankfurt 1986.

[11] L. H. Horsley: "Azeotropic Data III," in *Advances in Chemistry Series*, vol. **116**, American Chemical Society, Washington, D.C. 1973, pp. 74–83.

[12] Engineering Sciences Data, vol. **83 016**, ESDU International plc, London 1983.

[13] C. L. Yaws: *Physical Properties, a Guide to the Physical, Thermodynamic and Transport Property Data of Industrially Important Chemical Compounds*, McGraw-Hill, New York 1977, pp. 197–226.

[14] K. Noda et al., *J. Chem. Eng. Data* **27** (1982) 326.

[15] Th. W. Yergovich et al., *J. Chem. Eng. Data* **16** (1971) no. 2, 222.

[16] *Landolt-Börnstein*, new series, group 4, vol. **1**, part b, 117–118.

[17] *Landolt-Börnstein*, 6th ed., vol. **2**, part 5, 366.

[18] Y. Tanaka et al., *Int. J. Thermophys.* **8** (1987) no. 2, 147.

[19] A. E. Easteal, L. A. Woolf, *J. Chem. Thermodyn.* **17** (1985), 49–62, 69–82.

[20] J. Rex Goates et al., *J. Chem. Thermodyn.* **11** (1979) 739.

[21] *Engineering Sciences Data*, vol. 75 024, ESDU International plc, London 1975.

[22] *Landolt-Börnstein*, **7**, Elektrische Eigenschaften II, 18.

[23] *Landolt-Börnstein*, **6**, Elektrische Eigenschaften I, 632, 711–712.

[24] D. W. Nageli et al.: *Practical Ignition Limits for Low Molecular Weight Alcohols*, The Combustion Institute, March 1984, pp. 18–20.

[25] N. J. Sax, R. J. Lewis: *Dangerous Properties of Industrial Materials*, 7th ed., vol. **3**, Van Norstrand Reinhold, New York 1989, p. 2217.

[26] J. A. Monick: *Alcohols, Their Chemistry, Properties and Manufacture*, Reinhold Publ. Co., New York 1968.

[27] E. J. Wickson: "Monohydric Alcohols," *ACS Symp. Ser.* 1981, no. 159 (March 25–26, 1980).

[28] C. Ferri: *Reaktionen der organischen Synthese*, Thieme Verlag, Stuttgart 1978.

[29] P. Sykes: *A Guidebook to Mechanisms in Organic Chemistry*, 6th ed., Longman Group, London 1986.
[30] D. R. Stull, H. Prophet et al. (eds.): *JANAF Thermochemical Tables*, 2nd ed., NSDRS, NBS-37, U.S. Government Printing Office, Washington, D.C. 1971.
[31] W. J. Thomas, S. Portalski, *Ind. Eng. Chem.* **50** (1958) 967.
[32] R. H. Newton, B. F. Dodge, *J. Am. Chem. Soc.* **56** (1934) 1287.
[33] R. M. Ewell, *Ind. Eng. Chem.* **32** (1940) 149.
[34] V. M. Cherednichenko, Ph. D. Thesis, Korpova, Physico-Chemical Institute, Moscow 1953.
[35] T. Chang, R. W. Rousseau, P. K. Kilpatrick, *Ind. Eng. Chem. Process Des. Dev.* **25** (1986) 477.
[36] G. H. Graaf, P. J. J. M. Sijtsema, E. J. Stamhuis, G. E. H. Joosten, *Chem. Eng. Sci.* **41** (1986) no. 11, 2883.
[37] W. Kotowski, *Przem. Chem.* **44** (1965) 66.
[38] *Kirk-Othmer*, **15**, 398–415.
[39] L. Bisset, *Chem. Eng. (N.Y.)* **84** (1977) no. 21, 155.
[40] G. C. Chinchen et al., *Appl. Catal.* **36** (1988) 1–65.
[41] O. A. Hougen, K. M. Watson, R. A. Ragatz: *Chemical Process Principles*, part 2, Wiley-Interscience, New York 1959.
[42] G. Soave, *Chem. Eng. Sci.* **27** (1972) 1197.
[43] D. Y. Peng, D. B. Robinson, *Ind. Eng. Chem. Fundam.* **15** (1976) 59.
[44] F. Marschner, F. W. Moeller *Appl. Ind. Catal.* **2** (1983) 349–411.
[45] K. Klier, V. Chatikavanij, R. G. Herman, G. W. Simmons, *J. Catal.* **74** (1982) 343.
[46] K. Klier, *Adv. Catal.* **31** (1982) 243–313.
[47] J. C. J. Bart, R. C. P. Sneeden, *Catal. Today* **2** (1987) 1–124.
[48] G. C. Chinchen et al.: *ACS Symposium on Methanol and Synthetic Fuels*, Philadelphia, August 1984.
[49] M. Bowker et al., *J. Catal.* **109** (1988) 263.
[50] M. A. McNeil, C. J. Schack, R. G. Rinker, *Appl. Catal.* **50** (1989) 265.
[51] A. Fischer et al., *Proc. Int. Congr. Catal. 7th, 1980*, 1981, 341.
[52] P. J. Denny, D. A. Whan, *Catalysis (London)* **2** (1978) no. 3, 46–86.
[53] H. H. Kung, *Catal. Rev. Sci. Eng.* **22** (1980) no. 2, 235.
[54] R. J. Hawkins, R. J. Kane, W. E. Slinkard, J. L. Trumbley: "Methanol" in J. J. Ketta, W. A. Cunningham (eds.): *Encyclopedia of Chemical Processing and Design*, Marcel Dekker, New York 1988, pp. 418–483.
[55] A. Deluzarche, R. Kieffer, A. Muth, *Tetrahedron Lett.* **38** (1977) 3357.
[56] R. G. Herman et al., *J. Catal.* **56** (1979) 407.
[57] G. Henrici-Olive, S. Olive, *J. Mol. Catal.* **17** (1982) 89.
[58] W. Seyfert, G. Luft, *Chem. Ing. Tech.* **57** (1985) 482.
[59] R. M. Agny, C. G. Takoudis, *Ind. Eng. Chem. Prod. Res. Dev.* **24** (1985) 50.
[60] Y. B. Kagan et al., *Dokl. Akad. Nauk. SSSR* **224** (1975) 1081.
[61] G. Liu, D. Willcox, M. Garland, H. H. Kung, *J. Catal.* **96** (1985) 251.
[62] G. C. Chinchen et al., *Appl. Catal.* **30** (1987) 333.
[63] M. Bowker, H. Houghton, K. C. Waugh, *J. Chem. Soc. Faraday Trans 1*, **77** (1981) 3023.
[64] V. D. Kuznetsov, F. S. Shub, M. I. Temkin, *Kinet. Katal.* **25** (1984) no. 3, 606.
[65] G. H. Graaf, E. J. Stamhuis, A. Beenackers, *Chem. Eng. Sci.* **43** (1988) no. 12, 3185.
[66] K. J. Smith, R. B. Anderson, *J. Catal.* **85** (1984) 428.
[67] K. Klier in S. Kaliaguine, A. Mahay (eds.): *Catalysis on the Energy Scene, Studies in Surface Science and Catalysis*, vol. **19**, Elsevier, Amsterdam 1984, p. 439.

[68] B. Denise, R. P. A. Sneeden, C. Hamon, *J. Mol. Catal.* **17** (1982) 359.
[69] R. Bardet, J. Thivolle-Cazat, Y. Trambouze, $C_1$ *Mol. Chem.* **1** (1985) no. 3, 201.
[70] E. Ramaroson, R. Kieffer, A. Kiennermann, *Appl. Catal.* **4** (1982) 281.
[71] B. Denise, R. P. A. Sneeden, $C_1$ *Mol. Chem.* **1** (1985) 307.
[72] E. R. A. Matulewicz, Ph. D. Thesis, University of Amsterdam, 1984.
[73] D. J. Elliott, F. Pennella, *J. Catal.* **119** (1989) 359.
[74] G. Natta: *Catalysis*, vol. **3**, Reinhold Publ. Co., New York 1955, pp. 349 ff.
[75] BASF, DE 441433, 1923 (A. Mittasch, M. Pier, K. Winkler).
[76] BASF, DE 109495, 1923.
[77] BASF, GB 237030, 1925.
[78] C. Lormand, *Ind. Eng. Chem.* **17** (1925) 430.
[79] E. Blasiak, PL 340000, 1947.
[80] ICI, GB 1159035, 1965 (J. Gallagher, Y. H. Kiold).
[81] *Crit. Report Appl. Chem.* **13** (1985) 102.
[82] Ph. Courty, Ch. Marceilly: *Preparation of Catalysts III*, Elsevier, Amsterdam 1983, pp. 485 ff.
[83] IFP, EP 0152314, 1987 (P. Courty et al.).
[84] Süd Chemie, EP 0125689, 1987 (M. Schneider, K. Kochloefl, J. Ladebeck).
[85] Shell, US 3709919, 1973 (E. F. Magoon).
[86] ICI, GB 1159035, 1965 (J. Gallagher, Y. H. Kiold).
[87] BASF, DE 2846614, 1978 (F. J. Bröcker et al.).
[88] R. H. Höppener, E. B. M. Doesburg, J. J. F. Scholten, *Appl. Catal.* **25** (1986) 109–194.
[89] P. Gherardi, O. Ruggeri, F. Trifiro, A. Vaccari: *Preparation of Catalysts III*, Elsevier, Amsterdam 1983, pp. 723 ff.
[90] M. V. Twigg (ed.): *Catalyst Handbook*, 2nd ed., Wolfe Publishing Ltd., London 1989, pp. 441–468.
[91] S. Lee, A. Sawant, K. Rodrigues, *Energy Fuels* **3** (1989) 2–7.
[92] Mitsubishi Gas Chemical, GB 2047556, 1980 (K. Asakawa, Y. Yamamoto, S. Ebata, T. Nakamura).
[93] Shell, GB 1371638, 1974 (E. F. Magoon).
[94] Institut Ciezkiý Syntezy, GB 2025252, 1980 (W. Kotowski et al.).
[95] R. G. Herman, G. W. Simmons, K. Klier: *New Horizons in Catalysis, Studies in Surface Science and Catalysis*, vol. **7**, Elsevier, Amsterdam 1981, p. 475.
[96] IFP, GB 2037176, 1980 (A. Sugier, P. Courty, E. Freund).
[97] Shell, DP 2154074, 1972 (E. F. Magoon, L. H. Slaugh).
[98] P. G. Bondar et al., US 4107089, 1978.
[99] Metallgesellschaft, GB 1287327, 1971 (R. Herbert, H. Liebgott).
[100] Mitsubishi Gas Chemical, GB 2095233, 1982 (K. Asakawa).
[101] W. L. Marsden, M. S. Wainwright, J. B. Friedrich, *Ind. Eng. Chem. Prod. Res. Dev.* **19** (1980) 551.
[102] Unisearch, GB 2066856, 1981 (M. S. Wainwright, W. L. Marsden, J. B. Friedrich).
[103] H. E. Curry-Hyde, D. J. Young, M. S. Wainwright, *Applied Catalysis* **29** (1987) 31–41.
[104] H. E. Curry-Hyde, M. S. Wainwright, D. J. Young: *Methane Conversion*, Elsevier, Amsterdam 1988, p. 239.
[105] W. G. Baglin, G. B. Atkinson, L. J. Nicks, *Ind. Eng. Chem. Prod. Res. Dev.* **20** (1981) 87.
[106] J. E. France, W. E. Wallace, *Paper to the 12th Regional ACS Meeting*, Pittsburgh 1980.
[107] J. B. Friedrich, M. S. Wainwright, D. J. Young, *J. Catal.* **81** (1983) 14.
[108] E. Supp, *Energy Prog.* **5** (1985) no. 3, 127.

[109] U. Zardi, *Hydrocarbon Process.* **61** (1982) no. 8, 129.
[110] A. Pinto, P. L. Rogerson, *Chem. Eng. Prog.* **73** (1977) no. 7, 95.
[111] L. R. LeBlanc, J. M. Rovner, H. Roos: *Advanced Methanol Plant Design,* 1989 World Methanol Conference, Houston, Crocco & Associates, 1989.
[112] *Hydrocarbon Process.* **62** (1983) no. 11, 111.
[113] R. E. Smith, G. C. Humphreys, G. W. Griffiths, *Hydrocarbon Process.* **63** (1984) no. 5, 95.
[114] E. Supp, W. Hilsebein, *CEER Chem. Econ. Eng. Rev.* **17** (1985) no. 5, 18.
[115] *CEER Chem. Econ. Eng. Rev.* **17** (1985) no. 5, 23.
[116] K. Ohsaki et al., *CEER Chem. Econ. Eng. Rev.* **17** (1985) no. 5, 31.
[117] I. Takase, K. Niva, *CEER Chem. Econ. Eng. Rev.* **17** (1985) no. 5, 24. *Hydrocarbon Process.* **64** (1985) no. 11, 146.
[118] Manufacturing Chemists Association, Washington, D. C.: Safety Guide SG-3, Flammable Liquids Storage and Handling Drum Lots and Smaller Quantities. Chemical Safety Data Sheet SD-22, 1970. Manual TC-8, Recommended Practices for Bulk Loading and Unloading Flammable Liquid Chemicals to and from Tank Trucks. Manual TC-29, Loading and Unloading Flammable Liquid Chemical, Tank Cars.
[119] Kühn/Birett: *Merkblätter gefährlicher Arbeitsstoffe,* 42. Erg.-Lfg. 2/89, Blatt M 10, Ecomed-Verlag, München.
[120] Hommel: *Handbuch der gefährlichen Güter,* 4th ed., Merkblatt 123, Springer Verlag, Berlin – Heidelberg 1987.
[121] P. J. Jackson, N. White: "Technologies for the Conversion of Natural Gas," *Austr. Inst. Energy Conference,* 1985.
[122] F. Asinger: *Methanol – Chemie- und Energierohstoff,* Springer Verlag, Berlin – Heidelberg 1986.
[123] R. Hagar, *Oil Gas J.,* December 19th, 1988, p. 23.
[124] F. Pischinger, C. Havenith, G. Finsterwalder, *VDI-Ber.* **370** (1980) 331–338.
[125] H. Hiller, E. Supp, *Erdöl Kohle Erdgas Petrochem.* **38** (1985) no. 1, 19–22.
[126] *Linde Ber. Tech. Wiss.* **33** (1973) 6–12.
[127] *Ullmann,* 4th ed., **16**, 621–633.
[128] ChemFacts: *Methanol 1989,* Chemical Intelligence Service, London 1989.
[129] J. F. McFarlan, *Pharm. J. Trans.* **15** (1855) 310.
[130] G. E. De Schweinitz, *Ophthalmic. Rec.* **10** (1901) 289.
[131] M. W. Haggard, L. A. Greenberg, *J. Pharmacol. Exp. Ther.* **66** (1939) 479.
[132] B. Dutkiewicz, J. Konczalik, W. Karwacki: *Int. Arch. Arbeitsmed.* **47** (1980) 81.
[133] G. Leaf, L. J. Zatmann, *Br. J. Ind. Med.* **9** (1952) 19.
[134] J. Pohl, *Arch. Exp. Pathol. Pharmakol.* **31** (1893) 281.
[135] R. Heinrich, J. Angerer, *Int. Arch. Arbeitsmed.* **50** (1982) 341.
[136] US Department of Health Education and Welfare, NIOSH, *Criteria for a recommended Standard Occupational Exposure to Methyl Alcohol,* 1976.
[137] W. Wirth, G. Hecht, C. Gloxhuber: *Toxikologie Fibel,* 2nd ed., Thieme Verlag, Stuttgart 1971, p. 200.
[138] S. Moeschlin: *Klinik und Therapie der Vergiftungen,* 6th ed., Thieme Verlag, Stuttgart 1980, p. 259.
[139] W. Forth, D. Henschler, W. Rummel: *Pharmakologie und Toxikologie,* 5th ed., Wissenschaftsverlag, Mannheim 1987.
[140] I. L. Bennett, T. C. Nation, J. F. Olley, *J. Lab. Clin. Med.* **40** (1952) 405.
[141] W. S. Duke-Elder: *Textbook of Ophthalmology,* vol. **3**, The C. V. Mosby Co., St. Louis 1945, p. 3021.

[142] M. Burk, *Mbl. Augenheilk.* **130** (1957) 845.
[143] A. Dreyfus, *Z. Unfallmed. Berufskrankh.* **1** (1946) 84.
[144] B. K. Nelson et al., *Fundam. Appl. Toxicol.* **5** (1985) 727.
[145] V. F. Simmon, K. Kauhanen, R. G. Tardiff: "Mutagenic activity of chemicals identified in drinking water," in D. Scott, B. A. Bridges, F. H. Sobels, *Progress in Genetic Toxicology*, Elsevier, Amsterdam 1977.
[146] E. Gocke et al., *Mutat. Res.* **90** (1981) 91.
[147] *MAK-Werte 1986/Technische Regeln für Gefahrstoffe TRGS 900 (UVV Anlage 4)*, Jedermann Verlag, Heidelberg 1986.

# Methylamines

AUGUST B. VAN GYSEL, UCB, Brussels, Belgium
WILLY MUSIN, UCB, Brussels, Belgium

| | | | | |
|---|---|---|---|---|
| 1. | Introduction ............ 3325 | 6. | Quality Specifications....... 3331 |
| 2. | Physical Properties ........ 3326 | 7. | Storage and Transportation .. 3331 |
| 3. | Chemical Properties ....... 3326 | 8. | Uses .................. 3332 |
| 4. | Production ............. 3327 | 9. | Economic Aspects ......... 3333 |
| 4.1. | Synthesis............... 3327 | 10. | Toxicology and Occupational Health.................. 3334 |
| 4.2. | Purification............. 3329 | | |
| 5. | Environmental Protection ... 3330 | 11. | References............... 3334 |

## 1. Introduction

Monomethylamine [74-89-5], methanamine, aminomethane, $CH_3NH_2$, $M_r$ 31.06; dimethylamine [124-40-3], N-methylmethanamine, $(CH_3)_2NH$, $M_r$ 45.08; and trimethylamine [75-50-3], N,N-dimethylmethanamine, $(CH_3)_3N$, $M_r$ 59.11, are colorless gases at room temperature. All three are produced by the reaction of methanol and ammonia over solid acid catalysts.

Methylamines are the simplest members of the amine family and constitute an economical source of reactive amino nitrogen. Methylamines are valuable for the synthesis of a broad range of products applicable in such diverse fields as medicine, agriculture, rubber, plastics, and synthetic fibers.

Methylamines display chemical reactivity typical of small aliphatic amines. They bear a strong resemblance to ammonia from which they are derived, and their physical properties largely parallel those of ammonia.

Even at concentrations as low as 0.1 ppm in air, methylamines have a distinct, offensive, fishy odor. At higher concentrations, their odor is more ammoniacal.

Table 1. Physical properties of methylamines

| Property | Monomethyl-amine | Dimethylamine | Trimethylamine |
|---|---|---|---|
| bp (101.33 kPa), °C | −6.33 | 6.88 | 2.87 |
| mp, °C | −93.46 | −92.19 | −117.3 |
| Density ($d^{20}$), g/cm$^3$ | | | |
|   at 101.33 kPa (gas) | 0.0014 | 0.0020 | 0.0027 |
|   at 20 °C (liquid) | 0.6624 | 0.6556 | 0.6331 |
| Surface tension (25 °C), 10$^{-3}$ N/m | 19.19 | 16.33 | 13.47 |
| Heat of vaporization, kJ/mol | | | |
|   at 25 °C | 24.249 | 23.663 | 22.864 |
|   at bp | 26.0 | 24.626 | 27.708 |
| Heat of fusion, kJ/mol | 6.054 | 5.945 | 6.548 |
| Standard heat of formation, kJ/mol | | | |
|   at 25 °C (liquid) | −47.31 | 43.96 | −45.80 |
|   at 25 °C (gas) | −22.98 | −18.46 | −23.86 |
| Heat of combustion, standard state at 25 °C, liquid, kJ/mol | −1061.35 | −1744.63 | −2422.60 |
| Heat capacity, ideal gas (25 °C), J K$^{-1}$ mol$^{-1}$ | 49.27 | 70.84 | 90.97 |
| $t_{crit}$, °C | 156.9 | 164.5 | 160.1 |
| $p_{crit}$, MPa | 7.46 | 5.31 | 4.07 |
| Dielectric constant (25 °C), liquid | 9.4 | 5.26 | 2.44 |
| Dipole moment in benzene (25 °C), D | 1.47 | 1.18 | 0.87 |
| Flash point (closed cup), °C | 0 | −18 | −7 |
| Ignition temperature in air, °C | 430 | 400 | 190 |
| Lower explosion limit in air, vol% | 4.95 | 2.8 | 2 |
| Upper explosion limit in air, vol% | 20.75 | 14.4 | 11.6 |

## 2. Physical Properties

The most important physical properties of methylamines are listed in Table 1. Table 2 summarizes the physical properties of aqueous methylamine solutions [1], [2].

## 3. Chemical Properties (→ Amines, Aliphatic)

All three methylamines readily form salts with acids (e.g., mono-, di-, or trimethylammonium halides—also known as the corresponding amine hydrohalides—with the hydrohalic acids). Reaction with alkyl halides leads to further alkylation, up to and including the formation of quaternary ammonium salts, $R_{4-n}(CH_3N)_n^+ X^-$ (e.g., methylethylamine, methyldiethylamine, and methyltriethylammonium halide in the reaction of ethyl halide with monomethylamine).

Reaction of mono- or dimethylamines with carboxylic acids and their derivatives results in *N*-methylamides. Similarly, substituted mono- and dimethylsulfonamides can be prepared by treatment with a sulfonyl chloride.

*Monomethylamine* (MMA) reacts with phosgene to produce methyl isocyanate, with carbon disulfide in the presence of sodium hydroxide to give sodium monomethyl-

Table 2. Physical properties of aqueous solutions of methylamines

| Property | 40% Mono-methylamine | 40% Dimethyl-amine | 60% Dimethyl-amine | 40% Trimethyl-amine |
|---|---|---|---|---|
| bp (101.33 kPa), °C | 49.4 | 51.5 | 36.0 | 30.8 |
| mp, °C | −38 | ca. −38 | −60 | 1.7 |
| Flash point, °C | −10 | −10 | −30 | ca. −20 |
| Density (15.5 °C), g/cm$^3$ | 0.904 | 0.898 | 0.827 | 0.880 |
| Vapor pressure (total), kPa | | | | |
| 20 °C | 34 | 27 | 57 | 61 |
| 30 °C | 53 | 42 | 80 | 90 |
| 40 °C | 80 | 65 | 115 | 131 |
| 50 °C | 120 | 97 | 158 | 185 |
| Viscosity (15.5 °C), mPa · s | 1.5 | 1.7 | | 1.8 |

dithiocarbamate, and with urea to give *N*-methylurea via the intermediate ammonium cyanate. A mixture of chloroform and alkali converts monomethylamine to methyl isonitrile. Alkaline hypochlorites lead to a mixture of methylmono- and dichloroamines. Monomethylamine is decomposed by nitrous acid to methanol and nitrogen, and by nitrosyl chloride to methyl chloride and nitrogen. Ethylene oxide reacts with monomethylamine to produce methylethanolamine and methyldiethanolamine.

*Dimethylamine* (DMA) and phosgene react to produce tetramethylurea. Reaction with carbon disulfide and alkali in this case gives an alkali dimethyldithiocarbamate, whereas urea leads to the expected *N,N*-dimethylurea. Both nitrous acid and nitrosyl chloride convert dimethylamine to the corresponding nitrosamine. Ethylene oxide reacts with dimethylamine to give dimethylethanolamine.

Oxidation of *trimethylamine* (TMA) with hydrogen peroxide or a peracid gives trimethylamine oxide. The reaction with ethylene oxide in this case yields choline. Trimethylamine can form quaternary ammonium compounds by N-alkylation.

# 4. Production

## 4.1. Synthesis

Methylamines are prepared commercially by the exothermic reaction of methanol with ammonia over an amorphous silica–alumina catalyst at 390–430 °C. All three possible methylamines are produced. The reaction proceeds to thermodynamic equilibrium, whose position is governed by the temperature and the nitrogen–carbon ratio as shown in Figure 1 [3], [4].

The equilibrium composition favors the formation of trimethylamine, whereas market demand is greater for mono- and especially dimethylamine. Excess mono- and trimethylamines are recycled over the same catalyst, leading to additional dimethylamine by dismutation or disproportionation.

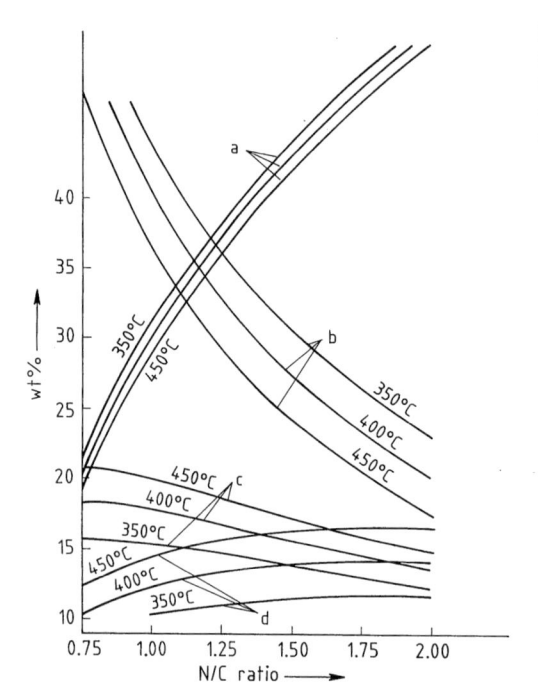

**Figure 1.** Composition of an equilibrium mixture of methylamines as a function of the nitrogen–carbon ratio at 350, 400, and 450 °C
a) Ammonia; b) Trimethylamine; c) Dimethylamine; d) Monomethylamine

The kinetics of the overall reaction have been studied by using several amorphous solid acid catalysts [5], [6]. Amine synthesis can be modeled by six bimolecular equilibrium reactions, three involving amine synthesis and three describing dismutation:

Amination reactions

$NH_3 + CH_3OH \rightleftharpoons CH_3NH_2 + H_2O$
$CH_3NH_2 + CH_3OH \rightleftharpoons (CH_3)_2NH + H_2O$
$(CH_3)_2NH + CH_3OH \rightleftharpoons (CH_3)_3N + H_2O$

Dismutation reactions

$CH_3NH_2 + CH_3NH_2 \rightleftharpoons (CH_3)_2NH + NH_3$
$CH_3NH_2 + (CH_3)_2NH \rightleftharpoons (CH_3)_3N + NH_3$
$(CH_3)_2NH + (CH_3)_2NH \rightleftharpoons (CH_3)_3N + CH_3NH_2$

All six equilibrium constants are known [3], [7]. A parallel reaction leads to the formation of dimethylether, which may also be an intermediate in the synthesis of the amines:

$2\,CH_3OH \rightleftharpoons CH_3\text{–O–}CH_3 + H_2O$

Recent research has been directed toward improving the selectivity for dimethylamine through the use of shape-selective zeolite catalysts [8]–[16].

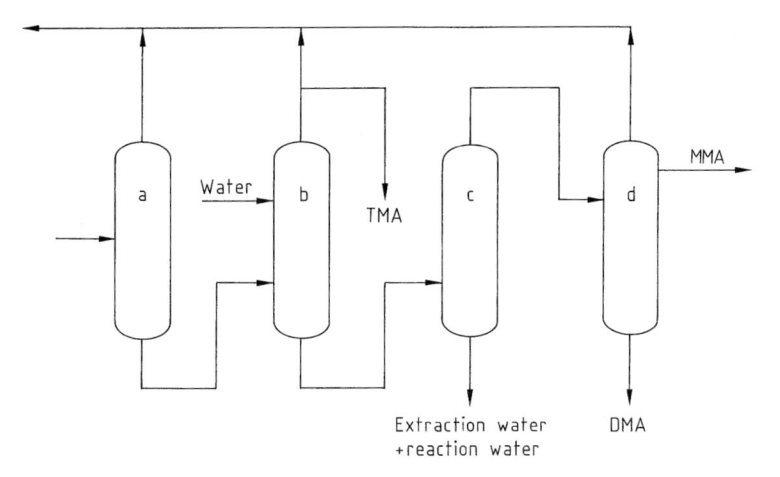

**Figure 2.** Schematic of methylamine purification
a) Ammonia column; b) Extractive distillation column; c) Column for water removal; d) MMA and DMA product

*Reactant selectivity* is a consequence of one of the reactants being too large to pass through the zeolite channels, whereas *product selectivity* is observed when only certain products are of the proper size and shape to diffuse out of the channels [17]. The activity of zeolites with respect to the conversion of methanol is much higher than that of amorphous silica–alumina, but coke formation within the catalyst pores limits catalyst life even at 300 °C (instead of 400–430 °C used with amorphous silica–alumina catalysts) [18].

## 4.2. Purification

The crude reaction mixture consists essentially of excess ammonia; mono-, di-, and trimethylamines; reaction water; and unconverted methanol. Purification is generally effected in a train of four to five distillation columns.

In one technique (Fig. 2) the operating pressures of the columns are chosen in such a way that the top condensors can be cooled by standard 30 °C cooling water, and that the bottom reboilers are fed with saturated steam of about 195 °C. The first column (a) (known as the ammonia column), which operates at 1.7 MPa, removes excess ammonia. The latter is recycled to the synthesis section. In the second column (b), pure trimethylamine is recovered overhead by using extractive distillation with water and a pressure of 1.0 MPa. Extraction water, reaction water, and unconverted methanol are removed from the bottom of the third column (c); the mixture of mono- and dimethylamines is distilled overhead. This mixture is separated in the fourth distillation column (d) at 0.7 MPa, MMA being the overhead and DMA the bottom product.

In another technique, (Fig. 3), a supplementary column (b) operating at 0.7–0.9 MPa is inserted after the ammonia column (a) for bottoms removal of reaction water,

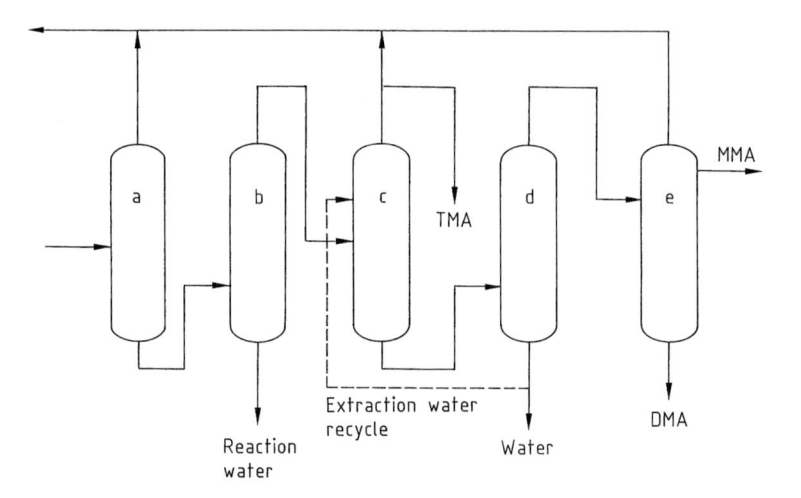

**Figure 3.** Methylamine purification with supplementary column
a) Ammonia column; b) Supplementary column; c) Extractive distillation column; d) Column for water removal; e) MMA and DMA product column

certain impurities, and unconverted methanol. Most of the extraction water may be recycled; the small remaining quantity of bottom water, containing some impurities and methanol, is disposed of easily either by incineration or biological treatment.

# 5. Environmental Protection

Analysis of intermediate byproducts and final products occurs automatically in the operations control room with the aid of an on-line chromatographic analyzer coupled to an on-line computer. Centralizing the data handling in this way assures safe plant operation.

Two effluents are formed in the process:

1) gas containing amines and various byproducts ($H_2$, CO, $N_2$, $O_2$, ethylene), which is incinerated at high temperature to avoid formation of nitrosoamines; and
2) aqueous wastes, whose nature depends on the selectivity and activity of the catalysts. These aqueous solutions contain unconverted methanol and longer chain alkylamines, and can be treated by biological degradation with activated sludge.

**Table 3.** Specifications for monomethylamine

|  | Anhydrous | Aqueous solution |
|---|---|---|
| Monomethylamine, % | min. 99.2 | min. 40 |
| Di- and trimethylamine, % | max. 0.5 | max. 0.2 |
| Ammonia, % | max. 0.05 | max. 0.02 |
| Methanol, % | max. 0.1 | max. 0.05 |
| Water, % | max. 0.3 | balance |

**Table 4.** Specifications for dimethylamine

|  | Anhydrous | Aqueous solutions | |
|---|---|---|---|
|  |  | 40% | 60% |
| Dimethylamine, % | min. 99.5 | min. 40.0 | min. 60.0 |
| Mono- and trimethylamine, % | max. 0.3 | max. 0.15 | max. 0.2 |
| Ammonia, % | max. 0.02 | max. 0.01 | max. 0.01 |
| Methanol, % | max. 0.1 | max. 0.05 | max. 0.05 |
| Water, % | max. 0.2 | balance | balance |

**Table 5.** Specifications for trimethylamine

|  | Anhydrous | Aqueous solution |
|---|---|---|
| Trimethylamine, % | min. 99.0 | min. 45 |
| Mono- and dimethylamine, % | max. 0.3 | max. 0.15 |
| Ammonia, % | max. 0.1 | max. 0.05 |
| Methanol, % | max. 0.1 | max. 0.05 |
| Water, % | max. 0.5 | max. 0.25 |

# 6. Quality Specifications

Product purity as determined by gas chromatography is high (> 99%). Commercial methylamines must be free of higher chain alkylamines. Tables 3, 4, 5 provide typical commercial specifications.

# 7. Storage and Transportation

Methylamines are sold as anhydrous compressed liquefied gases and as aqueous solutions. They are extremely flammable products that should be stored in a well-ventilated area and protected from fire risk.

**Monomethylamine.** The hazard classifications for anhydrous monomethylamine are as follows:

| | |
|---|---|
| ADR/RID | Class 2, no. 3 bt; labels no. 3 + 6.1, flammable; orange panel 236/1061 |
|  | Class 2, D 2200 |
| UN no. | 1061 |

whereas 40% aqueous solutions of monomethylamine are classified as

| | |
|---|---|
| ADR/RID | Class 3, no. 22 a/b; labels no. 3 + 8, flammable, caustic; orange panel 338/1235 |
| IMDG Code | Class 3.1, D 3152 |
| UN no. | 1235 |

**Dimethylamine.** Anhydrous dimethylamine must be classified according to

| | |
|---|---|
| ADR/RID | Class 2, no. 3 bt; labels no. 3 + 6.1, flammable; orange panel 236/1032 |
| IMDG Code | Class 3.1, D 3152 |
| UN no. | 1235 |

The hazard classification for aqueous solutions containing > 40% DMA are

| | |
|---|---|
| ADR/RID | Class 3, no. 22 a/b; labels no. 3 + 8, flammable, caustic; orange panel 338/1160 |
| IMDG Code | Class 3.1, 3.2, D 3275 |
| UN no. | 1160 |

**Trimethylamine.** Hazard classifications for anhydrous trimethylamine are as follows:

| | |
|---|---|
| ADR/RID | Class 2, no. 3 bt; labels no. 3 + 6.1, flammable; orange panel 236/1083 |
| IMDG Code | Class 2, D 2256 |
| UN no. | 1083 |

For an aqueous solution of $\leq 30\%$ trimethylamine, the following hazard classifications are used:

| | |
|---|---|
| ADR/RID | Class 3, no. 22 a; flammable, caustic |
| IMDG Code | Class 3.2, D 3413 |
| UN no. | 1297 |

# 8. Uses

**Monomethylamine** is an intermediate in the synthesis of pharmaceuticals (e.g., theophylline), pesticides (carbaryl, sodium metham, carbofuran), surfactants, photographic developers, explosives, and solvents such as *N*-methyl-2-pyrrolidone.

**Dimethylamine** is also a synthetic intermediate. Examples of products based on dimethylamine include fungicides and rubber accelerators [zinc bis(dimethyldithiocarbamate) (Ziram), tetramethyl thioperoxydicarbonic diamide (TMTD), tetramethyl thiodicarbonic diamide (MTMT)], the propellant 1,1-dimethylhydrazine, various pharmaceuticals, monomers such as dimethylaminoethyl methacrylate, solvents (*N,N*-dimethyl-

**Table 6.** Methylamine production capacity

| Location | Company | Capacity, $10^3$ t |
|---|---|---|
| North America | | 170 |
|   United States | Du Pont | |
| | Air Products | |
| | Pitman-Moore | |
| | GAF | |
|   Canada | Chinook | |
|   Mexico | Celanese | |
| South America | | 10 |
|   Brazil | | |
| Western Europe | | 210 |
|   Fed. Rep. of Germany | BASF | |
|   Netherlands | Akzo | |
|   Belgium | UCB | |
|   United Kingdom | ICI | |
|   Italy | ICT | |
|   Spain | Ertisa | |
| Eastern Europe | | 30 |
|   German Democratic Republic | VEB Leuna | |
|   Soviet Union | | |
|   Rumania | | |
| Asia | | 60 |
|   Japan | Nitto | |
| | Mitsubishi Gas | |
|   India | Fertilizers Chem. | |
|   Korea | Korea Fertilizer | |
|   China | (various plants) | |

formamide, *N,N*-dimethyl-acetamide), catalysts [e.g., 2,4,6-tris[(dimethylamino)-methyl]phenol (DMP 30)], the insecticide dimefox, surfactants, and ion-exchange resins.

**Trimethylamine** is used in the manufacture of choline salts, cationic starches, disinfectants, flotation agents, intense sweeteners, and ion-exchange resins.

# 9. Economic Aspects

Table 6 lists the principal manufacturers of methylamines by country, together with summary data on production capacity. The global market breakdown for methylamines is estimated to be 24% MMA, 57% DMA, and 19% TMA [19].

## 10. Toxicology and Occupational Health

All gaseous amines are regarded as hazardous materials. At high airborne concentration, their odor resembles that of ammonia and the methylamines display an asphyxiating character. At lower concentration, they induce severe irritation and damage to the eyes and respiratory tract; skin contact results in burns. The slightest leak is easily recognized due to the characteristic fishy odor of the compounds, which is detectable above 0.1 ppm by volume. The TLV and the MAK value are 10 ppm by volume.

**Monomethylamine** (RTECS 1.1.78–31.1.87, NIOSH PF 6 300 000)
$LD_{50}$ (mouse, inhalation): 2400 mg/m$^3$ over 2 h
Effects on skin (guinea pig): severe irritant at a dose of 100 mg (Draize test)

**Dimethylamine** (RTECS 1.1.78–31.1.87, NIOSH IP 8 750 000)
$LD_{50}$ (rat, oral): 698 mg/kg
$LC_{50}$ (rat, inhalation): 4540 ppm over 6 h
Toxic effects on eyes, respiratory system, and pulmonary system

**Trimethylamine** (RTECS 1.1.78–31.1.87, NIOSH PA 0 350 000)
$LD_{50}$ (rat, inhalation): 3500 ppm over 4 h

## 11. References

[1] J. A. Riddick, W. B. Bunger, T. K. Sakano: "Organic Solvents," in A. Weissberger (ed.): *Techniques of Chemistry*, vol. II, 4th ed., Wiley-Interscience, New York 1986.
[2] N. I. Sax, R. J. Lewis, Sr.: *Dangerous Properties of Industrial Materials*, 7th ed., Van Nostrand Reinhold, New York 1989.
[3] J. Issoire, C. Van Long, *Bull. Soc. Chim. Fr.* (1960) 2004; cited in D. R. Stull, E. F. Westrum, G. C. Sinke:
*The Chemical Thermodynamics of Organic Compounds*, Wiley, New York 1969.
[4] J. Ramioulle, A. David, *Hydrocarbon Process. Int. Ed.* **60** (1981) no. 7, 113–117.
[5] F. J. Weigert, *J. Catal.* **103** (1987) 20–29.
[6] M. Keane et al., *Appl. Catal.* **32** (1987) no. 1–2, 361–366.
[7] G. Schmitz, *J. Chim. Phys.* **77** (1980) 393; **72** (1975) 579.
[8] D. G. Parker, A. J. Tompsett, GB 2 019 394 B, 1982.
[9] F. J. Weigert, US 4 313 003, 1982.
[10] I. Mochida, JP 49 340, 1983.
[11] M. Deeba, R. N. Cochran, EP-A 85 408, 1983.
[12] R. N. Cochran, M. Deeba, US 4 398 041, 1983.

[13] A. J. Tompsett, T. V. Whittam, EP-A 107 457, 1984.
[14] A. J. Tompsett, US 4 436 938, 1984.
[15] Y. Ashina, T. Fujita, M. Fukatsu, J. Yagi, EP-A 125 616, 1984; EP-A 130 407, 1985.
[16] M. Deeba, W. J. Ambs, R. N. Cochran, US 4 434 300, 1984.
[17] P. B. Weisz, *Pure Appl. Chem.* **52** (1980) 2091.
[18] I. Mochida, A. Yasutake, H. Fujitsu, K. Takeshita, *J. Catal.* **82** (1983) no. 2, 313–321.
[19] T. Gibson et al.: "Alkylamines," *Chemical Economics Handbook,* SRI International 611.5030, 1984.

# Methyl *Tert*-Butyl Ether

BERNHARD SCHOLZ, Hüls Aktiengesellschaft, Marl, Federal Republic of Germany
HARTMUT BUTZERT, Hüls Aktiengesellschaft, Marl, Federal Republic of Germany
JOACHIM NEUMEISTER, Hüls Aktiengesellschaft, Marl, Federal Republic of Germany
FRANZ NIERLICH, Hüls Aktiengesellschaft, Marl, Federal Republic of Germany

| | | | | |
|---|---|---|---|---|
| 1. | Introduction .............. 3337 | 8. | Storage and Transportation .. | 3343 |
| 2. | Physical and Chemical Properties ............... 3338 | 9. | Legal Aspects............. | 3344 |
| | | 10. | Uses ................... | 3344 |
| 3. | Resources and Raw Materials . 3339 | 10.1. | As an Octane Enhancer ..... | 3344 |
| 4. | Production .............. 3340 | 10.2. | Other Uses ............. | 3346 |
| 5. | Environmental Protection ... 3342 | 11. | Toxicology and Occupational Health.................. | 3346 |
| 6. | Quality Specifications....... 3343 | | | |
| 7. | Chemical Analysis ......... 3343 | 12. | References............... | 3347 |

# 1. Introduction

Methyl *tert*-butyl ether, 2-methoxy-2-methylpropane (MTBE) [*1634-04-4*], $M_r$ 88.15, was first synthesized (by the classical Williamson ether synthesis) and characterized in 1904 [1].

$$CH_3-O-\underset{\underset{CH_3}{|}}{\overset{\overset{CH_3}{|}}{C}}-CH_3$$

Methyl *tert*-butyl ether (MTBE)

Extensive U.S. studies during World War II demonstrated the outstanding qualities of MTBE as a high-octane fuel component [2]. Even so, it was not until 1973 that the first commercial plant went on stream in Italy.

Reduction of the lead content in gasoline in the mid-1970s led to a drastic increase in demand for octane enhancers, with MTBE being used increasingly in this role. At present, MTBE is produced from isobutene and methanol. In 1987, MTBE production of $1.6 \times 10^6$ t ranked 32nd among chemicals in the United States [3]. Political decisions about gasoline quality (e.g., lower aromatics content and lower vapor pressure) may result in increased demand for MTBE.

3337

**Table 1.** Vapor pressure, density, and water miscibility of MTBE

| Temperature, °C | Vapor pressure, kPa | Density, g/cm³ | Miscibility | |
|---|---|---|---|---|
| | | | Water in MTBE, wt % | MTBE in water, wt % |
| 0  | 10.8 | 0.7613 | 1.19 | 7.3 |
| 10 | 17.4 | 0.7510 | 1.22 | 5.0 |
| 12 |      | 0.7489 |      |     |
| 15 |      | 0.7458 |      |     |
| 20 | 26.8 | 0.7407 | 1.28 | 3.3 |
| 30 | 40.6 | 0.7304 | 1.36 | 2.2 |
| 40 | 60.5 |        | 1.47 | 1.5 |

**Table 2.** Binary azeotropes with MTBE

| Azeotrope | bp, °C | MTBE content, wt % |
|---|---|---|
| MTBE – water | 52.6 | 96* |
| MTBE – methanol | 51.6 | 86 |
| MTBE – methanol (1.0 MPa) | 130 | 68 |
| MTBE – methanol (2.5 MPa) | 175 | 54 |

* Condensate separates into two phases.

# 2. Physical and Chemical Properties

**Physical Properties.** Methyl *tert*-butyl ether is a colorless, readily mobile liquid with a characteristic terpene-like odor. Its most important physical properties are as follows:

| | |
|---|---|
| mp | – 108.6 °C |
| bp | 55.3 °C |
| $n_D^{20}$ | 1.3692 |
| Dielectric constant (20 °C) | 4.5 |
| Viscosity (20 °C) | 0.36 mPa · s |
| Surface tension (20 °C) | 20 mN/m |
| Specific heat (20 °C) | 2.18 kJ kg$^{-1}$ K$^{-1}$ |
| Heat of vaporization (at *bp*) | 337 kJ/kg |
| Heat of formation (25 °C) | – 314 kJ/mol |
| Heat of combustion | – 34.88 MJ/kg |
| Flash point (Abel – Pensky) | – 28 °C |
| Ignition temperature (DIN 51794) | 460 °C |
| Explosion limits in air | 1.65 – 8.4 vol% |
| Critical data | |
| $t_{cr}$ | 224.0 °C |
| $p_{cr}$ | 3.43 MPa |

Vapor pressure, density, and solubility in water, as well as the composition and boiling points of azeotropes with water and methanol, are given in Tables 1 and 2 [4]. Methyl *tert*-butyl ether has unlimited miscibility with all ordinary organic solvents and all hydrocarbons.

**Chemical Properties.** Methyl *tert*-butyl ether is very stable under alkaline, neutral, and weakly acidic conditions. In the presence of strong acids, it is cleaved to methanol and isobutene. The latter reacts to form isobutene oligomers.

# 3. Resources and Raw Materials

At present, *isobutene* from the following sources is used as feedstock for MTBE production [5]:

1) isobutene in the form of raffinate I (see Table 3 for typical composition) from steam crackers. 40% of this stream is employed as MTBE feedstock worldwide. (This corresponds to 24% of the isobutene used for present MTBE production.) [6];
2) isobutene in the form of butene–butane (BB) fractions from fluid catalytic crackers (FCC–BB; 28% of MTBE feedstocks; for typical composition, see Table 3);
3) isobutene from dehydrogenation of isobutane, which is obtained both from refineries and from field butanes (12% of MTBE feedstocks); and
4) isobutene by dehydration of *tert*-butanol, a coproduct of propylene oxide synthesis (Halcon/Arco process) (36% of MTBE feedstocks).

**Table 3.** Typical composition of $C_4$ hydrocarbon streams from steam crackers (raffinate I) and fluid catalytic crackers (FCC–BB)

| Compound | Raffinate I, wt % | FCC–BB, wt % |
|---|---|---|
| Isobutane | 4 | 36 |
| *n*-Butane | 12 | 13 |
| Isobutene | 44 | 15 |
| 1-Butene | 24 | 12 |
| *cis*-2-Butene | 6 | 9 |
| *trans*-2-Butene | 9 | 14 |
| 1,3-Butadiene | 0.5 | 0.3 |
| Balance | 0.5 | 0.7 |

As demand for MTBE increases, the first source to be exploited will probably be free FCC–BB reserves. Any further expansion—and this also holds for the raffinate I route—is thus tied to new cracker construction. Field butanes will grow more than proportionally in importance. The relative share of *tert*-butanol as an isobutene source for MTBE production, on the other hand, is expected to decline because it is formed as a coproduct and thus linked to future propylene oxide demand. Raffinate I and FCC–BBs can be utilized directly in MTBE synthesis. By contrast, isobutane must be dehydrogenated. The same holds true for field butanes after isomerization of the *n*-butane fraction to isobutane. A number of industrial processes can be used for this [7]–[13]. For the primary butane isomerization, the Butamer process is most commonly used. For the isobutane dehydrogenation, the Catofin and Olefex processes, and in the Soviet Union the LIT/SP process, are at present of industrial importance.

To produce MTBE from *tert*-butanol [14], isobutene must first be obtained by elimination of water from the alcohol before the olefin can be used for ether production. Synthesis of MTBE from isobutanol [15], which is available from synthesis gas, is primarily of academic interest.

*Methanol* (→ Methanol), the second reactant in MTBE synthesis, is produced at a typical purity of >99.9% and is used directly for ether synthesis without further purification. The methanol capacities available today can support an increase in MTBE output to ca. $14 \times 10^6$ t/a, a level that will presumably be reached in 1993 [16].

## 4. Production

Methyl *tert*-butyl ether can be obtained by the acid-catalyzed addition of methanol to isobutene [17]. Suitable catalysts are solid acids such as bentonites [18] and, especially, macroporous acidic ion-exchange resins [19]. The reaction is weakly exothermic with a heat of reaction of −37 kJ/mol [20].

The reaction occurs in either a mixed gas–liquid or an exclusively liquid phase. In the mixed phase, formation of MTBE can be described by the Langmuir–Hinshelwood model [21]; in the pure liquid phase, the Rideal–Eley model gives better results [22].

Equilibrium constants at various temperatures have also been determined [23], [24]. For example, only 92% conversion can be achieved with equimolar amounts of isobutene and methanol. An excess of methanol not only increases the conversion of isobutene but also suppresses its dimerization. At a molar methanol excess as low as 10%, the selectivity for MTBE is practically 100%.

In recent years, in addition to the Snamprogetti [25] and Hüls processes [26], [27], the Arco process [28] and the IFP/CR & L-process (Institut Franais du Petrole/Chemical Research & Licensing, the latter nowadays the Nova Corp Subsidiary CD TECH) have been established [29].

Other industrial processes have been developed by DEA (formerly Deutsche Texaco) [30], Shell (Netherlands) [31], Phillips Petroleum [32], and Sumitomo [33]. A total of 54 MTBE plants with a total installed capacity of $7.3 \times 10^6$ t are currently on stream [34].

**Process Description.** *Reaction Section.* All processes have in common the reaction of isobutene with a certain molar excess of methanol on a macroporous acidic ion exchanger at 50–90 °C. In the Snamprogetti, Hüls, and Arco processes, the pressure in the reaction section is chosen so that the reactants over the catalyst are entirely in the liquid phase (1.0–1.5 MPa). This promotes the selectivity for MTBE formation and enhances catalyst service life. For example, only ca. 100–200 ppm by weight of isobutene is converted to diisobutene in the "Hüls-MTB-Process". At the plant located in Marl, the first catalyst charge dating from 1978 is still in service.

Design of the reaction section is dictated by the heat of reaction to be removed and thus by the isobutene concentration in the feedstock. In the first three processes mentioned (Snamprogetti, Hüls, Arco) adiabatic reactors are always employed with a *FCC–BB feedstock*. If *raffinate I* is used as feedstock, Snamprogetti and Hüls prefer tubular reactors for reasons of process engineering, whereas Arco employs recycle reactors. In the adiabatic reactors employed in the IFP/CR&L process (which also require recycle at higher isobutene concentration) an isobutene conversion of approximately 85% is achieved. The heat of reaction is removed by partial evaporation of the $C_4$ hydrocarbons in the reactor. This process design, however, gives rise to a certain loss in selectivity for MTBE. A final isobutene conversion of ca. 97% is achieved by a subsequent *catalytic distillation stage*. (In catalytic distillation a certain part of a distillation tower is filled with catalyst to enforce additional conversion of the reactants.) The spatial configuration of the catalyst in the tower was developed by CR&L [35]. A similar catalytic distillation step is employed in an early MTBE process developed by Chevron. This process has, however, not yet been commercialized [36].

*Refining Section.* In the removal of nonconverted $C_4$ hydrocarbons from MTBE by distillation (debutenizing), excess *methanol* appears in the distillate as an azeotrope depending on the pressure (at 0.6 MPa the azeotrope contains 2–4 wt % methanol, depending on the $C_4$ composition). This methanol can be completely removed and recovered by a water wash.

If methanol losses of ca. 200 ppm by weight are acceptable, methanol recovery from the distillate with molecular sieves is also appropriate [37]. This 200 ppm residual methanol together with dimethyl ether, also formed in trace amounts, can be separated from the butene fraction in a subsequent molecular sieve adsorption unit [38].

A greater excess of methanol naturally increases isobutene conversion, however the excess methanol unavoidably remains in the MTBE. It can, nevertheless, be recovered by distillation, for example, as a methanol-poor azeotrope with MTBE [39]. Not only does this give MTBE with purities of > 99.7% (balance chiefly *tert*-butanol produced by reaction of isobutene with water dissolved in the starting materials), it also results in isobutene conversions of ca. 98%.

In most industrial plants, isobutene conversion of 95–97% is sufficient. *Residual butenes* are mainly used for the manufacture of alkylate gasoline, recycled to the cracker, or simply burned. If they are to be utilized for other chemical purposes such as the production of polymer-grade 1-butene, the degree of isobutene conversion must be significantly increased. In the Hüls- MTB-Process (see Fig. 1), this is carried out in two stages with typical isobutene conversions of > 99.9%, corresponding to ca. 300 ppm by weight of residual isobutene in the spent $C_4$ fraction. (If raffinate I is used as a feedstock, this fraction is called raffinate II.)

*Construction Materials.* Because corrosive media are not used anywhere and do not result from catalyst disintegration, the entire plant is fabricated from standard steel. Moreover under normal operating conditions, residues, process wastewater, and gaseous emissions are not formed.

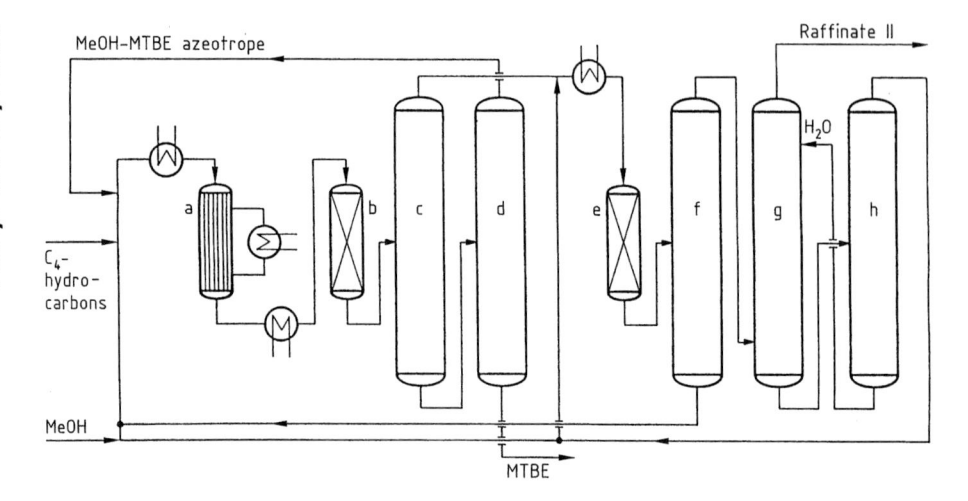

**Figure 1.** Two-stage Hüls-MTB-Process
First stage: a) Tubular reactor; b) Adiabatic reactor; c) First $C_4$ distillation tower; d) Methanol-MTBE azeotrope distillation tower
Second stage: e) Secondary reactor; f) Second $C_4$ distillation towerMethanol recovery: g) Methanol extraction; h) Methanol tower

Investment costs (inside battery limits, IBL) for a $4 \times 10^5$ t/a MTBE plant (Hüls-MTB-Process) in the Federal Republic of Germany, running on raffinate I feedstock, amount to $25 \times 10^6$ DM.

# 5. Environmental Protection

The small amounts of byproducts (see Chap. 7) in MTBE need not be removed if the product is to be used for gasoline. The catalyst is regenerable, and no environmental problems arise from its disposal.

Because MTBE is soluble in water, suitable precautions against groundwater contamination must be taken in loading and storage. For the environmental significance of MTBE, see Section 10.1. For emission control in the storage and handling of MTBE, see Chapter 9.

# 6. Quality Specifications

The usual purity of commercial MTBE is 98–99 wt%. The byproducts, *tert*-butanol and diisobutenes, as well as excess, residual methanol, do not have a detrimental effect on octane number improvement through the use of MTBE. Depending on the quality of the $C_4$ feedstock mixture, the MTBE product may also contain $C_5$ and $C_6$ hydrocarbons; removal of these by distillation is generally not necessary.

A typical MTBE composition adopted worldwide for use in the fuel sector is given below:

| | |
|---|---|
| MTBE | 98–99 wt% |
| Alcohols (methanol, *tert*-butanol) | 0.5–1.5 wt% |
| Hydrocarbons ($C_5$, $C_6$- hydrocarbons, diisobutenes) | 0.1–1.0 wt% |
| Water | 50–1500 ppm by weight |
| Total sulfur | max. 10 ppm by weight |
| Residue on evaporation | max. 10 ppm by weight |

For special applications (see Section 10.2), a special quality with 99.95% MTBE is marketed under the trade name Driveron S (Hüls).

# 7. Chemical Analysis

Pure MTBE is analyzed by gas chromatography, preferably in capillary columns with a highly polar stationary phase, e.g., TCEP [1,2,3-tris-(2-cyanoethoxy)propane], Carbowax 20 M, or DX-1. For GC analysis of MTBE-containing fuels, an oxygen-specific detector (O-FID) [40] or a column combination technique [41], [42] can be employed. An IR spectroscopic method also exists [43]. For other references on analysis of MTBE, see [44]–[47].

# 8. Storage and Transportation

Being noncorrosive, MTBE can be handled in the same way as fuel, and the existing distribution system can also be used for fuels containing MTBE. The ether has an indefinite storage life even in the presence of air. Because water is miscible with MTBE (though only slightly), dry storage is required. The product can be stored in nonpressurized containers. Carbon steel can be used as container material, as well as aluminum, brass, copper, polyethylene, or polypropylene. Teflon, Buna-N, and other fuel-resistant plastics and rubbers can be used for seals; the use of vinylidene fluoride-hexafluoropropene copolymer (Viton) is not recommended.

Methyl *tert*-butyl ether has a vapor pressure of 62.0 kPa at 40 °C. Emission from storage facilities can be controlled or prevented by ordinary measures. The usual safety precautions for highly inflammable liquids must be employed.

Extinguishing agents for firefighting are powder, carbon dioxide, and alcohol-resistant foam such as Tutogen L. With alcohol-resistant foams, a higher application rate is necessary for MTBE than for pure hydrocarbons.

Transport classifications are as follows:

| | |
|---|---|
| ADR/RID II | class 3, no. 3 b |
| ADNR II | class 3, no. 1 a, category K 1 n |
| GGVSee/IMDG | class 3.1 |
| UN no. | 2398 |
| ICAO Code/IATA-DGR II | class 3 |
| UN no. | 2398 |

## 9. Legal Aspects

The principal reason for the current importance of MTBE is environmental legislation (especially in the United States, Japan, and Western Europe). Although laws do not require the use of MTBE, it is often the best way to satisfy their provisions.

In the *European Community*, the Council Directive 85/536/EEC, December 5, 1985 [48], does not require special labeling at filling stations for fuels containing up to 15 vol% MTBE.

In the *United States*, the use of oxygenates in gasoline is governed by EPA waivers from the Clear Air Act. Sun Refining received a provisional waiver in 1988, permitting the use of up to 15 vol% MTBE (2.7 wt% oxygen) [49].

Legislation has developed since oxygenates in fuel have become generally recognized. To reduce carbon monoxide emission and comply with provisions of the Clean Air Act, some U.S. cities and states mandate the use of ca. 2 wt% oxygen in fuel (chiefly as MTBE) in winter months.

## 10. Uses

### 10.1. As an Octane Enhancer

More than 95% of MTBE produced goes into the gasoline pool. The market for fuels, amounting to ca. $100 \times 10^6$ t/a in Western Europe and ca. $320 \times 10^6$ t/a in the United States, has a capacity for MTBE that will exceed production for some time to come. In addition to many other favorable characteristics, the current importance of MTBE is based primarily on its exceptionally good octane-enhancing properties when used as a gasoline blendstock [50], [54]. These antiknock properties are especially important because the use of cheap but toxic alkyllead compounds has been restricted or banned

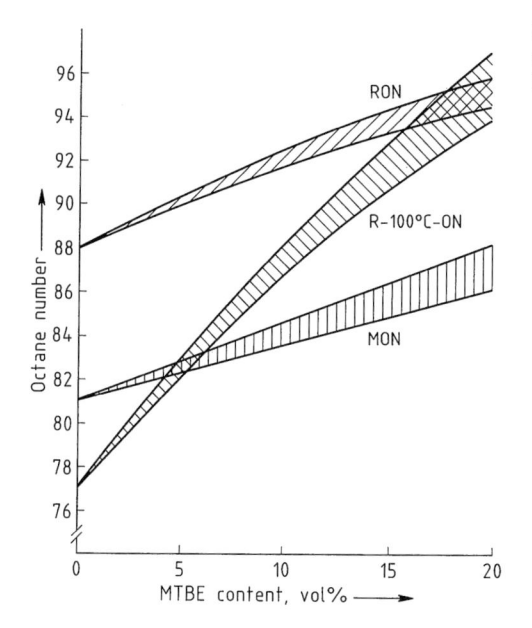

**Figure 2.** Ranges of octane improvement by addition of MTBE to an unleaded gasoline with RON = 88, MON = 81, and R-100 °C-ON = 77

by law, both on environmental grounds and to permit the use of exhaust catalytic converters. Depending on the composition of the base gasoline, blend octane numbers of 115–135 (research octane number, RON) and 98–120 (motor octane number, MON) can be achieved. The highest blend values are obtained with saturated, paraffin-rich gasolines; the lowest, with olefin-rich gasolines. The lead and aromatics content also influences the blend octane numbers of MTBE. Because of the relatively low boiling point of MTBE, the effect on the front-end octane number (FON or R-100 °C-ON, i.e., the research octane number of the fraction with a boiling range < 100 °C) is especially pronounced (see Fig. 2) [51].

Besides increasing the octane number, addition of MTBE to fuel has other positive effects. It does not necessitate any modifications to existing vehicles. The fuel vapor pressure (Reid vapor pressure, RVP) is decreased [52], so that vapor emissions during automobile fueling and operation are reduced. Addition of MTBE cuts down exhaust emissions, particularly carbon monoxide, unburned hydrocarbons, polycyclic aromatics, and particulate carbon. Although MTBE has a somewhat lower heat of combustion than gasoline, addition of up to 20 vol% neither impairs motor power nor increases fuel consumption. Easier cold starting and prevention of carburetor icing are other advantages. With regard to the hot-weather drivability of modern automobiles, no difference exists between fuels containing MTBE and those without it [53]. Because of the miscibility of water with MTBE, the fuel's cloud point is significantly lowered. The MTBE-blended fuels are compatible with all materials used in automobile manufacture (e.g., gaskets; lacquers; metals in the carburetor, injection pump, or elsewhere; and elastomers).

## 10.2. Other Uses

Methyl *tert*-butyl ether is also used in the petrochemical industry. Production of isobutene by splitting MTBE is the only application that has been used on an industrial scale (for applications of isobutene, → Butenes).

By reversal of its formation reaction, MTBE can be cracked to isobutene and methanol on acidic catalysts at > 100 °C [55]. The methanol obtained as a coproduct is recycled to MTBE synthesis. For economic and ecological reasons, the MTBE-cracking route and that involving the splitting of *tert*-butanol obtained by direct hydration of isobutene [57] will continue to displace the conventional sulfuric acid extraction process for the isolation (recovery) of pure isobutene from $C_4$ streams [58]. At present, isobutene production with MTBE as feedstock is employed by Exxon Chemical and Sumitomo. These processes consume ca. 3% of world MTBE output [56]. Three more plants are in the design stage, one of them using Hüls technology.

Methyl *tert*-butyl ether itself can be used in a number of chemical reactions, for example, the production of methacrolein and methycrylic acid [59] and of isoprene [60]. For other possible applications, see [61]–[65], [66].

The lack of acidic hydrogen atoms makes MTBE a suitable *solvent* for chemical reactions such as Grignard reactions. The preferred grade for this purpose is high-purity MTBE with a residual alcohol content < 500 ppm by weight, such as the product marketed by Hüls under the trade name Driveron S. Because of its negligible tendency to form peroxides, high ignition temperature, and narrow explosion limits, MTBE is a good solvent for analytical use [66]. It is also used as an *extractant*, for example, in solvent dewaxing of hydrocarbon oils [66], [67].

# 11. Toxicology and Occupational Health

**Toxicology.** $LD_{50}$ rat, oral 4 g/kg; $LD_{50}$ rat, dermal 4 g/kg; $LD_{50}$ rabbit, dermal 10 g/kg; $LC_{50}$ rat, inhalation (4 h) 85 000 mg/m$^3$ (STP).

The ether MTBE is weakly narcotic and irritates skin and eyes; it does not sensitize the skin. No neurotoxicological effects (e.g., morphological changes of the central nervous system) have been observed.

In a *subchronic* inhalation study, rats were exposed to MTBE concentrations of up to 1000 ppm (90 d, 6 h/d, 5 d/week). Exposure did not produce any effects. Subchronic toxicity tests on mice (50 000 ppm, 10 min/d, 30 d) and monkeys (3000 ppm, 6 h/d, 5 d/week, 14 d) also proved to be negative. On the basis of these and other studies the NOEL (no observable effect level) is assumed to be in the range 100–300 ppm.

Most of the inhaled MTBE is exhaled unchanged within 6 h. Bioaccumulation does not occur. *tert*-Butanol and formic acid are the main metabolites [68].

When tested in bacteria and mammalian cells, MTBE does not produce any *mutagenic* effects. Concentrations of up to 2500 ppm in inhalation tests on rats and mice do not produce any malformations in newborns or affect the fertility of the adults.

Due to the widespread use of MTBE as a motor fuel additive, discussions are being held in the United States concerning *chronic* inhalative toxicity tests. A comprehensive collection of toxicity data has been compiled by the American Petroleum Institute and the Oxygenated Fuels Association [69].

**Occupational Health.** Neither a MAK nor a TLV value has been established for MTBE.

Respirators should be worn by personnel working in an atmosphere containing MTBE. Fresh air should be supplied after inhalation; in severe cases, oxygen should be administered.

# 12. References

[1]  C. Henry, *Chem. Zentralbl.* (1904) 1065.
[2]  H. C. Barnett et al., *Natl. Advis. Comm. Aeronaut.*, Wartime Report ACR No. E 4 H 03 (1944).
[3]  *Chem. Eng. News* April 10 (1989) 12.
[4]  *Ullmann*, 4th ed., **17**, 58.
[5]  W. I. Ludlow: "Future Sources of MTBE,"*Dewitt 1989 Petrochemical Review*, Houston, March 28 – 30, 1989.
[6]  E. Debreczeni: *1987 – 1988 Butylenes Annual*, Debreczeni & Associates, Houston, January 1989.
[7]  K. C. Rock, R. O. Dunn: "Economics of Producing Butene Intermediate Feedstocks from Butanes," *Energ. Prog.* **8** (1988) no. 4, 191 – 195.
[8]  R. S. Andre, R. G. Clark: *Butane-derived MTBE can Fill the Octane Gap*, Air Products and Chemicals Inc., company brochure 1986.
[9]  G. C. Sturtevant et al.: *Selective Production of Light Olefins*, UOP-Technology Conference 1988,company brochure.
[10] Snamprogetti and Niimsk Nauchno Issledovatelskii Institut Monomerov dlia Sinteticheskogo Kauchuka a Yaroslavl, URSS, IT 120 142 1, 1985 (F. Buomo et al.).
[11] *Eur. Chem. News*, June 20 (1988) 22.
[12] G. L. Frishkorn et al.: "Butane Isomerisation Technology: Now and for the Future," *AIChE Meeting*, Minneapolis, Minn., Aug. 16 – 19, 1987.
[13] B. W. Burbridge et al., *NPRA annual meeting*, San Antonio 1975.
[14] D. J. Miller: "Arco's Processes for Oxygenated Fuels – MTBE/TAME and ETBE," *Dewitt 1989 Petrochemical Review*, Houston, March 28 – 30, 1989.
[15] W. Keim, W. Falter, *Catal. Lett.* **3** (1989) 59 – 64.
[16] D. Simons: "Supply Outlook for Methanol and Isobutylene for Production of Ethers," *National Conference on Octane and Oxygenates*, San Francisco, March 21 – 23, 1989.
[17] F. Ancilotti et al., *J. Mol. Catal.* **4** (1978) 37 – 48.
[18] A. Gicquel, B. Torck, *J. Catal.* **31** (1983) 27 – 31.
[19] J. M. Adams et al., *Clays Clay Miner.* **34** (1986) no. 5, 597 – 609.

[20]  F. Ancilotti et al., *J. Catal.* **46** (1977) 49–57.
[21]  C. Subramaniam, S. Bhatia, *Can. J. Chem.* **65** (1987) no. 8, 613–620.
[22]  Adnan M. Al-Jarallah et al., *Can. J. Chem.* **66** (1988) no. 10, 802–807.
[23]  F. Colombo et al., *Ind. Eng. Chem. Fundam.* **22** (1983) 219–223.
[24]  Snamprogetti, DE-OS 2 521 1963, 1975 (F. Ancilotti).
[25]  R. Trotta: "The Snamprogetti MTBE-Technology," *Dewitt 1988 Petrochemical Review,* Houston, March 23–25, 1988.
[26]  F. Obenaus: "Wege zur besseren Nutzung der $C_4$-Kohlenwasserstoffe," *DGMK-Meeting Karlsruhe,* February 8, 1982.
[27]  J. Neumeister, *Der Hüls MTB-Prozeß als Schlüssel für eine $C_4$-Chemie,* lecture held at the DGMK meeting, Karlsruhe, April 6–7, 1989.
[28]  Th. A. Beck, D. J. Miller: *Arco MTBE – The Maximum in Flexibility,* Dewitt 1988, Petrochemical Review, Houston, TX., March 23–25, 1988.
[29]  P. H. O. Dixon et al., *Catalytic Distillation Technology and MTBE-Production,* 1989 NPRA Annual Meeting, San Francisco, March 19–21, 1989.
[30]  DEA (formerly Deutsche Texaco), DE 27 06 465, 1987 (H. Humberg, H. Wegner).
[31]  Shell NL, DE-OS 29 11 077, 1979 (H. Groeneveld, A. Kwantes).
[32]  *Phillips Methyl Tertiary Butyl Ether Process,* Phillips Sales Brochure 899-85 TL.
[33]  Sumitomo Chemical Company, EP-A 0 206 594, 1986 (S. Arakawa et al.).
[34]  European Fuel Oxygenates Association: *EFOA Newsletter* **4** (1989).
[35]  CR & L, EP-A 0 008 860, 1979 (L. A. Smith Jr.).
[36]  Chevron Research, US 3 634 935, 1972 (W. M. Haunshild).
[37]  E. Trubac: "Oxygenate Removal from MTBE Raffinate," lecture presented at the *AIChE Meeting,* Minneapolis, Minn., Aug. 16–19, 1987.
[38]  Union Carbide Corp., EP-A 0 229 994, 1987 (M. Nagji).
[39]  Hüls AG, DE 2 629 769, 1976 (F. Obenaus et al.).
[40]  W. Schneider, J.-Ch. Frohne, H. Bruderreck, *J. Chromatogr.* **245** (1982) 71–83.
[41]  J. Winskowski, *Erdöl Kohle Erdgas Petrochem.* **41** (1988) no. 1, 30–36.
[42]  J. M. Levy, J. A. Yancey, *J. High Resolut. Chromatogr. Commun. HRC CC J. High Resolut. Chromatogr.* **9** (1986) no. 7, 383–387.
[43]  F. Luigart, *Erdöl Kohle Erdgas Petrochem.* **37** (1984) no. 3, 127.
[44]  AFNOR (Association Franaise de Normalisation) M 07-084, Dec. 1984, Norme experimentale.
[45]  DIN 51 413, Parts 4 and 5.
[46]  ASTM D 2 (proposal).
[47]  BSI (British Standards Institution) proposed method JP-STG-6 C (The Institute of Petroleum).
[48]  EEC Council Directive 85/536/EEC, Dec. 5, 1985 (Column 3).
[49]  *Chem. Eng. News,* Oct. 24, (1988) 12.
[50]  G. H. Unzelmann, *Oil Gas J.* (1989, April 10) 33–37.
[51]  F. Obenaus, W. Droste, *Erdöl Kohle Erdgas Petrochem.* **33** (1980) no. 6, 271–275.
[52]  EFOA, European Fuel Oxygenates Association, Brüssel, Belgium 250, Avenue Louise, *A Laboratory Assessment of the Technical Blending Properties of Oxygenates as Motor Gasoline Components,* 1988.
[53]  M. Denham, M. Mays, K. Schug, M. Sposini, lecture presented at the *8th International Symposium on Alcohol Fuels,* Nov. 13–16, Tokyo 1988.
[54]  W. J. Piel, *Energ. Prog.* **8** (1988) no. 4, 201–204.
[55]  Fattore et al., *Hydrocarbon Process.* August (1981) 101.
[56]  V. Guercio (ed.): *$C_4$-Monitor,* CTC International, Montclair N.J., May 1989.

[57] *Chem. Eng. N.Y.* **90** (1983) no. 25, 60.
[58] H. Kröper, K. Schlömer, H. M. Weitz, *Hydrocarbon Process* **48** (1969) no. 9, 195–198.
[59] Nippon Shokubai Kagaku Kogyo Co., EP-A 0 304 867, 1988 (K. Kinumi et al.).
[60] Kuraray Co, JP-Kokai 60 193 932 <85 193 932>, 1985 (Y. Ninagawa et al.). Sumitomo Chem., US 3 574 780, 1971 (Y. Watanabe et al.).
[61] Degussa AG, EP 0 050 869, 1981 (B. Lehmann).
[62] Anic S.p.A., DE-OS 3 130 428, 1982 (F. Ancillotti et al.). BASF AG, DE-OS 27 40 590, 1979 (F. Merger et al.).
[63] BASF AG, DE-OS 29 11 466, 1980 (H. Naarmann et al.).
[64] BASF AG, DE-OS 29 28 098, 1981 (W. Hoffmann).
[65] Bayer AG, DE-OS 30 02 203, 1981 (G. Bonse et al.).
[66] H. Grangette, C. Bernasconi, S. Randriamahefa, R. Gallo, *Inf. Chim.* **246** (1984) Jan./Feb., 135–140.
[67] Exxon Research and Engineering Co., EP-A 00 88 603, 1983 (T. H. West).
[68] H. Savolainen et al., *Arch. Toxicol.* **57** (1985) 285–288.
[69] American Petroleum Institute 2101 L Street NW., Washington D.C. 20037, United States. Oxygenated Fuels Association, *MTBE Committee*, 1330 Connecticut Avenue, NW, Suite 300, Washington, DC 20036, United States.

# Naphthalene and Hydronaphthalenes

GERD COLLIN, Rütgerswerke AG, Duisburg-Meiderich, Federal Republic of Germany (Chaps. 1–6)
HARTMUT HÖKE, Rütgers-BioTech, Mannheim, Federal Republic of Germany (Chap. 7)

1. Physical and Chemical Properties ............... 3351
2. Production .............. 3352
3. Uses ................... 3354
4. Alkylnaphthalenes ........ 3355
5. Hydronaphthalenes ........ 3358
6. Economic Aspects ......... 3358
7. Toxicology ............... 3359
8. References ............... 3360

Naphthalene was discovered in coal tar by A. GARDEN in 1819.

## 1. Physical and Chemical Properties

Naphthalene [91-20-3], $C_{10}H_8$, $M_r$ 128.18, *mp* 80.29 °C, *bp* 217.95 °C (101.3 kPa), forms colorless flakes or monoclinic crystals with a characteristic odor. It is readily soluble in benzene, diethyl ether, chloroform and carbon disulfide, soluble in ethanol, and insoluble in water. Naphthalene is volatile in steam and readily sublimable. It forms azeotropic mixtures with, for example, ethylene glycol, acetamide, *m*-cresol, and benzyl alcohol. Some important physical properties are as follows:

| | |
|---|---|
| Density (20 °C) | 1.1789 g/cm³ |
| Refractive index (99.5 °C) | 1.5829 |
| Heat capacity (25 °C) | 1.294 kJ/kg⁻¹ K⁻¹ |
| Heat of fusion | 148 kJ/kg |
| Heat of vaporization | 352 kJ/kg |
| Vapor pressure at 70 °C | 0.525 kPa |
| at 88 °C | 1.33 kPa |
| at 181 °C | 40 kPa |
| at 270 °C | 295 kPa |
| at 349 °C | 1.01 MPa |
| at 479 °C | 4.13 MPa |
| Critical temperature | 475 °C |
| Flash point (closed cup) | 80 °C |
| Ignition temperature | 540 °C |

| | |
|---|---|
| Explosion limits in air | |
| upper | 5.9 vol% |
| lower | 0.88 vol% |
| Dielectric constant (20 °C) | 2.47 |
| Odor threshold | 0.004 mg/m$^3$ |

Naphthalene can be catalytically hydrogenated in the liquid phase, preferably in the presence of nickel, with high selectivity to give tetra- and decahydronaphthalene, and in the gas phase in the presence of oxidic catalysts less selectively to give mixtures of tetra-, hexa-, octa-, and decahydronaphthalene. Liquid-phase oxidation with various oxidants or gas-phase oxidation with air in the presence of vanadium pentoxide catalysts gives (via 1,4-naphthoquinone) phthalic acid, which is readily dehydrated to phthalic anhydride (→Phthalic Acid and Derivatives). The 1-position of naphthalene is the most nucleophilic. Consequently, electrophilic substitution reactions, such as nitration, halogenation, sulfonation, and alkylation at lower temperature, preferentially give the corresponding 1-substituted products; at higher temperature, isomerization occurs to give the thermodynamically more stable 2-substituted products. Nitration with nitric acid at low temperature produces 1-nitronaphthalene, and with $H_2SO_4$–$HNO_3$ at high temperature, 1,5- and 1,8-dinitronaphthalene (→Nitro Compounds, Aromatic). Halogenation forms 1-halo- and/or 1,4- and 1,5-dihalonaphthalenes, depending on the reaction conditions. Polychloronaphthalenes are obtained by catalytic chlorination at higher temperature (→Chlorinated Hydrocarbons). Sulfonation with concentrated sulfuric acid gives 1-naphthalenesulfonic acid, and at higher temperature, 2-naphthalenesulfonic acid; oleum gives 1,5-, 1,6-, and 2,7-naphthalenedisulfonic acids and trisulfonic acids, depending on the $SO_3$ content (→Naphthalene Derivatives). Naphthalene can be alkylated at higher temperature with alkyl halides, alcohols, and olefins in the presence of strong acids or Friedel-Crafts catalysts, with preferential formation of 2-alkyl derivatives. Dehydrogenating condensation of naphthalene (Scholl reaction) gives perylene [198-55-0]. Acylation of naphthalene leads to mixtures of the 1- and 2-acyl derivatives, with the ratio depending on the solvent used. 1-Chloromethylation with paraformaldehyde and hydrochloric acid affords 1-chloromethylnaphthalene.

# 2. Production

The main feedstock for naphthalene production is coal tar. At an average content of around 10%, naphthalene represents the most important compound in high-temperature coal tar (coke-oven tar) in terms of quantity. It can already be concentrated to over 90% in primary tar distillation, in the naphthalene fraction boiling between 210 and 220 °C, with a yield in excess of 90%. Further refining requires separation of the co-boiling compounds (phenols, nitrogen bases, benzo[b]thiophene, and other aromatic hydrocarbons). For production of technical naphthalene (e.g., mp 78.5 °C), redistillation of the naphthalene fraction is sufficient. Generally, phenols are extracted with sodium

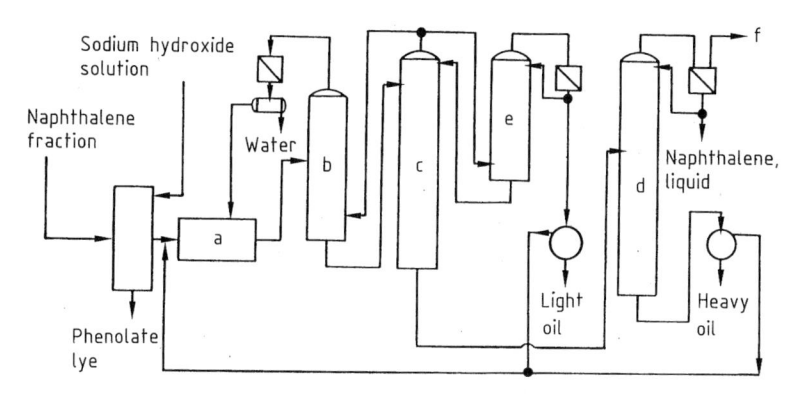

**Figure 1.** Naphthalene recovery from coal tar by extraction and distillation
a) Dephenolated naphthalene fraction; b) Dewatering column; c) Light oil column; d) Naphthalene column; e) Cooling; f) To the vacuum unit

hydroxide solution before distillation (Fig. 1). Pure naphthalene ($mp \geq 79.6\,°C$) is produced by, e.g., crystallization or hydrorefining.

Crystallization especially separates benzo[*b*]-thiophene, which boils at only 2 °C above naphthalene. Benzo[*b*]thiophene forms a eutectic mixture with naphthalene at a benzo[*b*]thiophene content of 93%, and mixed crystals elsewhere in the phase diagram. Whereas until recently, naphthalene was predominantly recovered by crystallization from the melt, with formation of suspensions, newer processes of crystallization from the melt result in the formation of layers or blocks. Benzo[*b*]thiophene and other impurities can be separated from naphthalene by multistage countercurrent crystallization from the melt without adding selective solvents. Four to six crystallization stages are needed to produce pure naphthalene with a melting point of 80 °C from a coal tar primary fraction containing 90% naphthalene.

Hydrorefining of naphthalene is carried out at 400 °C and ca. 1.4 MPa in the presence of cobalt–molybdenum catalysts, whereby benzo[*b*]-thiophene is converted to hydrogen sulfide and ethylbenzene. Other co-boiling compounds are broken down to lower-boiling hydrocarbons and can be separated by distillation. Naphthalene produced by this method has a sulfur content of $\leq 100$ ppm and contains tetralin as the main by-product (ca. 1%).

Other possible methods for separating benzo[*b*]thiophene include selective sulfonation, polycondensation with formaldehyde and acid, reaction with sodium metal, and azeotropic distillation with glycols.

Naphthalene can also be recovered from pyrolysis residue oils (from the pyrolysis of hydrocarbon fractions to olefins) by distillation and crystallization, in a manner similar to the recovery from coal tar. These pyrolysis residue oils normally contain 10–16% naphthalene [6].

In addition to recovery from coal tar, naphthalene is also produced in the United States by hydrodealkylation of aromatized petroleum-derived fractions (Fig. 2) [7], [8]. Typical feedstocks are residue oils from catalytic naphtha reforming, from naphtha

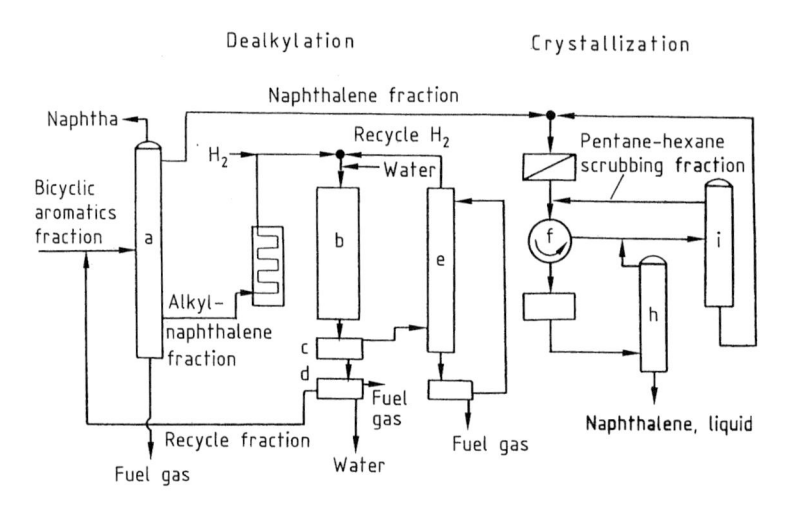

**Figure 2.** Naphthalene recovery by catalytic hydrodealkylation of alkylnaphthalene fractions (Unidak process)
a) Naphthalene column; b) Reactor; c) High-pressure gas separator; d) Low-pressure gas separator; e) Methane scrubber; f) Centrifuge; g) Melting vessel; h) Stripping column; i) Solvent recovery column

pyrolysis in olefin plants, or catcracker recycle oils. These aromatized oils are distilled to give a fraction containing naphthalene and alkylnaphthalenes and, in the case of catcracker recycle oils, subjected to extraction of the aromatics. The bicyclic aromatic fraction, with a boiling range of, e.g., 220–270 °C, is dealkylated under hydrogen pressure, either thermally above 700 °C, or catalytically in the presence of a chromium oxide/aluminum oxide or cobalt oxide/molybdenum oxide catalyst at 550–650 °C. Methyl- and dimethylnaphthalene fractions from coal tar can also be hydrodealkylated. The crude dealkylation product is processed to a low-sulfur pure naphthalene (*mp* 80 °C) by distillation.

# 3. Uses

Worldwide, the main derivative of naphthalene is still phthalic anhydride (from catalytic gas-phase oxidation of naphthalene), an intermediate for PVC plasticizers (→Phthalic Acid and Derivatives). Other important applications for naphthalene are dyes, particularly azo dyes produced from 2-naphthol [135-19-3] and naphthalene sulfonic acids. Naphthalene and alkylnaphthalene sulfonates are used as surfactants, and naphthalene sulfonate – formaldehyde condensates as tanning agents (syntans) and dispersants, e.g., as superplasticizers for concrete. Naphthalene is used in the production of moth repellents and fumigants and the insecticide carbaryl (1-naphthyl-*N*-methylcarbamate). It is also used for the synthesis of tetrahydroanthraquinone from naphthoquinone and butadiene (→Anthraquinone) and for alkylnaphthalene solvents for carbonless copy paper (Kureha Micro Capsule Oil, KMC, a mixture of diisopropylnaphthalene isomers), the solvents tetralin and decalin, as well as chlorinated naphthalenes (→Chlorinated Hydrocarbons), of which only the monochloronaphthalenes retain commercial significance, e.g., as dye dispersants and as fungicides and insecticides for preservatives. Polychlorinated naphthalenes are rarely produced due to their toxicity and resistance to environmental degradation. 1-Methyl- and 1-chloromethylnaphthalene are used in the production of the plant growth regulators 1-naphthaleneacetamide [86-86-2] and 1-naphthaleneacetic acid [86-87-3]. The alkaline-earth salts of sulfonated dinonylnaphthalenes are important lube-oil additives. New applications are opening up for 2,6-naphthalenedicarboxylic acid and 6-hydroxy-2-naphthoic acid as intermediates for liquid crystal polymers and heat-resistant polyester fibers and films. For further details on the uses of naphthalene, see Figure 3.

# 4. Alkylnaphthalenes

**1-Methylnaphthalene** [90-12-0], $C_{11}H_{10}$, $M_r$ 142.20, bp 244.4 °C (101.3 kPa), mp −30.5 °C, $d^{20}$ 1.0203, is a colorless, blue-fluorescing liquid which is readily soluble in benzene, ethanol, and diethyl ether, and insoluble in water. 1-Methylnaphthalene is present in high-temperature coal tar in a concentration of 0.5% and is produced industrially from the methylnaphthalene fraction, which boils between 240 and 245 °C, by redistillation of the 2-methylnaphthalene filtrate following crystallization and separation of 2-methylnaphthalene. 1-Methylnaphthalene can be isomerized to 2-methylnaphthalene in the presence of, e.g., $BF_3$, phosphoric acid, and phase-transfer catalysts [9], or on zeolite catalysts [10]. It is used for the synthesis of 1-naphthaleneacetic acid and, in mixtures with 2-methylnaphthalene, as a solvent and heattransfer oil.

**2-Methylnaphthalene** [91-57-6], $C_{11}H_{10}$, $M_r$ 142.20, bp 241.1 °C (101.3 kPa), mp 34.6 °C, $d^{20}$ 1.029, forms colorless crystals. The compound is readily soluble in benzene, ethanol and carbon disulfide, but insoluble in water. It is recovered industrially by crystallization from the methylnaphthalene fraction of high-temperature coal tar, which contains 1.5% 2-methylnaphthalene. 2-Methylnaphthalene can also be produced by isomerization of 1-methylnaphthalene (see above). 2-Methylnaphthalene is the feedstock for the production of 2-methyl-1,4-naphthoquinone [58-27-5] (menadione, vitamin $K_3$), and mixtures with 1-methylnaphthalene are used as solvents and heat-transfer oils. It is used in small quantities for the production of alkylmethylnaphthalene sulfonates as textile auxiliaries, surfactants, and emulsifiers. 2-Methylnaphthalene can be used via 2-methyl-6-acetylnaphthalene [11], [12], or in an older process (also as a mixture with 1-methylnaphthalene) by oxidation with subsequent Henkel rearrangement, to give 2,6-naphthalenedicarboxylic acid [1141-38-4]. This compound is used in the production of special heat-resistant polyester fibers and films, and of liquid crystal polymers. 2-Methylnaphthalene can be acylated in 1,2-dichlorbenzene

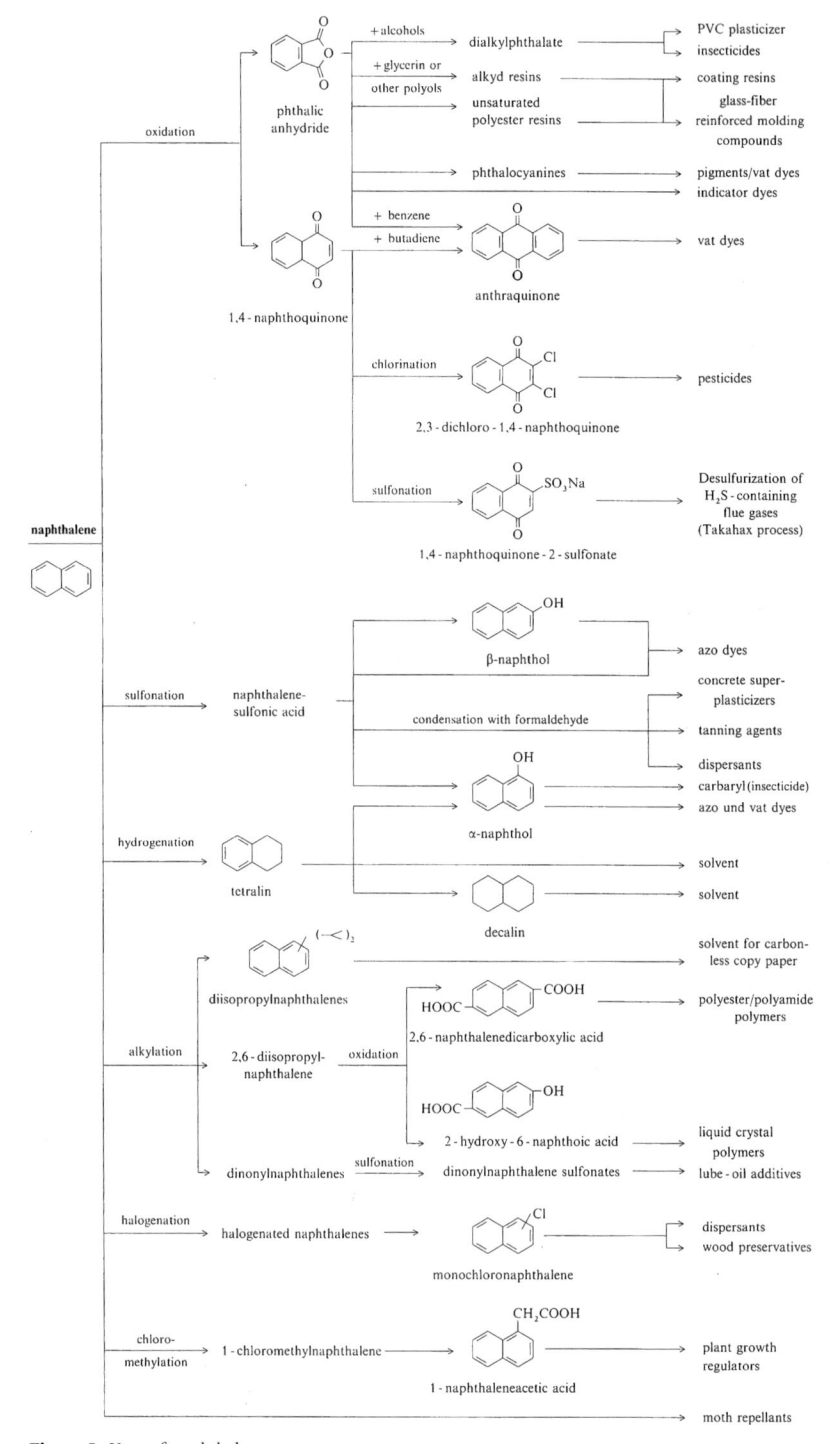

**Figure 3.** Uses of naphthalene

to give 1-acyl-7-methylnaphthalene [13], which can be used as an intermediate for pharmaceuticals [14].

**2,6-Dimethylnaphthalene** [*581-42-0*], $C_{12}H_{12}$, $M_r$ 156.23, *bp* 261 °C (101.3 kPa), *mp* 110 °C, forms colorless crystals which are slightly soluble in ethanol and insoluble in water. It can be recovered from the dimethylnaphthalene fraction of coal tar by redistillation and crystallization; from aromatized catcracker recycle oil by aromatics extraction, redistillation, hydrogenation of the isomer mixture to dimethyltetralins, isomerization, and dehydrogenation; or by disproportionation of methylnaphthalenes to give naphthalene and dimethylnaphthalenes, from which the 2,6-isomer can be obtained analogously to the production from aromatized catcracker recycle oil. 2,6-Dimethylnaphthalene is also obtained by alkylation of 2-methylnaphthalene or naphthalene on zeolite catalysts [15], [16].

2,6-Dimethylnaphthalene can be converted to 2,6-naphthalene dicarboxylic acid by liquid-phase oxidation with oxygen in acetic acid as solvent in the presence of a cobalt acetate/manganese bromide catalyst and a co-catalyst, e.g., ruthenium chloride [17].

**2-Isopropylnaphthalene** [*2027-17-0*], $C_{13}H_{14}$, $M_r$ 170.25, *bp* 268.9 °C (101.3 kPa), *mp* 15.1 °C, $d^{20}$ 0.9762, is a colorless liquid. It can be used to synthesize 1-naphthol [*90-15-3*] by oxidation via the hydroperoxide (Hock synthesis) (→Naphthalene Derivatives). Catalytic gas-phase dehydrogenation produces polymerizable 2-isopropenylnaphthalene [*3710-23-4*] [18].

**Diisopropylnaphthalenes.** Selective Friedel-Crafts alkylation of naphthalene with propene in the presence of an aluminum silicate catalyst yields as the main product a mixture of diisopropylnaphthalene isomers that is liquid over a wide temperature range. This isomer mixture is used as a solvent for carbonless copy paper [19], [20]. The solvent, trade name KMC (Kureha Micro Capsule Oil), is odorless, colorless, and environmentally harmless [21]. KMC can also be used as a solvent for scintillation measurements and as a heat transfer oil. Some physical properties of KMC are as follows:

| | |
|---|---|
| Boiling range (101.3 kPa) | 290 – 299 °C |
| Pour point | < – 40 °C |
| Flash point | > 130 °C |
| Density (15 °C) | 0.96 g/cm$^3$ |
| Refractive index (25 °C) | 1.565 |
| Specific heat (20 °C) | 1.71 kJ/kg K |
| Thermal conductivity (20 °C) | 0.12 W/Km |
| Dielectric constant (90 °C) | 2.46 |

Technically pure *2,6-diisopropylnaphthalene* [*24157-81-1*] (*mp* 70 °C) can be isolated from the isomer mixture by crystallization [22]. It can be converted to 2,6-naphthalene dicarboxylic acid [23] – [25] or to 6-hydroxy-2-naphthoic acid [*16712-64-4*] by liquid-phase oxidation [26]. These compounds are used as intermediates for heat-resistant liquid crystal polyesters [27].

# 5. Hydronaphthalenes

**1,2,3,4-Tetrahydronaphthalene** [*119-64-2*], tetralin, $C_{10}H_{12}$, $M_r$ 132.21, $bp$ 207.3 °C (101.3 kPa), $mp$ −35 °C, $d^{20}$ 0.9729, $n_D^{20}$ 1.5461, is a colorless liquid which is infinitely soluble in chloroform, petroleum ether (ligroin), diethyl ether, and ethanol, but insoluble in water. Tetralin is produced industrially by selective hydrogenation of low-sulfur naphthalene in the presence of nickel catalysts at 180–260 °C and 1–1.5 MPa.

Tetralin is used as a solvent for fats, resins, and paints; for the production of detergents; and for the synthesis of carbaryl via 1-naphthol. It can be applied as a hydrogen-donor solvent in coal extraction.

**Decahydronaphthalene** [*91-17-8*], decalin, $C_{10}H_{18}$, $M_r$ 138.25, exists in two stereoisomeric forms: *trans*-decalin, $bp$ 185.5 °C (101.3 kPa), $mp$ −32.5 °C, $d^{20}$ 0.8700, $n_D^{20}$ 1.4696; and *cis*-decalin, $bp$ 195 °C, $mp$ −45 °C (101.3 kPa), $d^{20}$ 0.8967, $n_D^{20}$ 1.4811.

The isomer mixture is a colorless liquid with a camphor-like odor. Technical decalin is infinitely soluble in butanol, soluble in chloroform, acetone, carbon disulfide and benzene, and slightly soluble in ethanol and methanol.

Decalin is produced by liquid-phase hydrogenation of naphthalene under pressure in the presence of nickel catalysts. The proportion of the two stereoisomers formed depends on the reaction conditions.

Like tetralin, decalin is used as a solvent for fats, resins, waxes, and paints. It can also be used to make cyclodecanone, an intermediate in the production of polyamide 10.

# 6. Economic Aspects

World production of naphthalene in 1987 was ca. $10^6$ t. About one-fourth of this came from Western Europe, one-fifth each from Japan and Eastern Europe, and one-eighth from the United States. In addition to coal-tar based naphthalene, petroleum-derived naphthalene has been recovered since 1961, predominantly in the United States, by dealkylation of aromatics-rich fractions from reforming and catalytic cracking. The proportion of petroleum-based naphthalene has decreased during the last decade from ca. 15% to ca. 5%, because coal-tar naphthalene is available in sufficient quantity. Table 1 summarizes the main uses of naphthalene as percentage of consumption in Western Europe, Japan, and the United States.

A new application for alkylnaphthalenes is the use of diisopropylnaphthalene isomer mixtures as solvents for carbonless copy paper. Production capacities for these mix-

**Table 1.** Uses of naphthalene in % (1987)

| Use | Western Europe | United States | Japan |
| --- | --- | --- | --- |
| Phthalic anhydride | 40 | 65 | 70 |
| Dyes | 29 |  | 10 |
| Surfactants, tanning agents, dispersants | 15 | 14 | 5 |
| Moth repellants, fumigants | 9 | 5 | 2 |
| Insecticides |  | 14 |  |
| Solvents and other uses | 7 | 2 | 12 |

tures, trade name KMC, total 10 000 t/a each in Japan and the Federal Republic of Germany. World production of 2-methylnaphthalene and 1-/2-methylnaphthalene mixtures is estimated at 1500 t/a each. For hydronaphthalenes, tetralin production is estimated to be 25 000 t/a worldwide. Naphthalenesulfonate–formaldehyde condensates are produced worldwide in quantities of ca. 40 000 t/a, predominantly for concrete plasticizers. Although polyester products derived from 2,6-naphthalenedicarboxylic acid or 6-hydroxy-2-naphthoic acid have been on the market for some time, they have not yet gained major commercial importance.

# 7. Toxicology

**Naphthalene** can be absorbed orally, by inhalation, or through the skin. The oral $LD_{50}$ values range from 1100 to 2400 mg/kg in rats, and from 350 to 710 mg/kg in mice. Although naphthalene is generally of low toxicity, intoxication after ingestion of large quantities, or owing to individual hypersensitivity, may be considerable. Headache and nausea may occur after inhalation of vapors. Metabolites formed from naphthalene in the organism cause hemolytic anemia; infants and fetuses, and also certain ethnic groups in which a high proportion of the population suffers from hereditary deficiency of glucose-6-phosphate dehydrogenase, are particularly at risk [28]. In isolated cases naphthalene workers have been found to suffer from cataracts after many years of exposure [29]. This causal relationship has been supported by positive experimental findings with rabbits [30]. Allergic reactions in humans are rare [31]. Epithelial necrosis of the lungs, detected after intraperitoneal administration to mice [32], [33], has not been reported in humans. Naphthalene has no mutagenic and cancerogenic potential in experimental models; marginal results that have been interpreted as positive are probably based on experimental inadequacies [34] or erroneous interpretations [35]. For reviews and further information, see [36], [37], [38].

**1-Methylnaphthalene and 2-methylnaphthalene** show only low toxicity; the oral $LD_{50}$ values in rats were determined at 2800 and 3850 mg/kg, respectively [39]. At high oral doses in rats, acute symptoms are unspecific and indicate impairment of the nervous system and hemorrhagic edema of inner organs. After intraperitoneal injection into mice, epithelial necrosis of the lung was induced, morphologically similar to that found with naphthalene [40], [41]. Other specific phenomena that are characteristic of naphthalene exposure, i.e., hemolytic anemia and formation of cataracts, are unknown for methylnaphthalenes. In humans, no intoxications related to these aromatic compounds have been reported.

**Diisopropylnaphthalenes** (mixture of isomers, KMC) cause no toxic symptoms in rats up to high dosage levels: in the range from about 3.0 g/kg under acute conditions to 30 mg kg$^{-1}$ d$^{-1}$ in a long-term study for 24 months. There are generally involved unspecific, mostly reversible phenomena such as loss of body weight, digestive dysfunction, enlargement of the liver, and local disturbed circulation (hyperemia) [42]. In skin and eye irritation tests conducted on rabbits, diisopropylnapthalenes exhibited slightly irritant, but reversible effects, which allows the product to be classified as nonirritant. Likewise, there is no evidence of sensitization [42]. No adverse effects were observed in a group of 80 workers who had had contact with diisopropylnapthalenes during production of carbonless copy paper [42]. Mutagenicity and cancerogenicity tests (24 months, oral, rat) gave no evidence of a cancerogenic potential, nor were teratogenic malformations found (mice, oral) [42]. Diisopropylnaphthalenes were rapidly released form the body, i.e., >80% within 15 h (mice) [42], and within 10 d (carp) [43], and they proved to be biodegradable [44].

**Tetralin and decalin** are irritant to the skin, eye, and to the mucous membranes on inhalation. They may cause nervous disturbances, headache, or numbness. The lowest concentration of decalin in air that exhibited an effect on humans was ca. 100 ppm; a 4-h exposure to 500 ppm was lethal to 4 from 6 rats [45]. In guinea pigs, morphological changes of the kidney, liver, and lung were observed after subacute inhalation of decalin [45]. Methemoglobinuria was induced by tetralin in cats; the same effect was found in infants after accidental exposure to tetralin [46]. The color of human urine may change to brownish green or green-gray after incorporation of decalin or tetralin [45]. Acute range finding studies revealed an oral LD$_{50}$ (rat) of 4.17 g/kg for decalin and 2.68 g/kg for tetralin [45]. A TLV of 25 ppm has been suggested for decalin [47].

# 8. References

General References

[1] *Beilstein*, naphthalene **5** 531, **5**(1), 257, **5**(3) 432, **5**(3) 1549, **5**(4) 1640; 1-methylnaphthalene **5** 566, **5**(1) 265, **5**(2) 460, **5**(3) 1617, **5**(4) 1687; 2-methylnaphthalene **5** 567, **5**(1) 266, **5**(2) 460, **5**(3) 1627, **5**(4) 1693; 2,6-dimethylnaphthalene **5** 570, **5**(1) 268, **5**(2) 468, **5**(3) 1651, **5**(4) 1714; 2-isopropylnaphthalene **5** 571, **5**(2) 470, **5**(3) 1657, **5**(4) 1723; 2,6-diisopropylnaphthalene **5**(4) 1768; tetralin **5** 491, **5**(1) 236, **5**(2) 382, **5**(3) 1219, **5**(4) 1388; decalin **5** 92, **5**(1) 46, **5**(2) 56, **5**(3) 242, **5**(4) 310.

[2] H.-G. Franck, G. Collin: *Steinkohlenteer*, Springer Verlag, Heidelberg/Berlin/New York 1968, pp. 60–61, 180–181.

[3] H.-G. Franck, J. W. Stadelhofer: *Industrial Aromatic Chemistry*, Springer Verlag, Berlin 1988, pp. 298–308, 328–333, 336–339.

[4] G.-P. Blümer, G. Collin, *Erdöl, Kohle, Erdgas, Petrochem.* **36** (1983) 22–27.

[5] G. Collin, M. Zander, *Nat. Resour. Dev.* **27**, 61–81 (1988).

Specific References

[6] Rütgerswerke AG, DE 1 815 568, 1968 (O. Wegener, R. Oberkobusch, G. Collin, M. Zander, H. Buffleb); DE 3 442 275, 1984 (J. Talbiersky, N. Drescher, H. E. Carl).

[7] F. Asinger: Die petrolchemische Industrie, vol. 1, Akademie-Verlag, Berlin 1971, pp. 603–609.

[8] G. F. Asselin, R. A. Erickson, *Chem. Eng. Prog.* **58** (1962) 47–52.

[9] Rütgerswerke AG, DE 3 028 199, 1980 (G.-P. Blümer).

[10] Rütgerswerke AG, DE 3 723 104, 1987 (J. Weitkamp, M. Neuber, W. Höltmann, G. Collin).

[11] Rütgerswerke AG, DE 3 519 009, 1985 (F. Kajetanczyk, R. Steinbach, I. Ruppert, K. Schlich); DE 3 519 009, 1985 (R. Steinbach, I. Ruppert, K. Schlich).

[12] Mitsubishi Gas, JP 87 273 260, 1987 (T. Hirai, S. Kitaoka).

[13] Rütgerswerke AG, DE 3 701 960, 1987 (R. Steinbach, F. Kajetanczyk).

[14] Max-Planck-Gesellschaft, DE 3 120 099, 1981 (H. G. Schlossberger).

[15] Hoechst, DE 3 334 084, 1983 (K. Eichler, E. I. Leupold).

[16] Rütgerswerke AG, EP 280 055, 1988 (J. Weitkamp, M. Neuber, W. Höltmann, G. Collin, H. Spengler).

[17] Rütgerswerke AG, DE 3 520 841, 1985 (G. Schmitt, K.-R. Kurtz).

[18] K. Handrick, *Proc. Neue Verfahren Kohleveredlung*, Luxemburg, November 26–28, 1979, pp. 27–37. Bergwerksverband, DE 2 644 624, 1976 (R. C. Schulz, D. Engel).

[19] Kureha, GB 1 359 512, 1974 (A. Konishi, M. Takahashi, F. Kimura, T. Toguchi).

[20] Rütgerswerke AG, DE 3 735 976, 1987 (R. Zellerhoff, M. Gürtler).

[21] J. W. Stadelhofer, R. B. Zellerhoff, *Chem. Ind. (London)*, 1989, April 3, 208–211.

[22] Rütgerswerke AG, EP 216 009, 1986 (W. Höltmann, R. Zellerhoff, R. Oberkobusch, P. Stäglich, B. Charpey).

[23] Teijin, JP 88 63 150, 1986 (I. Hirose).

[24] Kureha, DE 3 531 982, 1986 (T. Yamauchi, S. Hayashi, A. Sasakawa).

[25] Amoco, EP 329 273, 1989 (P. A. Sanchez, D. A. Young, G. E. Kuhlmann, W. Partenheimer, W. P. Schammel).

[26] Kureha, DE 3 517 158, 1985 (A. Iizuka, Y. Konai, T. Yamauchi, S. Hayashi).

[27] Celanese, EP 172 012, 1986 (D. E. Stuetz).

[28] W. W. Zuelzer, L. Apt, *J. Am. Med. Assoc.* **141** (1949) 185.

[29] H. W. Gerarde (ed.): *Toxicology and Biochemistry of Aromatic Hydrocarbons*, Elsevier, Amsterdam/London/New York/Princeton 1960, pp. 225–232.

[30] R. van Heyningen, A. Pirie, *Biochem. J.* **102**, (1967) 842.

[31] S. J. Fanbury, *Arch. Dermatol. Syphil.* **42** (1940) 53–55.

[32] D. Mahvi et al., *Am. J. Pathol.* **86** (1977) 559.

[33] D. L. Warren et al., *Chem. Biol. Interact.* **40** (1982) 287.

[34] E. Knake, *Virchows Arch.* **329** (1956) 141.

[35] B. Adkins et al., *J. Toxicol. Environ. Health* **17** (1986) 311.

[36] United States Environmental Protection Agency (EPA), Rep. No. 600/8-87/-055 F, Washington D.C., 1987.

[37] R. B. Franklin in R. Snyder (ed.): *Ethel Browning's Toxicity and Metabolism of Industrial Solvents,* Elsevier, Amsterdam/New York/Oxford 1987.
[38] Beratergremium für umweltrelevante Altstoffe der Gesellschaft Deutscher Chemiker: *Naphthalin, BUA-Stoffbericht 39,* VCH Verlagsgesellschaft, Weinheim 1989.
[39] Huntingdon Research Centre (by order of Rütgerswerke AG), Münster, Federal Republic of Germany, 1979 (unpublished results).
[40] In [36] p. 176.
[41] A. R. Buchpitt, R. B. Franklin, *Pharmacol. Ther.* **41** (1989) 393.
[42] Rütgers Kureha Solvents: *Toxicological and Physiochemical Studies on KMC,* Duisburg 1985 (unpublished results).
[43] T. Yoshida, H. Kojima, *Chemosphere* (1978) no. 6, 491–496.
[44] T. Yoshida, H. Kojima, *Chemosphere* (1978) no. 6, 497–501.
[45] *Patty,* vol. **2B,** pp. 3233, 3240–3242.
[46] *Ullmann,* 3rd ed., **12,** p. 589.
[47] D. J. de Renzo (ed.): *Solvents Safety Handbook,* Noyes Data, Park Ridge, N.J. 1986, pp. 174, 626.

# Naphthalene Derivatives

*Individual keywords: Alkylnaphthalenes →Naphthalene and Hydronaphthalenes; Chloronaphthalenes →Chlorinated Hydrocarbons; Cyanonaphthalenes →Nitriles; Hydronaphthalenes →Naphthalene and Hydronaphthalenes; Naphthalenedicarboxylic Acids →Carboxylic Acids, Aromatic*

GERALD BOOTH, Booth Consultancy Services, Thorpe House, Uppermill, Oldham OL3 6DP, United Kingdom

| | | |
|---|---|---|
| 1. | Introduction | 3364 |
| 2. | Naphthalenesulfonic Acids | 3365 |
| 2.1. | Production and Properties | 3365 |
| 2.2. | Monosulfonic Acids | 3367 |
| 2.3. | Disulfonic Acids | 3369 |
| 2.4. | Tri- and Tetrasulfonic Acids | 3371 |
| 2.5. | Alkylnaphthalenesulfonic Acids | 3372 |
| 3. | Naphthols | 3372 |
| 3.1. | 1-Naphthol | 3373 |
| 3.2. | 2-Naphthol | 3377 |
| 3.3. | Naphthalenediols | 3379 |
| 3.4. | Hydroxynaphthoic Acids | 3381 |
| 4. | Hydroxynaphthalenesulfonic Acids | 3384 |
| 4.1. | Production and Properties | 3385 |
| 4.2. | 1-Hydroxynaphthalenesulfonic Acids | 3389 |
| 4.3. | 2-Hydroxynaphthalenesulfonic Acids | 3392 |
| 4.4. | 1-Hydroxynaphthalenedisulfonic Acids | 3394 |
| 4.5. | 2-Hydroxynaphthalenedisulfonic Acids | 3396 |
| 4.6. | Hydroxynaphthalenetrisulfonic Acids | 3398 |
| 4.7. | Dihydroxynaphthalenesulfonic Acids | 3399 |
| 4.8. | Dihydroxynaphthalenedisulfonic Acids | 3400 |
| 5. | Aminonaphthalenes | 3401 |
| 5.1. | Naphthylamines | 3401 |
| 5.2. | Naphthalenediamines | 3404 |
| 6. | Aminonaphthalenesulfonic Acids | 3406 |
| 6.1. | Production and Properties | 3406 |
| 6.2. | 1-Aminonaphthalenesulfonic Acids | 3410 |
| 6.3. | 2-Aminonaphthalenesulfonic Acids | 3414 |
| 6.4. | 1-Aminonaphthalenedisulfonic Acids | 3416 |
| 6.5. | 2-Aminonaphthalenedisulfonic Acids | 3418 |
| 6.6. | Aminonaphthalenetrisulfonic Acids | 3420 |
| 6.7. | Diaminonaphthalenesulfonic Acids | 3423 |
| 6.8. | Diaminonaphthalenedisulfonic Acids | 3424 |
| 6.9. | Toxicity | 3425 |
| 7. | Aminonaphthols | 3425 |
| 8. | Aminohydroxynaphthalenesulfonic Acids | 3427 |
| 8.1. | Production and Properties | 3427 |
| 8.2. | Aminohydroxynaphthalenemonosulfonic Acids | 3429 |

8.3. Aminohydroxynaphthalenedisulfonic Acids ............ 3439

9. References ............... 3443

# 1. Introduction

Although the basis of naphthalene chemistry can be said to have started with ERLENMEYER in 1866, conclusive chemical evidence for the structure of naphthalene and its early derivatives (e.g., mono- and dinitroderivatives and α-naphthylamine) was published in 1888 by REVERDIN and NOELTING [5].

By 1900, hundreds of new naphthalene derivatives had been prepared as components for azo dyes. Not only was this early work, reported in German patents and journals, remarkably accurate in view of the comparatively primitive analytical techniques available, but the key compounds that emerged at that time remain dominant to this day. The trivial names given to these dye intermediates (Table 1) have also survived and are commonly used in the industry.

The important Armstrong – Wynne rules for polysubstitution of naphthalene and its derivatives by nitration and sulfonation were also formulated in this early period. Although empirical, they have stood the test of time against increasingly sophisticated theories and calculations.

Improved processes based on the key unit processes of sulfonation, nitration, reduction, hydroxylation, and amination (Bucherer reaction) for a wide range of naphthalene derivatives were developed during the next 40 years [6]. Although these were not published [7] until after 1945 with the end of the I.G. Farbenindustrie era, all the important subsequent reviews [5], [8] – [10] were based largely on these comprehensive data.

The only significant production development work published in the last 40 years is related to the project for the new Schelde-Chemie plant (Bayer Ciba-Geigy joint venture) designed to manufacture 14 000 t/a of naphthalene intermediates at Brunsbüttel [16]. The process development on important letter acids (H, J, γ, C, Peri, and Laurent's) required a major change from the traditional processes, optimized over 80 years, to meet present-day energy and environmental requirements [17].

Azo dyes and pigments continue to be major outlets for naphthalene intermediates. The Colour Index lists some 270 different naphthalene intermediates as precursors to many more colorants [18]. This represents about 20% of the total list of intermediates. Supplementary updating volumes list only seven new naphthalene intermediates and relatively few new outlets for existing intermediates [19]. From this one may infer that very little new colorant research is being carried out based on novel naphthalene derivatives. In contrast, other areas such as agrochemicals and pharmaceuticals have been most active in exploiting new naphthalene derivatives over the last 20 years [11].

## 2. Naphthalenesulfonic Acids

### 2.1. Production and Properties

Controlled sulfonation using a range of sulfuric acid and oleum strengths under a variety of reaction conditions leads to formation of mono-, di-, tri-, and tetrasulfonic acids, whose separation is frequently complicated by desulfonation (i.e., reverse sulfonation) or isomerization.

A common method for isolating naphthalenesulfonic acids and substituted derivatives obtained by sulfonation is the liming-out process. This consists of neutralizing the quenched sulfonation mass (i.e., after addition to excess water) with lime and filtering off the precipitated calcium sulfate while still hot. The solution of the calcium salt of the product is then titrated with sodium carbonate to form the sodium salt. Precipitated calcium carbonate is removed by filtration, and the solution is either evaporated to give the solid sodium salt or used to precipitate a less soluble salt. This technique may be employed in the laboratory to give pure products by using barium instead of calcium, provided the barium salt of the product has sufficient solubility in hot water.

The twelve sulfonated products most readily obtained from naphthalene are shown in Figure 1.

Separation of naphthalenesulfonic acids and downstream derivatives for analysis may be accomplished on small-scale columns by ion-exchange [20], ion-pair [21], or high-performance liquid [17] chromatography.

The naphthalenesulfonic acids are all extremely water soluble and are strong acids in solution. Alkali-metal and alkaline-earth salts are also water soluble but with a progression toward sparingly soluble barium salts, as shown in Table 2. The moderately soluble salts are readily precipitated from the mineral acid solution, with variations in crystal form according to conditions.

Naphthalene-2-sulfonic acid is isolated for conversion to 2-naphthol, but the sulfonation mass is further sulfonated in situ for the production of di- and trisulfonic acids. Naphthalene-1-sulfonic acid is not usually isolated on a large scale because 1-naphthol is produced advantageously by other routes. The 1-isomer is, however, nitrated in situ for the manufacture of Peri and Laurent's acids (see Chap. 6).

Naphthalenedisulfonic acids are converted into naphthol sulfonic acids (Chap. 4) and dihydroxynaphthalenes (Section 3.3), or are nitrated to form intermediates for aminonaphthalenedisulfonic acids and aminonaphtholsulfonic acids (Chap. 8). Naphthalenetrisulfonic acids similarly afford aminohydroxynaphthalenedisulfonic acids (e.g., H acid). The one readily accessible naphthalenetetrasulfonic acid has no large-scale application.

**Table 1.** Letter acids and other code-named naphthalene intermediates

| | |
|---|---|
| A acid | 3,5-dihydroxynaphthalene-2,7-disulfonic acid |
| Amino epsilon acid | 1-aminonaphthalene-3,8-disulfonic acid |
| Amino F acid | 2-aminonaphthalene-7-sulfonic acid |
| Amino G acid | 2-aminonaphthalene-6,8-disulfonic acid |
| Amino J acid | 2-aminonaphthalene-5,7-disulfonic acid |
| Amino R acid | 2-aminonaphthalene-3,6-disulfonic acid |
| Armstrong acid | naphthalene-1,5-disulfonic acid |
| B acid | 1-aminonaphthalene-4,6,8-trisulfonic acid |
| Badische acid | 2-aminonaphthalene-8-sulfonic acid |
| BON acid | 3-hydroxy-2-naphthoic acid |
| Böniger acid | 1-amino-2-hydroxynaphthalene-4-sulfonic acid |
| Bronner acid | 2-aminonaphthalene-6-sulfonic acid |
| C (Cassella) acid | 2-aminonaphthalene-4,8-disulfonic acid |
| Chicago (2 S) acid | 1-amino-8-hydroxynaphthalene-2,4-disulfonic acid |
| Chromotropic acid | 1,8-dihydroxynaphthalene-3,6-disulfonic acid |
| 1,6-Cleve's acid | 1-aminonaphthalene-6-sulfonic acid |
| 1,7-Cleve's acid | 1-aminonaphthalene-7-sulfonic acid |
| Crocein (Bayer) acid | 2-hydroxynaphthalene-8-sulfonic acid |
| Cyanol | 1-amino-7-naphthol |
| D (Dahl's) acid | 2-aminonaphthalene-5-sulfonic acid |
| Dahl's acid II | 1-aminonaphthalene-4,6-disulfonic acid |
| Dahl's acid III | 1-aminonaphthalene-4,7-disulfonic acid |
| Delta (δ) acid | 1-hydroxynaphthalene-4,8-disulfonic acid |
| Epsilon (ε) acid | 1-hydroxynaphthalene-3,8-disulfonic acid |
| F acid | 2-hydroxynaphthalene-7-sulfonic acid |
| Freund's acid (1,3,6) | 1-aminonaphthalene-3,6-disulfonic acid |
| Freund's acid (1,3,7) | 1-aminonaphthalene-3,7-disulfonic acid |
| G acid | 2-hydroxynaphthalene-6,8-disulfonic acid |
| Gamma (γ) acid | 2-amino-8-hydroxynaphthalene-6-sulfonic acid |
| H acid | 1-amino-8-hydroxynaphthalene-3,6-disulfonic acid |
| J acid | 2-amino-5-hydroxynaphthalene-7-sulfonic acid |
| K acid | 1-amino-8-hydroxynaphthalene-4,6-disulfonic acid |
| Kalle's acid | 1-aminonaphthalene-2,7-disulfonic acid |
| Koch acid | 1-aminonaphthalene-3,6,8-trisulfonic acid |
| Laurent's (L) acid | 1-aminonaphthalene-5-sulfonic acid |
| M acid | 1-amino-5-hydroxynaphthalene-7-sulfonic acid |
| Naphthionic acid | 1-aminonaphthalene-4-sulfonic acid |
| NW (Nevile and Winther) acid | 1-hydroxynaphthalene-4-sulfonic acid |
| Oxy Chicago acid | 1-hydroxynaphthalene-4,8-disulfonic acid |
| Oxy Koch acid | 1-hydroxynaphthalene-3,6,8-trisulfonic acid |
| Oxy L acid | 1-hydroxynaphthalene-5-sulfonic acid |
| Oxy Tobias acid | 2-hydroxynaphthalene-1-sulfonic acid |
| Peri acid | 1-aminonaphthalene-8-sulfonic acid |
| Purpurol | 1-amino-5-naphthol |
| R acid | 2-hydroxynaphthalene-3,6-disulfonic acid |
| 2 R (Columbia) acid | 2-amino-8-hydroxynaphthalene-3,6-di-sulfonic acid |
| RM acid | 2-amino-3-hydroxynaphthalene-6-sulfonic acid |
| S acid | 1-amino-8-hydroxynaphthalene-4-sulfonic acid |
| Schaeffer acid | 2-hydroxynaphthalene-6-sulfonic acid |
| T acid | 1-aminonaphthalene-3,6,8-trisulfonic acid |
| Tobias acid | 2-aminonaphthalene-1-sulfonic acid |
| Violet (RG) acid | 1-hydroxynaphthalene-3,6-disulfonic acid |

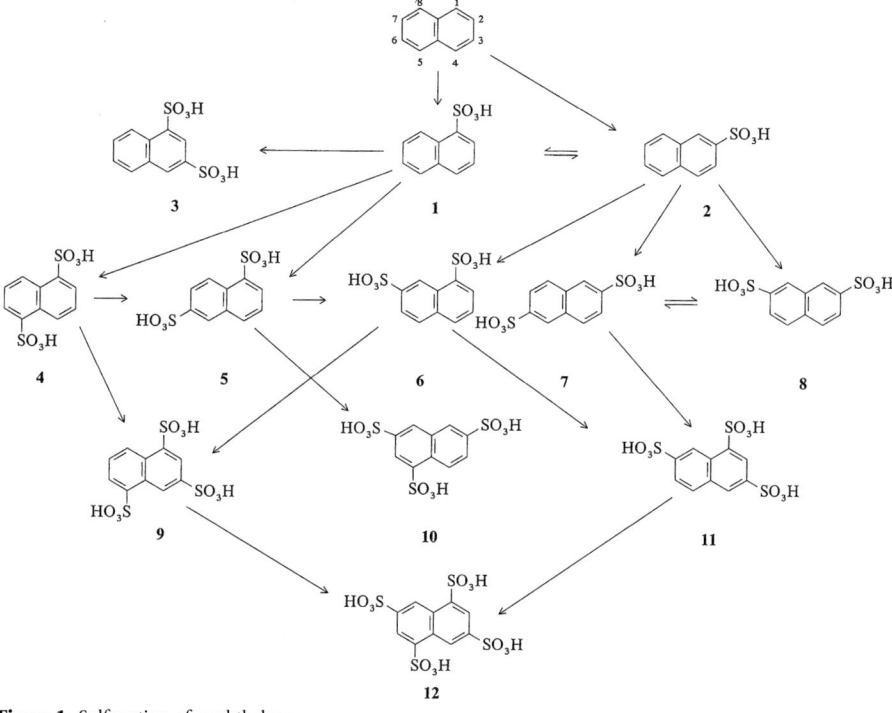

**Figure 1.** Sulfonation of naphthalene

## 2.2. Monosulfonic Acids

**Naphthalene-1-sulfonic acid** [*85-47-2*] (**1**), naphthalene-α-sulfonic acid, $C_{10}H_8O_3S$, $M_r$ 208.23, crystallizes from aqueous solutions as the dihydrate, *mp* 90 °C; dehydration ($P_2O_5$, 1.33 kPa) gives the anhydrous acid, *mp* 139 – 140 °C. Aqueous solubilities of the alkali-metal salts are given in Table 2.

Naphthalene-1-sulfonic acid is desulfonated to naphthalene by heating with dilute mineral acid. Halogenation also removes the sulfonic acid group, mainly with formation of 1,5-dihalonaphthalenes. Naphthalene-1-sulfonyl chloride, *mp* 68 °C, is prepared by treatment of the anhydrous acid with phosphorus pentachloride at 100 °C.

*Production.* Naphthalene-1-sulfonic acid is produced by adding naphthalene to 96 % sulfuric acid at 20 °C and slowly raising the temperature to 70 °C [7]. After 3 h at 70 – 75 °C the reaction mass is poured into water and limed out. A technical-grade product is obtained by evaporation and a purer product by precipitating the aniline salt.

An alternative sulfonation process uses sulfur trioxide in a solvent such as tetrachloroethane.

3367

**Table 2.** Salts of key naphthalenesulfonic acids

| Salt | Hydrate | Solubility, g in 100 mL H$_2$O |
|---|---|---|
| *1-Sulfonic acid* | | |
| Sodium | ½ H$_2$O | 9.1 (10 °C) |
| Potassium | ½ H$_2$O | 7.7 (10 °C) |
| Calcium | 2 H$_2$O | 6.1 (10 °C) |
| Barium | 1 H$_2$O | 1.2 (10 °C) |
| *2-Sulfonic acid* | | |
| Sodium | | 5.9 (25 °C) |
| Potassium | ½ H$_2$O | 6.7 (10 °C) |
| Calcium | 1 H$_2$O | 1.3 (10 °C) |
| Barium | 1 H$_2$O | 0.35 (10 °C) |
| *1,5-Disulfonic acid* | | |
| Sodium | 2 H$_2$O | 11.1 (18 °C) |
| Potassium | 2 H$_2$O | 6.7 (18 °C) |
| Calcium | 2 H$_2$O | 2.5 (18 °C) |
| Barium | 1 H$_2$O | 0.21 (18 °C) |
| *1,6-Disulfonic acid* | | |
| Sodium | 7 H$_2$O | 33.3 (18 °C) |
| Potassium | | 20.0 (18 °C) |
| Calcium | 4 H$_2$O | 10.0 (18 °C) |
| Barium | 3½ H$_2$O | 6.2 (100 °C) |
| *2,6-Disulfonic acid* | | |
| Sodium | 1 H$_2$O | 11.9 (18 °C) |
| Potassium | | 5.2 (18 °C) |
| Calcium | | 6.2 (18 °C) |
| Barium | 1 H$_2$O | <0.1 |
| *2,7-Disulfonic acid* | | |
| Sodium | 6 H$_2$O | 45 (18 °C) |
| Potassium | 2 H$_2$O | 71 (18 °C) |
| Calcium | 6 H$_2$O | 16 (18 °C) |
| Barium | 2 H$_2$O | 1.2 (19 °C) |

*Uses.* Naphthalene-1-sulfonic acid is further sulfonated or nitrated without isolation. At lower temperature (35 °C), sulfuric acid with oleum yields the 1,5-disulfonic acid, whereas at higher temperature (100 °C) the 1,6-disulfonic acid is predominantly obtained. Nitration gives mainly the 5-nitro and 8-nitro derivatives (see Section 6.1). Production of 1-naphthol by caustic fusion (hydroxylation) has long since been superseded by alternatives for reasons of product quality (Section 3.1).

**Naphthalene-2-sulfonic acid** [*120-18-3*] (**2**), naphthalene-$\beta$-sulfonic acid, C$_{10}$H$_8$O$_3$S, $M_r$ 208.23, crystallizes from aqueous solutions as the hydrate (*mp* 124 °C) or trihydrate (*mp* 83 °C). Aqueous solubilities of the free acid and its salts are somewhat lower than those of the corresponding $\alpha$-isomers (Table 2). The isolated sodium salt is known as $\beta$ salt.

Naphthalene-2-sulfonic acid is desulfonated by hot aqueous mineral acids with much more difficulty than naphthalene-1-sulfonic acid; the rate has been measured at 50 times slower [22]. Halogenation attacks either or both the 5- and the 8-positions. Bromination may be used to separate a mixture of naphthalene-1-sulfonic acid from naphthalene-2-sulfonic acid because the latter forms a soluble product whereas the former is desulfonated.

*Production.* Molten naphthalene is added to 96% sulfuric acid in an iron vessel and the mixture is agitated at 163 °C for 2 h. A complex work-up consisting of gradual dilution, heating, and neutralization with caustic soda and sodium sulfite ensures an 88% yield of isolated $\beta$ salt and desulfonation of the $\alpha$-isomer, which is formed as a coproduct in 5 – 10% yield [7].

*Uses.* The major outlet for $\beta$ salt is the production of 2-naphthol (Section 3.2). The 2-sulfonic acid is a stage in the production of the 1,6-, 2,6-, and 2,7-disulfonic acids and the 1,3,6-trisulfonic acid. Nitration gives primarily the 5- and 8-nitro derivatives as intermediates for Cleve's acids (Section 6.1). The 2-sulfonic acid condenses with formaldehyde or alcohols to form surface-active agents (see Section 2.5).

The intermediate 2-thionaphthol can be obtained by catalytic hydrogenation of naphthalene-2-sulfonic acid, but is traditionally made by zinc reduction of naphthalene-2-sulfonyl chloride.

## 2.3. Disulfonic Acids

**Naphthalene-1,3-disulfonic acid** [*6094-26-4*] (**3**), $C_{10}H_8O_6S_2$, $M_r$ 288.28, is produced only in small proportion on disulfonation of naphthalene and cannot be separated. It is prepared in good yield in the laboratory by diazotization and copper-catalyzed deamination of 2-aminonaphthalene-6,8-disulfonic acid [23].

**Naphthalene-1,5-disulfonic acid** [*81-04-9*] (**4**), Armstrong acid, $C_{10}H_8O_6S_2$, $M_r$ 288.28, crystallizes as the tetrahydrate but dehydrates at 125 °C to give the anhydrous acid (*mp* 240 – 245 °C). Solubilities of the alkali-metal salts are given in Table 2.

The disulfonic acid forms an anhydride when treated with oleum below 50 °C. This anhydride reacts with ammonia to give a mixture of naphthalene-1,5-disulfonamide and naphthalene-1-sulfonamide-5-sulfonic acid. Treatment of the disodium salt of naphthalene-1,5-disulfonic acid with phosphorus pentachloride at 110 °C gives naphthalene-1,5-disulfonyl chloride (*mp* 184 °C), which may also be prepared by chlorosulfonation of naphthalene. Reaction of an aqueous solution of naphthalene-1,5-disulfonic acid with bromide – bromate gives 1,5-dibromonaphthalene so readily that the yellow precipitate may be used to identify very low concentrations of the disulfonic acid.

*Production.* Naphthalene is mixed with 20% oleum at 20 – 35 °C followed by gradual addition of 65% oleum and further naphthalene alternately. After heating for 6 h at 55 °C the reaction mixture is added to water, and the product is precipitated as the free acid by cooling or as the disodium salt by the addition of alkaline sodium sulfate. In each case, the isolated yield is ca. 53%. The 1,6-disulfonic acid can be recovered from the filtrate.

Higher yields of **4** have been reported for alternative processes; for example, solvent sulfonation with sulfur trioxide and chlorosulfonic acid in a solvent such as tetrachloroethylene [24] or sulfonation with oleum through a double screw-feed apparatus [25].

*Uses.* The disodium salt of naphthalene-1,5-disulfonic acid gives 1-hydroxynaphthalene-5-sulfonic acid or 1,5-dihydroxynaphthalene (see Section 3.3) by fusion processes. Further sulfonation of the disulfonated reaction mixture gives naphthalene-1,3,5-trisulfonic acid (**9**), and nitration leads to 1- and 2-nitronaphthalene-4,8-disulfonic acids as intermediates to the corresponding aminonaphthalene disulfonic acids (Chap. 6).

Naphthalene-1,5-disulfonic acid is an important stabilizer for diazo compounds [26].

**Naphthalene-1,6-disulfonic acid** [525-37-1] (**5**), $C_{10}H_8O_6S_2$, $M_r$ 288.28, crystallizes as the tetrahydrate, *mp* 125 °C (decomp.). Solubilities of the alkali-metal salts are given in Table 2.

*Production.* Separation from the 1,5-disulfonic acid process (see above) usually meets requirements for the preparation of **5**. It can also be prepared by sulfonation of naphthalene-2-sulfonic acid with oleum, which gives naphthalene-1,7-disulfonic acid (**6**) as a byproduct [12].

*Uses.* Alkali fusion of **5** yields 1,6-dihydroxynaphthalene (Section 3.3), and sulfonation with oleum produces naphthalene-1,3,6-trisulfonic acid (**10**). However, the main outlet for naphthalene-1,6-disulfonic acid is via nitration to 1-nitronaphthalene-3,8-disulfonic acid, which is the intermediate for 1-aminonaphthalene-3,8-disulfonic acid (Section 6.4).

**Naphthalene-1,7-disulfonic acid** [5724-16-3] (**6**), $C_{10}H_8O_6S_2$, $M_r$ 288.28, is produced by sulfonation of naphthalene-2-sulfonic acid (**2**) with cold oleum to give a mixture of 1,6- (80%) and 1,7- (20%) acids. Although these are difficult to separate, the corresponding disulfonyl chlorides are separable because of their differing solubilities in, for example, benzene.

**Naphthalene-2,6-disulfonic acid** [581-75-9] (**7**), $C_{10}H_8O_6S_2$, $M_r$ 288.28, crystallizes as the dihydrate, *mp* 129 °C. The alkali-metal salts are much less soluble than those of the 2,7-isomer (Table 2).

*Production.* Naphthalene is added to sulfuric acid–monohydrate ($SO_3 \cdot H_2O$) mixture at 135 °C, and the reaction is completed by heating at 170–175 °C for 5 h. After quenching in water, the disodium salt of **7** is precipitated at 90 °C by adding sodium chloride and sodium sulfate. Filtration at 95 °C followed by washing with sodium sulfate solution gives the required product, free from the 2,7-isomer, in 21% yield. The 2,7-isomer is then isolated from the filtrates.

*Uses.* The disodium salt of **7** gives 2-hydroxynaphthalene-6-sulfonic acid or 2,6-dihydroxynaphthalene (Section 3.3) by fusion processes. Nitration of the crude sulfonation product containing 2,6- and 2,7-disulfonic acids gives the 4-nitro derivatives, which, on reduction, give the mixed 1-aminonaphthalene-3,6(3,7)-disulfonic acid (Freund's acid).

**Naphthalene-2,7-disulfonic acid** [92-41-1] (**8**), $C_{10}H_8O_6S_2$, $M_r$ 288.28; the sodium salt of **8** is much more soluble than that of the 2,6-isomer (Table 2) but can be isolated and purified by precipitation of the calcium salt at lower temperature followed by recrystallization.

*Production.* The 2,7-disulfonic acid is always the main product of high-temperature sulfonation of naphthalene, but at 160 °C it converts slowly to an equilibrium mixture containing ca. 30 % of the 2,6-isomer [12]. Crystallization of the pure product from conventional sulfonation is difficult, and alternative processes, (for example, reaction of naphthalene vapor with sulfur trioxide in the gas phase at 220 °C) are said to give a purer product.

*Uses.* The disodium salt of **8** gives 2-hydroxynaphthalene-7-sulfonic acid or 2,7-dihydroxynaphthalene (Section 3.3) by fusion processes. Sulfonation by oleum gives the 1,3,6-trisulfonic acid, and nitration gives 1-nitronaphthalene-3,6-disulfonic acid.

In sulfuric acid solution, the disodium salt of naphthalene-2,7-disulfonic acid undergoes an unusual reaction at the 4-position with 4,4'-bis-(dimethylamino)benzhydrol. The resulting condensation product is oxidized with lead dioxide to give the acid basic dye C.I. Acid Green 16.

## 2.4. Tri- and Tetrasulfonic Acids

**Naphthalene-1,3,5-trisulfonic acid** [6654-64-4] (**9**), $C_{10}H_8O_9S_3$, $M_r$ 368.34, is extremely soluble in water; the sodium salt crystallizes as the tetrahydrate.

*Production.* Naphthalene and 65 % oleum are added simultaneously to sulfuric acid monohydrate over a period of 32 h at 30 – 35 °C. The mixture is then heated at 50 °C for 1 h, 70 °C for 1 h, and finally 90 °C for 7 h. This procedure minimizes formation of the 1,3,6-isomer, but a purer product may be obtained by using isolated naphthalene-1,5-disulfonic acid as starting material.

*Uses.* The total reaction mixture is nitrated and the resulting 8-nitro derivative is reduced to give a mixture of 1-aminonaphthalene-3,6,8- and 4,6,8-trisulfonic acids.

**Naphthalene-1,3,6-trisulfonic acid** [86-66-8] (**10**), $C_{10}H_8O_9S_3$, $M_r$ 368.34, crystallizes as the hexahydrate, *mp* 170 °C.

*Production.* Naphthalene is initially sulfonated with sulfuric acid monohydrate under programmed temperature control between 80 and 145 °C. Then, 65 % oleum is added gradually at 40 °C and heating is continued for 2.5 h at 145 °C. Trisulfonation is complete after the further addition of 65 % oleum and heating for 3 h at 150 °C. This complex procedure [7] maximizes the formation of **10** with a conversion of up to 75 %.

*Uses.* The total reaction mixture is nitrated, and the resulting 8-nitro derivative is reduced to give 1-aminonaphthalene-3,6,8-trisulfonic acid (Koch acid), the key intermediate in the production of the important H acid. The crude trisodium salt may be precipitated from the quenched mixture after dilution to 75% acid strength to give a product (Azoguard) used as a diazo stabilizer.

**Naphthalene-1,3,7-trisulfonic acid** [*85-49-4*] (**11**), $C_{10}H_8O_9S_3$, $M_r$ 368.34, is much less important than the 1,3,5- and 1,3,6-isomers. It is produced by sulfonation of naphthalene-2,6-disulfonic acid (**7**) with 65% oleum in sulfuric acid monohydrate [7].

**Naphthalene-1,3,5,7-tetrasulfonic acid** [*6654-67-7*] (**12**), $C_{10}H_8O_{12}S_4$, $M_r$ 448.4, has no technical importance although it is easily obtained as the end product of the oleum sulfonation of naphthalene, along with the 1,3,7-trisulfonic acid (**11**). In the latter, unlike the 1,3,5- and 1,3,6-trisulfonic acids, direct entry of a fourth sulfonic acid group does not occur.

## 2.5. Alkylnaphthalenesulfonic Acids

A wide range of mixed alkylnaphthalenesulfonic acids of undefined or undisclosed constitution are produced as surface-active wetting agents and dispersants. For example, methylnaphthalenes are sulfonated with sulfuric acid at 160 °C, or naphthalenesulfonic acids are alkylated with alkenes, higher alcohols, or formaldehyde. Propene or butene produce isopropyl- or butylnaphthalene sulfonates, respectively. Formaldehyde gives the dimer **13** as a simple product from β salt, or polymeric condensation products (**14**), which are important as synthetic tanning agents or as plasticizers for concrete. Dinonylnaphthalene, prepared from naphthalene and nonene, is sulfonated to give derivatives that are important as lubricating oil additives [13].

## 3. Naphthols

Naphthols are the most important naphthalene derivatives because they are key intermediates in the production of many chemicals other than dyes and pigments. World production is estimated at 40 000 t/a of 1-naphthol and 100 000 t/a of 2-naphthol.

**Table 3.** Physical properties of naphthols

| Property | 1-Naphthol | 2-Naphthol |
|---|---|---|
| $mp$, °C | 96 | 123 |
| $bp$, °C | 280 | 295 |
| $d_4$ | 1.224 | 1.217 |
| Solubility in water, % | 0.03 (25 °C) | 0.075 (25 °C) |
| Dissociation constant ($H_2O$) | $1.4 \times 10^{-10}$ (20 °C) | $1.14 \times 10^{-10}$ (20 °C) |

The production methods for naphthols are as follows (the first four methods are analogous to those for phenols):

1) Alkali fusion of naphthalenesulfonic acids
2) Alkaline hydrolysis of chloronaphthalenes
3) Hydroperoxidation of 2-isopropylnaphthalene to form 2-naphthol and acetone (not applicable to 1-naphthol)
4) Dehydrogenation of tetralones
5) Direct, single-step replacement of the amino group of naphthylamine by a hydroxyl group

The last method is not applicable to phenols, for which the corresponding reaction requires two steps, diazotization of the aminobenzene followed by hydrolysis.

Naphthols resemble phenols in their chemical properties, but their hydroxyl groups are more reactive. For example, they are readily converted to ethers by reaction with alcohols, and to amines by heating with ammonia and bisulfite (Bucherer reaction).

## 3.1. 1-Naphthol

1-Naphthol [90-15-3] (**15**), α-naphthol, 1-naphthalenol, 1-hydroxynaphthalene, $C_{10}H_8O$, $M_r$ 144.16, forms colorless prisms (from toluene) which darken on exposure to air or light. The compound is steam volatile and sublimable. Other physical properties are listed in Table 3.

1-Naphthol is reduced by sodium in liquid ammonia to give 5,6,7,8-tetrahydro-1-naphthol and is oxidized by $NaOCl-FeCl_3$ or $I_2-KI$ to give a violet color. It is chlorinated by phosphorus pentachloride at 150 °C to give 1-chloronaphthalene, by sulfuryl chloride to give 4-chloro-1-naphthol, or by $Cl_2-CH_3COOH$ to give 2,4-dichloro-1-naphthol. Similarly bromine forms 2,4-dibromo-1-naphthol, and this reaction may be used quantitatively for titration.

Nitration of 1-naphthol gives complex mixtures. The crude 2,4-dinitro derivative was used in the mid-1800s as an acid yellow dye comparable to picric acid. Sulfonation at 50 °C readily yields 1-naphthol-2,4-disulfonic acid. Nitrous acid gives mainly the 2-nitroso derivative, contaminated with the 4-nitroso isomer.

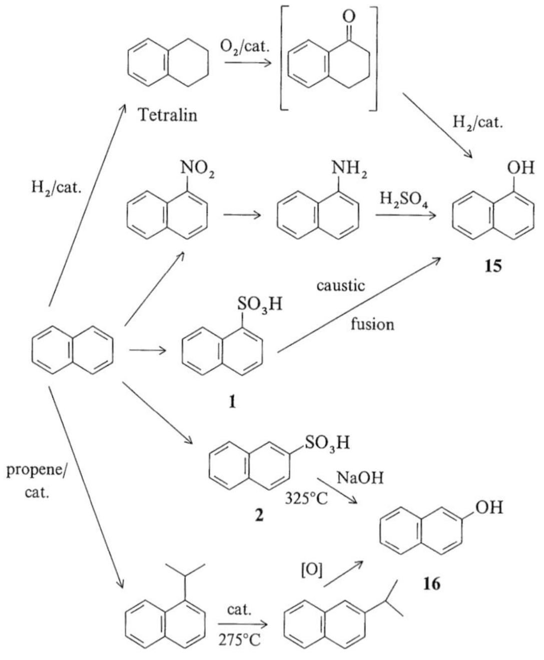

**Figure 2.** Industrial routes to 1-naphthol and 2-naphthol

*Production.* Historically 1-naphthol was produced via caustic fusion of naphthalene-1-sulfonic acid, but this was superseded by the I.G. Farbenindustrie process involving hydrolysis of 1-naphthylamine (Section 5.1) with aqueous 22% sulfuric acid at 200 °C under pressure in a lead-lined autoclave [7]. To obtain a purer product, Union Carbide developed a process based on catalytic oxidation of tetralin to 1-tetralol and 1-tetralone followed by dehydrogenation (see Fig. 2) [27]. The two-stage catalytic process is claimed to give an overall yield of 72% 1-naphthol, with an overall efficiency of 97%.

*Uses.* Several important agrochemicals are derived from 1-naphthol. 1-Naphthoxyacetic acid (**17**, R = H) is prepared by aqueous alkylation with chloroacetic acid. Although 2-naphthoxyacetic acid (**31**) appears to be more widely used, recent work [28] suggests that the 1-isomers, especially (**17**, R = Me), have greater herbicidal activity. Devrinol [*15299-99-7*] (**18**), Stauffer's 2-(1-naphthoxy)-*N,N*-diethylpropionamide herbicide, is obtained by the reaction of 1-naphthol and *N,N*-diethyl-α-bromopropionamide in MeOH – NaOMe at reflux [29]. Sevin is Union Carbide's insecticide based on 1-naphthyl methylcarbamate [*63-25-2*], carbaryl (**19**), which is produced from 1-naphthyl chloroformate and methylamine [30] or by reaction of 1-naphthol with methyl isocyanate. Its main use lies in the control of earthworms and certain caterpillars. Many analogues and derivatives of **19** have been patented [11].

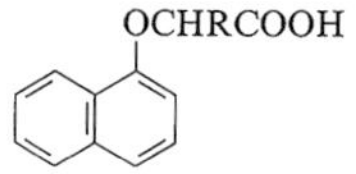

**17** OCHRCOOH-naphthalene

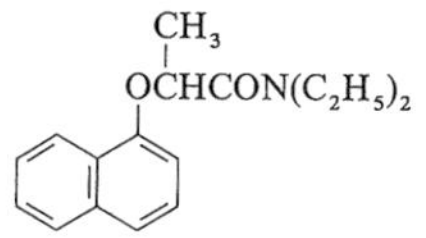

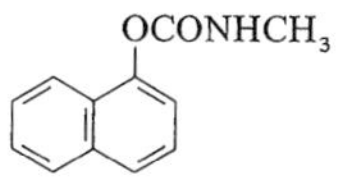

**18** OCH(CH₃)CON(C₂H₅)₂-naphthalene (shown as OCHCON(C₂H₅)₂ with CH₃)

**19** OCONHCH₃-naphthalene

Important drugs are also derived from 1-naphthol. The ester 1-naphthyl salicylate [550-97-0], Alphol (**20**), has been used as an antiseptic and antirheumatic, and ICI's 1-isopropylamino-3-(1-naphthoxy)-2-propanol [3506-09-0], Inderal (**21**), is an important member of a new class of adrenergic β-blockers for the treatment of heart disease [31]. Production of **21** involves reaction of **15** with epichlorohydrin followed by isopropylamine.

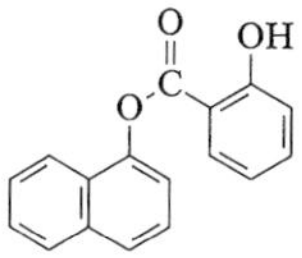

**20** 1-naphthyl salicylate

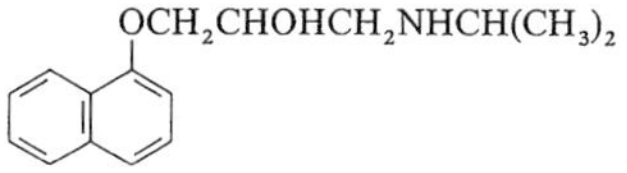

**21** OCH₂CHOHCH₂NHCH(CH₃)₂-naphthalene

Related products used as β-blockers or antiarrhythmics include nadoxolol (**22**) and nadolol (**23**). Many analogues and alternative syntheses of this type of compound have been patented [11], [12].

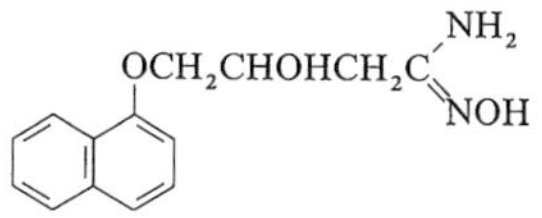

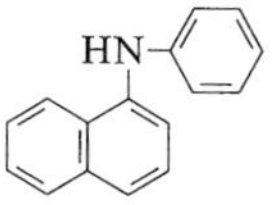

**22** OCH₂CHOHCH₂C(NH₂)=NOH-naphthalene

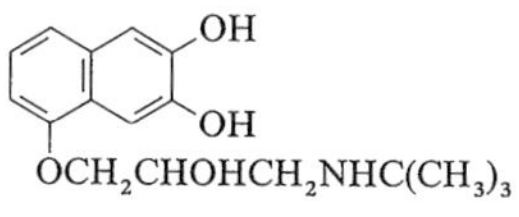

**23** naphthalene-2,3-diol with OCH₂CHOHCH₂NHC(CH₃)₃

**24** HN-phenyl-naphthalene

Condensation of 1-naphthol with aniline at 180 °C gives *N*-phenyl-1-naphthylamine (**24**) which is used as a rubber antioxidant.

1-Naphthol is much less important than 2-naphthol as a dye intermediate. Apart from the historical interest in simple azo derivatives such as Orange I (**25**) (GRIESS, 1876; C.I. Acid Orange 20), the equilibrium with the keto form (**26**) is an example of the prevalence of keto forms in many azo naphthol derivatives [32].

Certain diazo components, for example, 1-amino-2-hydroxynaphthalene-4-sulfonic acid, under strongly alkaline conditions, couple with 1-naphthol in the 2-position rather than the conventional 4-position. This leads to compounds such as C.I. Mordant Black 3 (**27**) which form stable chromium complexes. Excess diazo compound can lead to disazo acid dyes such as C.I. Acid Brown 43 (**28**).

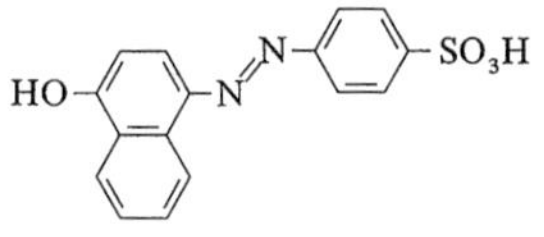

**25**

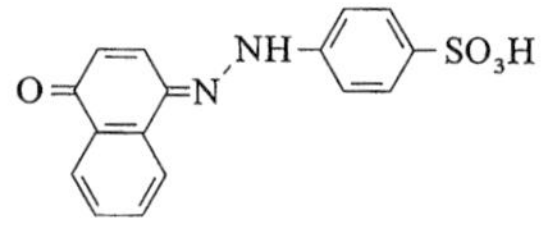

**26**

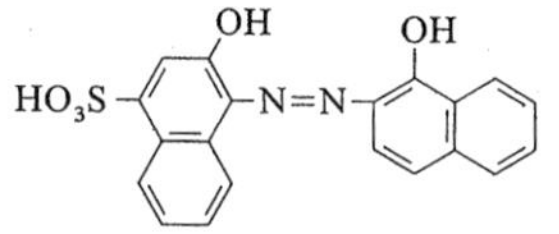

**27**

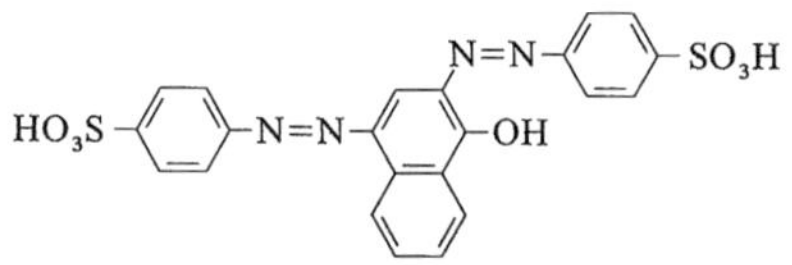

**28**

Nitroso compounds can condense with 1-naphthol for example, *N,N*-dimethyl-4-nitrosoaniline reacts to give indophenol (**29**).

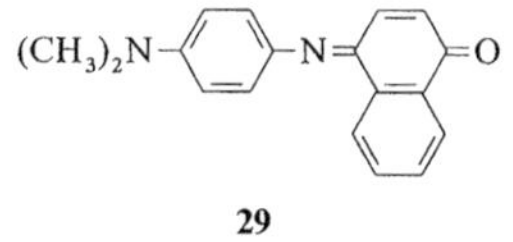

**29**

Halogenated derivatives of 1-naphthol have minor uses as shown by the reaction of 4-chloro-1-naphthol and 4-bromo-1-naphthol with substituted isatin-α-chlorides to form blue indigoid vat dyes [18]

*Toxicity.* 1-Naphthol is moderately toxic; the dust is liable to cause skin irritation and eye injury [14].

## 3.2. 2-Naphthol

2-Naphthol [*135-19-3*] (**16**), β-naphthol, 2-naphthalenol, 2-hydroxynaphthalene, $C_{10}H_8O$, $M_r$ 144.16, forms colorless plates upon sublimation, which darken on exposure to air or light. Unlike 1-naphthol it is nonvolatile in steam. Other physical properties are presented in Table 3.

2-Naphthol is reduced with sodium or with hydrogen in the presence of a catalyst to give mainly 1,2,3,4-tetrahydro-2-naphthol (unlike 1-naphthol which gives the arylphenol under the same conditions). Oxidation with ferric chloride forms 2,2′-dihydroxy-1,1′-binaphthyl, with any blue coloration indicating the presence of 1-naphthol. Air oxidation at 300 °C with a vanadium pentoxide catalyst also yields 1,1′-bi-2-naphthol, which dehydrates at higher temperature to the oxide and finally decomposes to give benzoic acid. Treatment with sulfuryl chloride in carbon disulfide or with NaOCl–NaOH leads to 1-chloro-2-naphthol, whereas chlorine in sodium carbonate gives 8-chloro-2-naphthol (plus 1, 1′-bi-2-naphthol). Reaction with phosphorus pentachloride at 150 °C gives 2-chloronaphthalene or at lower temperature, tri-2-naphthyl phosphate (*mp* 111 °C), which can be isolated after treatment of the reaction mixture with alkali. Reaction with one equivalent of bromine in acetic acid gives 1-bromo-2-naphthol, whereas excess bromine yields 1,6-dibromo-2-naphthol. The latter is readily debrominated in dilute mineral acid to yield 6-bromo-2-naphthol. Excess bromine reacts with 2-naphthol at 100 °C to form 1,5,6,-tribromo- and 1,3,5,6-tetrabromo-2-naphthol.

Careful nitration with strong nitric acid in acetic acid gives 1,6-dinitro-2-naphthol, whereas nitrous acid readily forms 1-nitroso-2-naphthol [*131-91-9*]. Sulfonation leads to a series of commercially important products (see Chap. 4).

*Production.* 2-Naphthol is produced by caustic fusion of naphthalene-2-sulfonic acid (Section 2.2). Typically, the sodium salt of the sulfonic acid is added gradually to 50 % sodium hydroxide liquor at 300 °C; the melt is then heated further at 320 °C in a gas-fired iron vessel with vigorous agitation. After completion of the reaction, the melt is run into excess water, possibly including filtrate from the previous batch at a proven tolerable level, and the naphtholate solution is neutralized to pH 8 with dilute sulfuric acid. If the temperature is maintained at >100 °C during neutralization, the crude product comes out of solution as an oil, which is separated, washed with hot water, and distilled under vacuum to give pure 2-naphthol. The molten material is processed through a flaker to give the final product for packaging. The fusion yield is about 80 % of the theoretical value, resulting in an overall yield of 70 % based on naphthalene. Typical specifications for 2-naphthol are (1) clear solution in dilute caustic soda, (2) *mp* 120.5 °C, and (3) 1-naphthol content <0.3 %.

The newer method of manufacture (see Fig. 2) is economically and environmentally favored in the United States, because despite requiring three stages, it is more amenable to continuous operation with recycle streams. The alkylation and isomerization are carried out up to 240 °C with a phosphoric acid catalyst [33]. Final catalytic oxidation at

90–110 °C gives the hydroperoxide, which is cleaved with dilute sulfuric acid to give 2-naphthol in high overall yield in spite of modest oxidation conversion.

*Uses.* The sulfonated and carboxylated derivatives are described in Chapter 4 and Section 3.4.

The simple ether derivatives, 2-methoxynaphthalene [*93-04-9*], nerolin, and 2-ethoxynaphthalene [*93-18-5*], nerolin "new", are readily prepared by conventional alcohol esterification processes for use in perfume formulation and, more recently as intermediates for drugs. The $\beta$-hydroxyethyl ether, anavenol (**30**), prepared by condensation of 2-naphthol with ethylene chlorohydrin or ethylene oxide, has been used as an animal sedative. Reaction of 2-naphthol with chloroacetic acid in aqueous alkaline medium gives 2-naphthoxyacetic acid (**31**), which is used as a growth promotor for fruit.

The lactate, benzoate, and salicylate esters have been used as antiseptics and are improvements of the naphthols themselves.

Reaction of 2-naphthol with aniline at 180 °C gives *N*-phenyl-2-naphthylamine (PBN, **32**) an antioxidant for rubber processing which is gradually being displaced by improved products.

The major pharmaceutical products based on 2-naphthol are the antifungal tolnaftate (**33**), produced by reaction with thiophosgene and *N*-methyl-*m*-toluidine [34]; the semisynthetic penicillin nafcillin (**34**), produced via 2-ethoxynaphthalene; and the antirheumatic naproxen (**35**), produced via 2-methoxynaphthalene.

The C-alkylation of 2-naphthol by using 2-chlorocyclohexanone in boiling dioxane in the presence of potassium carbonate has been exploited in the preparation of the antitussive naphthanone (**36**).

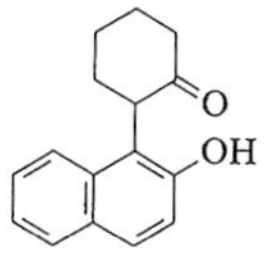

**36**

2-Naphthol is used in solvent dyes, e.g., C.I. Solvent Yellow 14 (**37**, R = H); in acid dyes, e.g., C.I. Acid Orange 7 (**37**, R = SO$_3$H); in metallizable dyes, e.g., C.I. Mordant Black 15 (**38**); and in pigments, e.g., C.I. Pigment Red 3 (**39**).

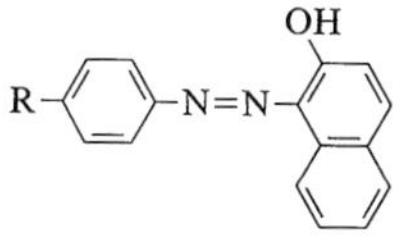

**37**

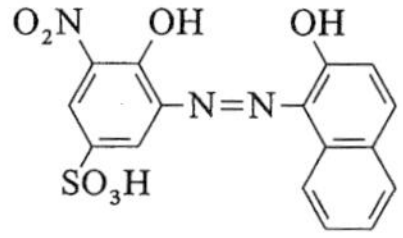

**38**

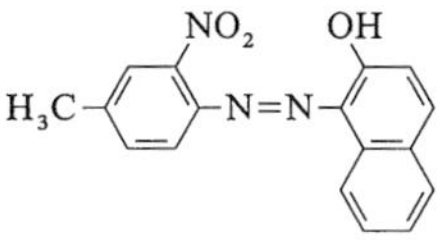

**39**

However, 3-hydroxy-2-naphthoic acid and derived arylamides (Section 3.4) are more important coupling components for pigments than is 2-naphthol.

*Toxicity.* 2-Naphthol is moderately toxic but less of a skin irritant than 1-naphthol [15].

## 3.3. Naphthalenediols

All ten isomeric naphthalenediols are known. They are usually prepared by alkali fusion of sulfonic acids or by acid hydrolysis of aminonaphthalenes, provided appropriately substituted compounds are available as precursors. With α,β-disubstituted naphthalenes, hydroxylation takes place more readily in the α-position; this is particularly important if hydroxynaphthalenesulfonic acids (Chap. 4) are required [12].

Conversely, with amination of dihydroxynaphthalenes by the Bucherer reaction, the β-position is more readily attacked. Thus, 1,7-dihydroxynaphthalene gives 7-amino-1-naphthol.

Of the ten dihydroxynaphthalenes, only 1,5-naphthalenediol has significant applications.

**1,5-Naphthalenediol** [*83-56-7*], 1,5-dihydroxynaphthalene, Azurol, C$_{10}$H$_8$O$_2$, $M_r$ 160.16, forms colorless needles (from aqueous methyl cellosolve). The compound is sublimable, readily soluble in ether and acetone, but only sparingly soluble in water and hydrocarbons. It is oxidized to 5-hydroxy-1,4-naphthoquinone (juglone) by $CrO_3$. Amination gives 1,5-naphthalenediamine; coupling with diazotized anilines occurs in the 2- or 4-position.

*Production* [7]. The sodium salt of naphthalene-1,5-disulfonic acid (Section 2.3) is added gradually to excess 75% sodium hydroxide. The mixture is heated under pressure to 276 °C over the course of 7 h and the reaction is completed by heating for a further 5 h at this temperature. After precipitation with sulfuric acid, the product is isolated in 90% yield.

*Uses.* 1,5-Naphthalenediol is used as a coupling component in azo dyes as exemplified in the metallizable (Cr) wool dye, C.I. Mordant Black 9 (**40**).

**1,2-Naphthalenediol** [*574-00-5*], *mp* 108 °C, is prepared by reduction of 1,2-naphthoquinone with hydrazine hydrate or sodium dithionite. It is oxidized under mild conditions to 1,2-naphthoquinone and couples with diazotized anilines in the 4-position.

**1,3-Naphthalenediol** [*132-86-5*], naphthoresorcinol, *mp* 124 °C, is synthesized by hydrolysis of 1,3-naphthalenediamine with sulfuric acid; by heating 1-amino-3-hydroxy-naphthalene-4-sulfonic acid with acid; or by cyclization of ethyl phenylacetoacetate. It is used as an analytical reagent for sugars and glucuronic acid in urine. It is oxidized in alkaline solution to 2-hydroxy-1,4-naphthoquinone and reacts with ammonia at 140 °C to give 3-amino-1-naphthol and 1,3-naphthalenediamine.

**1,4-Naphthalenediol** [*571-60-8*], *mp* 195 °C, is synthesized by reduction of 1,4-naphthoquinone with sodium dithionite, or by hydrolysis of 4-amino-1-naphthol with 20% sulfuric acid. It undergoes oxidation to give 1,4-naphthoquinone; amination to give 4-amino-1-naphthol; and methylation (MeOH – HCl) to give 4-methoxy-1-naphthol.

**1,6-Naphthalenediol** [*575-44-0*], *mp* 138 °C, is synthesized by caustic fusion of naphthalene-1,6-disulfonic acid at 330 °C. It couples with diazotized anilines to form disazo and trisazo derivatives.

**1,7-Naphthalenediol** [*575-38-2*], *mp* 181 °C, is prepared by caustic fusion of 2-hydroxy-naphthalene-8-sulfonic acid at 300 °C or by acid desulfonation of 2,8-dihydroxynaphthalene-6-sulfonic acid. It couples with diazotized anilines to form monoazo and disazo derivatives.

**1,8-Naphthalenediol** [*569-42-6*], *mp* 144 °C, is produced by caustic fusion of naphthosultone (Section 4.2) or naphthosultam (Section 6.1), and by acid desulfonation of 1,8-dihydroxynaphthalene-4-sulfonic acid. It is oxidized to juglone by chromium(VI) oxide; gives a dark green color with iron(III) chloride; undergoes amination at 300 °C to give 1,8-naphthalenediamine; and couples with diazotized anilines in the 4-position.

**2,3-Naphthalenediol** [*92-44-4*], *mp* 161 °C, is prepared by caustic fusion of 3-hydroxynaphthalene-2,6-disulfonic acid followed by treatment with dilute sulfuric acid under pressure, or by acid desulfonation of 2,3-dihydroxynaphthalene-6-sulfonic acid. It gives a dark blue color with iron(III) chloride; is aminated to 2-amino-3-naphthol at 140 °C and 2,3-naphthalenediamine at 240 °C; and couples with diazotized anilines in the 1- or 1,4-positions.

**2,6-Naphthalenediol** [*581-43-1*], 2,6-naphthohydroquinone, *mp* 222 °C, is synthesized by potassium hydroxide fusion of 2-hydroxynaphthalene-6-sulfonic acid at 295 °C. It undergoes amination at 250 °C to give 2,6-naphthalenediamine and couples with diazotized anilines in the 1-position under acid conditions or the 1,5-positions under alkaline conditions.

**2,7-Naphthalenediol** [*582-17-2*], *mp* 194 °C, is produced by caustic fusion of naphthalene-2,7-disulfonic acid or 2-hydroxynaphthalene-7-sulfonic acid at 300 °C. It undergoes amination at 250 °C to give 2,7-naphthalenediamine and couples with diazotized anilines in the 1-position or in the 1,8-positions. It forms a resin with formaldehyde.

## 3.4. Hydroxynaphthoic Acids

The industrially important hydroxynaphthoic acids are all produced by carboxylation (with $CO_2$) of the corresponding sodium or potassium naphtholate (Kolbe–Schmitt reaction). 3-Hydroxy-2-naphthoic acid (BON acid) is one of the most important naphthalene-based colorant intermediates. Arylamide derivatives were introduced in 1912 as improved azoic dyes, replacing 2-naphthol which had been in use since 1880 as a component in ice colors. Apart from their use in azoic dyes for cotton, these arylamides are intermediates for important pigments.

For isomers that are not accessible by carboxylation, the precursor cyano group is introduced by Sandmeyer reaction of the appropriate aminonaphthalenesulfonic acid, followed by hydrolysis and caustic fusion.

Although carboxylation of naphtholates may be carried out in a solvent whereby the polarity of the solvent has a marked effect on the reaction [35], large-scale operation is usually effected by using a dry melt in a rotating baker or specially designed agitated vessel. Whereas sodium α-naphtholate yields 1-hydroxy-2-naphthoic acid at 130 °C and sodium β-naphtholate yields 2-hydroxy-1-naphthoic acid at 120 °C; somewhat surprisingly, 3-hydroxy-2-naphthoic acid is produced from β-naphthol above 200 °C. The mechanism is not clear, and the course of the reaction is affected by temperature, pressure, alkali metal, and the presence of water [36]. However, the identified intermediate stages from sodium β-naphtholate are the formation of **41** at 120 °C and its conversion to equimolar quantities of **42** and 2-naphthol at 145 °C, which explains why the maximum yield is only 50%. At 200 °C, **42** rearranges to **43**, which also has a maximum yield of 50%.

                    COOH        COONa
                      \ ONa       \ ONa              ONa
                                                      \
                                                       COONa
                     41           42            43

The Marasse modification (1893) of the Kolbe–Schmitt reaction uses potassium carbonate with carbon dioxide and in certain cases gives better yields [36]. This has been confirmed for hydroxynaphthoic acids at 300 °C [37]. However, with potassium β-naphtholate an alternative rearrangement can take place, in which 6-hydroxy-2-naphthoic acid is coproduced at 220 °C.

**2-Hydroxy-1-naphthoic acid** [*2283-08-1*], $C_{11}H_8O_3$, $M_r$ 188.17, has a *mp* of 156–157 °C (decomp. to 2-naphthol and $CO_2$). The monosodium salt is unstable in hot aqueous solution. 2-Hydroxy-1-naphthoic acid is decarboxylated by nitrous acid, forming 1-nitroso-2-naphthol, and by diazo salts, forming a 1-azo-2-naphthol. An ethanolic solution of 2-hydroxy-1-naphthoic acid gives a dark blue color with iron(III) chloride. Chlorosulfonic acid at 30 °C gives the 6-sulfonyl chloride derivative.

*Production.* [7] Treatment of the dried sodium or potassium salt of 2-naphthol with carbon dioxide at 120 °C under pressure gives an 80 % yield of product after careful work-up. Solvents such as toluene, tetralin, or diphenyl oxide are patented alternatives [15].

*Uses.* 2-Hydroxy-1-naphthoic acid is an intermediate of minor importance in the synthesis of azo dyes.

**3-Hydroxy-1-naphthoic acid** [*19700-42-6*], *mp* 248–249 °C, is prepared by hydrolysis of diazotized 3-amino-1-naphthoic acid with boiling aqueous copper sulfate. It gives a reddish brown color with iron(III) chloride. Coupling with diazo compounds occurs in the 4-position.

**4-Hydroxy-1-naphthoic acid** [*7474-97-7*], *mp* 188–189 °C, is synthesized by potassium hydroxide fusion (200 °C) of 4-sulfo-1-naphthoic acid or 4-cyanonaphthalene-1-sulfonic acid (from 1-aminonaphthalene-4-sulfonic acid). It undergoes decarboxylation with bromine, nitrous acid, or diazo salts to form the corresponding 4-derivative.

**5-Hydroxy-1-naphthoic acid** [*2437-16-3*], *mp* 236 °C, is prepared by potassium hydroxide fusion (245 °C) of 1-cyanonaphthalene-5-sulfonic acid. It couples with diazo compounds in the 8-position.

**6-Hydroxy-1-naphthoic acid** [*2437-17-4*], *mp* 213 °C, is produced by hydrolysis (10 % KOH) and potassium hydroxide fusion (260 °C) of 1-cyanonaphthalene-6-sulfonic acid (from 1-aminonaphthalene-6-sulfonic acid and cyano-Sandmeyer reaction).

**7-Hydroxy-1-naphthoic acid** [*2623-37-2*], *mp* 256 °C, is synthesized by hydrolysis (50 % KOH) of 8-cyano-2-naphthol (from 8-amino-2-naphthol and cyano-Sandmeyer reaction). It couples in the 8-position with diazo compounds.

**8-Hydroxy-1-naphthoic acid** [1769-88-6], mp 169 °C, is prepared by hydrolysis (60% $H_2SO_4$) of 8-cyanonaphthalene-1-sulfonic acid and caustic fusion of the resulting anhydride of 8-sulfo-1-naphthoic acid. It couples with diazo compounds in the 5-position.

**1-Hydroxy-2-naphthoic acid** [86-48-6], $C_{11}H_8O_3$, $M_r$ 188.17, mp 200 °C, is only sparingly soluble in hot water but readily soluble in alkali and ethanol. It couples with diazo compounds in the 4-position and, with excess, in the 4,7-positions. It is sulfonated by oleum to give the 4-sulfonic and then the 4,7-disulfonic acids. Halogenation occurs in the 4-position, and nitrosation gives 2-nitroso-1-naphthol.

*Production.* [7] Treatment of the dried sodium salt of 1-naphthol with carbon dioxide at 135 °C under pressure gives an 80% yield of product.

*Uses.* 1-Hydroxy-2-naphthoic acid is used as an intermediate for azo and triphenylmethane dyes and, more recently, for color-film dyes [38].

**3-Hydroxy-2-naphthoic acid** [92-70-6], BON acid, $C_{11}H_8O_3$, $M_r$ 188.17, mp 222 °C, forms pale yellow crystals from aqueous ethanol. Alkaline solutions have a green fluorescence. Aqueous ferric chloride produces a blue color, and warming with $CHCl_3$–NaOH results in a green color. Oxidation with alkaline permanganate gives 2-carboxyphenylglyoxylic acid and then phthalic acid. Dry distillation produces dibenzoxanthone. Bromination gives the 4,7-dibromo derivative; nitration, the 4-nitro derivative; and sulfonation with oleum, the 5,7-disulfonic acid derivative.

*Production* is invariably by the Kolbe–Schmitt reaction of carbon dioxide with the dried sodium salt of 2-naphthol at 220–260 °C [7]. Many improvements in both process and equipment have been developed. The main problems are removal of the 50% unreacted 2-naphthol and avoidance of a tarry byproduct. Yields of 80% based on the amount of 2-naphthol consumed are possible despite the low conversion.

*Uses.* The major use of 3-hydroxy-2-naphthoic acid as a dye intermediate is reflected by the quoted production figure of 21 000 t/a in Japan alone. With annual imports of 2500 t/a in the United States, BON acid is the largest imported naphthalene intermediate and one of the largest of all dye and pigment intermediates. Naphtol AS (Naphtol Anilid-Säure) (**44**) represents the most important arylamide derivative. It is produced by forming the acid chloride of 3-hydroxy-2-naphthoic acid with phosphorus trichloride in a solvent such as toluene and then adding aniline for condensation in situ. Many other arylamides are manufactured similarly by using a range of substituted anilines and other arylamines. These arylamides are used in azoic dyes or as intermediates for azo pigments.

Sulfonation of 3-hydroxy-2-naphthoic acid with 98% sulfuric acid at 30 °C gives a mixture of the 5- and 7-sulfonic acid derivatives, which can be separated via their potassium salts. After isolation, the more soluble 5-isomer can be fused with NaOH–KOH at 260 °C to give 3,5-dihydroxy-2-naphthoic acid (**45**).

                    44                              45

**4-Hydroxy-2-naphthoic acid** [*1573-91-7*], *mp* 225–226 °C, is produced by air oxidation of 1-tetralone-3-carboxylic acid.

**5-Hydroxy-2-naphthoic acid** [*2437-18-5*], *mp* 215–216 °C, is prepared by potassium hydroxide fusion (260 °C) of 5-sulfo-2-naphthoic acid.

**6-Hydroxy-2-naphthoic acid** [*16712-64-4*], *mp* 245–248 °C, is synthesized by carboxylation of the potassium salt of 2-naphthol at 170–230 °C. It couples with diazo compounds in the 5-position and undergoes sulfonation in the 4- and 7-positions.

**7-Hydroxy-2-naphthoic acid** [*613-17-2*], *mp* 274 °C, is prepared by sulfonation (160 °C) of 2-naphthoic acid to give the 7-sulfo derivative, followed by potassium hydroxide fusion at 260 °C. It couples with diazo compounds in the 8-position.

**8-Hydroxy-2-naphthoic acid** [*5776-28-3*], *mp* 229 °C, is produced by the cyano-Sandmeyer reaction of 2-aminonaphthalene-8-sulfonic acid, followed by hydrolysis and potassium hydroxide fusion.

Di- and trihydroxynaphthoic acids and hydroxy- and dihydroxynaphthalenedicarboxylic acids are also obtainable, but are of only academic interest at present [5].

# 4. Hydroxynaphthalenesulfonic Acids

The importance of the various mono- and dihydroxynaphthalene mono- and disulfonic acids is demonstrated by the fact that about 25 products in this group have been produced over the years with many of them still available and referred to by code names (see Table 1). Most of the products have been used as azo dye coupling components, but they are equally important as intermediates for aminonaphthalenesulfonic acids (Chap. 6) and aminohydroxynaphthalenesulfonic acids (Chap. 8). Only those isomers that have a history of commercial use are dealt with on an individual basis. Many of the other isomers are also obtainable [5] and may be of present or future interest.

Because in usage this group of products is considered to be derived from 1-naphthol or 2-naphthol they are dealt with on this basis rather than as hydroxy derivatives of naphthalenesulfonic acids. The latter alternative is the IUPAC convention, and these more formal names are given in parentheses alongside alternative trivial names.

## 4.1. Production and Properties

Production depends on four main routes to achieve the required substitution patterns. All routes involve selectivity and separation at some stage, usually the sulfonation stage (cf. Section 2.1). The use of HPLC is essential for following separations and assessing the purity of products.

**Direct Sulfonation of Naphthols.** The reaction of 1-naphthol with sulfuric acid at 60 °C gives mainly 1-hydroxynaphthalene-2-sulfonic acid (**46**) with a small proportion of 1-hydroxynaphthalene-4-sulfonic acid (**47**), from which it can be separated. Increasing the sulfuric acid–naphthol ratio and raising the temperature produces first 1-hydroxynaphthalene-2,4-disulfonic acid (**48**) and then 1-hydroxynaphthalene-2,4,7-trisulfonic acid (**49**).

Sulfonation of 2-naphthol is more complex because concentration and amount of sulfonating agent, temperature, and reaction time all significantly influence the nature and yield of product. At 0–20 °C, sulfuric acid rapidly forms 2-hydroxynaphthalene-1-sulfonic acid (**50**); with longer reaction times a second sulfonic acid group is introduced to form 2-hydroxynaphthalene-1,6-disulfonic acid (**51**). At 50–55 °C, sulfonation occurs in the 6- and 8-positions giving **52** and **53**. More vigorous conditions lead to the disulfonic acids **54** and **55** with some 1- and, to a lesser extent, 8-desulfonation being involved to affect isomer yields. Finally, oleum at higher temperature sulfonates the disulfonic acids to give 2-hydroxynaphthalene-3,6,8-trisulfonic acid (**56**).

**Alkali Fusion of Naphthalenedisulfonic Acids or Hydroxynaphthalenedi- and Trisulfonic Acids** (cf. Chap. 3). Hydrolysis of sulfonic acid substituents to form substituted naphthols is a key process with selective monoreaction able to be controlled in disulfonic acids. This is easier with α,β-disulfonic acids, in which reaction at the α-group is much more facile.

The following ten derivatives (**57–66**) prepared by this route are all commercially important. The asterisks indicate hydroxyl groups introduced by the fusion process.

**Hydrolysis of Aminonaphthalenesulfonic Acids.** Certain naphthylaminesulfonic acids resemble 1-naphthylamine in hydrolyzing to the corresponding naphthol on prolonged heating with dilute sulfuric acid at 180–200 °C under pressure. This process is preferred for the synthesis of 1-hydroxynaphthalene-3-sulfonic acid (**67**), 2-hydroxynaphthalene-5,7-disulfonic acid (**68**), and 1-hydroxynaphthalene-3,6,8-trisulfonic acid (**69**).

An alternative to sulfuric acid hydrolysis is the use of sodium bisulfite in a so-called reverse-Bucherer reaction. This is the preferred process for 1-hydroxynaphthalene-4-sulfonic acid (**47**).

Finally, where neither sulfuric acid nor bisulfite hydrolysis is applicable, indirect hydrolysis via diazotization of the aminonaphthalenesulfonic acid and treatment with dilute sulfuric acid are employed. Although this method is generally applicable, it is more expensive to operate on a large scale, and its use is confined to derivatives of 1-hydroxynaphthalene-8-sulfonic acid.

**Acid Desulfonation of Hydroxynaphthalenedi- or Trisulfonic Acids.** With certain substitution patterns, sulfonic acid groups may be removed by heating with 40% sulfuric acid. This reaction should be regarded as reverse sulfonation rather than hydrolysis, and includes the examples already given in Section 3.3 for the formation of naphthalenediols. An example is the selective desulfonation of 2-hydroxynaphthalene-4,8-disulfonic acid to give 2-hydroxynaphthalene-4-sulfonic acid (**70**).

Although hydroxynaphthalenesulfonic acids are usually soluble in water and more soluble than their alkali-metal salts, they can be isolated by salting out from solution.

Table 4. Dissociation constants for hydroxyl groups of hydroxynaphthalenesulfonic acids in water at 20 °C

| Hydroxynaphthalenesulfonic acid | $pK_2$ | $pK_3$ | $pK_4$ |
|---|---|---|---|
| 1-Hydroxynaphthalene-3-sulfonic acid | 8.7 | | |
| 1-Hydroxynaphthalene-4-sulfonic acid | 8.2 | | |
| 1-Hydroxynaphthalene-5-sulfonic acid | 9.0 | | |
| 2-Hydroxynaphthalene-1-sulfonic acid | 10.96 | | |
| 2-Hydroxynaphthalene-6-sulfonic acid | 9.1 | | |
| 2-Hydroxynaphthalene-7-sulfonic acid | 9.2 | | |
| 1,8-Hydroxynaphthalene-4-sulfonic acid | 5.33 | >12.7 | |
| 2,3-Hydroxynaphthalene-6-sulfonic acid | 8.24 | 12.13 | |
| 2,8-Hydroxynaphthalene-6-sulfonic acid | 8.61 | 10.21 | |
| 1,8-Hydroxynaphthalene-3,6-sulfonic acid | | 5.53 | >12.75 |

The hydroxyl group is dissociated in aqueous solution as shown by the dissociation constants (Table 4) [15].

Purification and characterization are difficult unless chromatographic standards have been obtained [20]. The simplest form of analysis is to determine strength by coupling via titration of an alkaline solution with a standard solution of a diazonium salt.

Many of the hydroxynaphthalenesulfonic acids are used as azo coupling components. Where they serve as precursors in the synthesis of downstream intermediates the most common reactions are amination of the hydroxyl group by the Bucherer reaction to form aminonaphthalenesulfonic acids (Chap. 6) and caustic fusion of the sulfonic acid group to form naphthalenediols (Section 3.3) or their sulfonic acid derivatives.

## 4.2. 1-Hydroxynaphthalenesulfonic Acids

**1-Hydroxynaphthalene-2-sulfonic acid** [567-18-0] (**46**), Baum's acid, $C_{10}H_8O_4S$, $M_r$ 224.23, is sparingly soluble in water and is precipitated from aqueous solution by hydrochloric acid. Coupling with diazo compounds and nitrosation both take place in the 4-position. Sulfonation gives 1-naphthol-2,4-disulfonic acid and nitration gives 2,4-dinitro-1-naphthol.

*Production.* The reaction of 1-naphthol with oleum in the cold yields a mixture of the 2- and 4-sulfonic acids; the former may be separated from the aqueous quenched reaction mixture by salting out the potassium salt after removing unreacted 1-naphthol.

**1-Hydroxynaphthalene-3-sulfonic acid** [3771-14-0] (**67**) (4-hydroxynaphthalene-2-sulfonic acid), $C_{10}H_8O_4S$, $M_r$ 224.2, and its salts are very soluble in water. Coupling with diazo compounds occurs in the 2-position with weak diazo compounds, but in the 4-position with strong diazo compounds such as diazotized nitroanilines. Heating with ammonia at 170 °C gives 1,3-naphthalenediamine.

*Production.* 1-Aminonaphthalene-3-sulfonic acid (Section 6.1) is heated with dilute sulfuric acid by using high-pressure steam to maintain a pressure of 1 MPa for 20 h [7].

After cooling, screening, and adding salt to the hot liquor, the sodium salt crystallizes on cooling and is isolated in 80% yield.

*Uses.* 1-Hydroxynaphthalene-3-sulfonic acid is used as a coupling component for the production of direct dyes.

**1-Hydroxynaphthalene-4-sulfonic acid** [*84-87-7*] (**47**) (4-hydroxynaphthalene-1-sulfonic acid), Nevile and Winther's acid, NW acid, $C_{10}H_8O_4S$, $M_r$ 224.23, and its salts are water soluble, but the sodium salt may be precipitated from solution by salting out. Coupling with diazo compounds takes place in the 2-position. Nitration yields 2,4-dinitro-1-naphthol, whereas nitrosation gives 2-nitroso-1-hydroxynaphthalene-4-sulfonic acid. Sulfonation leads to a mixture of 1-hydroxynaphthalene-2,4-di- and 2,4,7-trisulfonic acids.

*Production.* Sodium naphthionate (Section 6.1) in aqueous alkaline solution is heated with sodium bisulfite and liquid sulfur dioxide for 24 h at 95 °C [7]. The batch is heated with excess lime to hydrolyze the bisulfite compound and boil off ammonia. After filtration the solution of the calcium salt of the product is neutralized with sodium carbonate, and precipitated calcium carbonate is filtered off. The solution of the sodium salt can be used directly as a coupling component, or salt can be added to precipitate and isolate the product.

An alternative synthesis involves selective sulfonation of 1-naphthol in the 4-position by using liquid sulfur trioxide in solvents such as trichloroethane or tetrachloroethane.

*Uses.* 1-Hydroxynaphthalene-4-sulfonic acid is used as a coupling component for a wide range of acid azo dyes, e.g., C.I. Acid Orange 19, C.I. Mordant Brown 35, and C.I. Food Red 3 (**71**).

**71**

**1-Hydroxynaphthalene-5-sulfonic acid** [*117-59-9*] (**57**) (5-hydroxynaphthalene-1-sulfonic acid), oxy L acid, Azurin acid, $C_{10}H_8O_4S$, $M_r$ 224.2, and its salts are very soluble in water, but the sodium salt may be precipitated from solution by salting out. Coupling with diazo compounds normally occurs in the 2-position but takes place in the 4-position with stronger diazo compounds such as diazotized nitroanilines. Sulfonation yields 1-hydroxynaphthalene-2,5-disulfonic acid; vigorous alkali fusion gives 1,5-dihydroxynaphthalene. Bromination in acetic acid yields 2,3-dibromo-1,4-naphthoquinone-5-sulfonic acid.

*Production.* The disodium salt of naphthalene-1,5-disulfonic acid (Section 2.3) is heated in an autoclave with 20% sodium hydroxide at 230 °C for 24 h to selectively hydrolyze one sulfonic acid group [7]. Dilution, salting out, and acidification give a technical product in 90% yield. A purer product, in lower yield, is obtained by acidification of the diluted melt with sulfuric acid, followed by liming and salting out.

*Uses.* 1-Hydroxynaphthalene-5-sulfonic acid is used as a coupling component for a range of azo colorants, for example, C.I. Pigment Red 54 and C.I. Mordant Black 29.

**1-Hydroxynaphthalene-7-sulfonic acid** [*20191-62-2*] (**72**) (8-hydroxynaphthalene-2-sulfonic acid), $C_{10}H_8O_4S$, $M_r$ 224.2, is very soluble in water, and most of its metal salts are also readily soluble. It couples with diazotized 4-nitroaniline to give the 2,4-disazo compound. Nitration gives 2,4-dinitro-1-naphthol-7-sulfonic acid, which was known as Flavianic acid when used as a yellow dye for wool.

*Production.* 1-Naphthol is sulfonated with concentrated sulfuric acid at 130 °C to give a mixture of 2,7- and 4,7-disulfonic acids. This mixture is difficult to separate, but dilution with water and heating the resulting dilute sulfuric acid solution at 120 °C for 30 min result in selective desulfonation in the 2- and 4-positions [39].

**1-Hydroxynaphthalene-8-sulfonic acid** [*117-22-6*] (**74**) (8-hydroxynaphthalene-1-sulfonic acid), $C_{10}H_8O_4S$, $M_r$ 224.2, crystallizes as the monohydrate with *mp* 106 °C. Sulfonation gives 1-hydroxynaphthalene-4,8-disulfonic acid, and alkali fusion with potassium hydroxide at 230 °C yields 1,8-dihydroxynaphthalene. Coupling with diazo compounds occurs in the 2-position.

*Production.* 1-Aminonaphthalene-8-sulfonic acid (Section 6.1) is diazotized over 16 h at 50 °C and then heated to 80 °C [7]. After cooling, the precipitated naphthosultone (**73**) is filtered off and washed. The sultone is heated with dilute sodium hydroxide at 100 °C, which causes ring opening; the product is isolated after salting out.

*Uses.* The naphthosultone (**73**) is usually satisfactory as an intermediate for further processing; however, for azo dye preparation the free naphthol (**74**) must be used, for example, in C.I. Acid Black 54 and C.I. Acid Blue 158 (the chromium complex of **75**).

## 4.3. 2-Hydroxynaphthalenesulfonic Acids

**2-Hydroxynaphthalene-1-sulfonic acid,** [567-47-5] (**50**), oxy Tobias acid, $C_{10}H_8O_4S$, $M_r$ 224.2, and its salts are very soluble in water; neutral solutions give a strong blue color with iron(III) chloride. The sulfonic acid group is sufficiently labile for desulfonation to take place slowly in aqueous solution and rapidly with diazo compounds to form 1-azo-2-hydroxynaphthalene derivatives (i.e., the same product as from 2-naphthol), and with bromine to form 1-bromo-2-hydroxynaphthalene. Low-temperature sulfonation gives 2-hydroxynaphthalene-1,6-disulfonic acid, whereas higher temperatures result in 1-desulfonation and formation of 2-hydroxynaphthalene-6- and 8-sulfonic acids. Nitration affords 1,6-dinitro-2-hydroxynaphthalene.

*Production.* 2-Naphthol is dissolved in 1,2-dichloroethane and chlorosulfonic acid is added at 0 °C; the reaction is allowed to go to completion at low temperature. During work-up and isolation, desulfonation is avoided by careful extraction of the product into aqueous alkali and salting out of the sodium salt. Alternative solvents, such as dichloromethane, tetrachloroethane, or 1,2-dichlorobenzene, and other sulfonating agents, (e.g., liquid sulfur trioxide) can be used.

*Uses.* Virtually all the 2-hydroxynaphthalene-1-sulfonic acid produced is converted to 2-aminonaphthalene-1-sulfonic acid (Tobias acid) by amination (Section 6.1).

The calcium salt Asaprol is soluble in aqueous ethanol and has been used in place of gypsum in surgical plaster.


**2-Hydroxynaphthalene-4-sulfonic acid** [6357-85-3] (**70**) (3-hydroxynaphthalene-1-sulfonic acid), $C_{10}H_8O_4S$, $M_r$ 224.2, and its salts are extremely soluble in water and difficult to isolate. It undergoes nitrosation and couples with diazo compounds in the 1-position. Sulfonation with oleum gives 2-hydroxynaphthalene-4,8-disulfonic acid. Amination with ammonia under pressure gives 2-aminonaphthalene-4-sulfonic acid.

*Production.* 2-Hydroxynaphthalene-4,8-disulfonic acid (Section 4.5) is heated with dilute sulfuric acid at 190 °C for 10 h in an autoclave under pressure [7]. After cooling, a small quantity of 2-naphthol is filtered off and sulfuric acid is removed by liming. The resulting solution of the calcium salt of the product is converted to the sodium salt, which is used directly as a solution or evaporated to isolate the product.


**2-Hydroxynaphthalene-6-sulfonic acid** [93-01-6] (**52**) (6-hydroxynaphthalene-2-sulfonic acid), Schaeffer acid, $C_{10}H_8O_4S$, $M_r$ 224.23, is readily soluble in water, from which it can be recrystallized as the monohydrate (*mp* 129 °C). The sodium salt is sparingly soluble in cold water (1.7%) but forms a 30% solution at 80 °C. Coupling with diazo compounds results in substitution in the 1-position as does treatment with bromine, nitrous acid, or oleum. Fusion with potassium hydroxide gives 2,6-dihydroxynaphthalene (Section 3.3), amination gives 2-aminonaphthalene-6-sulfonic acid (Section 6.1), and sulfonation (20% oleum, 25 °C) results in 2-hydroxynaphthalene-1,6-disulfonic acid.

*Production.* 2-Naphthol is added to a mixture of sulfuric acid and sodium sulfate, and the reaction is continued at 90 °C for 12 h [7]. After pouring the reaction mixture into water, the sodium salt of the product is precipitated by addition of salt and filtered at 60 °C to separate the isomers.

*Uses.* 2-Hydroxynaphthalene-6-sulfonic acid is used as a coupling component for a wide range of azo dyes, e.g., C.I. Acid Orange 12, C.I. Food Yellow 3, and C.I. Food Orange 2 (**76**). It is also used as an intermediate for more highly substituted dye components and synthetic tanning agents. The 1-nitroso derivative forms an iron complex (C.I. Acid Green 1) which was formerly used as a wool dye.

76

**2-Hydroxynaphthalene-7-sulfonic acid** [*92-40-0*] (**58**) (7-hydroxynaphthalene-2-sulfonic acid), F acid, Cassella acid, $C_{10}H_8O_4S$, $M_r$ 224.23, is soluble in water, from which it can be recrystallized as the monohydrate (*mp* 89 °C). The sodium and potassium salts are also readily soluble, but the barium salt is sparingly soluble. Iron(III) chloride gives a dark blue color with neutral solutions of **58**. Azo coupling and nitrosation occurs in the 1-position. Sulfonation gives a variety of products depending on the conditions.

*Production.* Naphthalene-2,7-disulfonic acid (Section 2.3) is heated with aqueous sodium hydroxide under pressure at 230 °C for 5 h and then at 260 °C for another 5 h. After neutralizing and screening, the liquor is acidified to precipitate the product, which is isolated after boiling off the sulfur dioxide and cooling to 30 °C.

*Uses.* 2-Hydroxynaphthalene-7-sulfonic acid is used as an intermediate in the production of 2-aminonaphthalene-7-sulfonic acid (Section 6.1), 2-hydroxynaphthalene-3,7-disulfonic acid, and 2,7-dihydroxynaphthalene (Section 3.3). It has minor uses as an azo coupling component.

**2-Hydroxynaphthalene-8-sulfonic acid** [*132-57-0*] (**53**) (7-hydroxynaphthalene-1-sulfonic acid), crocein acid, Bayer's acid, $C_{10}H_8O_4S$, $M_r$ 224.23, is more soluble in water than 2-hydroxynaphthalene-6-sulfonic acid and cannot readily be isolated. Its salts are also more soluble than their 2,6-analogues and are difficult to isolate. On evaporation of its solutions, the free acid (**53**) decomposes to 2-naphthol. Coupling with diazo compounds takes place in the 1-position, but reaction is hindered by the peri-sulfonic acid group so that concentrated conditions and sometimes catalysts are required. Above 60 °C, sulfuric acid converts **53** into 2-hydroxynaphthalene-6-sulfonic acid; sulfonation then proceeds further to give the 6,8-disulfonic acid. Amination (Bucherer reaction) gives 2-aminonaphthalene-8-sulfonic acid (Section 6.1).

*Production.* 2-Naphthol reacts with excess 98% sulfuric acid at 20 – 35 °C to form 2-hydroxynaphthalene-1-sulfonic acid. Further reaction at 55 – 60 °C produces a mixture of 6- and 8-sulfonic acids, from which the less soluble 6-isomer is separated before and after liming to give a crude crocein acid liquor containing 16% Schaeffer acid. If pure **53** is required, the latter is separated by selective coupling with diazotized *o*-toluidine and removal of the precipitated azo dye.

*Uses.* Crocein acid is used as a coupling component for azo dyes e.g., C.I. Acid Red 25 (**77**).

HO<sub>3</sub>S—[naphthyl]—N=N—[naphthyl with OH]<br>
HO<sub>3</sub>S—<br>
77

## 4.4. 1-Hydroxynaphthalenedisulfonic Acids

**1-Hydroxynaphthalene-3,6-disulfonic acid** [*578-85-8*] (**59**) (4-hydroxynaphthalene-2,7-disulfonic acid), RG acid, Violet acid, $C_{10}H_8O_7S_2$, $M_r$ 304.3: the free acid ($pK_a$ 8.56) and its alkalimetal salts are readily soluble in water, but the acid sodium salt can be precipitated from concentrated solution with salt. Sulfonation with oleum at 125 °C gives a tetrasulfonic acid, and amination with ammonia and ammonium chloride at 160 °C gives a mixture of 3-amino-1-hydroxynaphthalene-6-sulfonic acid and 6-amino-1-hydroxynaphthalene-3-sulfonic acid. Alkali fusion at 170 °C gives mainly 1,6-dihydroxynaphthalene-3-sulfonic acid, and at 250 °C yields 1,3,6-trihydroxynaphthalene. Diazo coupling takes place in the 2-position.

*Production.* Naphthalene-1,3,6-trisulfonic acid (Section 2.4) is heated with 50% sodium hydroxide at 180 °C for 15 h in an autoclave [7]. After dilution with water, the product is salted out as the monosodium salt and purified by washing with brine.

*Uses.* 1-Hydroxynaphthalene-3,6-disulfonic acid is used as a coupling component for azo dyes, e.g., C.I. Acid Red 7 (**78**).

CH<sub>3</sub>COHN—[phenyl]—N=N—[naphthyl with OH]<br>
HO<sub>3</sub>S      SO<sub>3</sub>H<br>
78

**1-Hydroxynaphthalene-3,8-disulfonic acid** [*117-43-1*] (**79**) (8-hydroxynaphthalene-1,6-disulfonic acid), epsilon acid, $C_{10}H_8O_7S_2$, $M_r$ 304.3: the sodium salt crystallizes as the hexahydrate and is readily soluble in water. Alkali fusion at 200 °C gives 1,8-dihydroxynaphthalene-3-sulfonic acid and amination with ammonia and ammonium chloride at 170 °C gives 6,8-diaminonaphthalene-1-sulfonic acid. Diazo coupling takes place in the 2-position and bromination in the 4-position. The 1,8-sultone is formed on dehydration with concentrated sulfuric acid.

*Production.* The acid sodium salt of 1-aminonaphthalene-3,8-disulfonic acid (Section 6.2) is hydrolyzed by heating with water under pressure at 180 °C.

*Uses.* Epsilon acid is used as a coupling component for azo dyes, e.g., C.I. Direct Red 47 (**80**).

**1-Hydroxynaphthalene-4,7-disulfonic acid** [*6361-37-1*] (**81**) (4-hydroxynaphthalene-1,6-disulfonic acid), D (Dahl's) acid, $C_{10}H_8O_7S_2$, $M_r$ 304.3: its alkali-metal salts are readily soluble in water. Boiling with 65% sulfuric acid results in desulfonation and formation of 1-hydroxynaphthalene-7-sulfonic acid (**72**). Diazo coupling and nitrosation take place in the 2-position.

*Production.* 1-Aminonaphthalene-4,7-disulfonic acid (Section 6.2) is heated with sodium bisulfite under conditions similar to those described for the production of 1-hydroxynaphthalene-4-sulfonic acid (**47**) from sodium naphthionate (see Section 4.2). High yields of **81** have been claimed via sulfonation of 1-(dichlorophosphinyloxy)-naphthalene (1-naphthol + $POCl_3$) [40].

**1-Hydroxynaphthalene-4,8-disulfonic acid** [*117-56-6*] (**83**) (4-hydroxynaphthalene-1,5-disulfonic acid), Schollkopf's acid, delta acid, oxy Chicago acid, $C_{10}H_8O_7S_2$, $M_r$ 304.3: the sodium salt is readily soluble in water, but the barium salt is only sparingly soluble. Cold 5% oleum dehydrates **83** to the sultone **82**, whereas with hot 25% oleum it undergoes sulfonation to give 1-hydroxynaphthalene-2,4,8-trisulfonic acid. Amination (Bucherer) gives 1-aminonaphthalene-4,8-disulfonic acid and caustic fusion gives 1,8-dihydroxynaphthalene-4-sulfonic acid. Diazo coupling and nitrosation take place in the 2-position.

*Production.* 1-Hydroxynaphthalene-8-sulfonic acid-1,8-sultone (**73**) (Section 4.2) is heated with sulfuric acid at 80–90 °C. The reaction mixture is then quenched by pouring into concentrated brine, whereby the sodium salt of **82** crystallizes. After isolation, this intermediate is hydrolyzed by heating with aqueous sodium carbonate and the product is isolated as its sodium salt after cooling.

*Uses.* Delta acid is used as a coupling component for azo dyes, e.g., C.I. Acid Blue 169.

## 4.5. 2-Hydroxynaphthalenedisulfonic Acids

**2-Hydroxynaphthalene-3,6-disulfonic acid** [*148-75-4*] (**54**) (3-hydroxynaphthalene-2,7-disulfonic acid), R acid, $C_{10}H_8O_7S_2$, $M_r$ 304.3, is readily soluble in water and ethanol. Although the sodium salt (R salt) is soluble in water, it can be precipitated by sodium chloride. The barium salt is sparingly soluble in cold water but is 8% soluble in boiling water. Whereas sulfonation with 20% oleum yields 2-hydroxynaphthalene-3,6,8-trisulfonic acid (**56**), the reaction with $Na_2SO_3$–$MnO_2$ gives 2-hydroxy-1,3,6-trisulfonic acid. Alkali fusion with sodium hydroxide gives 2,3-dihydroxynaphthalene-6-sulfonic acid and some 2,3-dihydroxynaphthalene. Amination (Bucherer) produces 2-aminonaphthalene-3,6-disulfonic acid. Diazo coupling and nitrosation take place in the 1-position.

*Production.* 2-Naphthol is heated in excess 98% sulfuric acid to 60 °C, and anhydrous sodium sulfate is added [7]. The mixture is then heated for 24 h at 105 °C, 12 h at 110 °C, and 12 h at 120 °C before being diluted with more sulfuric acid and quenched in water. Sodium chloride is added at 60 °C, and the mixture is cooled slowly to 30 °C with crystallization of the product as its sodium salt. The isolated yield is 68%, and the product contains about 3% of Schaeffer's salt as the major impurity. Alternatively, R salt may be recovered as a byproduct from the G acid process (see p. 3397). Further purification may be effected via the aniline salt [41].

*Uses.* The so-called R acid is used as a coupling component for a wide range of azo colorants, e.g., C.I. Mordant Red 9, C.I. Pigment Red 60, C.I. Acid Red 27, and C.I. Food Red 9 (**84**).

The aluminum salt of R acid has been used as an astringent and antiseptic (Alumnol).

This acid is an intermediate in the manufacture of 2-aminonaphthalene-3,6-disulfonic acid and 2,3-dihydroxynaphthalene-6-sulfonic acid.

**2-Hydroxynaphthalene-3,7-disulfonic acid** [6361-38-2] (**85**) (3-hydroxynaphthalene-2,6-disulfonic acid), $C_{10}H_8O_7S_2$, $M_r$ 304.3: the alkali-metal salts of **85** are readily soluble in water, but its barium salt is only sparingly soluble even at 100 °C. Sulfonation with 20% oleum at 85 °C gives mainly 2-hydroxynaphthalene-1,3,7-trisulfonic acid. Amination yields 2-aminonaphthalene-3,7-disulfonic acid, and caustic fusion yields 2,7-dihydroxynaphthalene-3-sulfonic acid. Diazo coupling takes place in the 1-position.

*Production.* 2-Hydroxynaphthalene-7-sulfonic acid (Section 4.3) is heated at 120 °C with concentrated sulfuric acid for 12 h, the mixture is poured into water, and the product is separated after liming out.

HO₃S — [naphthalene] — OH, SO₃H

**85**

**2-Hydroxynaphthalene-5,7-disulfonic acid** [575-05-3] (**68**) (6-hydroxynaphthalene-1,3-disulfonic acid), $C_{10}H_8O_7S_2$, $M_r$ 304.3, and its alkali-metal salts are readily soluble in water. Diazo coupling takes place in the 1-position.

*Production.* 2-Aminonaphthalene-5,7-disulfonic acid (Section 6.2) is heated with 5% sulfuric acid for 15 h at 185 °C under pressure [7]. After liming out and removal of ammonia and calcium ions, the product is obtained as a solution of the sodium salt for further use.

*Uses.* The acid has some use as an azo coupling component and as an intermediate in the production of 2,5-dihydroxynaphthalene-7-sulfonic acid.

**2-Hydroxynaphthalene-6,8-disulfonic acid** [118-32-1] (**55**) (7-hydroxynaphthalene-1,3-disulfonic acid), G acid, $C_{10}H_8O_7S_2$, $M_r$ 304.3: most salts of **55** are more readily soluble in water than those of R acid, with the notable exception of the potassium salt. Sulfonation with 20% oleum at 100 °C yields 2-hydroxynaphthalene-3,6,8-trisulfonic acid (**56**). Diazo coupling takes place in the 1-position, although the reaction is sluggish, as in the case of 2-hydroxynaphthalene-8-sulfonic acid.

*Production.* 2-Naphthol is added to 98% sulfuric acid at 40 °C. Then, 20% oleum is added over 6 h and the temperature is allowed to rise to 60 °C. The reaction is completed by heating for a further 16 h at 60 °C followed by 15 h at 80 °C. After the reaction mixture has been poured into water, potassium chloride is added to the hot solution and the batch is cooled slowly to 35 °C before filtering off the crystalline dipotassium salt of G acid and carefully washing out any residual R salt; the yield is 60%. In addition, R salt can be recovered from the filtrate in 12% yield after conversion to its sodium salt.

*Uses.* The G acid is aminated (Bucherer reaction) to give 2-aminonaphthalene-6,8-disulfonic acid and subjected to caustic fusion at 200 °C to yield 2,8-dihydroxynaphthalene-6-sulfonic acid (a precursor of γ acid). It is also used as a coupling component for a wide range of azo colorants, e.g., C.I. Acid Orange 10, C.I. Acid Red 187, and C.I. Food Red 7 (**86**).

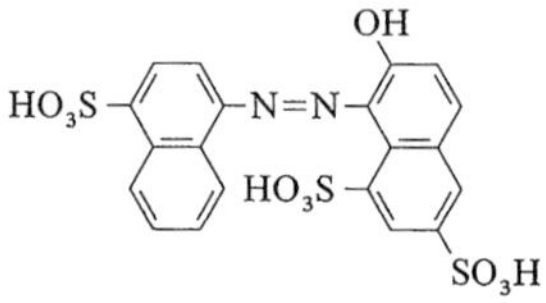

**86**

## 4.6. Hydroxynaphthalenetrisulfonic Acids

**1-Hydroxynaphthalene-3,6,8-trisulfonic acid** [*3316-02-7*] (**69**) (8-hydroxynaphthalene-1,3,6-trisulfonic acid), oxy Koch acid, $C_{10}H_8O_{10}S_3$, $M_r$ 384.3, and its alkali-metal salts are readily soluble in water. Amination at 170 °C gives 1-aminonaphthalene-3,6,8-trisulfonic acid. Diazo coupling takes place in the 2-position.

*Production.* 1-Aminonaphthalene-3,6,8-trisulfonic acid (Section 6.3) is heated with dilute sulfuric acid at 180 °C under pressure for 20 h [7]. After the mixture is run into excess sodium hydroxide solution, ammonia is removed by heating and the total liquor is used directly for the production of chromotropic acid (Section 4.7) by caustic fusion.

**2-Hydroxynaphthalene-3,6,8-trisulfonic acid** [*6259-66-1*] (**56**) (7-hydroxynaphthalene-1,3,6-trisulfonic acid), $C_{10}H_8O_{10}S_3$, $M_r$ 384.3, and its alkali-metal salts are readily soluble in water; nevertheless, the sodium salt is obtainable by salting out. Amination at 240 °C gives 2-aminonaphthalene-3,6,8-trisulfonic acid. Alkali fusion gives a mixture of dihydroxynaphthalenedisulfonic acids and finally 4,6,7-trihydroxynaphthalene-2-sulfonic acid. Diazo coupling occurs slowly in the 1-position.

*Production.* 2-Naphthol is heated with sulfuric acid monohydrate at 115 °C for 3 h. The reaction mixture is cooled and 65 % oleum is added, followed by heating at 115 °C for 12 h [7], after which it is poured into water. Salt is then added to the hot solution and the sodium salt of the product crystallizes. Further purification is effected by redissolving in water and salting out.

*Uses.* The acid is used as an azo coupling component in, e.g., C.I. Solvent Red 31 (the dicyclohexylamine salt of **87**).

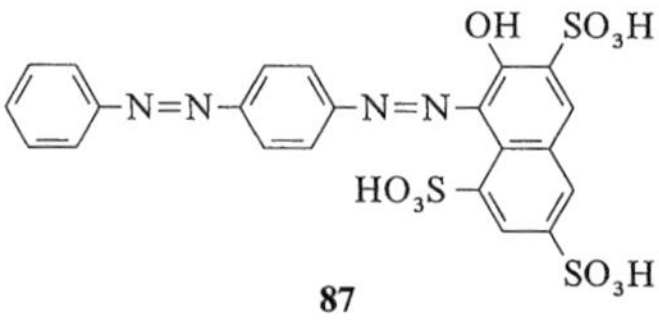

**87**

## 4.7. Dihydroxynaphthalenesulfonic Acids

Although many isomers are known [5], only the four compounds that have been used industrially (mainly as azo coupling components) are described. All are obtained in high yield by selective fusion processes.

**4,5-Dihydroxynaphthalene-1-sulfonic acid** [83-65-8] (**60**) (1,8-dihydroxynaphthalene-4-sulfonic acid), dioxy Chicago acid, dioxy S acid, $C_{10}H_8O_5S$, $M_r$ 240.2, is sparingly soluble in water, but the sodium salt is readily soluble. Heating with dilute sulfuric acid at 150 °C gives 1,8-dihydroxynaphthalene (Section 3.3).

*Production.* 1-Hydroxynaphthalene-4,8-disulfonic acid (Section 4.4) is heated with 82% caustic soda liquor for 3 h at 235 °C [7]. The melt is poured into excess hydrochloric acid and sulfur dioxide is expelled. After cooling, the product is filtered off and washed with brine; the yield is 83%.

**4,6-Dihydroxynaphthalene-2-sulfonic acid** [6357-93-3] (**61**) (2,8-dihydroxynaphthalene-6-sulfonic acid), dioxy G acid, $C_{10}H_8O_5S$, $M_r$ 240.2, is readily soluble in water, but the sodium salt is less so and can be salted out.

*Production.* The potassium salt of 2-hydroxynaphthalene-6,8-disulfonic acid (Section 4.4) is heated with 70% caustic liquor for 3 h at 230 °C and then at 240 °C for completion of the reaction. The melt is diluted with water and poured into excess hydrochloric acid. After sulfur dioxide has been expelled, salt is added and the batch is cooled to precipitate the product, which is filtered off and washed with brine; the yield is 80%.

*Uses.* Dioxy G acid is used as an intermediate in the production of 2-amino-8-hydroxynaphthalene-6-sulfonic acid (γ acid) and its N-substituted derivatives by means of the Bucherer reaction with ammonia or primary amines. It is also used as a coupling component for azo dyes such as C.I. Mordant Black 56 (**88**).

**4,7-Dihydroxynaphthalene-2-sulfonic acid** [6357-94-4] (**62**) (2,5-dihydroxynaphthalene-7-sulfonic acid), dioxy J acid, $C_{10}H_8O_5S$, $M_r$ 240.2, is readily soluble in water; its sodium salt is much less soluble and is easily salted out. Amination with $NH_3$–$NaHSO_3$ yields J acid (Section 8.2).

*Production.* 2-Hydroxynaphthalene-5,7-disulfonic acid (Section 4.4) is heated with aqueous NaOH–KOH for 5 h at 175 °C [7]. The melt is poured into water and acidified with hydrochloric acid, and the precipitated solid is filtered off and washed with brine; the yield is 80%.

*Uses.* Although not used as a precursor for J acid itself, **62** can be used advantageously for the production of some N-substituted J acids.

**6,7-Dihydroxynaphthalene-2-sulfonic acid** [*92-27-3*] (**63**) (2,3-dihydroxynaphthalene-6-sulfonic acid), dioxy R acid, $C_{10}H_8O_5S$, $M_r$ 240.2, is readily soluble in water. The sodium salt is sparingly soluble in cold water but readily soluble in hot water. Heating with dilute sulfuric acid at 190 °C gives 2,3-dihydroxynaphthalene (Section 3.3).

*Production.* 2-Hydroxynaphthalene-3,6-disulfonic acid (Section 4.4) is heated with 70 % caustic soda liquor for 4 h at 270 °C, after which the reaction mixture is diluted with water and poured into excess hydrochloric acid [7]. The precipitated product is filtered off and washed with brine; the yield is 90 %.

## 4.8. Dihydroxynaphthalenedisulfonic Acids

Although many isomers are known [5], only the three significant ones are described here. All are obtained by selective fusion processes, although yields are lower than for the monosulfonic acids (Section 4.7).

**3,5-Dihydroxynaphthalene-2,7-sulfonic acid** [*81344-22-1*] (**64**), A acid, $C_{10}H_8O_8S_2$, $M_r$ 320.3.

*Production.* 2-Hydroxynaphthalene-3,6,8-trisulfonic acid (Section 4.6) is heated with caustic soda liquor at 240 °C, and the melt is poured into water and neutralized. Careful addition of salt precipitates the product and leaves the more soluble 2,3-dihydroxy byproduct in solution.

**3,6-Dihydroxynaphthalene-2,7-disulfonic acid** [*23894-07-7*] (**65**), $C_{10}H_8O_8S_2$, $M_r$ 320.3.

*Production.* 2-Hydroxynaphthalene-3,6,7-trisulfonic acid is heated with 66 % caustic soda liquor at 200 – 300 °C and the product worked up by dilution, neutralization, and salting out.

**4,5-Dihydroxynaphthalene-2,7-disulfonic acid** [*148-25-4*] (**66**) (1,8-dihydroxynaphthalene-3,6-disulfonic acid), chromotropic acid, $C_{10}H_8O_8S_2$, $M_r$ 320.3: the disodium salt is readily soluble in water. Alkali fusion yields 1,3,8-trihydroxynaphthalene-6-sulfonic acid. Diazo coupling occurs in the 3- and 6-positions.

*Production.* 2-Hydroxynaphthalene-3,6,8-trisulfonic acid total liquor (Section 4.6) is concentrated and heated with caustic soda liquor to 158 °C for 20 – 30 h [7]. The melt is diluted with water and poured into excess hydrochloric acid. Salt is added, sulfur dioxide expelled, and the mixture cooled to precipitate the product as the disodium salt in 88% yield.

An alternative process is based on direct fusion (caustic liquor at 228 °C under pressure) of 1-aminonaphthalene-3,6,8-trisulfonic acid (Section 6.3) to chromotropic acid without isolating the intermediate oxy Koch acid [7].

*Uses.* Chromotropic acid is used as an analytical reagent and as a coupling component for a wide range of azo dyes, e.g., C.I. Acid Violet 6, C.I. Mordant Blue 13, and C.I. Direct Blue 84 (the dicopper complex of demethylated **89**).

## 5. Aminonaphthalenes

The two general methods for the synthesis of monoamino- and diaminonaphthalenes are reduction or hydrogenation of the corresponding nitronaphthalene derivative, and amination (Bucherer reaction) of the corresponding hydroxynaphthalene derivative. Although the former method is analogous to the general method for substituted anilines, the Bucherer reaction is not applicable to phenols.

## 5.1. Naphthylamines

**1-Naphthylamine** [*134-32-7*] (**90**), 1-naphthaleneamine, 1-aminonaphthalene, α-naphthylamine, $C_{10}H_9N$, $M_r$ 143.18, mp 50 °C, bp 301 °C, $pK_a$ 10.0, $d_4^{20}$ 1.131, sublimes and is steam volatile. It crystallizes from aqueous ethanol as colorless needles, which slowly turn red on exposure to air. Solubility in water is 0.16% at 20 °C. It behaves as a typical primary aromatic amine in forming salts with strong acids (but not with acetic or benzoic acid) and readily forming *N*-acyl derivatives.

1-Naphthylamine couples with diazo compounds in the 4-position, with up to 10% byproduct being formed by coupling in the 2-position.

Oxidation of 1-naphthylamine salts with iron(III) chloride in aqueous solution gives an insoluble blue-violet compound, whereas oxidation with chromic acid yields 1,4-naphthoquinone and phthalic acid. Reduction with Na–NaOEt leads to the 5,6,7,8-tetrahydro compound, which can also be formed together with the 1,2,3,4-tetrahydro derivative by catalytic hydrogenation. Sulfonation gives a complex mixture of mono-, di-, and trisulfonic acids.

*Production.* Naphthalene is nitrated with $H_2SO_4$–$HNO_3$ at 50–60 °C [7]. The crude product (*mp* 52 °C) is separated, washed with hot water, and further purified by crystallization or sweating (draining off the lower melting isomer) to remove 2-nitronaphthalene (3%) and traces of 2,4-dinitronaphthalene. Purified 1-nitronaphthalene was traditionally reduced with iron in boiling dilute hydrochloric acid, but modern plants use hydrogenation with a nickel catalyst. The 1-naphthylamine produced is further purified by distillation under vacuum. The content of 2-naphthylamine in the commercial product is specified at <10 ppm.

*Uses.* A large proportion of the 1-naphthylamine produced is converted to 1-naphthol (Section 3.1). Other major uses are in the production of aminonaphthalenesulfonic acid dye intermediates (Chap. 6) and as a component of azo dyes. In the latter, use as a coupling component, followed by diazotization and further coupling of the resulting aminoazo compound, predominates, e.g., for C.I. Acid Blue 113 (**92**).

**92**

Other important derivatives are C.I. Solvent Brown 1, C.I. Acid Red 20, C.I. Direct Brown 208, and C.I. Mordant Black 68.

The rubber antioxidant 1-phenylaminonaphthalene (**24**) can be obtained from 1-naphthol (Section 3.1). However, because the latter is now manufactured from 1-naphthylamine, it is more convenient to manufacture **24** directly from 1-naphthylamine and aniline by a vapor-phase catalytic process (alumina gel, 300 °C). *N*-Alkyl-1-naphthylamines are usually manufactured from 1-naphthol; for example, *N*-ethyl-1-naphthylamine is of particular importance as an intermediate for Victoria Blues.

The selective preemergence herbicide naptalam [*132-66-1*] (**93**), *N*-1-naphthylphthalamic acid, is produced by acylation of **90** by heating with an equimolar quantity of phthalic anhydride in a solvent such as xylene [42].

The rodenticide ANTU [*86-88-4*] (**94**), 1-naphthalenethiourea, is manufactured from 1-naphthylamine by reaction with thiocyanate ion in aqueous acid solution. The elimination of ammonia from **94** by heating in chlorobenzene is a convenient method for producing 1-naphthyl isothiocyanate.

Acylation of *N*-methyl-1-naphthylamine with fluoroacetyl chloride yields the insecticide and acaricide Nissol [*5903-13-9*] (**95**). An improved process is based on reaction of the corresponding chloroacetamido derivative with potassium fluoride in a glycol solvent [43].

**93** (2-carboxyphenyl-CONH-naphthalene), **94** (NHCSNH₂-naphthalene), **95** (N(CH₃)(COCH₂F)-naphthalene)

*Toxicology* [14], [44]. 1-Naphthylamine is toxic to humans by inhalation or skin absorption and is classed as a potential carcinogen. Samples are often contaminated with highly carcinogenic 2-naphthylamine, which confused many earlier results. Although frequently quoted, data on toxicity to animals are inconclusive compared with those available for the 2-isomer.

The metabolism of 1-naphthylamine in animals and humans has been studied, with a variety of results. *N*-(1-naphthyl)hydroxylamine and *N*-hydroxy-1-naphthylacetamide appear to be key metabolic products, in contrast to the behavior of 2-naphthylamine.

Use in the United Kingdom is controlled by the Carcinogenic Substances Regulations of 1967 and in the United States by the Federal Register of Carcinogens (1973).

**2-Naphthylamine** [91-59-8] (**91**), 2-aminonaphthalene, 2-naphthaleneneamine, β-naphthylamine, $C_{10}H_9N$, $M_r$ 143.18, mp 112 °C, bp 306 °C, $d_4^{98}$ 1.061 is steam volatile and is soluble in hot water, ethanol, and ether. It crystallizes from aqueous ethanol as white crystals that turn red on exposure to light and air.

Oxidation with hypochlorite, lead dioxide, or air in UV light gives dibenzophenazine. Reduction with sodium in amyl alcohol or catalytic hydrogenation gives mainly 2-aminotetralin. Substition reactions (e.g., halogenation) occur first in the 1-position unless the amino group is protonated, in which case substituents enter the 5- and 8-positions.

*Production.* 2-Naphthol is heated with an $NH_4OH-NH_4HSO_3$ mixture at 180 °C for 20 h, and the molten amine is separated after basification and removal of excess ammonia. After being dried, the oil is distilled under vacuum. Production of 2-naphthylamine ceased worldwide in the 1950s.

*Uses.* Historically, 2-naphthylamine was an important colorant intermediate. Two examples are solvent sulfonation to Tobias acid (Section 6.1) and formation of Naphthol BN (**96**). Until 1950 the latter was the most important azoic coupling component after Naphthol AS (**44**). After production of 2-naphthylamine ceased, Naphthol BN was prepared for a time via Tobias acid and desulfonation.

**96**

Although *N*-phenyl-2-aminonaphthalene (**32**) was produced originally from 2-naphthylamine, all production of this antioxidant is now based on 2-naphthol (Section 3.2).

*Toxicology* [14], [44]. 2-Naphthylamine is a human carcinogen that leads predominately to formation of tumors in the epithelium of the bladder. The conclusive epidemiological studies on humans are backed up by a wealth of animal data. However, although a proven carcinogen in mouse, hamster, dog, and monkey, 2-naphthylamine exhibits little, if any, carcinogenicity in rat or rabbit. Several metabolic pathways have been identified, but the main one involves N-hydroxylation followed by rearrangement to 2-amino-1-hydroxynaphthalene. Only the former is a proven animal carcinogen.

Measurement of 2-naphthylamine impurity levels requires sophisticated analysis. For example, 22 ng per cigarette was measured by GLC of the *N*-pentafluoropropionyl derivative. The levels found were considered to be biologically insignificant [45].

Use in the United Kingdom is prohibited under the Carcinogenic Substances Regulations of 1967; in the United States, by the Environmental Protection Agency; and in all industrialized nations, by comparable regulatory authorities.

Although *N*-phenyl-2-aminonaphthalene (PBN, **32**) has been classed as a potential carcinogen, its toxicity is more likely to be due to the presence of primary amines as impurities.

## 5.2. Naphthalenediamines

All ten of the isomeric diamines are known [5]. Table 5 lists their melting points, CAS registry numbers, and solubilities. Only the 1,5- and 1,8-isomers are readily available and of commercial significance.

**1,5-Naphthalenediamine** [2243-62-1] (**97**), 1,5-diaminonaphthalene, Alphamin, $C_{10}H_{10}N_2$, $M_r$ 158.2: oxidation of **97** with iron(III) chloride in water produces a blue-violet color. Treatment with boiling aqueous sodium bisulfite followed by addition of alkali gives a mixture of 1-amino-5-hydroxynaphthalene and 1,5-dihydroxynaphthalene. Sulfonation (5% oleum, 100 °C) gives 1,5-diaminonaphthalene-2-sulfonic acid, and nitration in acetic acid produces 2,4,6,8-tetranitro-1,5-diaminonaphthalene. Coupling with diazo compounds takes place in the 2-position; reduction of the resulting azo compound with $SnCl_2$–HCl produces 1,2,5-triaminonaphthalene.

*Production.* 1-Nitronaphthalene can be nitrated further to give a 40:60 mixture of 1,5- and 1,8-dinitronaphthalenes. Similar results are obtained by direct nitration of naphthalene with $H_2SO_4$–$HNO_3$ under careful control of temperature over the range 40–80 °C. Although separation of the isomers by fractional crystallization or solvent extraction [46] is usually carried out at this stage, the mixed isomers can also be reduced and the resulting diamines separated. Reduction of the dinitronaphthalenes is achieved by treatment of a nonaqueous solution with iron or hydrogen in the presence of a catalyst.

An alternative process for 1,5-naphthalenediamine involves amination of 1,5-dihydroxynaphthalene with ammonia and ammonium bisulfite. Although less efficient on a

**Table 5.** Physical properties of naphthalenediamines

| Isomer | CAS registry no. | mp, °C | Solubility |
|---|---|---|---|
| 1,2 | [938-25-0] | 98 | soluble in hot water, alcohol, and ether |
| 1,3 | [24824-28-0] | 96 | |
| 1,4 | [2243-61-0] | 120 | slightly soluble in hot water |
| 1,5 | [2243-62-1] | 189 | soluble in hot water and alcohol |
| 1,6 | [2243-63-2] | 78 | soluble in hot water and alcohol |
| 1,7 | [2243-64-3] | 117 | soluble in alcohol |
| 1,8 | [479-27-6] | 66 | soluble in alcohol and ether |
| 2,3 | [771-97-1] | 199 | soluble in alcohol and ether |
| 2,6 | [2243-67-6] | 220 | slightly soluble in alcohol and ether |
| 2,7 | [613-76-3] | 166 | |

stage basis it offers an economical alternative to nitration and reduction if the 1,8-naphthalenediamine (**98**) is not also required.

*Uses.* 1,5-Naphthalenediamine is an intermediate in the manufacture of naphthalene-1,5-diisocyanate [*3173-72-6*] (**99**), which is used for specialty polyurethane polymers. It has only limited use as a dye intermediate.

*Toxicity.* Like 1-naphthylamine, 1,5-naphthalenediamine is classed as having limited carcinogenicity in animals [14], [44].

**1,8-Naphthalenediamine** [*479-27-6*] (**98**), 1,8-diaminonaphthalene, Deltamin, $C_{10}H_{10}N_2$, $M_r$ 158.2, turns brown in air even in the absence of light. Its solutions give a brown precipitate on addition of iron(III) chloride. Treatment with boiling aqueous sodium bisulfite followed by addition of alkali gives 8-amino-1-hydroxynaphthalene. Sulfonation takes place in the 4-position. Coupling with azo compounds also occurs in the 4-position, but in almost neutral solution 4,5-bis coupling can occur. The resulting bisazo compound is reduced with SnCl$_2$–HCl to give 1,4,5,8-tetraaminonaphthalene.

*Production.* Reduction of 1,8-dinitronaphthalene is carried out with iron and acetic acid in xylene or by catalytic hydrogenation.

*Uses.* 1,8-Naphthalenediamine has minor use as an antioxidant for lubricating oils and as an analytical reagent for the detection of selenium and nitrites.

Treatment of a solution of the hydrochloride of **98** with acetic anhydride yields methylperimidine (**100**), a colorant intermediate. Most important are the phthaloperinone pigments obtained by reaction of **98** with substituted phthalic anhydrides [47]. The parent product (**101**) is a lightfast, heat-stable orange dye for poly(methyl methacrylate) and is used to color the covers for automobile indicator lights.

# 6. Aminonaphthalenesulfonic Acids

Many aminonaphthalenesulfonic acids are important intermediates either for the direct production of azo dyes or for conversion to more highly substituted intermediates for azo dyes. About 24 products in this group have been made industrially, and many of them are still commercially available and referred to by code names (Table 1). The most important are the C, Peri, and Laurent's acids [17]. Only those isomers that have a history of large-scale use are dealt with here on an individual basis. Many other isomers are also obtainable and may be of potential present or future interest [5].

Because in usage this group of products is considered to be derived from 1- and 2-naphthylamines they are dealt with on this basis rather than as amino derivatives of naphthalenesulfonic acids. The latter is the IUPAC convention, and these names are given in parentheses alongside alternative trivial names.

## 6.1. Production and Properties

Production depends on four main routes to achieve the required substitution patterns. As with most naphthalene substitution reactions, isomeric mixtures often arise so that isomer separation or purification to remove unwanted byproducts is required. High-performance liquid chromatography is an essential analytical tool for following separations and assessing the purity of products.

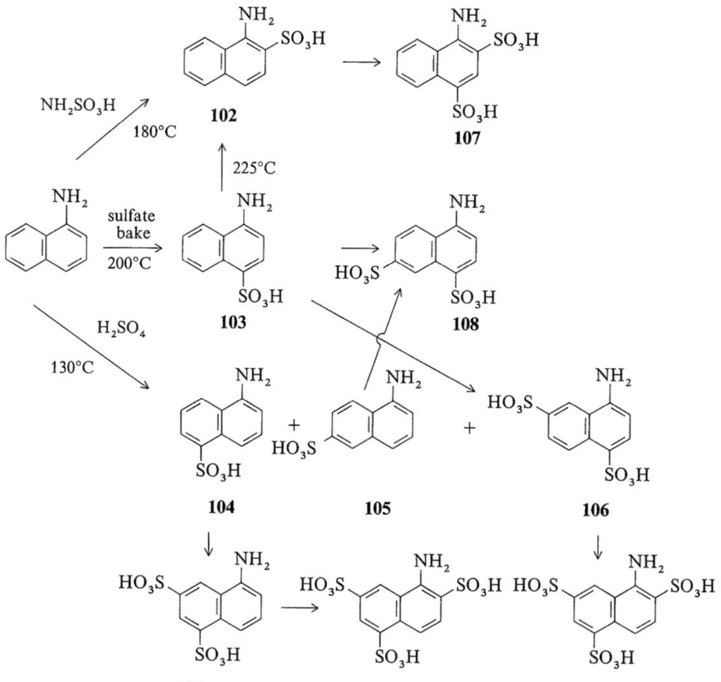

**Figure 3.** Sulfonation of 1-naphthylamine

**Sulfonation of Naphthylamines.** Sulfonation of 1-naphthylamine gives a variety of products depending on conditions (Fig. 3). Further sulfonation of the separated and isolated primary products leads to di- and trisulfonic acids.

Although 2-naphthylamine can be sulfonated to give a range of isomeric mono- and disulfonic acids, this method is not discussed here because use of this starting material is now prohibited.

**Nitration–Reduction of Naphthalenesulfonic Acids.** Both naphthalene-1-sulfonic acid and 2-sulfonic acid undergo nitration to give separable mixtures of 5- and 8-nitro compounds, which are isolated and reduced to the corresponding aminonaphthalenesulfonic acids (Fig. 4).

Naphthalene-1,5-disulfonic acid is nitrated to give the 3-nitro derivative, which is reduced to 2-aminonaphthalene-4,8-disulfonic acid (**118**). Other available disulfonic acid isomers are similarly mononitrated and reduced to aminonaphthalenedisulfonic acids (see. Fig. 5).

**Amination of Hydroxynaphthalenesulfonic Acids.** The Bucherer reaction is most readily applicable to 2-hydroxynaphthalenemono- and disulfonic acids, as shown in the following examples.

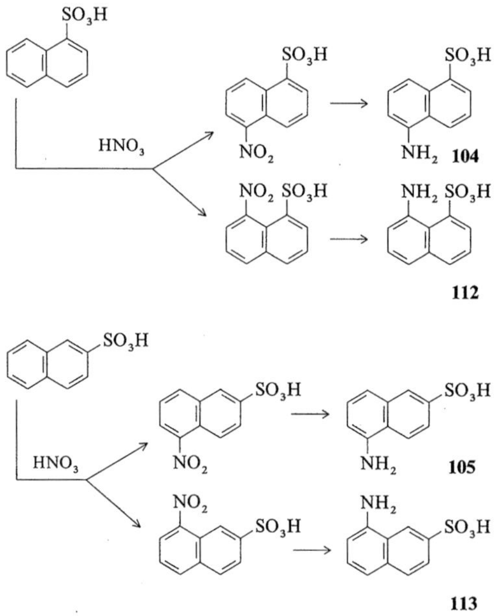

**Figure 4.** Nitration–reduction of naphthalenesulfonic acids

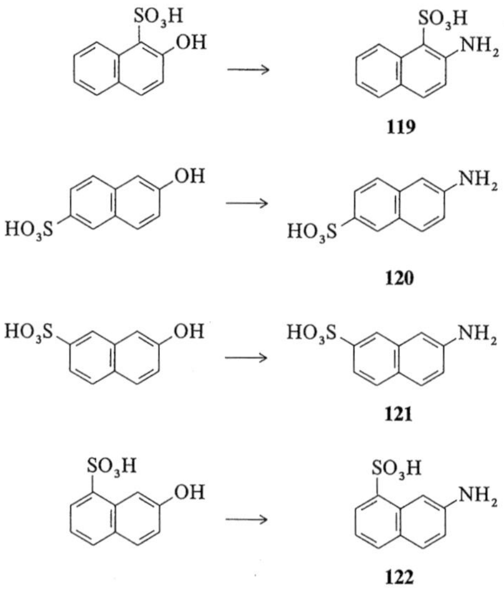

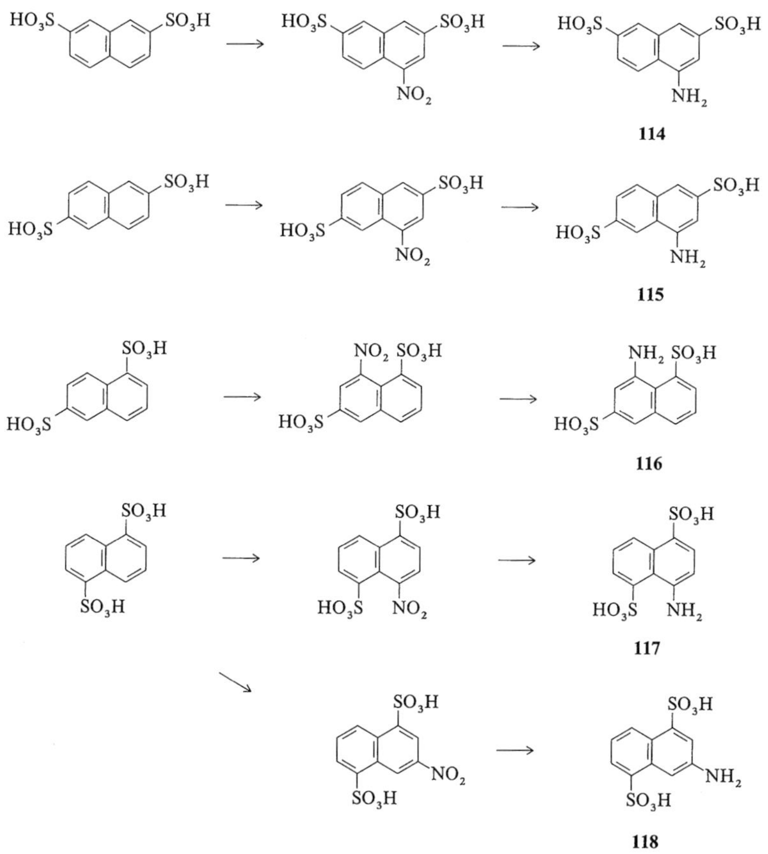

**Figure 5.** Nitration–reduction of naphthalenedisulfonic acids

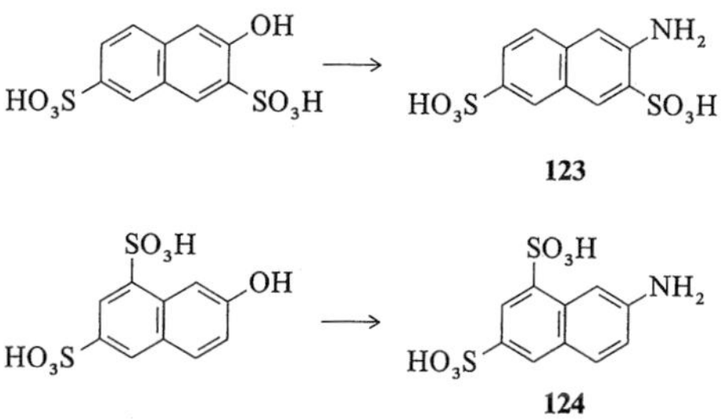

**Desulfonation.** As with certain hydroxynaphthalenedi- and trisulfonic acids (Section 4.1), selective desulfonation is exploited with aminonaphthalenedi- and trisulfonic acids. Heating with dilute sulfuric acid is the standard process; sulfonic acid groups adjacent to the amino group are usually labile. A different mechanism may apply when sodium amalgam is required in the case of peri-desulfonation, e.g., with **116**.

Although many naphthalene derivatives exhibit UV fluorescence in solution, aminonaphthalenesulfonic acids are notable in commonly showing solution fluorescence in daylight. The fluorescent color is in the green–blue–violet area [15], with no obvious correlation between color and structure. Nevertheless, an attempt has been made to systematize the spectral identification of substituted naphthalenesulfonic acids as a contribution to the structural determination of azo dyes after reductive scission [48].

The fluorescence has been exploited with 5-dimethylamino-1-naphthalenesulfonyl chloride [605-65-2] (dansyl chloride), which is reacted with amino acids for labeling purposes in protein sequencing.

# 6.2. 1-Aminonaphthalenesulfonic Acids

**1-Aminonaphthalene-2-sulfonic acid** [81-06-1] (**102**), *ortho*-naphthionic acid, $C_{10}H_9NO_3S$, $M_r$ 223.24, mp 262–265 °C (decomp.), is sparingly soluble in cold water but dissolves in hot water to give a 3% solution. The sodium salt is more readily soluble in water (1.7% at 20 °C and 10% at 100 °C). The calcium salt is similarly soluble, but the barium salt is almost insoluble.

*Production.* 1-Naphthylamine (Section 5.1) can be sulfonated by heating with sulfuric acid at 180–185 °C, but the transient intermediate naphthalene-1-sulfamic acid is best prepared from 1-aminonaphthalene-4-sulfonic acid (naphthionic acid) [7]. When sodium naphthionate is heated in a solvent such as *o*-dichlorobenzene or xylene at 225 °C, thermal rearrangement occurs to give the 1,2-isomer in high yield.

**1-Aminonaphthalene-3-sulfonic acid** [*134-54-3*] (**125**) (4-aminonaphthalene-2-sulfonic acid), $C_{10}H_9NO_3S$, $M_r$ 223.24, is only sparingly soluble in water, but its alkali-metal salts are readily soluble. Sulfonation with 20% oleum yields 1-aminonaphthalene-3,5-disulfonic acid, and heating with water under pressure gives 1-hydroxynaphthalene-3-sulfonic acid, which can also be obtained by diazotization and acid hydrolysis of **125**.

*Production.* 1-Aminonaphthalene-3,8-disulfonic acid (Section 6.4) in dilute sodium hydroxide solution is pumped at 40 °C through a special reactor that effects contact with continuously produced sodium amalgam [7]. The resulting liquor is used directly for production of 1-hydroxynaphthalene-3-sulfonic acid (Section 4.2).

**1-Aminonaphthalene-4-sulfonic acid** [*84-86-6*] (**103**) (4-aminonaphthalene-1-sulfonic acid), naphthionic acid, Piria's acid, $C_{10}H_9NO_3S$, $M_r$ 223.24, is sparingly soluble in water (0.25% at 100 °C), but the sodium, potassium, calcium, and barium salts are all readily soluble.

Chlorination of **103** with copper(II) chloride produces 4-chloro-1-aminonaphthalene, whereas the action of chlorine on an aqueous sulfuric acid solution gives 2,3-dichloro-1,4-naphthoquinone. High-temperature bromination gives 2,4-dibromo-1-aminonaphthalene, whereas vigorous nitration yields 2,4-dinitro-1-hydroxynaphthalene. Sulfonation with 25% oleum gives 1-aminonaphthalene-4,6- and 4,7-disulfonic acids, whereas prolonged heating with sulfuric acid at 130 °C leads to 1-aminonaphthalene-5- and 6-sulfonic acids. Hydrolysis of the 1-amino group to give 1-hydroxynaphthalene-4-sulfonic acid is best achieved by a reverse-Bucherer reaction but also results from prolonged heating with 50% sodium hydroxide.

*Production.* Naphthionic acid was first prepared by PIRIA by heating 1-nitronaphthalene with ammonium sulfite; however, the formation of 1-aminonaphthalene-2,4-disulfonic acid and other byproducts makes this route unattractive for large-scale production. Although it can also be prepared by direct sulfonation of 1-naphthylamine, it is more conveniently produced by a solvent-bake process [6]. 1-Naphthylamine is dissolved in *o*-dichlorobenzene and $H_2SO_4$ (molar ratio 1:1) added to form a suspension of the sulfate. After being heated gradually to 180 °C, the reaction is completed by maintaining this temperature until all the water of reaction has been distilled off with solvent recycle. The product is extracted from the final mixture by addition to aqueous sodium carbonate solution. After separation, the aqueous layer is steam distilled to remove residual solvent, and the product is salted out, filtered, and washed with brine to give sodium naphthionate as a paste in 90% yield [7].

*Uses.* Naphthionic acid is used as an intermediate in the production of 1-hydroxynaphthalene-4-sulfonic acid, 1-aminonaphthalene-2-sulfonic acid, and 1-aminonaphthalene-4,6- and 4,7-disulfonic acids. Use as a diazo component has been previously described, e.g., C.I. Food Red 3 (**71**), C.I. Food Red 7 (**86**), C.I. Food Red 9 (**84**), and C.I. Acid Red 25 (**77**).

**1-Aminonaphthalene-5-sulfonic acid** [84-89-9] (**104**) (5-aminonaphthalene-1-sulfonic acid), Laurent's acid, L acid, Purpurin acid, $C_{10}H_9NO_3S$, $M_r$ 223.24, crystallizes from hot water as needles of the monohydrate. The alkali-metal salts are all readily soluble in water.

Bromination yields the 2,4-dibromo derivative, vigorous nitration gives 2,4-dinitro-1-hydroxynaphthalene, and sulfonation results in 1-aminonaphthalene-2,5,7-trisulfonic acid. 1-Hydroxynaphthalene-5-sulfonic acid is obtained by heating **104** with aqueous bisulfite (reverse-Bucherer reaction) or via the diazo compound. Caustic fusion produces 5-amino-1-hydroxynaphthalene (Chap. 7). Coupling with diazo compounds takes place in the 2-position.

*Production.* Laurent's acid is coproduced with Peri acid as the minor component of a 5:2 isolated product ratio.

*Uses.* Laurent's acid is used as an intermediate for 1-hydroxynaphthalene-5-sulfonic acid, 5-amino-1-naphthol, 1-aminonaphthalene-5,7-disulfonic acid, and 8-amino-4-hydroxynaphthalene-2-sulfonic acid.

**1-Aminonaphthalene-6-sulfonic acid** [119-79-9] (**105**) (5-aminonaphthalene-2-sulfonic acid), 1,6-Cleve's acid, $C_{10}H_9NO_3S$, $M_r$ 223.24, crystallizes from boiling water as plates of the dihydrate. Its sodium, potassium, and calcium salts are all readily soluble in water, but the barium salt is only sparingly soluble. Sulfonation with 10% oleum yields 1-aminonaphthalene-4,6-disulfonic acid. Caustic fusion gives 5-amino-2-hydroxynaphthalene, and Bucherer-type hydrolysis gives 1-hydroxynaphthalene-6-sulfonic acid.

*Production.* Sulfonation of 1-naphthylamine with concentrated sulfuric acid at 130 °C gives mainly 1-aminonaphthalene-6-sulfonic acid. However, the latter is always made, together with the 1,7-isomer, by nitration of naphthalene-2-sulfonic acid to give the mixed 5- and 8-nitronaphthalene-2-sulfonic acids follwed by reduction to "mixed Cleve's acids." A β-sulfonation mass (Section 2.2) is diluted with sulfuric acid, then 67.5% nitric acid is added slowly at 34 °C. After being quenched and neutralized with calcium carbonate, the filtered liquor is reduced by boiling with iron powder, slight acidity being maintained with sulfuric acid. After basification with magnesia the iron residues are filtered off and the filtrate is evaporated until the magnesium salt of the product crystallizes on cooling. Filtration followed by washing with magnesium sulfate gives a 34% yield of the magnesium salt of 1,6-Cleve's acid. Purification by dissolution in hot water, acidification, and filtration of the free acid while hot reduce the yield to 25% based on naphthalene.

**1-Aminonaphthalene-7-sulfonic acid** [119-28-8] (**113**) (8-aminonaphthalene-2-sulfonic acid), 1,7-Cleve's acid, $C_{10}H_9NO_3S$, $M_r$ 223.24, is 0.5% soluble in water at 25 °C and 3% soluble at 100 °C. The sodium, potassium, calcium, and magnesium salts are readily soluble in water but the barium salt is only sparingly soluble. Sulfonation with oleum at 50 °C gives 1-aminonaphthalene-4,7-disulfonic acid. Alkali fusion gives 8-amino-2-naphthol, and Bucherer-type hydrolysis gives 1-hydroxynaphthalene-7-sulfonic acid.

*Production.* The filtrate from the production of 1,6-Cleve's acid is acidified to precipitate crude 1,7-Cleve's acid (34% yield), which can be purified by recrystallization of the sodium salt in an overall yield of 29% based on naphthalene.

*Uses.* For many azo dye outlets, isolated mixed Cleve's acid from acidification of the reduction liquor is adequate as an economical component in the production, for example, of C.I. Acid Black 36 (**131**).

**131**

**1-Aminonaphthalene-8-sulfonic acid** [*82-75-7*] (**112**) (8-aminonaphthalene-1-sulfonic acid), Peri acid, $C_{10}H_9NO_3S$, $M_r$ 223.24, crystallizes as the monohydrate and is sparingly soluble even in boiling water. The sodium salt is also sparingly soluble in water.

Caustic fusion at 200 °C produces 8-amino-1-naphthol, whereas slow hydrolysis with 10% sodium hydroxide at 220–260 °C yields 1,8-dihydroxynaphthalene. Sulfonation with cold 10% oleum produces 1-aminonaphthalene-4,8-disulfonic acid, whereas hot concentrated sulfuric acid results in slow isomerization into naphthionic acid.

*Production* [7]. Naphthalene is converted to naphthalene-1-sulfonic acid (see Section 2.2). The crude sulfonation product is diluted with sulfuric acid, and 67.5% nitric acid is added slowly at 30 °C to give the 5- and 8-nitro derivatives. The products are converted to the readily soluble magnesium salts by addition of a slurry of dolomite. The reaction mixture is neutralized with chalk and filtered hot. The filtrate is run into a suspension of iron powder in boiling water, which is kept slightly acidic with sulfuric acid. When reduction is complete, dissolved iron is precipitated by addition of magnesia, the iron residues are filtered off while hot, and the resulting liquor is acidified to pH 4.5 with sulfuric acid to give 99% pure Peri acid in 48% yield. The filtrates are further acidified to precipitate crude Laurent's acid. This may be purified via its magnesium salt to give a 17% yield of pure 1,5-isomer.

*Uses.* Dehydration with phosphorus oxychloride yields naphthosultam (**132**), and diazotization followed by hydrolysis results in naphthosultone (**133**), the intermediate for 1-hydroxynaphthalene-8-sulfonic acid. Diazotization followed by a Sandmeyer cyanide reaction gives 8-cyanonaphthalene-1-sulfonic acid, one of the most industrially important Sandmeyer reaction products because the nitrile forms the lactam, naphthostyril [*130-00-7*] (**134**), upon treatment with concentrated alkali at 185 °C. Hydrolysis of naphthostyril with dilute alkali yields 8-aminonaphthalene-1-carboxylic acid (**135**), which is the key intermediate for production of anthanthrone vat dyes via 1,1'-binaphthyl-8,8'-dicarboxylic acid (**136**).

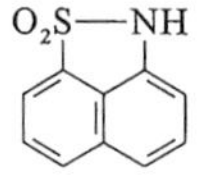

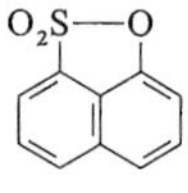

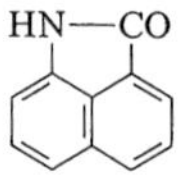

132 133 134

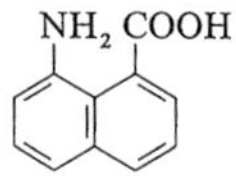

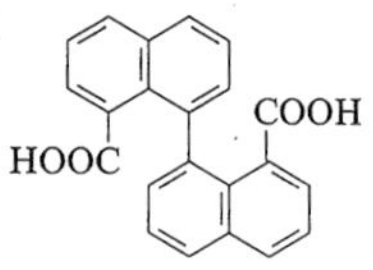

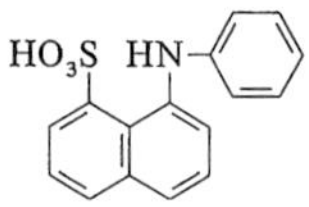

135 136 137

Although Peri acid couples in the 4-position to form azo dyes, its N-aryl derivatives [e.g., *N*-phenyl (**137**) or *N*-*p*-tolyl] are generally used. These are produced by heating Peri acid with excess amine at 170 °C in the presence of amine hydrochloride. A typical and important acid dye for wool based on **137** is C.I. Acid Blue 113 (**92**). The monoazo compound derived from H acid (C.I. Acid Blue 92) is purified to anazolene sodium (**138**) for diagnostic use as a plasma volume indicator.

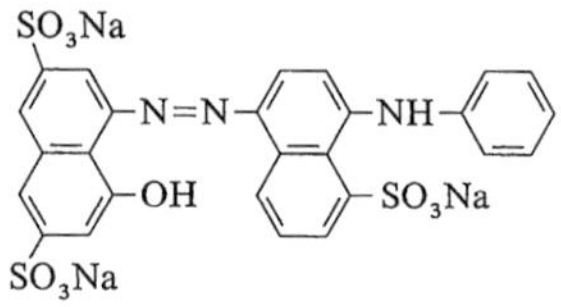

138

## 6.3. 2-Aminonaphthalenesulfonic Acids

**2-Aminonaphthalene-1-sulfonic acid** [*81-16-3*] (**119**), Tobias acid, $C_{10}H_9NO_3S$, $M_r$ 223.24, is sparingly soluble in water but the sodium salt is readily soluble.

Coupling with diazo compounds takes place in the 1-position with elimination of the sulfonic acid group.

*Production.* 2-Hydroxynaphthalene-1-sulfonic acid (Section 4.3) is heated with a mixture of ammonia and ammonium sulfite in an autoclave at 150 °C and 1 MPa for 30 h [7]. The reaction mixture is added to a slurry of lime, and excess ammonia is driven off. Inorganic calcium salts are filtered off at 80 °C, and the filtrate is acidified with hydrochloric acid at 40 °C to precipitate Tobias acid in 95 % yield. Care must be taken throughout the process to avoid desulfonation and formation of 2-naphthylamine.

*Uses.* Sulfonation with oleum gives 2-aminonaphthalene-1,5-disulfonic acid, which can be desulfonated to produce 2-aminonaphthalene-5-sulfonic acid.

Tobias acid is used as a diazo component in acid dyes (e.g., C.I. Acid Yellow 19), red reactive dyes, and pigments such as the important Lithol Reds, [e.g., C.I. Pigment Red 49 (**139**)]. The free acid of **139** is also remarkably insoluble and can be used as a pigment.

**139**

**2-Aminonaphthalene-5-sulfonic acid** [*81-05-0*] (**127**) (6-aminonaphthalene-1-sulfonic acid), Dahl's acid, Dressel acid, D acid, $C_{10}H_9NO_3S$, $M_r$ 223.24, crystallizes from water as needles although only 0.4% soluble at 100 °C. The sodium, calcium, and barium salts are moderately soluble in water.

Bromination in acetic acid gives the 1-bromo derivative. Sulfonation with oleum gives a mixture of 2-aminonaphthalene-5,7-disulfonic acid together with some 1,5-disulfonic acid. Fusion with potassium hydroxide at 260 °C yields 6-amino-1-naphthol.

*Production.* Tobias acid is added to a mixture of sulfuric acid and oleum, and the reaction is completed by stirring for 12 h at 25 °C. The resulting sulfonation mass, containing mainly 2-aminonaphthalene-1,5-disulfonic acid (**126**), is poured into water and heated at 105 °C for 3 h to desulfonate and precipitate the product which, after cooling, is filtered off; the yield is 95%.

**2-Aminonaphthalene-6-sulfonic acid** [*93-00-5*] (**120**) (6-aminonaphthalene-2-sulfonic acid), Bronner acid, $C_{10}H_9NO_3S$, $M_r$ 223.24, crystallizes from water as the monohydrate in plates although only 0.2% soluble at 100 °C. Sulfonation with 20% oleum at 20 °C gives a mixture of 1,6- and 6,8-disulfonic acids.

*Production.* 2-Hydroxynaphthalene-6-sulfonic acid (Section 4.3) is aminated by a Bucherer reaction under conditions similar to those used for Tobias acid.

**2-Aminonaphthalene-7-sulfonic acid** [*494-44-0*] (**121**) (7-aminonaphthalene-2-sulfonic acid), amido F acid, $C_{10}H_9NO_3S$, $M_r$ 223.24, crystallizes as the monohydrate in needles or prisms although only 0.3% soluble at 100 °C.

Bromination gives the 1-bromo derivative, caustic fusion results in 7-amino-2-naphthol, and sulfonation with oleum gives a mixture of disulfonic acids.

*Production.* 2-Hydroxynaphthalene-7-sulfonic acid (Section 4.3) is aminated by a Bucherer reaction under conditions similar to those used for Tobias acid.

**2-Aminonaphthalene-8-sulfonic acid** [*86-60-2*] (**122**) (7-aminonaphthalene-1-sulfonic acid), Badische acid, $C_{10}H_9NO_3S$, $M_r$ 223.24, crystallizes from water as needles or prisms although only 0.5% soluble at 100 °C. Alkali-metal salts are quite soluble in water. Caustic fusion yields 7-amino-1-naphthol, sulfonation in the cold with 20% oleum gives 2-aminonaphthalene-6,8-disulfonic acid, but nitration gives a mixture of products. 2-Aminonaphthalene-8-sulfonic acid reacts with diazo compounds to form diazoamino compounds rather than coupling to give azo derivatives.

*Production.* 2-Hydroxynaphthalene-8-sulfonic acid (Section 4.3) is aminated by a Bucherer reaction under conditions similar to those used for Tobias acid.

## 6.4. 1-Aminonaphthalenedisulfonic Acids

Although preparation of fourteen of the possible isomers has been reported [5], only processes for those seven isomers that have been used as intermediates or dye components are described here.

**1-Aminonaphthalene-3,6-sulfonic acid** [*6251-07-6*] (**114**) (4-aminonaphthalene-2,7-disulfonic acid), Freund's acid(1,3,6), $C_{10}H_9NO_6S_2$, $M_r$ 303.3, and its salts are readily soluble in water. Caustic fusion results in a mixture of aminonaphtholsulfonic acids, but amination occurs selectively, producing 1,3-diaminonaphthalene-6-sulfonic acid.

*Production.* Naphthalene is sulfonated with sulfuric acid at 160 °C for 14 h to give a mixture of mainly 2,6- and 2,7-disulfonic acids [7]. The reaction mass is nitrated with $H_2SO_4$–$HNO_3$ at 30 °C, poured into water, and limed out. The filtrate is then reduced with iron and hydrochloric acid, and the reduction liquor is worked up to selectively salt out the sodium salt of 1-aminonaphthalene-3,7-disulfonic acid in 17.5 % yield. The filtrates and brine wash are then acidified to precipitate crude 1-aminonaphthalene-3,6-disulfonic acid, which is purified by dissolution in aqueous potassium hydroxide and precipitation with acid; the yield is 35 %.

**1-Aminonaphthalene-3,7-disulfonic acid** [*6362-05-6*] (**115**) (4-aminonaphthalene-2,6-disulfonic acid), Freund's acid(1,3,7), $C_{10}H_9NO_6S_2$, $M_r$ 303.3, is readily soluble in water, but its salts are less soluble than those of the 1,3,6-isomer. The amino group is hydrolyzed by water at 180 °C to give 1-hydroxynaphthalene-3,7-disulfonic acid. Caustic fusion results in formation of 1-amino-7-hydroxynaphthalene-3-sulfonic acid, and amination gives 1,3-diaminonaphthalene-7-sulfonic acid.

*Production.* The 1,3,7-acid is produced together with the 1,3,6-isomer [7].

**1-Aminonaphthalene-3,8-disulfonic acid** [*129-91-9*] (**116**) (8-Aminonaphthalene-1,6-disulfonic acid), amino epsilon acid, $C_{10}H_9NO_6S_2$, $M_r$ 303.3, and its sodium salt are very soluble in water, but the barium salt is only sparingly soluble. Reduction with zinc and alkali (or with sodium amalgam) desulfonates **116** to give 1-aminonaphthalene-3-sulfonic acid. The amino group is hydrolyzed by water at 180 °C to give 1-hydroxynaphthalene-3,8-disulfonic acid. Caustic fusion results in 1-amino-8-hydroxynaphthalene-3-sulfonic acid, whereas reaction with 10 % sodium hydroxide at 250 °C produces 1,8-dihydroxynaphthalene-3-sulfonic acid. Amination produces 1,3-diaminonaphthalene-8-sulfonic acid (**149**), but the production of 6,8-di(phenylamino)naphthalene-1-sulfonic acid (diphenyl epsilon acid; **140**) by reaction with aniline is more important.

*Production.* Naphthalene is sulfonated under conditions that give predominately the 1,6-disulfonic acid. The sulfonation mass is nitrated with $H_2SO_4$–$HNO_3$ at 30–40 °C [7]. After addition of dolomite and neutralization with lime, the liquor is reduced with iron and basified with magnesia, and the iron residues are removed. Careful adjustment of temperature and acidity precipitates the product as the acid magnesium salt, which is purified by dissolution in aqueous sodium carbonate and removal of magnesium carbonate; the diacid is precipitated with acid in 32% overall yield.

*Uses.* Amino ε acid is used mainly as an intermediate rather than as a color component, but the derived diphenyl ε acid (**140**) is used for safranine dyes such as C.I. Acid Blue 61 (**141**).

**1-Aminonaphthalene-4,6-disulfonic acid** [*85-74-5*] (**108**) (4-aminonaphthalene-1,7-disulfonic acid), Dahl's acid II, $C_{10}H_9NO_6S_2$, $M_r$ 303.3, and its alkali-metal salts are soluble in water; the calcium salt is soluble in aqueous ethanol. Caustic fusion at 180 °C yields 1-amino-6-hydroxynaphthalene-4-sulfonic acid, but at higher temperature 4,7-dihydroxynaphthalene-1-sulfonic acid is formed. Sulfonation with 35% oleum at 80 °C gives 1-aminonaphthalene-2,4,6-trisulfonic acid.

*Production.* Naphthionic acid is sulfonated with 25% oleum at 20 °C to give a 70:30 mixture of 1,4,7- and 1,4,6-trisulfonic acids, which is used as mixed Dahl's acids.

**1-Aminonaphthalene-4,7-disulfonic acid** [*85-75-6*] (**106**) (4-aminonaphthalene-1,6-disulfonic acid), Dahl's acid III, $C_{10}H_9NO_6S_2$, $M_r$ 303.3, and its sodium salt are readily soluble in water, but the calcium and barium salts are only sparingly soluble. Conversion to 1-hydroxynaphthalene-4,7-disulfonic acid is achieved by reverse-Bucherer reaction or diazotization and hydrolysis. Sulfonation (35% oleum, 90 °C) yields 1-aminonaphthalene-2,4,7-trisulfonic acid.

*Production.* Although separation of **106** from the mixture obtained by sulfonation of naphthionic acid (see above) is not practicable on a large scale, the 1,4,7-isomer can be obtained directly in high yield by sulfonation of 1-aminonaphthalene-7-sulfonic acid with oleum.

**1-Aminonaphthalene-4,8-disulfonic acid** [*117-55-5*] (**117**) (4-aminonaphthalene-1,5-disulfonic acid), $C_{10}H_9NO_6S_2$, $M_r$ 303.3, and its acid monosodium salt are sparingly soluble in cold water, but the disodium salt is readily soluble. Hydrolysis with sodium hydroxide at 200 °C yields 1-amino-8-hydroxynaphthalene-4-sulfonic acid; at higher temperature, 1,8-dihydroxynaphthalene-4-sulfonic acid is formed. Heating with bisulfite produces 1-hydroxynaphthalene-4,8-disulfonic acid. Diazo compounds couple with 1-aminonaphthalene-4,8-disulfonic acid to form azo compounds, in contrast to 2-aminonaphthalene-4,8-disulfonic acid, which reacts to form diazoamino compounds.

*Production.* Naphthalene is sulfonated under conditions that give predominantly the 1,5-disulfonic acid, and the sulfonation mass is nitrated at 20 °C [7]. 3-Nitronaphthalene-1,5-disulfonic acid is precipitated from the quenched reaction mixture as its iron(II) salt and removed. After dilution and liming out, the remainder of the mixed nitro solution is reduced with iron and dilute acid, and the 1,4,8-isomer is salted out preferentially in ca. 40% yield. A small proportion (ca. 10% of 1-aminonaphthalene-3,8-disulfonic acid) is then salted out, after which the more soluble 1-aminonaphthalene-5,7-disulfonic acid crystallizes on concentration.

**1-Aminonaphthalene-5,7-disulfonic acid** [*13306-42-8*] (**130**) (5-aminonaphthalene-1,3-disulfonic acid), $C_{10}H_9NO_6S_2$, $M_r$ 303.3, and its sodium salt are very soluble in water. Caustic fusion produces 1-amino-5-hydroxynaphthalene-7-sulfonic acid, and sulfonation (40% oleum, 130 °C) yields 1-aminonaphthalene-2,5,7-trisulfonic acid.

*Production.* 1-Aminonaphthalene-5-sulfonic acid (Section 6.2) in sulfuric acid solution is treated with 65% oleum, and the mixture is heated at 120 °C for 4 h. The sulfonation mass, which contains mainly the 2,5,7-trisulfonic acid, is diluted with water, and the batch is heated at 130 °C for 10 h to cause desulfonation. The product crystallizes on cooling and, after isolation, is deacidified by treatment of the solution with lime. The final sodium salt solution is used directly for production of 1-amino-5-hydroxynaphthalene-7-sulfonic acid.

## 6.5. 2-Aminonaphthalenedisulfonic Acids

Although preparation of ten of the possible isomers has been reported [5], only processes for those five isomers that have been used as intermediates or dye components are described here.

**2-Aminonaphthalene-1,5-disulfonic acid** [*117-62-4*] (**126**), 5-sulfo-Tobias acid, $C_{10}H_9NO_6S_2$, $M_r$ 303.32, is purified by crystallization of the dipotassium salt from water. Alkali fusion results in formation of 2-amino-5-hydroxynaphthalene-1-sulfonic acid. Sulfonation (40% oleum, 100 °C) gives 2-aminonaphthalene-1,5,7-trisulfonic acid. Heating with aqueous sulfuric acid produces 2-aminonaphthalene-5-sulfonic acid. The diazo compound is converted to 2-diazo-1-hydroxynaphthalene-5-sulfonic acid on treatment with alkaline hypochlorite.

*Production.* Sulfonation of Tobias acid to produce 2-aminonaphthalene-5-sulfonic acid (Section 6.3) proceeds via **126**, which is isolated from the quenched reaction mixture with careful avoidance of desulfonation.

*Uses.* As a diazo component in orange and red azo reactive dyes, **126** gives more bathochromic shades than Tobias acid.

**2-Aminonaphthalene-3,6-disulfonic acid** [*92-28-4*] (**123**) (3-aminonaphthalene-2,7-disulfonic acid), amino R acid, $C_{10}H_9NO_6S_2$, $M_r$ 303.3.

*Production.* The sodium salt of 2-hydroxynaphthalene-3,6-disulfonic acid (Section 4.5) is aminated by heating with ammonia and ammonium bisulfite for 24 h at 160 °C. The charge is added to excess lime and heated to remove excess ammonia, and calcium sulfite is filtered off. The product is precipitated by acidification of the filtrate and purified by redissolving the paste in alkali and reprecipitating with acid. The yield is 90 %.

*Uses.* Amino R acid is used mainly as an intermediate for conversion to 2-amino-3-hydroxynaphthalene-6-sulfonic acid by caustic fusion at 250 °C.

**2-Aminonaphthalene-4,8-disulfonic acid** [*131-27-1*] (**118**) (3-aminonaphthalene-1,5-disulfonic acid), Cassella acid, C acid, $C_{10}H_9NO_6S_2$, $M_r$ 303.32, and its salts are readily soluble in water.

Reduction with sodium amalgam desulfonates **118** to give 2-aminonaphthalene-8-sulfonic acid, whereas boiling with dilute sulfuric acid results in desulfonation and hydrolysis to give 2-hydroxynaphthalene-4-sulfonic acid. Alkali fusion (KOH, 215 °C) produces 2-amino-4-hydroxynaphthalene-8-sulfonic acid.

*Production.* The iron(II) salt of 3-nitronaphthalene-1,5-disulfonic acid (nitro-Armstrong acid) (Section 6.4) is reduced with iron in dilute acid [7].

A claimed improvement involves isolation of nitro-Armstrong acid from the nitration products of naphthalene-1,5-disulfonic acid as the crystalline magnesium salt prior to catalytic reduction to C acid [49].

*Uses.* Cassella acid is an important diazo component for yellow direct and reactive dyes.

**2-Aminonaphthalene-5,7-disulfonic acid** [*118-33-2*] (**129**) (6-aminonaphthalene-1,3-disulfonic acid), amino J acid, $C_{10}H_9NO_6S_2$, $M_r$ 303.32, and the bis-alkali-metal salts are very soluble in water, but the mono-alkali-metal salts, especially the acid potassium salt, are sparingly soluble in cold water. Reduction with sodium amalgam produces 2-aminonaphthalene-7-sulfonic acid, and heating with ammonium bisulfite gives 2-hydroxynaphthalene-5,7-disulfonic acid.

*Production.* 2-Aminonaphthalene-1-sulfonic acid is sulfonated by heating with excess oleum for 12 h at 100 °C. This produces 2-aminonaphthalene-1,5,7-trisulfonic acid (**128**), which is desulfonated by heating the quenched acid mixture at 105 °C for 3 h. The product is filtered off after cooling and dissolved in hot aqueous alkali, and excess sulfate is removed by precipitation with lime. The resulting sodium salt liquor (80% yield) is used directly for caustic fusion to give J acid, the major outlet.

**2-Aminonaphthalene-6,8-disulfonic acid** [86-65-7] (**124**) (7-aminonaphthalene-1,3-disulfonic acid), amino G acid, $C_{10}H_9NO_6S_2$, $M_r$ 303.3, and its bis-alkali-metal salts are very soluble in water, but the acid salts, especially the acid potassium salt, are much less soluble.

*Production.* The disodium or dipotassium salt of 2-hydroxynaphthalene-6,8-disulfonic acid (Section 4.5) is heated with excess aqueous ammonia and ammonium bisulfite at 185 °C for 18 h. The reaction mixture is basified, heated to remove excess ammonia, neutralized, and evaporated while any residual sulfite is removed. The resulting amino G acid liquor (95% yield) is used directly for caustic fusion to give γ acid, the major outlet.

## 6.6. Aminonaphthalenetrisulfonic Acids

Only those isomers that have been used as intermediates or dye components are described here. Many other isomers have been reported, as have aminonaphthalenetetrasulfonic acids [5], but they are of only academic interest. The major outlet for aminonaphthalenetrisulfonic acids is in caustic fusion reactions to produce important aminohydroxynaphthalenedisulfonic acids.

**1-Aminonaphthalene-2,4,8-trisulfonic acid** [76530-15-9] (4-aminonaphthalene-1,3,5-trisulfonic acid), $C_{10}H_9NO_9S_3$, $M_r$ 383.36, is not isolable as the free acid and always reverts to the sultam (**142**) with elimination of water between the peri-substituents. The trisodium salt is also water soluble but can be crystallized in the form of yellow plates. Desulfonation occurs with 20% hydrochloric acid at 150 °C to form 1-aminonaphthalene-8-sulfonic acid and with 40% sulfuric acid at 110 °C to give 1-aminonaphthalene-2,8-disulfonic acid.

*Production.* A sulfuric acid solution of 1-aminonaphthalene-8-sulfonic acid (Section 6.2) is treated with 65% oleum in the presence of anhydrous sodium sulfate as a moderator, and the temperature is slowly increased to 85 °C. The quenched solution is neutralized with chalk, calcium sulfate is removed, and the calcium salt solution is basified with sodium carbonate. After removal of calcium carbonate, the resulting solution of the disodium salt is evaporated for subsequent caustic fusion to produce Chicago acid.

**1-Aminonaphthalene-3,6,8-trisulfonic acid** [*117-42-0*] (**143**) (8-aminonaphthalene-1,3,6-trisulfonic acid), Koch acid, T acid, $C_{10}H_9NO_9S_3$, $M_r$ 383.36, and its alkali-metal salts are readily soluble in water. The solutions are unusual for this class of compound in that they exhibit no fluorescence. Hydrolysis with dilute sulfuric acid produces 1-hydroxynaphthalene-3,6,8-trisulfonic acid (Section 4.6), whereas diazotization followed by acid hydrolysis results in formation of the corresponding sultone. Reduction with zinc and alkali desulfonates **143** to give 1-amino-naphthalene-3,6-disulfonic acid. Amination gives 1,3-diaminonaphthalene-6,8-disulfonic acid.

*Production.* Naphthalene in sulfuric acid solution is treated with gradual addition of 65 % oleum and temperature profiling (to 160 °C) to optimize formation of the 1,3,6-trisulfonic acid. A mixture of sulfuric acid and nitric acids is added slowly at 40 °C to form 1-nitronaphthalene-3,6,8-trisulfonic acid (nitro-Koch acid) which is converted to neutral liquor by liming out. The liquor is reduced with iron powder in the presence of acetic acid and, after removal of iron residues, the crude product is isolated, as the acid calcium – sodium salt. This is purified by dissolution with sodium carbonate and removal of calcium carbonate. The trisodium salt liquor is concentrated before direct use in H acid manufacture. To obtain pure Koch acid, the 1-nitro-3,6,8-trisulfonic acid can be isolated before reduction. The reduction step can alternatively be carried out by hydrogenation of the aqueous solution of nitro-Koch acid in the presence of a nickel catalyst [50].

*Uses.* Apart from caustic fusion to give H acid, the other main use is the production of chromotropic acid (**66**) by treatment with strong caustic under pressure.

**1-Aminonaphthalene-4,6,8-trisulfonic acid** [*17894-99-4*] (**144**) (8-aminonaphthalene-1,3,5-trisulfonic acid), B acid, $C_{10}H_9NO_9S_3$, $M_r$ 383.36, is reduced with zinc and alkali to 1-aminonaphthalene-4,6-disulfonic acid. Heating with aqueous sulfuric acid causes desulfonation to form 1-aminonaphthalene-6,8-disulfonic acid. Sulfonation (25 % oleum, 90 °C) gives the sultam of 1-aminonaphthalene-2,4,6,8-tetrasulfonic acid.

*Production.* Naphthalene is sulfonated under conditions that maximize formation of the 1,3,5-trisulfonic acid (Section 2.4), followed by nitration with $H_2SO_4$–$HNO_3$ in situ at 35 °C. After quenching, neutralization with calcium carbonate, and reduction of the nitro-derivative with iron under acid conditions, the byproduct 1-aminonaphthalene-3,6,8-trisulfonic acid (18 % yield) is selectively precipitated by acidification. The desired isomer is then precipitated by further acidification and salting of the filtrates, and finally purified by reprecipitation. Overall yield is 25 – 30 %.

*Uses.* Alkali fusion of B acid yields K acid. A special outlet for purified B acid is in the synthesis of the drug suramin sodium (Bayer 205) (**145**). 1-Aminonaphthalene-4,6,8-trisulfonic acid is acylated with 2-nitrotoluyl chloride, and this product is reduced and acylated with *m*-nitrobenzoyl chloride. This intermediate is reduced and finally phosgenated to give **145**, which has been used to treat trypanosomiasis and onchocerciasis.

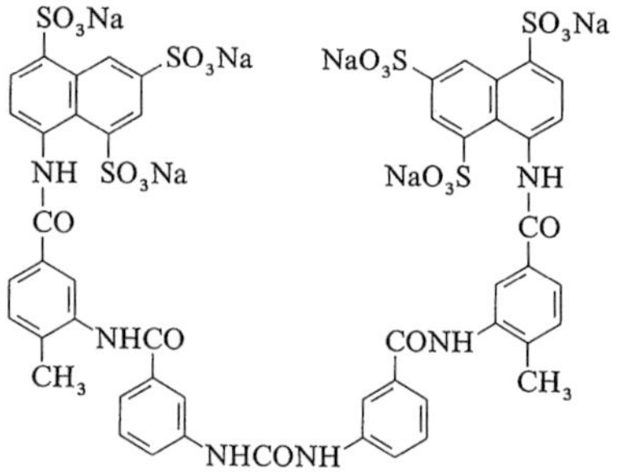

**145**

**2-Aminonaphthalene-1,5,7-trisulfonic acid** [*55524-84-0*] (**128**) (6-aminonaphthalene-1,3,5-trisulfonic acid), $C_{10}H_9NO_9S_3$, $M_r$ 383.36, and its salts are readily soluble in water. Alkali fusion (65% NaOH, 220 °C) forms 2-amino-5-hydroxynaphthalene-1,7-disulfonic acid. Desulfonation to give 2-aminonaphthalene-5,7-disulfonic acid occurs on heating with dilute sulfuric acid. Diazotization and treatment with sodium carbonate yield the diazo oxide of 2-diazo-1-hydroxynaphthalene-5,7-disulfonic acid.

*Production.* Vigorous sulfonation of Tobias acid to produce 2-aminonaphthalene-5,7-disulfonic acid is described in Section 6.5. The intermediate **128** is obtained by quenching the sulfonation mass and liming out carefully to avoid desulfonation. The calcium salt solution is converted to the sodium salt with sodium carbonate and used as a total liquor for conversion to 2-amino-5-hydroxynaphthalene-1,7-disulfonic acid.

**2-Aminonaphthalene-3,6,8-trisulfonic acid** [*118-03-6*] (**146**) (7-aminonaphthalene-1,3,6-trisulfonic acid), $C_{10}H_9NO_9S_3$, $M_r$ 383.36, and its disodium salts are soluble in water, but the dipotassium salt is much less soluble. Reduction with zinc dust and alkali results in 2-aminonaphthalene-3,6-disulfonic acid, and caustic fusion at 240 °C gives 2-amino-8-hydroxynaphthalene-3,6-disulfonic acid.

*Production.* 2-Aminonaphthalene-6,8-disulfonic acid (Section 6.5) is sulfonated with 65% oleum at 125 °C for 12 h in the presence of anhydrous sodium sulfate. After quenching and liming out, the hot solution of calcium salt is basified with sodium carbonate and the resulting solution of the trisodium salt is used directly for caustic fusion to give 2-amino-8-hydroxynaphthalene-3,6-disulfonic acid.

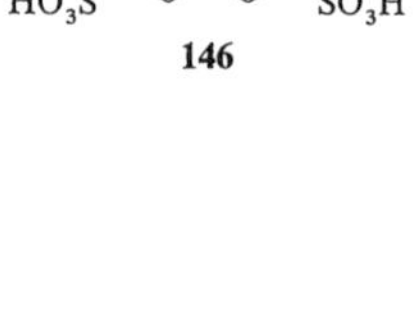

**146**

## 6.7. Diaminonaphthalenesulfonic Acids

**4,5-Diaminonaphthalene-1-sulfonic acid** [6362-18-1] (**147**) 1,8-diaminonaphthalene-4-sulfonic acid, $C_{10}H_{10}N_2O_3S$, $M_r$ 238.27.

*Production.* 1-Nitronaphthalene (Section 5.1) is sulfonated with oleum at 20 °C, and the sulfonation mass is nitrated at 35 °C [7]. After quenching and liming out, the liquors are reduced with iron powder under slightly acidic conditions. After removal of iron residues, the product **147** is isolated in 40% yield by acidifying and salting.

*Uses.* 4,5-Diaminonaphthalene-1-sulfonic acid is used as an intermediate for 4,5-dihydroxynaphthalene-1-sulfonic acid (Section 4.7) and perimidine derivatives.

**5,6-Diaminonaphthalene-1-sulfonic acid** [84-92-4] (**148**) 1,2-diaminonaphthalene-5-sulfonic acid, $C_{10}H_{10}N_2O_3S$, $M_r$ 238.27.

*Production.* 1-Aminonaphthalene-5-sulfonic acid (Section 6.2) is coupled with diazotized aniline, and the resulting isolated azo compound (purpurin base) is reduced by heating with zinc and ammonia at 100 °C and allowing ammonia and aniline to distill off [7]. The product (**148**) is isolated in 76% yield by acidification of the screened liquors.

*Uses.* 5,6-Diaminonaphthalene-1-sulfonic acid is used as an intermediate for naphthotriazoles. A green pigment (C.I. Pigment Green 9) is obtained by treatment with iron(III) salts.

**6,8-Diaminonaphthalene-1-sulfonic acid** (**149**) 1,3-diaminonaphthalene-8-sulfonic acid, $M_r$ 238.27, is prepared by amination of 1-aminonaphthalene-3,8-disulfonic acid (Section 6.4) or 1-hydroxynaphthalene-3,8-disulfonic acid but has no significant use.

**6,8-Dianilinonaphthalene-1-sulfonic acid** [129-93-1] (**140**), diphenyl epsilon acid, $C_{22}H_{18}N_2O_3S$, $M_r$ 390.38.

*Production.* 1-Aminonaphthalene-3,8-disulfonic acid (Section 6.4) is heated with excess aniline in the presence of aniline sulfate at 130 °C.

*Uses.* A dye component as exemplified in **141**.

**1,4-Diaminonaphthalene-2-sulfonic acid** [6357-89-7] (**150**), $C_{10}H_{10}N_2O_3S$, $M_r$ 238.27.

*Production.* 1-Aminonaphthalene-2-sulfonic acid (Section 6.2) is coupled with diazotized aniline, and the resulting azo compound is reduced in a process analogous to that used for **148**.

## 6.8. Diaminonaphthalenedisulfonic Acids

**3,8-Diaminonaphthalene-1,5-disulfonic acid** [19659-81-5] (**151**), amino C acid, $C_{10}H_{10}N_2O_6S_2$, $M_r$ 318.32.

*Production.* Naphthalene-1,5-disulfonic acid sulfonation mass is nitrated to give the 3,8-dinitro derivative [7]. The quenched and limed-out liquor is reduced with iron to give **151**.

**4,8-Diaminonaphthalene-2,6-disulfonic acid** [6362-06-7] (**152**), $C_{10}H_{10}N_2O_6S_2$, $M_r$ 318.32.

*Production.* Naphthalene-2,6-disulfonic acid is dinitrated and reduced in a similar way to the 1,5-isomer [7].

**4,5-Diaminonaphthalene-2,7-sulfonic acid** [6362-11-4] (**153**), Alen acid, $C_{10}H_{10}N_2O_6S_2$, $M_r$ 318.32.

*Production.* Naphthalene-2,7-disulfonic acid is dinitrated and reduced in a similar way to the 1,5-isomer [7].

## 6.9. Toxicity

Aminonaphthalenesulfonic acids were once regarded as innocuous compared to the parent amines. Later, however, some risk was recognized to be involved [9]. Although several aminonaphthalene mono- and disulfonic acids are listed in the EPA inventory [14] as having mildly toxic and irritant properties, no evidence of carcinogenic properties exists.

The hypothesis that water-soluble agents are less likely to be absorbed by bodily organs is supported by the full and rapid excretion of $^{35}$S-labeled Tobias acid in experiments with rats [51]. Neither 2-naphthylamine nor its derivatives were detected in these experiments, indicating that, given the short dwell time of water-soluble compounds in the body, desulfonation is unlikely.

# 7. Aminonaphthols

Although all 14 isomeric aminonaphthols have been described [5], they are less useful than their sulfonic acids (Chap. 8) and only those of commercial significance will be identified. As a class they are amphoteric and dissolve in both acid and alkali. One reason for their lack of use is that their alkaline solutions are very readily oxidized.

There are four main methods of preparation:

1) Caustic fusion of aminonaphthalenesulfonic acids as previously mentioned under properties of 1-aminonaphthalene-5-sulfonic acid, 1-aminonaphthalene-6-sulfonic acid, 1-aminonaphthalene-7-sulfonic acid, and 1-aminonaphthalene-8-sulfonic acid (Section 6.2), as well as 2-aminonaphthalene-5-sulfonic acid, 2-aminonaphthalene-7-sulfonic acid, and 2-aminonaphthalene-8-sulfonic acid (Section 6.3). With some isomers (e.g., 1,4- and 2,6-), hydroxynaphthalenesulfonic acids are formed by replacement of the amino group rather than the sulfonic acid group.
2) Bucherer amination of dihydroxynaphthalenes with ammonia and bisulfite. Alternatively, the reverse-Bucherer reaction can be applied to certain naphthalenediamines.
3) Desulfonation of certain aminohydroxynaphthalenesulfonic acids by heating with dilute acid or by treatment with sodium amalgam.
4) Reduction of arylazo, nitroso, or nitro derivatives of 1-naphthol or 2-naphthol by zinc or sodium. A more recent process uses aqueous hydrazine as the reducing agent [52].

**4-Amino-1-naphthol** [2834-90-4] (**154**), 4-hydroxy-1-aminonaphthalene, 4-amino-1-naphthalenol, $C_{10}H_9NO$, $M_r$ 159.2.

*Production.* Orange I (**25**) in aqueous alkaline solution is treated with sodium dithionite at 40–50 °C, and the resulting suspension is heated to 70 °C to complete the reaction. The crude product is collected by filtration and dissolved in hot aqueous hydrochloric acid in the presence of a reducing agent. On cooling, needles of the hydrochloride salt crystallize and can be isolated in 70% yield. Alternatively, the crude paste may be dissolved in dilute alkali in the presence of a reducing agent and acylated with acetic anhydride or acryloyl chloride, for example, to give 4-acetylamino-1-naphthol or 4-acryloylamino-1-naphthol, which have been used as azo coupling components.

*Uses.* 4-Amino-1-naphthol is used as a polymerization inhibitor.

**5-Amino-1-naphthol** [83-55-6] (**155**), 5-hydroxy-1-aminonaphthalene, 5-amino-1-naphthalenol, Purpurol, $C_{10}H_9NO$, $M_r$ 159.2.

*Production.* 1-Aminonaphthalene-5-sulfonic acid paste (Section 6.2) is added to molten sodium hydroxide at 200 °C, and the reaction is completed by heating slowly to 250 °C and maintaining this temperature for 3 h [7]. After dilution and careful neutralization with hydrochloric acid, 5-amino-1-naphthol precipitates in 85% yield (*mp* 192 °C).

*Uses.* 5-Amino-1-naphthol is used as a coupling component for azo dyes (coupling takes place ortho to the amino or hydroxyl group, or para to the hydroxyl group depending on the pH), and as an intermediate for sulfur dyes.

**7-Amino-1-naphthol** [4384-92-3] (**156**), 8-hydroxy-2-aminonaphthalene, 7-amino-1-naphthalenol, $C_{10}H_9NO$, $M_r$ 159.2.

*Production.* 2-Aminonaphthalene-8-sulfonic acid (Section 6.3) is heated with molten potassium hydroxide for 2–3 h at 260 °C in an autoclave. The product is obtained in 74% yield after dilution and acidification of the melt.

**1-Amino-1-naphthol** [2834-92-6] (**157**), 1-amino-2-naphthalenol, is prepared by reduction of Orange II (1-*p*-sulfobenzeneazo-2-naphthol) by a process analogous to that described for 4-amino-1-naphthol, but it has no significant use.

**1-Amino-2-ethoxynaphthalene** [118-30-9] (**158**), $C_{12}H_{13}NO$, $M_r$ 188.

*Production.* A solution of 2-ethoxynaphthalene (Section 3.2) in 1,2-dichlorobenzene is reacted with nitric acid at 45 °C. The organic layer is separated and washed at 60 °C before cooling to crystallize 1-nitro-2-ethoxynaphthalene (*mp* 95–100 °C) in 80% yield. Reduction with iron gives **158** in 74% yield.

*Uses.* Sulfonation ($H_2SO_4$, 60 °C) of **158** gives 1-amino-2-ethoxynaphthalene-6-sulfonic acid for use principally as an intermediate component in azo direct green dyes.

**8-Amino-2-naphthol** [118-46-7] (**159**), 1-amino-7-naphthol, 8-amino-2-naphthalenol, Cyanol, $C_{10}H_9NO$, $M_r$ 159.2.

*Production.* The sodium salt of 1-aminonaphthalene-7-sulfonic acid (Section 6.2) is added to 80% potassium hydroxide at 210 °C, and the reaction is completed by heating for 6 h at 230 °C [7]. After dilution and careful acidification with hydrochloric acid, the acid solution is screened to remove impurities and neutralized to precipitate **159** in 80% yield (*mp* 207 °C).

*Uses.* Sulfonation gives 1-amino-7-hydroxynaphthalene-4-sulfonic acid. The *N*-acetyl derivative is an important coupling component for after-chrome and metal-complex dyes, e.g., C.I. Mordant Black 38 and C.I. Acid Black 60.

## 8. Aminohydroxynaphthalene-sulfonic Acids

Many of the letter acids listed in Table 1 are aminohydroxynaphthalenesulfonic acids, (aminonaphtholsulfonic acids), the most important of which are probably the H, J, and γ acids [17]. Most of the large-scale use of this class involves the production of water-soluble azo dyes. Only those products of known use are dealt with on an individual basis. Registration of compounds in the European core inventory (ECOIN) in 1981 has been used as a guide to current commercial applications. Thirteen aminonaphthol monosulfonic acids and seven aminonaphthol disulfonic acids are listed therein.

Nomenclature is used as in Table 1, based on tradition (and current common usage) with precedence given to the amino and then the hydroxyl group. The IUPAC convention, based on derivatives of naphthalenesulfonic acids, is added in parentheses for completion.

A number of aminohydroxynaphthalenetrisulfonic acids and diaminohydroxynaphthalenesulfonic acids have been prepared [5] by routes analogous to these described here but they have no technical advantage over the simpler derivatives.

### 8.1. Production and Properties

Four main routes are available depending on the substitution pattern required (route 1 predominates as the basis for all major aminonaphthol sulfonic acids).

1) Caustic fusion hydrolysis of aminonaphthalenedi- and trisulfonic acids.

One sulfonic acid group of aminonaphthalenedi- and trisulfonic acids (see Chap. 6) can be hydrolyzed in a way similar to the selective hydrolysis described for naphthalenedi- and trisulfonic acids (Chap. 4).

The following ten derivatives (**160–169**) manufactured by this route are all commercially important and are described individually later. With α,β-disulfonic acids, the α-group is the more mobile; with α,α-configurations, reaction occurs with the sulfonic acid group in the ring not containing the amino group. Hydrolysis of a sulfonic acid group in the peri-position to the amino group is particularly facile.

2) Reduction of nitroso or arylazo derivatives of hydroxynaphthalenesulfonic acids.

Hydroxynaphthalenesulfonic acids react readily with nitrous acid or diazo compounds with the introduction of substituents ortho to the hydroxyl group. Reduction of the nitroso or azo substituent leads to an ortho arrangement of the amino and hydroxyl groups that is difficult to achieve by other routes.

The nitrosation route can be extended to include 1-naphthol or 2-naphthol, because simultaneous reduction and sulfonation can be achieved to obtain, for example, 1-amino-2-hydroxynaphthalene-4-sulfonic acid (**170**) directly from 2-naphthol.

3) Sulfonation of aminonaphthols.

When an aminonaphthol is readily accessible (Chap. 7), sulfonation may lead to useful aminohydroxynaphthalenesulfonic acids that are not otherwise readily obtainable. However, isomer formation and the need for separation may be a problem.

4) Selective desulfonation of aminohydroxynaphthalenedisulfonic acids.

Facile removal of certain α-sulfonic acid groups may occasionally make an aminohydroxynaphthalenedisulfonic acid (e.g., K acid) a viable starting material.

Aminohydroxynaphthalenesulfonic acids resemble aminonaphthalenesulfonic acids in forming internal salts. The monosulfonic acids are therefore sparingly soluble in water and are isolated as the free acid. The disulfonic acids are much more soluble in water and are often isolated as the monosodium or monopotassium salts even under acidic conditions.

Coupling with diazo compounds is the most important property of aminohydroxynaphthalenesulfonic acids, and many of them couple under both acidic and alkaline conditions. The hues and properties of the resulting $o$-aminoazo or $o$-hydroxyazo dyes vary widely. The most useful products are derived from alkaline-coupled J acid (monoazo oranges) or H acid (monoazo reds), acid-coupled γ acid (monoazo reds), and acid- and alkali-twice-coupled H acid (disazo blues and blacks), even though all of these letter acids undergo both acid and alkaline coupling.

Although all aminohydroxynaphthalenesulfonic acids are diazotizable, only 1-amino-2-hydroxynaphthalene-4-sulfonic acid and its derivatives are industrially important as diazo components.

## 8.2. Aminohydroxynaphthalenemono-sulfonic Acids

**1-Amino-2-hydroxynaphthalene-4-sulfonic acid** [*116-63-2*] (**170**) (4-amino-3-hydroxynaphthalene-1-sulfonic acid), Boeniger acid, $C_{10}H_9NO_4S$, $M_r$ 239.2, is obtained as white or gray needles of the hemihydrate, which is only sparingly soluble in hot water. The alkali-metal salts are soluble in water, but the solutions slowly oxidize in air. Oxidation with nitric acid gives 1,2-naphthoquinone-4-sulfonic acid.

*Production.* 2-Naphthol is dissolved in dilute aqueous sodium hydroxide and precipitated in finely divided form by addition of sulfuric acid at 5 °C [7]. Sodium nitrite solution is added to the slightly alkaline suspension, followed by sulfuric acid to effect the nitrosation while the temperature is held at 5 – 8 °C. The suspension of 1-nitroso-2-naphthol is neutralized with aqueous sodium hydroxide; a solution of sodium bisulfite is then added. The temperature is allowed to rise to 20 °C, and the soluble bisulfite addition complex (**171**) is formed. Excess sulfuric acid is then added and the temperature is raised to 40 °C, whereupon reductive rearrangement takes place and the product slowly crystallizes at 50 °C. Filtration and washing give the product as a 25 % paste in

80% yield. All plant items and reactants must be free from iron to avoid complex formation.

*Uses.* The 1,2,4-acid paste is usually converted directly to the diazo oxide (**172**) by addition to excess sodium nitrite solution in the presence of copper(II) sulfate and in the absence of mineral acid [7]. After completion of the reaction, the product is precipitated by careful addition of excess sulfuric acid and isolated in 95% yield as an orange-gray sandy paste. This can be dried if required for further nonaqueous processing at a temperature not exceeding 50 °C to avoid decomposition.

The so-called diazo-1,2,4-acid (**172**) is usually thought to exist as the naphthoquinone–diazide (**173**; 4-diazo-3,4-dihydro-3-oxonaphthalene-1-sulfonic acid) [26]. It is an important component in azo dyes for after-chroming, e.g., C.I. Mordant Black 3 (**27**), C.I. Mordant Black 17, and C.I. Mordant Red 7 (**174**), and for producing preformed chromium complex dyes such as C.I. Acid Violet 90 (the 2:1 chromium complex of **174**).

The quinone–diazide **173** [*4857-47-0*], is remarkably stable, which allows direct nitration in sulfuric acid solution by treatment with mixed acid over 8 h at 10–15 °C to give the so-called 6-nitrodiazo-1,2,4-acid [*5366-84-7*] (**175**) as a paste in 85% yield after isolation and slurry washing. This nitroderivative is also considered to exist in the stable quinone–diazide form (4-diazo-3,4-dihydro-7-nitro-3-oxonaphthalene-1-sulfonic acid) and is an even more important component for metallized and metallizable dyes than is **173**; typical examples are C.I. Acid Black 52 (chromium complex of **176**), C.I. Direct Red 173, C.I. Reactive Black 1, C.I. Mordant Black 1 (**176**), and C.I. Mordant Black 11.

Chlorination of dry **173** in chlorosulfonic acid solution under a positive pressure of chlorine results in formation (50% yield) of the diazo oxide of 6-chloro-1-amino-2-hydroxynaphthalene-4-sulfonic acid, which is, however, of lesser importance as a diazo component.

**1-Amino-2-hydroxynaphthalene-6-sulfonic acid** [*5639-34-9*] (**177**) (5-amino-6-hydroxynaphthalene-2-sulfonic acid), amino-Schaeffer acid, $C_{10}H_9NO_4S$, $M_r$ 239.2, is obtained as needles or prisms that are only slightly soluble in boiling water. The sodium salt is soluble in water but can be isolated as the hemipentahydrate. Diazotization does not occur in mineral acid medium. The diazo oxide is formed by reaction with sodium nitrite in the presence of copper(II) sulfate.

*Production.* 2-Hydroxynaphthalene-6-sulfonic acid is nitrosated in aqueous solution, and the resulting nitroso derivative is reduced in situ by addition of zinc dust to the acidified batch, with the temperature allowed to rise to 40 °C [7]. The precipitated product is isolated in 70% yield.

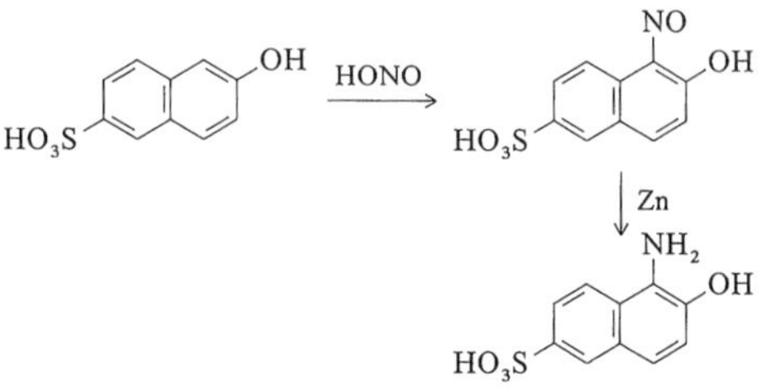

[Structure: Schaeffer acid → nitrosation (HONO) → nitroso-naphthol-sulfonic acid → Zn reduction → **177** (1-amino-2-hydroxynaphthalene-6-sulfonic acid)]

*Uses.* The sodium salt of **177** has found some use in the detection of potassium and as a photographic developer (Eikonogen). The *O*-ethyl derivative of **177** is an important middle component for dis- and trisazo green dyes.

1-Amino-2-ethoxynaphthalene-6-sulfonic acid [*118-28-5*] (**178**) can be obtained by ethylation of **177** with diethyl sulfate in methanol [53] or by sulfonation of 1-amino-2-ethoxynaphthalene (**158**; Chap. 7). The older, established route is based on Schaeffer acid and a three-step process involving ethylation, nitration, and reduction.

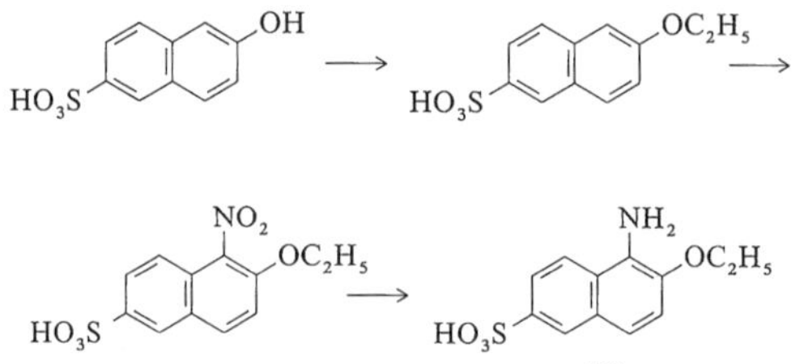

[Scheme: Schaeffer acid → 2-ethoxy derivative → nitration → **178** (1-amino-2-ethoxynaphthalene-6-sulfonic acid)]

Important products derived from **178** are C.I. Direct Greens 23, 33 (**179**), 34, and 38.

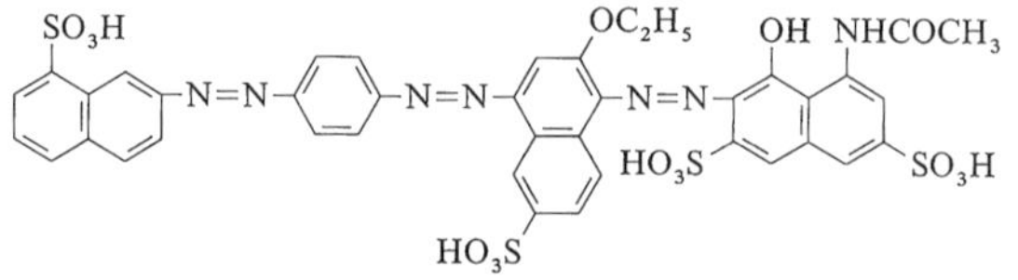

[Structure of **179**: trisazo dye]

**1-Amino-5-hydroxynaphthalene-6-sulfonic acid** [*58596-07-9*] (**180**) (5-amino-1-hydroxynaphthalene-2-sulfonic acid), $C_{10}H_9NO_4S$, $M_r$ 239.2.

*Production.* 5-amino-1-naphthol (Chap. 7) is sulfonated with 96% sulfuric acid at 20 °C, and the sparingly soluble product is isolated after quenching.

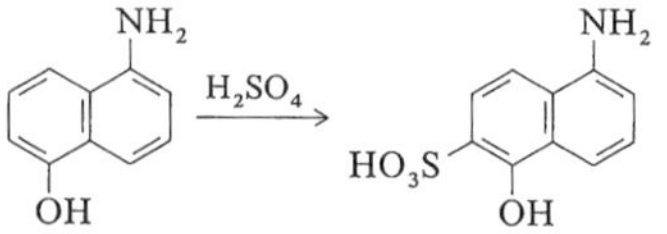

[Scheme: **155** (5-amino-1-naphthol) + $H_2SO_4$ → **180**]

**1-Amino-5-hydroxynaphthalene-7-sulfonic acid** [489-78-1] (**160**) (8-amino-4-hydroxynaphthalene-2-sulfonic acid), M acid, $C_{10}H_9NO_4S$, $M_r$ 239.2, and its sodium salt are sparingly soluble in water. Coupling with diazo compounds takes place under acidic and alkaline conditions, but a bisazo derivative cannot be obtained. Reaction with bisulfite gives 4,8-dihydroxynaphthalene-2-sulfonic acid.

*Production.* 1-Aminonaphthalene-5,7-disulfonic acid (Section 6.4) is heated with concentrated sodium hydroxide for 16 h at 172–176 °C. After the reaction mixture is quenched, acidified, and purged of sulfur dioxide, the product is isolated in 80% yield.

**1-Amino-8-hydroxynaphthalene-4-sulfonic acid** [83-64-7] (**161**) (4-amino-5-hydroxynaphthalene-1-sulfonic acid), S acid, $C_{10}H_9NO_4S$, $M_r$ 239.2, is only sparingly soluble in hot water. Reaction with bisulfite gives 4,5-dihydroxynaphthalene-1-sulfonic acid. Diazo compounds couple under acid conditions α to the amino group and under alkaline conditions α to the hydroxy group; bisazo derivatives can be obtained.

*Production.* 1-Aminonaphthalene-4,8-disulfonic acid (Section 6.4) is heated with concentrated potassium hydroxide at 200–230 °C.

*Uses.* The S acid and its acetyl derivative are used as azo coupling components in e.g., C.I. Mordant Green 14, C.I. Mordant Blue 7, and C.I. Direct Green 39.

**1-Amino-8-hydroxynaphthalene-6-sulfonic acid** [35400-55-6] (**181**) (5-amino-4-hydroxynaphthalene-2-sulfonic acid), $C_{10}H_9NO_4S$, $M_r$ 239.2.

*Production.* 1-Aminonaphthalene-6,8-disulfonic acid gives **181** on fusion with 50% potassium hydroxide at 200 °C, but because the former is not readily available, a more convenient process is the desulfonation of K acid (**167**) by boiling with zinc dust in aqueous sodium hydroxide solution.

**2-Amino-1-hydroxynaphthalene-4-sulfonic acid** [567-13-5] (**182**) (3-amino-4-hydroxynaphthalene-1-sulfonic acid), $C_{10}H_9NO_4S$, $M_r$ 239.2.

*Production.* Although **182** may be obtained from 2-nitroso-1-naphthol by heating with bisulfite, this is less efficient than the corresponding reaction with 1-nitroso-2-naphthol. The preferred method is reduction of the azo dye derived from aniline and 1-hydroxynaphthalene-4-sulfonic acid with sodium dithionite or hydrazine [52].

*Uses.* 2-Amino-1-hydroxynaphthalene-4-sulfonic acid forms the stable quinone–diazide, 3-diazo-3,4-dihydro-4-oxonaphthalene-1-sulfonic acid [20680-48-2] (**183**) when diazotized in the presence of copper(II) sulfate. The stability of **183** and **173**, together with their moderate photosensitivity, led to their use in the earliest diazotype reprographic processes.

**2-Amino-3-hydroxynaphthalene-6-sulfonic acid** [*6399-72-0*] (**162**) (6-amino-7-hydroxynaphthalene-2-sulfonic acid), RM acid, $C_{10}H_9NO_4S$, $M_r$ 239.2, is only sparingly soluble in hot water.

*Production.* 2-Aminonaphthalene-3,6-disulfonic acid (Section 6.5) is fused with 85% potassium hydroxide at 180 °C for 20 h [7]. After quenching and acidification, the product is isolated in 75% yield.

**2-Amino-5-hydroxynaphthalene-7-sulfonic acid** [*87-02-5*] (**163**) (7-amino-4-hydroxynaphthalene-2-sulfonic acid), J acid, $C_{10}H_9NO_4S$, $M_r$ 239.2, is only slightly soluble (0.1%) in cold water but readily soluble at 100 °C. The alkali-metal salts are readily soluble in water. Diazo compounds are normally coupled with J acid under alkaline conditions with reaction in the 6-position (ortho to hydroxyl); however, under acid conditions, coupling can occur in the 1-position. Reaction with bisulfite gives 4,7-dihydroxynaphthalene-2-sulfonic acid. The N-acylation, N-alkylation, and N-arylation reactions are important.

*Production.* A concentrated disodium salt solution of 2-aminonaphthalene-5,7-disulfonic acid (Section 6.5) is added to 72% sodium hydroxide at 150 °C [7]. After heating to 187 °C over 10 h reaction is continued at this temperature for a further 8 h until completion of the reaction is indicated by giving equivalent diazotization and coupling titers for a worked-up sample. Quenching, acidification, purging of sulfur dioxide, and filtration at 55 °C followed by washing with water at 50 °C result in an 88% yield of paste. Purification by salting out the sodium salt from a hot alkaline solution gives a 90% recovery of the product as gray crystals.

An improved process with 90% yield is claimed by feeding the concentrated 2-aminonaphthalene-5,7-disulfonic acid liquor and the caustic liquor simultaneously into the fusion reactor to maintain more stable conditions [54].

*Uses.* The J acid is an important coupling component in a wide range of direct dyes [e.g., C.I. Direct Blue 71 (**184**)] and reactive dyes [e.g., C.I. Reactive Orange 1 and C.I. Reactive Red 9 (**185**)].

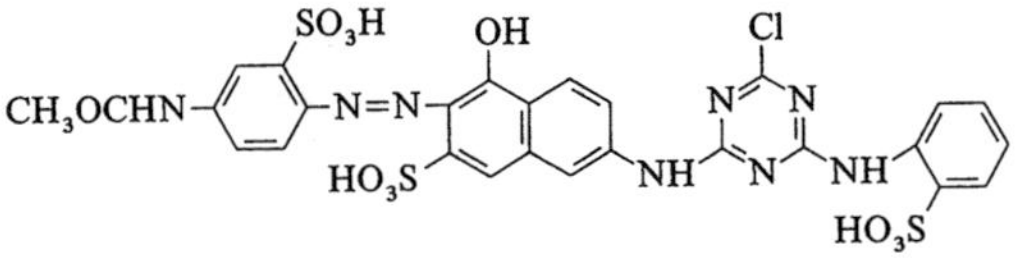

**185**

Many J acid derivatives, e.g., *N*-acetyl, *N*-methyl, *N*-phenyl, di-J acid, and the urea carbonyl J acid, are of comparable importance to J acid itself.

*7-Acetamido-4-hydroxynaphthalene-2-sulfonic acid* [6334-97-0] (**186**), *N*-acetyl J acid, is produced by reaction of the sodium salt of J acid with acetic anhydride in aqueous solution at 20–30 °C, followed by heating at 95 °C with sodium carbonate to hydrolyze any *O*-acetyl derivative present (yield 85%). It is used to produce C.I. Acid Orange 27, C.I. Acid Red 137, and C.I. Direct Red 77.

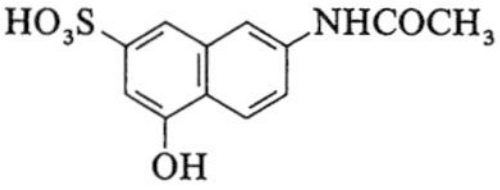

**186**

*4-Hydroxy-7-methylaminonaphthalene-2-sulfonic acid* [22346-43-6] (**187**), *N*-methyl J acid, is produced by reaction of 2,5-dihydroxynaphthalene-7-sulfonic acid (Section 4.7) with methylamine and aqueous sodium bisulfite at 96 °C for 36 h (yield 83%). It is used to produce C.I. Reactive Orange 4, C.I. Reactive Orange 13, and C.I. Reactive Red 33.

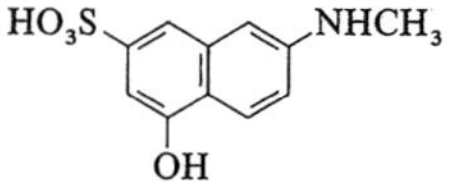

**187**

*4-Hydroxy-7-phenylaminonaphthalene-2-sulfonic acid* [119-40-4] (**188**), *N*-phenyl J acid, is produced by reaction of J acid with aniline in aqueous sodium bisulfite at reflux for 40 h (yield 85%). It is used to produce C.I. Direct Violet 7, C.I. Direct Violet 9, C.I. Direct Violet 51, C.I. Direct Blue 78, and C.I. Direct Blue 168.

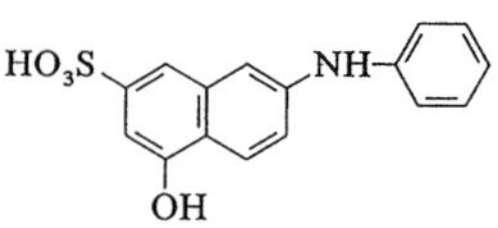

**188**

*7,7′-Iminobis(4-hydroxynaphthalene-2-sulfonic acid)* [87-03-6] (**189**), J acid imide, di-J acid, is produced by heating the sodium salt of J acid in aqueous sodium bisulfite at 106 °C for 36 h (yield 89%). It is used to produce C.I. Direct Reds 31 and 149, as well as C.I. Direct Violets 48, 66, and 102.

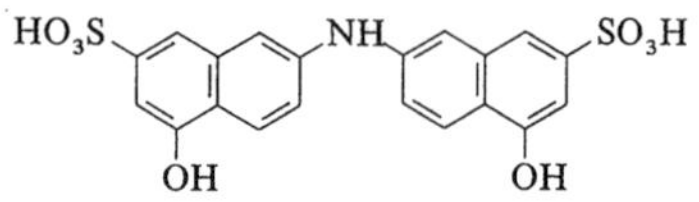

[Structure 189]

*7,7′-Urylenebis(4-hydroxynaphthalene-2-sulfonic acid)* [*134-47-4*] (**190**), J acid urea, carbonyl J acid, is produced by reacting the sodium salt of J acid with phosgene in aqueous solution at 30 °C and maintaining alkalinity with sodium carbonate. After being heated to 80 °C, the product is salted out with sodium chloride (yield ca. 100%). It is used to produce C.I. Direct Orange 26; C.I. Direct Reds 4, 14, 23, and 80; and C.I. Direct Brown 112.

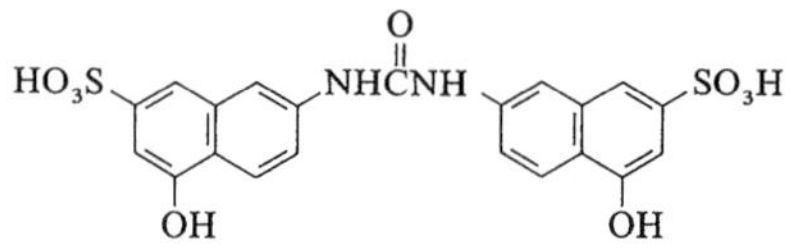

[Structure 190]

The related *N*-chloroacetyl, *N*-benzoyl, *N*-(4-aminobenzoyl), *N*-(3-aminobenzoyl), *N*-(3-methyl-4-aminobenzoyl), *N*-(4-methoxyphenyl), *N*-(3-carboxyphenyl), and *N*-(4-aminophenyl) derivatives are prepared by similar methods but have comparatively minor importance.

**2-Amino-6-hydroxynaphthalene-8-sulfonic acid** [*86-61-3*] (**191**), (7-amino-3-hydroxynaphthalene-1-sulfonic acid), $C_{10}H_9NO_4S$, $M_r$ 239.2.

*Production.* 6-Nitrodiazo-1,2,4-acid (**175**) is heated with aqueous sodium sulfite solution in the presence of copper(II) sulfate. The solution is then acidified and heated further to cause desulfonation. The resulting nitrohydroxynaphthalenesulfonic acid is reduced with iron in the presence of acid [55].

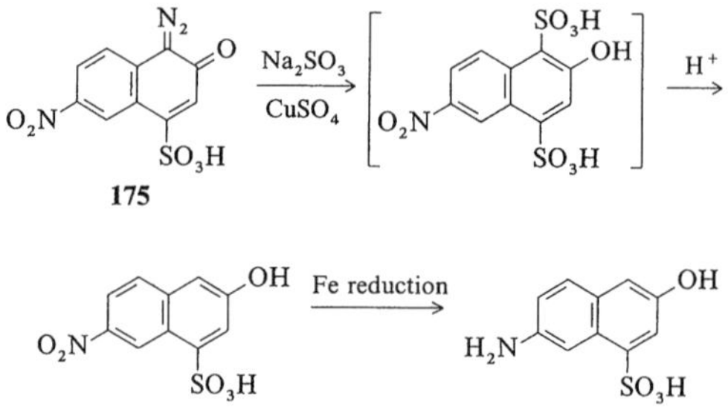

[Reaction scheme showing 175 → intermediate → 191]

This process is preferred to the traditional deamination of **175** with ethanol in the presence of a copper catalyst.

**2-Amino-8-hydroxynaphthalene-6-sulfonic acid** [*90-51-7*] (**164**) (6-amino-4-hydroxynaphthalene-2-sulfonic acid), γ acid, $C_{10}H_9NO_4S$, $M_r$ 239.2, is obtained as white or gray crystals which are only sparingly soluble in water even at 100 °C (0.4%). The sodium salt dissolves readily in water to form solutions that exhibit a blue fluorescence. Coupling with diazo compounds is usually carried out under acid conditions and occurs in the 1-position; coupling in the 7-position can be effected in alkaline solution.

*Production.* 2-Aminonaphthalene-6,8-disulfonic acid liquor (Section 6.5) is added over 16 h to 70% sodium hydroxide at 190 °C, while the water is allowed to evaporate to maintain the temperature [7]. The reaction mixture is then heated for a further 6 h. Completion of the reaction is indicated by a worked-up sample giving identical titers for both diazotization and coupling. The batch is diluted with water, acidified, and heated at 75 °C to expel sulfur dioxide. The product is filtered off at 55 °C and washed with warm water; yield is 84%.

An improved process with 90% yield is claimed by feeding concentrated 2-aminonaphthalene-6,8-disulfonic acid dipotassium salt liquor and caustic soda liquor simultaneously into the fusion reactor [54].

*Uses.* The so-called γ acid is an important component in azo dyes; coupling can be carried out under acid or alkaline conditions. The *o*-amino azo configuration obtained under acid conditions leads to high lightfastness of special importance in monoazo acid red dyes for wool and polyamide, such as C.I. Acid Reds 37, 57, and 266 (**192**). Disazo direct dyes are also obtained by acid coupling (e.g., C.I. Direct Red 75) but trisazo blacks, exemplified by C.I. Direct Black 80 (**193**), which use γ acid as an alkali-coupled end and middle component, are of more significance as replacements for benzidine-based black direct dyes.

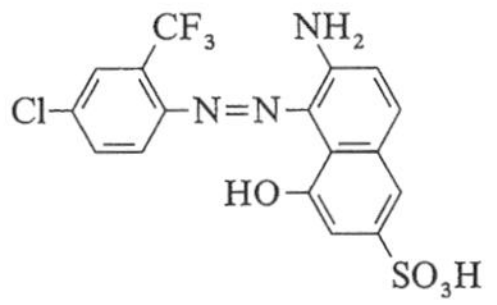

**192**

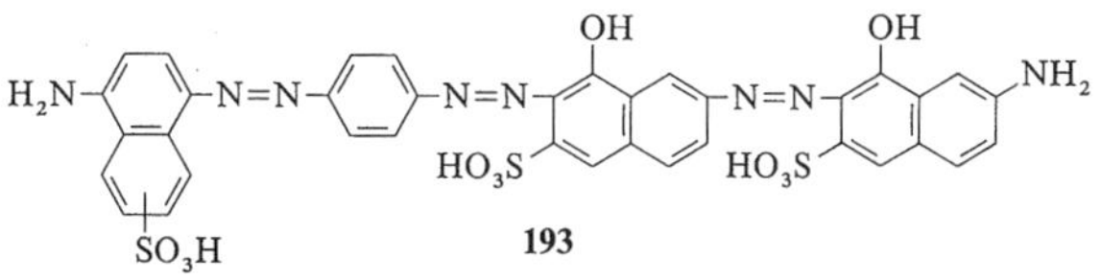

**193**

As with J acid, the *N*-acetyl and *N*-methyl derivatives are also important.

*6-Acetamido-4-hydroxynaphthalene-2-sulfonic acid* [6361-41-7] (**194**), *N*-acetyl γ acid, is produced by reaction of the sodium salt of γ acid with acetic anhydride in aqueous solution at 20 – 30 °C (yield 87%). It is used to produce C.I. Acid Red 68, C.I. Reactive Orange 7, and C.I. Reactive Orange 16.

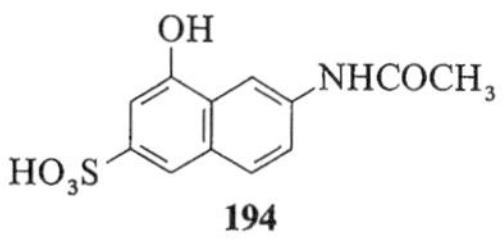

**194**

*4-Hydroxy-6-methylaminonaphthalene-2-sulfonic acid* [6259-53-6] (**195**), *N*-methyl γ acid, is produced by reaction of 2,8-dihydroxynaphthalene-6-sulfonic acid (Section 4.7) with methylamine in aqueous sodium bisulfite at 96 °C for 36 h (yield 95%). It is used to produce C.I. Acid Brown 91.

Other substituted γ acids that have been similarly used in specialized azo dyes include the *N*-β-hydroxymethyl, *N*-phenyl, *N*-(4-methoxyphenyl), *N*-(4-carboxyphenyl), *N*-(3-sulfophenyl), and *N*-(4-amino-3-sulfophenyl) derivatives.

**2-Amino-8-hydroxynaphthalene-7-sulfonic acid** (Na salt, [74525-31-8]) (**196**) (7-amino-1-hydroxynaphthalene-2-sulfonic acid), $C_{10}H_9NO_4S$, $M_r$ 239.2.

*Production.* 7-Amino-1-naphthol (Chap. 7) is sulfonated at 0 °C with 96% sulfuric acid to yield 65% of **196** together with 16% of the isomeric 2-amino-8-hydroxynaphthalene-5-sulfonic acid [21013-47-8] (**197**). Separation by fractional precipitation of the sodium salt is not easy, but should the latter isomer be required the sulfonation temperature is increased to 30 °C, where-upon the product ratio changes from 65:16 to 50:35.

## 8.3. Aminohydroxynaphthalenedisulfonic Acids

**1-Amino-2-hydroxynaphthalene-3,6-disulfonic acid** [2007-20-7] (**198**) (4-amino-3-hydroxynaphthalene-2,7-disulfonic acid), $C_{10}H_9NO_7S_2$, $M_r$ 319.3.

*Production.* R acid (Section 4.5) is coupled with diazotized aniline, and the isolated azo product is then reduced with sodium dithionite by a process similar to that employed for other *o*-aminonaphthols.

**1-Amino-8-hydroxynaphthalene-2,4-disulfonic acid** [82-47-3] (**165**) (4-amino-5-hydroxynaphthalene-1,3-disulfonic acid), Chicago acid, 2 S acid, $C_{10}H_9NO_7S_2$, $M_r$ 319.3, and its acid sodium salt are readily soluble in water; the latter can be salted out with sodium chloride. In alkaline solution, coupling with diazo compounds takes place ortho to the hydroxyl group. Alkaline hydrolysis at 170 °C gives 1,8-dihydroxynaphthalene-2,4-disulfonic acid. Heating with dilute sulfuric acid results in desulfonation to give 1-amino-8-hydroxynaphthalene-2-sulfonic acid.

*Production.* A solution of the disodium salt of naphthosultam-2,4-disulfonic acid (**142**; Section 6.6) is concentrated and mixed with potassium hydroxide liquor at 130 °C [7]. The temperature is raised to 155 °C over 20 h to complete the reaction. After being cooled and diluted with water, the melt is run into excess hydrochloric acid and the monopotassium salt is collected by filtration at 30 °C and slurry washed with recycling of the wash liquors. A yield of 87 % is obtained.

*Uses.* Chicago acid is used as a coupling component for C.I. Direct Blue 1 (**199**), C.I. Direct Blue 22, and C.I. Direct Blue 76 (the dicopper complex of demethylated **199**).

**1-Amino-8-hydroxynaphthalene-3,6-disulfonic acid** [90-20-0] (**166**) (4-amino-5-hydroxynaphthalene-2,7-disulfonic acid), H acid, $C_{10}H_9NO_7S_2$, $M_r$ 319.3, is usually isolated as its monosodium salt [5460-09-3], a gray powder that is slightly soluble in cold water and readily soluble in hot water; the disodium salt is very soluble in water. With diazo compounds, H acid couples ortho to the amino group under acid conditions and ortho to the hydroxyl group under alkaline conditions. Bisazo derivatives are formed by stepwise acid and alkaline coupling, with the much slower acid coupling being performed first.

*Production.* 1-Aminonaphthalene-3,6,8-trisulfonic acid (Section 6.6) as the trisodium salt liquor is heated with aqueous sodium hydroxide in an autoclave [7]. The temperature is raised gradually from 150 to 182 °C, and heating is continued until the test for completion (strength by coupling equals strength by diazotization) is satisfied. The melt is diluted, added to excess dilute sulfuric acid, heated at 95 °C to expel sulfur dioxide, cooled, and filtered. The paste is washed and dried for ease of handling. The yield obtained is 72 %.

Given a yield from naphthalene of 62 % for Koch acid, the maximum overall yield of H acid from naphthalene is thus 45 %. Large-scale H acid production uses naphthalene as starting material, and the through process to Koch acid involves the three process stages of sulfonation, nitration, and reduction. These all require optimization with a view to yield and purity of the ultimate H acid [56]. At the fusion stage, a higher temperature (200 °C) may be used to shorten the reaction period [57]. In a fully integrated process, many of the stage parameters are interrelated [58]. Equally important to the fusion reaction conditions are the work-up conditions to achieve good physical form, a major factor in final quality. The H acid is precipitated as the monosodium salt which has a strong tendency to be thixotropic and is hence difficult to wash.

*Uses.* Like certain other aminonaphthol derivatives, H acid has been claimed to be a useful corrosion inhibitor. The versatility of H acid as a coupling component is exemplified by alkaline coupling to give monoazo reds such as C.I. Acid Red 33 (**200**) and C.I. Reactive Red 12 (**201**). Even more important, on a volume basis, are disazo blues and blacks such as C.I. Direct Blue 15, C.I. Direct Blue 218, C.I. Acid Black 1, C.I. Direct Black 19, and C.I. Reactive Black 5 (**202**).

**202**, X = SO$_2$CH$_2$CH$_2$OSO$_3$H

Some N-substituted derivatives of H acid are isolated for more specialized uses. However, in situ N-acylation is more common with agents such as *p*-toluenesulfonyl chloride or fiber-reactive systems containing an acyl group or equivalent (e.g., as in **201**).

*4-Acetylamino-5-hydroxynaphthalene-2,7-disulfonic acid* [134-34-9] (**203**), *N*-acetyl H acid, is produced by reacting H acid in aqueous alkaline solution with acetic anhydride at 70 °C for 1 h, followed by heating at 95 °C for 2 h to hydrolyze the *O*-acetyl derivative (yield 88%). It is used to produce C.I. Acid Red 1, C.I. Acid Red 35, and C.I. Acid Violet 7.

$$\text{OH} \quad \text{NHCOCH}_3$$
$$\text{HO}_3\text{S} \quad\quad \text{SO}_3\text{H}$$
**203**

*4-Benzamido-5-hydroxynaphthalene-2,7-disulfonic acid* [117-46-4] (**204**), *N*-benzoyl H acid, is produced by reaction of H acid with benzoyl chloride in aqueous alkaline sodium carbonate solution at 35 °C for 5 h, followed by heating at 95 °C for 2 h to hydrolyze the *O*-benzoyl derivative (yield 80%). It is used to produce C.I. Reactive Red 4 and C.I. Direct Green 13.

$$\text{OH} \quad \text{NHOC-C}_6\text{H}_5$$
$$\text{HO}_3\text{S} \quad\quad \text{SO}_3\text{H}$$
**204**

**l-Amino-8-hydroxynaphthalene-4,6-disulfonic acid** [130-23-4] (**167**) (4-amino-5-hydroxynaphthalene-1,7-disulfonic acid), K acid, $C_{10}H_9NO_7S_2$, $M_r$ 319.3: the acid sodium salt of **167** is very soluble in water, but the acid potassium salt is less soluble and is more easily salted out. K acid couples under acid and alkaline conditions in a similar manner to H acid.

*Production.* 1-Aminonaphthalene-4,6,8-trisulfonic acid paste (Section 6.6) is heated with sodium hydroxide liquor in an autoclave at 170 °C for 12 h, after which the mixture is added to excess dilute hydrochloric acid. After heating to remove sulfur dioxide, the product (including ca. 8% of the isomeric 5-amino-2-hydroxynaphthalene-4,8-disulfonic acid) is completely precipitated by cooling and is isolated in 86% yield. Purification of crude K acid may be effected by dissolving it in aqueous alkali and filtering off the less soluble 5,2,4,8-isomer before reprecipitating.

*Uses.* The N-substituted derivatives of K acid, such as *N*-acetyl, *N*-benzoyl, *N*-(2,5-dichlorobenzoyl), and *N*-benzyl, are used as monoazo coupling components in special cases where an improvement in dye properties over the H acid analogue is demonstrable.

*4-Benzamido-5-hydroxynaphthalene-1,7-disulfonic acid* [6361-49-5] (**205**), *N*-benzoyl K acid, is produced by reacting purified K acid with benzoyl chloride in aqueous alkaline sodium carbonate solution at 35–40 °C for 18 h, followed by heating for 1 h at 90 °C to hydrolyze the *O*-benzoyl derivative (yield 80%). It is used to produce C.I. Acid Red 133, C.I. Acid Red 157, and C.I. Mordant Red 21.

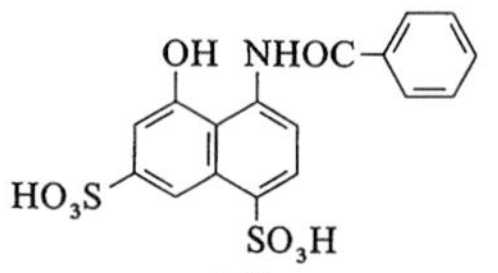

**205**

**2-Amino-5-hydroxynaphthalene-1,7-disulfonic acid** [*6535-70-2*] (**168**), sulfo J acid, $C_{10}H_9NO_7S_2$, $M_r$ 319.3.

*Production.* 2-Aminonaphthalene-1,5,7-trisulfonic acid (Section 6.6) is heated with sodium hydroxide liquor at 170 °C and worked up in a similar way to the K acid. Alternatively, J acid may be sulfonated with 65% oleum.

*Uses.* Sulfo J acid is used as a coupling component in azo reactive dyes, e.g., C.I. Reactive Red 6 (the copper complex of **206**).

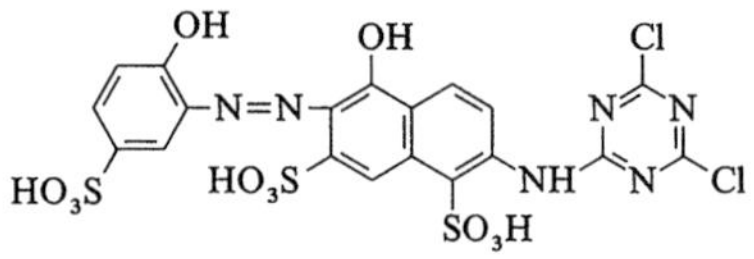

**206**

**2-Amino-5-hydroxynaphthalene-4,8-disulfonic acid** [*74832-35-2*] (**207**) (3-amino-8-hydroxynaphthalene-1,5-disulfonic acid), oxy C acid, $C_{10}H_9NO_7S_2$, $M_r$ 319.3.

*Production.* 3,8-Diaminonaphthalene-1,5-disulfonic acid (Section 6.8) is monoacetylated in aqueous solution, and the free amino group is diazotized. The diazo compound is added to dilute sulfuric acid at 100 °C to hydrolyze both the diazo and the acetylamino groups [59].

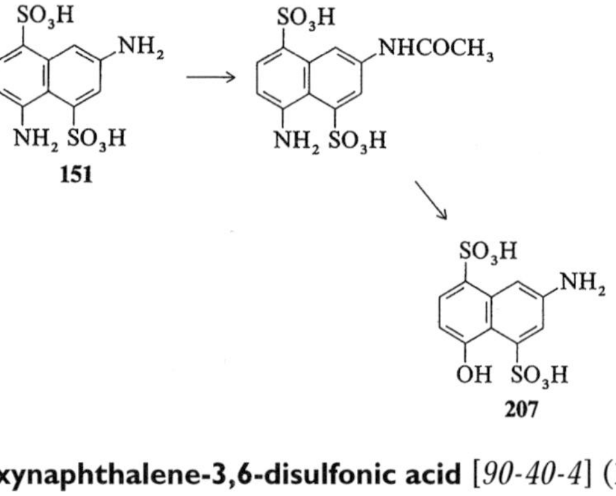

**151**            **207**

**2-Amino-8-hydroxynaphthalene-3,6-disulfonic acid** [*90-40-4*] (**169**) (3-amino-5-hydroxy-2,7-disulfonic acid), RR acid, 2 R acid, Columbia acid, $C_{10}H_9NO_7S_2$, $M_r$ 319.3, and its alkali salts are readily soluble in water. Coupling with diazo compounds takes place under alkaline conditions in the position ortho to the hydroxyl group.

*Production.* 2-Aminonaphthalene-3,6,8-trisulfonic acid liquor (Section 6.6) is concentrated by evaporation and heated with sodium hydroxide liquor in an autoclave [7]. The temperature is raised to 195 °C over 10 h and held at 195 °C for another 8–10 h until completion of the reaction is determined by working up a test sample and obtaining a strength by alkaline coupling equal to the diazotization titer. After dilution, acidification, and stirring while hot until all sulfur dioxide is expelled, the product is filtered off as its monosodium salt at 55 °C and washed with brine. Most of the isomeric 2-amino-3-hydroxynaphthalene-6,8-disulfonic acid impurity is removed in the filtrate. Crude RR acid is obtained in 65 % yield and is suitable for certain outlets. Further purification is necessary for critical derived dyes, and this is achieved by dissolving in dilute alkali, filtering at 80 °C, and reprecipitating by addition of sodium chloride and hydrochloric acid. Filtration at 55 °C gives 90 % recovery of product.

*Uses.* The RR acid is used as a coupling component in azo dyes, e.g., C.I. Food Black 2 and C.I. Direct Brown 151.

# 9. References

General References

[1] H. R. Schweizer: *Künstliche organische Farbstoffe und ihre Zwischenprodukte*, Springer Verlag, Berlin 1964.
[2] *Houben-Weyl*, 4th ed., **9**, 474–492.
[3] E. N. Abrahart: *Dyes and their Intermediates*, Arnold, London 1977.
[4] E. H. Rodd: *Chemistry of Carbon Compounds*, 2nd ed., vol. **3 G**, Elsevier, Amsterdam 1978.
[5] N. Donaldson: *The Chemistry and Technology of Naphthalene Compounds*, Arnold, London 1958.
[6] G. Booth: *The Manufacture of Organic Colorants and Intermediates*, Society of Dyers and Colourists, Bradford 1988.
[7] BIOS, FIAT reports: German dyestuffs and dyestuffs intermediates, including manufacturing processes, plant design and research data 1945–1948.
[8] H. E. Fierz-David, L. Blangey: *Fundamental Processes of Dye Chemistry*, Interscience, New York 1949.
[9] K. Venkataraman (ed.): *The Chemistry of Synthetic Dyes*, vol. 1, Academic Press, New York 1952.
[10] H. A. Lubs (ed.): *The Chemistry of Synthetic Dyes and Pigments*, Reinhold Publ. Co., New York 1954.
[11] H. Hiyama: "Medicines and Agricultural Chemicals Derived from Naphthalene," *Kagaku to Kogyo (Osaka)* **54** (1980) 238–247.
[12] M. Kodama, N. Sakota: "Recent Advances in Application and Synthesis of Naphthalene Compounds," *Yuki Gosei Kagaku Kyokaishi* **39** (1981) 960–972.
[13] H-G. Franck, J. W. Stadelhofer: *Industrial Aromatic Chemistry*, Springer-Verlag, Berlin 1988.
[14] N. I. Sax, R. J. Lewis: *Dangerous Properties of Industrial Materials*, 7th ed., vol. **3**, Van Nostrand Reinhold, New York 1989.
[15] *Ullmann*, 4th ed., **17**, 83. 3rd ed., **12**, 617.
*Ullmann*, 3th ed., **12**, 617.

Specific References

[16] G. Dittmar, *Chem. Rundsch.* **30** (1977) no. 46, 23.
[17] H. Bretscher, G. Eigenmann, E. Plattner, *Chimia* **32** (1978) 180.
[18] Colour Index, vol. 4, 3rd ed., Society of Dyers and Colourists, Bradford 1971.
[19] Colour Index, vol. 6 (1975); vol. **7** (1982), Society of Dyers and Colourists, Bradford.
[20] J. J. Kirkland: *Modern Practice of Liquid Chromatography*, Wiley-Interscience, New York 1971.
[21] L. R. Snyder, J. J. Kirkland: *Introduction to Modern Liquid Chromatography*, Wiley-Interscience, New York 1979.
[22] A. A. Spryskov, N. A. Orsyankina, *J. Gen. Chem. USSR (Engl. Transl.)* **20** (1950) 1043.
[23] H. E. Fierz-David, C. Richter, *Helv. Chim. Acta* **28** (1945) 257.
[24] Sumitomo Chem. Co. Ltd., JP 7 3 38 699, 1970 (T. Ikeda et al.).
[25] Bayer, DE 2 728 070, 1977 (H. U. Alles et al.).
[26] K. H. Saunders, R. L. M. Allen: *Aromatic Diazo Compounds*, Arnold, London 1985.
[27] Union Carbide, US 3 356 743, 1963; US 3 378 591, 1964 (B. T. Freure).
[28] F. A. Yukhro et al., Deposited Doc. 1981, SPSTL 566 Khp-D 81, *Chem. Abstr.* **98** 160 364.
[29] Stauffer, US 3 480 671, 1967 (H. Tilles et al.).
[30] Union Carbide, US 2 903 478, 1959 (J. A. Lambrech).
[31] ICI, GB 994 918; 995 800, 1962 (A. F. Crowther, L. H. Smith).
[32] R. Kuhn, F. Bar, *Justus Liebigs Ann. Chem.* **516** (1935) 143.
[33] Amer. Cyanamid, US 3 504 045; 3 504 046, 1968 (E. J. Scharf, H. R. Kemme).
[34] Japan Soda Co., FR 1 337 797, 1961; *Chem. Abstr.* **60** 2871.
[35] F. Seidel et al., *J. Prakt. Chem.* **2** (1954) 53. M. Scalera, H. Z. Lecher, *J. Prakt. Chem.* **3** (1956) 232.
[36] A. S. Lindsey, H. Jeskey, *Chem. Rev.* **57** (1957) 583.
[37] Henkel and Cie, DE 1 203 281, 1962.
[38] J. Bailey, L. A. Williams in K. Venkataraman (ed.): *Chemistry of Synthetic Dyes*, vol. **IV**, Academic Press, New York 1971.
[39] Sugai Chemical, JP 7 0 10 935, 1966 (H. Hiyama et al.).
[40] Bayer, DE 1 943 543, 1969 (C. P. Mertig).
[41] Amer. Cyanamid, DE 2 346 459, 1972 (M. L. Feldman et al.).
[42] U.S. Rubber, US 2 556 665, 1951 (A. E. Smith, O. L. Hoffman).
[43] Japan Soda, JP 6 7 17 016, 1964 (S. Kano); *Chem. Abstr.* **68** (1968) 29 479.
[44] *IARC Monograph on the Evaluation of Carcinogenic Risk of Chemicals to Man*, vol. **4**, IARC, Lyon 1974.
"Some Aromatic Amines and Azo Dyes in the General and Industrial Environment," *Environmental Carcinogens Selected Methods of Analysis*, vol. **4**, IARC, Lyon 1981.
[45] Y. Masuda, D. Hoffman, *Anal. Chem.* **41** (1969) 650.
[46] Bayer, DE 2 517 437-9, 1975 (H. U. Blank et al.).
[47] CFMC, FR 1 075 110, 1954 (J. Dassigny). CFMC, FR 1 108 109, 1956 (J. Robin).
[48] H. Koepernik, R. Borsdorf, *J. Prakt. Chem.* **325** (1983) 1002.
[49] Ciba-Geigy, DE 2 856 567, 1977 (B. Albrecht).
[50] Bayer, DE 2 703 076; 2 747 714, 1977 (O. Barth et al.).
[51] M. A. Marchisio et al., *Br. J. Ind. Med.* **33** (1976) 269.
[52] Taoka Chem. Co., JP 76 146 447, 1975 (S. Kawasaki et al.).
[53] Bayer, DE 3 010 372, 1980 (R. Puetter et al.).
[54] Ciba-Geigy, DE 2 813 570, 1978 (H. Breitschmid).

[55] Geigy, US 2 124 070, 1938 (A. Krebser, F. Vamotti).
[56] T. Hayashi, *Yamaguchi Daigaku Rigakkai Shi* **2** (1951) 67; *Chem. Abstr.* **49** (1955) 2390.
[57] Bayer, DE 2 716 030, 1977 (H. Behre et al.).
[58] Sumitomo, DE 2 727 345, 1977 (N. Kotera et al.).
[59] Bayer, DE 1 117 235, 1959 (K. H. Schuendehuette et al.).

# Naphthoquinones

JOHANN GROLIG, Bayer AG, Leverkusen, Federal Republic of Germany
RUDOLF WAGNER, Bayer AG, Leverkusen, Federal Republic of Germany

| | | | | | |
|---|---|---|---|---|---|
| 1. | Introduction | 3447 | 2.4. | Derivatives | 3450 |
| 2. | 1,4-Naphthoquinone | 3448 | 3. | Other Naphthoquinones | 3451 |
| 2.1. | Properties | 3448 | 4. | Naphthodiquinones | 3452 |
| 2.2. | Production | 3448 | 5. | Polynaphthoquinone | 3452 |
| 2.3. | Uses | 3449 | 6. | References | 3452 |

## 1. Introduction

Of the five possible naphthoquinones, the isomers (**1**)–(**3**) have been synthesized in a pure form. Of these, 1,4-naphthoquinone (**1**) is produced on an industrial scale as an intermediate for anthraquinone. 1,5-Naphthaquinone (**4**) is known only in the form of its derivatives. The existence of 2,3-naphthoquinone (**5**) was confirmed by trapping it as its Diels–Alder adduct with cyclopentadiene. The synthesis of the unstable 1,7-isomer or of its derivatives has to date not been possible. In contrast, polynaphthoquinone (**8**) is a stable derivative of 1,7-naphthoquinone. The naphthoquinones (**6**) and (**7**) have also been synthesized. The industrially most important compounds are 1,4-naphthoquinone and 2-methyl-1,4-naphthoquinone.

## 2. 1,4-Naphthoquinone

### 2.1. Properties

1,4-Naphthoquinone [*130-15-4*], α-naphthoquinone, 1,4-naphthalenedione, $M_r$ 158.16, *mp* 128.5 °C, $\varrho$ 1.42 g/cm$^3$, crystallizes as yellow needles. It has a penetrating odor, is steam volatile, and sublimes readily. It is insoluble in water but soluble in warm alcohols and hydrocarbons. Phenols and arylamines react at the 2-position. In acids and upon heating it condenses to form oligomers [1] and polymers. Its most important reaction is with butadiene to give tetrahydroanthraquinone, which is oxidized to anthraquinone (see Section 2.3).

### 2.2. Production

**Gas-Phase Oxidation of Naphthalene.** The most important process for the production of 1,4-naphthoquinone is the oxidation of naphthalene with air or gases containing molecular oxygen at 250–450 °C and 0.1–1 MPa, over a catalyst containing vanadium pentoxide.

Kawasaki Kasei Chemicals has been operating an anthraquinone plant with this process in Japan since 1980. Besides vanadium pentoxide, the catalyst also contains potassium sulfate, potassium pyrosulfate, and silicon dioxide [2], [3]. The latter also serves as a carrier and is obtained by treatment of aluminosilicates with acid [4]. As the naphthalene is converted nearly quantitatively, a recycling step is not necessary; 36 mol % of 1,4-naphthoquinone and 54 mol% of phthalic anhydride are obtained.

The reaction is best performed under a slightly positive pressure of 0.3 – 0.8 MPa [5]. Although the oxidation is normally carried out in a fixed bed, a flow-bed variant has also been described [6]. Lifetime and selectivity of the catalysts are increased by the addition of sulfur-containing compounds [7]. Phthalic anhydride is always obtained in substantial amounts as a byproduct.

In the Kawasaki process the oxidation product is first treated with water in order to separate the phthalic anhydride as phthalic acid from the reaction mixture [8]. The naphthoquinone is extracted with xylene and the extract washed with dilute alkali to remove residual phthalic acid [9]. Then the naphthoquinone is crystallized from the solvent. High-purity naphthoquinone can be obtained by recrystallizing from xylene [10], or by sublimation [11]. Purification of the quinone can also be achieved by treating the hot xylene solutions with lime [12]. In some applications it is possible to use the crude naphthoquinone in form of the oxidation mixture.

**Liquid-Phase Oxidation.** 1,4-Naphthoquinone can alternatively be obtained in the liquid phase in greater than 90% yield by oxidation of naphthalene or its derivatives. The following oxidizing agents have been used on a preparative scale: chromic acid [13], chromium trioxide [14], lead dioxide [15], manganese(II) sulfate [16], ammonium peroxodisulfate [17] – [19], tetrabutylammonium nitrate in nitric acid [20], and cerium(IV) compounds [21] – [27]. The oxidation with cerium(IV), in particular, gives high naphthoquinone selectivities, with yields as high as 98% [23]. Phase-transfer catalysts increase the rate of reaction [19], [20], [27]. Chemical oxidizing reagents can be regenerated electrochemically. The regeneration of cerium(IV) compounds has been extensively investigated in a pilot plant [26]. 1-Naphthol can be catalytically oxidized to 1,4-naphthoquinone with oxygen in the presence of cobalt complexes (Salcomin) in yields of 90% [28].

## 2.3. Uses

1,4-Naphthoquinone is industrially significant as an intermediate in the anthraquinone synthesis by Kawasaki (→ Anthraquinone) [29] – [30]. Starting from naphthalene, naphthoquinone is obtained by gas-phase oxidation, as described in Section 2.2. In the second step it is converted to 1,4,4a,9a tetrahydroanthraquinone [*56136-14-2*] with butadiene in a Diels – Alder reaction [29], [30]. The Diels – Alder adduct is extracted as its sodium salt with dilute alkali. Unconverted naphthoquinone in the organic phase is fed back into the Diels – Alder reaction. In the third step the basic tetrahydroanthraquinone solution is oxidized with air to anthraquinone, which forms a precipitate and is filtered off. Kawasaki produces 3000 t/a anthraquinone with this process (→ Anthraquinone).

In a process developed by Bayer, the 1,4-naphthoquinone is not isolated after the naphthalene oxidation step. Instead, it is reacted with butadiene and oxidized to anthraquinone in the mixture of naphthalene and phthalic anhydride. The products are then separated by distillation [31]–[34].

1,4-Naphthoquinone has been described as a polymerization regulator [35] and as a hardener for photochemically cross-linked polyesters [36]. It is used as a corrosion inhibitor [37] and as a stabilizer for transformer oils [38]. It reacts with cyclic aryl phosphinic acid esters to give 1,4-naphthalenediol adducts, which can be used for the production of thermally stable, flame-resistant polyesters [39].

## 2.4. Derivatives

**2-Methyl-1,4-naphthoquinone** [58-27-5], menadione 2-methyl-1,4-naphthalenedione, vitamin K3, $M_r$ 172.20, *mp* 105–106 °C is two to three times more effective as a blood coagulator than vitamin K1 (2-methyl-3-phytyl-1,4-naphthoquinone). It is obtained from 2-methylnaphthalene in 70–90% yield by oxidation with chromic acid [40], [41]. Slightly lower yields are obtained with chromium trioxide in conjunction with crown ethers [42]. The gas-phase oxidation of 2-methylnaphthalene results in maximally 40% yields of the quinone [43].

**2-Chloromethyl-1,4-naphthoquinone** [43027-41-4], 2-chloromethyl-1,4-naphthalenedione, and its derivatives are highly potent cytostatic compounds. It is obtained by the oxidation of 2-chloromethylnaphthalene with chromium trioxide [44].

**2,3-Dichloro-1,4-naphthoquinone** [117-80-6], 2,3-dichloro-1,4-naphthalenedione, dichlone, phygon, $M_r$ 227.05, *mp* 193 °C, is an important intermediate in the synthesis of dyes. It can also be converted to pigments of exceptional stability, which are suitable for integration into high performance membranes [45]. It is also used for conserving grain, as an algicide [46], and as an additive in dye binders [47]. This quinone is obtained by oxidation of 4-hydroxynaphthalene-1-sulfonic acid [84-87-7] with potassium chlorate in hydrochloric acid [48], or by direct chlorination of 1,4-naphthoquinone [49]–[51]. Yields of up to 96% can be obtained by the latter method if the chlorination is carried out in carbon tetrachloride at 50 °C [51].

**5-Nitro-1,4-naphthoquinone** [17788-47-5], 5-nitro-1,4-naphthalenedione, $M_r$ 203.15, *mp* 167 °C, is obtained in 70% yield by nitration of 1,4-naphthoquinone with a mixture of highly concentrated nitric and sulfuric acid [52]; 6-nitro-1,4-naphthoquinone [58200-82-1] is formed as a byproduct. Nitronium tetrafluoroborate ($NO_2^+BF_4^-$) is also a suitable nitrating reagent [53]. The 5-nitro-1,4-naphthoquinone is also accessible by the electrochemical oxidation of 1-nitronaphthalene in a mixture of a mineral acid and a water-soluble organic solvent [54]. Nearly quantitative yields can be obtained through the reduction of 5-nitro-1,4-

naphthoquinone monooxime with sulfur dioxide and phosphorus in the presence of hydrogen iodide, followed by oxidation of the resulting isomeric 1-aminonitronaphthols [55].

The reaction of 5-nitro-1,4-naphthoquinone with butadiene and subsequent oxydehydrogenation leads to 1-nitroanthraquinone [82-34-8] [56]–[58], which is the precursor of the important dye intermediate 1-aminoanthraquinone [25620-59-1].

**Hydroxy Derivatives.** A few monohydroxy derivatives of 1,4-naphthoquinone occur in nature as dyes in plants: for example, 5-hydroxy-1,4-naphthoquinone [481-39-0] (juglone) and 2-hydroxy-1,4-naphthoquinone [83-72-7] (lawsone). 2-Alkyl-3-hydroxy-1,4-naphthoquinones are used as antimalarial agents [59] and insecticides [60]. Derivatives of naphthazarin (5,8-dihydroxy-1,4-naphthoquinone) [475-38-7] are dyes (vat dyes, alkannins). Naphthazarin can be made from 1,5-dinitronaphthalene [605-71-0] by conversion with sulfur and sulfuric acid [61] via the intermediate 5-hydroxy-8-amino-1,4-naphthoquinone [68217-36-7]. Alternatively, it can be prepared from the reaction of dichloromaleic anhydride [1122-17-4] with 1,4-dimethoxybenzene [62]. The resulting 2,3-dichloronaphthazarin [14918-69-5] is converted to 2,3-dihydronaphthazarin [4988-51-6] with tin(II) chloride in hydrochloric acid in the second step and then converted to naphthazarin in alkaline solution with an overall yield of 66%.

# 3. Other Naphthoquinones

**1,2-Naphthoquinone** [524-42-5], α-naphthoquinone, 1,2-naphthalenedione, $M_r$ 158.16, mp 145–147 °C, crystallizes as yellow needles, is odorless, is involatile in steam, and is a relatively weak oxidizing agent. It is obtained in 80% yield when 2-naphthol is oxidized by air in the liquid phase with copper(I) chloride as catalyst [63]. Slightly lower yields are obtained when periodic acid and dimethyl formamide [64], iodosobenzene [65], or benzene seleninic anhydride is used as oxidant [66]. A 95% yield can, however, be obtained with tetramethyloxopiperidinium chloride [26864-01-7] at low temperature [67]. The quinone can also be obtained by the oxidation of 1-amino-2-hydroxynaphthalene hydrochloride [1198-27-2] with iron(III) chloride in hydrochloric acid [68].

Only the derivative 6-bromo-1,2-naphthoquinone [6954-48-9] (bonaphthon), which exhibits activity against type A influenza viruses, is of any practical interest [69].

**2,6-Naphthoquinone** [613-20-7], $M_r$ 158.16, mp 135 °C (decomp.), forms red prisms. It is a stronger oxidizing agent than the other naphthoquinones. It is obtained by the oxidation of 2,6-naphthalenediol [581-43-1] with lead dioxide in benzene [70].

**1,5-Naphthoquinone** [51583-62-1] is not known in the pure form. Its derivative, 3,7-di-*t*-butyl-1,5-naphthoquinone [54532-93-33], is obtained in virtually quantitative yield by oxidizing the corresponding hydroquinone with dichlorodicyano-*p*-benzoquinone

[71]. The naphthazarin isomer 4,8-dihydroxy-1,5-naphthoquinone [6251-03-2] is a starting material for thermally stable dyes [72].

**2,3-Naphthoquinone** [4939-92-8] has also not been isolated. The existence of this unstable quinone was verified by trapping it as the Diels–Alder adduct with cyclopentadiene during the oxidation of 2,3-naphthalenediol with potassium periodate [73].

**1,7-Naphthoquinone** [46001-16-5] is also very unstable and has not been isolated. The derivative, 3,6-di-*t*-butyl-8-methyl-1,7-naphthoquinone [83021-64-1], can be obtained in impure form in solution during the oxidation of the corresponding hydroquinone with dichlorodicyano-*p*-benzoquinine [74].

# 4.  Naphthodiquinones

**1,2,3,4-Naphthodiquinone** [30266-58-1], 1,2, 3,4-naphthalenetetrone, oxolene, $M_r$ 188.10, *mp* 131 °C, $\varrho$ 1.45 g/cm$^3$, forms colorless crystals. It is obtained from 2,3-dihydroxy-1,4-naphthoquinone through oxidation with nitric acid in glacial acetic acid [75].

**1,4,5,8-Naphthalenetetrone** [23077-93-2], $M_r$ 188.10, *mp* 220 °C, $\varrho$ 1.45 g/cm$^3$, is obtained as pale yellow prisms by the oxidation of 5,8-dihydroxy-1,4-naphthoquinone with lead tetraacetate [76] or with iodophenyl-bis(trifluoroacetate) in 81 % yield [77].

# 5.  Polynaphthoquinone

Polynaphthoquinone [38830-94-3], 1,7-naphthalenedione polymer, is a black powder which melts above 360 °C and sublimes at 450 °C. It is made by oxidizing 1,7-naphthalenediol [575-38-2] with dilute nitric acid in ethanol, followed by heating in a stream of air at 350 °C [78], [79]. The polyquinone is suitable as a catalyst for the oxydehydration of alkanes, alkenes, and alcohols. Its reactivity can be greatly enhanced through the addition of Lewis acids [78], [79].

# 6.  References

[1]   H. E. Höberg, *Acta Chem. Scand.* **26** (1972) 309.
[2]   Kawasaki, DE-OS 3 033 341, 1979 (R. Matsuura et al.).
[3]   Kawasaki, JP 8 259 827, 1980.
[4]   Kawasaki, JP 7 822 559, 1973 (R. Matsuura et al.).

[5] Bayer, DE-OS 2 437 221, 1974 (M. Martin, G. Scharfe); DE 2 453 232, 1974 (M. Martin, W. Schmidt, G. Scharfe).
[6] Sumikin Coke, JP 8 805 051, 1986 (T. Nishizaki, R. Minami).
[7] Bayer, DE-OS 2 234 306, 1972 (H. Dohm, K. Morgenstern); DE-OS 2 312 838, 1973 (R. Wiemers, K. Morgenstern, L. Müller, H. Dohm); Japan Distillation Co., DE-OS 2 506 809, 1974 (Y. Yokoyama, J. Yoshikawa, H. Ota, C. Ichikawa).
[8] Kawasaki, JP 7 783 443, 1975 (R. Matsuura et al.).
[9] Kawasaki, GB 2 039 897, 1978.
[10] Kawasaki, JP 8 385 835, 1981.
[11] Rütgerswerke, DE 1 232 943, 1965 (K. Rühl, L. Rappen).
[12] Kawasaki, JP 86 155 348, 1984 (A. Matsura, T. Komatsu, T. Sumino).
[13] Great Lakes Carbon Corp., US 3 681 401, 1966 (L. A. Joo, L. A. Bryan).
[14] Braude, Fawcett, *Org. Synth. Coll.*, vol. **4** (1963) 698.
[15] Nippon Kakayu Co., JP 7 375 555, 1972.
[16] M. Periasamy, M. V. Bhatt, *Tetrahedron Lett.* 1978, no. 46, 4561.
[17] ICI, DE-OS 2 301 803, 1972 (R. A. Rennie, R. Campbell).
[18] J. Skarzewsky, *Tetrahedron* **40** (1984) 4997.
[19] E. V. Dehmlow, J. K. Makrandi, *J. Chem. Res. Synop.* 1986, 32.
[20] D. Pletcher, E. M. Valdes, *J. Chem. Res. Synop.* 1987, 386.
[21] ICI, DE-OS 1 804 727, 1967 (R. A. Rennie, L. Campbell).
[22] M. Periasamy, M. V. Bhatt, *Synthesis* 1977, 330.
[23] Kawasaki, US 4 536 377, 1983 (T. Komatsu et al.).
[24] Diamond Shamrock Corp., EP-A 75 828, 1981 (E. A. Mayeda, D. W. Abrahamson).
[25] Grace, US 4 670 108, 1986 (R. P. Kreh, R. M. Spotnitz).
[26] I. M. Dalrymple, J. P. Millington, *J. Appl. Electrochem.* **16** (1986) 885–893.
[27] D. Pletcher, E. M. Valdes, *Electochimica Acta* no. 4, 509–515 (1988).
[28] Shell Int. Res., DE-OS 2 460 665, 1974 (A. J. de Jong, R. van Helden).
[29] Kawasaki, JP 5 108 256, 1974; JP 5 108 257, 1974; JP 5 322 559, 1978.
[30] Kawasaki, US 4 412 954, 1981 (T. Kamatsu, K. Usui).
[31] Bayer, DE-OS 2 532 388, 1975 (J. Priemer, N. Schenk, J. Krekel, W. Schwerdtel).
[32] Bayer, DE-OS 2 218 316, 1972 (J. Grolig, G. Scharfe).
[33] Bayer, DE-OS 2 245 555, 1972 (G. Scharfe, J. Grolig).
[34] Bayer, DE-OS 2 532 450, 1975 (J. Priemer, N. Schenk, M. Martin, W. Swodenk).
[35] US Rubber Co., US 2 457 701, 1945 (Ch. D. McCleary). F. A. Bovey, J. M. Kolthoff, *Chem. Rev.* **42** (1948) 491.
[36] Akzo GmbH, DE-OS 2 325 179, 1972 (L. Roskott, M. A. A. Groenendaal).
[37] Rohm & Haas, US 3 960 671, 1974 (J. S. Clovis, J. Dohling).
[38] S. N. Litvinenko, L. K. Mushkalo, A. V. Stetsenko, A. K. Tyltin, *Khim. Tekhnol. Topl. Masel* **18** (1973) no. 8, 41.
[39] Japan Exlan Co., JP 86 261 320 A2, 1985 (T. Imamura et al.); JP 86 268 691 A2, 1985 (T. Imamura, T. Matsumoto, K. Matsuzawa).
[40] Velsicol Corp., US 2 402 226, 1943 (J. Hyman, C. F. Peters).
[41] BASF, DE-OS 2 952 709, 1979 (H. Puetter, W. Bewert).
[42] M. Juaristi, J. M. Aizpurua, B. Lecea, C. Palomo, *Can. J. Chem.* **62** (1984) 2941.
[43] S. K. Ray, G. S. Murty, H. S. Rao, *Proc. Symp. Chem. Oil Coal* 1969, 320–329.
[44] A. J. Lin, A. S. Sartorelli, *Biochem. Pharmacol.* **2** (1976) no. 2, 206; *J. Med. Chem.* **1** (1976) no. 11, 1336; *J. Med. Chem.* **16** (1973) no. 11, 1268.

[45] NEC Corp., JP 8 515 458, 1983.
[46] US-Rubber Co., GB 657 977, 1949 (T. A. Clayton); *Chem. Eng. News* 1956, 3706.
[47] Asahi Chem. Ind. Co., JP 7 447 426, 1969.
[48] J. Ullmann, M. Ettich, *Ber. Dtsch. Chem. Ges.* **54** (1922) 259.
[49] sterr. Stickstoffwerke, AT 190 921, 1955 (A. Wagner).
[50] Kawasaki, JP 7 898 943, 1977 (M. Matsuura et al.).
[51] Seitetsu Chem. Ind. Co., JP 76 113 859, 1975 (H. Watanabe et al.).
[52] Mitsui Toatsu, DE-OS 2 509 819, 1974 (A. Iwamura et al.); Nippon Shokubai, DE-OS 2 436 157, 1974 (A. Fukui et al.).
[53] S. Chatterjee, W. D. Kwalwasser, *J. Chem. Soc. Perkin Trans. 2* 1974, no. 8, 873.
[54] Sumitomo Chem. Co., JP 8 076 083, 1978.
[55] Bayer, DE-OS 2 321 003, 1973 (F. Dürholz, R. Pütter, A. Vogel); DE-OS 2 359 950, 1973 (F. Dürholz, R. Pütter).
[56] Ciba Geigy, BE 8 22 754, 1973; BE 8 22 755, 1973.
[57] Mitsui Toatsu, DE-AS 2 539 631, 1975 (Y. Torisu et al.).
[58] Mitsui Toatsu, GB 1 468 480, 1974.
[59] G. Fawaz, L. F. Fieser, *J. Am. Chem. Soc.* **72** (1950) 996, 3999.
[60] Du Pont, US 2 572 946, 1949 (M. Paulshock).
[61] FIAT-Berichte 1313 II, 208, 1948.
[62] J. R. Lewis, J. Paul, *Z. Naturforsch. B: Anorg. Chem., Org. Chem.* **32 B** (1977) no. 12, 1473.
[63] G. Wurm, B. Goessler, *Arch. Pharm. (Weinheim, Ger.)* **320** (1987) no. 6, 564.
[64] A. V. Pinto, V. F. Ferreira, M. do. C. F. R. Pinto, *Synth.Commun.* **15** (1985) no. 13, 1177.
[65] D. H. R. Barton et al., *Tetrahedron Lett.* **23** (1982) no. 9, 957.
[66] D. H. R. Barton et al., *J. Chem. Soc. Perkin Trans 1* 1981, 1473.
[67] D. H. Hunter, D. H. R. Barton, W. S. Motherwell, *Tetrahedron Lett.* **25** (1984) no. 6, 603.
[68] L. F. Fieser, *Org. Synth. Coll.*, vol. **2** (1943) 430.
[69] M. P. Zykov et al., *Vopr. Virusol* 1975, no. 1, 58 – 62 .G. N. Pershin et al., *Farmakol. Toksikol. (Moscow)* **38** (1975) no. 1, 69 – 73.
[70] R. Kuhn, J. Hammer, *Chem. Ber.* **83** (1950) 413.
[71] H. L. K. Schmand, P. Boldt, *J. Am. Chem. Soc.* **97** (1975) no. 2, 447.
[72] Mitsubishi Chem. Ind. Co., JP 85 151 098, 1984 (T. Niwa, Y. Murata, S. Maeda).
[73] V. Horak et al., *Tetrahedron Lett.* **22** (1981) no. 37, 3577.
[74] P. Boldt et al., *J. Org. Chem.* **48** (1983) 2814.
[75] T. H. Zincke, A. Ossenbeck, *Justus Liebigs Ann. Chem.* **307** (1899) 1.
[76] K. Zahn, P. Ochwat, *Justus Liebigs Ann. Chem.* **462** (1928) 86.
[77] S. Yoshino, K. Hayakawa, K. Kanematsu, *J. Org. Chem.* **46** (1981) 3841.
[78] Y. Iwasawa et al., *J. Chem. Soc. Faraday Trans 1* **68** (1972) 1617; **70** (1974) 193; *J. Catal.* **31** (1973) 444.
[79] Y. Iwasawa, S. Ogasawara, *J. Catal.* **37** (1975) 148 – 157.

# Nitriles

*Individual keywords:* → Acrylonitrile; benzonitrile → Benzoic Acid; malononitrile → Malonic Acid and Derivatives

PETER POLLAK, Lonza AG, Basel, Switzerland (Chap. 2)

GÉRARD ROMEDER, Lonza AG, Basel, Switzerland (Chap. 2)

FERDINAND HAGEDORN, Bayer AG, Leverkusen, Federal Republic of Germany (Sections 3.1–3.3)

HEINZ-PETER GELBKE, BASF Aktiengesellschaft, Ludwigshafen, Federal Republic of Germany (Section 3.4)

| | | | | |
|---|---|---|---|---|
| 1. | Introduction . . . . . . . . . . . . . 3455 | 3. | Aromatic and Araliphatic Nitriles . . . . . . . . . . . . . . . . . | 3467 |
| 2. | Aliphatic Nitriles . . . . . . . . . 3456 | | | |
| 2.1. | Physical Properties . . . . . . . . 3456 | 3.1. | Properties . . . . . . . . . . . . . . | 3467 |
| 2.2. | Chemical Properties. . . . . . . . 3456 | 3.2. | General Production Methods . | 3468 |
| 2.3. | General Production Processes 3456 | 3.3. | Selected Araliphatic and Aromatic Nitriles. . . . . . . . . . | 3469 |
| 2.4. | Selected Aliphatic Nitriles . . . 3460 | 3.3.1. | Araliphatic Nitriles. . . . . . . . . | 3469 |
| 2.4.1. | Saturated Mono- and Dinitriles . 3460 | 3.3.2. | Aromatic Nitriles . . . . . . . . . . | 3471 |
| 2.4.2. | Unsaturated Mono- and Dinitriles 3462 | 3.3.2.1. | Cyanobenzenes . . . . . . . . . . . | 3471 |
| 2.4.3. | Substituted Nitriles . . . . . . . . . 3462 | 3.3.2.2. | Cyanonaphthalenes (Naphthalene Carboxylic Acid Nitriles) . . . . . . | 3472 |
| 2.4.4. | β-Iminonitriles . . . . . . . . . . . . 3465 | | | |
| 2.4.5. | β-Ketonitriles . . . . . . . . . . . . 3465 | 3.4. | Toxicology . . . . . . . . . . . . . . | 3473 |
| 2.5. | Toxicology and Occupational Health. . . . . . . . . . . . . . . . . 3466 | 4. | References. . . . . . . . . . . . . . | 3474 |

# 1. Introduction

Organic compounds containing the –CN group are generically called nitriles. The term "nitrile" refers to the triply bound nitrogen atom, ≡N, and not to the carbon atom attached to it; therefore numbering of the aliphatic chain starts with that carbon atom (e.g., 4-chlorobutyronitrile describes the chemical formula $ClCH_2CH_2CH_2CN$).

Nitriles R–CN are usually viewed as derivatives of the corresponding acids R–COOH. Consequently, they are also named by changing the ending "ic acid" or "oic acid" to "onitrile" (e.g., glutaric acid → glutaronitrile; dodecanoic acid → dodecanonitrile, propiononitrile, as an exception, is commonly referred to as propionitrile). Another way of naming nitriles is to state the name of the group R and use the prefix "cyano" or the suffix "cyanide" (e.g., cyanoethane and ethyl cyanide both describe $H_3CCH_2CN$). Nitriles that are obtained by addition of hydrogen cyanide to ketones or aldehydes and thus have both a hydroxy and a cyano group attached to the same carbon atom are commonly termed cyanohydrins (e.g., acetone cyanohydrin is identical to 2-hydroxy-2-methylpropionitrile). The same nomenclature applies to 3-hydroxypropionitrile which is commonly named ethylene cyanohydrin.

## 2. Aliphatic Nitriles

Aliphatic nitriles are important starting materials for polymers as well as for the synthesis of e.g., pharmaceuticals and pesticides.

### 2.1. Physical Properties

The carbon–nitrogen bond is extremely polar, which results in nitriles having a high dipole moment. This large dipole moment leads to intramolecular association; hence, nitriles have higher boiling points than would be expected from their molecular mass. Most of the lower molecular mass aliphatic nitriles (up to $C_{13}H_{27}CN$) are liquids at room temperature. Simple nitriles such as acetonitrile, propionitrile, glycolonitrile, and malononitrile are miscible with water; the latter two having a higher solubility because of the presence of two polar groups. Nitriles with higher molecular mass are sparingly water-soluble. Nitriles are good solvents for both polar and nonpolar solutes. The physical properties of selected nitriles are listed in Table 1.

### 2.2. Chemical Properties

The chemical properties of aliphatic nitriles are determined by two reactive centers:

1) The electrophilic cyano group which can react with nucleophiles
2) The activated carbon adjacent to the nitrile group which is prone to base-catalyzed substitution reactions

Typical reactions involving the cyano group and the activated carbon adjacent to it are illustrated in Figures 1 and 2, respectively.

### 2.3. General Production Processes

Of the many synthetic pathways that can be envisaged for producing aliphatic nitriles, the following four are widely used:

**Table 1.** Physical properties

| Nitrile | $M_r$ | mp, °C | bp (101.3 kPa), °C | $n_D^{20}$ | $d_4^{20}$ |
|---|---|---|---|---|---|
| Acetonitrile, $CH_3CN$ | 41.05 | − 45 | 81.6 | 1.3441 | 0.7138 |
| Propionitrile, $CH_3CH_2CN$ | 55.08 | − 93 | 97.2 | 1.3670 | 0.782 |
| Butyronitrile, $CH_3CH_2CH_2CN$ | 69.10 | −111.9 | 116 – 117 | 1.3838 | 0.7936 |
| Valeronitrile, $CH_3CH_2CH_2CH_2CN$ | 83.13 | − 96 | 141.3 | 1.3971 | 0.8008 |
| Tetradecanonitrile, $CH_3(CH_2)_{12}CN$ | 209.38 | 19.25 | 226 (10 kPa) | 1.4392 [a] | 0.8281 [d] |
| Malononitrile, $NCCH_2CN$ | 66.06 | 30 – 31 | 218 – 219 | 1.4146 [b] | 1.0494 [e] |
| Succinonitrile, $NCCH_2CH_2CN$ | 80.09 | 57 | 265 – 267 | 1.4173 [c] | 0.9867 [c] |
| Glutaronitrile, $NC(CH_2)_3CN$ | 94.12 | − 29 | 286 | 1.4295 | 0.9911 [f] |
| Allyl cyanide, $CH_2=CH-CH_2CN$ | 67.09 | − 84 | 119 | 1.4060 | 0.8329 |
| Methyleneglutaronitrile, $NCC_2H_4C(=CH_2)CN$ | 106.12 | − 9.0 | 113 (0.66 kPa) | 1.4558 | 0.9831 [f] |

Measuring temperature: [a] 25 °C; [b] 34 °C; [c] 60 °C; [d] 19 °C; [e] 35 °C; [f] 15 °C.

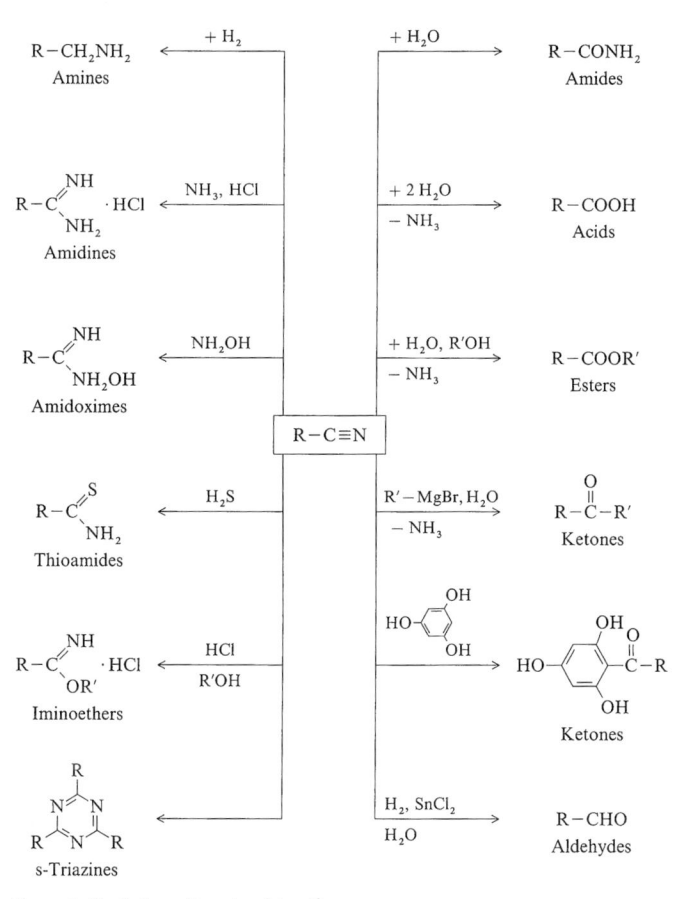

**Figure 1.** Typical reactions involving the cyano group

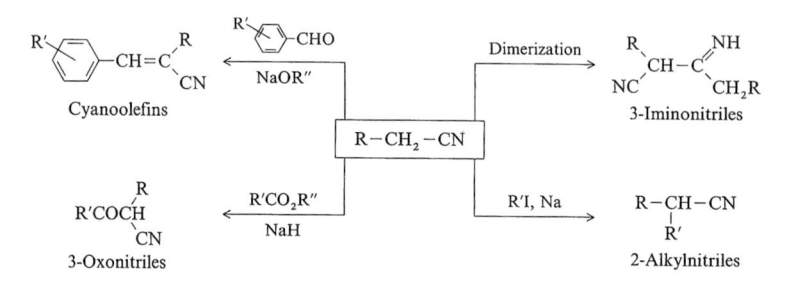

**Figure 2.** Typical reactions involving the activated carbon atom adjacent to the cyano group

**Reaction of Nitrogen-Free Precursors (such as Alkanes, Olefins, Alcohols, Aldehydes, or Acids) with Ammonia.** Gas phase reaction of *olefins with ammonia* in the presence of oxygen (ammoxidation) and oxidation catalysts (based on, e.g., vanadium or molybdenum) has attained the greatest industrial importance for the production of acrylonitrile from propene and methacrylonitrile from isobutene. This process is known as the *Sohio process* → Acrylonitrile.

$$H_2C=CH-CH_3 + NH_3 + 3/2\,O_2 \xrightarrow{\text{Catalyst}} H_2C=CH-CN + 3\,H_2O$$

Acetonitrile and hydrogen cyanide are formed as byproducts in typical quantities of 30–40 kg and 140–180 kg, respectively per 1000 kilograms of acrylonitrile [1]. Propane can also be used as feedstock (the first step being the in situ dehydrogenation to propene) as well as acrolein [2]. Similarly, the simplest aliphatic nitrile, hydrogen cyanide can be obtained from methane, ammonia, and oxygen (*Andrussow process*).

The ammoxidation process can also be advantageously carried out with *aldehydes* as starting materials, when they are readily available. This is especially the case for butyraldehyde and isobutyraldehyde which yield butyronitrile and isobutyronitrile, respectively [3]. The ammonolysis of *alcohols* or *aldehydes* can also be conducted under dehydrogenation conditions [4]:

$$(H_3C)_2CH-CH_2OH + NH_3 \xrightarrow[290\,°C]{Cu/SiO_2} (H_3C)_2CHCN + H_2O + 2\,H_2$$

*Acids* when reacted with ammonia over a dehydration catalyst such as $Al_2O_3$ at typical temperatures above 360 °C give nitriles without isolation of the intermediary amides [5].

$$(H_3C)_3CCOOH + NH_3 \xrightarrow[380\,°C]{Al_2O_3} (H_3C)_3CCN + 2\,H_2O$$

In addition to ammonia other nitrogen-containing reagents can be used, for example, hydroxylamine which easily yields nitriles upon reaction with aldehydes.

$$R-CHO + NH_2OH \xrightarrow{-H_2O} [R-CH=N-OH] \xrightarrow{-H_2O} R-CN$$

A multitude of dehydration agents is available.

Other nitrogen-containing reagents such as sulfonamides [6], hydroxylamine-*O*-sulfonic acid [7] and chlorosulfonylisocyanate [8] have also been investigated. However, none of those methods have gained industrial importance.

**Formation of the Cyano Moiety from Nitrogen-Containing Precursors such as Amines, Amides, or Formamides.** If an amine is employed as starting material, it is usually produced in situ from the corresponding alcohol and ammonia, the dehydrogenation step is then performed catalytically [4]. An electrochemical dehydrogenation with a nickel hydroxide anode has also been investigated [9].

$$R-CH_2-NH_2 \xrightarrow[KOH, H_2O]{Ni(OH)_2/Anode,} R-CN + 2H_2$$

Dehydration of *amides* with agents such as aluminum chloride or thionyl chloride is a widely applicable method, although not competitive in most cases. The direct catalytic conversion of acids with ammonia is preferred.

*Formamides* are also valuable precursors [10]:

$$HCONHC(CH_3)_3 \xrightarrow[t > 250\,°C]{Acetic\ anhydride} (H_3C)_3CCN + H_2O$$

**Reaction of Hydrogen Cyanide or Cyanide Ions with Double Bonds, Carbonyl Compounds, Hydrogen or Halogens.** A typical industrial example of the hydrogen cyanide addition to *double bonds* is the direct hydrocyanation of butadiene to adiponitrile [11].

Reaction with aldehydes and ketones yields the corresponding cyanohydrins.

$$\underset{R'}{\overset{R}{>}}C=O + HCN \rightleftharpoons NC-\underset{R'}{\overset{R}{\underset{|}{\overset{|}{C}}}}-OH$$

The reaction is usually carried out at 10–15 °C and a pH of 6.5–7.5. The cyanohydrin is continuously removed from the reaction mixture to shift the equilibrium toward cyanohydrin formation, which is usually stabilized with mineral acids or directly chemically converted to a more stable form [12].

A displacement reaction of hydrogen with a halogen cyanide, which has gained great industrial importance, is the synthesis of malononitrile from acetonitrile and cyanogen chloride above 800 °C (→ Malonic Acid and Derivatives):

$$H_3C-CN + ClCN \longrightarrow NC-CH_2-CN + HCl$$

The order of halogen reactivity towards cyanide displacement is Cl < Br < I, fluoride being essentially inert. Examples are the production of:

4-chlorobutyronitrile from 1-bromo-3-chloropropane [13]
adiponitrile via prior chlorination of butadiene [14]
cyanoacetates from chloroacetic acid (→ Malonic Acid and Derivatives)

Another important reaction is the synthesis of acyl cyanides in the presence of copper(I) cyanide [15]:

$$H_3C-\underset{O}{\overset{\|}{C}}-Cl + NaCN \xrightarrow{CuCN} H_3C-\underset{O}{\overset{\|}{C}}-CN + NaCl$$

In addition to the readily available hydrogen cyanide or cyanide salts (e.g., sodium cyanide) other cyanation agents such as tetraalkylammonium cyanides [16] or trimethylsilylcyanide [17] have been investigated.

**Reaction of the Activated Carbon in Nitrile-Containing Precursors.** In addition to the examples highlighted in Figure 2 the electrohydrodimerization of acrylonitrile to adiponitrile, known as "EHD process" is also significant [18]:

$$2\,H_2C=CH-CN \xrightarrow[-2\,OH^-]{2e,\,H_2O} NC-(CH_2)_4-CN$$

## 2.4. Selected Aliphatic Nitriles

### 2.4.1. Saturated Mono- and Dinitriles

**Acetonitrile** [75-05-8], methylcyanide, cyanomethane, ethanenitrile, $CH_3CN$, is a colorless liquid miscible with water, ethanol and many organic solvents but immiscible with many saturated hydrocarbons (e.g., petroleum fractions). The physical properties are listed in Table 1. Characteristics of acetonitrile are its high dipole moment (3.84 D) and dielectric constant (38.8 at 20 °C).

*Production.* Acetonitrile and hydrogen cyanide are the principal byproducts from the ammoxidation of propylene to acrylonitrile (Sohio process). Some acrylonitrile producers currently recover and purify acetonitrile, however, most companies burn the byproducts as plant fuel.

*Economic Aspects.* In 1986 the comsumption of acetonitrile amounted to ca. 13 000 t in the United States, Western Europe, and Japan [19]. Along with *N*-methylpyrrolidone and *N,N*-dimethylformamide, acetonitrile represents one of the major solvents for extraction of butadiene from crude $C_4$ streams. Acetonitrile is also a valuable reaction solvent, especially in the synthesis of antibiotics. Highly purified acetonitrile finds application as a solvent in high-performance liquid chromatography.

The major use of acetonitrile in *organic synthesis* is in the production of malononitrile through reaction with cyanogen chloride.

Other applications include the synthesis of trimethyl orthoacetate [H$_3$C – C(OCH$_3$)$_3$] and acetamidine hydrochloride [H$_3$C – C( = NH)NH$_2$ · HCl]. The former is used in the synthesis of 3-(2,2-dichlorovinyl)-2,2-dimethylcyclopropanecarboxylic acid (also known as DV-acid) which is a key precursor for pyrethroides [20]. The latter yields N-containing heterocycles such as 4-amino-5-cyano-2-methylpyrimidine used in the synthesis of thiamine [21].

*Transport Classification:* RID/ADR 3, no. 11B

**Propionitrile** [*107-12-0*], ethyl cyanide, CH$_3$CH$_2$CN, is a colorless liquid which is miscible with water, ethanol, dimethylformamide, and diethyl ether. The physical properties are listed in Table 1. Propionitrile is obtained either by hydrogenation of acrylonitrile [22] or by the gas-phase reaction of propanal or propanol with ammonia [4]. It is also obtained as a byproduct in the electrohydrodimerization of acrylonitrile to adiponitrile (EHD process) [18].

Hydrogenation of propionitrile yields *N*-propylamines [23]. It is also used as an organic intermediate, for example, in the synthesis of the pharmaceuticals flopropione (via Houben – Hoesch reaction) [24] and ketoprofen [25].

*Transport Classification:* RID/ADR 3, no. 11B

**Butyronitrile** [*109-74-0*], propyl cyanide, butanenitrile, CH$_3$(CH$_2$)$_2$CN, is a colorless liquid slightly miscible with water, miscible with ethanol and diethyl ether. The physical properties are listed in Table 1. Butyronitrile is usually obtained by the catalytic gase-phase reaction of butanol or butyraldehyde with ammonia [3], [4]. Its major use is the manufacture of the poultry drug amprolium [26].

*Transport Classification* RID/ADR 3, no. 11B

**Isobutyronitrile** [*78-82-0*], 2-methylpropanenitrile, isopropyl cyanide, 2-methylpropionitrile, (CH$_3$)$_2$CHCN, $M_r$ 69.11, mp −71.5 °C, bp 117 °C (101.3 kPa), $n_D^{20}$ 1.374, $d_4^{20}$ 0.770, is a colorless liquid miscible with ethanol and diethyl ether, slightly miscible with water.

Isobutyronitrile is usually obtained by the catalytic gas-phase reaction of isobutyraldehyde or isobutanol with ammonia [3], [4]. Its major use is the synthesis of the insecticide diazinon [27].

*Transport Classification:* RID/ADR 3, no. 11B

**Succinonitrile** [*110-61-2*], ethylene dicyanide, ethylene cyanide, dicyanoethane, butanedinitrile, NCCH$_2$CH$_2$CN, is a colorless, waxy solid slightly soluble in water and ethanol. The physical properties are listed in Table 1. Succinonitrile is obtained by addition of hydrogen cyanide to acrylonitrile [28]. Subsequent hydrogenation of succinonitrile yields 1,4-diaminobutane, which reacts with adipic acid to form the new polyamide 46. The polymer, developed by DSM (Netherlands), is marketed under the trade name Stanyl; the latter is reported to have excellent mechanical properties at high temperature [29].

*Transport Classification:* RID/ADR 6.1, no. 12C

## 2.4.2. Unsaturated Mono- and Dinitriles

**Methacrylonitrile** [*126-98-7*], 2-methyl-2-propenenitrile, $CH_2=C(CH_3)CN$, $M_r$ 67.09, mp −35.8 °C, bp 90.3 °C, $d_4^{20}$ 0.7998, $n_D^{20}$ 1.4003, is a liquid which is miscible with ethanol, diethyl ether and acetone; it is immiscible with water.

Methacrylonitrile can be produced by ammoxidation of isobutene. Its copolymerization with methacrylic acid gives poly(methacrylimide), an engineering plastic commercialized under the trade name Rohacell by Röhm (Federal Republic of Germany) [30].

**2-Methyleneglutaronitrile** [*1572-52-7*], acrylonitrile dimer, 2,4-dicyano-1-butene, 2-methylenepentanedinitrile, $NCC_2H_4C(=CH_2)CN$, is a colorless liquid slightly miscible with water, ethanol, acetone and benzene, immiscible with hexane. The physical properties are listed in Table 1.

2-Methyleneglutaronitrile is obtained by dimerization of acrylonitrile in the presence of a metal halide and a tertiary amine [31]. Its hydrogenation yields 2-methylglutaronitrile, which in turn can be converted into 3-methylpyridine [32].

**Tetracyanoethylene** [*670-54-2*], ethenetetracarbonitrile, TCNE, $(NC)_2C=C(CN)_2$, $M_r$ 128.09, mp 197−198 °C, bp 223 °C, $n_D^{25}$ 1.560. The preferred synthetic preparation of TCNE involves the debromination of the KBr complex of dibromomalononitrile. Tetracyanoethylene is a reactive compound that undergoes a variety of reactions including addition, replacement and cyclization. The chemistry of TCNE is exhaustively reviewed in [33].

## 2.4.3. Substituted Nitriles

**Chloroacetonitrile** [*107-14-2*], chloromethanenitrile, chloromethyl cyanide, $ClCH_2CN$, $M_r$ 75.50, bp 124−126 °C, $d_4^{20}$ 1.1896, $n_D^{20}$ 1.426, is a colorless liquid with pungent odor. It is miscible with hydrocarbons and ethanol; immiscible with water. Chloroacetonitrile can be selectively obtained by the photochemical chlorination of acetonitrile with chlorine in the presence of, for example, $SnCl_4$ [34]. Another method is based on the dehydration of chloroacetamide with, e.g., phosphorous pentoxide [35]. Chloroacetonitrile is used as an organic intermediate, e.g., in the synthesis of the cardiovascular drug guanethidine [36] and the insecticide fenoxycarb [37].

*Transport Classification:* RID/ADR 6.1, no. 11B

**Trichloroacetonitrile** [545-06-2], trichloroethanenitrile, trichloromethyl cyanide, $CCl_3CN$, $M_r$ 144.40, mp −42 °C, bp 85.7 °C (101.3 kPa), $d_4^{25}$ 1.4403, $n_D^{20}$ 1.4409, is a colorless liquid immiscible with water.

Trichloroacetonitrile can be obtained by dehydration of trichloroacetamide with phosphorous pentoxide [38] or by chlorination of acetonitrile with chlorine. Vapor phase chlorination in the presence of water [39] and photochemical chlorination in the presence of catalysts such as $HgCl_2$ or $AlCl_3$ have been reported [40]. Trichloroacetonitrile is an organic intermediate used, for example, in the synthesis of the fungicide etridiazole [41].

*Transport Classification* : RID/ADR 6.1, no. 11B

**Glycolonitrile** [107-16-4], hydroxyacetonitrile, formaldehyde cyanohydrin, $HOCH_2CN$, $M_r$ 57.05, bp 183 °C (slight decomp.), $d_4^{15}$ 1.104, $n_D^{20}$ 1.4117. Anhydrous glycolonitrile is a colorless liquid which is miscible with water, ethanol and diethyl ether.

Glycolonitrile is obtained by reaction of aqueous formaldehyde and aqueous hydrogen cyanide, in the presence of catalysts such as sodium hydroxide [42]. It is usually handled as an aqueous solution. Anhydrous glycolonitrile tends to be unstable at high pH values and therefore must be stabilized by addition of small amounts of, e.g., phosphoric acid.

The bulk of the glycolonitrile produced is not isolated but further upgraded by reaction with amines. Depending on the molecular ratio, reaction with ammonia yields aminoacetonitrile, iminodiacetonitrile or nitrilotriacetonitrile. Reaction with ethylenediamine gives ethylenediaminetetraacetonitrile which, in turn, is hydrolyzed to ethylenediaminetetraacetic acid (EDTA), a well known sequestrant.

**Aminoacetonitrile** [540-61-4], glycinonitrile, cyanomethylamine, $H_2NCH_2CN$, $M_r$ 56.07, bp 58 °C (20 kPa, partial decomp.).

Aminoacetonitrile is readily obtained as an aqueous solution by reacting an aqueous solution of glycolonitrile with liquid ammonia [43]. The solution is stabilized by lowering the pH with small amounts of sulfuric acid [44]. Aminoacetonitrile can also be isolated as the hydrochloride or sulfate.

Aminoacetonitrile is usually further hydrolyzed by sodium hydroxide to the sodium salt of glycine, which in turn is neutralized to glycine [45].

**Lactonitrile** [78-97-7], acetaldehyde cyanohydrin, 2-hydroxypropanenitrile, $H_3CCH(OH)CN$, $M_r$ 71.08, mp −40 °C, bp 182 – 184 °C (decomp.), $d_4^{20}$ 0.9877, $n_D^{18}$ 1.4058, is a colorless or straw-colored liquid, which is miscible with water and ethanol; it is miscible with diethyl ether and carbon disulfide.

Lactonitrile is obtained by reaction of acetaldehyde with hydrogen cyanide in the presence of sodium hydroxide. It is mainly used in the synthesis of lactic acid and lactates through hydrolysis with sulfuric acid and esterification [46].

**3-Aminopropionitrile,** [151-18-8], 3-aminopropanenitrile, β-aminopropionitrile, $H_2N-CH_2-CH_2CN$, $M_r$ 70.09, bp 185 °C (101.3 kPa), $n_D^{20}$ 1.4396, is a colorless liquid which tends to slowly polymerize if stored in the presence of air. 3-Aminopropionitrile is prepared by the reaction of acrylonitrile with ammonia [47] and is mainly used as an intermediate in the manufacture of β-alanine and pantothenic acid [48].

**3-Chloropropionitrile** [542-76-7], 3-chloropropanenitrile, β-chloropropionitrile, $ClCH_2CH_2CN$, $M_r$ 89.53, mp −51 °C, bp 175–176 °C (101.3 kPa), $n_D^{20}$ 1.4360, $d_4^{20}$ 1.1573, is a colorless liquid miscible with acetone, benzene, ethanol and diethyl ether. It can be obtained by hydrochlorination of acrylonitrile with hydrochloric acid [49] and is used as an organic intermediate in the synthesis of the $H_2$-receptor famotidine [50].
*Transport Classification* : RID/ADR 6.1, no. 11C

**3-Hydroxypropionitrile** [109-78-4], ethylene cyanohydrin, hydracrylonitrile, 3-hydroxypropanenitrile, $HOCH_2CH_2CN$, $M_r$ 71.08, mp −46 °C, bp 228 °C (101.3 kPa, slight decomp.), $d_4^{25}$ 1.0404, $n_D^{20}$ 1.4240, is a straw-colored liquid, miscible with water, acetone, and ethanol; immiscible with benzene, carbon tetrachloride, and naphtha. 3-Hydroxypropionitrile is obtained by reacting ethylene oxide with hydrogen cyanide or from the hydration of acrylonitrile [51]. It was used as feedstock in the production of acrylic acid and acrylates.

**4-Chlorobutyronitrile** [628-20-6], 4-chlorobutanenitrile, $Cl(CH_2)_3CN$, $M_r$ 103.55, bp 189–191 °C (101.3 kPa), $d_4^{15}$ 1.0934, $n_D^{20}$ 1.4413, is a liquid, miscible with ethanol and diethyl ether; it is immiscible with water.

4-Chlorobutyronitrile is readily obtained by reacting 1-bromo-3-chloropropane with sodium cyanide [13]. It is used in the synthesis of the pharmaceuticals buflomedil [52] and buspirone [53]. 4-Chlorobutyronitrile also yields cyclopropanecarbonitrile upon cyclization with base [54].

**Diaminomaleonitrile** [*1187-42-4*], HCN tetramer, "DAMN", 2,3-diamino-2-butenedinitrile, $NC(NH_2)C=C(NH_2)CN$, $M_r$ 108.10, $mp$ 183 °C (decomp.), $n_D^{20}$ 1.5898, forms pale-brown needle shaped crystals. They are soluble in methanol, ethanol, and N,N-dimethylformamide; slightly soluble in water and dioxane; and insoluble in acetone, benzene, xylene, and carbon disulfide.

Diaminomaleonitrile is obtained by tetramerization of hydrocyanic acid in the presence of catalysts such as alkylaluminum compounds [55]. Nippon Soda runs a 200 t/a plant in Japan and intends to expand the production capacity in the future [56]. Diaminomaleonitrile is a versatile starting material for synthesizing nitrogen-containing heterocycles [57].

## 2.4.4. β-Iminonitriles

**2-Amino-1-propene-1,1,3-tricarbonitrile** [*868-54-2*], 2-amino-1,1,3-tricyanopropene, malononitrile dimer, $(NC)_2C=C(NH_2)CH_2CN$, $M_r$ 132.13, $mp$ 171–173 °C, is a light brownish powder, insoluble in water.

Malononitrile dimer is obtained from the base-catalyzed dimerization of malononitrile [58]. It has been investigated as a potential precursor for new dyes but it has no current industrial significance [59].

**3-Amino-3-methylacrylonitrile** [*1118-61-2*], β-iminobutyronitrile, β-aminocrotononitrile, diacetonitrile, 3-amino-2-butenenitrile, $NCCH=C(NH_2)CH_3$, $M_r$ 82.11, $mp$ (cis-form) 79–83 °C, $mp$ (trans-form) 52–53 °C, $d_4^{77}$ 0.9519, is a yellowish solid soluble in ethanol and diethyl ether; it is sparingly soluble in benzene and water.

3-Amino-3-methylacrylonitrile is obtained by the base-catalyzed dimerization of acetonitrile [60]. It is mainly used for the synthesis of pyrazolimines.

## 2.4.5. β-Ketonitriles

**Acetylacetonitrile** [*2469-99-0*], β-ketobutyronitrile, 3-oxo-butanenitrile, $H_3CCOCCH_2CN$, $M_r$ 83.09, $bp$ 120–125 °C (101.3 kPa), can be obtained by hydrolysis of β-aminocrotononitrile [61]. A potential application is the synthesis of 3-amino-5-methylisoxazole, which is an intermediate for various drugs.

**3,3-Dimethoxypropanenitrile** [*57597-62-3*], cyanoacetaldehyde dimethyl acetal, 3,3-dimethoxy-propanenitrile, $(CH_3O)_2CHCH_2CN$, $M_r$ 115.15, $bp$ 195 °C, $d_4^{20}$ 1.001, is a colorless to slightly yellow liquid.

**Table 2.** Toxicological data

| Nitrile | $LD_{50}$, mg/kg (oral, rats) | Inhalation, rats | | $LD_{50}$, mg/kg (skin, rabbit) | Skin/eye irritation (rabbit) | Recommended workplace exposure limits, mg/m³/type of limit |
|---|---|---|---|---|---|---|
| Saturated nitriles | | | | | | |
| Acetonitrile | 3800 | $LC_{50}$: | 7551 ppm/8h | 1250 | mild/severe | 34/TWA; 70/MAK |
| Propionitrile | 230 | LCLo: | 500 ppm/4h | 210 | mild/moderate | 14/TWA |
| Butyronitrile | 50 | LCLo: | 1000 ppm/4h | 500 | mild/mild | 22/TWA |
| Isobutyronitrile | 102 | LCLo: | 1000 ppm/4h | 310 | mild/mild | 22/TWA |
| Saturated dinitriles | | | | | | |
| Malononitrile | 14 | | | | mild/mild | 8/TWA |
| Succinonitrile | 450 | | | | | 20/TWA |
| Adiponitrile | 300 | | | | | 18/TWA |
| Unsaturated nitriles | | | | | | |
| Acrylonitrile | 78 | LCLo: | 500 ppm/4h | 250 | mild/severe | 5/TWA |
| Allyl cyanide | 115 | LCLo: | 500 ppm/4h | LDLo: 1410 (rabbit) | mild/mild | |
| Substituted nitriles | | | | | | |
| Chloracetonitrile | 220 | LCLo: | 250 ppm/4h | 71 | mild/moderate | |
| Cyanhydrins | | | | | | |
| Acetoncyanhydrin | 17 | LCLo: | 63 ppm/4h | 17 | | 4/C * |
| Glycolonitrile | 16 | LCLo: | 250 ppm/4h | 5 | | 5/C * |
| Lactonitrile | 87 | LCLo: | 125 ppm/4h | 20 | | |
| 3-Hydroxypropionitrile | 3200 | | | 5000 | mild/mild | |

* C = Ceiling concentration not to be exceeded during a 15 min. sampling period.

3,3-Dimethoxypropanenitrile can be obtained from acrylonitrile by reaction with methanol in the presence of nitrite and platinum- or palladium-catalysts [62]. It is a valuable starting material for the synthesis of 4-aminopyrimidines such as cytosine.

## 2.5. Toxicology and Occupational Health

Occupational exposure to aliphatic nitriles mainly occurs by inhalation of vapors and/or by absorption through the skin. Pertinent toxicity and occupational health data relating to the most widely used aliphatic nitriles are summarized in Table 2.

The general toxicity of nitriles in humans includes gastric distress and vomiting, bronchial irritation, respiratory distress, convulsions, and coma. Animal toxicity studies also show gastrointestinal tract, liver, and pancreas irritations.

Whereas the old literature ascribes almost all the actions of nitriles to the liberation of cyanide ions, recent reports tend to confirm that both the cyanide group and the entire molecule are important for biologic actions. While many of the neural effects are probably due to the cyanide moiety, the irritating and necrotic effects on the skin, liver, and kidney may be associated with the nature of the alkyl group. Although most aliphatic nitriles have cyanide release in common, there are significant differences in the amounts and duration exposure necessary to cause poisoning and toxicity effects. Signs of toxicity following oral administration of aliphatic nitriles to rats have highlighted the fact, that saturated nitriles affect the central nervous system in the same way as potassium cyanide, whereas unsaturated nitriles cause cholinomimetic effects such as salivation, diarrhea, and vasodilatation. Thus, cyanide release from unsaturated nitriles seems to play a minimal role in causing toxicity. Nitriles have tentatively been divided into three groups, with respect to toxicity, namely:

1) Those extensively metabolized to cyanide ions such as malononitrile and α-cyanohydrins
2) Those moderately metabolized to cyanide ions such as propionitrile and butyronitrile
3) Those poorly metabolized to cyanide ions such as acetonitrile, tetramethylsuccinonitrile, aminonitriles and acrylonitrile.

# 3. Aromatic and Araliphatic Nitriles

*Aromatic* nitriles can be represented by the general formula $Ar–C\equiv N$. Further cyano groups or other substituents can be bonded to the aromatic nucleus. In *araliphatic* nitriles the aromatic nucleus is separated from the CN group by a $(CH_2)_n$ group, which can itself bear other substituents. Like the aliphatic nitriles the aromatic and araliphatic nitriles are important intermediates for organic synthesis.

## 3.1. Properties

**Physical Properties.** Aromatic and araliphatic nitriles are liquids or crystalline solids, mostly sparingly soluble in water (some heteroaromatic cyano-compounds have a higher water solubility) with sometimes considerable thermal stability. Benzonitrile and a range of substituted benzonitriles have a characteristic odor resembling bitter almonds. Some substituted benzonitriles sublime readily.

**Chemical Properties.** Reactions involving the cyano group are similar to the corresponding reactions of aliphatic nitriles (see Section 2.2). Ortho-substituents generally lower the reactivity of the cyano group. Functional groups attached to the aromatic ring normally react in the expected fashion. The bonds of halogens ortho and para to nitrile groups are weakened, i.e., these groups are prone to nucleophilic ipso-substitution. Ring-closure reactions involving the cyano-group are often possible.

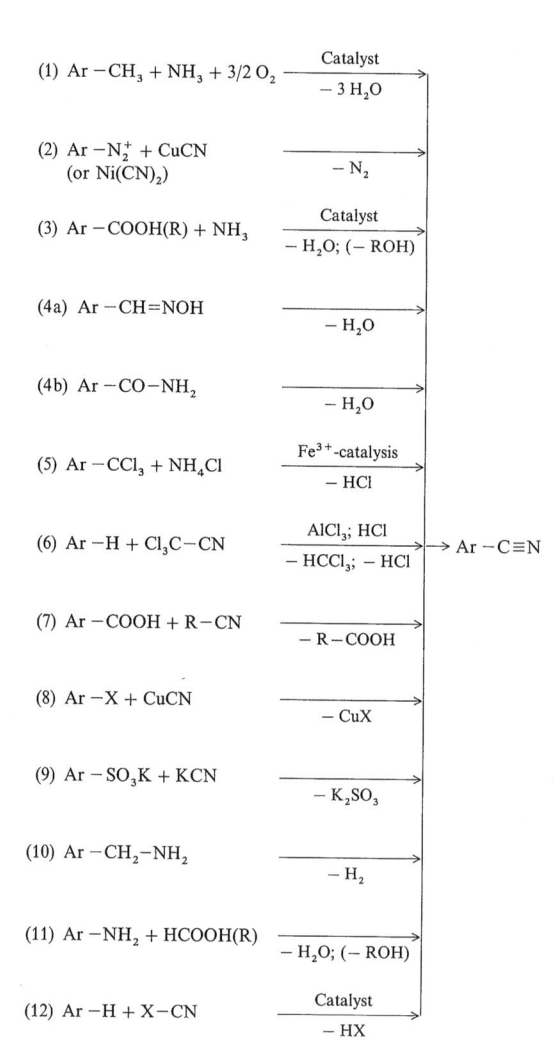

**Figure 3.** Synthetic routes to aromatic nitriles

(1) $Ar-CH_3 + NH_3 + 3/2\, O_2 \xrightarrow[-3\,H_2O]{\text{Catalyst}}$

(2) $Ar-N_2^+ + CuCN$ (or $Ni(CN)_2$) $\xrightarrow{-N_2}$

(3) $Ar-COOH(R) + NH_3 \xrightarrow[-H_2O;\,(-ROH)]{\text{Catalyst}}$

(4a) $Ar-CH=NOH \xrightarrow{-H_2O}$

(4b) $Ar-CO-NH_2 \xrightarrow{-H_2O}$

(5) $Ar-CCl_3 + NH_4Cl \xrightarrow[-HCl]{Fe^{3+}\text{-catalysis}}$

(6) $Ar-H + Cl_3C-CN \xrightarrow[-HCCl_3;\,-HCl]{AlCl_3;\,HCl}$ $\rightarrow Ar-C\equiv N$

(7) $Ar-COOH + R-CN \xrightarrow{-R-COOH}$

(8) $Ar-X + CuCN \xrightarrow{-CuX}$

(9) $Ar-SO_3K + KCN \xrightarrow{-K_2SO_3}$

(10) $Ar-CH_2-NH_2 \xrightarrow{-H_2}$

(11) $Ar-NH_2 + HCOOH(R) \xrightarrow{-H_2O;\,(-ROH)}$

(12) $Ar-H + X-CN \xrightarrow[-HX]{\text{Catalyst}}$

The nitrile group is a meta-directing substituent. The straight form of the cyano group allows even a hexacyano substitution at the benzene ring, in this case all six cyano groups being coplanar with the ring.

## 3.2. General Production Methods

*Araliphatic* nitriles can be produced by the methods similar to those described for aliphatic nitriles (see Section 2.3). The most important *cyanoaromatics*, e.g., benzonitrile, 2- and 4-chlorobenzonitrile, phthalonitrile, isophthalo- and terephthalonitrile, toluonitriles, and cyanopyridines are produced industrially by ammoxidation of the corresponding methylaromatics. Some newer special syntheses for aromatic nitriles also exist. The most important synthetic routes are shown schematically in Figure 3.

Method 1 represents the ammoxidation, by which mono- and dinitriles are obtained [72]. For this reaction mixtures of aromatics (e.g., xylenes) can be also used, in this case nitrile mixtures are formed [73]. Liquid-phase ammoxidations in the presence of manganese (II) bromide are also known [74].

The acid-catalyzed treatment of aromatic carboxylic acids or their esters with ammonia, (method 3) at elevated temperatures in the gas phase, leads to the corresponding nitriles, often with good yields [75].

Elimination of water from aromatic aldoximes (method 4 a) is also performed with high yield. Many variants of this reaction can be found in the literature [76]. The dehydration of aromatic carboxamides is carried out similarly to that of aliphatic amides [77].

Reaction 5 (trichloromethylaromatics with ammonia or ammonium chloride) is catalyzed by iron (III) chloride [78].

The Houben–Fischer method (method 6) known for a long time, uses trichloroacetonitrile (see p. 3463) and thus leads to the unavoidable production of trichloromethane. In reaction 7 the pK value of the aromatic carboxylic acid should be higher than that of the acid from which the nitrile R–CN was formed. The conversion requires a temperature of 200–300 °C (pressure vessel) and frequently gives the nitrile in good yield. Acetonitrile, for example, is a suitable starting nitrile [79]. In place of the carboxylic acids, their alkali salts or esters can be used in some cases [80].

Formylation of an aromatic amine, subsequent thermal dehydration of the formanilide (reaction 11), and reaction via the isonitrile yields nitriles (analogous to the corresponding conversion of aliphatic amines) [81].

Some ether-substituted aromatics (e.g., anisole) react with cyanogen bromide in the presence of catalysts to form nitriles [82]:

$$CH_3-O-\langle\phantom{x}\rangle + BrCN \xrightarrow[-HCl]{\text{Catalyst} \atop 100\,°C} CH_3-O-\langle\phantom{x}\rangle-CN$$

$+$ ortho-Isomer

Further special syntheses of aromatic nitriles include reactions of aromatic carboxylic acids with urea and amidosulfonic acid at 200–400 °C [12] or with chlorosulfonyl isocyanate and subsequent cleavage of the adduct with triethylamine [84].

## 3.3. Selected Araliphatic and Aromatic Nitriles

### 3.3.1. Araliphatic Nitriles

**Phenylacetonitrile** [*140-29-4*], α-cyanomethylbenzene, benzyl cyanide, $M_r$ 117.14, $mp$ −26 °C, $bp$ 233.5 °C, $d_4^{20}$ 1.0157, $n_D^{17}$ 1.5243, flash point 102 °C, is a colorless, oily, toxic liquid.

Phenylacetonitrile is produced by the reaction of benzyl chloride with alkali cyanide in alcohol or aqueous solution under phase transfer catalysis with N,N-dialkylbenzylammonium chloride [85].

Phenylacetonitrile is used as an intermediate in the production of synthetic penicillins or barbiturates, in the synthesis of optical bleaches for fibers, in the production of insecticides, and for perfumes and flavors (via phenylacetic acid to "honey-type" ester compounds).

*Labeling and transport:* RID/ADR 6.1, no. 12C

**1,2-Phenylenediacetonitrile** [*613-73-0*], 1,2-bis-(cyanomethyl)benzene, $M_r$ 156.18; mp 59–60 °C; forms colorless crystals.

The nitrile is used as an intermediate for the synthesis of optical brighteners and is produced by the reaction of 1,2-bis(chloromethyl)benzene with sodium cyanide in aqueous suspension under phase transfer catalysis [86].

**Phenylglyoxylonitrile** [*613-90-1*], benzoyl cyanide, $M_r$ 131.13, mp 34 °C, bp 206–208 °C (101.3 kPa), bp 96 °C (2 kPa), bp 82 °C (1.2 kPa), $n_D^{26}$ 1.5303, can be used for selective reactions at the carbonyl and cyano groups. It is produced from the reaction of benzoyl chloride or benzoic anhydride with sodium cyanide (or hydrogen cyanide in the presence of alkali cyanide) in practically quantitative yield with copper (I) cyanide as catalyst [87]. Benzoyl cyanide is used as an intermediate in the synthesis of plant protection agents.

*Labeling and transport:* RID/ADR 6.1, no. 24B

**D,L-Mandelonitrile** [*532-28-5*], benzaldehyde cyanohydrin, α-cyano-α-hydroxymethylbenzene, $C_6H_5-CH(OH)-CN$, $M_r$ 133.15, mp −10 °C, $d_4^{20}$ 1.1165 – 1.120, is miscible with ethanol and diethyl ether, immiscible with water.

Mandelonitrile occurs naturally as the β-glycoside of gentiobiose (amygdalin). It decomposes on heating into benzaldehyde and hydrogen cyanide. It is produced from benzaldehyde and alkali cyanide in the presence of mineral acid [88]. Mandelonitrile is used as an intermediate in the production of mandelic acid (→ Hydroxycarboxylic Acids, Aromatic).

**Benzothiazolyl-2-acetonitrile** [*56278-50-3*], 2-cyanomethylbenzothiazole, $M_r$ 174.21, mp 102 °C, forms colorless crystals.

This nitrile is produced by the reaction of 1-amino-2-mercaptobenzene (*o*-aminothiophenol) with the monoiminoether hydrochloride of malononitrile [89] and is used as a condensation component for methine dyes.

## 3.3.2. Aromatic Nitriles

### 3.3.2.1. Cyanobenzenes

**2-Chlorobenzonitrile** [*873-32-5*], 2-chloro-1-cyanobenzene, $M_r$ 137.57, *mp* 44.8 °C, *bp* (101.3 kPa) 232 °C, *bp* (1.6 kPa) 111.5 °C, forms colorless crystals that readily sublime. The crystals are readily soluble in organic solvents, only sparingly soluble in water, and steam volatile.

2-Chlorobenzonitrile is produced industrially by ammoxidation of 2-chlorotoluene [90]. It is used as an intermediate in the production of 2-amino-5-nitrobenzonitrile, which in turn is used for the synthesis of azo dyes (see below).

*Labeling and transport:* RID/ADR 6.1, no. 12C

**4-Chlorobenzonitrile** [*623-03-0*], 4-chloro-1-cyanobenzene, $Cl-C_6H_4-CN$, $M_r$ 137.6, *mp* 97 °C, forms colorless crystals that can be sublimed. The compound is steam volatile, readily soluble in organic solvents, and only sparingly soluble in water. 4-Chlorobenzonitrile is produced industrially by the ammoxidation of 4-chlorotoluene [91]. It is used in the synthesis of a red pigment for plastics [92].

**2,6-Dichlorobenzonitrile** [*1194-65-6*], 1-cyano-2,6-dichlorobenzene, $M_r$ 172.02, *mp* 144.5 – 146.5 °C, is a colorless, crystalline substance. Its solubility in water is 10 ppm at 25 °C. 2,6-Dichlorobenzonitrile is produced industrially by the ammoxidation of 2,6-dichlorotoluene [93]. In addition the nitrile is obtained by oxidation or side-chain-chlorination of 2,6-dichlorotoluene via 2,6-dichlorobenzoic acid and 2,6-dichlorobenzaldehyde, respectively. 2,6-Dichlorobenzonitrile shows herbicidal activity (Casoron, Duphar) and is used in fruit and vine cultivation. It is also used as an intermediate in the production of 2,6-difluorobenzonitrile and of 2,6-dichlorothiobenzamide (Prefix, Shell) which shows herbicidal activity.

**2,6-Difluorobenzonitrile** [*1897-52-5*], 1-cyano-2,6-difluorobenzene, $M_r$ 172.02, *mp* 29 – 30 °C, forms colorless crystals. The nitrile can be produced by treatment of 2,6-dichlorobenzonitrile with potassium fluoride in sulfolane [23] and is used as an intermediate in the synthesis of insecticides (e.g., Dimilin, Shell).

**2-Chloro-4-nitrobenzonitrile** [*28163-00-0*], 2-chloro-1-cyano-4-nitrobenzene, $M_r$ 182.57, *mp* 81 °C, is produced by the reaction of the diazonium salt derived from 2-chloro-4-nitroaniline with copper cyanide (Sandmeyer reaction) [95]. 4-Amino-2-chlorobenzonitrile can be obtained from 2-chloro-4-nitrobenzonitrile by reduction of the nitro group with iron (Béchamp method): it is used as an intermediate for the production of azo dyes.

**4-Chloro-2-nitrobenzonitrile** [*34662-32-3*], 1-chloro-4-cyano-3-nitrobenzene, $M_r$ 182.57, *mp* 100–101 °C, is obtained by the Sandmeyer reaction of 4-chloro-2-nitrobenzene diazonium salt with copper cyanide [95]. Reduction of the nitro group with iron yields 2-amino-4-chlorobenzonitrile: the latter is an intermediate for the production of azo dyes.

**2-Chloro-5-nitrobenzonitrile** [*16588-02-6*], 1-chloro-2-cyano-4-nitrobenzene, $M_r$ 182.57, *mp* 108–109 °C, forms yellow needle-shaped crystals. The chlorine atom ortho to the nitrile group is prone to a range of nucleophilic substitution reactions, of which that with ammonia has achieved industrial importance [96]. The nitrile is produced by nitration of 2-chlorobenzonitrile (see above) at 0 °C [97].

**2-Amino-5-nitrobenzonitrile** [*17420-30-3*], 2-cyano-4-nitroaniline, $M_r$ 163.13, *mp* 210 °C, is readily soluble in acetone, chloroform, and benzene; it is sparingly soluble in water and ethanol. As already mentioned, the reaction of 2-chloro-5-nitrobenzonitrile (see above) with ammonia gives 2-amino-5-nitrobenzonitrile [98], an intermediate for azo dyes. A newer synthetic route starts from 6-nitroisatoic anhydride [2H-3,1-benzoxazine-6-nitro-2,4(1H)-dione]: the latter is converted into 2-amino-5-nitrobenzonitrile by reaction with ammonia, subsequent treatment with phosgene in dimethylformamide or *N*-methylpyrrolidone, followed by hydrolysis of the formamidine [77].

**4-Hydroxybenzonitrile** [*767-00-0*], 4-cyanophenol, 1-cyano-4-hydroxybenzene, $M_r$ 119.2, forms colorless crystals of *mp* 113 °C. 4-Hydroxybenzonitrile is produced by reaction of 4-bromophenol with copper(I) cyanide [99], by ammoxidation of *p*-cresol [100], by reaction of 4-chlorobenzonitrile with sodium methoxide [101], or by the catalyzed gas phase reaction of ethyl 4-hydroxybenzoate with ammonia [75]. The nitrile is used for the production of the herbicides 3,5-dibromo- and 3,5-diiodo-4-hydroxybenzonitrile (Bromoxynil, Ioxynil, May & Baker).

### 3.3.2.2. Cyanonaphthalenes (Naphthalene Carboxylic Acid Nitriles)

In principle cyanonaphthalenes can be produced by the general methods given in Section 3.2. Additional production methods involving ring-closure condensation reactions are described under 1,4-dicyanonaphthalene. The industrial importance of cyanonaphthalenes is low.

**1-Cyanonaphthalene** [86-53-3], α-naphthonitrile, 1-naphthoic acid nitrile, $M_r$ 153.17, mp 35.5 – 36 °C, bp 296 – 299 °C, forms colorless crystals that are readily soluble in ethanol. It is used in organic synthesis.

**2-Cyanonaphthalene** [613-46-7], β-naphthonitrile, 2-naphthoic acid nitrile, $M_r$ 153.17, mp 66 °C, bp 303 °C, forms colorless crystals that are readily soluble in ethanol. It is used in organic synthesis and suitable as a reaction medium for obtaining propylene from ethylene and 2-butene by olefin metathesis [102].

**1,4-Dicyanonaphthalene** [3029-30-9], naphthalene-1,4-dicarboxylic acid dinitrile, $M_r$ 178.18, mp 208 °C, crystallizes in long, colorless needles. 1,4-Dicyanonaphthalene is produced by condensation of 1,2-phenylenediacetonitrile (see p. 3470) with glyoxal or its bisaldimines [103]. It is used as an intermediate for the synthesis of optical brighteners and dyes from naphtholactams.

**2,6-Dicyanonaphthalene** [31656-49-2], naphthalene-2,6-dicarboxylic acid dinitrile, $M_r$ 178.18, mp 294 – 297 °C, can be obtained by ammoxidation of 2,6-dimethylnaphthalene on sodium vanadate bronzes doped with $Fe_2O_3$ [104]. It is used for the production of 2,6-bis(aminomethyl)naphthalene, an intermediate for dyes.

**6-Cyano-β-naphthol** [52927-22-7], 6-cyano-2-hydroxynaphthalene, 6-hydroxy-2-naphthoic acid nitrile, $M_r$ 169.17, mp 166 – 167 °C. This nitrile is used as an intermediate for the production of liquid crystals [105] and monoamino-oxidase-inhibiting naphthalene-2-oxyalkylamines, which are used as antidepressive agents [106].

## 3.4. Toxicology

**Mandelonitrile** is rather toxic with an $LD_{50}$ value (rat, oral) of 116 mg/kg [107]. This high toxicity may be due to the equilibrium of the cyanohydrin with benzaldehyde and hydrogen cyanide.

**Phenylacetonitrile** has an oral $LD_{50}$ value (rat) of 270 mg/kg [108].

**Benzonitriles with Methyl Substituents.** The various methyl-substituted derivatives (2-, 3-, and 4-methylbenzonitrile and 2,3-, 2,4-, and 2,5-dimethylbenzonitrile) are less toxic than benzonitrile itself. Their $LD_{50}$ values (rat, oral) are in the range 3000 – 4500 mg/kg [109]. The monomethylbenzonitriles are reported to be metabolized by oxidation of the methyl group [109]. Mono- and dimethylbenzonitriles generally show a stronger irritating effect on the skin and mucous membrane than benzonitrile [109].

**Miscellaneous Substituted Benzonitriles.** The oral $LD_{50}$ values of some substituted benzonitriles are as follows:

| | |
|---|---|
| 2-Chlorobenzonitrile | 435 mg/kg [109] |
| 4-Chlorobenzonitrile | 887 mg/kg [107] |
| 2,6-Dichlorobenzonitrile | 2710 mg/kg [107] |
| 2-Chloro-5-nitrobenzonitrile | 3200 mg/kg [109] |
| 4-Nitrobenzonitrile | 30 mg/kg [109] |

The $LD_{50}$ values (rat, oral) for nitriles which are used as herbicides are as follows [110]:

| | |
|---|---|
| 3,5-Diiodo-4-hydroxybenzonitrile | 305 mg/kg |
| 3,5-Dibromo-4-octanoyloxybenzonitrile | 420 mg/kg |
| Trichlorophenylacetonitrile | 1050 mg/kg |
| Diphenylacetonitrile | 200 mg/kg |

# 4. References

[1] K. Weissermel, H.-J. Arpe: *Industrial Organic Chemistry*, Verlag Chemie, Weinheim 1978, pp. 268–269.
[2] Distillers Co., GB 723 003, 1955 (D. J. Hadley, C. A. Woodcock).
[3] Eastman Kodak Co., US 2 786 867, 1957 (H. J. Hagemeyer, B. Thompson, C. W. Hargis).
[4] Ruhrchemie AG, DE 3 014 729, 1981 (G. Horn, D. Frohning, H. Liebern).
[5] Bayer, DE 3 216 382, 1983 (F. Hagedorn et al.).
[6] A. Hulkenberg, J. J. Troost, *Tetrahedron Lett.* 1974, 3187.
[7] C. Fizet, J. Streith, *Tetrahedron Lett.* 1982, 1505.
[8] G. Lohaus, *Chem. Ber.* **100** (1967) 2719.
[9] U. Feldhues, H. J. Schäfer, *Synthesis* 1982, 145.
[10] Bayer, EP 58 333, 1982 (D. Arlt, G. Klein).
[11] in [1] pp. 217–218.
[12] W. Nagata, M. Yoshioka, *Org. React. (N.Y)* **25**, 1977 255–476.
[13] Ethyl Corp., DE 2 618 744, 1976 (E. G. Woods, L. H. Shephard, E. P. Breidenbach).
[14] In [1] pp. 216–217.
[15] Degussa, EP 36 441, 1981 (A. Kleeman, B. Lehmann, H. Klenk).
[16] M. Kobler, K. H. Schuster, G. Simchen, *Liebigs Ann. Chem.* 1978, 1946.
[17] T. M. Reetz, *Tetrahedron* **39** (1983) 961.
[18] Asahi Chemical Industry Co., JP 84 185 788, 1984.
[19] SRI International: *Chemical Economics Handbook*, Menlo Park, Feb. 1988.
[20] Sagami, BE 847 864, 1977 (K. Kondo et al.).
[21] Takeda, US 2 592 930, 1952.
[22] Lonza AG, CH 641 155, 1984 (H. Althaus).
[23] Mitsubishi Chemicals Ind. Co., Ltd., JP 74 47 303, 1974 (Y. Kageyama, Y. Fukai).
[24] Canter et al., *J. Chem. Soc.* 1931, 1245.
[25] E. R. Biehl, H. M. Li, *J. Org. Chem.* **31** (1966) 602.
[26] Merck and Co., Inc., US 3 020 200, 1962 (S. Rogers).

[27] J.R. Geigy AG, US 2 754 243, 1956 (M. Gysin).
[28] Standard Oil Co., Ohio, JP 82 67 551, 1982.
[29] *Chem. Eng. News* (1984, May 7) 19.
[30] H. G. Elias: *New Commercial Polymers 1969–1975*, Gordon and Breach Science Publishers, New York 1977, p. 21.
[31] Mitsubishi Petrochemicals Co., Ltd., JP 82 158 750, 1982.
[32] BASF, EP 57 890, 1982 (W. Rebafka, G. Heilen, K. Halbritter, W. Franzischka).
[33] A. J. Fatiadi, *Synthesis* 1987, 749, 959.
[34] Bayer, EP 9 788, 1980 (R. Schubart, R. Braden).
[35] Mitsubishi Chemical, JP 84 212 459, 1984.
[36] Ciba, US 2 928 829, 1960 (R. P. Mull).
[37] Hoffmann-La Roche, EP 4 334, 1979 (H. Fischer, F. Schneider, R. Zurflüh).
[38] Carpenter, *J. Org. Chem.* **27** (1962) 2085.
[39] Dow Chemical Co., US 3 923 860, 1975 (S. H. Ruetman).
[40] Du Pont, US 3 418 228, 1968 (P. L. Bartlett).
[41] Olin Mathieson, US 3 260 588, 1966.
[42] Dow Chemical, US 2 890 238, 1959 (A. R. Sexton).
[43] Mitsubishi Gas Chemical Co., Inc., JP 8 280 346, 1982.
[44] Showa Denko K.K., JP 79 46 720, 1979 (K. Nakayasu, O. Furuya).
[45] Showa Denko K.K., US 4 299 978, 1981 (K. Nakayasu, O. Furuya, Y. Hosaki).
[46] Asahi Chemical Ind. Co., JP 82 82 345, 1982.
[47] Merck and Co., US 2 742 491, 1956 (J. Weijlard, A. P. Sullivan).
[48] Pfizer, US 2 957 025, 1960 (J. O. Brooks).
[49] Hoechst, DE 2 555 043, 1977 (H. Erpenbach, K. Gehrmann, H. Joest, W. Lork).
[50] Yamanouchi, DE 2 951 675, 1979 (Y. Hirata, I. Yanagisawa, Y. Ishii, M. Takeda).
[51] Nitto Chemicals Ind. Co. Ltd., JP 84 196 851, 1984.
[52] Orsymonde, GB 1 325 192, 1971 (L. Lafon).
[53] Bristol-Myers Co., DE 2 057 845, 1971 (Y. H. Wu, J. W. Rayburn).
[54] Ciba-Geigy Co., US 3 853 942, 1974 (Y. S. Surey, H. C. Grace).
[55] Nippon Chemicals Co., Ltd., US 4 066 683, 1978 (S. Hideo).
[56] *Japan Chem. Week* Dec. 1 (1988).
[57] Nippon Soda, DAMN company brochure.
[58] M. Mittelback, *Monatsh. Chem.* **116** (1985) no. 5, 689–691.
[59] Nippon Kayaku, JP 62 220 557, 1987 (Tada, Shoji).
[60] Lentia GmbH, DE 3 231 052, 1984 (G. Stern, F. Grossgut, M. Joos, H. Weismann).
[61] Sogo Pharmaceuticals Corp., Ltd., US 4 152 336, 1979 (K. S. Masakata, K. M. Sadamitsu, S. Y. Kazuyuki).
[62] Ube Industries Ltd., EP 55 108, 1980 (K. Matsui, S. Uchimumi, A. Iwayama, T. Umeza).
[63] *NIOSH Criteria for a Recommended Standard*, US Department of Health, Education and Welfare, Washington 1978.
[64] G. D. Clayton, F. E. Clayton: *Patty's Industrial Hygiene and Toxicology*, 3rd revised ed., vol. **2 C**, pp. 4845–4900.
[65] N. I. Sax: *Dangerous Properties of Industrial Materials*, 6th ed., Van Nostrand Reinhold Comp., New York 1984.
[66] A. E. Ahmed, Y. M. Faroqui, *Toxicol. Lett.* **12** (1982) 157–163.
[67] E. H. Silver, S. H. Kuttab, T. Hasan, M. Hassan, *Drug. Metab. Dispos.* **10** (1982) 495–498.
[68] A. E. Ahmed, N. M. Trieff, *Prog. Drug Metab.* **7** (1983) 229–294.

[69]   S. Szabo, G. T. Gallagher, *Surv. Synth. Pathol. Res.* **3** (1984) 11–30.
[70]   H. Tanii, K. Hashimoto, *Arch. Toxicol.* **55** (1984) 47–54.
[71]   H. Tanii, K. Hashimoto, *Arch. Toxicol.* **57** (1985) 88–93.
[72]   BASF, DE-OS 1 643 722, 1967 (C. Palm et al.).BASF, EP 0 222 249, 1985 (H. Engelbach, R. Krokoszinski, W. Franzischka, M. Decker).
[73]   Showa Denko, *Hydrocarbon Process.* **45** (1969) no. 11, 252.
[74]   Rhône Poulenc, DE-OS 2 141 657, 1970 (Y. Colleuille, R. Gardon).
[75]   Ueno, EP 0 074 116, 1981 (R. Ueno, K. Sakota, K. Kawata, Y. Naito).
[76]   E. Vowinkel, J. Bartel, *Chem. Ber.* **107** (1974) 1221–1227.
[77]   BASF, DE-OS 2 628 055, 1976 (D. Schneider, H. Scheuermann).
[78]   Hoechst, DE-OS 1 668 068, 1967 (A. Hansel). Bayer, DE-OS 2 550 262, 1975 (F. Hagedorn, E. Klauke, K. Wedemeyer).
[79]   F. Becke, T. Burger, *Justus Liebigs Ann. Chem.* **716** (1968) 78–82.
[80]   Sumitomo, DE-OS 2 047 160, 1969 (S. Sagawa, T. Nakagawa, S. Okamoto).
[81]   BASF, EP 0 073 326, 1981 (P. Magnusson, W. Trauzettel).
[82]   Mobil Oil, US 3 433 821, 1965 (L. A. Hamilton, P. S. Landis).
[83]   J. Lücke, R. E. Winkler, *Chimia* **25** (1971) 94.
[84]   H. Vorbrüggen, *Tetrahedron Lett.* **13** (1968) 1631–1634. Schering, DE-OS 1 643 009, 1967 (H. Vorbrüggen).
[85]   K. Fukunaga, *Yuki Gosei Kagaku Kyokaishi* **33** (1975) no. 10, 774. Bayer, DE-OS 1 668 034, 1967 (H. Leuchs, K. H. Schmidt).
[86]   Ciba-Geigy, DE-OS 2 743 094, 1976 (L. Guglielmetti, A. C. Rochat, I. J. Fletcher).
[87]   Bayer, DE-OS 2 614 240, 2 614 241, 2 614 242, 1975 (K. Findeisen). Degussa, DE-OS 2 624 891, 1975 (H. Klenk et al.).
[88]   Bayer, DE-OS 2 914 091, 1980 (H. H. Schwarz, H. P. Wirges, A. Judat, G. Zumach).
[89]   Ciba-Geigy, DE-OS 2 632 402, 1975 (P. Loew, H. Schwander, H. Kristinsson).
[90]   Nippon Kayaku, DE-OS 2 711 332, 1978 (H. Hayami, H. Shimizu, G. Takasaki).
[91]   Nippon Shokubai, EP 0 290 996, 1987 (A. Inoue et al.).
[92]   Ciba-Geigy, EP 0 094 911, 1982 (A. C. Rochat, L. Caesar, A. Iqbal).
[93]   Nitto, EP 0 158 928, 1986 (Y. Kiyomiya, Y. Yamaguchi, M. Ushigome, H. Murata).Süddeutsche Kalkstickstoffwerke, EP 0 158 928, 1984 (H. Bayerl, W. Pollwein).
[94]   I.S.C. Chemicals, DE 2 902 877, 1978 (G. Fuller); *J. Agri. Food Chem.* **21** (1973) no. 6, 997.
[95]   C. W. N. Holmes, J. D. Loudon, *J. Chem. Soc.* 1940, 1521–1523.
[96]   J. Baudet, *Rec. Trav. Chim. Pays Bas* **43** (1924) 708–715.
[97]   B. B. Dey, Y. G. Doraiswami, *J. Indian Chem. Soc.* **10** (1933) 309.
[98]   B. Hartmanns, *Rec. Trav. Chim. Pays-Bas* **65** (1946) 468–469.
[99]   Sumitomo, DE-OS 2 047 161, 1969 (S. Sagawa, T. Nakagawa, S. Okamoto).
[100]  Akademie d. Wissenschaften d. DDR, DD 232 693, 1984 (A. Martin et al.).
[101]  J. Prachensky et al., CS 242 141, 1984.
[102]  Pittsburg Petroleum Gas, US 3 637 894, 1969 (J. C. Crano, E. K. Fleming).
[103]  Ciba-Geigy, DE-OS 2 649 167, 1975 (H. Schwander, C. Zickendraht, L. Guglielmetti).Hoechst, DE-OS 2 503 321, 1975 (L. Heiss).
[104]  Suntech, US 4 064 072, 1976 (R. D. Bushick).
[105]  RCA Corp., US 3 925 237, 1974 (D. L. Ross, D. M. Gavrilovic).
[106]  Kyowa Hakko, JP 52 083 533, 52 023 055, 1976 (N. Nakamizo, T. Hirata, T. Takahashi).
[107]  J. V. Marhold: *Sbornik Vysledku Toxikologickeho Vysetreni Latek A pripavku*, Institut pro Vychovu Veducicn Prakovniku Chemickeho Prumyclu, Praha, Czechoslovakia, 1972, p. 161.

[108] G. P. Galibin, V. J. Fedorova, N. M. Karamzina, *Gig. Sanit.* **32** (1967) no. 8, 20.
[109] H. Zeller, H. Th. Hoffmann, A. M. Thiess, W. Hey, *Zentralbl. Arbeitsmed. Arbeitsschutz* **19** (1969) 225.
[110] A. Fischer: BASF, *Herbizide,* 1971, p. 154.

# Nitrilotriacetic Acid

CHARALAMPOS GOUSETIS, BASF Aktiengesellschaft, Ludwigshafen, Federal Republic of Germany
HANS-JOACHIM OPGENORTH, BASF Aktiengesellschaft, Ludwigshafen, Federal Republic of Germany

| 1. | Introduction | 3479 | 5. | Uses | 3483 |
|---|---|---|---|---|---|
| 2. | Properties | 3479 | 6. | Economic Aspects | 3484 |
| 3. | Production | 3481 | 7. | Toxicology and Environmental Aspects | 3484 |
| 4. | Analysis | 3482 | 8. | References | 3485 |

## 1. Introduction

Nitrilotriacetic acid [139-13-9], NTA, $N,N$-bis(carboxymethyl)glycine, along with ethylenediaminetetraacetic acid (EDTA, → Ethylenediaminetetraacetic Acid and Related Chelating Agents) is one of the most important aminopolycarboxylate chelating agents.

$$N\begin{matrix}-CH_2COOH\\-CH_2COOH\\-CH_2COOH\end{matrix}$$

Its synthesis from ammonia and chloroacetic acid was accomplished by HEINTZ as early as 1861 [1]. Industrial production first took place in 1936 in Ludwigshafen. Since that time, NTA has become an established chelating agent and is used in many industries where multivalent metal ions, chiefly alkaline-earth ions, must be "masked" in order to modify their properties or prevent typical reactions such as the formation of insoluble precipitates.

## 2. Properties

**Physical Properties.** Nitrilotriacetic acid, $C_6H_9NO_6$, $M_r$ 191.14, is a tribasic acid, crystallizing from water as colorless needles. The most important physical data are as follows [2]:

| | |
|---|---|
| mp (decomp.) | 242 °C |
| Solubility in water, g/100 g | |
| 5 °C | 0.13 |
| 22.5 °C | 0.13 |
| 80 °C | 0.95 |
| 100 °C | 3.3 |
| Dissociation constants [3] (25 °C, ionic strength = 0.1 mol/L) | |
| p$K_1$ | 1.80 |
| p$K_2$ | 2.48 |
| p$K_3$ | 9.65 |

The pH of a saturated solution (ca. 0.13 g NTA per 100 g of solution at room temperature) is ca. 2.3. The alkali-metal salts are the most soluble NTA salts. For example, ca. 640 g of trisodium nitrilotriacetate [5064-31-3] dissolve in 1 L of water.

**Chemical Properties.** The most important property of NTA is its ability to form watersoluble chelates with multivalent metal ions. Not only the carboxyl groups but also the tertiary nitrogen atom function as ligands. Unoccupied coordination sites of the metal ion can be saturated by water molecules. Although the chelates of NTA with metal ions are often represented as octahedral (coordination number 6) for the sake of simplicity, recent structural studies (chiefly by X-ray methods) have shown that higher coordination numbers are also possible for some metal ions. For example, $Ca^{2+}$ in the chelate $CaNTA^-$ has a coordination number of 7 [4].

As a rule, complexes are 1:1, i.e., 1 mol of metal ions reacts with 1 mol of NTA. If NTA is present in excess, 1:2 complexes with 1 mol of metal ions complexed by 2 mol of NTA are formed.

The stability of the metal complex is described by stability constants ($K_1$ for 1:1 complexes, $K_2$ for 1:2 complexes; see Table 1). Further, because of the polarization of the H–O bond in the chelate, the 1:1 complexes behave as weak acids and dissociate. This effect is described by the dissociation constant $K_d$:

$$K_1 = \frac{[MNTA]}{[M][NTA]}$$

$$K_2 = \frac{[MNTA_2]}{[M][NTA]^2}$$

$$K_d = \frac{[MNTAOH][H^+]}{[MNTA]}$$

**Table 1.** Stability constants for NTA chelates*

| Metal ion | $\log K_1$ | $\log K_2$ | $pK_d$ |
|---|---|---|---|
| $Al^{3+}$ | 11.4 | | 5.09 |
| $Ca^{2+}$ | 6.39 | 8.76 | |
| $Cd^{2+}$ | 9.78 | 14.39 | 11.25 |
| $Co^{2+}$ | 10.38 | 14.33 | 10.80 |
| $Cu^{2+}$ | 12.94 | 17.42 | 9.14 |
| $Fe^{2+}$ | 8.33 | 12.80 | 10.60 |
| $Fe^{3+}$ | 15.90 | 24.30 | 4.1/7.8** |
| $Hg^{2+}$ | 14.60 | | |
| $Mg^{2+}$ | 5.47 | | |
| $Mn^{2+}$ | 7.46 | 10.94 | |
| $Ni^{2+}$ | 11.50 | 16.32 | 10.86 |
| $Pb^{2+}$ | 11.34 | | |
| $Zn^{2+}$ | 10.66 | 14.24 | 10.06 |

\* Ionic strength (25 °C) = 0.1 mol/L.
\*\* Behaves as a dibasic acid.

where M is a multivalent metal ion, NTA represents the anion $N(CH_2COO^-)_3$, and the charges have been omitted for simplicity except for $H^+$.

Complexing of metal ions with NTA competes with other secondary reactions. For example, metal ions react with anions such as carbonate, sulfide, sulfate, or oxalate to form sparingly soluble precipitates, and hydrogen ions compete with metal ions for sites on the NTA trianion. This means that several equilibria must be taken into consideration in the analysis of NTA systems.

Nitrilotriacetate chelates are stable over relatively wide pH ranges that depend on the concentration of the chelate and the excess of the complexing agent: $Ca^{2+}$, pH 9–12; $Mg^{2+}$, pH 7–10; $Cu^{2+}$, pH 3–12; $Fe^{3+}$, pH 1.5–3. Assuming a stoichiometric ratio for 1:1 complexing, in the optimal pH range 1 g of $Na_3NTA$ binds 156 mg of $Ca^{2+}$, 94 mg of $Mg^{2+}$, 247 mg of $Cu^{2+}$, or 217 mg of $Fe^{3+}$.

# 3. Production

Today the original synthesis of NTA from ammonia and chloroacetic acid has only historical significance [1]. The oxidation of triethanolamine is likewise of no industrial importance [5]. The one-stage alkaline and two-stage acid processes now in use are based on the cyanomethylation of ammonia (or ammonium sulfate) with formaldehyde and sodium cyanide (or hydrogen cyanide).

**Alkaline Process.** The alkaline process was long the established method for NTA production. Trisodium nitriloacetate is synthesized as follows:

$NH_3 + 3\ HCHO + 3\ NaCN \rightarrow N(CH_2CN)_3 + 3\ NaOH$
$N(CH_2CN)_3 + 3\ NaOH + 3\ H_2O \rightarrow N(CH_2COONa)_3 + 3\ NH_3$

The reaction can be carried out batchwise or continuously, but the continuous process is more economical.

In a typical process, aqueous NaCN solution is fed into a cascade reactor system along with formaldehyde solution [6]. Ammonia is liberated during the synthesis and need not be supplied. The reaction takes place at 80–100 °C. Because of the high pH (ca. 14), triscyanomethylamine $N(CH_2CN)_3$ [*7327-60-8*] is hydrolyzed in situ to $Na_3NTA$. This process generates three times as much ammonia as it consumes and the ammonia concentration must be limited to suppress the production of substances with low degrees of carboxymethylation (glycine, iminodiacetic acid). This is largely achieved by continuously distilling off the ammonia with steam or air throughout the process. However, formation of byproducts (chiefly glycolic acid, hexamethylenetetramine, and the above-mentioned amino acids) cannot be completely prevented.

The resulting solution is sold directly as a 40-wt % solution, or used in the production of $Na_3NTA \cdot H_2O$ in powder form, or acidified to pH 1–2 to yield the acid ($H_3NTA$).

**Acid Process** [7]. The significant yield of byproducts in the alkaline process has led in recent years to the construction of plants based on the acid process, which features much lower byproduct levels. The acid process is associated with stringent safety requirements due to the use of hydrogen cyanide; corrosion can also be a problem.

In the first stage, ammonia is reacted with formaldehyde to give hexamethylenetetramine, which is then reacted with hydrogen cyanide in sulfuric acid solution to yield triscyanomethylamine. The solid triscyanomethylamine is sparingly soluble in the acidic solution and is filtered off, washed, and saponified with NaOH to give $Na_3NTA$. The resulting solution has a far lower byproduct content than the solution from the alkaline method. It is also sold as 40 % product or used in the production of $Na_3NTA \cdot H_2O$ or $H_3NTA$ (see above).

**Trade Names.** The following trade names apply to the most important NTA products: Dissolvine A grades (Akzo), Hampshire NTA grades (W. R. Grace), Masquol NTA grades (Protex), Nervanaid NTA grades (ABM Chemicals), Rexene NTA grades (Rexolin Chemicals), Trilon A grades (BASF), and Versene NTA grades (Dow).

# 4. Analysis

A number of methods are suitable for the quantitative determination of NTA and its salts. One widely used technique is potentiometric titration with iron(III) chloride under acidic conditions [8], [13]. Ion-pair chromatography [9], GC analysis [10], and polarographic determinations [11] can also be employed in the trace range.

# 5. Uses

The uses of NTA and its salts are based on their complexing properties for metal ions [12], [13]. It is employed in many areas to eliminate the interfering effects of metal ions by complexing, to dissolve precipitates, to modify the redox potentials of metal ions, or to prepare metal-ion buffers. The most important application areas are described below.

**Water Softening.** Nitrilotriacetic acid softens water by complexing $Ca^{2+}$ and $Mg^{2+}$ ions; complexation occurs in the neutral to basic range. The softening of water with hardnesses of 1 °d, 1 °f, 1 °e, and 1 mmol $Ca^{2+}$/L requires the following amounts of chelating agent: 34, 19, 27, and 191 mg NTA, or 46, 26, 36, and 257 mg $Na_3NTA$, respectively. [1 °d (German hardness) corresponds to 0.01 g CaO/L water, 1 °e (English hardness) to 0.01 g $CaCO_3$/0.7 L water, and 1 °f (French hardness) to 0.01 g $CaCO_3$/L water.]

Water softening with NTA is used in the paper and textile industries, soap manufacture, cosmetics and toiletries, the production of detergents and cleansers, the treatment of boiler feedwater, and in many chemical process industries (e.g., photographic materials and electroplating industries).

**Phosphate Replacement in Detergents.** $Na_3NTA$ can be used as a substitute for pentasodium triphosphate in the formulation of low phosphate detergents or, with zeolite and sometimes polycarboxylates, in phosphate-free detergents. Substitution of phosphates can help reduce the eutrophication of surface waters.

In comparison with pentasodium triphosphate, $Na_3NTA$ has the advantage of resisting hydrolysis, so that it remains fully effective after spray drying of the detergent slurry and also during washing. The equivalence ratio of $Na_3NTA$ to pentasodium triphosphate is about 0.6 to 1 and there is no detrimental effect on primary and secondary detergency (anti-redeposition).

**Replacement of Tetrapotassium Diphosphate in Cleansers.** The equivalence ratio of $Na_3NTA$ to tetrapotassium diphosphate is favorable (ca. 1:3) and only low concentrations of $Na_3NTA$ are needed in cleansers. This often makes the use of hydrotropes or solubilizers superfluous. Synergetic builder effects occur with partial substitution of the phosphate.

**Masking of Heavy-Metal Ions.** Although inferior to EDTA in many applications, NTA is also frequently employed for masking heavy-metal ions because of its greater specific complexing power for interfering ions such as $Fe^{3+}$, $Cu^{2+}$, and $Mn^{2+}$. Examples of such applications are in soap manufacture, textile bleaching, the cleaning of hard surfaces, and in many areas of the chemical process industries.

**Special applications** include the separation of rare-earth elements, complexometric titrations, and the preparation of micronutrients for agriculture.

# 6. Economic Aspects

It is difficult to state figures for production capacity of NTA mainly because all existing plants can also be used for producing other aminopolycarboxylates. A worldwide capacity of ca. 100 000 t seems realistic however [15].

World consumption of Na$_3$NTA (as 100% product) was ca. 40 000 t in 1985 and about 70 000 t in 1989 [15]. There continues to be excess production capacity, built when a rapid increase in demand was expected for the formulation of phosphate-free and low-phosphate detergents, chiefly in the United States. The expected increase has only taken place in Canada, where ca. 40 000 t of Na$_3$NTA were used in 1988. In Western Europe, consumption is ca. 20 000 t/a; NTA is only used in detergents in a few European countries (Switzerland, The Netherlands). Consumption in the United States has stagnated at 5000–7000 t/a for industrial and institutional applications [14].

Important manufacturers are Monsanto (USA), W. R. Grace (USA, UK), BASF (FRG), ABM Chemicals (UK), Akzo (Netherlands), Aminkemi (Sweden), SA Dabeer (Spain), and Protex (France).

It is hard to predict market development for the coming years. Sizable increases should be expected in Western Europe if a number of countries (especially the FRG) adopt the use of NTA in detergents.

# 7. Toxicology and Environmental Aspects

The results of extensive studies on the environmental behavior and toxicology of NTA have been summarized [16]–[18].

After adaptation of relevant microorganisms (i.e., production of necessary enzymes), NTA is readily biodegradable. It is completely degraded to inorganic end products (i.e., mineralized) without any measurable concentration of organic metabolites being produced. Environmental studies performed after the introduction of NTA-containing detergents in Switzerland, also confirm the good biodegradability: > 98% of the NTA is biodegraded in biological water treatment plants [19]–[21]. The half-life in the receiving water is < 1 d [22]; virtually no NTA could be detected in ground water after infiltrating a few meters [23], [24].

As a powerful complexing agent, NTA can form heavy-metal complexes under certain conditions. Whether such an effect occurs in the environment depends on the NTA concentration. Since NTA is rapidly biodegraded, remobilization of heavy metals from the activated sludge of water treatment plants is negligible in comparison with the normal heavy-metal levels in the effluent [22], [23]. The same applies to the remobilization of heavy metals from river sediments, where under realistic conditions no measurable changes in the heavy-metal concentrations can be expected.

The acute toxicity to aquatic organisms is slight. Data for more than 50 freshwater and marine species are known; $LC_{50}$ values vary from ca. 100 to > 10 000 mg/L. The numerical values depend highly on the hardness of the test medium. Numerous studies have also been performed on the chronic toxicity of NTA in microorganisms, macro-invertebrates, amphibians, and fish. No toxic effects are seen at NTA concentrations below 1 mg/L. In most organisms, perceptible toxic effects only occur at very much higher levels.

The acute toxicity in mammals is species dependent. For example, $LD_{50}$ (oral) values of ca. 2000 mg/kg are found for the trisodium salt ($Na_3NTA \cdot H_2O$) in rodents; vomiting occurs in dogs and monkeys. After inhaling 5 mg of NTA dust per liter of respired air over 4 h, rats displayed no clinical symptoms. The irritant action of NTA on the skin, eyes, and respiratory tract is slight; indications of sensitization and allergies are not known. Nitrilotriacetic acid is not teratogenic either by itself or in combination with heavy metals such as cadmium and mercury. Numerous mutagenicity tests have not shown indications of genotoxicity.

Nitrilotriacetic acid is not metabolized by mammals; any absorbed NTA is excreted unaltered by the kidneys. Tissue concentrations are highest in the cells of the renal tubules and in the ureters. Subchronic and chronic studies therefore show toxic lesions of the kidneys, which can be attributed to disorders of electrolyte and iron metabolism. High doses (e.g., 2000 mg per kilogram body weight per day) result in injury to the epithelial cells of kidneys, ureters, and bladder, and as a consequence to tumor formation. There is, however, a threshold concentration below which tumors cannot develop. The threshold concentration is much higher than the NOEL (no adverse effect level) of 14 mg per kilogram body weight per day.

# 8. References

[1] W. Heintz, *Ann. Chem. Pharm.* **122** (1862) 257–294.
[2] *Beilstein*, **H** 369, **E I** 482, **E II** 801, **E III** 1180.
[3] A. E. Martell, R. M. Smith: *Critical Stability Constants*, vol. **1**, Plenum Press, New York 1974.
[4] G. Wilkinson (ed.): *Comprehensive Coordination Chemistry*, vol. **3**, Pergamon Press, Oxford 1987, p. 33.
[5] Carbide & Carbon Chemical, US 2 384 816, 1945 (G. Curme, Jr., H. Chitwood, J. Clark); US 2 384 818, 1945 (H. Chitwood).
[6] BASF, DE 837 999, 1952 (E. Ploetz); DE 2 062 435, 1970 (E. Hartert); DE 2 625 974, 1976 (K.-L. Hock).W. R. Grace, DE 1 813 718, 1969 (G. Busch, B. Elofsson).
[7] BASF, DE 3 029 205, 1980 (H. Distler, K.-L. Hock). W. R. Grace, US 3 463 811, 1967 (J. Goldfrey, J. Sykes, J. Harper, C. Morgan). Hampshire Chemical, US 2 855 428, 1958 (J. Singer,

M. Weisberg). Monsanto Co, EP 148 146, 1985 (C. Y. Shen). Stauffer Chemical US 3 984 453, 1966 (S. Chaberek).
- [8] G. Schwarzenbach, H. Flaschka: *Die komplexometrische Titration*, 5th ed., Enke Verlag, Stuttgart 1985. S. Siggia, D. Eichlin, R. Rheinhart, *Anal. Chem.* **27** (1955) 1745–1749.
- [9] J. Weiss: *Handbuch der Ionenchromatographie*, Dionex GmbH, Weiterstadt, 1985, p. 253.
- [10] N. T. de Oude, *Vom Wasser* **64** (1985) 283–292.
- [11] DIN 38 413,Part 5.
- [12] *Ullmann*, 4th ed., **17**, 340.
- [13] BASF, Technical Bulletin, *Trilon Chelating Agents*, Ludwigshafen 1989.
- [14] M. Salaices: "Chelating Agents," in *Chemical Economics Handbook Product Review*, SRI International, Menlo Park 1987.
- [15] BASF, personal communication, Ludwigshafen 1989.
- [16] H. Bernhardt (ed.): "NTA, Studie über die aquatische Umweltverträglichkeit von Nitrilotriacetat," Verlag H. Richarz, Sankt Augustin 1984.
- [17] R. L. Andersson, W. E. Bishop, R. L. Campbell, *CRC Crit. Rev. Toxicol.* **15** (1985) 1–102.
- [18] E. Bayer (ed.): "Nitrilotriessigsäure, BUA-Stoffbericht 5," Beratergremium für umweltrelevante Altstoffe (BUA) der Gesellschaft Deutscher Chemiker, VCH Verlagsgesellschaft, Weinheim – New York 1987.
- [19] W. Giger, M. Ahel, M. Koch, H. U. Laubscher, C. Schaffner, J. Schneider, *Water Sci. Technol.* **19** (1987) 449–460.
- [20] A. Alder et al. in *Eidgenössische Anstalt für Wasserversorgung, Abwasserreinigung und Gewässerschutz Jahresbericht 1987*, pp. 3.5–3.9.
- [21] H. Siegrist, A. Alder, W. Gujer, W. Giger, *Gas Wasser Abwasser* **68** (1988) 101–109.
- [22] P. Reichert et al., in *Eidgenössische Anstalt für Wasserversorgung, Abwasserreinigung und Gewässerschutz Jahresbericht 1987*, pp. 4.1–4.2.
- [23] E. Kuhn, M. van Loosdrecht, W. Giger, R. P. Schwarzenbach, *Water Res.* **21** (1987) 1237–1248.
- [24] L. Schaffner, M. Ahel, W. Giger, *Water Sci. Technol.* **19** (1987) 1195–1196.

# Nitro Compounds, Aliphatic

Sheldon B. Markofsky, W. G. Grace & Co., Columbia, Maryland 21044, United States

| | | | | |
|---|---|---|---|---|
| 1. | Introduction ............. 3487 | 6.1. | Nitromethane and Derivatives | 3494 |
| 2. | Properties ............... 3488 | 6.2. | Nitroethane and Derivatives.. | 3495 |
| 3. | Production .............. 3488 | 6.3. | 1-Nitropropane and Derivatives ............. | 3497 |
| 4. | Quality Specifications and Analysis ............... 3492 | 6.4. | 2-Nitropropane and Derivatives ............. | 3497 |
| 5. | Storage and Transportation .. 3492 | 7. | Toxicology and Occupational Health................. | 3498 |
| 6. | Uses of Nitroalkanes and their Derivatives ............. 3494 | 8. | References.............. | 3500 |

## 1. Introduction

Aliphatic nitro compounds ($RNO_2$), also called nitroalkanes and nitroparaffins, are isomeric with the chemically different alkyl nitrites (RONO). As shown by the following equilibrium involving a secondary nitroalkane, the nitroalkane and nitronic acid are tautomers, sharing a common nitronate anion.

$$\begin{matrix} R_2 \\ R_1 \end{matrix}\!\!\! CH\!-\!N\!\!\!\begin{matrix} O \\ O \end{matrix} \rightleftarrows \left[\begin{matrix} R_2 \\ R_1 \end{matrix}\!\!\! C\!=\!N^+\!\!\!\begin{matrix} O^- \\ O^- \end{matrix}\right] \rightleftarrows \begin{matrix} R_2 \\ R_1 \end{matrix}\!\!\! C\!=\!N\!\!\!\begin{matrix} O \\ OH \end{matrix}$$

The nitroalkane is less acidic than the corresponding nitronic acid and, in most cases, the equilibrium lies far to the left. The acidity of the primary and secondary nitroalkanes is due partly to the electron-withdrawing effect of the nitro group and the resonance stability of the nitronate anion.

## 2. Properties

**Physical Properties.** The four nitroalkanes of greatest industrial significance, nitromethane [75-52-5], $CH_3NO_2$; nitroethane [79-24-3], $CH_3CH_2NO_2$; 1-nitropropane [108-03-2], $CH_3CH_2CH_2NO_2$; and 2-nitropropane [79-46-9], $CH_3CH(NO_2)CH_3$ are all colorless liquids when pure. Nitroalkanes are only slightly soluble in water. Due in part to its polarity, the nitro group causes a large increase in flash point and boiling point compared to the corresponding hydrocarbon. Nitroalkanes are often useful because they readily form azeotropes with many organic solvents such as alcohols, ketones, and hydrocarbons [1]. The physical properties of the four basic nitroalkanes are listed in Table 1.

**Chemical Properties.** Primary and secondary nitroalkanes are excellent building blocks for the synthesis of more complex molecules, since the readily formed nitronate anion undergoes many useful condensation reactions. Especially important is the condensation of these anions with aldehydes and ketones (Henry reaction) [3]–[8], and the Michael reaction, in which the nitronate anion reacts with $\alpha,\beta$-unsaturated carbonyl compounds, nitriles, esters, and other activated alkenes [9]–[12]. The following is an example of the Henry reaction:

$$CH_3CH_2NO_2 + HCHO \xrightleftharpoons{:B} CH_3\overset{NO_2}{\underset{|}{C}H}CH_2OH$$

A typical Michael addition is shown as follows:

$$CH_3CH_2NO_2 + CH_2=CHCN \xrightarrow{:B} CH_3\overset{NO_2}{\underset{|}{C}H}CH_2CH_2CN$$

Figure 1 shows how nitroalkanes can be employed to make more complicated molecules.

The nitro group can subsequently be transformed into a variety of other functional groups, as illustrated in Figure 2. Thus, nitroalkanes allow the precise placement of these functional groups into complex structures. The chemistry depicted in Figures 1 and 2 is discussed in detail in [2], [13]–[18].

## 3. Production

For more than forty years, the four principal nitroalkanes have been produced by the high-temperature vapor-phase nitration of propane [19]. The process, which employs nitric acid as the nitrating agent, is based on a free radical reaction, in which the active species is the $NO_2$ radical. This procedure produces a nitroalkane mixture rich in nitropropanes. The process is used by Angus Chemical Company for bulk production of nitromethane, nitroethane, 1-nitropropane, and 2-nitropropane.

**Table 1.** Physical properties of the four common nitroalkanes

| Property | Nitromethane | Nitroethane | 1-Nitropropane | 2-Nitropropane |
|---|---|---|---|---|
| $M_r$ | 61.041 | 75.068 | 89.095 | 89.095 |
| bp (101.3 kPa), °C | 101.20 | 114.07 | 131.18 | 120.25 |
| *Aqueous azeotrope* | | | | |
|   bp (101.3 kPa), °C | 83.59 | 87.22 | 91.63 | 88.55 |
|   wt% nitroalkane | 76.4 | 71.0 | 63.5 | 70.6 |
| mp (101.3 kPa), °C | −28.55 | −89.52 | −103.99 | −91.32 |
| Density, g/cm$^3$ | | | | |
|   at 20 °C | 1.138 | 1.051 | 1.001 | 0.988 |
|   at 30 °C | 1.124 | 1.039 | 0.991 | 0.977 |
| $n_D^{20}$ | 1.38188 | 1.39193 | 1.40160 | 1.39439 |
| Vapor pressure at 25 °C, MPa | 4.89 | 2.79 | 1.36 | 2.29 |
| Vapor density (air = 1) | 2.1 | 2.58 | 3.06 | 3.06 |
| Evaporation rate (BuOAc = 1) | 1.39 | 1.21 | 0.88 | 1.10 |
| Evaporation number (diethyl ether = 1) | 9 | 11 | 16 | 10 |
| Heat of combustion, kJ/mol (liq.) at 25 °C | −709 | −1363 | −2017 | −2001 |
| Heat of formation, kJ/mol (liq.) at 25 °C | −113 | −142 | −168 | −181 |
| Heat of vaporization, kJ/mol (liq.) | | | | |
|   at 25 °C | 38.3 | 41.6 | 43.4 | 41.4 |
|   at bp | 34.4 | 38.0 | 38.5 | 36.8 |
| Specific heat capacity at 25 °C, Jmol$^{-1}$ K$^{-1}$ | 106 | 138.6 | 175.7 | 175.3 |
| Solubility in water, wt% | | | | |
|   at 25 °C | 11.1 | 4.7 | 1.5 | 1.7 |
|   at 70 °C | 19.3 | 6.6 | 2.2 | 2.3 |
| Solubility of water in nitroalkane, wt% | | | | |
|   at 25 °C | 2.1 | 1.1 | 0.6 | 0.5 |
|   at 70 °C | 7.6 | 3.0 | 1.7 | 1.6 |
| Lower flammability, wt% | 7.1 | 3.4 | 2.2 | 2.6 |
| Upper flammability, wt% | | | | 11 |
| Flash point, °C | | | | |
|   Tag open cup | 44.4 | 41.1 | 48.9 | 37.8 |
|   Tag closed cup | 35 | 30.6 | 35.6 | 27.8 |
| Ignition temperature, °C | 418 | 414 | 421 | 428 |
| Critical temperature $t_c$, °C (calculated) | 315 | 324 | 340 | 326 |
| Critical pressure $p_c$, MPa (calculated) | 6.30 | 4.98 | 4.33 | 4.49 |
| Critical density $\varrho_c$, g/cm$^3$ (calculated) | 0.353 | 0.329 | 0.314 | 0.318 |
| Surface tension, (N/cm) × 10$^{-5}$ at 20 °C | 37.48 | 32.66 | 30.64 | 29.87 |
| Viscosity, mPa · s | | | | |
|   at 10 °C | 0.731 | 0.769 | 0.972 | 0.883 |
|   at 20 °C | 0.647 | 0.677 | 0.844 | 0.770 |
|   at 30 °C | 0.576 | 0.602 | 0.740 | 0.677 |
| Dielectric constant, at 30 °C | 35.87 | 28.06 | 23.24 | 25.52 |
| pH of 0.01 M aqueous solution at 25 °C | 6.4 | 6.0 | 6.0 | 6.2 |

The reaction is carried out at 350 – 450 °C. At this temperature all participants in the reaction are gaseous. The reaction is fast but not too fast, so that good temperature control is still assured. The pressure is adjusted to 0.8 – 1.2 MPa, so that the reaction product leaving the reactor can be condensed without liquefying the hydrocarbon. The temperature of the exothermic reaction can be controlled in various ways:

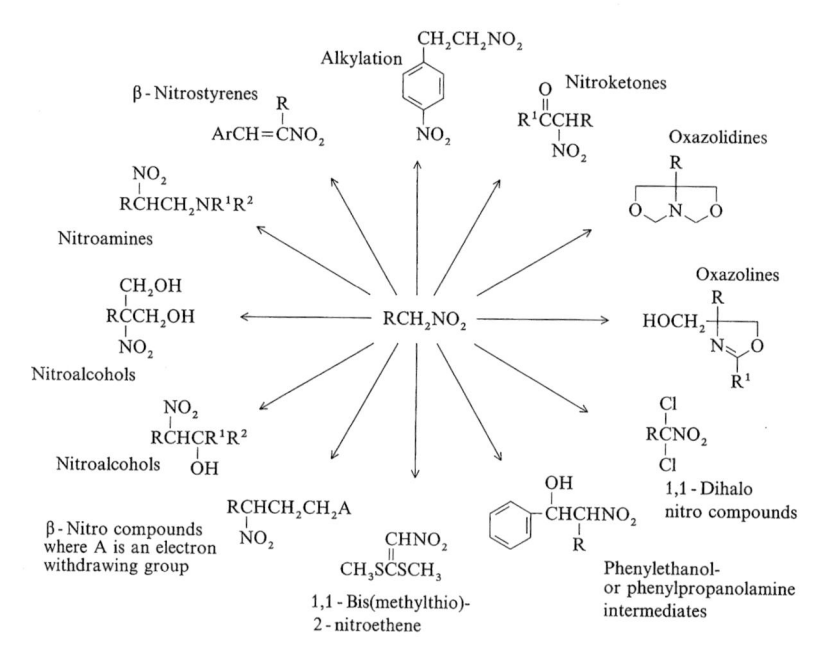

**Figure 1.** Building new chemicals with nitroalkanes

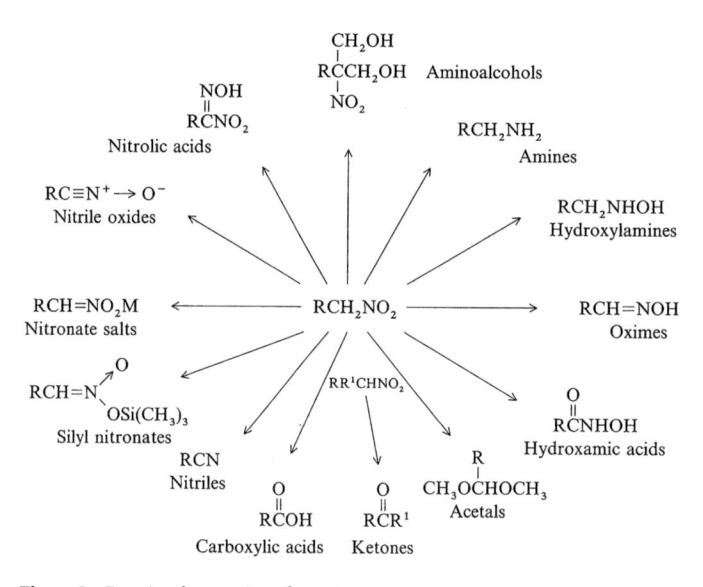

**Figure 2.** Functional group transformations

1) by using an excess of propane to remove the heat, the molar ratio of propane to nitric acid being at least 4:1;
2) by spraying liquid nitric acid into the heated propane, the heat of reaction being used to vaporize nitric acid and to produce nearly adiabatic conditions;

3) by using 60–70% nitric acid, thus producing a large volume of steam, which acts as an inert medium in the reactor; or
4) by controlling the temperature in the reactor via the residence time, the shorter the residence time the lower being the temperature. Short residence time is also important for maximum formation of nitroalkane and minimum formation of byproducts.

The vapor-phase nitration of propane by nitric acid proceeds mainly by a free radical mechanism. The decomposition of the nitric acid to ·OH and ·NO$_2$ or ·ONO, the postulated initiating step, requires temperatures above 350 °C.

By reaction of the OH radicals with alkanes, alkyl radicals can be formed, which then react further with NO$_2$ radicals to give nitroalkanes. However, the alkyl radicals can also react with HNO$_3$ to form nitroalkane and ·OH and in that way continue the chain. The most important side reaction of the alkyl radical is the formation of nitrite, which is unstable at the reaction temperature and decomposes to nitric oxide and an alkoxy radical.

$$R\cdot + \cdot ONO \longrightarrow RONO$$
$$RONO \longrightarrow RO\cdot + NO$$

Alkoxy radicals can decompose with cleavage of the carbon–carbon bond and formation of alkyl radicals of lower molecular weight.

$$R'CH_2O\cdot \longrightarrow R'\cdot + CH_2O$$

This can explain the formation of nitroethane and nitromethane during the nitration of propane. Other byproducts of this process are alcohols, aldehydes, ketones, and related oxygen-containing derivatives.

The conversion of nitric acid to nitroalkanes in vapor-phase nitration is less than 50%. An improved conversion can be achieved by adding a small amount of oxygen or halogen. This is not done in the industrial process, however, since addition of oxygen promotes the formation of large amounts of oxygen-containing byproducts, whereas with halogen addition, corrosion problems or difficulties with the processing of the reaction mixture can occur.

Most of the nitric acid is converted to NO, NO$_2$, N$_2$O, and N$_2$; NO and NO$_2$ are recovered. The overall loss of nitric acid is 20–40%.

Excess propane is sometimes also recovered from the reactor liquid and reused, so that about 60–80% of the propane reacts to give nitroalkanes.

Rapid cooling immediately after the nitration leads to liquefaction of the nitroalkanes and the oxygen-containing byproducts. The liquid and gas phases are separated, and after propane and nitrogen oxides have been recovered from the gas phase they are returned to the reactor. Some of the byproducts, e.g., formaldehyde, are removed by separation of the aqueous phase. The low-boiling oxygen-containing byproducts such as acetaldehyde or acetone, and nitrogen-containing byproducts such as acetonitrile, can

**Table 2.** Product specifications of the four common nitroalkanes (W. R. Grace & Co.)

| Specification | Nitromethane | Nitroethane | 1-Nitropropane | 2-Nitropropane |
|---|---|---|---|---|
| Purity, wt% (min.) | 97 | 97* | 94 | 94 |
| Total nitroalkanes, wt% (min.) | 99 | 99 | 99 | 99 |
| Acidity, wt% acetic acid (max.) | 0.1 | 0.1 | 0.2 | 0.1 |
| Water, wt% (max.) | 0.1 | 0.1 | 0.1 | 0.1 |
| Color, APHA (max.) | 20 | 20 | 20 | 20 |
| Relative density at 25 °C | 1.124 – 1.129 | 1.042 – 1.047 | 0.997 – 0.999 | 0.984 – 0.988 |

* A special grade of nitroethane is available with > 99% purity.

be separated as distillation first runnings from the organic phase. The nitroalkanes so obtained are subjected to a chemical washing process (to remove dissolved higher-boiling oxygen-containing impurities) and then washed with water. The remaining water is removed by passing through a drying tower, and the individual nitroalkanes are then obtained as pure products by fractional column distillation.

More recently, a number of publications have been issued describing improved methods for the syntheses of nitroalkanes [20]–[25]. However, in the past few years only one major new nitroalkane plant has been placed in operation, the W. R. Grace & Co. facility in Deer Park, Texas. The new Grace process employs a mixture of ethane and propane for the hot-tube, free-radical process, and uses $N_2O_4$ as the source of $NO_2$. This process produces a nitroalkane mixture richer in nitromethane and nitroethane.

# 4. Quality Specifications and Analysis

The specifications for the four nitroalkanes produced at the Grace plant in Texas are given in Table 2. Purity and total nitroalkane content are assayed by capillary gas chromatography; water content is measured by typical Karl Fischer procedures. The acidity is determined by dissolving the nitroalkane in methanol and titrating with dilute sodium hydroxide with bromocresol green as indicator.

# 5. Storage and Transportation

Nitroalkanes can be safely stored and transported, but certain precautions should be observed with respect to their shock sensitivity. It is necessary to protect nitroalkanes from extremes of temperature and pressure, e.g., fire and shock, and from chemical contamination which could further sensitize them to these extremes. Although the four commercial nitroalkanes are relatively insensitive to detonation by shock at ordinary temperatures, their sensitivity increases with increasing temperature. The tendency of the nitroalkanes toward detonation is inversely related to chain length. Nitromethane presents the most serious shock-sensitivity hazard of the four common nitroalkanes.

Adiabatic compression also poses a special problem in the handling of nitromethane. Thus, the transfer of nitromethane via pumps and piping requires special conditions designed to avoid adiabatic compression [1].

In addition, special care must be taken to avoid forming dry alkali-metal salts of the nitroalkanes, especially in the case of nitromethane. For example, in the presence of a strong base, such as sodium hydroxide, nitromethane can form the sodium salt of methazonic acid (2-nitroacetaldehyde oxime) which, when dry, is very shock sensitive and can explode. Mixtures of nitromethane and an amine and/or heavy metal oxides, such as those of silver, lead, and mercury, can lead to violent decompositions and should be avoided.

Nitromethane is packaged in 55-gallon drums to a net weight of 500 lb. The drum headspace is filled with nitrogen to exclude moisture and to reduce the hazard of adiabatic compression during transportation. The drums are thin-walled so that they rupture easily in the event of high energy impact. Nitromethane drums should not be stacked, but stored on-end in a single layer.

Nitroethane, 1-nitropropane, and 2-nitropropane are also transported in 55-gallon drums under nitrogen. However, because these nitroalkanes are not as shock sensitive as nitromethane, there are no special restrictions on stacking drums.

Nitroethane, 1-nitropropane, and 2-nitropropane are also available neat in bulk quantities transported in tank wagons or rail cars. Bulk nitromethane presents special hazards and, in the United States, may not be transported undiluted in containers having a capacity greater than 110 gallons (ca. 416 L). Nitromethane, however, can be transported in bulk when mixed with any of a variety of approved diluents. The following diluents, along with their minimum content in wt%, have been approved for the transportation of nitromethane: 1,2-butylene oxide (40%), cyclohexanone (25%), 1,4-dioxane (35%), methanol (45%), 1,1,1-trichloroethane (50%), 1-nitropropane (48%), and 2-nitropropane (47%). These bulk nitromethane mixtures are safely transported by road and ocean freight. All of the nitroalkanes should be kept away from oxidizing, corrosive, or sensitizing materials.

For more details on storage and transportation procedures, refer to a nitroalkane safety guide [1].

# 6. Uses of Nitroalkanes and their Derivatives

## 6.1. Nitromethane and Derivatives

One of the most important uses for nitromethane in which it is used without chemical modification is the stabilization of halogenated hydrocarbons. For example, small amounts of nitromethane (and sometimes nitroethane and/or 1-nitropropane) are widely used in industry to form stable noncorrosive mixtures with 1,1,1-trichloroethane that are used in vapor degreasing, dry cleaning, and for cleaning semiconductors and lenses. These nitroalkanes are useful for inhibiting corrosion on the interiors of tin-plated steel cans containing water-based aerosol formulations. Nitromethane is also employed to stabilize the halogenated propellants for aerosols.

Nitromethane is frequently employed as a polar solvent for cyanoacrylate adhesives and acrylic coatings. Nitromethane is also used for cleaning electronic circuit boards; nitroethane and nitropropanes are also used for this purpose. Nitromethane alone, and in mixtures with methanol and other nitroparaffins, is used as a fuel by professional drag racers and hobbyists. Hobbyists use the nitromethane-based fuel primarily for radio-controlled aircraft. The explosives industry utilizes nitromethane in a binary explosive formulation. The liquid (nitromethane) and solid (inorganic nitrate) ingredients are safely transported and stored separately as standard commercial products. When mixed, an explosive more energetic than 60% dynamite is formed. There are also commercial and military applications for nitromethane in shaped charges, which are often used for targeted undersea explosions and line trenching.

**Nitromethane Derivatives.** *Chloropicrin* [76-06-2], trichloronitromethane, $Cl_3CNO_2$ can be prepared by the reaction of nitromethane with sodium hypochlorite. This chlorinated nitro compound is an effective fungicide and nematocidal fumigant.

*Tris(hydroxymethyl)nitromethane* [126-11-4], [2-(hydroxymethyl)-2-nitro-1,3-propanediol], $(CH_2OH)_3CNO_2$ is obtained if three moles of formaldehyde react with nitromethane via the Henry reaction. This derivative is used as a biocide.

*Tris(hydroxymethyl)aminomethane* [77-86-1], [2-amino-2-(hydroxymethyl)-1,3-propanediol], $(CH_2OH)_3CNH_2$ is prepared by reduction of tris(hydroxymethyl)nitromethane. It is used industrially as a buffer and as a component in adhesives and resins.

The addition of two moles of formaldehyde to nitromethane by the Henry reaction gives di(hydroxymethyl)nitromethane $(CH_2OH)_2CHNO_2$, which can be converted to the corresponding amino compound, $(CH_2OH)_2CHNH_2$ (2-amino-1,3-propanediol) [1794-90-7]. This amine is employed in the synthesis of the X-ray contrast agent Iopamidol [26].

The brominated product $(CH_2OH)_2CBrNO_2$(2-bromo-2-nitro-1,3-propanediol) [*52-51-7*], which can be derived from di(hydroxymethyl)nitromethane is a widely used biocide (Bronopol).

The reaction between benzaldehyde and nitromethane does not stop at the corresponding nitroalcohol; instead, dehydration to β-nitrostyrene, $C_6H_5CH=CHNO_2$[(2-nitroethenyl)-benzene] [*102-96-5*], occurs. This nitroalkene has been used as a chain transfer agent, i.e., to lower the molecular weights of polymers in their free radical initiated synthesis.

Treatment of β-nitrostyrene with bromine, followed by dehydrobromination, gives bromonitrostyrene, $C_6H_5CH=CBrNO_2$ [(2-bromo-2-nitroethenyl)benzene] [*7166-19-0*], which is employed as a slimicide.

An important use of nitromethane is in the synthesis of the anti-ulcer drugs Nizatidine [27] and Ranitidine [28]. A commercial route to Ranitidine is shown below [29]:

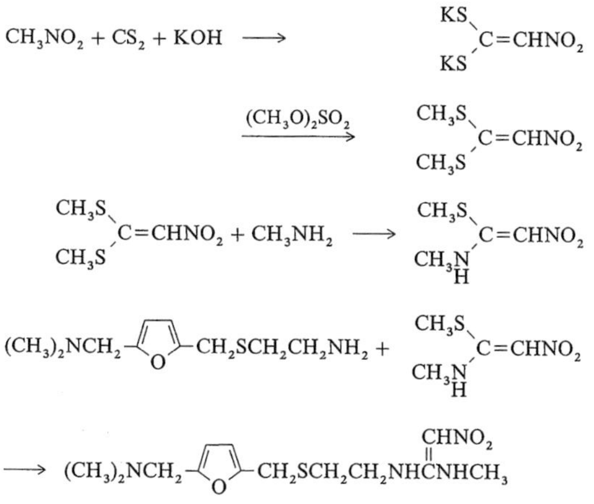

$$CH_3NO_2 + CS_2 + KOH \longrightarrow \begin{array}{c} KS \\ \phantom{KS}\diagdown \\ \phantom{KSKS}C=CHNO_2 \\ \phantom{KS}\diagup \\ KS \end{array}$$

$$\xrightarrow{(CH_3O)_2SO_2} \begin{array}{c} CH_3S \\ \phantom{CH_3S}\diagdown \\ \phantom{CH_3SCH}C=CHNO_2 \\ \phantom{CH_3S}\diagup \\ CH_3S \end{array}$$

$$\begin{array}{c} CH_3S \\ \phantom{CH_3S}\diagdown \\ \phantom{CH_3SCH}C=CHNO_2 + CH_3NH_2 \longrightarrow \\ \phantom{CH_3S}\diagup \\ CH_3S \end{array} \begin{array}{c} CH_3S \\ \phantom{CH_3S}\diagdown \\ \phantom{CH_3SCH}C=CHNO_2 \\ \phantom{CH_3S}\diagup \\ CH_3\underset{H}{N} \end{array}$$

$$(CH_3)_2NCH_2\text{-furan-}CH_2SCH_2CH_2NH_2 + \begin{array}{c} CH_3S \\ \phantom{CH_3S}\diagdown \\ \phantom{CH_3SCH}C=CHNO_2 \\ \phantom{CH_3S}\diagup \\ CH_3\underset{H}{N} \end{array}$$

$$\longrightarrow (CH_3)_2NCH_2\text{-furan-}CH_2SCH_2CH_2NHC(=CHNO_2)NHCH_3$$
Ranitidine

Finally, nitromethane is used in the preparation of the psychotropic agent Sulpiride [30].

## 6.2. Nitroethane and Derivatives

Nitroethane is employed as a solvent due to its excellent wetting properties and its ability to form azeotropes with many industrial solvents. For example, some printing inks contain nitroethane.

Nitroethane is used by the commercial blasting industry as either a fuel or sensitizer of water-based blasting agents. Nitroethane can be added to nitromethane to reduce its tendency to detonate when used as a fuel in internal combustion engines.

**Nitroethane Derivatives.** The Henry reaction between two moles of formaldehyde and nitroethane gives 2-methyl-2-nitro-1,3-propanediol [*77-49-6*], $CH_3CNO_2(CH_2OH)_2$ which is utilized as a biocide in cutting oils.

Catalytic hydrogenation of 2-methyl-2-nitro-1,3-propanediol affords the corresponding amine, 2-amino-2-methyl-1,3-propanediol [*115-69-5*]. After reaction with oleic acid, this aminoalcohol forms an oxazoline [31], which is used as a specialty cationic surfactant.

Nitroethane is used to make a plasticizer for solid rocket fuels [32]. The molecuar structure for this plasticizer is:

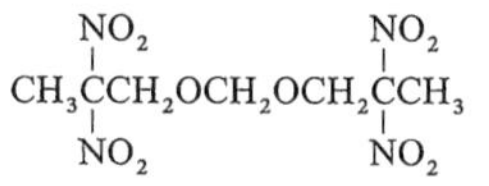

$$\underset{\underset{NO_2}{|}}{\overset{\overset{NO_2}{|}}{CH_3CCH_2OCH_2OCH_2CCH_3}}$$

The antihypertensive drug Aldomet (methyl DOPA) is made using nitroethane. The synthesis follows the general route [33], [34]:

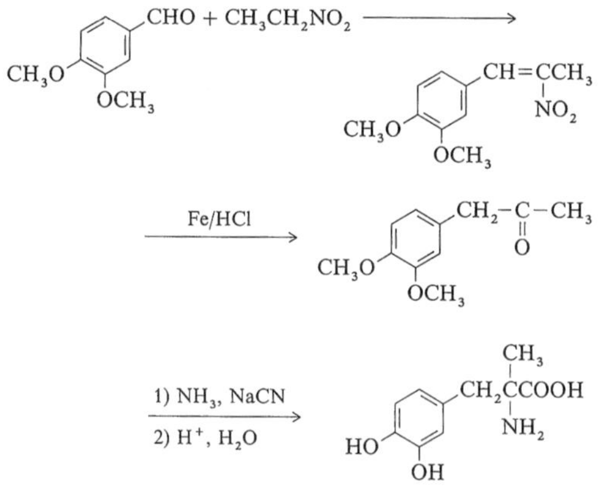

[Reaction scheme: 3,4-dimethoxybenzaldehyde + $CH_3CH_2NO_2$ → 3,4-dimethoxy-β-nitrostyrene (CH=CCH$_3$ with NO$_2$)]

$\xrightarrow{Fe/HCl}$ [3,4-dimethoxyphenyl-$CH_2$-C(=O)-$CH_3$]

$\xrightarrow[\text{2) } H^+, H_2O]{\text{1) } NH_3, NaCN}$ [3,4-dihydroxyphenyl-$CH_2$-C($CH_3$)($NH_2$)COOH]

Methyl DOPA

## 6.3. 1-Nitropropane and Derivatives

Like the other basic nitroalkanes, 1-nitropropane is employed as a solvent or cosolvent.

$$CH_3CH_2CH_2NO_2 + HCHO \longrightarrow CH_3CH_2\underset{NO_2}{\overset{|}{C}H}CH_2OH$$

$$\xrightarrow{H_2/Ni} CH_3CH_2\underset{NH_2}{\overset{|}{C}H}CH_2OH$$

$$2\,CH_3CH_2\underset{NH_2}{\overset{|}{C}H}CH_2OH + ClCH_2CH_2Cl \longrightarrow$$

$$\underset{\underset{CH_2OH}{|}}{CH_3CH_2CH}NHCH_2CH_2NH\underset{\underset{CH_2OH}{|}}{CH}CH_2CH_3$$
$$\text{Ethambutol}$$

The most important use for 1-nitropropane is in the synthesis of Ethambutol, an antituberculosis drug [35].

Only the D-enantiomer of 2-amino-1-butanol is used in the final synthetic step.

Relatively small quantities of 1-nitropropane are used in the preparation of biocides [36].

## 6.4. 2-Nitropropane and Derivatives

A major use of 2-nitropropane is as an industrial solvent. It is used in vinyl inks that are employed in printing, flexography, and photogravure. The 2-nitropropane, often mixed with alcohols, dissolves a large number of resins such as epoxy, polyurethane, polyester, vinyl, urea–formaldehyde, and phenolic. These solvent–resin mixtures are used for coatings. For example, can-coating varnishes contain the nitroalkane because of its excellent wetting properties. Other applications of these 2-nitropropane formulations are in adhesives and in electrostatic paints. Some of the properties of 2-nitropropane that make it so versatile in many of these applications are: high polarity and flash point, appropriate evaporation rate, good wetting and azeotropic properties, and satisfactory resistivity.

**2-Nitropropane Derivatives.** The Henry reaction of 2-nitropropane with formaldehyde yields 2-methyl-2-nitro-1-propanol (NMP) [*76-39-1*].

$$CH_3\underset{\underset{CH_3}{|}}{\overset{\overset{NO_2}{|}}{C}}CH_2OH$$

This nitroalcohol has been employed as an adhesive component to improve tire cord bonding [37].

Catalytic hydrogenation of NMP affords 2-amino-2-methyl-1-propanol (AMP) [124-68-5]. This aminoalcohol is a useful organic base for neutralizing and solubilizing applications. Other applications of AMP are:

1) AMP is used in toiletries, cosmetics, and in hair sprays, in which it neutralizes and solubilizes the carboxyl-containing polymers used in their formulation.
2) In conjunction with fatty acids, AMP is an excellent dispersing agent for powders and pigments, especially titanium dioxide.
3) AMP has been employed as a formaldehyde scavenger in melamine–formaldehyde, urea–formaldehyde, and phenol–formaldehyde resins.
4) AMP has also been used as a wetting agent, as an emulsifier in polishing waxes, and in textiles for permanent pleats.

Finally, 2-nitropropane has been utilized in the preparation of the experimental β-blocking drug, Bucindolol, [38] – [40], whose synthesis scheme shows how 2-nitropropane can be used to introduce a *tert*-butylamine functionality into a molecule:

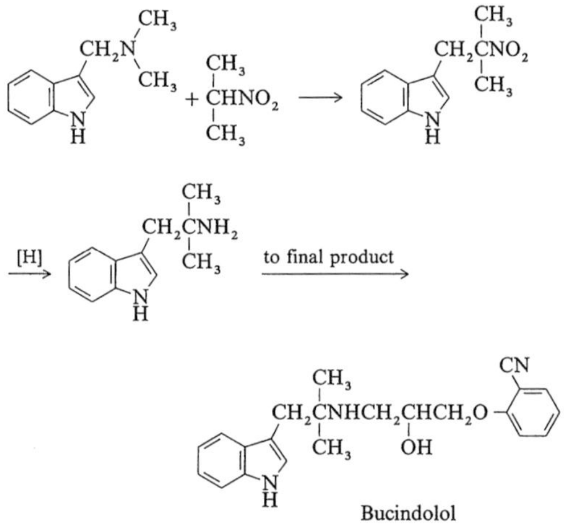

# 7. Toxicology and Occupational Health

The two most important health controls for nitroalkane exposure are adequate ventilation and prevention of skin contact.

In the United States, OSHA and ACGIH have set PE's and TLV's for the nitroalkanes [41], [42]. The OSHA and ACGIH limits are shown in Table 3, together with MAK values.

**Table 3.** Industrial exposure limits for nitroalkanes, mL/m$^3$ (ppm)

| Compound | PEL (OSHA) | TLV (ACGIH) | MAK |
|---|---|---|---|
| Nitromethane | 100 | 100 | 100 |
| Nitroethane | 100 | 100 | 100 |
| 1-Nitropropane | 25 | 25 | 25 |
| 2-Nitropropane | 25 | 10 | * |

*Identified as a suspected human carcinogen; safe exposure limit has not been established.

Inhalation is the major industrial hazard of nitroalkane exposure. The reported effects of overexposure to vapors are headache, nausea, vomiting, and convulsions. Although chronic exposure to animals indicates some liver and kidney injury, no such injury has been reported in humans when exposures were maintained below the TLV levels.

Nitroalkane vapors can cause eye irritation at levels above the recommended TLVs. Since the nitroalkanes' odor detectabilities are poor, odor detection does not serve as a warning for overexposure. Therefore, prolonged exposure to vapors above the TLV necessitates the use of respirators. Such exposure may occur in operations that require entry into tanks or closed vessels and in emergency situations. Acute exposure to extremely high concentrations of 2-nitropropane vapors has resulted in serious injuries and, in some cases, death when workers were exposed in enclosed spaces without using recommended respiratory protection or adequate ventilation. For respiratory protection, supplied-air or self-contained breathing apparatus with a full facepiece should be used.

Nitroalkanes are mild skin irritants due to their solvent action but are not absorbed through the skin. Although irritation can occur from prolonged or repeated skin contact, no allergic or sensitization reactions have been reported.

The acute LD$_{50}$ (rat, oral) of the nitroalkanes is as follows [43]:

| | |
|---|---|
| Nitromethane | $1210 \pm 322$ mg/kg |
| Nitroethane | $1620 \pm 193$ mg/kg |
| 1-Nitropropane | $455 \pm 75$ mg/kg |
| 2-Nitropropane | $725 \pm 160$ mg/kg |

The National Toxicology Program of the U.S. Department of Health and Human Services, The International Agency for Research on Cancer, and the American Conference of Governmental Industrial Hygienists list 2-nitropropane as a suspect carcinogen. The classification is based on studies in which prolonged exposure to 2-nitropropane was found to cause liver neoplasms in laboratory rats [44].

In 1979, an epidemiological study of workers exposed to 2-nitropropane was reported by the International Minerals and Chemical Corporation. The authors concluded that analysis of these data does not suggest any unusual cancer or other disease mortality pattern among this group of workers [45]. There is no evidence that 2-nitropropane causes cancer in humans.

# 8. References

[1] *Grace Nitroparaffins—Safety and Handling Guide*, W. R. Grace & Co., Lexington, MA.
[2] G. O'Neill, S. Markofsky: *Grace Nitroparaffins Chemistry Guide*, W. R. Grace & Co., Lexington, MA.
[3] S. Kanbe, H. Yasuda, *Bull. Chem. Soc. Jpn.* **41** (1968) 1444.
[4] R. H. Wollenberg, *Tetrahedron Lett.* 1978, 3219.
[5] T. I. Gubino, *Tr. Molodykh Uch. Sarat. Univ.* 1971, 177.
[6] T. D. Zheved, K. V. Altukov in G. V. Nekrasova (ed.): *Sint. Issled. Nitrosoedin. Aminokislot* 1983, 3–5.
[7] O. I. Rosumov et al., *Zh. Obsch. Khim.* **47** (1977) no. 3, 567.
[8] Henkel and Cie. GmbH, DE 1 954 173, 1969 (R. Wessendorf).
[9] D. A. White, M. M. Baizer, *Tetrahedron Lett.* 1973, 3597.
[10] E. D. Bergmann, R. Corett, *J. Org. Chem.* **21** (1956) 107; **23** (1958) 1507.
[11] E. A. Parfenov, A. R. Bekker, G. F. Kostereva, *Zh. Org. Khim.* **17** (1981) no. 8, 1591.
[12] G. A. Smirnov, T. A. Klimova, V. V. Sevost'yanova, *Izv. Akad. Nauk SSSR, Ser. Khim.* 1981, no. 11, 2624.
[13] D. Seebach, E. W. Colvin, F. Lehr, T. Weller, *Chimia* **33** (1979) 1.
[14] N. Ono, A. Kaji, *Yuki Gosei Kagaku Kyokaishi* **38** (1980) no. 2, 115.
[15] G. Rosini, R. Ballini, *Synthesis* 1988, 833.
[16] M. Braun, *Nachr. Chem. Tech. Lab.* **33** (1985) no. 7, 598.
[17] H. Feuer (ed.): *The Chemistry of the Nitro and Nitroso Groups*, Part 1, R. E. Krieger Co., New York 1981.
[18] K. Torssell: *Nitrile Oxides, Nitrones, and Nitronates in Organic Synthesis*, VCH Publishers, New York 1988.
[19] *Ullmann* 4th ed., **17**, 373.
[20] Société Chemique de la Grande Paroisse, FR 2 272 975, 1971 (M. Lucquin, J. Dechaux).
[21] Société Chemique de la Grande Paroisse, US 4 260 838, 1978; US 4 313 010, 1978 (P. Lhonore, B. Jacquinot, J. Quibel).
[22] W. R. Grace & Co., US 4 469 904, 1983; US 4 524 226, 1983; US 4 517 393, 1983 (M. Sherwin, P. Wang).
[23] W. R. Grace & Co., US 4 476 336, 1984; EP 85 328A, 1983 (M. Sherwin).
[24] W. R. Grace & Co., EP 174 600A, 1988 (P. Wang).
[25] Dow Chemical, US 4 421 940, 1982; US 4 431 842, 1982 (W. V. Hayes).
[26] Savac AG, US 4 001 323, 1975 (E. Felder, R. S. Vitale, D. E. Pitre).
[27] Eli Lilly and Co., US 4 375 547, 1980 (R. P. Pioch).
[28] Allen and Hansbury Ltd., GB 1 565 966, 1976 (B. Price, J. Clitherow, J. Bradshaw).
[29] Allen and Hansbury Ltd., US 4 128 658, 1977 (B. J. Price, J. W. Clitherow, J. Bradshaw).
[30] Etudes Scientifiques et Industrielle Fr., FR 2 019 350, 1969 (M. Hashimoto, T. Kamiya).
[31] J. A. Frump, *Chem. Rev.* **71** (1971) no. 5, 483.
[32] E. Hamel et al., *Ind. Eng. Chem. Prod. Res. Dev.* **1** (1962) no. 2, 108.
[33] Merck and Co. Inc., US 2 868 818, 1959 (K. Pfister, G. Stein).
[34] Merck and Co. Inc., US 3 158 648, 1964 (R. Jones, K. Krieger, J. Lago).
[35] R. G. Wilkinson, R. G. Shepherd, J. P. Thomas, C. Baughn, *J. Am. Chem. Soc.* **83** (1961) 2212.
[36] IMC Chemical Group, Inc., US 4 088 817, 1977 (J. Hunsucker, R. Shelton).
[37] Uniroyal Inc., US 3 598 690, 1967 (A. Danielson).

[38] H. R. Snyder, L. Katz, *J. Am. Chem. Soc.* **69** (1947) 3140.
[39] Mead Johnson & Co., US 4 234 595, 1979 (W. Kreighbaum, W. Comer).
[40] Bristol Myers Co., DE 3 421 252, 1984 (W. Kreighbaum).
[41] General Industry Safety and Health Standards in Occupational Safety and Health Act (OSHA), 29 CFR 1910.1000, 1989.
[42] Threshold Limit Values and Biological Exposure Indices for 1986–87, American Conference of Governmental Industrial Hygienists.
[43] E. Bingham, A. Robbins: "Health... 2-NP..." *DHHS (NIOSH) Publ. (U.S.)* 80-142 (1980) Oct.
[44] Huntington Research Center, *HEW Publ. (NIOSH) (U.S.)* 210-75-0039 (1975).
[45] M. Miller, G. Temple: *2-NP Mortality Epidemiological Study of the Sterlington, LA, Employees*, International Minerals and Chemical Corporation, Mundelein, IL 1979.

# Nitro Compounds, Aromatic

*Individual keywords: Nitrobenzaldehydes → Benzaldehyde; Nitrobenzoic acids → Benzoic Acid and Derivatives; Nitronaphthalenesulfonic acids → Naphthalene Derivatives*

GERALD BOOTH, Booth Consultancy Services, Thorpe House, Uppermill, Oldham OL3 6DP, United Kingdom

| | | | | | |
|---|---|---|---|---|---|
| 1. | Introduction | 3504 | 6. | Nitroaromatic Sulfonic Acids and Derivatives | 3548 |
| 2. | Nitration | 3505 | 6.1. | Nitrobenzenesulfonic Acids and Derivatives | 3549 |
| 2.1. | Nitrating Agents | 3506 | 6.2. | Nitrotoluenesulfonic Acids and Derivatives | 3552 |
| 2.2. | Reaction Mechanisms | 3507 | 6.3. | Chloronitrobenzenesulfonic Acids and Derivatives | 3555 |
| 3. | Nitro Aromatics | 3509 | 6.4. | Chloronitrotoluenesulfonic Acids | 3559 |
| 3.1. | Nitrobenzenes | 3509 | 7. | Nitrohydroxy and Alkoxy Aromatics | 3560 |
| 3.2. | Nitrotoluenes | 3515 | 7.1. | Nitrophenols and Derived Ethers | 3560 |
| 3.3. | Nitroxylenes | 3521 | 7.2. | Nitroalkylphenols | 3565 |
| 3.4. | Nitronaphthalenes | 3522 | 8. | Nitroketones | 3567 |
| 3.5. | Other Nitro Aromatics | 3525 | 9. | Nitroheterocycles | 3568 |
| 4. | Nitrohalo Aromatics | 3526 | 9.1. | Pyridine Derivatives | 3569 |
| 4.1. | Chloronitrobenzenes | 3526 | 9.2. | Quinoline Derivatives | 3570 |
| 4.2. | Dichloro- and Polychloronitrobenzenes | 3532 | 9.3. | Imidazole Derivatives | 3570 |
| 4.3. | Chloronitrotoluenes | 3534 | 9.4. | Furan Derivatives | 3571 |
| 4.4. | Fluoronitrobenzenes and Fluoronitrotoluenes | 3536 | 9.5. | Thiophene Derivatives | 3571 |
| 5. | Nitroamino Aromatics | 3538 | 9.6. | Thiazole Derivatives | 3572 |
| 5.1. | Nitroanilines | 3540 | 10. | References | 3574 |
| 5.2. | Nitrotoluidines | 3544 | | | |
| 5.3. | Halogenonitroanilines | 3546 | | | |
| 5.4. | Cyanonitroanilines | 3547 | | | |

# 1. Introduction

The earliest aromatic nitro compounds were obtained by MITSCHERLICH in 1834 by treating hydrocarbons derived from coal tar with fuming nitric acid. By 1835 LAURENT was working on the nitration of naphthalene, the most readily available pure aromatic hydrocarbon at that time. DALE reported on mixed nitro compounds derived from crude benzene at the 1838 annual meeting of the British Association for the Advancement of Science. Not until 1845, however, did HOFMANN and MUSPRATT report their systematic work on the nitration of benzene to give mono- and dinitrobenzenes by using a mixture of nitric and sulfuric acids.

The first small-scale production of nitrobenzene was carefully distilled to give a yellow liquid with a smell of bitter almonds for sale to soap and perfume manufacturers as "essence of mirbane."

Bechamp's iron reduction process, which made aniline more readily available, was published in 1854, and the discovery of aniline mauve by PERKIN in 1856 started the European aniline dye industry that became the basis for a worldwide synthetic colorant industry estimated to have had sales of $ $6 \times 10^9$ in 1988.

Process development and scaleup of the nitration and reduction processes, begun by PERKIN and continued by many others, resulted in the trade price of aniline dropping from 50 Fr/kg in 1858 to 10 Fr/kg in 1863; by 1871, the European production of aniline had reached 3500 t/a [23].

In 1985 European production of aniline was 500 000 t/a, and it is still the single largest product based on a nitration process. Its use in colorants now accounts for only 4% of aniline production because most of the growth has been due to rubber chemicals and isocyanates, the latter now consuming well over 50% of nitrobenzene production for the manufacture of methylene diphenyldiisocyanate.

The number of naturally occurring nitroaromatic compounds is small; the first to be recognized was chloramphenicol (**1**), an important compound extracted from cultures of a soil mold *Streptomyces venezuelae* and identified in 1949.

$$O_2N-\underset{}{\bigcirc}-\underset{\underset{OH}{|}}{\overset{\overset{H}{|}}{C}}-\underset{\underset{H}{|}}{\overset{\overset{NHCOCHCl_2}{|}}{C}}-CH_2OH$$

**1**

This discovery stimulated investigations into the role of the nitro group in pharmacological activity, following the earlier (1943) discovery of the antibacterial activity of nitrofuran derivatives. Many synthetic pharmaceuticals and agrochemicals contain nitroaromatic groups, although the function of the nitro group is often obscure.

Nitration is the dominant method for introducing the nitro group into aromatic systems. Several so-called indirect methods also exist:

1) oxidation of nitroso or amino compounds,
2) replacement of diazonium groups (nitro-Sandmeyer reaction),

3) rearrangement of nitramines, and
4) nucleophilic displacement reactions.

Although useful in overcoming problems of substituent orientation or substrate sensitivity to nitration conditions [14], they have very little industrial significance. Of more than 200 nitroaromatic compounds described in this article, only two require synthesis by indirect methods. Many important nitro compounds are produced by applying unit processes (e.g., sulfonation, halogenation, or amination) to primary nitro starting materials, most of which are derived from the key primaries nitrobenzene, nitrotoluenes, and nitrochlorobenzenes.

Nucleophilic displacement of activated nitro groups is of minor industrial significance but has considerable synthetic potential. The nucleofugicity may rival that of the corresponding fluorine substituent [24], and the nitro precursor may be more readily available than its halogen analogue. For example, the readily available 1,3-dinitrobenzene reacts with sodium methoxide to form 3-nitroanisole or with potassium fluoride (180 °C, DMF) to form 3-fluoronitrobenzene.

The choice of nitro compounds covered here is influenced strongly by their commercial application, and a good guide to this is the registration of compounds in the 1981 European Core Inventory (ECOIN). Most nitro compounds, or their derivatives, are intermediates for colorants, agrochemicals, pharmaceuticals, or other fine chemicals with a few major volume outlets for synthetic materials and explosives.

For many compounds, CAS nomenclature conventions differ from the more important names in common usage. The latter are given here, with the CAS name and number where meaningful.

## 2. Nitration

The unit process of nitration can be defined simply as the irreversible introduction of one or more nitro ($NO_2$) groups into an aromatic nucleus by replacement of a hydrogen atom. O-Nitration to give nitrates and N-nitration to give nitramines are much far less important commercially for aromatic compounds and are not dealt with here.

Nitration, an electrophilic substitution reaction, is represented by the equation:

$$R\text{-}C_6H_5 + X\text{-}NO_2 \longrightarrow R\text{-}C_6H_4\text{-}NO_2 + HX$$

Substrate  Nitrating agent  Product

Introduction of the nitro group so deactivates the ring to further electrophilic substitution that dinitration rarely occurs under the conditions used for mononitration. The more vigorous conditions required for dinitration (i.e., excess, stronger acid and higher temperature) must usually be applied to isolated mononitro compounds, rather than carrying out stepwise reaction in situ.

The nitration reaction is always strongly exothermic, as exemplified by the mononitration of benzene ($\Delta H = -117$ kJ/mol) and naphthalene ($\Delta H = -209$ kJ/mol) and is probably the most potentially hazardous industrially operated unit process. This is due fundamentally to the heat generated, which can trigger the power of nitric acid to degrade organic materials exothermically to gaseous products with explosive violence [19]. Nitroaryl compounds, especially those with more than one nitro group, are potentially hazardous due to their very high oxygen contents. Some polynitro compounds (e.g., trinitrotoluene and picric acid) are detonable and have a long history of use as explosives [8]. A less publicized hazard with nitroaromatic compounds is occasioned by the violent decomposition reactions that may occur on heating with alkali [19].

The nitration process may be carried out on either a batch (discontinuous) or a continuous basis. Lower tonnage requirements are met by batch reaction in reactors (nitrators) designed to accommodate a variety of products. Maximum flexibility is obtained by using the more expensive enamel-lined mild-steel reactors in place of the traditional cast iron or stainless steel. The exothermic process requires efficient cooling of the reactor by jacket, internal cooling, or both. For safety reasons, the vessel is usually limited to about 6000-L capacity and is fitted with an efficient agitator.

Continuous reaction for large-tonnage intermediates (e.g., nitrobenzene and nitrotoluenes) is attractive for reasons of safety as well as the obvious economies of a dedicated unit. Whether continuous operation is achieved by a continuous mixing device or by a cascade of small agitated reactors, the material inventory is much lower than for batch reactors so the reaction temperature is easier to control. The continuous reactor requires closely defined operating parameters and management.

Typical nitroaromatic production is based on high-yield processes, with more than 80% of the total cost being represented by the raw materials. This means that only a small proportion of the cost is available for operational savings. Two major areas that have yielded improvements recently are (1) the sulfuric acid recycle that has been an integral requirement of all efficientnitration processes and (2) isomer control and separation. Both these points are highlighted in more detail, together with information on plant design, in the descriptions of processes for individual products.

## 2.1. Nitrating Agents

For reasons of practicability and economics, industrial-scale nitration is usually carried out with a mixture of nitric and sulfuric acids (mixed acid), and occasionally with aqueous nitric acid, nitric acid in acetic acid, or nitric acid in acetic anhydride. Use of alternative component acids such as perchloric acid, hydrofluoric acid, or boron trifluoride is limited to important supporting studies. These are sometimes carried out in inert organic solvents such as chlorohydrocarbons or sulfolane to give homogeneous reaction mixtures.

The strength of the nitrating agent (X–NO$_2$) decreases with decreasing electronegativity of X [25]: nitronium ion (e.g., BF$_4^-$NO$_2^+$) > nitracidium ion (OH$_2^+$–NO$_2$) > nitronium chloride (Cl–NO$_2$) > acetyl nitrate (AcO–NO$_2$) > nitric acid (HO–NO$_2$) > ethyl nitrate (C$_2$H$_5$O–NO$_2$).

The nitronium ion, NO$_2^+$, is considered to be the active species in all of these systems. In the most common system the overall equation

$$HNO_3 + 2\,H_2SO_4 \rightleftharpoons NO_2^+ + H_3O^+ + 2\,HSO_4^-$$

is a composite of many equilibria present in HNO$_3$–H$_2$SO$_4$–H$_2$O mixtures. These must all be taken into account when considering the reactivity of the substrate and the extent of nitration required.

A typical nitrating agent for large-scale aromatic mononitration consists of 20% nitric acid, 60% sulfuric acid, and 20% water; this is referred to as 20/60/20 mixed acid. An alternative definition of the same acid, used in some situations, is 15 mol% nitric acid, 30 mol% sulfuric acid, and 55 mol% water. Commonly, the liquid aromatic substrate and the nitrated product together form a separate phase from the aqueous mixed acid. Efficient agitation is therefore required to maximize contact with the organic phase and minimize resistance to mass transfer. Solid substrates are best dissolved in the sulfuric acid phase. Stronger nitric acid leads to oxidative side reactions, whereas higher temperature leads to decreased nitronium ion concentration. Much detailed developmental work is required for each individual nitration to optimize these and other variables, to maximize formation of the required isomer, and to minimize side reactions.

Free-radical reactions (typically in the vapor phase with nitric acid or nitrogen dioxide as nitrating agent) tend not to be employed in aromatic nitration, being more appropriate for aliphatic nitration (→ Nitro Compounds, Aliphatic). However, claims have been made that a radical mechanism makes some contribution, even when ionic mechanisms clearly predominate [26].

## 2.2. Reaction Mechanisms

Much experimentation has been applied to aromatic nitration since the nitronium ion, NO$_2^+$, was confirmed as the active species by the Ingold–Hughes school in 1950 through study of Raman spectra. Temperature and degree of mixing are important parameters for all nitrating agents, but other variables often lead to inconsistent results [27], [28], [16]. In commercial nitration with HNO$_3$–H$_2$SO$_4$ the relative rate constants cover a wide range, with reaction times varying from several seconds for active substrates to many hours for inactive substrates. It has been argued that under the nonideal conditions of large-scale production, nitration is controlled more by mass transfer than by kinetics, and that many laboratory studies wrongly assumed that mass-

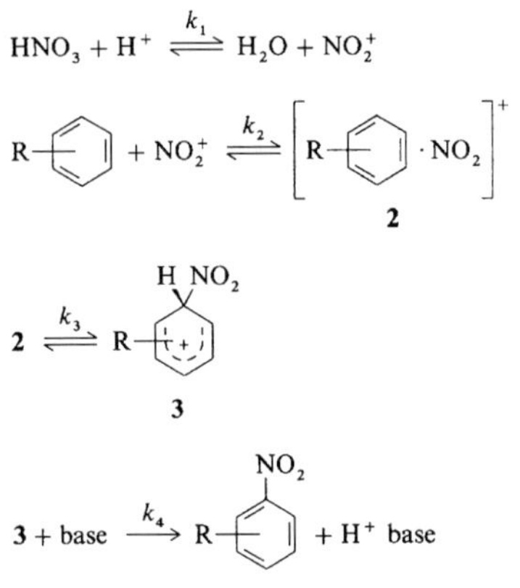

**Figure 1.** Basic mechanism of aromatic nitration

$$HNO_3 + H^+ \underset{}{\overset{k_1}{\rightleftharpoons}} H_2O + NO_2^+$$

$$R-C_6H_5 + NO_2^+ \underset{}{\overset{k_2}{\rightleftharpoons}} [R-C_6H_5 \cdot NO_2]^+$$
$$\mathbf{2}$$

$$\mathbf{2} \overset{k_3}{\rightleftharpoons} R-C_6H_5(H)(NO_2)^+$$
$$\mathbf{3}$$

$$\mathbf{3} + base \overset{k_4}{\longrightarrow} R-C_6H_5-NO_2 + H^+ base$$

transfer resistance is negligible [29]. Another early assumption that the rate of nitration parallels the equilibrium formation of the nitronium ion was also challenged because of the realization that different mechanisms may apply at differing acid strengths [30].

The mechanism that can accommodate all the facts available on the nitration of simple substrates (i.e., benzene and toluene, but not phenols or polyalkylbenzenes) is shown in Figure 1 [28]. The possible involvement of four rate constants and three significant reversible reactions makes the kinetics extremely complex even if the reaction is totally kinetically controlled.

The encounter complex, or encounter pair (**2**) [27], is the novel feature of this mechanism; it is a molecular complex of less specific nature than the earlier concept of an intermediate π-complex. The important point is that **2** has a kinetic role and may indeed be rate determining. Conversion of the σ-complex (**3**), sometimes called a Wheland intermediate, to the product is not rate limiting because the nitration reaction does not exhibit a primary kinetic isotope effect. Although under certain conditions formation of the nitronium ion may be rate determining, for reactive substrates in aqueous sulfuric acid the formation of **2** is predominately the rate-determining step, whereas nitration of toluene, benzene, and other less reactive aromatic compounds obeys an overall rate law, for which formation of **3** is predominately the rate-determining step [28]; that is

$$\text{rate} = k_{obsd}[HNO_3][ArH]\,a_{H^+}$$

where $k_{obsd}$ depends on the structure of the aromatic compound.

When the ring is highly activated towards electrophilic attack, as in the case of phenol, the nitrosonium ion ($NO^+$), arising from a catalytic amount of nitrous acid, comes into play, with formation of an intermediate nitroso compound, which is oxidized by nitric acid to the nitro derivative with regeneration of nitrous acid.

$$\text{ArH} + \text{NO}^+ \rightleftharpoons \left[\text{Ar}^+\begin{smallmatrix}\text{H}\\ \text{NO}\end{smallmatrix}\right] \longrightarrow \text{ArNO} + \text{H}^+$$

$$\text{ArNO} + \text{HNO}_3 \longrightarrow \text{ArNO}_2 + \text{HONO}$$

Ipso-electrophilic attack at a substituent position represents another exception to the general mechanism. With di- and polyalkylbenzenes the *ipso*-nitroarenium ion (**4**) may be formed instead of **3**. Although rearrangement of **4** to the conventional Wheland intermediate does occur, it is accompanied by side reactions such as nitrodealkylation, thus giving polyalkylbenzene nitration potentially more variable mixture products than other nitrations [31]. Increased understanding of

the formation of ipso-Wheland intermediates has led to their exploitation in other reactions, especially with nucleophiles [16].

# 3. Nitro Aromatics

## 3.1. Nitrobenzenes

Mono-, di-, and symmetrical trinitrobenzenes are readily available by sequential nitration of benzene. Neither *o*- and *p*-dinitrobenzene nor unsymmetrical trinitrobenzenes have any significant industrial importance.

**Nitrobenzene** [*98-95-3*] (**5**), oil of mirbane, $C_6H_5NO_2$, $M_r$ 123.1, is a pale yellow liquid with an odor of bitter almonds, which is readily soluble in most organic solvents and miscible with benzene in all proportions. It is a good solvent for aluminum chloride and is therefore used as a solvent in Friedel–Crafts reactions. It is only slightly soluble in water (0.19 % at 20 °C; 0.8 % at 80 °C) and is steam volatile. Binary azeotropes of nitrobenzene are listed in Table 1. Some other physical properties of nitrobenzene are as follows:

| | |
|---|---|
| *mp* | 5.85 °C |
| *mp* (tech. spec.) | 5.5 °C (min.) |
| *bp* | |
| at 101 kPa | 210.9 °C |
| at 13 kPa | 139.9 °C |
| at 4 kPa | 108.2 °C |
| at 0.13 kPa | 53.1 °C |

**Table 1.** Nitrobenzene-containing binary azeotropes

| Second component | $bp$ (101 kPa), °C | Nitrobenzene, wt% |
|---|---|---|
| Acetamide | 202.0 | 76 |
| Benzotrichloride | 210.7 | 98.5 |
| Benzyl alcohol | 204.2 | 38 |
| 4-Chlorophenol | 219.9 | 92 |
| N,N-Diethylaniline | 210.7 | 97 |
| Ethyl benzoate | 210.6 | 81 |
| Ethylene glycol | 185.9 | 41 |
| Propionamide | 205.4 | 76 |

| | |
|---|---|
| $d_4^{1.5}$ (solid) | 1.344 |
| $d_4^{10}$ | 1.213 |
| $d_4^{15.5}$ (tech. spec.) | 1.208–1.211 |
| $d_4^{25}$ | 1.199 |
| Viscosity (15 °C) | $2.17 \times 10^{-2}$ mPa · s |
| Surface tension (20 °C) | 43.35 mN/m |
| Dielectric constant | |
|   at 20 °C | 35.97 |
|   at 170 °C | 18.15 |
| Specific heat (30 °C) | 1.418 J/g |
| Latent heat of fusion | 94.1 J/g |
| Latent heat of vaporization | 331 J/g |
| Flash point (closed cup) | 88 °C |
| Autoignition temperature | 482 °C |
| Explosive limit in air (93 °C) | 1.8 vol% |
| Vapor density | 4.1 |
| Refractive index $n^{15}$ | 1.55457 |

Reactions of nitrobenzene involve either electrophilic meta substitution of the aromatic ring by nitration, sulfonation, or halogenation (Fig. 2), for example, or reduction of the nitro group to yield a variety of products, as shown in Table 2. The primary reduction products are nitrosobenzene, N-phenylhydroxylamine, and aniline, with the other products being formed by interactions and rearrangements, as shown in Figure 3. Very specific reduction conditions are required to obtain pure products, and intermediate stages (e.g., nitrosobenzene) are not usually isolable.

*Production.* The traditional batchwise nitration process, in which mixed acid (27–32/56–60/8–17 wt%) is added to a slight excess of benzene (to avoid nitric acid in the spent acid), with the temperature controlled at 50–55 °C and the reaction completed by heating to 80–90 °C, has been replaced by continuous processes that operate under similar conditions. Both economy and safety, as summarized in Section 2.1, result from the more efficient mixing and higher reaction rates possible in smaller reactors. An 120-L continuous reactor has been reported to give the same output of nitrobenzene as a 6000-L batch reactor. The variety of reactor configurations employed is typified by a stirred cylindrical reactor operating at 50–100 °C. The critical reuse of spent acid is achieved by continuous concentration and addition of fresh acid, as shown in Figure 4. This differs from the concept of batchwise reuse although, in each case,

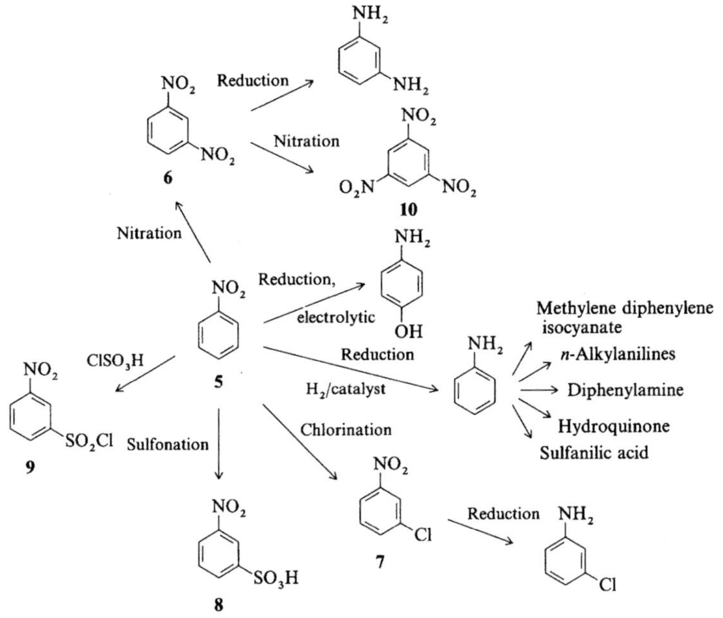

**Figure 2.** Key intermediates derived from nitrobenzene

**Table 2.** Nitrobenzene reduction products

| Reducing agent | Product |
| --- | --- |
| Fe, Zn, or Sn/HCl, $H_2$– catalyst | aniline |
| Zn – $H_2O$ | N-phenylhydroxylamine (rearranges with acid to 4-aminophenol) |
| $Na_3AsO_3$ | azoxybenzene |
| Zn – NaOH | azobenzene, hydrazobenzene (rearranges with acid to benzidine) |
| $LiAlH_4$ | azobenzene |
| $Na_2S_2O_3$– $Na_3PO_4$ | sodium phenylsulfamate ($C_6H_5NHSO_3Na$) |
| Electrolytic reduction | 4-aminophenol |

addition of fresh acid or $SO_3$ results in the disadvantage of an equivalent amount of spent acid for disposal.

Many of the world's current production facilities are package units based on years of design experience. For example, Meissner units, with capacities up to 12 t of nitroaromatics per hour, have built-in nitrogen blanketing for additional safety. Each output stream passes through purging steps; thus, spent acid is extracted with incoming benzene to remove both residual nitrobenzene and nitric acid, while residual waste gases are scrubbed by a mixed acid loop to meet environmental regulations [32].

An alternative development from the turbulent flow tube reactor is the pump nitration circuit of Nobel Chematur. When nitration actually takes place in the pump itself, reaction times can be less than a second due to the intensive mixing, and many consequential advantages are claimed [33], [34].

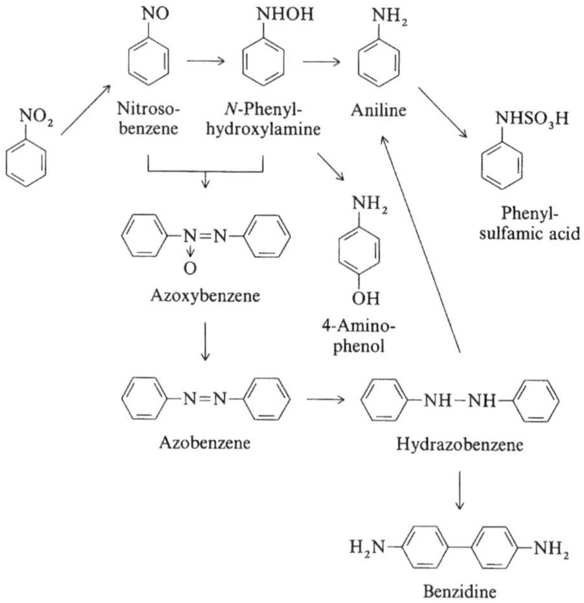

**Figure 3.** Nitrobenzene reduction products

The concept of an adiabatic process that uses the heat of reaction for concentration of spent acid has been around since 1941 [35]. The American Cyanamid adiabatic process [36], jointly developed with CLR, forms the basis for one of the world's largest nitrobenzene plants, rated at 159 000 t/a. This process takes cocurrent streams of benzene (1.1 mol per mole of $HNO_3$) and mixed acid (6–8/62–68/24–32 wt%) through a vigorously agitated tubular reactor, with entry at 60–80 °C (below the boiling point of benzene) and exit at ca. 120 °C, so engineered that the residence time is around 4 min. The requirement for much weaker sulfuric acid makes acid recycle much more economical, especially because process heat is virtually sufficient for evaporation if this is carried out under vacuum.

Another option for eliminating the need to reconcentrate the sulfuric acid in a separate step is to carry out the nitration at higher temperature (120–160 °C), with excess water being distilled from the nitrator as an azeotrope with benzene. In a duplex process the benzene azeotrope resulting from a high-temperature first-stage partial reaction is used in a lower temperature second stage to complete the reaction [37]. Azeotropic nitration offers considerable energy savings, but it is not thought to be competitive with adiabatic nitration.

In all these processes the work-up streams are very similar (cf. Fig. 4). The reaction mixture passes from the nitrator to a separator or centrifuge, and aqueous spent acid is fed to the recycle loop via a concentrator or strength adjustment stage as appropriate. Crude nitrobenzene, as the top layer, is passed through a series of washer–separators in which residual acid is removed by washing first with dilute alkali and then with water. The washed product is topped in a still to remove water and benzene (recycled) and, if required, finally vacuum distilled to give pure product in ca. 96% overall yield.

**Figure 4.** Production of nitrobenzene—continuous process

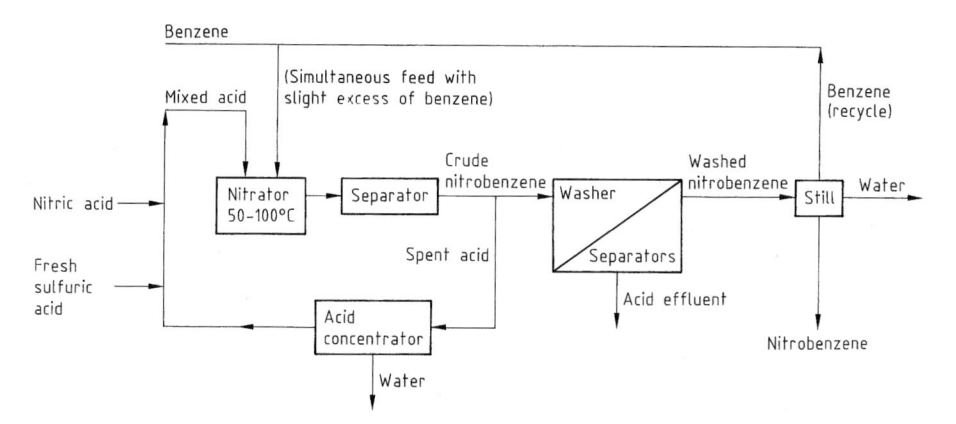

The washing stages give rise to an effluent problem, and wastewater treatment is necessary, principally to remove nitrobenzene.

World capacity for nitrobenzene in 1985 was ca. $1.7 \times 10^6$ t/a, with about one-third located in Western Europe and one-third in the United States [20].

*Uses.* The use of nitrobenzene as a processing solvent in specific chemical reactions is minor but important. Most (95% or more) nitrobenzene produced is converted to aniline, which has hundreds of downstream products. Lower volume, but nevertheless important, industrial outlets include electrolytic reduction to 4-aminophenol, nitration to give 1,3-dinitrobenzene (**6**), chlorination to give 3-chloronitrobenzene (**7**), sulfonation to give 3-nitrobenzenesulfonic acid (**8**), and chlorosulfonation to give 3-nitrobenzenesulfonyl chloride (**9**). The last three products are consumed mainly as their reduction products, 3-chloroaniline, metanilic acid, and 3-aminobenzenesulfonamide, respectively.

Nigrosin (C.I. Solvent Black 5) still survives as the crude mixture obtained by reacting nitrobenzene with aniline and aniline hydrochloride at 200 °C in the presence of iron or copper.

*Toxicity.* Nitrobenzene (TLV 5 mg/m$^3$) is very toxic and can cause acute poisoning because it is readily absorbed via the skin or respiratory tract. The primary effects are lowering of the hemoglobin level, methemoglobinemia, cyanosis, and breathlessness. Chronic exposure can lead to spleen and liver damage, jaundice, and anemia. Ingestion of alcohol may speed up and exaggerate the effects. Working conditions must be tightly specified and controlled [22].

**1,3-Dinitrobenzene** [*99-65-0*] (**6**), *m*-dinitrobenzene, $C_6H_4N_2O_4$, $M_r$ 168.1, crystallizes from ethanol as pale yellow rhombic–bipyramidal plates; the compound is steam volatile. Physical properties are listed in Table 3, together with those of the 1,2- and 1,4-isomers for comparison.

**Table 3.** Physical properties of the three dinitrobenzenes

| Property | Isomer | | |
| --- | --- | --- | --- |
| | Ortho | Meta | Para |
| *mp*, °C | 118 | 91 | 178 * |
| *bp*, (101 kPa) °C | 318 | 307 | 298 |
| Density, g/cm$^3$; (*t*, °C) | 1.3119 (120) | 1.3729 (100) | 1.617 (20) |
| Dipole moment, D | | 3.88 | |
| Solubility in H$_2$O (25 °C), % | 0.013 | 0.081 | 0.065 |
| Solubility in C$_2$H$_5$OH (20 °C), % | 1.9 | 3.5 | 0.4 |
| Solubility in benzene (18 °C), % | 5.66 | 39.45 | 2.56 |
| Solubility in chloroform (18 °C), % | 27.1 | 32.4 | 1.82 |

* After sublimation.

Chemical reduction of 1,3-dinitrobenzene gives 3-nitroaniline or *m*-phenylenediamine, depending on conditions. Under mild conditions, 1,3-dinitrobenzene does not react with aqueous sodium hydroxide or sodium sulfite solution, whereas the 1,2- and 1,4-isomers react to form alkali-soluble nitrophenols or nitrobenzene sulfonic acids. These reactions provide the basis for the removal of small amounts of isomeric impurities from 1,3-dinitrobenzene.

*Production.* Nitrobenzene is further nitrated with strong mixed acid (HNO$_3$ 33 wt%, H$_2$SO$_4$ 67 wt%) by controlled addition and reaction at 60–90 °C. After completion of the reaction and addition to water the crude oil is separated while still hot. The ortho isomer (10%), the para isomer (2%), and nitrophenol (2%) byproducts are removed by treatment with an aqueous NaOH – NaHSO$_3$ mixture. The final yield is ca. 82% of 99% pure product (*mp* 88–89 °C).

*Uses.* Catalytic hydrogenation of 1,3-dinitrobenzene yields *m*-phenylenediamine; smaller quantities are reduced with aqueous sodium sulfide to give 3-nitroaniline (Section 5.1). The I.G. Farbenindustrie developed an interesting alternative isomer-directing process for 1,3-dichlorobenzene by passing chlorine into boiling 1,3-dinitrobenzene; a 79% yield of distilled product was reported [1]. Historically, 1,3-dinitrobenzene has been used in the formulation of explosives, lubricants, polymerization inhibitors, and corrosion inhibitors.

*Toxicity.* 1,3-Dinitrobenzene and its isomers are highly toxic [8] by inhalation or skin absorption. The TLV dermal is 1 mg/m$^3$. Prolonged exposure to small amounts can lead to accumulation in fatty tissue, with the additional danger that ingestion of alcohol can flush out these residues to produce the same severe toxic symptoms as massive exposure. Progressive symptoms of cyanosis, headaches, and vomiting can lead to severe poisoning which is signaled by yellowing of the hair and nails in prolonged cases.

**1,3,5-Trinitrobenzene** [*99-35-4*] (**10**), *s*-trinitrobenzene, C$_6$H$_3$O$_6$N$_3$, $M_r$ 213.1, crystallizes as yellow dimorphic crystals from ethanol or nitric acid (*mp* 61 and 122.5 °C, respectively). 1,3,5,-Trinitrobenzene is inert to further ring substitution. Reduction with sodium sulfide gives 3,5-dinitroaniline and with Fe – HCl, 1,3,5-triaminobenzene. The latter is an intermediate in the production of phloroglucinol. Reaction with potassium methoxide results in the formation of 3,5-dinitroanisole via a classical Meisenheimer complex. An important property is the formation of π-complexes with hydrocarbons and other donors, which are useful in their characterization or isolation. Although trinitrobenzene has more explosive power than trinitrotoluene (TNT), it is too expensive in comparison [8].

*Production.* Although 1,3-dinitrobenzene may be further nitrated with strong mixed acid, extensive (dangerous) side reactions occur and the yield of 1,3,5-trinitrobenzene is low. The preferred synthesis is by decarboxylation of 2,4,6-trinitrobenzoic acid (obtained from TNT by oxidation with chromic acid) by heating in aqueous medium.

## 3.2. Nitrotoluenes

All three isomeric mononitrotoluenes, 2,4-dinitrotoluene (together with mixed isomeric dinitrotoluenes), and 2,4,6-trinitrotoluene are industrially important products obtained from the sequential nitration of toluene. The main outlets for mono- and dinitrotoluenes depend on their catalytic hydrogenation to the derived amines, but minor outlets also exist for sulfonated and chlorinated derivatives. Figure 5 shows key intermediates derived from nitrotoluenes.

**2-Nitrotoluene** [88-72-2] (**11**), *o*-nitrotoluene, ONT, 1-methyl-2-nitrobenzene, $C_7H_7NO_2$, $M_r$ 137.1, is a clear yellow liquid that crystallizes at lower temperatures to solid α- or β-forms, depending on conditions. The compound is steam distillable at 100 g of ONT per kilogram of distillate. 2-Nitrotoluene is soluble in most organic solvents but only slightly soluble in water (0.065 % at 30 °C). Some physical properties are listed in Table 4, together with those of 3- and 4-nitrotoluenes for comparison.

Reduction of 2-nitrotoluene with (1) hydrogen or iron borings and acid gives *o*-toluidine; (2) zinc dust and ammonia in aqueous ethanol gives *o*-tolylhydroxylamine, (3) iron powder and aqueous caustic soda gives *o*-azoxy- and *o*-azotoluene; and (4) zinc dust and alcoholic sodium hydroxide yield *o*-hydrazotoluene (cf. Table 2).

Oxidation of 2-nitrotoluene with manganese dioxide in sulfuric acid gives 2-nitrobenzaldehyde [552-89-6] (**14**) or 2-nitrobenzoic acid [552-16-9] (**15**), depending on the strength of the sulfuric acid. When potassium permanganate, potassium dichromate, or nitric acid is used as oxidizing agent, only 2-nitrobenzoic acid is obtained. Treatment of 2-nitrotoluene with boiling aqueous sodium hydroxide produces anthranilic acid through the unusual combination of autoxidation and reduction. However, the yield is too low for industrial use.

Reaction of 2-nitrotoluene with diethyl oxalate gives 2-nitrophenylpyruvic acid [5461-32-5] (**27**), which on reductive cyclization and decarboxylation

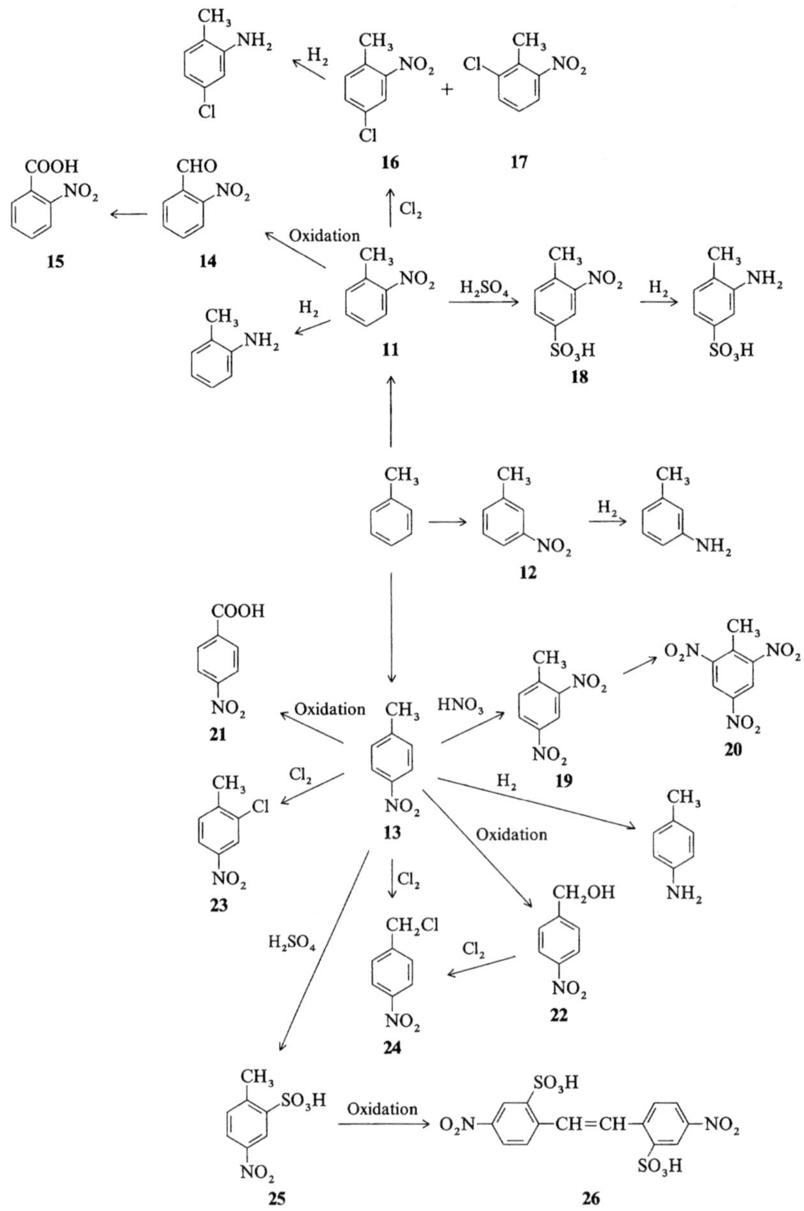

**Figure 5.** Key intermediates derived from nitrotoluenes

Nitro Aromaticsconverts to indole—the model example of the Reissert indole synthesis. This is one of the many reactions of 2-nitrotoluenes with neighboring group interaction to give cyclized products [38].

**Table 4.** Physical properties of the three nitrotoluenes

| Property | Isomer | | |
|---|---|---|---|
| | Ortho | Meta | Para |
| $mp$, °C | −9.27 (α-form) | 16.1 | 44.5 (unstable) |
| | −3.17 (β-form) | | 51.9 (stable form) |
| $bp$, °C | | | |
|   at 101 kPa | 222.3 | 231.9 | 238.3 |
|   at 0.13 kPa | 50.0 | 50.2 | 53.7 |
| $d_4^{20}$ | 1.163 | 1.1571 | 1.286 |
| Vapor density (air = 1) | 4.72 | 4.72 | 4.72 |
| Refractive index $n_D^{20}$ | 1.5474 | 1.5470 | 1.5460 |
| Surface tension, mN/m | 42.3 (15 °C) | 39.9 (30 °C) | 36.8 (60 °C) |
| Dielectric constant (58 °C) | 21.61 | 22.2 | 22.2 |
| Flash point (closed cup), °C | 91 | 102 | 106 |

Nitration of 2-nitrotoluene yields a mixture of 2,4-dinitrotoluene and 2,6-dinitrotoluene. Chlorination in the presence of iron or Friedel–Crafts catalysts yields a mixture of 4-chloro-2-nitrotoluene (**16**) and 2-chloro-6-nitrotoluene (**17**), whereas in the absence of iron, 2-nitrobenzyl chloride [612-23-7] (**28**) or 2-chlorotoluene is formed, depending on reaction conditions. 2-Nitrotoluene reacts with formaldehyde in concentrated sulfuric acid to give dinitroditolylmethane.

*Production.* 2-Nitrotoluene is the dominant isomer in conventional mixed acid nitration of toluene. Table 5 shows a $60 \pm 5\%$ ortho, $38 \pm 5\%$ para mix at different sulfuric acid strengths, as well as the products obtained by using aromatic sulfonic acid or phosphoric acid, which contain lower ortho–para ratios. This is a desirable feature because the ratio of ca. 1.6 obtained under practical mixed acid conditions does not normally match the requirements and leads to a surplus of 2-nitrotoluene. However, the phosphoric acid alternative with the desirable ratio [39] shown in Table 5 has, apart from the problem of phosphoric acid recovery, a tendency to increase the amount of dinitrotoluene formed, which leads to an apparent reduction in the amount of ortho product. Under conditions in which dinitrotoluene formation is eliminated, one evaluation of the phosphoric acid process could only achieve a reduction in the ortho–para ratio to 1.35 [40]. Many claims for alternative processes and reaction media have been patented [17] without becoming established.

Large-scale nitration is still carried out under the usual conditions of mixed acid (17–32/67–52/ca. 16%) at 25–40 °C and a nitric acid to toluene molar ratio close to one. The inherent speed of reaction (17 times faster than with benzene) makes continuous reaction with low residence times most suitable. An ortho–para ratio of 1.6 and an isolated yield of 96% total isomers are typical in a plant similar to that described for nitrobenzene but operating at as low a temperature as possible to avoid the more readily formed byproducts. The completed nitration mix is fed to a continuous, usually centrifugal, separator while hot, and the crude isomer mix is washed with dilute alkali and water (cf. nitrobenzene production). The washed product is transferred to a still,

**Table 5.** Nitration products of toluene with various agents

| Nitrating agent | Temperature, °C | 2-Nitrotoluene, % | 3-Nitrotoluene, % | 4-Nitrotoluene, % | Ortho–para ratio |
|---|---|---|---|---|---|
| 85% HNO$_3$ | −30 | 55.6 | 2.7 | 41.7 | 1.33 |
| 85% HNO$_3$ | 60 | 57.5 | 4.0 | 38.5 | 1.49 |
| 23.8% HNO$_3$, 58.7% H$_2$SO$_4$, 17.5% H$_2$O | 50 | 62.2 | 4.3 | 33.5 | 1.86 |
| 5.4% HNO$_3$, 71.4% H$_2$SO$_4$, 21.5% H$_2$O | 60 | 58.2 | 5.5 | 36.2 | 1.61 |
| 90% HNO$_3$ + aromatic sulfonic acid | 50 | 51.0 | 3.6 | 45.4 | 1.12 |
| 10% HNO$_3$, 90% H$_3$PO$_4$ | 30 | 45.1 | 4.3 | 50.2 | 0.9 |

and the water and residual toluene are topped off before fractional distillation under vacuum (96–97 °C, 1.6 kPa) to give fairly pure 2-nitrotoluene [8].

The 1984 capacity for mononitrotoluene (all isomers) production in the Western world was ca. 200 000 t/a [20].

*Uses.* 2-Nitrotoluene derivatives are used principally as colorant intermediates. For example, 2-toluidine, 2-amino-4-chlorotoluene [Fast Scarlet TR Base, by reduction of 4-chloro-2-nitrotoluene (**16**)], 2-amino-6-chlorotoluene [Fast Red KB Base, by reduction of 2-chloro-6-nitrotoluene (**17**)], and 2-toluidine-4-sulfonic acid [by reduction of 2-nitrotoluene-4-sulfonic acid (**18**)] are all diazo components of azo dyes.

A more recent outlet for 2-toluidine (which is important in consuming the normal surplus of 2-nitrotoluene) is its conversion to 2-ethyl-6-methylaniline, an agrochemical intermediate.

Reduction to *o*-hydrazotoluene followed by rearrangement to 2-tolidine has virtually been phased out based on analogy with the proven carcinogen benzidine.

**3-Nitrotoluene** [99-08-1] (**12**), *m*-nitrotoluene, 1-methyl-3-nitrobenzene, C$_7$H$_7$NO$_2$, $M_r$ 137.1, is a yellow liquid at ambient temperature; it is soluble in most organic solvents, but only slightly soluble in water (0.05% at 30 °C). Physical properties are listed in Table 4.

Reduction of 3-nitrotoluene with hydrogen or strong reducing agents produces 3-toluidine, but treatment with Zn–H$_2$O or Zn–EtOH–KOH yields *m*-tolylhydroxylamine and *m*-azotoluene, respectively. Oxidation with chromic acid gives 3-nitrobenzoic acid [121-92-6]. Further nitration leads to a mixture of 2,3-, 2,5-, and 3,4-dinitrotoluenes (Table 6).

*Production.* The middle "meta fraction" distilled from the mononitrotoluene isomer mix after 2-nitrotoluene has been removed is predominantly 3-nitrotoluene, but with a large 4-nitrotoluene content. This is accumulated to form the basis for more rigorous fractional distillation to give technical quality 3-nitrotoluene.

*Uses.* Catalytic hydrogenation or reduction with iron converts 3-nitrotoluene to *m*-toluidine, which provides important coupling components for azo dyes, usually after N-alkylation.

**Table 6.** Dinitrotoluene isomeric mixtures obtained by nitration

| Starting material | Dinitrotoluene, wt% | | | | | cp*, °C |
|---|---|---|---|---|---|---|
| | 2,3- | 2,4- | 2,5- | 2,6- | 3,4- | |
| Toluene | 1.3 | 78 | 0.5 | 18 | 2.4 | 55–57 |
| 2-Nitrotoluene | | 67 | | 33 | < 0.5 | 50–51 |
| 3-Nitrotoluene | 25 | | 20 | | 55 | |
| 4-Nitrotoluene | | 99 | | | | 69 |

* Crystallization point of the isomeric mixture.

**4-Nitrotoluene** [99-99-0] (**13**), *p*-nitrotoluene, PNT, 4-nitro-1-methylbenzene, $C_7H_7NO_2$, $M_r$ 137.1, forms colorless to light yellow rhombic–bipyramidal crystals, which are soluble in most organic solvents but only slightly soluble in water (0.044% at 30 °C). Physical properties are listed in Table 4.

Reduction of 4-nitrotoluene with hydrogen, iron, and acid, or electrolytically in aqueous hydrochloric acid produces *p*-toluidine. Treatment with zinc dust in hot aqueous–alcoholic calcium chloride gives *p*-tolylhydroxylamine. Oxidation with chromic oxide in acetic acid produces 4-nitrobenzaldehyde [555-16-8]. Electrolytic oxidation in the presence of manganese salts or treatment with other strong oxidizing agents yields 4-nitrobenzoic acid, whereas controlled electrolytic oxidation in acetic acid–sulfuric acid results in 4-nitrobenzyl alcohol (**22**). Air oxidation (MeOH–KOH medium) leads to 4,4′-dinitrodibenzyl and 4,4′-dinitrostilbene.

Chlorination in the presence of Friedel–Crafts catalysts gives 2-chloro-4-nitrotoluene (**23**), whereas without catalyst 4-nitrobenzyl chloride (**24**) is produced. Further nitration or sulfonation of 4-nitrotoluene gives highly selective 2-substitution.

*Production.* After distillation of the meta fraction from the isomer mixture, the still residues are cooled in a crystallizer to separate technical quality 4-nitrotoluene. Further distillation of nitrotoluene residues and fractions should be implemented with great care because this has been reported to cause explosions. Holding residues at 150–200 °C results in an undefined "aging" process that can lead to unpredictable evolution of heat, especially if air is introduced [41].

*Uses.* 4-Nitrotoluene derivatives are used principally as intermediates for colorants and related products; for example, *p*-toluidine, 4-nitrobenzoic acid [62-23-7] (by oxidation of 4-nitrotoluene with 15% $HNO_3$ at 175 °C), 4-amino-2-chlorotoluene (by reduction of 2-chloro-4-nitrotoluene), and 4-nitrotoluene-2-sulfonic acid (**25**), which is of great importance in forming stilbene intermediates for fluorescent whitening agents (see Section 6.2).

**2,4-Dinitrotoluene** [121-14-2] (**19**), DNT, 2,4-dinitro-1-methylbenzene, $C_7H_6N_2O_4$, $M_r$ 182.1, *mp* 71 °C, *bp* 300 °C (decomp., 101 kPa), $d_4^{15}$ 1.521, $d_4^{71}$ 1.321, vapor density 6.27, forms yellow needles from ethanol or carbon disulfide, which are moderately soluble in most organic solvents but only sparingly soluble in water (0.03% at 22 °C).

Reduction of 2,4-dinitrotoluene with iron–acid or $H_2$–catalyst gives 2,4-diaminotoluene smoothly. Selective reduction of either nitro group may be achieved under carefully controlled conditions by using the stoichiometric amount of hydrogen with a platinum black catalyst to obtain 4-amino-2-nitrotoluene [*119-32-4*], or an equimolar quantity of tin(II) chloride in alcoholic hydrochloric acid to obtain the isomeric 2-amino-4-nitrotoluene [*99-55-8*].

Oxidation of 2,4-dinitrotoluene with aqueous nitric acid or potassium permanganate yields 2,4-dinitrobenzoic acid [*610-30-0*].

*Production.* 4-Nitrotoluene is nitrated with mixed acid (containing equimolar $HNO_3$) under controlled conditions and continuous operation to give a 96% yield of 2,4-dinitrotoluene. In Meissner units, for tighter control of the whole process, cocurrent flow is used in each step to separate mono- and dinitration, but countercurrent flow is used between them to prevent under- or overnitration; this is also done in one-step nitration plants [32]. After nitration, if separation of the resulting emulsion is difficult, the introduction of pure product can give surprisingly rapid separation [42].

Alternative processes yield mixed isomer products. When toluene is nitrated directly under similar conditions with 2.1 equivalents of nitric acid, the product is a ca. 80:20 mixture of 2,4-dinitrotoluene and 2,6-dinitrotoluene [*606-20-2*]. Nitration of 2-nitrotoluene, which is sometimes present in surplus, gives a ca. 67:33 mixture of 2,4- and 2,6-dinitrotoluene (Table 6).

*Uses.* Most of the 2,4-dinitrotoluene produced is hydrogenated (nickel catalyst) to 2,4-diaminotoluene (*m*-tolylenediamine) [*95-80-7*] for conversion to toluene diisocyanate (TDI), which is a component of a polyurethane. A much smaller amount is used in explosives and for further nitration to TNT (**20**).

Crude mixtures of 2,4- and 2,6-dinitrotoluene are also used to produce mixed diamines (80:20 or 67:33, depending on the nitration process used), which in turn are converted to mixed isocyanates for polyurethane production, the use of isomer mixtures being considerably cheaper.

**Toxicity of Nitrotoluenes.** The toxic effects of mononitrotoluenes are similar to those described for nitrobenzene. Generally, the toxicity of alkylnitrobenzenes decreases with increasing number of alkyl groups and increases with increasing number of nitro groups. Mononitrotoluenes (TLV 30 mg/m$^3$) are methemoglobin formers and may be absorbed through the skin or the respiratory tract. Dinitrotoluenes (TLV 1.5 mg/m$^3$) and TNT (TLV 1.3 mg/m$^3$) are progressively more potent.

## 3.3. Nitroxylenes

All three xylenes can be nitrated readily, with *o*- and *m*-xylene giving mixtures of products. Even under simple conditions with mixed acid, isomer ratios can be changed significantly by varying the reaction conditions.

o-xylene → HNO₃/H₂SO₄ →
**29** (55–31 %) + **30** (45–69 %)

m-xylene → HNO₃/H₂SO₄ →
**31** (25–16 %) + **32** (75–84 %)

p-xylene → HNO₃/H₂SO₄ →
**33** (89 %)

All five important mononitroxylenes are pale yellow oils or low-melting solids, $C_8H_9NO_2$, $M_r$ 151.17, which are soluble in most organic solvents but only slightly soluble in water. Some physical properties are listed in Table 7.

The nitroxylenes (**29–33**) are all readily reduced to the corresponding aminoxylenes (xylidines). Oxidation can result in conversion of one or both of the methyl groups to a carboxylic acid group, depending on the reagent and reaction conditions. For example, **30** with aqueous nitric acid yields 2-methyl-4-nitrobenzoic acid and **31** with potassium permanganate gives 2-nitroisophthalic acid, whereas **32** results in a mixture of 4-nitroisophthalic acid and 3-methyl-4-nitrobenzoic acid under similar conditions. Further nitration of nitroxylenes in mixed acid at 35–50 °C with a slight excess of nitric acid gives high yields of dinitroxylenes.

*Production.* Batchwise nitration of *p*-xylene in mixed acid at 25–35 °C with a slight excess of nitric acid gives yields of 85–95 %. Conventional separation and workup lead to fairly pure 1,4-dimethyl-2-nitrobenzene (**33**). The mixture of **31** and **32** obtained from *m*-xylene is distilled to remove **31** and leave the main product 1,3-dimethyl-4-nitrobenzene (**32**). The mixture of **29** and **30** obtained from *o*-xylene is much more difficult to separate because the ratio is nearly 1:1 and their boiling points are closer. Initial fractionation by vacuum distillation is followed by a sweating process (dependent on melting point differences) to complete the purification [1].

Table 7. Physical properties of nitroxylenes

| Compound | CAS registry number | mp, °C | bp (101 kPa), °C | $\varrho$, g/cm$^3$ |
|---|---|---|---|---|
| 1,2-Dimethyl-3-nitrobenzene (**29**) | [83-41-0] | 15 | 246 | 1.1402 |
| 1,2-Dimethyl-4-nitrobenzene (**30**) | [99-51-4] | 28–30 | 258 | 1.139 |
| 1,3-Dimethyl-2-nitrobenzene (**31**) | [81-20-9] | 15 | 222 | 1.112 |
| 1,3-Dimethyl-4-nitrobenzene (**32**) | [89-87-2] | 9 | 244 | 1.126 |
| 1,4-Dimethyl-2-nitrobenzene (**33**) | [89-58-7] | 25 | 240 | 1.132 |

*Uses.* The sole outlet for nitroxylenes is via their reduction to the corresponding xylidines. 3,4-Dimethylaniline [95-64-7] (3,4-xylidine, **34**), and 2,6-dimethylaniline [87-62-7] (2,6-xylidine, **35**) are starting materials for the production of riboflavin and agrochemicals, respectively. Other isomers are used as colorant intermediates; the most important of these are 2,4-dimethylaniline [95-68-1] (2,4-xylidine, **36**), used as a diazo component (e.g., in C.I. Solvent Orange 7 and C.I. Acid Red 26) and 2,5-dimethylaniline [95-78-3] (2,5-xylidine, **37**) used as a coupling component (e.g., in Dispersol Orange 7 and C.I. Solvent Red 26). Conveniently, mixed xylidines can be used for some colorants.

## 3.4. Nitronaphthalenes

Under moderate conditions, naphthalene is nitrated to give 95 % 1-nitronaphthalene (**38**) and 5 % 2-nitronaphthalene (**39**). Fierz – David [43] obtained this ratio, together with traces of dinitronaphthalene and dinitronaphthol, under a variety of conditions: (1) 20 % nitric acid at 95 – 98 °C, (2) 95 % nitric acid in acetic acid and acetic anhydride at 50 – 70 °C, and (3) HNO$_3$ – H$_2$SO$_4$ at 60 °C. The last conditions form the basis of large-scale production. Mixed acid nitration at higher temperature (80 – 100 °C) results in a mixture of 1,5-dinitronaphthalene (**40**) and 1,8-dinitronaphthalene (**41**) in ca. 2:3 ratio. This demonstrates deactivation of the ring containing the first-entering group, with the peri position then being favored in the second ring. Unlike the sulfonic acid substituent, the nitro group does not undergo migration in the naphthalene ring, and isomer ratios are therefore more constant and controllable.

**38** 1-nitronaphthalene

**39** 2-nitronaphthalene

**40** 1,5-dinitronaphthalene

**41** 1,8-dinitronaphthalene

All ten dinitronaphthalene isomers and many trinitro and tetranitro derivatives have been described in the literature, but only the commercially important 1,5- and 1,8-isomers are described here, together with 1- and 2-nitronaphthalenes.

Nitration of naphthalene mono- and disulfonic acids, leading to an important series of aminonaphthalenesulfonic acids, is usually performed with the appropriately formed sulfonation mass (→ Naphthalene Derivatives).

**1-Nitronaphthalene** [86-57-7] (**38**), α-nitronaphthalene, $C_{10}H_7NO_2$, $M_r$ 173.17, mp 52 (metastable) – 61 °C, bp 304 °C (101 kPa), forms pale yellow, odorless needles that sublime at 30 – 50 °C (1.33 Pa). It is soluble in ethanol, chloroform, or ether, but insoluble in water.

Halogenation of 1-nitronaphthalene occurs in the 5- and 8-positions, and sulfonation results in 5-nitronaphthalene-1-sulfonic acid, although the latter is produced on a large scale by nitration of naphthalene-1-sulfonic acid. When chlorine gas is passed into molten 1-nitronaphthalene, the nitro group is replaced, and a mixture of chloronaphthalenes is formed.

Hydrogenation or iron reduction results in 1-aminonaphthalene; intermediate azoxy, azo, and hydrazo stages can be obtained by reduction with Zn – NaOH in alcohol. Controlled reduction with aqueous ammonium sulfide under mild conditions yields 1-naphthylhydroxylamine. Oxidation with air in aqueous $Ca(OH)_2$ – KOH at 140 °C produces 1-hydroxy-4-nitronaphthalene. Reaction with $NH_2OH$ – NaOEt yields 2-amino-4-nitronaphthalene.

*Production.* Naphthalene is charged to sulfuric acid alternately with mixed acid at 40 – 50 °C over 8 h, with the final temperature being 55 °C. The acid strength used allows the molten product to be separated as an oil and washed. Continuous operation at 50 – 60 °C with mixed acid (33/48/19) gives a comparable product. The crude reaction product is obtained in 90 – 95 % yield and is purified by vacuum distillation.

*Uses.* Virtually all of the 1-nitronaphthalene produced is catalytically reduced to 1-naphthylamine. Historically, it has also been used as a deblooming agent for petroleum and oils and as a component in the formulation of explosives.

*Toxicology* [22]. Vapors of 1-nitronaphthalene are poisonous and lacrimatory, being skin, eye, and mucous membrane irritants.

**2-Nitronaphthalene** [*581-89-5*] (**39**), β-nitronaphthalene, $C_{10}H_7NO_2$, $M_r$ 173.17, *mp* 79 °C, *bp* 312 °C (97.4 kPa), *bp* 165 °C (2 kPa), does not sublime and is volatile in steam, in contrast to 1-nitronaphthalene. It is soluble in ethanol, ether, or chloroform, but insoluble in water.

Halogenation and sulfonation take place mainly in the 5-position, and further nitration gives a mixture of 1,6- and 1,7-dinitro- and 1,3,8-trinitronaphthalenes.

Preparation is carried out by indirect methods, for example, a nitro-Sandmeyer reaction with 2-naphthylamine.

*Toxicology.* 2-Nitronaphthalene is moderately toxic by ingestion and intraperitoneal routes; it is also a skin and lung irritant. 2-Nitronaphthalene is treated as a human carcinogen because it is thought to be metabolized to 2-aminonaphthalene in the human body. This is most relevant at the purification stage in the production of 1-nitronaphthalene.

**1,5-Dinitronaphthalene** [*605-71-0*] (**40**), $C_{10}H_6N_2O_4$, $M_r$ 218.17, *mp* 219 °C, forms colorless needles from benzene, acetic acid, or acetone; the compound is potentially explosive [19], [22].

Reduction with iron – acid or hydrogen – catalyst yields 1,5-naphthalenediamine, but $Zn – NH_4Cl$ in ethanol produces 5,5′-dinitro-1,1′-azoxynaphthalene, and reduction with ammonium sulfide gives 5-nitro-1-aminonaphthalene. Nitration of **40** produces 1,4,5-trinitronaphthalene, whereas sulfonation with 20% oleum results in 1,5-dinitronaphthalene-3-sulfonic acid. Vigorous oxidation with nitric acid at 150 °C gives a mixture of 3,5-dinitrobenzoic acid, 3-nitrophthalic acid, and picric acid.

*Production.* Naphthalene is added slowly to mixed acid (22/58/20) at 40 °C, and the temperature is raised to 80 °C over 4 h to give a mixture of 1,5- and 1,8-dinitronaphthalenes. The isomers are separated by fractional crystallization (e.g., from ethylene dichloride) or, preferably, by solvent extraction [44]. The 1,5-isomer can be extracted with toluene, leaving 99% pure 1,8-dinitronaphthalene. After evaporation of the toluene, the residue is extracted with a strongly polar solvent (e.g., sulfolane) to leave 99% pure 1,5-isomer.

World production of dinitronaphthalenes occurs mainly in the Federal Republic of Germany and Japan, with the capacity of the latter reported to be 1200 t/a.

*Uses.* 1,5-Dinitronaphthalene is an intermediate in the production of naphthazarin (5,8-dihydroxy-1,4-naphthoquinone) and 1,5-naphthalenediamine, which is mainly converted to naphthalene 1,5-diisocyanate. As a sensitizing agent for ammonium nitrate explosives the use of mixed isomers is adequate.

**1,8-Dinitronaphthalene** [*602-38-0*] (**41**), $C_{10}H_6N_2O_4$, $M_r$ 218.17, *mp* 172 °C, forms colorless or pale yellow plates from ethanol or chloroform; the compound is potentially explosive. It is sparingly soluble in organic solvents and insoluble in water.

Table 8. Isomer ratios obtained from nitration of monoalkylbenzenes with mixed acid

| Starting material | Nitro isomers, wt% | | | Ortho–para ratio |
|---|---|---|---|---|
| | Ortho | Meta | Para | |
| Toluene | 58.5 | 4.5 | 37 | 1.58 |
| Ethylbenzene | 45.0 | 6.5 | 48.5 | 0.93 |
| Cumene | 30.0 | 7.7 | 62.3 | 0.48 |
| tert-Butylbenzene | 15.8 | 11.5 | 72.7 | 0.22 |

Chlorination yields 1,4,5-trichloronaphthalene, and nitration gives 1,3,8-tri- and 1,3,6,8-tetranitronaphthalenes. Sulfonation with sulfuric acid yields 4,5-dinitronaphthalene-2-sulfonic acid, but treatment with oleum at 40–50 °C results in 4-nitroso-5-nitro-1-naphthol. Selective hydrogenation to give 8-nitro-1-aminonaphthalene can be achieved with a platinum black catalyst.

*Production.* 1,8-Dinitronaphthalene is produced by separation from the mixed isomers obtained on nitration of naphthalene (see above).

*Uses.* 1,8-Dinitronaphthalene is catalytically hydrogenated to 1,8-naphthalenediamine for use mainly as a colorant intermediate for naphthperinones. If demand for the 1,5-isomer as an isocyanate precursor increases, one outlet for the surplus 1,8-isomer would be direct conversion to sulfur dyes.

## 3.5. Other Nitro Aromatics

Although ethylbenzene and cumene are readily available and can be nitrated under similar conditions to toluene, their nitro derivatives very rarely supplant the nitrotoluenes as precursors. A startling technical advantage of, for example, an ethyl homologue over a toluene derivative would be required to justify the establishment of large-scale nitro isomer separation. Table 8 records the isomer trends on nitration of higher alkylbenzenes under comparable conditions. The ortho–para ratio decreases incrementally due to steric hindrance [45].

Nitration of polyalkylbenzenes (e.g., mesitylene and durene) requires extra care because of oxidation and other side reactions that can easily result from nitrosation and ipso substitution. Trinitromesitylene is prepared readily from mesitylene and mixed acid, and has been used as an explosive [8]. Two nitro derivatives of mesitylene are, however, of more recent importance as intermediates for colorants. Mesidine [88-05-1] (2,4,6-trimethylaniline, **42**) is produced by reduction of 1,3,5-trimethyl-2-nitrobenzene [603-71-4], which is obtained by nitration of mesitylene under special conditions, e.g., low temperature in the presence of acetonitrile and sulfamic acid [46]. 3,5-Diamino 2,4,6-trimethyl-benzenesulfonic acid [32432-55-6] (**43**) is produced by dinitration of the reaction mixture from the monosulfonation of mesitylene followed by reduction of the dinitro mesitylenesulfonic acid.

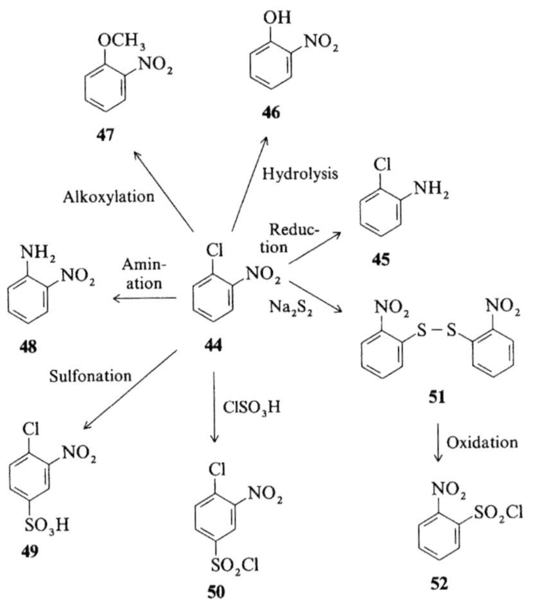

**Figure 6.** Key intermediates derived from 2-chloronitrobenzene

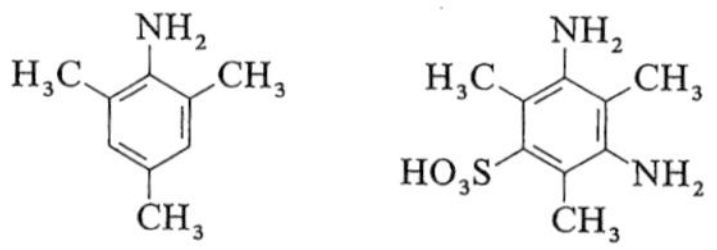

# 4. Nitrohalo Aromatics

## 4.1. Chloronitrobenzenes

The three isomeric monochloronitrobenzenes and 2,4-dinitrochlorobenzene are industrially important; all except 3-chloronitrobenzene are obtained by nitration of chlorobenzene. The primary outlets for chloronitrobenzenes depend on nucleophilic reactions (e.g., amination and alkoxylation), reduction of the nitro group, and electrophilic ring substitution (e.g., sulfonation). This makes them highly versatile primary intermediates with a wide range of derived products, as shown in Figures 6 and 7, for use in the production of colorants and other effect chemicals [47].

**2-Chloronitrobenzene** [88-73-3] (**44**), OCNB, 2-nitrochlorobenzene, $C_6H_4ClNO_2$, $M_r$ 157.56, crystallizes as light yellow monoclinic needles. The compound is very soluble in ether, benzene, or hot ethanol, but insoluble in water. Physical properties

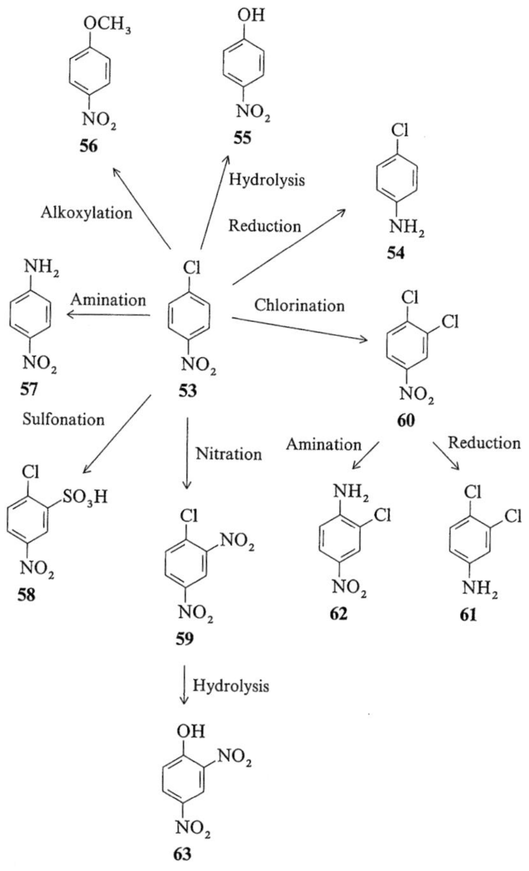

**Figure 7.** Key intermediates derived from 4-chloronitrobenzene

are listed in Table 9, together with those of 3- and 4-chloronitrobenzene for comparison.

Reduction of the nitro group with Fe–HCl gives 2-chloroaniline, with alternative reducing agents giving intermediate reduction products analogously to nitrobenzene (Fig. 3). The chlorine atom is readily replaced by nucleophilic attack to give hydroxy, alkoxy, and amino derivatives (Fig. 6). Electrophilic ring substitution (e.g., by halogenation, nitration, or sulfonation) takes place mainly in the position meta to the nitro group.

*Production.* Nitration of chlorobenzene with mixed acid (30/56/14) typically gives an isomer mix in 98% yield consisting of 34–36% 2-chloronitrobenzene, 63–65% 4-chloronitrobenzene, and only ca. 1% 3-chloronitrobenzene. The ortho–para ratio of about 0.55 is in sharp contrast to the 1.6 obtained with nitrotoluene isomers (Section 3.2). As with the nitration of toluene, much work has been done on isomer control so that the producer might have some flexibility towards the balance of isomer demand. Although no major change in ratio has been achieved, the situation is better for nitrochlorobenzene than nitrotoluenes, because the favored isomer is in much greater

**Table 9.** Physical properties of the three monochloronitrobenzenes

| Property | Isomer | | |
|---|---|---|---|
| | Ortho | Meta | Para |
| mp, °C | 33 | 46 (stable) | 83 |
| | | 24 (labile) | |
| bp, °C (kPa) | 245.5 (100) | 235.5 (101) | 242 (101) |
| | 119 (1.1) | | 113 (1.1) |
| Density, g/cm$^3$ (°C) | 1.368 (22) | 1.534 (20) | 1.520 (22) |
| | 1.305 (80) | 1.310 (80) | 1.303 (85) |
| Dielectric constant (°C) | 37.7 (50) | 20.9 (50) | 12.7 (83.5) |
| Flash point (closed cup), °C | 127 | 127 | |

demand. In further contrast to the nitration of toluene, the nitration of chlorobenzene in the presence of phosphoric acid decreases the proportion of 4-chloronitrobenzene [40], and the use of phosphoric acid in the presence of a transition-metal catalyst is said to increase the ortho–para ratio to ca. 0.8 [48].

Even though the rate of nitration of chlorobenzene is an order of magnitude slower than that of benzene, comparable temperatures (40–70 °C) are adequate, and the techniques and equipment are very similar to those described for benzene (Section 3.1). In Meissner units the slower reaction rate is compensated by placing additional reactors in series, thereby facilitating flexible operation of this plant type as a basis for multipurpose installations [32].

The mixed product output stream is the same whether produced by a batch or continuous process, and the isomers are separated by a combination of fractional crystallization and distillation. For a simple first separation the isomer mixture is held at a temperature slightly above its crystallization point (15 °C), whereby much of the 4-chloronitrobenzene crystallizes and can be separated. Fractional distillation gives a para-rich distillate containing all the meta isomer and an ortho-rich still residue. Each of these is crystallized, separated, and the liquid component is refractionated to gradually accumulate high-purity ortho and para products, together with intermediate fractions for continual recycle.

Estimated 1985 product figures for 2- and 4-chloronitrobenzene are for the Federal Republic of Germany 60 000 t/a, the United States 40 000 t/a, and Japan 30 000 t/a.

*Uses.* 2-Chloronitrobenzene derivatives (Fig. 6) find many outlets in the synthesis of colorants and effect chemicals. Reduction with iron produces 2-chloroaniline [95-51-2] (**45**, Fast Yellow G Base), and electrolytic reduction followed by rearrangement of the resulting hydrazo derivative leads to 3,3′-dichlorobenzidine [91-94-1], both of which are important diazo components. In the alternative, more economical hydrogenation process, a modified catalyst is required to inhibit dechlorination as a side reaction. Typically, a platinum on carbon catalyst, together with a small quantity of an inorganic acid acceptor such as magnesium oxide, is used; however, up to 2 % aniline may still be formed. The alternative use of morpholine as a dechlorination suppressor, rather than just an acid acceptor, is recommended for 2- and 4-chloronitrobenzene, this reduces the extent of dechlorination to 0.5% [49]. The use of modified catalyst systems based on platinum, rather than the conventional nickel catalyst used in the production of most aniline derivatives, is essential for hydrogenation of all nitrohalo aromatics.

Treatment of 2-chloronitrobenzene with aqueous sodium hydroxide at 130 °C produces 2-nitrophenol (**46**; cf. Chap. 7), whereas treatment with MeOH–NaOH gives 2-nitroanisole (**47**) or with EtOH–NaOH 2-nitrophenetole (Chap. 7), all of which are used as precursors of the derived amines and many other products. Treatment with aqueous ammonia at 175 °C under pressure yields 2-nitroaniline (**48**; cf. Chap. 5). Sulfonation yields 4-chloro-3-nitrobenzenesulfonic acid (**49**), and chlorosulfonation gives the corresponding 4-chloro-3-nitrobenzenesulfonyl chloride (**50**). Reaction with sodium disulfide gives 2,2′-dinitrodiphenyl disulfide (**51**), which on oxidation with chlorine gives 2-nitrobenzenesulfonyl chloride (**52**). The three derivatives **49**, **50**, and **52** are all precursors to further series of intermediates (cf. Chaps. 6 and 7).

**3-Chloronitrobenzene** [*121-73-3*] (**7**), 3-nitrochlorobenzene $C_6H_4ClNO_2$, $M_r$ 157.56, forms pale yellow prisms (from ethanol), which can exist as a stable or a labile modification in the solid state. The compound is insoluble in water, readily soluble in benzene or ether, and soluble in acetone or hot ethanol. Physical properties are listed in Table 9 together with those of 2- and 4-chloronitrobenzenes for comparison.

Reduction of the nitro group with Fe–HCl gives 3-chloroaniline, and ring substitution with electrophiles yields mixtures of products. Unlike the 2- and 4-isomers, the chlorine atom of 3-chloronitrobenzene is not activated towards nucleophilic substitution.

*Production.* Nitrobenzene is chlorinated at 35–45 °C in the presence of sublimed iron(III) chloride to give an isomer mixture containing 86% of the desired 3-isomer. A continuous process has been described that uses a series of reactors operating at 35–55 °C, with a residence time of 5 h [50]. Purification is achieved by a combination of distillation and crystallization [51]. Disposal of the byproducts, 10% 2-chloronitrobenzene and 4% 4-chloronitrobenzene, can be achieved by feeding them to the mixed isomer stream in the chlorobenzene nitration plant. The 1% crude 3-chloronitrobenzene from this plant may similarly be fed to the purification stream of the chlorination plant. Final purification of 3-chloronitrobenzene may be achieved chemically by caustic hydrolysis of the residual 2- and 4-chloronitrobenzenes and washing them out as nitrophenols.

*Uses.* Reduction of 3-chloronitrobenzene to 3-chloroaniline [*108-42-9*] (Orange GC base) is its primary outlet, with minor uses in other fields [51]. Crude 3-chloronitrobenzene can be used for exhaustive chlorination to give pentachloronitrobenzene (Section 4.2).

**4-Chloronitrobenzene** [*100-00-5*] (**53**), PCNB, 4-nitrochlorobenzene, $C_6H_4ClNO_2$, $M_r$ 157.56, crystallizes as light yellow monoclinic prisms, which are insoluble in water and very soluble in toluene, ether, acetone, or hot ethanol. Physical properties are listed in Table 9 together with those of 2- and 3-chloronitrobenzenes for comparison.

Reactions of 4-chloronitrobenzene are similar to those of 2-chloronitrobenzene: nitration gives 2,4-dinitrochlorobenzene, chlorination gives 3,4-dichloronitrobenzene, and sulfonation gives 6-chloro-3-nitrobenzenesulfonic acid.

*Production.* Nitration of chlorobenzene and separation of the isomers have already been described for 2-chloronitrobenzene.

*Uses.* 4-Chloronitrobenzene derivatives find important outlets in many key intermediate chains (Fig. 7). The first stages in these chains are 4-chloroaniline [106-47-8] (**54**), 4-nitrophenol (**55**), 4-nitroanisole (**56**) (cf. Chap. 7), 4-nitroaniline (**57**; cf. Chap. 5), 6-chloro-3-nitrobenzenesulfonic acid (**58**; cf. Chap. 6), 2,4-dinitrochlorobenzene (**59**), and 3,4-dichloronitrobenzene (**60**).

Reaction with sodium sulfide can be controlled to give 4-amino-4'-nitrodiphenyl sulfide (**64**), which after N-acetylation and oxidation with hydrogen peroxide forms **65**: subsequent reduction and hydrolysis yields 4,4'-diaminodiphenyl sulfone [80-08-0] (dapsone, **66**), a valuable drug in the treatment of leprosy. Dapsone is also used as an analytic reagent and can alternatively be obtained via the reaction of 4-chloronitrobenzene with 4-acetylaminobenzenesulfinic acid to give (**65**).

$$O_2N-\text{C}_6H_4-S-\text{C}_6H_4-NH_2 \longrightarrow$$
**64**

$$O_2N-\text{C}_6H_4-SO_2-\text{C}_6H_4-NHCOCH_3$$
**65**

$$\downarrow$$

$$H_2N-\text{C}_6H_4-SO_2-\text{C}_6H_4-NH_2$$
**66**

Condensation of 4-chloronitrobenzene with 2,4-dichlorophenol gives 2,4-dichloro-4'-nitrodiphenyl ether [1836-75-5] (**67**) which is used as a herbicide (nitrofen). There are many related products of this type; for example, the more recently developed preemergent herbicide Fluoronitrofor [13738-63-1] (**68**) — derived by reaction of 4-chloronitrobenzene with 2,4-dichloro-6-fluorophenol.

**67**    **68**

A major outlet for 4-chloronitrobenzene exploits the condensation with aniline to give 4-nitrodiphenylamine [836-30-6] (PNDPA, **69**), which on reductive N-alkylation gives important antioxidants for rubber, e.g., IPPD [101-72-4] (**70**).

⟨⟩—NH—⟨⟩—NO₂ ⟶

**69**

⟨⟩—NH—⟨⟩—NHCH(CH₃)₂

**70**

**1-Chloro-2,4-dinitrobenzene** [*97-00-7*] (**59**), 2,4-dinitrochlorobenzene, DNCB, $C_6H_3ClN_2O_4$, $M_r$ 202.56, *mp* 53.4 °C (α), 43 °C (β), 27 °C (γ), *bp* 315 °C (101 kPa, decomp.), $d_4^{75}$ 1.4982, flash point (closed cup) 194 °C. The stable α-form is obtained as yellow rhombic crystals from ether; two other unstable crystal modifications are also known. The compound is soluble in benzene or ether, and insoluble in water.

Reduction of **59** with Fe – HCl gives 2,4-diaminochlorobenzene, but selective reduction is difficult because of the high reactivity of the chlorine atom toward basic reagents. Pyridine reacts readily to form a stable pyridinium salt. Nucleophilic substitution reactions are similar to those of 2-chloronitrobenzene, except that the greatly increased activity of the chlorine atom (usually several thousandfold) means that much milder reaction conditions suffice. Further substitution on the aromatic ring is difficult, although nitration under forcing conditions yields 2,4,6-trinitrochlorobenzene [*88-88-0*] (picryl chloride).

*Production.* Although chlorobenzene can be dinitrated directly, this results in unnecessary isomer problems. 4-Chloronitrobenzene is usually nitrated with mixed acid (35/65) at 60 °C to give the pure dinitro isomer. However, 2-chloronitrobenzene can be nitrated to produce 1-chloro-2,4-dinitrobenzene, together with ca. 10 wt% of the isomeric 2-chloro-1,3-dinitrobenzene. This may be separated for disposal, but the mixed isomers are preferably used directly if tolerated by the end product (e.g., sulfur dyes).

*Uses.* Industrially, the most important derivatives of 1-chloro-2,4-dinitrobenzene are obtained by nucleophilic reactions in aqueous media at moderate temperature. Ammonia gives 2,4-dinitroaniline (Chap. 5), alkali gives 2,4-dinitrophenol (Chap. 7), and methanolic sodium hydroxide gives 2,4-dinitroanisole (Chap. 7). Refluxing with hydrazine in ethanol yields 2,4-dinitrophenylhydrazine [*119-26-6*], a reagent used for the characterization of carbonyl compounds. Reaction with ammonium thiocyanate in aqueous medium at 80 °C gives dinitrophenylrhodanate, which was used as an insecticide. Reaction with substituted anilines gives 2,4-dinitrodiphenylamine derivatives that are used as yellow disperse dyes. Reaction with pyridine gives the reactive (2,4-dinitrophenyl)pyridinium chloride [*4185-69-7*], an intermediate in the preparation of pentamethine dyes.

Nitrochlorobenzenes, especially dinitrochlorobenzene, have traditionally been used to produce sulfur dyes; for example, C.I. Sulfur Black 1 is obtained from 2,4-dinitrochlorobenzene by prolonged refluxing with sodium polysulfide liquor.

**Toxicity of Nitrochlorobenzenes.** The mononitrochlorobenzenes (TLV 1 mg/m$^3$) are considered more toxic than nitrobenzene. They are absorbed through the skin and lungs, with lowering of the hemoglobin level and its consequential effects. 2,4-Dinitrochlorobenzene is similarly toxic and is also an extreme irritant to the skin, frequently leading to skin eruptions and allergic dermatitis [22]. These allergic properties are probably due to the antigen–antibody reaction resulting from bonding with natural proteins (cf. the Sanger test with 2,4-dinitrofluorobenzene).

## 4.2. Dichloro- and Polychloronitrobenzenes

Both 1,2- and 1,4-dichlorobenzene are nitrated on a large scale to give important intermediates for colorants via reduction or nucleophilic substitution reactions analogous to those used for 2- and 4-chloronitrobenzene and shown in Figures 6 and 7.

**1,2-Dichloro-4-nitrobenzene** [99-54-7] (**60**), DCNB, 3,4-dichloronitrobenzene, $C_6H_3Cl_2NO_2$, $M_r$ 192.01, crystallizes as yellow needles from ethanol. It is insoluble in water, and soluble in benzene, ether, or hot ethanol. The stable α-form has a *mp* of 43 °C; the unstable β-form is a liquid that reverts to the α-form at 15 °C, *bp* 263 °C (101.3 kPa), $d_4^{75}$ 1.4588.

*Production.* Nitration of 1,2-dichlorobenzene with mixed acid at 35–60 °C results in a mixture of 3-nitro (10%) and 4-nitro (90%) isomers, which are separated by crystallization. A more expensive process (two stages from chlorobenzene) is based on chlorination of molten 4-chloronitrobenzene (90–100 °C) by a process similar to that used for 3-chloronitrobenzene production. Chlorination has the advantage of giving a pure product, but it cannot be used if the isomeric 1,2-dichloro-3-nitrobenzene is also required.

*Uses.* Iron reduction of **60** yields 3,4-dichloroaniline [95-76-1], an important agrochemical intermediate. Amination produces 2-chloro-4-nitroaniline, which is an important diazo component and intermediate (Section 5.3).

Reaction of 1,2-dichloro-4-nitrobenzene with potassium fluoride in an aprotic solvent gives 3-chloro-4-fluoronitrobenzene (**74**); the derived amine is used as a herbicide intermediate.

**1,2-Dichloro-3-nitrobenzene** [3209-22-1] (**71**), 2,3-dichloronitrobenzene, 3-nitro-1,2-dichlorobenzene, $C_6H_3Cl_2NO_2$, $M_r$ 192.01, mp 60 °C, bp > 230 °C (decomp.), $d_4^{80}$ 1.449, crystallizes as yellow monoclinic needles, which are moderately soluble in most organic solvents and insoluble in water.

*Production.* 1,2-Dichloro-3-nitrobenzene is a byproduct of the nitration of 1,2-dichlorobenzene.

**1,3-Dichloro-4-nitrobenzene** [611-06-3] (**72**), 2,4-dichloronitrobenzene, $C_6H_3Cl_2NO_2$, $M_r$ 192.01, mp 34 °C, bp 101 °C (0.5 kPa), $d_4^{78}$ 1.551, crystallizes as pale yellow needles from ethanol.

*Production.* 1,3-Dichlorobenzene is nitrated with mixed acid, and the product is separated from the resulting isomer mixture.

*Uses.* 1,3-Dichloro-4-nitrobenzene is reduced to the intermediate 2,4-dichloroaniline [554-00-7] (**75**). Preferential amination in the ortho position can be effected to give 5-chloro-2-nitroaniline [1635-61-6] which is, however, much less important than the other chloronitroanilines (see Section 5.3).

**1,4-Dichloro-2-nitrobenzene** [89-61-2] (**73**), 2,5-dichloronitrobenzene, 2-nitro-1,4-dichlorobenzene, $C_6H_3Cl_2NO_2$, $M_r$ 192.01, mp 56 °C, bp 266 °C (101 kPa), $d_4^{75}$ 1.4390, crystallizes as pale yellow prisms or plates from ethanol. The compound is insoluble in water and soluble in benzene, ether, or hot ethanol.

*Production.* Nitration of 1,4-dichlorobenzene with mixed acid at 35 – 65 °C results in a 98 % yield of essentially pure 1,4-dichloro-2-nitrobenzene.

*Uses.* Derivatives of **73** are used mainly as colorant intermediates. Reduction yields 2,5-dichloroaniline [95-82-9] (Fast Scarlet GG Base, **76**), which is sulfonated to give the equally important 2,5-dichloroaniline-4-sulfonic acid [88-50-6] (Fast Red FR base). Amination produces 4-chloro-2-nitroaniline (Section 5.3); hydrolysis, 4-chloro-2-nitrophenol (Section 7.1); and methoxylation, 4-chloro-2-nitroanisole (Section 7.1). Condensation with phenol gives 4-chloro-2-nitrodiphenyl ether (**77**), which is reduced to 4-chloro-2-aminodiphenyl ether (Red FG Base); a series of similarly prepared aminodiphenyl ethers are used as diazo components.

Condensation with arylamines gives nitrodiphenylamines that are used as yellow to orange disperse dyes.

**1,2,4-Trichloro-5-nitrobenzene** [*89-69-0*] (**78**), 1-nitro-2,4,5-trichlorobenzene, $C_6H_2Cl_3NO_2$, $M_r$ 226.46, *mp* 57 °C, *bp* 288 °C (101 kPa), $d_4^{80}$ 1.554, crystallizes as yellow prisms from ethanol.

*Production.* 1,2,4-Trichlorobenzene is nitrated to give mainly the 5-nitro compound, which is purified by fractional crystallization.

*Uses.* Reduction of **78** gives 2,4,5-trichloroaniline [*636-30-6*], and methoxylation yields 5-chloro-2,4-dimethoxynitrobenzene (**79**) [*119-21-1*]; the latter is reduced to 5-chloro-2,4-dimethoxyaniline [*97-50-7*].

**Pentachloronitrobenzene** [*82-68-8*] (**80**), $C_6Cl_5NO_2$, $M_r$ 295.36, *mp* 144 °C, *bp* 328 °C (101 kPa, decomp.), $d_4^{20}$ 1.718, forms yellow crystals, which are moderately soluble in organic solvents and insoluble in water.

*Production.* Pentachloronitrobenzene is produced by nitration of pentachlorobenzene.

*Uses.* The early use of **80** as a fungicide (terrachlor) has led to several series of nitro-containing agrochemicals. Reduction gives pentachloroaniline [*527-20-8*], which has limited application. Reaction with ethanol – potassium hydroxide yields pentachlorophenetole, exemplifying an easy route to pentachlorophenyl ethers.

**Toxicity of Di- and Polychloronitrobenzenes.** All di- and polychloronitrobenzenes are moderately toxic [22] and should be handled as potential eye, nose, and skin irritants.

## 4.3. Chloronitrotoluenes

Of the ten chloronitrotoluene isomers available, only five are of potential technical interest. Both nitration of chlorotoluenes and chlorination of nitrotoluenes result in isomer problems, but the latter is more substituent directive.

Side-chain chlorination of nitrotoluenes is mentioned in Section 3.2, but only 4-nitrobenzyl chloride (**24**) is technically important.

**4-Chloro-2-nitrotoluene** [*89-59-8*] (**16**), 2-nitro-4-chloro-1-methylbenzene, $C_7H_6ClNO_2$, $M_r$ 171.58, mp 38.5 °C, bp 240 °C (96 kPa), $d_4^{80}$ 1.2559, crystallizes as yellow monoclinic needles.

*Production.* 4-Chlorotoluene is nitrated with mixed acid (39/59/2) at 25 °C to give a mixture of 2-nitro (65%) and 3-nitro (35%) isomers. The major component (**16**) is separated as the lower boiling fraction on vacuum distillation. 4-Chloro-3-nitrotoluene [*89-60-1*] (**81**) can be obtained, if required, from the residue by distillation and sweating.

*Uses.* Reduction of **16** gives 4-chloro-2-toluidine [*95-81-8*] (Fast Red KB Base, **82**). The byproduct **81** can be methoxylated and reduced to produce cresidine [*120-71-8*] (**83**), which can, however, be obtained by several alternative routes.

**2-Chloro-6-nitrotoluene** [*83-42-1*] (**17**), 6-chloro-2-nitrotoluene, 2-nitro-6-chloro-1-methylbenzene, $C_7H_6ClNO_2$, $M_r$ 171.58, mp 37.5 °C, bp 238 °C (101 kPa).

*Production.* Chlorination of 2-nitrotoluene, in a process similar to the chlorination of nitrobenzene, gives a mixture of 6-chloro (80%) and 4-chloro (20%) isomers. The major component **17** is separated by vacuum distillation.

*Uses.* Reduction of **17** gives 6-chloro-*o*-toluidine [*87-63-8*] (Fast Scarlet TR Base, **84**) which, apart from forming azo dyes, is used to produce 2,6-dichlorotoluene, by Sandmeyer reaction, for conversion to 2,6-dichlorobenzaldehyde.

**2-Chloro-4-nitrotoluene** [*121-86-8*] (**23**), 4-nitro-2-chloro-1-methylbenzene, $C_7H_6ClNO_2$, $M_r$ 171.58, mp 68 °C, bp 264 °C (101 kPa), crystallizes as yellow needles from ethanol. The compound is soluble in most organic solvents, only slightly soluble in hot water, and steam volatile.

*Production.* Chlorination of 4-nitrotoluene at 50 °C in the presence of iron(III) chloride catalyst gives good conversion to 2-chloro-4-nitrotoluene, the most readily obtained of all the chloronitrotoluene isomers.

*Uses.* Reduction of **23** gives 2-chloro-4-toluidine [*95-74-9*] (**85**), which undergoes sulfonation to give 4-amino-6-chlorotoluene-3-sulfonic acid [*88-51-7*]. Further chlorination of **23** gives 2,6-dichloro-4-nitrotoluene [*7149-69-1*] (**86**) for reduction to 2,6-dichloro-*p*-toluidine [*56461-98-4*].

**4-Nitrobenzyl chloride** [*100-14-1*] (**24**), 1-chloromethyl-4-nitrobenzene, $C_7H_6ClNO_2$, $M_r$ 171.58, mp 71 °C, forms pale yellow plates or needles from ethanol.

*Production.* Chlorination of 4-nitrotoluene at 190 °C without catalyst gives **24**. The alternative process of nitration of benzylchloride coproduces 2-nitrobenzyl chloride (**28**) and requires a difficult separation stage.

*Uses.* 4-Nitrobenzyl chloride is used as an alkylating agent (e.g., in a biosynthetic penicillin) and in the production of 4-nitrobenzyl cyanide [*555-21-5*] (**87**), which is hydrolyzed to 4-nitrophenylacetic acid [*104-03-0*] (**88**), an acylating agent.

**Toxicity of Chloronitrotoluenes.** In the absence of detailed information on individual products, all chloronitrotoluenes should be assumed to have a toxicity similar to that of chloronitrobenzenes and should be handled accordingly.

## 4.4. Fluoronitrobenzenes and Fluoronitrotoluenes

The two most technically important classes of fluoronitrobenzenes and fluoronitrotoluenes are (1) those fluoronitrobenzenes that can be manufactured conveniently by halogen-exchange reactions, with the readily available chloronitrobenzenes and (2) nitration products from benzotrifluoride and chlorobenzotrifluorides.

Nitration of fluorobenzene with mixed acid gives a 92:8 mixture of 4-fluoronitrobenzene [*350-46-9*] (**89**) and 2-fluoronitrobenzene [*1493-27-2*] (**90**). However, production of these two compounds and of 2,4-dinitrofluorobenzene [*70-34-8*] (**91**) is always based on fluorine-exchange

reaction of the corresponding chloro compound by treatment with potassium fluoride in a polar solvent such as dimethylformamide or dimethyl sulfoxide [52]. These compounds and their homologues form the basis for many substituted fluoro aromatics by further substitution of the nitro compounds or their derived amines. Fluoro aromatics often justify the additional costs over the corresponding chloro compounds in agrochemicals because of the lower dosage required and their greater selectivity and safety. Examples are the fungicide fluoroimide [*41205-21-4*] (**92**), derived from 4-

fluoroaniline, and the herbicide flampropisopropyl [52756-22-6] (**93**), from 3-chloro-4-fluoroaniline (the reduction product of **74**).

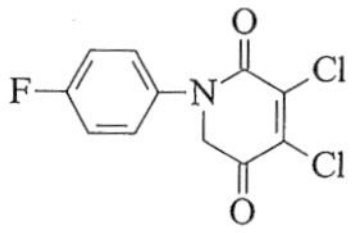

**92** 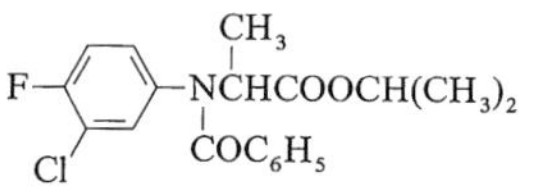  **93**

Ring-substituted fluoronitrotoluenes have no particular advantage over the available fluoronitrobenzenes, but side-chain fluorinated toluenes (i.e., substituted benzotrifluorides) are of great importance. The I.G. Farbenindustrie produced a large range of substituted aminobenzotrifluorides as speciality fast salts that give Naphtol azo derivatives of high lightfastness. Most of these amines were derived from the corresponding nitro compounds, and some products have survived as speciality dye components. More important, the chemistry of these substituted benzotrifluorides has been exploited in the production of agrochemicals containing the trifluoromethyl group.

Benzotrifluoride is nitrated to give 3-nitrobenzotrifluoride [98-46-4] (**94**) in high yield, even after separation of the coproduced 6% ortho and 3% para isomers. Under similar conditions, 2-chlorobenzotrifluoride yields 2-chloro-5-nitrobenzotrifluoride [777-37-7] (**95**), and 4-chlorobenzotrifluoride yields 4-chloro-3-nitrobenzotrifluoride [121-17-5] (**96**). Forcing nitration of **94**, **95**, or **96** at 100 °C gives the 3,5-dinitro derivatives, with 4-chloro-3,5-dinitrobenzotrifluoride [393-75-9] (**97**) being the most important example.

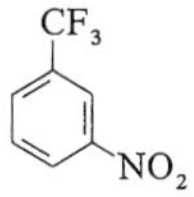

**94** 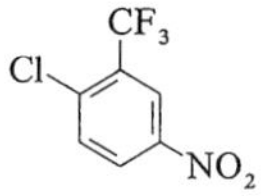  **95** 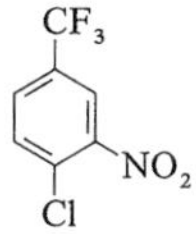  **96**

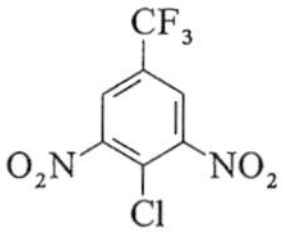

**97**

Nitration of 3-chlorobenzotrifluoride gives a mixture of isomers from which the desired 5-chloro-2-nitrobenzotrifluoride [118-83-2] (**98**) is separated by crystallizing it at 5 °C and centrifuging off the unwanted isomer as an oil.

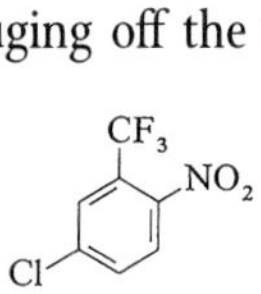

**98** 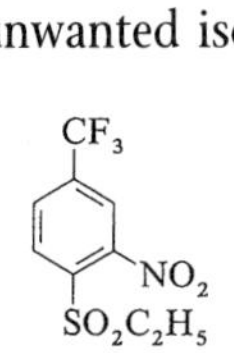  **99** 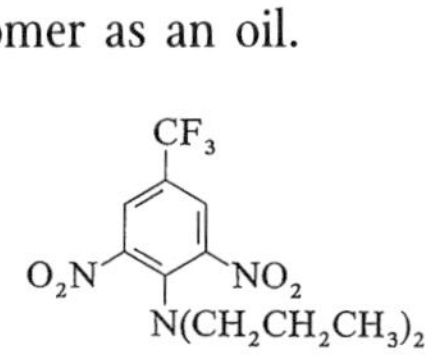  **100**

Iron reduction of the mononitro compounds **94**, **95**, **96**, and **98** gives a series of

Table 10. p$K_a$ values for the ionization of primary nitroanilines in aqueous $H_2SO_4$

| Compound | 25 °C | 90 °C |
|---|---|---|
| 4-Nitroaniline | 1.00 | 0.60 |
| 2-Nitroaniline | − 0.30 | − 0.51 |
| 4-Chloro-2-nitroaniline | − 1.06 | − 1.15 |
| 2,5-Dichloro-4-nitroaniline | − 1.75 | − 1.76 |
| 2-Chloro-6-nitroaniline | − 2.38 | − 2.35 |
| 2,6-Dichloro-4-nitroaniline | − 3.27 | − 3.12 |
| 2,4-Dinitroaniline | − 4.27 | − 3.81 |
| 2,6-Dinitroaniline | − 5.39 | − 4.72 |
| 2-Bromo-4,6-dinitroaniline | − 6.69 | − 5.81 |
| 2,4,6-Trinitroaniline | − 10.03 | − 8.67 |

trifluoromethylanilines used as diazo compounds, e.g., 2- chloro-5-trifluoromethylaniline [121-50-6] from **96** as Fast Orange RD Base and 4- chloro-2-trifluoromethylaniline [445-03-4] from **98** as Scarlet VD Base. Reaction of **96** with sodium ethyl sulfinate gives the nitrosulfone (**99**), which on reduction gives 3-amino-4-ethylsulfonylbenzotrifluoride [382-85-4]; used as Fast Golden Orange GR Base. 3-Aminobenzotrifluoride [98-16-8], obtained by hydrogenation of **94**, is more important as an intermediate for drugs and crop protection chemicals than for colorants.

Reaction of **97** with dialkylamines produces a range of 2,6-dinitro-4-trifluoromethyl-N,N-dialkylanilines used as preemergent herbicides, typified by trifluralin [1582-09-8] (**100**).

The fluorine atom in dinitrofluorobenzenes is readily displaced, as evidenced by the use of 2,4-dinitrofluorobenzene (**91**)in the Sanger method for determination of the N-terminal amino group of a peptide chain.

# 5. Nitroamino Aromatics

All the isomeric mononitroanilines and many of their derivatives are industrially important. Production as colorant intermediates has made them available as starting materials for many other outlets. The most important additional substituents are nitro (i.e., dinitroanilines), chloro (i.e., chloronitroanilines), and methyl (i.e., nitrotoluidines).

The aromatic nitroamino compounds are weakly basic amines due to the strongly electron-withdrawing nature of the nitro group, especially when it is in a position ortho or para to the amino group. Widely varying p$K$ values have been reported, and a comparison of their ionization constants relative to 4-nitroaniline (Table 10) is useful to indicate the extent of protonation in different strengths of sulfuric acid at different temperatures [53]. This is important when nitroanilines undergo reaction (e.g., nitration or diazotization) in sulfuric acid, although solubility must also be considered.

An alternative measure of the basicity of nitroanilines is to consider their polarity as exhibited by the main absorption maximum in a comparable series of monoazo dyes, for example, as exemplified by **101**.

**Table 11.** Substituent effect on absorption in a series of nitroaniline-based dyes (**101**)

| Example | X | Y | Z | $\lambda_{max}$ (in MeOH), nm |
|---|---|---|---|---|
| 1 | NO$_2$ | H | H | 425 |
| 2 | H | H | NO$_2$ | 453 |
| 3 | Cl | H | NO$_2$ | 475 |
| 4 | Cl | Cl | NO$_2$ | 417 |
| 5 | NO$_2$ | H | NO$_2$ | 491 |
| 6 | CN | H | NO$_2$ | 504 |
| 7 | NO$_2$ | NO$_2$ | NO$_2$ | 520 |
| 8 | CN | CN | NO$_2$ | 549 |

**101**

Table 11 shows the trend of bathochromic shift with increasing polarity [54]. These absorption maxima are not necessarily related directly to the p$K$ values of the nitroanilines because the steric effects of the ortho substituents may differ in each series. Crowding due to the two chlorine substituents in example 4 results in loss of molecular planarity and broadening of the absorption band. The *o*-cyano group (examples 6 and 8) does not exert steric effects in the same way and therefore has a more powerful effect than a nitro group in the same position. 2,6-Dicyano-4-nitroaniline (example 8) is the most bathochromic nitroaniline in this series for comparison with nitro heterocyclic amines used in monoazo dyes (cf. Chap. 9).

Nearly all the primary nitroanilines have found use as diazo components, mainly in pigments or disperse dyes, and the most weakly basic of them (e.g., examples 4–8 in Table 11) require diazotization with nitrosylsulfuric acid. Isolated diazo compounds have long been known to be unstable, and often detonatable. However, not until a serious accident in 1969 during diazotization of 6-chloro-2,4-dinitroaniline with nitrosylsulfuric acid were concentrated solutions of certain diazotized nitroanilines in sulfuric acid found to be detonatable [55].

Although aniline and *C*-alkylanilines can be nitrated, this is not a viable route for production of the more important nitroanilines. However, nitrations of this type are important in studying the effects of cationic species (positive poles) on aromatic nitration. Careful analysis of the classical work claiming complete meta nitration of PhN$^+$Me$_3$ showed that 89% of the meta isomer and 11% of the para isomer are produced, and that the proportion of para substitution increases for PhN$^+$HMe$_2$ (22%), PhN$^+$H$_2$Me (30%), and PhN$^+$H$_3$ (38%), all nitrated at 25 °C in 98% sulfuric acid. Direct nitration of aniline is sensitive to sulfuric acid concentration, with para nitration increasing to 52% in 89% sulfuric acid, but this does not correlate with the calculated amount of protonated amine, so both meta and para substitution must result from nitration of the anilinium ion [56].

Table 12. Physical properties of the three mononitroanilines

| Property | Isomer | | |
|---|---|---|---|
| | Ortho | Meta | Para |
| *mp*, °C | 71.5 | 114 | 148 |
| *bp*, °C | 284 (decomp.) | 307 (decomp.) | 331 (decomp.) |
| Density (15 °C), g/cm³ | 1.442 | 1.398 | 1.437 |
| Dipole moment (in benzene), D | 4.23 | 4.68 | 6.2 |
| Flash point, °C | 167 | | 199 |
| Refractive index (°C) | 1.5362 (100) | 1.595 (111) | 1.5401 (160) |

Nitration of *o*- and *p*-toluidines at low temperature gives desirable isomers by more controlled meta substitution than with aniline. Nitrochlorotoluenes with the required substitution pattern are less readily available than nitrochlorobenzenes or nitrodichlorobenzenes, which are used predominantly as precursors in the manufacture of nitroanilines and chloronitroanilines by nucleophilic amination of the active chlorine atom. When 3-nitroaniline configurations are required, partial reduction of the dinitro compound is an option, but difficulties arise if the compound is unsymmetrically substituted.

The primary reactions undergone by nitroamino aromatics are diazotization, reduction to the corresponding diamine, and various N-acylation reactions. As diazo components, the nitroanilines and many of their derivatives found early use as fast bases for azoic dyes. The fast base name indicates the hue derived on diazotization and coupling with Naphtol AS.

## 5.1. Nitroanilines

**2-Nitroaniline** [*88-74-4*] (**48**), *o*-nitroaniline, ONA, $C_6H_6N_2O_2$, $M_r$ 138.13, crystallizes as orange-yellow plates from water. The compound is moderately soluble in organic solvents and sparingly soluble in water (0.13%). Physical properties are listed in Table 12 alongside those of *m*- and *p*-nitroanilines for comparison.

*Production.* 2-Chloronitrobenzene (Section 4.1) is heated with excess (10 mol/mol) strong aqueous ammonia in an autoclave. The temperature is gradually increased to 180 °C over 4 h and held there for 5 h more. The pressure builds up to around 4 MPa and is released to an ammonia recycle loop before the product is isolated by filtration and washing. The reaction is extremely exothermic ($\Delta H = -168$ kJ/mol), and too rapid heating or inadequate temperature control can result in a runaway reaction. Because of this hazard, I.G. Farbenindustrie [1] developed a continuous amination unit [17] for amination of chloronitrobenzenes; the process is summarized under 4-nitroaniline.

*Uses.* 2-Nitroaniline is reduced with Fe – acid or $H_2$ – catalyst to *o*-phenylenediamine [*95-54-5*]. Acetylation followed by reduction gives 2-aminoacetanilide [*555-48-6*]. 2-Nitroaniline is used as a diazo component (Fast Orange GR Base) in azo dyes and pigments (e.g., C.I. Disperse Yellow 10 and C.I. Pigment Yellow 5) and as an intermediate for vat dyes (e.g., C.I. Vat Red 14).

48      102      55

**3-Nitroaniline** [99-09-2] (**102**), *m*-nitroaniline, MNA, $C_6H_6N_2O_2$, $M_r$ 138.13, crystallizes as yellow needles from water. It is moderately soluble in organic solvents and sparingly soluble in water (0.11%). Its physical properties are listed in Table 12.

*Production.* 1,3-Dinitrobenzene (Section 3.1) is added to warm water containing magnesium sulfate. An aqueous solution of sodium hydrogen sulfide (6 molar equivalents) is added gradually to the vigorously stirred emulsion, and reduction is completed by heating to 90 °C. The 3-nitroaniline produced solidifies on cooling and is separated by filtration.

*Uses.* Acetylation of 3-nitroaniline followed by reduction gives 3-aminoacetanilide [102-28-3]. Diazotization, followed by reduction of the diazosulfonate with ammonium bisulfite and subsequent hydrolysis, gives (3-nitrophenyl)hydrazine [619-27-2]; an intermediate in the production of pyrazolone azo coupling components. 3-Nitroaniline is used as a diazo component (Fast Orange R Base) in azo dyes (e.g., C.I. Disperse Yellow 5 and C.I. Acid Orange 18).

**4-Nitroaniline** [100-01-6] (**55**), *p*-nitroaniline, PNA, $C_6H_6N_2O_2$, $M_r$ 138.13, crystallizes as yellow needles from water. 4-Nitroaniline is moderately soluble in organic solvents and sparingly soluble in water (0.08%). Its physical properties are listed in Table 12.

*Production.* Batchwise amination of 4-chloronitrobenzene (Section 4.1) is carried out at 195 °C (4.5 MPa) in a process similar to that for 2-nitroaniline.

In the alternative continuous process [1], molten 4-chloronitrobenzene and 40% ammonia liquor in a molar ratio of 1:17 are pumped at 20 MPa through a steel reaction coil held at 237–240 °C by oil bath heating. Inflow is controlled to give a contact time of ca. 10 min, and outflow is directed into an expansion chamber with a tangential flow of hot water to give, in quantitative yield, a suspension of product that is readily isolated and washed.

The alternative historical process based on nitration of acetanilide is not economical compared with amination, but nitration of formanilide followed by alcoholysis and recycle of ethyl formate brings the costs closer [57].

*Uses.* The volume outlet for 4-nitroaniline is in iron reduction or catalytic hydrogenation to produce *p*-phenylenediamine (PPD) [106-50-3]. This is still the major route to *p*-phenylenediamine despite more recent alternatives [21]. Reaction of 4-nitroaniline with acetic anhydride gives 4-nitroacetanilide [104-04-1] (**103**) which, on reduction, yields 4-aminoacetanilide [122-80-5], an important colorant intermediate. Chlorination (aqueous HCl–$H_2O_2$) gives 2,6-dichloro-4-nitroaniline [99-30-9] (**104**), a diazo component specifically for yellow-brown disperse dyes (e.g., C.I. Disperse Brown 1). Bromination (aqueous HBr – $H_2O_2$) gives 2-bromo-4-nitroaniline [*13296-94-1*] (**105**), which is occasionally used as an alternative diazo component to 2- chloro-4-nitroaniline.

**103** 4-acetamidonitrobenzene (NHCOCH$_3$, NO$_2$)
**104** 2,6-dichloro-4-nitroaniline
**105** 2-bromo-4-nitroaniline

Diazotization of 4-nitroaniline followed by reduction with $(NH_4)_2SO_3$–$NH_4HSO_3$ and hydrolysis gives (4-nitrophenyl)hydrazine [*100-16-3*], an intermediate in the production of pyrazolone azo coupling components. Reaction of 4- chloro-1-nitrobenzene with hydrazine is not a viable alternative.

4-Nitroaniline is used as a diazo component (Fast Red GG Base) in the production of disperse monoazo dyes (e.g., C.I. Disperse Orange 25, example 2 in Table 11) and pigments (e.g., C.I. Pigment Red 1). It is also used in dyes such as C.I. Direct Black 19, where the 4-nitroaniline monoazo dye is reduced with sodium sulfide to the aminoazo compound for further diazotization and coupling. Coupling of diazotized 4-nitroaniline with 2,5-dimethoxyaniline produces Black K Base (**106**).

**106** Black K Base
**107** 4-nitrophenyl isocyanate

Phosgenation of 4-nitroaniline in aqueous medium gives *N,N′*-bis(4-nitrophenyl)urea [*587-90-6*], whereas phosgenation in toluene yields 4-nitrophenyl isocyanate [*100-28-7*] (**107**). The latter is used for the production of unsymmetrical ureas by reaction with amines. For example, with 3-picolylamine the urea [*53558-25-1*] (**108**) is formed, which has been used as a rodenticide [58]. An analogous process with guanidine hydrochloride leads to Nitroquanil [*51-58-1*] (**109**), an antimalarial drug which, although it is less active against malaria than its analogue Paludrine (derived from 4- chloroaniline), is also less toxic.

**108**
**109**

**2,4-Dinitroaniline** [*97-02-9*] (**110**), 2,4 DNA, $C_6H_5N_3O_4$, $M_r$ 183.12, *mp* 188 °C, $d_4^{15}$

1.615, flash point 224 °C, forms yellow needles from aqueous acetic acid. The compound is sparingly soluble in ethanol and insoluble in water.

*Production.* 2,4-Dinitrochlorobenzene (Section 4.1) is aminated with aqueous ammonia initially at 70 °C; then the temperature is allowed to rise gradually to 120 °C to complete the reaction. Only a low-pressure autoclave is required, compared with the amination of 2-chloro- and 4-chloronitrobenzene.

*Uses.* Reduction with iron–acid gives 1,2,4-triaminobenzene [615-71-4] and with sodium sulfide, 3,4-diaminonitrobenzene [99-56-9] (**111**), an intermediate for the preparation of nitroheterocyclics. Acetylation followed by reduction gives 2,4-diaminoacetanilide [6373-15-5]. Bromination of a sulfuric acid solution yields 6-bromo-2,4-dinitroaniline [1817-73-8] (**112**).

As a diazo component, 2,4-dinitroaniline typically gives violet monoazo disperse dyes when coupled with arylamines (e.g., C.I. Disperse Violet 12, **113**). The derived 6-chloro- and 6-bromo-2,4-dinitroanilines are much more important in the production of navy disperse dyes for polyester, (e.g., C.I. Disperse Blue 79, **114**) [59].

**2,6-Dinitroaniline** [606-22-4] (**117**), 2,6 DNA, $C_6H_5N_3O_4$, $M_r$ 183.12, *mp* 142 °C, forms yellow needles from ethanol.

*Production.* Chlorobenzene-4-sulfonic acid is dinitrated to give 4-chloro-3,5-dinitrobenzenesulfonic acid [88-91-5] (**115**), which is desulfonated by heating the diluted sulfuric acid solution from the nitration mix to give 2-chloro-1,3-dinitrobenzene [606-21-3] (**116**); the latter is aminated to **117**.

*Uses.* 2,6-Dinitroaniline and its derivatives are reported to have algicidal and herbicidal activity; for example, heating with phthalic anhydride yields *N*-(2,6-dinitrophenyl)phthalimide [57491-99-3] (**118**) for use as a herbicide.

**2,4,6-Trinitroaniline** [489-98-5] (**119**), picramide, $C_6H_4N_4O_6$, $M_r$ 228.12, crystallizes as yellow prisms from aqueous acetic acid; these melt at 192–195 °C and explode on further heating.

*Production.* Picric acid (Section 7.1) is converted to 2,4,6-trinitrochlorobenzene by treatment with $PCl_5$, or a similar chlorinating agent; this is then aminated under mild reaction conditions.

*Uses.* 2,4,6-Trinitroaniline is used in explosives [8] and as a diazo component for disperse dyes; in the latter, 2,4-dinitroaniline is nitrated to give picramide, and the resulting sulfuric acid solution is diazotized in situ [60]. Dyes based on picramide are reportedly improved by introduction of the nitro group by a final nucleophilic displacement reaction [$NaNO_2$ – $Cu(OAc)_2$ – DMF] [61] on the dye derived from 6-bromo-2,4-dinitroaniline (e.g., **114**).

**Toxicity of Nitroamino Aromatics.** All of the readily available nitroanilines are classified as highly toxic on inhalation or skin absorption. The TLV for 4-nitroaniline is 1 ppm [22].

## 5.2. Nitrotoluidines

The commercially available nitrotoluidine isomers, historically important as diazo bases, are obtained by low-temperature nitration of toluidines or conventional nitration of acylated toluidines. Certain isomers can be obtained by mono reduction of dinitrotoluenes, but mixtures arise unless the molecule is symmetrical (e.g., 2,6-dinitrotoluene). Nomenclature is a problem because the old usage, which still prevails in much of industry, numbers toluidines with 1-methyl, whereas many reference books adopt a convention of 1-amino.

**4-Nitro-o-toluidine** [*99-55-8*] (**120**), 2-methyl-5-nitroaniline, $C_7H_8N_2O_2$, $M_r$ 152.15, mp 107–109 °C, $d_4^{15}$ 1.365, crystallizes as yellow prisms from ethanol.

*Production.* o-Toluidine is nitrated at −10 °C by slow addition of nitric acid to a solution of the amine in 98 % sulfuric acid. On quenching the nitration mixture in water and allowing the temperature to rise, a solution is obtained from which the sulfate of the product crystallizes on cooling, leaving most of the 6-nitro-o-toluidine byproduct in solution.

*Uses.* 4-Nitro-o-toluidine is used as a diazo component (Fast Red RL Base) in C.I. Pigment Red 12, for example.

**5-Nitro-o-toluidine** [*99-52-5*] (**121**), 2-methyl-4-nitroaniline, $C_7H_8N_2O_2$, $M_r$ 152.15, mp 132–133 °C, $d_4^{15}$ 1.366, crystallizes from ethanol as yellow needles.

*Production.* N-Benzenesulfonyl-o-toluidine is dissolved in chlorobenzene at 40–50 °C, and 62 % nitric acid is added gradually to give 5-nitro-N-benzenesulfonyl-o-toluidine, which is isolated and hydrolyzed in sulfuric acid to give 5-nitro-o-toluidine in 80 % yield.

*Uses.* 5-Nitro-o-toluidine is used as a diazo component (Fast Scarlet G Base), for example, in C.I. Pigment Red 162 and C.I. Acid Black 28.

**6-Nitro-o-toluidine** [*603-83-8*] (**122**), 2-methyl-3-nitroaniline, $C_7H_8N_2O_2$, $M_r$ 152.15, mp 92 °C, bp 305 °C, $d_4^{15}$ 1.378, crystallizes from ethanol as yellow leaflets.

*Production.* 6-Nitro-o-toluidine is produced by sulfide reduction of 2,6-dinitrotoluene in a process similar to that for 3-nitroaniline (Section 5.1) or by controlled hydrogenation of 2,6-dinitrotoluene with palladium on a carbon catalyst [62].

**3-Nitro-p-toluidine** [*89-62-3*] (**123**), m-nitro-p-toluidine, MNPT, 4-methyl-2-nitroaniline, $C_7H_8N_2O_2$, $M_r$ 152.15, mp 117 °C, bp 134 °C (0.26 kPa), $d^{15}$ 1.31, forms orange leaflets from aqueous ethanol.

*Production.* N-Acetyl-p-toluidine is nitrated with mixed acid, and the quenched reaction mixture is heated and hydrolyzed to 3-nitro-p-toluidine.

*Uses.* 3-Nitro-p-toluidine is used as a diazo component (Red GL Base) in C.I. Pigment Yellow 1 and C.I. Pigment Red 3 (**124**), among others.

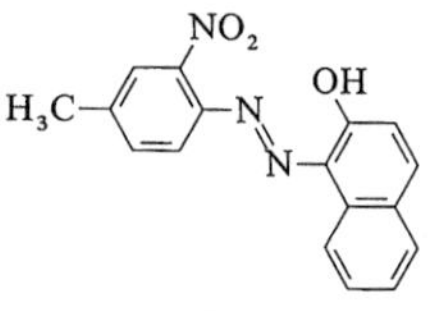

**124**

## 5.3. Halogenonitroanilines

Many halogenonitroaniline isomers exist [17] but only those obtained from the readily accessible dichloronitrobenzenes have large-scale use.

**4-Chloro-2-nitroaniline** [*89-63-4*] (**125**), $C_6H_5ClN_2O_2$, $M_r$ 172.57, *mp* 116–117 °C, flash point 191 °C, forms orange needles from water, in which it is sparingly soluble (0.05 % at 20 °C, 0.65 % at 100 °C).

*Production.* 1,4-Dichloro-2-nitrobenzene (Section 4.2) is aminated at 165 °C by using a process similar to that described for 2-nitroaniline.

*Uses.* 4-Chloro-2-nitroaniline is used as a diazo component (Red 3GL Base) in C.I. Pigment Yellow 3 and C.I. Pigment Red 6, for example.

**2-Chloro-4-nitroaniline** [*121-87-9*] (**62**), $C_6H_5ClN_2O_2$, $M_r$ 172.57, *mp* 108 °C, flash point 205 °C, crystallizes from water as yellow needles.

*Production.* 1,2-Dichloro-4-nitrobenzene (Section 4.2) is aminated by using a similar process to that described for 4-nitroaniline.

*Uses.* 2-Chloro-4-nitroaniline is used as a diazo component in disperse dyes (e.g., C.I. Disperse Red 65, **126**, example 3, Table 11) for polyester or in modified basic dyes (e.g., C.I. Basic Red 18, **127**) for acrylic fibers.

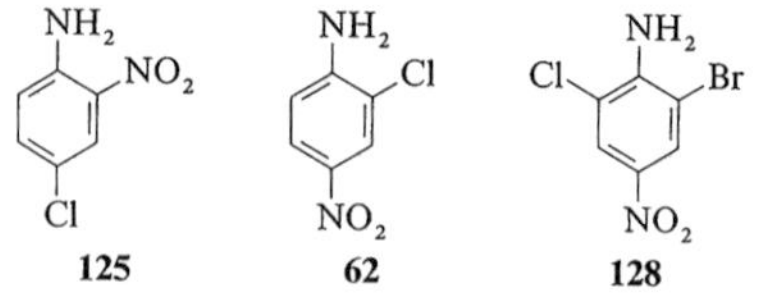

**125**     **62**     **128**

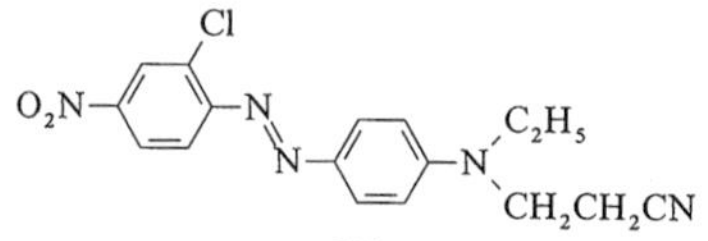

**126**

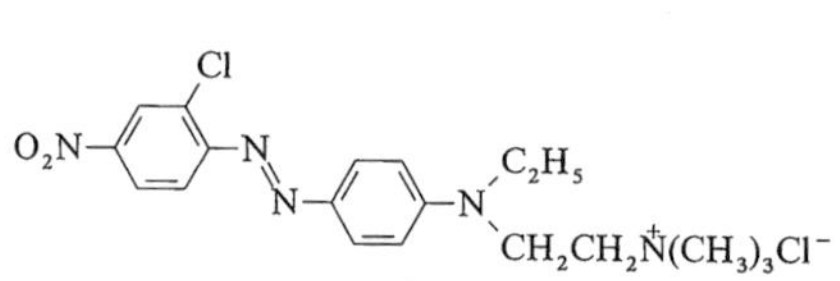

**127**

Bromination of (62) by treatment of a sulfuric acid solution with bromine or by treatment of a hydrobromic acid solution with hydrogen peroxide gives 2-bromo-6-chloro-4-nitroaniline [99-29-6] (128), which is used occasionally as an alternative to 2,6-dichloro-4-nitroaniline (104) in yellow-brown dyes for synthetic fibers.

Acylation of 62 with the acid chloride derived from 5-chlorosalicylic acid gives N-(2-chloro-4-nitrophenyl)-5-chlorosalicylanilide (niclosamid) [50-65-7] (129), which is a molluscicide and anthelminthic and typifies a series of related nitrophenylsalicylanilides used for similar outlets [63].

$$\underset{129}{\text{HO-C}_6\text{H}_3(\text{Cl})\text{-CONH-C}_6\text{H}_3(\text{Cl})\text{-NO}_2}$$

## 5.4. Cyanonitroanilines

The cyano group as a more bathochromic (Table 11) pseudohalogen substitutent in nitromonoazodisperse dyes has found increasing use in dyeing polyester because it also confers greater lightfastness.

**2-Cyano-4-nitroaniline** [17420-30-3] (130) is produced by nitration of 2-chlorobenzonitrile, separation of the required 2-chloro-5-nitrobenzonitrile [16588-02-6], and amination of this with ammonia.

Of the many monoazo disperse dyes derived from 130, C.I. Disperse Red 73 (134, example 6 of Table 11) is typical [64]. Bromination of 130 gives 6-bromo-2-cyano-4-nitroaniline (131), which is used in blue dyes (e.g., C.I. Disperse Blue 183, 135) [65].

2-Cyano-4,6-dinitroaniline (132) and 2,6-dicyano-4-nitroaniline (133) are difficult to prepare and to diazotize. The derived bright blue dyes (e.g., 136 and 137) are best prepared by introduction of the cyano group (cuprous cyanide in dimethylformamide) into the preformed monoazo dyes obtained from 2-bromo-4,6-dinitroaniline [66] and 6-bromo-2-cyano-4-nitroaniline [67], respectively.

**130**, **131**, **132**, **133**, **134**, **135**, **136**

**137**
C. I. Disperse Blue 165

# 6. Nitroaromatic Sulfonic Acids and Derivatives

Most of the sulfonic acids derived from nitrobenzene, nitrotoluenes, chloronitrobenzenes, and related products are reduced to yield anilinesulfonic acids, for use in the synthesis of water-soluble dyes or pigments insolubilized as lakes. Some of the derived sulfonamides and sulfones are also important intermediates for colorants and fine chemicals; they are produced from the corresponding sulfonyl chlorides. The sulfonyl chloride group is more economically introduced by direct chlorosulfonation of the nitroaromatic rather than by conversion of the sulfonic acid.

## 6.1. Nitrobenzenesulfonic Acids and Derivatives

**2-Nitrobenzenesulfonic acid** [80-82-0] (**138**), *o*-nitrobenzenesulfonic acid, $C_6H_5NO_5S$, $M_r$ 203.18, *mp* (anhydride) 85 °C, crystallizes as very hygroscopic pale yellow crystals that form several hydrates.

*Production.* 2-Chloronitrobenzene is reacted with sodium disulfide at 80 °C in aqueous ethanol to give 2,2′-dinitrodiphenyl disulfide (**51**), which is isolated and treated with chlorine in mineral acid medium in the presence of nitric acid to produce 2-nitrobenzenesulfonyl chloride (**52**) by oxidative chlorination. This is hydrolyzed in dilute alkali, and the liquor of the resulting **138** is reduced with Fe – acid or $H_2$ – catalyst to orthanilic acid [88-21-1] (**139**), an important diazo component in reactive dyes for cotton.

**2-Nitrobenzenesulfonyl chloride** [1694-92-4] (**52**), $C_6H_4ClNO_4S$, $M_r$ 221.62, crystallizes from ether, *mp* 68 – 69 °C.

*Production.* 2-Nitrobenzenesulfonyl chloride is isolated as an intermediate in the production of 2-nitrobenzenesulfonic acid. When 2,2′-dinitrodiphenyl disulfide is chlorinated in an organic solvent the product is 2-nitrophenylsulfenyl chloride [7669-54-7] (**140**).

*Uses.* Reduction with zinc – acid gives 2-aminothiophenol [137-07-5], although this is more readily produced from aniline. The main use of **52** is via derived sulfonamides; reaction with ammonia produces 2-nitrobenzenesulfonamide [5455-59-4] (**141**), and methylamine gives the corresponding *N*-methylsulfonamide [23530-40-7], both of which have been patented for use as herbicides. Reaction with *N*-ethylaniline produces a nitrosulfonamide that, on reduction, gives 2-amino-*N*-ethylbenzenesulfonanilide [81-10-7] (**142**), a valuable diazo component for acid wool dyes.

Many other sulfonamides derived from aminoheterocyclics or *o*-aminobenzoic esters can be ring-closed after reduction to give compounds with a variety of reported pharmacological properties.

**3-Nitrobenzenesulfonic acid** [98-47-5] (**8**), m-nitrobenzenesulfonic acid, $C_6H_5NO_5S$, $M_r$ 203.18, crystallizes in yellow plates that are slightly deliquescent in air. Both the free acid and the sodium salt are potentially unstable to heat.

*Production* [1]. 3-Nitrobenzenesulfonic acid is produced by dissolving nitrobenzene in 98 % sulfuric acid and heating to 80 °C. Then 65 % oleum is added at this temperature, and the reaction is completed by heating for 9 h at 105 °C (higher temperatures are unsafe). After quenching the sulfonation mass in water, neutralizing with lime, and filtering off gypsum, the calcium salt of the product is converted to the sodium salt with sodium carbonate. After calcium carbonate has been removed by filtration, the sodium salt solution can be used directly or evaporated.

The formation of 3,3′-dinitrodiphenylsulfone [1228-53-1] as a byproduct is claimed to be reduced from 8 to 4 % by addition of sulfur trioxide [68].

Although benzenesulfonic acid can be nitrated to give mainly 3-nitrobenzenesulfonic acid (54 %), the process is not economical.

*Uses.* 3-Nitrobenzenesulfonic acid is used primarily as a mild oxidizing agent or as a precursor for colorant intermediates. In the former case it is used in processing certain anthraquinone intermediates (e.g., amination of anthraquinone-1-sulfonic acid to give 1-aminoanthraquinone) or as a dye-printing auxiliary to obtain resist effects with, for example, vat dyes. Reduction of 3-nitrobenzenesulfonic acid (usually as a liquor) to the important metanilic acid is achieved with iron or by catalytic hydrogenation. The alternative reduction with zinc and alkali or with sodium amalgam yields hydrazobenzene-3,3′-disulfonic acid, which rearranges to benzidine-2,2′-disulfonic acid [117-61-3] on treatment with acid.

**3-Nitrobenzenesulfonyl chloride** [121-51-7] (**9**), $C_6H_4ClNO_4S$, $M_r$ 221.62, mp 64 °C, is crystallized from petroleum ether.

*Production.* Nitrobenzene is added to excess chlorosulfonic acid, and the temperature is raised gradually to 100 °C and held there for 6 h. After cooling, the mixture is poured into ice and water, and 3-nitrobenzenesulfonyl chloride is filtered off in 75 % yield for use as a paste, after being washed (to remove acid) with cold water.

*Uses.* Halide exchange gives 3-nitrobenzenesulfonyl fluoride [349-78-0]. Reduction with aqueous sodium sulfite yields 3-nitrobenzenesulfinic acid [13257-95-9] (**143**), which is readily alkylated to form sulfones. When alkylation is carried out with ethylene chlorohydrin or ethylene oxide, the product is 3-nitrophenyl-2-hydroxyethylsulfone [41687-30-3], which is reduced to 3-aminophenyl-2-hydroxyethylsulfone [5246-57-1] (**144**), a valuable intermediate in the production of dyes (e.g., C.I. Reactive Blue 19).

3-Nitrobenzenesulfonyl chloride reacts with amines forming 3-nitrobenzenesulfonamide [121-52-8] with ammonia, and 3-nitrobenzenesulfonanilide [80-37-5] (**145**) with aniline. Reduction of these nitrosulfonamides yields the corresponding amines for use as diazo components.

**4-Nitrobenzenesulfonic acid** [138-42-1] (**146**), *p*-nitrobenzenesulfonic acid, $C_6H_5NO_5S$, $M_r$ 203.18, forms hygroscopic crystals from ethyl acetate–benzene, *mp* (anhydrous) 106–108 °C.

*Production.* 4-Nitrobenzenesulfonic acid is produced by hydrolysis of 4-nitrobenzenesulfonyl chloride.

*Uses.* 4-Nitrobenzenesulfonic acid is less important than the ortho and meta isomers because sulfanilic acid and its derivatives are readily available from aniline and acetanilide.

**4-Nitrobenzenesulfonyl chloride** [98-74-8] (**147**), $C_6H_4ClNO_4S$, $M_r$ 221.62, *mp* 80.5 °C, crystallizes from petroleum ether as colorless needles.

*Production.* 4,4′-Dinitrodiphenyldisulfide [100-32-3] (**148**), produced by reaction of 4-chloronitrobenzene with sodium disulfide, is treated with chlorine in HCl–HNO₃ by a process similar to that used for 2-nitrobenzenesulfonyl chloride.

*Uses.* Nitrosulfonamides derived from **147** have been used as herbicides. Sulfa drugs are rarely prepared from **147** because the route via *N*-acetylsulfanilyl chloride is much more economical. However, when *N*-acylsulfonamides are required, the nitro route must be used, e.g., reduction of **149** [derived from reaction of **147** with a heterocyclic amine (RNH₂), followed by acylation with acetic anhydride] under mild conditions to give the required product.

**2,4-Dinitrobenzenesulfonic acid** [89-02-1] (**150**), $C_6H_4N_2O_7S$, $M_r$ 248.17, crystallizes from water as colorless needles of the trihydrate (*mp* 108 °C), which dehydrate above 130 °C.

*Production.* 2,4-Dinitrochlorobenzene is warmed with aqueous sodium sulfite to give an 80% yield of the sodium salt of **150** [*885-62-1*] as yellow needles after salting out. The lower reactivity of 2- and 4- chloronitrobenzene requires much higher temperature for this nucleophilic sulfitation, with too many side reactions to be practicable.

*Uses.* Reduction of **150** gives the colorant intermediate 1,3-diaminobenzene-4-sulfonic acid [*88-63-1*]as an alternative to the sulfonation of

SO₃H
NO₂
NO₂
**150**

*m*-phenylenediamine. The sulfonic acid group of **150** can be replaced by amino and hydroxyl groups in the same way as the chlorine atom of 2,4-dinitrochlorobenzene, and sometimes the use of a water-soluble alternative may be advantageous. Treatment of **150** with chlorosulfonic acid gives 2,4-dinitrobenzenesulfonyl chloride [*1656-44-6*] which is used as a photographic chemical together with the parent sulfonic acid.

## 6.2. Nitrotoluenesulfonic Acids and Derivatives

Nitrotoluenesulfonic acids have properties similar to 3-nitrobenzenesulfonic acids and are reduced to toluidinesulfonic acids, which sometimes offer advantages over anilinesulfonic acids as colorant intermediates. The methyl group introduces an additional functionality, leading to alternative derivatives, mainly via oxidation reactions.

**2-Nitrotoluene-4-sulfonic acid** [*97-06-3*] (**18**), 3-nitro-4-methylbenzenesulfonic acid $C_7H_7NO_5S$, $M_r$ 217.20, crystallizes from water as pale yellow hygroscopic needles of the dihydrate (*mp* 92 °C, decomp. < 245 °C).

*Production.* 2-Nitrotoluene is sulfonated by heating with 25% oleum at 80 °C for 3 h, then quenching and liming to give a neutral solution of the sodium salt of **18** [1].

*Uses.* The total liquor of **18** is hydrogenated in situ to give *o*-toluidine-4-sulfonic acid [*618-03-1*] (**151**). Isolated 2-nitrotoluene-4-sulfonic acid may be converted to 2-nitro-*p*-toluenesulfonyl chloride [*54090-41-4*] (**152**), but the preferred process for **152** is chlorosulfonation of 2-nitrotoluene. Reaction of **152** with dimethylamine followed by reduction gives 3-amino-4-methyl-*N*,*N*-dimethylbenzenesulfonamide [*6331-68-6*] for use as an azoic diazo component.

**151** (structure: toluene with NH₂ ortho to CH₃ and SO₃H)
**152** (structure: toluene with NO₂ ortho to CH₃ and SO₂Cl)

Heating a solution of 2-nitrotoluene-4-sulfonic acid with sodium hydroxide produces the diazo component 4-sulfoanthranilic acid [98-43-1], and the autoxidation can be better controlled than with 2-nitrotoluene (cf. Section 3.2) to give a viable manufacturing process.

**4-Nitrotoluene-2-sulfonic acid** [121-03-9] (**25**), 5-nitro-2-methylbenzenesulfonic acid, *p*-nitrotoluene-*o*-sulfonic acid, PNTOS, $C_7H_7NO_5S$, $M_r$ 217.20, crystallizes from water as plates of the dihydrate (*mp* 133.5).

*Production.* 4-Nitrotoluene is sulfonated with 25 % oleum at 60 °C and worked up to give a solution of the sodium salt of **25**.

*Uses.* The most important outlets for 4-nitro-toluene-2-sulfonic acid exploit the readiness with which it forms stilbene derivatives. Selfcondensation products obtained by heating in aqueous sodium hydroxide are used as yellow water-soluble cotton dyes and consist of mixtures containing 4,4′-dinitroso-2,2′-stilbenedisulfonic acid [58058-72-3] (**153**), 4,4′-dinitro-2,2′-stilbenedisulfonic acid [128-42-7] (**154**), and 4,4′-dinitrodibenzyl-2,2′-disulfonic acid [6268-17-3] (**155**).

The entire class of stilbene dyes is based on this chemistry via further reaction of the condensation products, especially **154**, with aromatic amines or, more usually, with aminoazo compounds to give cheap, water-soluble orange and brown dyes.

Manufacture of pure **154** is achieved by controlled air oxidation of **25** in the presence of a catalyst. I.G. Farbenindustrie [1] used manganese sulfate and a specially designed unit to optimize air distribution and agitation. A simpler alternative is the use of sodium hypochlorite as oxidant in a process involving the simultaneous addition of oxidant and sodium hydroxide to an aqueous solution of **25** at 75 – 80 °C [69].

**153**: ON–C₆H₃(SO₃H)–CH=CH–C₆H₃(SO₃H)–NO

$$\text{ON-}\underset{}{\text{C}_6\text{H}_3(\text{SO}_3\text{H})}\text{-CH=CH-}\underset{}{\text{C}_6\text{H}_3(\text{HO}_3\text{S})}\text{-NO} \quad \mathbf{153}$$

$$\text{O}_2\text{N-}\underset{}{\text{C}_6\text{H}_3(\text{SO}_3\text{H})}\text{-CH=CH-}\underset{}{\text{C}_6\text{H}_3(\text{HO}_3\text{S})}\text{-NO}_2 \quad \mathbf{154}$$

$$\text{O}_2\text{N-}\underset{}{\text{C}_6\text{H}_3(\text{SO}_3\text{H})}\text{-CH}_2\text{-CH}_2\text{-}\underset{}{\text{C}_6\text{H}_3(\text{HO}_3\text{S})}\text{-NO}_2 \quad \mathbf{155}$$

$$\text{O}_2\text{N-}\underset{}{\text{C}_6\text{H}_3(\text{SO}_3\text{H})}\text{-CH=CH-}\underset{}{\text{C}_6\text{H}_3(\text{HO}_3\text{S})}\text{-NH}_2 \quad \mathbf{156}$$

Apart from its use in the production of stilbene dyes, **154** is even more important as the precursor, by Fe–HCl reduction at 100 °C, of 4,4′-diamino-2,2′-stilbenedisulfonic acid [*81-11-8*], which is the basis for 80% of all fluorescent whitening agents.

Reduction of **154** with sodium sulfide results in production of the disodium salt of 4′-nitro-4-aminostilbene-2,2′-disulfonic acid [*6634-82-8*] (**156**) for use as a diazo component; for example, in polyazo dyes for which the nitro group of **156** is reduced to an azoxy group by treatment with glucose and sodium hydroxide [70].

Hydrogenation of **25** gives *p*-toluidine-2-sulfonic acid (5-amino-2-methylbenzenesulfonic acid) [*118-88-7*], which can be used as an alternative to metanilic acid.

4-Nitrotoluene-2-sulfonyl chloride [*121-02-8*] (**157**), the acid chloride of **25**, is best obtained by chlorosulfonation of 4-nitrotoluene. Reaction of **157** with aniline followed by reduction gives 4-aminotoluene-2-sulfonanilide [*79-72-1*] (**158**) for use as a diazo component. The sulfonamide derived from **157** is oxidized and ring closed in one step to give 6-nitrosaccharin (**159**) by treatment with $CrO_3$–$H_2SO_4$ at 65–70 °C [71].

**157**: 2-CH₃-5-NO₂-C₆H₃-SO₂Cl

**158**: 2-CH₃-5-NH₂-C₆H₃-SO₂NH-C₆H₅

**159**: 6-nitrosaccharin (O₂N-C₆H₃ fused with -SO₂-NH-C(=O)-)

**2,6-Dinitrotoluene-4-sulfonic acid** [*88-90-4*] (**160**), 3,5-dinitro-4-methylbenzenesulfonic acid, $C_7H_6N_2O_7S$, $M_r$ 262.20, crystallizes as a hydrate in the form of pale yellow crystals, which soften at 110 °C, dehydrate at 140 °C, and melt at 165 °C.

*Production.* 2-Nitrotoluene is sulfonated to form 2-nitrotoluene-4-sulfonic acid; the reaction mixture is diluted with 90 % sulfuric acid, and sodium nitrate is added over 6 h at 80 °C. After a further 1 h at this temperature the reaction mass is poured into water and the sodium salt of the product is salted out by addition of sodium sulfate. After being washed acid-free with brine, the yield of **160** is ca. 80 %.

*Uses.* Reduction of **160** with sulfide gives 2-amino-6-nitrotoluene-4-sulfonic acid (**161**), and reduction with iron–acid gives 2,6-diaminotoluene-4-sulfonic acid [*98-25-9*]. 2,6-Dinitrotoluene-4-sulfonyl chloride [*80198-19-2*] is obtained from **160** by treatment with thionyl chloride. Reaction of this sulfonyl chloride with an alkylamine followed by hydrogenation leads to a readily accessible series of symmetrical diaminotoluenesulfonamides [72].

## 6.3. Chloronitrobenzenesulfonic Acids and Derivatives

The commercially available chloronitrobenzenesulfonic acids are usually obtained by sulfonation or chlorosulfonation of chloronitrobenzenes or by sulfition of dichloronitrobenzenes. Although the corresponding substituted anilines can be obtained by reduction, the main outlets for this group of products are via nucleophilic displacement reactions of the chlorine substituent.

**2-Chloro-5-nitrobenzenesulfonic acid** [*96-73-1*] (**58**), 4-chloronitrobenzene-3-sulfonic acid, PN salt, $C_6H_4ClNO_5S$, $M_r$ 237.62, crystallizes as the dihydrate from water (*mp* 169 °C, decomp.). The compound undergoes explosive decomposition when heated above 200 °C, and all reactions, including its production, must be undertaken with great caution.

*Production.* 4-Chloronitrobenzene is sulfonated with 65 % oleum at 115 °C; special precaution must be taken to control temperature [41]. Work-up is by quenching and isolation of the sodium salt after it has been salted out.

*Uses.* Amination of **58** gives 4-nitroaniline-2-sulfonic acid [*96-75-3*] (**162**), which is used as a diazo component or reduced to the important colorant intermediate 2,5-diaminobenzenesulfonic acid [*88-45-9*]. Chlorination of **162** in aqueous solution is a more facile route to 2,6-dichloro-4-nitroaniline (**104**) than chlorination of 4-nitroaniline.

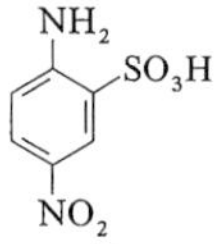

**162**

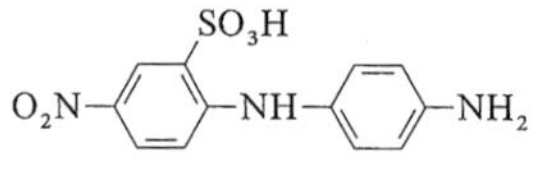

**163**

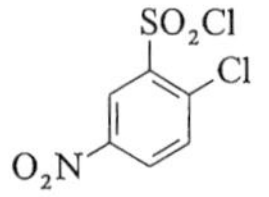

**164**

Monocondensation of **58** with *p*-phenylenediamine gives 4′-amino-4-nitrodiphenylamine-2-sulfonic acid [*91-29-2*] (**163**), which can be reduced to 4,4′-diaminodiphenylamine-2-sulfonic acid [*119-70-0*], a tetrazo component in direct cotton dyes. Monoaminodiphenylaminesulfonic acid diazo components are obtained by condensation of **58** with aniline or toluidine, followed by reduction.

2-Chloro-5-nitrobenzenesulfonyl chloride [*4533-95-3*] (**164**) may be obtained from **58** but is best prepared by direct chlorosulfonation of 4-chloronitrobenzene. Amination of **164** gives 4-nitroaniline-2-sulfonamide [*54734-85-9*].


**4-Chloro-2-nitrobenzenesulfonic acid** (**165**), $C_6H_4ClNO_5S$, $M_r$ 237.62, *mp* 114–115 °C (anhydrous), forms hygroscopic crystals. The sodium salt is water soluble; the potassium salt, less so (ca. 1% at 25 °C).

*Production.* 4-Chloro-2-nitrobenzenesulfonic acid is obtained by alkaline hydrolysis of 4-chloro-2-nitrobenzenesulfonyl chloride [*4533-96-4*] (**166**), which is obtained from bis(4-chloro-2-nitrophenyl) disulfide (the latter is formed from 2-nitro-1,4-dichlorobenzene) by analogy with the process for 2-nitrobenzenesulfonyl chloride. The primary outlets for **166** are amination to 4-chloro-2-nitrobenzenesulfonamide [*13852-81-8*] (**167**) and further chlorosulfonation to the disulfonyl dichloride (**168**), an intermediate for a series of 4-amino-6-chlorobenzene-1,3-disulfonamides that are useful diuretics.

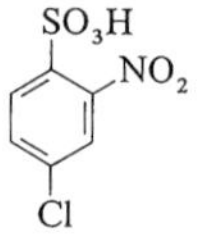

**165** 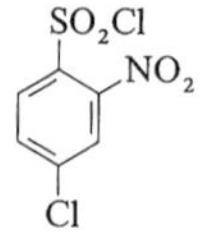   **166** 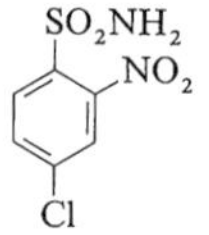   **167**

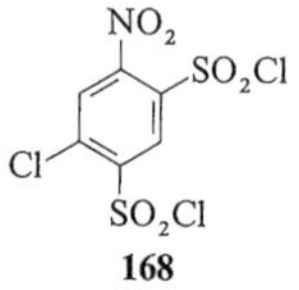

**168**

**4-Chloro-3-nitrobenzenesulfonic acid** [*121-18-6*] (**49**), $C_6H_4ClNO_5S$, $M_r$ 237.62, crystallizes as needles from water (decomp. > 200 °C). The potassium salt forms pale yellow crystals in water, *mp* 325 °C (decomp.). The solubility of the potassium salt in water is 2% at 30 °C versus 19% for the sodium salt.

*Production.* 2-Chloronitrobenzene can be sulfonated with 65% oleum analogously to 4-chloronitrobenzene, but the preferred process is to sulfonate chlorobenzene at 100 °C and nitrate the sulfonation mass in situ by the addition of nitric acid or potassium nitrate at 20–30 °C. The product is isolated in 90% yield as its sodium or potassium salt by salting out from the quenched reaction mixture after addition of lime and removal of the precipitated calcium sulfate. Purification, if required, is achieved by recrystallization of the potassium salt.

*Uses.* Hydrolysis of **49** gives 2-nitrophenol-4-sulfonic acid [616-85-3] (**169**) for reduction to the diazo component 2-aminophenol-4-sulfonic acid [98-37-3]. Amination yields 2-nitroaniline-4-sulfonic acid [616-84-2] (**170**), and iron reduction produces 2-chloroaniline-5-sulfonic acid [98-36-2], both of which are used as diazo components. Heating **49** with aqueous sodium sulfite results in nitrobenzene-2,5-disulfonic acid [119-00-6] (**171**), an intermediate for an alternative route to aniline-2,5-disulfonic acid [98-44-2].

The acid chloride of **49** is best produced by direct chlorosulfonation of 2-chloronitrobenzene. 4-Chloro-3-nitrobenzenesulfonyl chloride [97-08-5] (**50**) is especially important as the intermediate for 4-chloro-3-nitrobenzenesulfonamide [97-09-6] (**172**); the latter leads to the diazo component 2-aminophenol-4-sulfonamide [98-32-8]. 4-Chloro-3-nitrobenzenesulfonyl chloride is also used for the production of nitro dyes, e.g., Disperse Yellow 42 (**173**) by condensation of **50** with two equivalents of aniline.

**5-Chloro-2-nitrobenzenesulfonic acid** [54481-12-8] (**174**), $C_6H_4ClNO_5S$, $M_r$ 237.62, forms hygroscopic crystals of the monohydrate (*mp* 93 °C).

*Production.* 5,5′-Dichloro-2,2′-dinitrodiphenyldisulfide (obtained from the reaction of 2,4-dichloronitrobenzene with $Na_2S_2$) is oxidatively chlorinated with chlorine in $HCl-HNO_3$ to give the sulfonyl chloride, which hydrolyzes to **174**.

*Uses.* Reduction of **174** gives 4-chloroaniline-2-sulfonic acid, and amination results in 4-nitroaniline-3-sulfonic acid, both of minor importance as intermediates. The derived sulfonyl chloride leads to 5-chloro-2-nitrobenzenesulfonamide (**175**), an intermediate in the production of the benzothiadiazine derivative diazoxide, used as a vasodilator.

### 2-Chloro-3,5-dinitrobenzenesulfonic acid [4515-26-8] (**176**), $C_6H_3ClN_2O_7S$, $M_r$ 282.62, crystallizes from water as needles.

*Production.* 4-Chloronitrobenzene is sulfonated by gradual addition of 25% oleum at 60 °C; the temperature is raised slowly to 125 °C, and held there for 8 h. The sulfonation mixture is cooled to 70 °C, nitric acid is added over 5 h, and the reaction is completed by heating for 12 h at 70 °C followed by 4 h at 100 °C. After cooling and quenching, the sodium salt of the product is salted out by addition of sodium sulfate and isolated in 80% yield.

*Uses.* Hydrolysis of **176** with aqueous alkali readily yields 2,4-dinitrophenol-6-sulfonic acid, which is reduced with sodium sulfide to 4-nitro-2-aminophenol-6-sulfonic acid [96-67-3] (**177**), a diazo component of metallizable azo dyes. Under very mild conditions, **176** is aminated to 2,4-dinitroaniline-6-sulfonic acid, and **176** reacts with sodium sulfite to give 3,5-dinitrobenzene-1,2-disulfonic acid.

As a reagent for the identification of α-amino acids ($RCHNH_2COOH$), **176** forms derivatives (**178**) that are more readily hydrolyzed in acid solution to give the parent α-amino acid than the corresponding 2,4-dinitrophenyl derivatives [73].

### 4-Chloro-3,5-dinitrobenzenesulfonic acid [88-91-5] (**115**), $C_6H_3ClN_2O_7S$, $M_r$ 282.62, forms hygroscopic crystals in water (*mp* 293 °C).

*Production.* 4-Chloro-3,5-dinitrobenzenesulfonic acid is produced by a process based on the sulfonation and nitration of chlorobenzene that uses 2 mol/mol of potassium nitrate, in place of the 1 mol/mol employed in the production of 4-chloro-3-nitrobenzenesulfonic acid. The product is isolated from the quenched reaction mixture as the potassium salt, after addition of potassium chloride.

*Uses.* Conversion of **115** to 2,6-dinitroaniline is described in Section 5.1. More important is the use of the derived sulfonyl chloride to form the sulfonamide (**179**) (a fungicide) and the sulfone (**180**). Reaction of **179** with di-*n*-propylamine forms oryzalin [*19044-88-3*], and reaction of **180** with di-*n*-propylamine forms nitralin [*4726-14-1*], both of which are important members of the large class of substituted 2,6-dinitroaniline herbicides.

## 6.4. Chloronitrotoluenesulfonic Acids

**6-Chloro-2-nitrotoluene-4-sulfonic acid** [*68189-28-6*] (**181**), 3-chloro-4-methyl-5-nitrobenzenesulfonic acid, $C_7H_6ClNO_5S$, $M_r$ 251.65.

*Production.* 6-Chloro-2-nitrotoluene is sulfonated with 20% oleum at 60 °C with careful temperature control to avoid a runaway reaction. Work-up is by conventional processes.

*Uses.* Reduction of **181** gives 2-amino-6-chlorotoluene-4-sulfonic acid [*6387-27-5*], which is used as a diazo component.

**6-Chloro-4-nitrotoluene-2-sulfonic acid** (**182**), 3-chloro-2-methyl-5-nitrobenzenesulfonic acid, $C_7H_6ClNO_5S$, $M_r$ 251.65, crystallizes from water as hygroscopic needles.

*Production.* 2-Chloro-4-nitrotoluene is sulfonated with 30% oleum at 75 °C to produce 6-chloro-4-nitrotoluene-2-sulfonic acid.

*Uses.* Oxidation of **182** with sodium hypochlorite gives 6,6′-dichloro-4,4′-dinitro-2,2′-stilbene disulfonic acid (**183**), which is reduced to the corresponding diamine; the latter is used in experimental fluorescent whitening agents, analogous to those obtained from 4,4′-diamino-2,2′-stilbene disulfonic acid.

**Toxicology of Nitrohydrocarbon Sulfonic Acids.** As a general rule the water solubility conferred by the sulfonic acid group, as well as the hydrophilicity conferred by the sulfonamide and sulfone groups, renders this group of products less toxic than their unsulfonated parent compounds. The products are less volatile, penetrate the skin less readily, and are more readily excreted if ingested. This is demonstrated by the sodium salt of 3-nitrobenzenesulfonic acid, which has an oral $LD_{50}$ (rat) of 11 000 mg/kg, compared to 640 mg/kg for nitrobenzene.

# 7. Nitrohydroxy and Alkoxy Aromatics

Phenol can be nitrated with aqueous nitric acid in the presence of nitrous acid (cf. Section 2.2) to give a mixture consisting mainly of 2- and 4-nitrophenol. Although these can be separated (the ortho isomer is steam volatile), the economics of this route compares unfavorably with hydrolysis of 2- and 4-nitrochlorobenzene. Hydrolysis is also used for 2,4-dinitrophenol. Nitrocresols are more frequently obtained by nitrosation or nitration, although mixtures of isomers often result.

Nitrophenyl ethers are also produced from reactive nitrochloro compounds by reaction with alcohols or phenols. The less common *meta*-nitrophenyl ethers are prepared by methylation of the 3-nitrophenol precursor. Nomenclature with substituted anisidines and their homologues is confusing. Common practice (used here) maintains the methoxy group at the 1-position, in line with substituted phenols, but many references use the convention of the amino group in the 1-position. Only the formula or the CAS name is unambiguous for compounds such as 4-nitro-*o*-anisidine.

The most general use of both nitrophenols and nitro ethers involves reduction to the corresponding hydroxy- and alkoxyanilines.

## 7.1. Nitrophenols and Derived Ethers

**2-Nitrophenol** [88-75-5] (**46**), *o*-nitrophenol, $C_6H_5NO_3$, $M_r$ 139.11, *mp* 45 °C, *bp* 216 °C (101 kPa), forms yellow needles from ethanol. It is slightly soluble in water (1 % at 100 °C) and steam volatile (intramolecular hydrogen bonding, unlike the para isomer).

*Production.* 2-Chloronitrobenzene in 8.5 % sodium hydroxide solution is heated gradually (exothermic) to 170 °C in an autoclave and held there under pressure for 8 h. The resulting solution is cooled and acidified to give the product in 95 % yield.

*Uses.* Catalytic hydrogenation (Pd–C) of 2-nitrophenol gives 2-aminophenol [95-55-6], which is used as a photographic developer and, in larger amounts, as a versatile intermediate for dyes and fine chemicals. 2-Aminophenol reacts with acetic anhydride to give 2-methylbenzoxazole, which is nitrated with mixed acid to give the 6-nitro derivative for subsequent hydrolysis to 5-nitro-2-aminophenol [121-880-0] (**184**).

**2-Nitroanisole** [91-23-6] (**47**), 2-methoxynitrobenzene, 2-nitrophenyl methyl ether, $C_7H_7NO_3$, $M_r$ 153.13, $mp$ 10 °C, $bp$ 272 °C (101 kPa), is steam volatile. The compound is sparingly soluble in water, but readily soluble in ethanol, ether, or benzene.

*Production.* Methanolic sodium hydroxide is added slowly to a solution of 2-chloronitrobenzene in methanol at 70 °C; the mixture is then heated gradually to 95 °C under pressure to complete the reaction. After dilution with water the product is separated as an oil, in 90 % yield, and methanol is recovered from the aqueous layer.

*Uses.* 2-Nitroanisole is reduced ($H_2$– catalyst or iron – formic acid) to *o*-anisidine [90-04-0] or (benzidine-type reaction) to *o*-dianisidine [119-90-4], both of which are important as dye intermediates. Low-temperature nitration of *o*-anisidine, either directly or by addition of *o*-anisidine nitrate to sulfuric acid, gives 4-nitro-*o*-anisidine (Fast Scarlet base) [99-59-2] (**185**) in good yield. Nitration of *N*-acetyl-*o*-anisidine with mixed acid at 25 °C gives mainly the 5-nitro derivative, which can be separated from the 4-nitro isomer after hydrolysis to the amine because the more weakly basic 5-nitro-*o*-anisidine (Fast Red B base) [97-52-9] (**186**) is precipitated first on dilution of the sulfuric acid hydrolysate. Chlorosulfonation of **47** yields 2-nitroanisole-4-sulfonyl chloride [22117-79-9], from which *o*-anisidine-4-sulfonamide derivatives are obtained.

**2-Ethoxynitrobenzene** [610-67-3] (**187**), 2-nitrophenyl ethyl ether, 2-nitrophenetole, $C_8H_9NO_3$, $M_r$ 167.16, $cp$ 4.5 °C, $bp$ 275 °C (101 kPa), is a greenish yellow oil.

*Production.* 2-Chloronitrobenzene in ethanol solution is treated with ethanolic sodium or potassium hydroxide in a process similar to that for 2-nitroanisole, except that the reaction mixture is aerated to avoid reduction to the potentially explosive azoxy derivative. The nitrophenol byproduct is formed in larger amounts than with nitroanisole and requires washing with alkali or post-ethylation with ethyl chloride under pressure [1].

*Uses.* Hydrogenation of **187** gives *o*-phenetidine [94-70-2] (2-ethoxyaniline), which is nitrated to 4-nitro-*o*-phenetidine.

**4-Chloro-2-nitrophenol** [89-64-5] (**188**), $C_6H_4ClNO_3$, $M_r$ 173.56, $mp$ 87 °C, forms yellow prisms from ethanol. The compound is sparingly soluble in water and is steam volatile.

*Production.* 2,5-Dichloronitrobenzene is hydrolyzed with 8 % aqueous sodium hydroxide at 120 – 130 °C under pressure [1].

*Uses.* Reduction of **188** gives 2-amino-4-chlorophenol [95-85-2], and chlorination yields 2,4-dichloro-6-nitrophenol [609-89-2] (**189**), which is reduced to 2-amino-4,6-dichlorophenol [527-62-8]. These substituted aminophenols are members of a large class of diazo components used for hydroxyazo dyes, which are important for wool and polyamide when converted to their chromium complexes. For the amine derived from the methyl ether of **188**, 4-chloro-*o*-anisidine [*93-50-5*], the preferred route is methoxylation of 2,5-dichloronitrobenzene followed by reduction. 2-amino-4-chlorophenol reacts with phosgene to give 5-chlorobenzoxolinone which nitrates to the 6-nitro compound (**190**); the latter undergoes alkaline hydrolysis to 2-amino-4-chloro-5-nitrophenol.

**3-Nitrophenol** [*554-84-7*] (**191**), *m*-nitrophenol, $C_6H_5NO_3$, $M_r$ 139.11, *mp* 97 °C, *bp* 194 °C, forms yellowish crystals from dilute hydrochloric acid. The compound is moderately soluble in water (1 % at 20 °C, 13 % at 90 °C).

*Production.* 3-Nitroaniline is diazotized in aqueous sulfuric acid and then hydrolyzed by being added gradually to boiling dilute sulfuric acid. The crude product solidifies on cooling and is filtered off in 90 % yield.

*Uses.* Reduction of **191** gives 3-aminophenol [*591-27-5*], and methylation to 3-nitroanisole [*555-03-3*] followed by reduction gives *m*-anisidine [*536-90-3*]. Carboxylation of 3-aminophenol gives the tuberculostatic 4-aminosalicylic acid [*65-49-6*].

**4-Nitrophenol** [*100-02-7*] (**55**), *p*-nitrophenol, $C_6H_5NO_3$, $M_r$ 139.11, forms yellow needles (mixed α- and β-forms), *mp* 114 °C, *bp* 216 °C (101 kPa). The (metastable) α-form crystallizes from toluene above 63 °C, and the yellow, prismatic β-form crystallizes from toluene below 63 °C. 4-Nitrophenol is not steam volatile and is much more soluble in water (30 % at 100 °C) than the ortho isomer. Its flash point is 105 °C, and the decomposition point (explosive) of the sodium salt is 156 °C.

*Production.* 4-Chloronitrobenzene is hydrolyzed by a process identical to that described for 2-nitrophenol.

*Uses.* Reduction of **55** to 4-aminophenol [*123-30-8*] has largely been superseded by processes based on direct reduction of nitrobenzene. Reaction of sodium 4-nitrophenoxide with dialkylthiophosphoric chlorides (125 °C in chlorobenzene) gives a series of insecticides typified by parathion [*56-38-2*] (**192**; R = ethyl), which has largely been displaced by the less toxic methylparathion [*298-00-0*] (**192**; R = methyl) and related products. 4-Nitrophenol reacts with 4-chloro-3-nitrobenzotrifluoride (**96**) to form fluorodifen (**193**)—related to nitrofen (**67**), mentioned in Section 4.1; it is used as a herbicide specifically on drilled rice crops.

**4-Nitroanisole** [100-17-4] (**56**), 4-methoxynitrobenzene, 4-nitrophenyl methyl ether, $C_7H_7NO_3$, $M_r$ 153.13, mp 54 °C, bp 274 °C (101 kPa), forms prisms from ethanol. 4-Nitroanisole is insoluble in water, moderately soluble in ethanol or benzene and steam volatile.

*Production.* 4-Chloronitrobenzene is dissolved in methanol and treated with methanolic sodium hydroxide, which is added slowly at 80 – 90 °C. After completion of reaction and neutralization, the excess methanol is removed by distillation, and after cooling, the product is filtered off in 95 % yield.

*Uses.* Reduction of **56** with sulfide gives *p*-anisidine [104-94-9]. *N*-Acetyl-*p*-anisidine is nitrated with 62 % nitric acid in chlorobenzene solution to give the 3-nitro derivative for hydrolysis to 3-nitro-*p*-anisidine (Fast Bordeaux GP base) [96-96-8] (**194**).

**4-Ethoxynitrobenzene** [100-29-8] (**195**), 4-nitrophenyl ethyl ether, 4-nitrophenetole, $C_8H_9NO_3$, $M_r$ 167.16, mp 60 °C, bp 283 °C (101 kPa), crystallizes from ethanol as prisms, which are insoluble in water and moderately soluble in ethanol or ether.

*Production.* 4-Ethoxynitrobenzene is obtained from 4-chloronitrobenzene, by a process analogous to that described for 2-ethoxynitrobenzene.

*Uses.* Reduction of **195** gives *p*-phenetidine [156-43-4], which undergoes similar reactions to *p*-anisidine and is also the intermediate for the analgesic phenacetin (4-ethoxyacetanilide [62-44-2]) and the sweetener dulcin (4-ethoxyphenylurea [150-69-6]).

**2,4-Dinitrophenol** [51-28-5] (**63**), $C_6H_4N_2O_5$, $M_r$ 184.11, mp 114 °C (decomposes explosively at higher temperature), $pK_a$ 4.89, forms yellow needles. 2,4-Dinitrophenol is sublimable, steam volatile, sparingly soluble in water (1.3% at 100 °C), and readily soluble in acetone, ethanol, or benzene.

*Production.* 2,4-Dinitrochlorobenzene is hydrolyzed by heating with 6% aqueous sodium hydroxide at 95–100 °C for 4 h. The product is precipitated by addition of acid, filtered off, and washed to remove acid and also a small quantity of the more soluble 2,6-isomer. The yield is 95%.

*Uses.* 2,4-Dinitrophenol is used in antiseptics, as an explosives additive, as a pest control agent, and as an intermediate for black sulfur dyes. Sulfide reduction gives 4-nitro-2-aminophenol [*99-57-0*] (**196**).

**2,4-Dinitroanisole** [*119-27-7*] (**197**), 1-methoxy-2,4-dinitrobenzene, $C_7H_6N_2O_5$, $M_r$ 198.13, mp 83 °C, forms yellow crystals from ethanol. 2,4-Dinitroanisole is potentially explosive on heating.

*Production.* 2,4-Dinitrochlorobenzene is treated with methanolic sodium hydroxide in a process similar to that described for 4-nitroanisole. 2,4-Dinitrophenetole [*610-54-8*] is similarly manufactured by using ethanol.

*Uses.* Catalytic hydrogenation of **197** yields 2,4-diaminoanisole [*615-05-4*], which is converted by N-acylation and N'-alkylation to azo coupling components. Sulfide reduction gives 4-nitro-*o*-anisidine (**185**) as an alternative route to nitration of *o*-anisidine, but the product from this selective reduction is less pure.

**2,5-Dimethoxynitrobenzene** [*89-39-4*] (**198**), 2-nitrohydroquinone dimethyl ether, 1,4-dimethoxy-2-nitrobenzene, $C_8H_9NO_4$, $M_r$ 183.13, mp 72–73 °C, forms yellow needles from aqueous ethanol.

*Production.* 1,4-Dimethoxybenzene is nitrated with 34% nitric acid at 35–85 °C to give **198**.

*Uses.* 2,5-Dimethoxynitrobenzene is reduced to 2,5-dimethoxyaniline [*102-56-7*] for use as an azo coupling component or as an intermediate for Fast Blue RR Base by benzoylation, nitration, and reduction.

2-Chlorohydroquinone dimethyl ether is similarly nitrated to give 4-chloro-2,5-dimethoxynitrobenzene [*6940-53-0*] (**199**) for reduction to 4-chloro-2,5-dimethoxyaniline [*6358-64-1*].

**2,5-Diethoxynitrobenzene** [*119-23-3*] (**200**), 2-nitrohydroquinone diethyl ether, 1,4-diethoxy-2-nitrobenzene, $C_{10}H_{13}NO_4$, $M_r$ 211.16.

*Production.* 1,4-Diethoxybenzene is nitrated in a manner similar to the dimethyl ether to give **200**.

*Uses.* Reduction of **200** yields 2,5-diethoxyaniline [*94-85-9*], which is used as an azo coupling component or as the intermediate for Fast Blue BB Base by consecutive benzoylation, nitration, and reduction.

## 7.2. Nitroalkylphenols

Many mono- and dinitrocresols are available, but only those that can be produced economically by nitrosation or nitration of readily available cresols are significant. The next most accessible group of nitrocresols is obtained by diazotization of nitrotoluidines. Phenols substituted with higher alkyl groups are readily available, and certain of their nitro derivatives are used as agrochemicals.

**4-Nitro-*m*-cresol** [*2581-34-2*] (**201**), 1-hydroxy-3-methyl-4-nitrobenzene, $C_7H_7NO_3$, $M_r$ 153.13, mp 128–129 °C, $bp$ > 200 °C (101 kPa, decomp.), flash point ca. 110 °C, ignition temperature 455 °C, crystallizes as yellow needles, which are sparingly soluble in water and readily soluble in alcohol, chloroform, or benzene.

*Production.* *m*-Cresol is nitrosated in isopropanol solution and the isolated 4-nitroso-*m*-cresol is treated with aqueous nitric acid to oxidize it to **201**.

*Uses.* The derived insecticide metathion (fenitrothion) [*122-14-5*] (**202**), an analogue of methylparathion, is less toxic than parathion (**192**).

**2-Nitro-*p*-cresol** [*119-33-5*] (**203**), 1-hydroxy-4-methyl-2-nitrobenzene, $C_7H_7NO_3$, $M_r$ 153.13, mp 33 °C, $bp$ 125 °C (2.9 kPa), forms yellow needles, which are sparingly soluble in water and readily soluble in alcohol or benzene.

*Production.* *p*-Toluidine is diazotized and hydrolyzed to *p*-cresol, which in the presence of nitrous acid forms 2-nitroso-*p*-cresol; the latter is readily oxidized in situ to **203** by addition of aqueous nitric acid [1].

*Uses.* Treatment of an aqueous alkaline solution of 2-nitro-*p*-cresol with methyl chloride under pressure gives the methyl ether (**204**), which is reduced (Fe – formic acid) to cresidine (**83**). This azo coupling component is also available by an alternative process from the less readily available 4-chloro-3-nitrotoluene (**81**; see Section 4.3).

**4,6-Dinitro-o-cresol** [534-52-1] (**205**), DNOC, 2-methyl-4,6-dinitrophenol, 2-hydroxy-1-methyl-3,5-dinitrobenzene, $C_7H_6N_2O_5$, $M_r$ 198.13, mp 86.5 °C, forms yellow crystals. The compound is sparingly soluble in water and readily soluble in alcohol, ether, or benzene.

*Production.* o-Cresol is sulfonated in excess 75% sulfuric acid to give the disulfonic acid. The sulfonation mass is diluted with water, and 2 equivalents of nitric acid are added at 70 °C to form the dinitro derivative. The product is separated while molten and washed with hot water.

*Uses.* 4,6-Dinitro-o-cresol has been used since the 1930s as a contact herbicide, being gradually displaced by improved products, some of which have related structures. Examples are dinoseb (2-sec-butyl-4,6-dinitrophenol) [88-85-7] (**206**; R = sec-butyl) and dinoterb (2-tert-butyl-4,6-dinitrophenol) [1420-07-1] (**206**; R = tert-butyl), which are produced by controlled nitration of the corresponding alkylphenol. Members of this class of dinitro compounds have also been used as insecticides and acaricides, with dinocap [39300-45-3] (**206**, crotonate, R = 1-methylheptyl) specifically claimed to be effective against fruit mildew.

**2,6-Dinitro-p-cresol** [609-93-8] (**207**), 4-methyl-2,6-dinitrophenol, 4-hydroxy-1-methyl-3,5-dinitrobenzene, $C_7H_6N_2O_5$, $M_r$ 198.13, mp 82 °C, forms yellow crystals.

*Production.* p-Cresol is dinitrated directly in dilute sulfuric acid solution by careful addition of nitric acid to yield 2,6-dinitro-p-cresol.

*Uses.* 2,6-Dinitro-p-cresol and analogous alkyl compounds have been used as agrochemicals similarly to the 4,6-dinitro-p-cresol.

**Toxicity of Nitrophenols and Derivatives.** Nitrophenol compounds are all at least moderately toxic by inhalation and skin absorption. Nitro- and, especially, dinitrophenols and cresols are pseudoacids and must be handled accordingly. 2,4-Dinitrophenol and the dinitrocresols are more poisonous by ingestion than mononitrophenols. Detailed toxicity data are available on most of the compounds, especially the known active agents (e.g., pesticides) [22].

## 8. Nitroketones

Acyl substituents on the benzene ring make nitration difficult, and the isomer ratio is very sensitive to temperature and nitration conditions. Introduction of an alkyl or alkoxyl group makes nitration easier, and para substitution leads to uniform products. For example, nitration of 4-methylacetophenone gives 4-methyl-3-nitroacetophenone (**208**) in good yield. Similarly, 4-methoxypropiophenone gives the 3-nitro derivative (**209**), provided that nitration is carried out at low temperature (–5 °C). Above 0 °C, ipso nitration takes place with the formation of 2,4-dinitroanisole.

Difficult accessibility of nitroaromatic ketones together with the problem of selfcondensation of derived anilines has eliminated this class of compound from the historical colorant usage common to all other classes. Special requirements, such as pharmaceutical precursors, have a small but important place.

**2-Nitroacetophenone** [577-59-3] (**210**), $C_8H_7NO_3$, $M_r$ 165.13, mp 24.5 °C, bp 134 °C (0.5 kPa).

*Production.* O-Acetyl-1-phenylethanol is nitrated to give a mixture of o- and p-nitrophenylethyl acetates, which are hydrolyzed with acid and fractionally distilled to yield approximately equal amounts of 1-(2-nitrophenyl)ethanol (**211**) and the corresponding 4-nitro isomer. Oxidation of **211** with dichromate–$H_2SO_4$ gives an 85 % yield of **210**, which can be further purified by vacuum distillation.

An alternative process involves oximination of 2-nitroethylbenzene (Section 3.5) followed by acid hydrolysis to the required product [74].

**3-Nitroacetophenone** [*121-89-1*] (**212**), 1-(3-nitrophenyl)ethanone, $C_8H_7NO_3$, $M_r$ 165.13, mp 81 °C, bp 167 °C (2.4 kPa), crystallizes as needles from ethanol. The compound is steam volatile.

*Production.* Acetophenone is carefully dissolved in concentrated sulfuric acid at 0 °C, and mixed acid is added slowly below 0 °C. The crude product is precipitated by quenching into ice, filtered, and washed acid free. Purification, mainly from the ortho isomer, is achieved by recrystallization from ethanol with significant lowering of the overall yield to ca. 55%.

**4-Nitroacetophenone** [*100-19-6*] (**213**), 4-nitrophenyl methyl ketone, $C_8H_7NO_3$, $M_r$ 165.13, mp 80 – 81 °C, bp 138 °C (0.2 kPa), forms yellow prisms.

*Production.* 1-(4-Nitrophenyl)ethanol, coproduced with the 2-nitro compound, is oxidized with dichromate to give **213**, which is purified, if required, by recrystallization from methanol.

*Uses.* One of several alternative routes to synthetic chloroamphenicol (**1**) uses **213** as starting material with a first-stage bromination to α-bromo-4-nitroacetophenone [*99-81-0*] (**214**) followed by buildup of the side chain. The intermediate **214** is the key precursor in the synthesis of the bronchodilator clenbuterol [*37148-27-9*].

# 9. Nitroheterocycles

The chemistry of nitro derivatives of many heterocyclic systems has long been established, but their commercial exploitation has developed only over the last 40 years. The more important of the diverse outlets are summarized here. Many of the compounds mentioned are likely to be covered in more detail under alternative headings (e.g., antibiotics or dyes) but not, however, compared in either synthesis or mode of action. Given the large number of nitroaryl derivatives that are used as agrochemicals, surprisingly few examples of nitroheterocyclics are reported in this field. Availability in bulk for use in one area often leads to outlets in other fine chemicals.

The nitrofuran group of broad-spectrum antibacterial drugs, based on 5-nitro-2-furaldehyde (**215**), was developed in the 1940s [75]. The less important nitroimidazole group of antibiotics was studied in the 1950s, following isolation of the naturally occurring azomycin (**216**). Also about this time, synthetic nitrothiazoles, typified by 2-acetylamino-5-nitrothiazole [*140-40-9*] (**217**), were found to have antitrichomonal activity.

$$\underset{\mathbf{215}}{O_2N-\text{furan}-CHO} \quad \underset{\mathbf{216}}{\text{imidazole}-NO_2} \quad \underset{\mathbf{217}}{O_2N-\text{thiazole}-NHCOCH_3}$$

The 1950s witnessed a change in the field of colorants with the introduction by Eastman Kodak [76] of the nitroheterocyclic-based blue disperse dye (**218**) for cellulose acetate fibers. Since then, a succession of useful nitroheterocyclic diazo components have been developed in the dyemakers continual search for high strength and more bathochromic azo dyes [77].

$$\underset{\mathbf{218}}{O_2N-\text{thiazole}-N=N-\text{Ar}(CH_3)-N(C_2H_5)(CH_2CHOHCH_2OH)}$$

In this brief review, only those nitroheterocycles of known commercial interest are considered. This limits the systems to pyridine, quinoline, imidazole, furan, thiophene, and thiazole.

## 9.1. Pyridine Derivatives

Pyridine, in common with many other nitrogen heterocycles, is not rapidly nitrated because of the deactivation deriving from the protonated ring nitrogen. High-temperature nitration in oleum gives a low yield of 3-nitropyridine [*2530-26-9*]. However, pyridine *N*-oxide is nitrated under moderate conditions (mixed acid at 100 °C) [78] to give 4-nitropyridine *N*-oxide [*1124-33-0*] (**219**), which requires purification from a small quantity of the 2-isomer.

Hydrogenation of **219** under different conditions yields either the amine oxide or 4-aminopyridine [*504-24-5*]. Compound **219** is also a useful intermediate in the synthesis of many 4-substituted pyridines by facile nucleophilic displacement of the nitro group.

**219**    **220**    **221**

## 9.2. Quinoline Derivatives

Quinoline is readily nitrated in the isocyclic ring with mixed acid at 0 °C, yielding a 1:1 mixture of 5-nitroquinoline [607-34-1] (**220**) and 8-nitroquinoline [607-35-2] (**221**). By analogy with pyridine N-oxide, quinoline N-oxide nitrates to give the 4-nitro derivative, which is reduced to 4-nitroquinoline [3741-15-9] or 4-aminoquinoline [578-68-7], depending on the reducing agent. The 5- and 8-nitroquinolines are reduced to 5-aminoquinoline [611-34-7] and 8-aminoquinoline [578-66-5], respectively. All of these derivatives are patented in a variety of uses, many of which depend on the modified metal–ligand behavior compared to quinoline and 8-hydroxy quinoline in, for example, corrosion inhibition applications.

## 9.3. Imidazole Derivatives

2-Nitroimidazole [527-73-1] (**216**) has low toxicity and exhibits broad antibiotic activity. Many substituted nitroimidazoles have been synthesized in an attempt to clarify the structure–activity relationships of such apparently simple molecules. Synthesis of **216** is difficult and requires reaction of 2-aminoimidazole sulfate with sodium nitrite in the presence of excess copper sulfate [79]. 4-Nitroimidazole [3034-38-6] (4-nitroglyoxaline, **222**), containing a proportion of the 5-nitrotautomer, is more readily obtained by direct nitration of imidazole [80].

Alkylation of the sodium derivative of **222** with 2-morpholinoethyl chloride gives nimorazole [6506-37-2] (**223**). Similarly, 2-methylimidazole is nitrated [81] to give 2-methyl-5-nitroimidazole [696-23-1] (**224**), which on alkylation with 2-chloroethanol gives metronidazole [443-48-1] (**225**). Both of these drugs are used in the treatment of amoebiasis and trichomoniasis.

The nitro group is essential for the activity of imidazole compounds but may, however, be the cause of recently suspected carcinogenic potential of certain derivatives.

## 9.4. Furan Derivatives

The use of nitrofurans as antibacterial agents was established in 1944 for the treatment of war wounds. In 1947 the same drugs were found to be effective in the prevention of coccidiosis in chickens and enteritis in swine. These and other veterinary applications led to large-volume usage in animal feed supplements.

**5-Nitro-2-furaldehyde** [698-63-5] (**215**), 2-formyl-5-nitrofuran, $C_5H_3NO_4$, $M_r$ 141, mp 35–36 °C, bp 159–161 °C (1.3 kPa), forms crystals in petroleum ether; it is moderately soluble in water.

*Production.* Furfural is nitrated in acetic anhydride solution at 25–40 °C to give the diacetate [92-55-7] (**226**), which is purified by recrystallization from isopropanol and then used directly for the production of derived drugs [82].

*Uses.* Derivatives of **215** used as antibacterial agents include the semicarbazone nitrofurazone [59-87-0] (**227**) and the other hydrazones nitrofurantoin [67-20-9] (**228**) and furazolidone [67-45-8] (**229**), all containing the same active center. The mode of action is not known but probably involves an unstable reduction product. High activity is offset by the general toxicity of the class, and recent reports of mutagenic and carcinogenic activity have discouraged their use.

## 9.5. Thiophene Derivatives

Thiophenes are used less commonly than furans as chemotherapeutic agents, although nifurzide [39978-42-2] (**230**), obtained by condensation of 5-nitrothiophene-2-carboxylic acid hydrazide with 3-(2-nitro-5-furyl) acrolein, contains both a nitrofuran and a nitrothiophene residue.

Thiophene itself is nitrated to 2-nitrothiophene by gradual addition of a solution of thiophene in acetic anhydride to a mixture of nitric and glacial acetic acids at 10 °C. Attempts to circumvent acetyl nitrate processes are reported to be both difficult and dangerous. Substituted thiophenes more amenable to nitration are limited. 2-Acetylthiophene is nitrated with acetyl nitrate to give 2-nitro-5-acetylthiophene (**231**), the derived amine of which has been compared with 2-amino-3-nitro-5-acetylthiophene (**232**) as a diazo component in azo disperse dyes [83]. The latter is prepared by amination of 2-chloro-3-nitro-5-acetylthiophene; the derived dyes would have been of commercial interest if a more economical route had existed [77].

The Gewald synthesis of substituted 2-aminothiophenes [84] made 2-acetylamino-3-thiophenecarboxylic acid ethyl ester readily available for mononitration, with mixed acid at 0–5 °C, to give 2-amino-3-ethoxycarbonyl-5-nitrothiophene [*42783-04-0*] (**233**) [85] after deacylation. Similarly, the *N*-acetyl derivative of the readily available 2-amino-3-thiophenecarboxylic acid is dinitrated with mixed acid at 0 °C, with decarboxylation and deacylation to give 2-amino-3,5-dinitrothiophene [*2045-70-7*] (**234**) [86]. Monoazo dyes derived from **234** yield important blue and green shades on polyester, e.g., C.I. Disperse Green 9 (**235**) with $\lambda_{max}$ 614 nm (cf. Table 11).

## 9.6. Thiazole Derivatives

**2-Amino-5-nitrothiazole** [*121-66-4*] (**236**), 5-nitro-2-thiazolamine, Enheptin, Entramin, $C_3H_3N_3O_2S$, $M_r$ 145, *mp* 200 °C.

*Production.* Early production of **236** was based on nitration of 2-acetylaminothiazole and careful hydrolysis [87]. Direct nitration of 2-aminothiazole gives a cleaner product and is best achieved by adding 2-aminothiazole nitrate [*57530-25-3*] to concentrated sulfuric acid at 0–10 °C [88]. This process and the nitration of *o*-anisidine are the only large-scale processes based on rearrangement of an amine nitrate, and both are potentially hazardous due to the possibility of runaway exothermic reactions [89].

*Uses.* Although **236** has been used as a selective chemotherapeutic agent, the *N*-formyl derivative, forminitrazole [*500-08-3*] and the *N*-acetyl derivative, acinitrazole [*140-40-9*] (**217**) are more important, being used in the treatment of trichomoniasis and as a veterinary antibacterial, respectively. The derivative niridazole [*61-57-4*] (**237**), used in the treatment of schistosomiasis, is obtained by reaction of **236** with 2-chloroethyl isocyanate followed by ring closure [88]. The antiparasiticthenitrazole [*3810-35-3*] (**238**) is obtained by acylation of **236** with 2-thenoyl chloride in pyridine.

O₂N—[thiazole]—NH₂
**236**

O₂N—[thiazole]—N(CO)NH (imidazolidinone)
**237**

O₂N—[thiazole]—NHCO—[thienyl]
**238**

Production of dye **218** (C.I. Disperse Blue 106) has resulted in the synthesis of many analogous dyes [76], [77], [90], but with only marginal improvement in specific properties. Diazotization of **236** with nitrosylsulfuric acid, followed by coupling with tertiary arylamines, gives azo products in yields of 50–60%. Improvements in both yield and purity are claimed to result from using displacement coupling, for example, with 4-formyl- or 4-carboxylarylamine coupling components [91].

**Benzothiazole and Benzoisothiazole Derivatives.** Nitroamino derivatives of benzothiazole and benzoisothiazole have both been used as diazo components. For example, **239**, derived from 2-amino-6-nitrobenzothiazole [*6285-57-0*], is a valuable bluish red dye for polyester. It is bathochromic to the extent of 68 nm in the visible spectrum, compared with the corresponding dye obtained from 4-nitroaniline [92], thus demonstrating the comparatively low polarizing effect of a nitro substituent in an aromatic ring compared with a nitro substituent in a heterocyclic ring, e.g., in dyes derived from **236**. The acylic substituted products are, however, more thermally stable and have superior lightfastness. Blue and greenish blue dyes based on 7-amino-5-nitrobenz [3,4-c] isothiazole (**240**) are more important because the *o*-quinonoid structure leads to more bathochromic colors than those based on **236**, while having good fastness properties [93]. Synthesis of **240** occurs by treatment of 2-cyano-4-nitroaniline (**130**) with sulfide, followed by oxidative cyclization of the derived thioamide with hydrogen peroxide [94].

O₂N—[benzothiazole]—N=N—[phenyl]—N(C₂H₅)(CH₂CH₂CN)
**239**

O₂N—[benzisothiazole]—NH₂
**240**

# 10. References

General References

[1] B.I.O.S., F.I.A.T. reports: *German Dyestuffs and Dyestuffs Intermediates, Including Manufacturing Processes, Plant Design and Research Data 1945–1948.*
[2] H. E. Fierz-David, L. Blangey: *Fundamental Processes of Dye Chemistry,* Interscience, New York 1949.
[3] K. Venkataraman (ed.): *The Chemistry of Synthetic Dyes,* vol. **1**, Academic Press, New York 1952.
[4] P. H. Groggins (ed.): *Unit Processes in Organic Synthesis,* McGraw Hill, New York 1958.
[5] P. B. P. De La Mare, I. H. Ridd: *Aromatic Substitution, Nitration and Halogenation,* Butterworths, London 1959.
[6] A. V. Topchiev: *Nitration of Hydrocarbons,* Pergamon Press, London 1959.
[7] M. J. Astle: *Industrial Organic Nitrogen Compounds,* Reinhold Publ. Co., New York 1961.
[8] T. Urbanski: *Chemistry and Technology of Explosives,* vol. **1**, Pergamon Press, Oxford 1964.
[9] H. R. Schweizer: *Künstliche organische Farbstoffe und ihre Zwischenprodukte,* Springer Verlag, Berlin 1964.
[10] P. A. S. Smith: *The Chemistry of Open-Chain Organic Nitrogen Compounds* vol. **2**, Benjamin, New York 1966.
[11] N. V. Sidgwick, I. T. Millar, H. D. Springall: *Organic Chemistry of Nitrogen,* 3rd ed., Oxford 1966.
[12] R. O. C. Norman, R. Taylor: *Electrophilic Substitution in Benzeneoid Compounds,* Elsevier, Amsterdam 1969.
[13] H. Feuer (ed.): *The Chemistry of the Nitro and Nitroso Groups,* part 1, Interscience, New York 1969, part 2, 1970.
[14] *Houben-Weyl,* **X/1,** 463–889.
[15] E. G. Hancock (ed.): *Benzene and its Industrial Derivatives,* Benn, London 1975.
[16] K. Schofield: *Nitration and Aromatic Reactivity,* University Press, Cambridge 1979.
[17] *Ullmann,* 4th ed., **17,** 383–416.
[18] S. Patai (ed.): *The Chemistry of Amino, Nitroso and Nitro Compounds and their Derivatives,* parts 1, 2, Wiley-Interscience, New York 1982.
[19] L. Bretherick: *Handbook of Reactive Chemical Hazards,* 3rd ed., Butterworths, London 1985.
[20] H.-G. Franck, J. W. Stadelhofer: *Industrial Aromatic Chemistry,* Springer-Verlag, Berlin 1988.
[21] G. Booth: *The Manufacture of Organic Colorants and Intermediates,* Society of Dyers and Colourists, Bradford 1988.
[22] N. I. Sax, R. J. Lewis: *Dangerous Properties of Industrial Materials,* 7th ed. vol. **3,** Van Nostrand Reinhold, New York 1989.

Specific References

[23] T. Travis: "Early Intermediates for the Synthetic Dyestuffs Industry," *Chem. Ind. London* 1988, 508–514.
[24] J. R. Beck: "Nucleophilic Displacement of the Aromatic Nitro Group," *Tetrahedron* **34** (1978) 2057–2068.
[25] R. J. Gillespie, D. J. Millen, *Rev. Chem. Soc.* **2** (1948) 277–306.
[26] A. I. Titov, *Tetrahedron* **19** (1963) 557–580.

[27]  G. A. Olah: "Preparative and Mechanistic Aspects of Electrophilic Nitration in Industrial and Laboratory Nitrations," *ACS Symp. Ser.* **22** (1976).
[28]  L. M. Stock: "Classic Mechanism for Organic Nitration," *Prog. Phys. Org. Chem.* **12** (1976) 21–47.
[29]  C. Hanson, J. G. Marsland, K. G. Wilson: "Macro-kinetics Applied to Large-Scale Nitrations," *Chem. Ind. (London)* 1966, 675–683.
[30]  R. C. Miller, D. S. Noyce, T. Vermeulen, *Ind. Eng. Chem.* **56** (1964) no. 6, 43–53.
[31]  R. C. Hahn, H. Shosenji, D. L. Strack: "Recent Developments in *ipso*-Nitration" in *Industrial and Laboratory Nitrations, ACS Symp. Ser.* **22** (1976).
[32]  Josef Meissner GmbH & Co., DE 2 921 487, 1979.
[33]  Nobel Chematur AB (Sweden); personal communication.
[34]  G. Erlandsson, *Chem. Ztg.* **104** (1980) 353–358.
[35]  DuPont, US 2 256 999, 1941 (J. B. Castner).
[36]  Am. Cyanamid, US 4 021 498, 1975 (V. Alexanderson, J. B. Trecek, C. M. Vanderwaart).
[37]  DuPont, US 3 981 935, 1975 (R. McCall).
[38]  P. N. Preston, G. Tennant, *Chem. Rev.* **72** (1976) 627–678.
[39]  Bofors, GB 1 207 384, 1966 (C. L. Hakansson et al.).
[40]  G. F. P. Harris: "Isomer Control in the Mononitration of Toluene" in *Industrial and Laboratory Nitrations, ACS Symp. Ser.* **22** (1976).
[41]  G. Booth, *A. I. Chem. Eng. Progress Manual,* vol. **10**, (1976).
[42]  Mobay, US 3 350 466, 1966 (W. K. Menke).
[43]  H. E. Fierz-David, R. Sponagel, *Helv. Chim. Acta* **26** (1943) 98–111.
[44]  Bayer, DE 2 517 437-9, 1975 (H. U. Blank et al.).
[45]  H. C. Brown, W. G. Bonner, *J. Am. Chem. Soc.* **76** (1954) 605–606.
[46]  Sumitomo, JP 7 505 339, 1973 (T. Ikeda, T. Hadano).
[47]  N. R. Ayyanger, A. G. Lugade, *Colourage* **29** (1982) no. 13, 3–9.
[48]  Monsanto, GB 1 420 733; 1 421 107, 1973 (I. Schumacher).
[49]  J. R. Kosak: "Catalytic Hydrogenation of Aromatic Halonitro Compounds," *Ann. N. Y. Acad. Sci.* **172** (1970) 175–186.
[50]  Hoechst, DE 2 156 285, 1971 (K. Baessler).
[51]  N. R. Ayyanger, A. G. Lugade, *Colourage* **29** (1982) no. 26, 3–9.
[52]  G. C. Finger, C. W. Kruse, *J. Am. Chem. Soc.* **78** (1956) 6034–6037.
[53]  C. D. Johnson, A. R. Katritsky, S. A. Shapiro, *J. Am. Chem. Soc.* **91** (1969) 6654–6662.
[54]  P. F. Gordon, P. Gregory: *Organic Chemistry in Colour,* Springer-Verlag, Berlin 1983.
[55]  P. Bersier, L. Valpiana, H. Zubler, *Chem. Ing. Tech.* **43** (1971) 1311–1315.
[56]  J. H. Ridd in T. Urbanski (ed.): *Nitro Compounds,* Pergamon Press, Warsaw-Oxford, 1964.
[57]  C. S. Rondestvedt, *Ind. Eng. Chem. Prod. Res. Dev.* **16** (1977) 177–179.
[58]  Röhm and Haas, DE 2 409 686, 1973 (J. E. Ware, D. L. Pearden, E. E. Kilbourn).
[59]  Sandoz, GB 952 468, 1959.
[60]  ICI, GB 1 220 448, 1967 (T. D. Baron, B. R. Fishwick).
[61]  ICI, GB 1 226 950, 1967 (T. D. Baron, B. R. Fishwick).
[62]  M. Lounasmaa, *Acta Chem. Scand.* **22** (1968) 2388–2390.
[63]  Bayer, US 3 079 297, 1960 (E. Schraufstaetter, R. Goennert).
[64]  Sandoz, GB 855 488, 1958.
[65]  Sandoz, GB 1 051 264, 1963.
[66]  Kuhlmann, FR 1 458 333, 1965 (M. Jirou, J. Leroy).
[67]  Bayer, GB 1 125 683–4, 1966 (A. Gottschlich et al.).

[68] Witco Chem., DE 2 328 574, 1972 (A. Benson, M. L. Mausner).
[69] M. Zahradnik: *Production and Application of Fluorescent Brightening Agents,* Wiley-Interscience, Chichester 1982.
[70] Bayer, DE 2 910 458, 1979 (H. Nickel).
[71] N. C. Rose, *J. Heterocycl. Chem.* **6** (1969) 745–746.
[72] Bayer, DE 3 012 800, 1980 (R. Kopp et al.).
[73] R. J. Pollitt, *J. Chem. Soc.* 1965, 6198–6201.
[74] A. H. Ford-Moore, H. N. Rydon, *J. Chem. Soc.* 1946, 679–681.
[75] K. Miura, H. K. Reckendorf, *Prog. Med. Chem.* **5** (1967) 320–381.
[76] Eastman Kodak, US 2 659 719, 1953; US 2 683 708-9, 1954; US 2 730 523, 1956; US 2 746 953, 1956 (J. B. Dickey, E. B. Towne).
[77] M. A. Weaver, L. Shuttleworth, *Dyes and Pigm.* **3** (1982) 81–121.
[78] H. J. Denttertog, J. Overhoff, *Rec. Trav. Chim. Pays Bas* **69** (1950) 468–473.
[79] A. G. Beaman et al., *J. Am. Chem. Soc.* **87** (1965) 389–90.
[80] BASF, DE 2 208 924, 1972 (H. Spaenig et al.).
[81] Zaklady Chem., GB 1 418 538, 1973.
[82] H. J. Sanders et al., *Ind. Eng. Chem.* **47** (1955) 358–367.
[83] J. B. Dickey et al., *J. Soc. Dyers Colour.* **74** (1958) 123–132.
[84] K. Gewald, *Chem. Ber.* **98** (1965) 3571–3577.
[85] ICI, GB 1 394 365, 1972.
[86] ICI, GB 1 394 367-8, 1972 (D. B. Baird et al.).
[87] Monsanto, US 2 573 641, 1951; US 2 573 656-7, 1951 (H. L. Hubbard, G. W. Steahley).
[88] M. Wilhelm et al., *Helv. Chim. Acta* **49** (1966) 2443–2452.
[89] L. Silver, *Chem. Eng. Prog.* **63** (1967) no. 8, 43–49.
[90] J. B. Dickey et al., *J. Org. Chem.* **24** (1959) 187–196.
[91] Eastman Kodak, US 4 247 458, 1981 (L. Shuttleworth).
[92] M. F. Satori, *J. Soc. Dyers Colour.* **83** (1967) 144–146.
[93] BASF, GB 1 112 146, 1965 (M. Seefelder et al.).
[94] H. G. Wippel, *Melliand Textilber.,* **50** (1969) 1090–1096.

# Oxalic Acid

WILHELM RIEMENSCHNEIDER, Hoechst AG, Frankfurt, Federal Republic of Germany

MINORU TANIFUJI, Ube Industries, Tokyo, Japan (Section 4.4, Chap. 7 in part)

| | | | | |
|---|---|---|---|---|
| 1. | Introduction ............ 3577 | 6. | Uses ................ 3590 |
| 2. | Physical Properties ....... 3578 | 6.1. | Metal Treatment ......... 3591 |
| 2.1. | Anhydrous Oxalic Acid..... 3578 | 6.2. | Textile Treatment......... 3591 |
| 2.2. | Oxalic Acid Dihydrate ..... 3579 | 6.3. | Bleaching Agents......... 3591 |
| 3. | Chemical Properties....... 3580 | 6.4. | Chemical Uses .......... 3591 |
| 4. | Production Processes and Raw Materials.............. 3581 | 7. | Economic Aspects ........ 3592 |
| 4.1. | Oxidation of Carbohydrates .. 3582 | 8. | Storage, Handling, Transportation, Waste Disposal ............... 3593 |
| 4.2. | Oxidation of Ethylene Glycol . 3583 |
| 4.3. | Oxidation of Propene....... 3585 | 9. | Derivatives ............ 3593 |
| 4.4. | Production from Carbon Monoxide .............. 3586 | 9.1. | Salts ................ 3593 |
| 4.5. | Production of Anhydrous Oxalic Acid ............. 3589 | 9.2. | Organic Derivatives ....... 3595 |
| | | 10. | Toxicology ............ 3597 |
| 5. | Chemical Analysis ........ 3590 | 11. | References............. 3597 |

# 1. Introduction

Oxalic acid, ethanedioic acid, acidum oxalicum, has the structural formula

$$\text{HO}-\overset{\overset{\displaystyle O}{\|}}{C}-\overset{\overset{\displaystyle O}{\|}}{C}-\text{OH}$$

and is the simplest saturated dicarboxylic acid [1]. The compound exists in anhydrous form [144-62-7] or as a dihydrate [6153-56-6]. The anhydrous acid is not found in nature and must be prepared from the dihydrate even when produced industrially. Oxalic acid is widely distributed in the plant and animal kingdom (nearly always in the form of its salts) and has various industrial applications.

The acidic potassium salt of oxalic acid is found in common sorrel (Latin: oxalis acetosella) and the name oxalic acid is derived from that plant. Examples of plants in which oxalic acid occurs (in the form of potassium, sodium, calcium, magnesium salts, or iron complex salts) are given below (oxalic acid content in milligrams per 100 g dry weight):

| | |
|---|---|
| Spinach | 460–3200 |
| Rhubarb | 500–2400 |
| Chard | 690 |
| Parsley | 190 |
| Beets | 340 |
| Cocoa | 4500 |
| Tea | 3700 |
| Beet leaves | up to 12 000 |

Oxalic acid is formed in plants through incomplete oxidation of carbohydrates, e.g., by fungi (aspergillus niger) or bacteria (acetobacter) and in the animal kingdom through carbohydrate metabolism via the tricarboxylic acid cycle. The urine of humans and of most mammals also contains a small amount of calcium oxalate. In pathological cases, an increased calcium oxalate content in urine leads to the formation of kidney stones [2]. Calcium and iron(II) oxalates are also found as minerals.

Oxalic acid is one of the oldest known acids and was already identified in the potassium salt of sorrel by WIEGLEB in 1769. The compound also has historical significance in chemistry because WOEHLER prepared oxalic acid by hydrolysis of cyanogen [460-19-5] in 1824, thereby synthesizing the first natural product, even before his famous urea synthesis in 1828.

# 2. Physical Properties

Both the anhydrous and dihydrated forms of oxalic acid form colorless and odorless crystals.

## 2.1. Anhydrous Oxalic Acid

Anhydrous oxalic acid [144-62-7] exists as rhombic crystals in the α-form and as monoclinic crystals in the β-form [3]. These forms differ mainly in their melting points. The slightly stable β-form changes into the α-form at 97 °C and 0.2 bar. Anhydrous oxalic acid is prepared by dehydration of the dihydrate through careful heating to 100 °C. It is then sublimated in a dry air stream. The sublimation is fast at 125 °C and can be carried out at temperatures up to 157 °C without decomposition. The dehydration can also be accomplished by azeotropic distillation with benzene or toluene [4]. Anhydrous oxalic acid is slightly hygroscopic, it absorbs water from moist air ("weathers") to form the dihydrate again. The hydration occurs very slowly because of surface caking. The physical and thermal chemical properties of anhydrous oxalic acid may be summarized as follows:

| | |
|---|---|
| $M_r$ | 90.04 |
| mp | |
| $\alpha$ | 189.5 °C |
| $\beta$ | 182 °C |
| Decomposition temperature | 187 °C |
| Density $d_4^{20}$ | |
| $\alpha$ | 1.900 g/cm$^3$ |
| $\beta$ | 1.895 g/cm$^3$ |
| Refractive index $\beta$, $n_4^{20}$ | 1.540 |
| Vapor pressure (solid, 57–107 °C) | log $P = -(4726.95/T) + 11.3478$ kPa |
| Specific heat (solid, −200 to 50 °C) | $c_p = 1.084 + 0.0318\,t$ J/g |
| Heat of combustion $\Delta H_c°$ (25 °C) | −245.61 kJ/mol |
| Standard enthalpy of formation $\Delta H_f°$ (25 °C) | −826.78 kJ/mol |
| Standard free energy of formation $\Delta G_f°$ (25 °C) | −697.91 kJ/mol |
| Heat of solution (in water) | −9.58 kJ/mol |
| Heat of sublimation | 90.58 kJ/mol |
| Heat of decomposition | 826.78 kJ/mol |
| Entropy, $S°$ (25 °C) | 120.08 J mol$^{-1}$ K$^{-1}$ |
| Thermal conductivity (0 °C) | 0.9 W m$^{-1}$ K$^{-1}$ |
| Ionization constant | |
| $K_1$ | $6.5 \times 10^{-2}$ |
| $K_2$ | $6.0 \times 10^{-5}$ |

## 2.2. Oxalic Acid Dihydrate

Oxalic acid dihydrate [6153-56-6], HOOC–COOH · 2 H$_2$O is the industrially produced and usual commercial form of oxalic acid. The compound forms colorless and odorless prisms or granules that contain 71.42 wt% oxalic acid and 28.58 wt% water. Oxalic acid dihydrate is stable at room temperature and under normal storage conditions. The most important physical properties are as follows:

| | |
|---|---|
| $M_r$ | 126.07 |
| mp | 101.5 °C |
| Density $d_4^{20}$ | 1.653 g/cm$^3$ |
| Refractive index $n_4^{20}$ | 1.475 |
| Standard enthalpy of formation $\Delta H_f°$ (18 °C) | −1422 kJ/mol |
| Enthalpy of solution (in water) | −35.5 kJ/mol |
| pH (0.1 M solution) | 1.3 |
| Water vapor partial pressure | |
| at 20 °C | 2.5 kPa |
| at 50 °C | 20 kPa |
| Bulk density, regular product | 0.977 g/cm$^3$ |
| Bulk density, coarse product | 0.881 g/cm$^3$ |

The solubility in water and the density of these solutions are presented in Table 1 [5]. Oxalic acid is readily soluble in polar solvents such as alcohols (although partial

**Table 1.** Solubility of oxalic acid in water at different temperatures and corresponding densities

| Temperature, °C | Solubility, g/100 g* | Relative density $d_4^{17.5}$ |
|---|---|---|
| 0    | 3.5   | 1.016 |
| 10   | 5.5   | 1.025 |
| 17.5 | 8.5   | 1.038 |
| 20   | 9.5   |       |
| 30   | 14.5  |       |
| 40   | 22    |       |
| 50   | 32    |       |
| 60   | 46    |       |
| 80   | 85    |       |
| 90   | 120   |       |

\* Anhydrous basis.

esterification occurs), acetone, dioxane, tetrahydrofuran, and furfural. Oxalic acid is sparingly soluble in diethyl ether (1.5 g oxalic acid dihydrate in 100 g ether at 25 °C), and insoluble in benzene, chloroform, and petroleum ether. The ionization constants show that oxalic acid is a strong acid. The value of $K_1$ is comparable to that of mineral acids and the value of $K_2$ corresponds to ionization constants of strong organic acids, for example, benzoic acid.

## 3. Chemical Properties

In the homologous series of dicarboxylic acids, oxalic acid, the first member, shows unique behavior because of the interaction of the neighboring carboxylate groups. This results in an increase in the value of the dissociation constant and in the ease of decarboxylation: Upon rapid heating to 100 °C oxalic acid decomposes into carbon monoxide, carbon dioxide, and water with formic acid as an isolable intermediate. In aqueous solution decomposition is induced by light, and to a much greater extent by $\gamma$- or X-rays (to carbon monoxide, carbon dioxide, formic acid, and occasionally hydrogen). This decomposition is catalyzed by the salts of heavy metals, for example, by uranyl salts. Oxalic acid cannot form an intramolecular anhydride. Upon heating to over 190 °C or warming in concentrated sulfuric or phosphoric acid, oxalic acid decomposes to carbon monoxide, carbon dioxide, and water: this decomposition is not exothermic.

Ease of reaction with oxygen is also a typical characteristic of oxalic acid:

$$HOOC-COOH + 1/2\ O_2 \longrightarrow 2\ CO_2 + H_2O$$

This reaction can be exploited in quantitative analysis (Manganometry, see Chap. 5).

The reducing properties of oxalic acid (which itself is oxidized to the harmless end products carbon dioxide and water) form the basis for the variety of practical applications (see Chap. 6). Oxalic acid is also oxidized relatively easily to carbon dioxide by many other oxidizing agents in addition to air, especially in the presence of the salts of heavy metals. Oxalic acid is easily esterified, whereby two types, the acidic mono or neutral diesters can result. These esters are applied as intermediates in chemical syntheses. They react relatively easy with water, ammonia, or amines to afford the corresponding acyl derivatives.

Important chemical characteristics are also demonstrated by the metal salts of oxalic acid. These exist in two types—the acidic and neutral salts. The alkali metal and iron (III) salts are readily soluble in water. All other salts are sparingly soluble in water. The near complete insolubility of the alkaline-earth salts of oxalic acid, especially of calcium oxalate, finds some applications in quantitative analysis. When heated all these metal salts lose carbon monoxide. Other salts which are easier decomposable lose carbon dioxide in addition. The alkali and alkaline-earth salts form carbonates under these conditions. Manganese, zinc, and tin salts form oxides; iron, cadmium, mercury, and copper salts form mixtures of oxides and metals. Nickel, cobalt, and silver salts afford pure metals. Anhydrous fusion of oxalates with alkali yield carbonates and hydrogen. For review see [6].

# 4. Production Processes and Raw Materials

There are many industrial processes for the production of oxalic acid, which have been carried out by a large number of companies and are in part still used today. In recent years both the number of firms producing oxalic acid and the number of starting materials used for its production have been reduced. Generally, only three classes of compounds are currently employed as raw materials for the production of oxalic acid. These are:

carbohydrates (including ethylene glycol and molasses)
olefins
carbon monoxide

Nevertheless, three outdated methods, which have been abandoned in important industrial nations for economic reasons, are described briefly [7], because in the third world small firms might exist that still employ these production methods:

1) The sodium formate method, which is also based on carbon monoxide as the raw material, but involves very cumbersome procedures of questionable technical safety that are no longer economically feasible (heating to 400 °C, elimination of hydrogen, calcium precipitation, and precipitation with sulfuric acid). Nevertheless, this process is probably still being used in China.
2) The alkali fusion of cellulose, which is the oldest production method [8]. Although cellulose-containing material from every source can be used in this method, this

process has been discontinued because of the low yields, large amounts of required alkali and sulfuric acid, and the large amounts of wastewater.

3) The isolation of oxalic acid as a byproduct of carbohydrate fermentation, for example, in the synthesis of citric acid [9]. The process involves the biotechnical exploitation of the tricarboxylic acid cycle through enzymatic cleavage of sugars. The intermediary formed oxalacetic acid esters and oxalosuccinic acid (1-oxopropane-1,2,3-tricarboxylic acid) esters are cleaved by hydrolases. Especially because of the handling of large amounts of water, this process is no longer economically feasible.

In all industrial processes, oxalic acid is produced as the crystalline dihydrate. In the following sections four methods are described that are currently used industrially. Three of them are based on a nitric acid oxidation and one is a synthesis using carbon monoxide. A new review of oxalic acid production from various sources is given in [10].

## 4.1. Oxidation of Carbohydrates

This method, the oldest of chemical productions, was in principal discovered by SCHEELE more than 200 years ago (1776). He oxidized sugar with concentrated nitric acid to oxalic acid. However, it was not until ca. 1940 that this process gained industrial importance when the nitrogen oxides produced in the reaction could be recovered and recycled. In Germany, I.G. Farben produced 2000 t of oxalic acid per year by this method until 1944 [11]. Currently this process is employed in Brazil, Czechoslovakia, China, and in several Eastern European countries. Raw materials employed in this process include sugar, glucose, fructose, corn starch, wheat starch, reclaimed starch, potato starch, corncobs, tapioca, and molasses. Choice of starting material depends on availability and price and on optimization of the reaction process involved. For example, depending on the starting materials foam, greases, slime, and varying yields can be expected and must be dealt with accordingly [1]. Recently molasses and other agricultural waste have gained increasing importance as raw materials. These materials are not only inexpensive, but must be further processed in any case due to environmental reasons. Molasses contains, as a production residue of sugar refinement, many nitrogen-containing compounds, which are, however, for the most part removed during the nitric acid oxidation. This explains the difficulties encountered with excessive foam formation during production that must be overcome. If starch products are used as raw materials they must be hydrolyzed to glucose in an extra batchwise step prior to the actual oxidation [12], [13]. This is carried out by boiling the starch suspension in the presence of oxalic acid or sulfuric acid for about six hours. The oxidation of the resulting glucose and the following steps are then essentially continuous operations. The oxidation occurs according to the following idealized equation:

$$C_6H_{12}O_6 + 6\,HNO_3 \xrightarrow{V_2O_5,\,Fe^{3+}} 3\,(COOH)_2 + 6\,NO + 6\,H_2O$$

In fact besides NO other nitrogen oxides such as $NO_2$, $N_2O$, $N_2O_5$ are formed.

**Allied Chemical Process.** The Allied Chemical process is schematically presented in Figure 1. The concentration of the glucose solution obtained by hydrolysis of starch is approximately 50–60 wt%. This material, along with the recycled supernatent from the crude centrifugation, is placed in a reactor with ca. 50% sulfuric acid, vanadium pentoxide (0.001 – 0.05 wt%) and iron(III) sulfate (0.39–0.8 wt%), based on the total mass of the mixture. Nitric acid (65%) is then added slowly under vigorous stirring in a tightly controlled temperature range around 70°C— the temperature range differs according to the raw material being used. The strongly exothermic reaction must be well cooled. Air is simultaneously blown into the reactor to support the reaction and to remove NO and other nitrogen oxides of low oxidation number. The nitrogen oxides expelled are collected in an adsorption system and are recycled after oxidation. Crude oxalic acid is obtained after cooling and centrifugation of the reaction mixture. The crude acid is again dissolved in hot water, put through a grease separator and recrystallized. After a second centrifugation and drying, oxalic acid dihydrate is obtained with a purity >99%. Precise control of the reaction parameters is critical for a good yield [14]. For good quality raw materials ca. 65% of the theoretical yield is obtained. The absent ca. 35% of the yield mainly results from losses through $CO_2$ formation. This process is employed today in Spain, Brazil, The People's Republic of China, Taiwan, Korea, and India.

## 4.2. Oxidation of Ethylene Glycol

The oxidation of ethylene glycol with nitric acid is a one-step process. A flow chart is shown in Figure 2. An oxidizing mixture of 30–40% sulfuric acid and 20–25% nitric acid is used. The oxidation is carried out in the presence of vanadium pentoxide and iron (III) salts, at 50–70°C and atmospheric pressure. Only $CO_2$ is formed as a byproduct. This original process has been improved by the Mitsubishi Gas Company and in Japan most of the oxalic acid is produced this way [15]. The improved process operates at about 10 bar in a pressure reactor, in which ca. 60% nitric acid and possibly also sulfuric acid at 80°C, is pressurized with oxygen at 3–10 bar. Ethylene glycol is then continually pumped in and the temperature and pressure are maintained by addition of oxygen. The exothermic reaction must be held at a temperature of 50–70°C by effective cooling. The yield of oxalic acid, based on ethylene glycol, is 90–94% of the theoretical value. The nitric acid used can be recovered almost completely. This indicates that the nitrogen oxides and nitric acid only act as catalysts and the actual oxidation has been effected by oxygen. The idealized overall equation for this reaction is:

$$\begin{array}{c} CH_2OH \\ | \\ CH_2OH \end{array} + 2\,O_2 \xrightarrow[V_2O_5,\,Fe^{3+}]{NO_2} \begin{array}{c} COOH \\ | \\ COOH \end{array} + 2\,H_2O$$

Other suitable starting materials are propylene glycol, acetaldehyde, or glycolic acid.

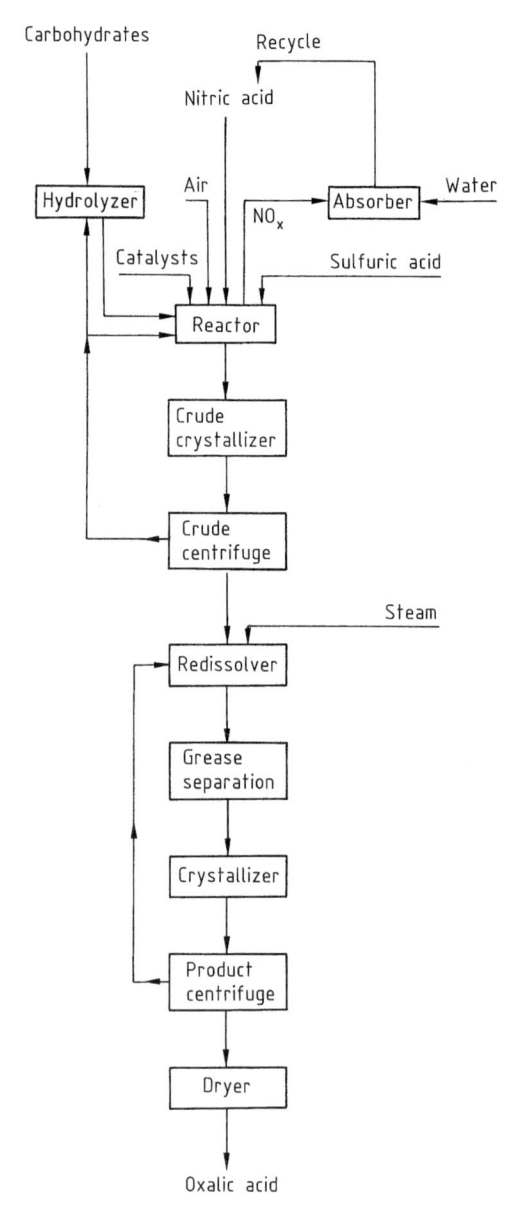

**Figure 1.** Production of oxalic acid by oxidation of carbohydrates (Allied Chemical process)

Vanadium compounds (0.001 – 0.1 wt %) are recommended as promoters. Sodium nitrite, formic acid, or formaldehyde are recommended as initiators [16]. Ube Industries in Japan have further developed the process so that it operates only with nitric acid (no sulfuric acid), however, pressure is required. In an oxalic acid process developed in the Soviet Union a mixture of nitric and sulfuric acids is used. The reaction is performed at 1.3 – 4.0 kPa (the pressure is less than that employed in the Mitsubishi process) to remove $NO_x$ more effectively.

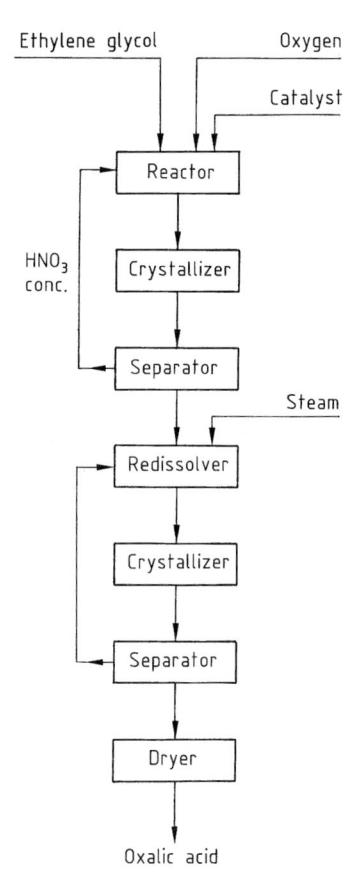

**Figure 2.** Production of oxalic acid from ethylene glycol

Although the ethylene glycol process clearly affords the best yields by a wide margin, the corn starch and molasses oxidations are still more advantageous from an economic point of view. Ethylene glycol oxidation is carried out only by Mitsubishi in Japan.

## 4.3. Oxidation of Propene

The second most important industrial process for the production of oxalic acid (after carbohydrate oxidation) is the reaction of propene with nitric acid. The worldwide production of oxalic acid by this method in 1978 was 65 000 t/a. Pure propene or propene rich fractions from petrochemical refineries can be used as feedstocks. This earlier developed reaction was significantly improved by Rhône-Poulenc in France by technical engineering and construction improvements and strict observance of the reaction conditions. These refinements increased the yield and solved the serious safety problems [17]. Rhône-Poulenc is reputed to be the largest oxalic acid producer in the world. The two-step process can be represented by the following equations approximate stoichiometry:

Initial stage:

$$CH_3CH=CH_2 + 3\,HNO_3 \longrightarrow CH_3\underset{ONO_2}{CHCOOH} + 2\,NO + 2\,H_2O$$

Propene

α-Nitratolactic acid

Final stage:

$$CH_3\underset{ONO_2}{CHCOOH} + 5/2\,O_2 \longrightarrow HOOC-COOH + CO_2 + HNO_3 + H_2O$$

Oxalic acid

**Process Description** (see Fig. 3). Propene is introduced into a solution of nitric acid (50–75 wt%) in the first reactor at 10–40 °C, while keeping the molar ratio of propene to nitric acid at 0.01/1.0 – 0.5/1.0 to produce watersoluble intermediates of α-nitratolactic acid and lactic acid. In the second step the solution of these partially oxidized products is treated with oxygen at 45 – 100 °C in the presence of a catalyst. Oxalic acid is formed, crystallized, and filtered. The overall yield based on propene is claimed to be ca. 90% and the conversion of propene is ca. 100%. The nitrogen oxides are removed from the second reactor and recovered. The consumption of nitric acid is 1.2 kg per kg of oxalic acid dihydrate produced. The process can also be employed for the production of lactic acid instead of oxalic acid. An improvement in the yield is achieved by addition of catalysts (salts or compounds containing Fe, Al, Cr, Sn, Bi, or I). In the chemical industry chromium(III) nitrate, iron(III) nitrate or tin(II) chloride are employed in quantities of 0.4 – 0.5 wt% (based on the metal) and dissolved in 65% nitric acid. The scope and limitations of this process are restricted by safety problems because the nitrates of α-hydroxycarboxylic acids that are formed as intermediates from α-olefins and $NO_2$ are unstable and lead to uncontrollable decomposition and violent explosions [18]. As a result many technical difficulties need to be overcome to carry out these processes. Today this process is only operational by Rhône-Poulenc. Analogous to propene, ethylene can also be oxidized to oxalic acid.

## 4.4. Production from Carbon Monoxide
[19], [20]

The newest process for the industrial production of oxalic acid was developed by Ube Industries. Their synthesis is performed in two steps. In the first step CO and a lower alcohol are reacted under pressure and in the presence of a catalyst to form the corresponding diester of oxalic acid. This diester is hydrolyzed in the second step to oxalic acid and the original alcohol which is then recycled. Palladium on charcoal and alkyl nitrites are employed as catalysts and the reaction is carried out at 10 – 11 MPa. *n*-Butanol is the preferred alcohol and is simultaneously used as the solvent. This process has the advantage of inexpensive starting materials, however, it is disadvantageous with regard to the much higher capital investment (high pressure) and utility costs.

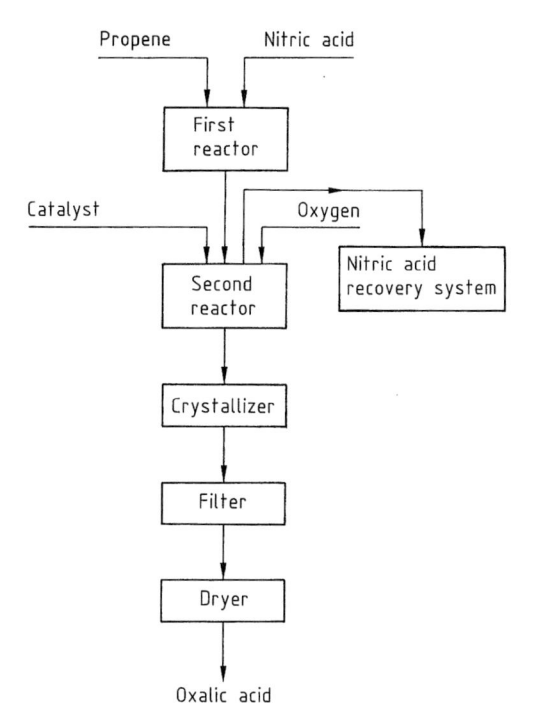

**Figure 3.** Production of oxalic acid from propene

$$2\,CO + 2\,ROH + 1/2\,O_2 \xrightarrow{Pd/C,\ RONO} \begin{array}{c} COOR \\ | \\ COOR \end{array} + H_2O \quad (1)$$

$$\Delta H = -52.8 \text{ kJ/mol}$$
Oxalic acid

$$\begin{array}{c} COOR \\ | \\ COOR \end{array} + 2\,H_2O \longrightarrow \begin{array}{c} COOH \\ | \\ COOH \end{array} + 2\,ROH \quad (2)$$

New perspectives and developments are given in [21]. The use of palladium catalysts is discussed in [22].

The catalytic action of the alkyl nitrite on the first reaction is believed to be the following:

$$2\,RONO + 2\,CO \longrightarrow (COOR)_2 + 2\,NO \quad (3)$$
$$2\,NO + 1/2\,O_2 + 2\,ROH \longrightarrow RONO + H_2O \quad (4)$$

Butyl nitrite and methyl nitrite are the preferred alkyl nitrites for industrial purposes. The first plant using butyl nitrite in the production of oxalic acid was constructed in 1978 (6000 t/a oxalic acid); this plant still operates satisfactorily. The advantages of using butyl nitrite are as follows:

1) It functions not only as reaction component but also as a dehydrating agent. Butyl nitrite has a higher boiling point than water and butyl nitrite and water form an azeotropic mixture with a lower boiling point than butyl nitrite. Thus, the water formed can be removed as an azeotropic mixture and most of the butyl nitrite remains in the system. Butyl nitrite is recovered from the azeotropic mixture and reused.
2) Butyl nitrite is finally hydrolyzed to butanol which can easily be separated from oxalic acid solution by phase separation.

A flow chart of the process using butyl nitrite is given in Figure 4. A circulation liquid containing butyl nitrite, butanol, and the palladium catalyst and a circulation gas containing carbon monoxide and oxygen are pressurized and fed to a reaction tower. Here dibutyl oxalate is formed and butyl nitrite is consumed (Eq. 3). Butyl nitrite is regenerated simultaneously (Eq. 4). Both reactions are exothermic. The selectivity for dibutyl oxalate increases with increasing reaction pressure and decreasing reaction temperature; however, the velocity of the reaction increases with increasing temperature. Thus, the reaction is conducted industrially at 10–11 MPa and 90–100 °C.

Gas circulation is carried out in order to hold the oxygen concentration below the explosion limit during the reaction and to improve the carbon monoxide utilization and gas–liquid contact rate.

The solution released from the tower is flashed to evaporate the carbon dioxide gas formed as a byproduct. The reaction water is removed by azeotropic distillation with butyl nitrite and most of the solution is recycled to the reaction tower. Part of the solution is further processed: it is first filtered to separate the catalysts; then butyl nitrite, and butanol are removed to yield pure dibutyl oxalate. The palladium catalyst, butyl nitrite, and butanol are recovered and returned to the circulation liquid, which is also supplied with make-up butanol and nitric acid. The circulation liquid is then pressurized and fed back to the reaction tower.

The refined dibutyl oxalate is mixed with the mother liquor from the oxalic acid crystallization and hydrolyzed at about 80 °C into oxalic acid and butanol. The oxalic acid solution and butanol are separated by phase separation. Then oxalic acid dihydrate is crystallized from the oxalic acid solution, and butanol is recycled to the process after refining. Recently, a gas-phase process using methyl nitrite has been established [21], [22]. This process employs a lower reaction pressure (390 kPa) and is more economic than the liquid-phase method because of the following reasons:

1) The consumption of electric power is greatly reduced owing to the low pressure operation (about 60 % saving).
2) Dimethyl oxalate formation and methyl nitrite regeneration reaction are separated; hence, side-reactions are decreased and the yield of dimethyl oxalate is increased (98 %).
3) Inexpensive methanol is used instead of butanol.

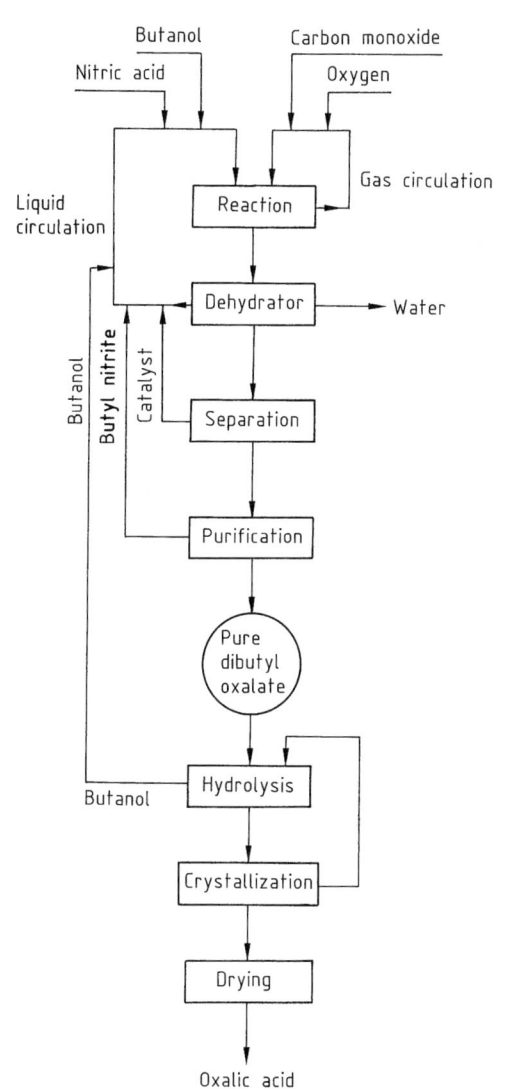

**Figure 4.** Production of oxalic acid from carbon monoxide an butanol using butyl nitrite as catalyst

## 4.5. Production of Anhydrous Oxalic Acid

In all processes oxalic acid is obtained as the dihydrate. Anhydrous oxalic acid can be obtained by heating oxalic acid dihydrate to ca. 3 °C below its melting point or by azeotropic distillation with benzene or toluene. The demand for anhydrous oxalic acid is, however, very low.

# 5. Chemical Analysis

**Qualitative Analysis.** Upon heating oxalic acid with diphenylamine a blue color results, the so-called "aniline blue". Analysis limit: 5 µg [23].

**Quantitative Analysis.** In addition to simple acid–base titration, which of course can only be conducted in the absence of other acids, four important analytical methods exist.

*Manganometry.* As the reaction of oxalic acid and $KMnO_4$ is quantitative, oxalic acid can be easily titrated with a standard solution of $KMnO_4$.

$$5\,(COOH)_2 + 2\,KMnO_4 + 3\,H_2SO_4 \longrightarrow K_2SO_4 + 2\,MnSO_4 + 10\,CO_2 + 8\,H_2O$$

This method can only be used in the absence of other reducing agents.

*Gravimetry.* Because of the insolubility of calcium oxalate in water (100 g of water at 18 °C dissolves only 0.0006 g of calcium oxalate), oxalic acid can be very easily and quantitatively precipitated out of aqueous solutions, e.g., with calcium chloride. The oxalic acid content of the precipitate is determined either gravimetrically (after ashing as CaO) or manganometrically.

*Determination by HPLC.* In this determination the oxalic acid must be esterified, separated by HPLC and analyzed by gas chromatography subsequently [24].

*Enzymatic Determination.* This method is usually used for analysis of oxalic acid in biological material, such as in fruit or vegetables and in body fluids. Either oxalate oxidase [25], [26], oxalate decarboxylase, or formate dehydrogenase are employed [27].

# 6. Uses

Apart from the application of oxalic acid as a synthetic intermediate, the properties of oxalic acid or its salts have generally been exploited as reducing agents (in which case the harmless final products of carbon dioxide and water are formed), as precipitants for calcium ions, or as complexing agents for the salts of heavy metals. The principal applications worldwide can be divided into the following groups in percents of total use:

| | |
|---|---|
| Metal treatment | ca. 28% |
| Textile treatment | ca. 25% |
| Bleaching agents | ca. 20% |
| Chemical uses | ca. 27% |

## 6.1. Metal Treatment

Sodium, potassium, or ammonium oxalates form soluble double salts with heavy metals, especially iron. Iron (III) oxalate is the only readily soluble heavy metal salt. Oxalic acid salts are therefore used (1) for rust removal, e.g., from cooling systems (automobile radiators), boilers, or from steel plates before phosphating, (2) for removal of iron veins in marble, (3) as a constituent of metal cleaners with rust protection, especially for use on copper, silver, or aluminum, (4) in anodizing of aluminum, for the formation of very hard, abrasion and corrosion protective coatings, where through the use of additives different surface colors can be imparted to aluminum [28], [29]. Oxalic acid is also used for the cleaning of materials for electronic devices, and as coatings for protection of stainless steel, nickel alloys, chromium, alloy steels, and titanium [30], [31].

## 6.2. Textile Treatment

Oxalic acid is used as mordant for the printing and dyeing of wool and cotton. It serves as a stripping agent for wool colors for special pattern affects, as auxilliary in indigo caustic discharge printing, and is used in vat dyeing as a reducing agent for potassium dichromate. In laundries oxalic acid is used for the removal of rust stains and the neutralization of alkalinity. It also has an antibacterial effect [32]. Oxalic acid can be employed as a catalyst for cross linking of textile finishing agents in the permanent pressing of cellulose fabrics and for flameproofing [33].

## 6.3. Bleaching Agents

Oxalic acid is used in the tanning and bleaching of leather, and in bleaching of cork, wood (especially veneered wood), straw, cane, feathers as well as natural and synthetic waxes. The majority of oxalic acid is used for the bleaching of pulpwood. In leather processing oxalic acid is used in whitening instead of sodium dithionite.

## 6.4. Chemical Uses

**Preparation of Esters and Salts.** Because of the two carboxyl groups oxalic acid forms acid esters and salts as well as normal esters and salts.

**Reagent in Chemical Synthesis.** Since oxalic acid is a strong dehydrating agent it can be used in condensation reactions, in the esterification of secondary alcohols to esters, or in cyclization reactions [34].

**Concentration of Rare Earth Elements.** Oxalic acid is used for the concentration and isolation of rare earth elements from aqueous solutions that are formed during the digestion of lanthanide ores (for example, monazite) with sulfuric, hydrochloric, or phosphoric acids. The precipitation of oxalates is almost quantitative even from strongly acidic solutions. After precipitation the oxalates are converted to their oxides by heating and processes further in this form [35], [36].

**Other Uses.** Small particles or powders, which are prepared from oxalates, are used for magnetic recording tapes. Oxalic acid is used for the preparation of polyester oxalates and copolyoxamides [37], [38]; for the production of inks, pigments, and paints [39]; and for the preparation of catalysts especially those which contain precious metals and as redox initiators.

# 7. Economic Aspects

Specification for commercial-grade oxalic acid dihydrate is a purity of $>99.5\%$. Supply and demand in the world market is presented in Table 2. No significant increase in demand is anticipated since no new application areas are indicated. The average prices per kilogram in the United States (truckload quantities) in 1988 were [40]:

| | |
|---|---|
| Oxalic acid | $1.10/kg |
| Ammonium oxalate | $3.80/kg |
| Ammonium iron (III) oxalate | $4.40/kg |
| Potassium oxalate | $4.60/kg |
| Sodium oxalate | $5.50/kg |

The most important producers of oxalic acid are:

| | |
|---|---|
| Western Europe: | Rhône-Poulenc (Chalampé, France) |
| | Soc. Franc. Hoechst (Trosly–Breuil, France) |
| | EniChem Sintesi (Pieve-Vergonte, Spain) |
| | DAVSA (Vimbodi, Spain) |
| Japan: | Mitsubishi Gas (Yokkaichi, Mie) |
| | Ube Industries (Ube, Yamaguchi) |
| Taiwan: | Uranus Chemicals (Hsinchu, Hsien) |

# 8. Storage, Handling, Transportation, Waste Disposal

Dry oxalic acid is packed and sold in polyethylene lined multilayered bags or in polyethylene or poly(vinyl chloride) vessels or drums. Reinforced fiberglass plastics are also used commonly and are recommended for storage.

Oxalic acid is not flammable and also not self inflammatory up to 400 °C. It is not shock sensitive and does not cause dust explosions. Oxalic acid is a weakly dangerous substance for water (hazard class 1) and exhibits low toxicity for fish and bacteria. Labelling according to GefStoffV: $X_n$ (less poisonous). R-phrase: 21/22 (dangerous to health through contact with the skin and through ingestion); S-phrase: 2, 24/25. For inhalation protection masks with fine particle filters (filter class P 2) and protective gloves made from PVC or rubber are recommended [41].

Neither oxalic acid nor its solutions should be discarded in the environment before proper treatment. Common treatments include incineration or heating with sulfuric acid to convert it to carbon monoxide, carbon dioxide, and water. Small amounts of oxalic acid can be treated with potassium permanganate to give carbon dioxide. Neutralization with lime results in precipitation of insoluble calcium oxalate. The latter can be safely deposited.

# 9. Derivatives

## 9.1. Salts

In addition to neutral and acid salts, oxalic acid also forms many complex compounds that easily crystallize. Simple oxalates of nearly all organic and inorganic bases are known. The three most important salts of oxalic acid are potassium hydrogen oxalate (salt of sorrel), ammonium oxalate, and ammonium iron (III) oxalate. All other salts of oxalic acid are less important industrially.

**Potassium oxalate** [583-52-8], $K_2C_2O_4 \cdot H_2O$, $M_r$ 184.24, has the highest water solubility of any neutral oxalate (at 20 °C, 25 wt%). It was formerly used for the preparation of photographic developers and is employed today in the production of uranium in the extraction of the so-called yellow cake and in metal and textile treatment.

**Potassium hydrogen oxalate** [127-95-7], $KHC_2O_4 \cdot H_2O$, $M_r$ 146.15, is found in many plants and was the first oxalate discovered in nature. LDLo, oral, human: 660 mg/kg.

**Table 2.** World supply–demand balance of oxalic acid *

| Country | Production capacity | Production | Consumption | Export | Import |
|---|---|---|---|---|---|
| *North America* | | | | | |
| United States | 0.0 | 0.0 | 9.0 | 0.0 | 9.0 |
| Canada, Mexico | 0.0 | 0.0 | 0.9 | 0.0 | 0.9 |
| Total | 0.0 | 0.0 | 9.9 | 0.0 | 9.9 |
| *South America* | | | | | |
| Brazil | 7.0 | 5.0 | 1.0 | 4.0 | 0.0 |
| Venezuela | 0.0 | 0.0 | 0.1 | 0.0 | 0.1 |
| Total | 7.0 | 5.0 | 1.1 | 4.0 | 0.1 |
| *Western Europe* | | | | | |
| Spain | 10.0 | 8.5 | 7.0 | 1.5 | 0.0 |
| France | 8.0 | 8.0 | 5.0 | 4.0 | 1.0 |
| Others | 4.0 | 3.0 | 15.0 | 1.0 | 13.0 |
| Total | 22.0 | 19.5 | 27.0 | 6.5 | 14.0 |
| *Eastern Europe* | 13.0 | 8.0 | 10.0 | 3.0 | 5.0 |
| *Asia* | | | | | |
| China | 100.0 | 64.0 | 50.0 | 14.0 | 0.0 |
| Japan | 18.0 | 12.4 | 9.9 | 4.8 | 2.3 |
| Korea | 4.8 | 4.5 | 2.5 | 2.0 | 0.0 |
| Taiwan | 4.9 | 4.8 | 2.2 | 2.5 | 0.0 |
| India | 6.5 | 5.5 | 9.5 | 0.0 | 4.0 |
| Others | 0.0 | 0.0 | 1.6 | 0.0 | 1.6 |
| Total | 134.2 | 91.2 | 75.7 | 23.4 | 7.9 |
| World total | 176.2 | 123.7 | 123.7 | 36.9 | 36.9 |

* Data are from an Ube survey, 1. Feb. 1990; data are given in 1000 t.

**Ammonium oxalate** [1113-38-8], $(NH_4)_2C_2O_4 \cdot H_2O$, $M_r$ 142.08. The anhydrous salt dissolves in water at 0 °C up to 2.27 wt% and at 100 °C up to 25.8 wt%. Ammonium oxalate is used in textile and leather processing as a replacement for oxalic acid, for the precipitation of rare earth elements, and in ore beneficiation (precipitation of heavy metals from acidic extracts).

**Ammonium iron (III) oxalate** [29696-35-3], $(NH_4)_3[Fe(C_2O_4)_3] \cdot 3H_2O$, $M_r$ 428.08, forms emerald green crystals and is readily soluble in water and sensitive to light. It is used in the anodizing of aluminum and imparts, dependant on concentration, either yellow, gold, or brown colorations to aluminum. It is also employed in the treatment of other metals.

**Sodium oxalate** [62-76-0], $Na_2C_2O_4$, $M_r$ 134.01, is easily prepared in high purity from oxalic acid and sodium hydroxide. Therefore, it is used as a titrimetric standard substance. The solubility in water at 0 °C is 2.6 wt% and at 100 °C is 6.3 wt%. Sodium oxalate is used for the production of super white pulp.

**Table 3.** Physical properties of selected oxalates

| Ester | $M_r$ | mp, °C | bp, °C | $d_4^{20}$ | CAS registry no. |
|---|---|---|---|---|---|
| Dimethyl | 118.09 | 54 | 163.5 | 1.148 (54 °C) | [553-90-2] |
| Diethyl | 146.15 | −40.6 | 184.4 | 1.0785 | [95-92-1] |
| Di-$n$-butyl | 202.15 | −29.6 | 245.5 | 0.9873 | [2050-60-4] |

## 9.2. Organic Derivatives

**Esters.** Oxalic acid can be easily esterified by the usual methods. It forms both neutral (ROOC–COOR) and acidic esters (ROOC–COOH). The inherent acidity of oxalic acid is sufficient to catalyze the esterification in most instances. In some cases the esterification stops at the monoester stage. In this case anhydrous oxalic acid is used in preparation of the diester. Newer methods include the preparation from carbon monoxide and alcohols according to the Ube process (see Section 4.4) and the reaction of the easily accessible oxamide (see below) with alcohols in the presence of concentrated sulfuric acid [42].

Of industrial interest are the neutral esters, especially dimethyl, diethyl, and di-$n$-butyl oxalates. Except for the dimethyl ester which is a solid at room temperature they are all oily liquids with a faint odor, and thermally stable. Some esters can be distilled at atmospheric pressure without decomposition. The esters are readily soluble in alcohols, benzene, diethyl ether, acetone, and dioxane. Selected physical properties are presented in Table 3. The neutral esters are good solvents for ethers and esters of cellulose (e.g., nitrocellulose) and for resins. They are therefore used for the production of special lacquers. Of greatest significance is the diethyl ester. The di-$n$-butyl ester is also used as a plasticizer. A further application results from the ability of oxalic acid esters to easily undergo condensation reactions. Especially diethyl oxalate plays an important role in organic synthesis.

**Oxamide** [471-46-5], $H_2N$-OC-CO-$NH_2$, $M_r$ 88.07, mp ca. 350 °C (decomp.), is sparingly soluble is water (0.3 g/L at 20 °C; 6 g/L at 100 °C).

The current most economic method for the production of oxamide is a one-step process by Hoechst that is based on a catalytic oxidation of hydrogen cyanide [43]:

$$2\,HCN + 1/2\,O_2 + H_2O \longrightarrow \begin{array}{c} CONH_2 \\ | \\ CONH_2 \end{array}$$
$$\Delta H = -440 \text{ kJ/mol}$$
$$\text{Oxamide}$$

The flow chart is shown in Figure 5. A solution of copper nitrate in aqueous acetic acid is employed as the catalyst. The reaction is run at 65 °C and the yield is nearly quantitative. The oxamide that is insoluble in the catalytic solution precipitates at the bottom of the reactor and is pumped from there to a centrifuge, washed, dried and, if desired, granulated. Without further treatment it has a purity of >99% [44]. Using

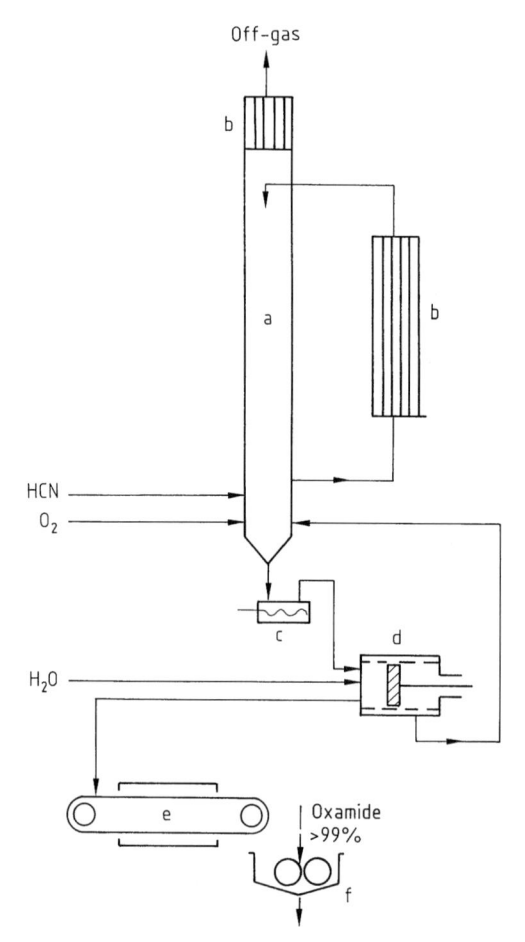

**Figure 5.** Continuous production of oxamide
a) Reactor; b) Cooler; c) Sludge pump; d) Centrifuge; e) Dryer; f) Granulator

this method Enimont in Gela/Sicily, began a production plant in 1990 with a capacity of 10 000 t/a [45].

Oxamide can be converted to ammonium oxalate and oxalic acid by heating it for long periods with water; best results are obtained by the application of pressure. Oxamide has proved to be of great interest as a long-term fertilizer. A single application is sufficient because the nitrogen is released gradually throughout the vegetation period in a form that the plant can easily use. A granular application form of oxamide is required. The poor aqueous solubility of oxamide guarantees that the fertilizer will not be washed away by rainfall and that it can be used in rice plantations. Oxamide can also be used as a starting material for chemical syntheses.

# 10. Toxicology [46], [47]

Like other strong acids, oxalic acid has a local caustic effect. The compound is absorbed relatively well through the skin and very well through mucous membranes, so that contact with larger amounts can quickly lead to absorptive poisoning. Oral ingestion causes acute irritation and pain in the mouth, esophagus, and stomach. Nausea, vomiting of blood, collapse, convulsions, irritability, and cramps (calcium precipitation) are known to occur. Premature coronary and/or circulatory arrest, and/or a kidney malfunction which results in anuria and uremia are possible. Eventual transition to polyuremia can occur. Death occurs in the coma or through kidney failure. Workers who had contact with a 5% solution of oxalic acid without wearing protective gloves, complained of pains in their hands within a few days. Blood clots and damaged blood vessels and finally a very painful gangrene made amputation of the fingers necessary. With solutions of higher concentrations the superficial caustic effects become increasingly significant. Inhaling the vapors and/or dust evokes responses that range from irritation of the mucous membranes to rashes, headaches, overexcitability, and albuminuria. Persistent coughing, vomiting, and general weakness are signs of chronic effects. Skin rashes and brittle yellow fingernails are also known.

Intake of food containing higher quantities of oxalic acid should be avoided because of an impaired calcium absorption through the formation of insoluble calcium oxalate: 4–5 g oxalic acid can be deadly for humans because the precipitated calcium oxalate plugs the kidney channels. The normal excretion of oxalic acid from humans is 20 mg per day. Two-thirds of all kidney stones are composed of calcium oxalate. The toxicological levels are [48]:

| | |
|---|---|
| $LD_{50}$, oral, rat | 375 mg/kg |
| LDLo, oral, human | 71 mg/kg |
| RTECS-no. | Ro 245000 |
| EC-no. | 607-006-00-8 |

# 11. References

[1] *Kirk-Othmer*, 3rd ed., **16**, 618–636.
[2] R. Schröder, *DMW Dtsch. Med. Wochenschr.* **105** (1980) 997.
[3] R. C. West (ed.): *CRC Handbook of Chemistry and Physics*, 61st ed., CRC Press, Boca Raton, FL, 1980.
[4] H. T. Clarke, A. W. Davis: *Organic Syntheses*, 2nd ed., coll. vol. **I**, J. Wiley and Sons, New York 1946, pp. 421–425.
[5] W. Riemenschneider: unpublished results.
[6] D. Dollimore, *Thermochim. Acta* **117** (1987) 331–363.
[7] *Ullmann*, 4th ed., **17**, 476, 477.
[8] *Ullmann*, 4th ed., **17**, 478, 479.
[9] *Ullmann*, 4th ed., **17**, 479.

[10]  S. K. Dube et al., *J. Chem. Technol. Biotechnol.* **32** (1982) no. 10, 909–919.
[11]  BIOS final report no. 663, 1946.
[12]  Allied Chem., US 2 057 119, 1936 (G. S. Simpson).
[13]  Allied Chem., US 2 322 915, 1943 (M. J. Brooks).
[14]  Allied Chem., US 3 536 754, 1970.
[15]  Mitsubishi Gas, US 3 691 232, 1972 (E. Yonemitsu et al.).
[16]  Mitsubishi Gas, DE 2 011 998, 1969.
[17]  Allied Chem., US 3 081 345, 1963 (E. J. Carlson, E. E. Gilbert).
[18]  H. Saito, *Shokubai* **23** (1981) no. 2, 139–142; *Chem. Abstr.* **95** (1981): 79382u.
[19]  S. Uchiumi, Y. Masayosi, *J. Japan Petrol. Inst.* **25** (1982) no. 4, 197–204.
[20]  D. M. Fenton, P. J. Steinwand, *J. Org. Chem.* **39** (1974) 701.
[21]  Rhône-Poulenc, DE 1 618 824, 1967.
[22]  A. M. Gaffney et al., *Chem. Ind. (Dekker)* **33** (1988) 87–100.
[23]  F. Feigl et al., *Mikrochemie* **18** (1935) 272.
[24]  S. M. Steinberg et al., *Int. J. Environ. Anal. Chem.* **19** (1985) 251–260.
[25]  U. Rehmert et al., *Lab. Med.* **7** (1983) 29–32.
[26]  J. E. Buttery et al., *Clin. Chem. (Winston-Salem N.C.)* **29** (1983) 700–702.
[27]  H.-O. Beutler et al., *Z. Anal. Chem.* **301** (1980) 186.
[28]  S. Werneck, R. Pinner: *The Surface Treatment and Finishing of Aluminum and its Alloys*, R. Draper Ltd., Teddington 1972.
[29]  S. John, B. A. Shenoi, *Met. Finish* **74** (1976) no. 9, 48, 57.
[30]  P. E. Morris, *Corrosion (Houston)* **33** (1977) 17.
[31]  W. Widerholt: *The Chemical Surface Treatment of Metals*, R. Draper Ltd., Teddington 1965.
[32]  H. Cohen, G. E. Linton: *Chemistry and Textiles for the Laundry Industry*, Textile Book Publishers Inc., New York 1961.
[33]  N. A. Cashen et al., US 3 811 210, 1974.
[34]  L. Re, B. Maurer, T. Ohloff, *Helv. Chim. Acta* **56** (1973) 1882.
[35]  L. A. Sarrer et al., *J. Am. Chem. Soc.* **49** (1972) 943.
[36]  G. M. Varshal, *Metody Khim. Anal. Khim. Sostav Miner.*, 1971, 18–33; *Chem. Abstr.* **75** (1971): 133488k.
[37]  A. Alsnis et al., *Latv. PSR Zinat. Akad. Vestis* **5** (1976) 81–90; *Chem. Abstr.* **85** (1976): 78382a.
[38]  O. Vogl et al., *Polym. Prepr. (Am. Chem. Soc. Div. Polym. Chem.)* **19** (1978) no. 2, 75–80.
[39]  N. H. Andersen et al., *Synth. Commun.* **3** (1973) 125.
[40]  SRI International: *Chem. Econ. Handbook*, June 1988, 682 1000 A-P.
[41]  GefStoffV of Aug. 28th, 1986.
[42]  Röhm GmbH, DE 1 668 639, 1967 (W. Gruber, J. Dehler).
[43]  Hoechst AG, DE 2 308 941, 1973 (W. Riemenschneider).
[44]  W. Riemenschneider, *Chem. Ing. Tech.* **50** (1978) 55.
[45]  *Chem. Ind.* **7** (1989) 18.
[46]  N. I. Sax: *Dangerous Properties of Industrial Material*, 6th ed., Van Nostrand Reinhold, New York 1984, p. 2098.
[47]  Kühn-Berett: *Merkblätter Gefährliche Arbeitsstoffe* **12/1982**, Verlag Moderne Industrie, München 0-03-1/2.
[48]  *Registry of Toxic Effects of Chemical Substances*, US Department of Health and Human Services, Cincinnati, Ohio 1983/1984.

# Oxocarboxylic Acids

*Individual keyword:* → *Glyoxylic Acid.*

FRANZ DIETRICH KLINGLER, Boehringer Ingelheim, Federal Republic of Germany (Chaps. 1, 2, 4–6)

WOLFGANG EBERTZ, Hoechst Aktiengesellschaft, Frankfurt, Federal Republic of Germany (Chap. 3)

| | | | | |
|---|---|---|---|---|
| 1. | Introduction . . . . . . . . . . . . . 3599 | 3.4. | Toxicology and Environmental Protection . . . . . . . . . . . . . . . | 3606 |
| 2. | Pyruvic Acid . . . . . . . . . . . . . 3600 | 4. | Levulinic Acid . . . . . . . . . . . . | 3606 |
| 3. | Acetoacetic Acid and Derivatives . . . . . . . . . . . . . . . 3602 | 5. | Acetonedicarboxylic Acid . . . . | 3607 |
| 3.1. | Properties . . . . . . . . . . . . . . . 3602 | 6. | Other Oxocarboxylic Acids . . . | 3608 |
| 3.2. | Production . . . . . . . . . . . . . 3604 | 7. | References . . . . . . . . . . . . . . | 3609 |
| 3.3. | Uses and Economic Aspects . . 3605 | | | |

# 1. Introduction

Oxocarboxylic acids are classified according to the relative positions of the oxo and carboxyl groups as α-, β-, γ-, and δ-oxocarboxylic acids. Glyoxylic acid [298-12-4] is the simplest α-oxocarboxylic acid (an aldehyde carboxylic acid) (→ Glyoxylic Acid).

Oxocarboxylic acids, particularly α-oxocarboxylic acids [1], are important metabolites in biochemical cycles. They form the link between α-amino acid, carbohydrate, and fat metabolism [2], [3]. Whereas the free α-oxocarboxylic acids, the most important example of which is pyruvic acid, are relatively stable, the β-oxocarboxylic acids can be isolated in the free state only by taking special precautions. They decompose by decarboxylation to give the corresponding ketones. The γ-oxocarboxylic acids, with levulinic acid as the simplest example, are again stable and do not decarboxylate.

Very specific processes are required for the industrial production of oxocarboxylic acids. A general method for the production of oxocarboxylic acids is the hydrolysis of fatty acids with concentrated sulfuric acid or other hydrolytic reagents. α-Oxocarboxylic acids can be produced by treatment of α-amino acids with trifluoro acetic anhydride [407-25-0]. 2-Trifluorooxazolones are formed as intermediates which hydrolyze to the corresponding α-oxocarboxylic acids [4]. The carbonylation of alkyl halides with cobalt carbonyls (typically carried out at ca. 5 MPa and 100 °C) leads to α-oxocarboxylic acids.

This method has recently come into use [5]. General methods for the synthesis of oxocarboxylic acids have been described by ROBINSON [6] and STETTER [7].

## 2. Pyruvic Acid

Pyruvic acid [127-17-3], 2-oxopropanoic acid, pyroracemic acid, α-ketopropionic acid, $H_3C-CO-COOH$, $M_r$ 88.06, is the most important α-oxocarboxylic acid. It plays a central role in energy metabolism in living organisms [8]. During exertion, pyruvic acid is formed from glycogen in the muscle and reduced to lactic acid [79-33-4]. In the liver, pyruvic acid can be converted into alanine [56-41-7] by reductive amination. Pyruvic acid was discovered and first described in 1835 by BERZELIUS [9].

**Physical Properties.** Pyruvic acid is a colorless liquid with an odor similar to that of acetic acid. Its most important physical data may be summarized as follows:

| | |
|---|---|
| Density (20 °C) | 1.268 g/cm$^3$ |
| mp | 13.6 °C |
| bp | |
| 101.31 kPa | 165 °C (decomp.) |
| 1.33 kPa | 57.9 °C |
| 0.13 kPa | 21.4 °C |
| Refractive index (25 °C) | 1.4259 |
| Flash point (DIN 51758) | 91 °C |
| Ignition temperature (DIN 51794) | 305 °C |

Pyruvic acid is totally miscible with water, ethanol, and ether. Pyruvic acid exists in the keto form; the enol form cannot be detected [10].

**Chemical Properties.** Pyruvic acid reacts as both an acid and a ketone. It forms, for example, oximes, hydrazones, and salts. 4,5-Dioxo-2-methyltetrahydrofuran-2-carboxylic acid [24891-71-2] (**1**) is formed from pyruvic acid either slowly on standing or more quickly under acid catalysis [11].

On standing in aqueous solution, pyruvic acid polymerizes to higher molecular mass products via the dimeric ketoglutaric acid (**2**) [19071-44-4] and the trimeric aldol product [12], [13].

Like all 2-oxo acids, pyruvic acid eliminates carbon monoxide on treatment with concentrated sulfuric acid [14].

Oxidation of pyruvic acid gives acetic or oxalic acid [144-62-7] and carbon dioxide, depending on the conditions [15]. Lactic acid is obtained by reaction with reducing agents [1].

Reaction of α-amino acids with pyruvic acid gives, besides carbon dioxide, alanine [56-41-7] (transamination reaction) and the corresponding aldehyde with one carbon atom less [16]. Alanine is also obtained by reductive amination of pyruvic acid [1]. Phenylethylamines react with pyruvic acid to form the corresponding tetrahydroisoquinolines via the Bischler–Napieralski reaction [17]. Reaction with o-phenylenediamines gives quinoxalinols [18]. In a similar reaction the corresponding hydroxypteridines are obtained from 4,5-diaminopyrimidines and pyruvic acid [19]. Pyruvic acid reacts with aldehydes to form the corresponding α-keto-γ-hydroxy acids, which then cyclize to butyrolactone derivatives [1]. Friedel–Crafts type reactions of aromatic compounds with pyruvic acid yield diarylpropionic acids. These compounds have achieved a certain degree of importance because they provide a good route to 1,1-diarylethylenes by dehydration and decarbonylation [15], [20].

**Production.** On an industrial scale, pyruvic acid is produced by *dehydration and decarboxylation of tartaric acid* [87-69-4] [21]. In this process, pyruvic acid is distilled from a mixture of tartaric acid and potassium and sodium hydrogen sulfates at 220 °C. The crude acid obtained (ca. 60%) is then distilled in vacuum. The reaction temperature can be lowered to 160 °C by adding ethylene glycol [107-21-1] [22]. Pyruvic acid can also be obtained by the *gas-phase oxidation of lactic acid* [23], but this process has not been successful industrially. In contrast, *microbial oxidation of* D-*lactic acid* by a new process results in high yields [24]. Microbial oxidation of 1,2-propanediol [57-55-6] to pyruvic acid has also been described [25]. Another process describes the hydrolysis of 2,2-dihalopropionic acids to pyruvic acid [26]. A process for the oxidation of methylglyoxal [78-98-8] with halogens has recently been published [27].

**Uses.** Pyruvic acid is used mainly as an intermediate in the synthesis of pharmaceuticals. It is also employed in the production of crop protection agents, polymers, cosmetics, and foods.

**Storage and Quality Specifications.** Pyruvic acid is stored and transported in tightly closed polyethylene containers. It can be kept for only a limited period and must therefore be stored in refrigerated areas at a maximum of 10 °C. At higher temperature, explosion can occur through spontaneous self-condensation [28]. The concentration of the commercial product is determined acidimetrically and decreases by ca. 1% per month during storage.

**Toxicology.** Pyruvic acid has a corrosive effect and irritates the eyes, skin, and respiratory passages.

# 3. Acetoacetic Acid and Derivatives

## 3.1. Properties

The physical properties of the most important acetoacetic esters and aromatic amides are listed in Table 1.

**Chemical Properties.** Acetoacetic acid and its derivatives exist in the tautomeric keto–enol forms.

$$CH_3-\overset{O}{\underset{\|}{C}}-CH_2-\overset{O}{\underset{\|}{C}}- \rightleftharpoons CH_3-\overset{OH}{\underset{|}{C}}=CH-\overset{O}{\underset{\|}{C}}-$$

The enol content depends on the type of derivative and the polarity of the solvent. In nonpolar solvents, stabilization of the enol form occurs by formation of a hydrogen bond [29].

Acetoacetic acid is a strong organic acid that decomposes rapidly to acetone and carbon dioxide when heated above its melting point of 35–37 °C.

$$CH_3-\overset{O}{\underset{\|}{C}}-CH_2-COOH \longrightarrow CH_3-\overset{O}{\underset{\|}{C}}-CH_3 + CO_2$$

Because of this instability, acetoacetic acid as such is not used in organic synthesis. Its derivatives, however, in particular esters and amides, are important intermediates because of their higher stability in comparison to the free acid.

$\beta$-Ketoacids and their derivatives have three reactive functional groups, which allow the synthesis of many important compounds. The carbonyl, methylene, and carboxylic groups take different reaction paths.

The tendency of $\beta$-dicarbonyl compounds to form the enol results in high reactivity of the C–H bond of the *methylene group*. Hydrogen can be abstracted easily from the methylene group by strong alkali. This reaction is usually carried out with acetoacetic esters and alkoxides. The resulting salt of the ester can be alkylated or acylated.

**Table 1.** Physical properties of important acetoacetic esters and amides*

| | R | CAS registry number | $M_r$ | mp, °C | bp, °C | $d^{20}$, kg/m³ | Flash point, °C |
|---|---|---|---|---|---|---|---|
| 1 | —OCH₃ | [105-45-3] | 116.1 | −35 | 170 | 1075 | 67 |
| 2 | —OC₂H₅ | [141-97-9] | 130.1 | −43 | 180 | 1030 | 71 |
| 3 | —NH—⟨⟩ | [102-01-2] | 177.2 | 82 | | 1260 | 185 |
| 4 | —NH—⟨⟩—CH₃ (CH₃) | [97-36-9] | 205.0 | 88 | | 1240 | 171 |
| 5 | —NH—⟨⟩ (CH₃O) | [92-15-9] | 207.2 | 86 | | 1130 | 162 |
| 6 | —NH—⟨⟩ (Cl) | [93-70-9] | 211.6 | 104 | | 1440 | 176 |
| 7** | —NH—⟨⟩—SO₃⁻ (CH₃, CH₃O, NH₄⁺) | [72705-22-7] | 318.3 | | | | |

\*
$$CH_3-\overset{O}{\underset{\|}{C}}-CH_2-\overset{O}{\underset{\|}{C}}-R$$

\*\* Hygroscopic.

$$CH_3-\overset{O}{\underset{\|}{C}}-CH_2-COOR + M-OR^1 \longrightarrow$$

$$CH_3-\overset{O}{\underset{\|}{C}}-\overset{-}{C}H-COOR + R^1-OH$$
$$M^+$$
$$\downarrow + XR^2$$
$$CH_3-\overset{O}{\underset{\|}{C}}-\underset{R^2}{\overset{|}{C}H}-COOR + MX$$

The reactivity of the remaining C—H bond is still high enough to undergo a second substitution to form disubstituted acetoacetic esters. When treated with diluted alkali these substituted esters are decarboxylated to form substituted acetones [30].

$$CH_3-\overset{O}{\underset{\|}{C}}-\underset{R^2}{\overset{R^3}{\underset{|}{C}}}-COOR \xrightarrow{OH^-/H_2O}$$
$$CH_3-\overset{O}{\underset{\|}{C}}-\underset{R^2}{\overset{R^3}{\underset{|}{C}H}} + CO_2 + R-OH$$

The cleavage of substituted acetoacetic esters under strongly basic conditions leads to carboxylic acids (retro-Claisen condensation) [30].

$$CH_3-\underset{\underset{O}{\|}}{C}-\underset{\underset{R^2}{|}}{\overset{\overset{R^3}{|}}{C}}-COOR \xrightarrow{OH^-/ROH} \xrightarrow{H_2O}$$

$$CH_3COOH + H\underset{\underset{R^2}{|}}{\overset{\overset{R^3}{|}}{C}}-COOH + R-OH$$

The *carbonyl group* is the second reactive group in acetoacetic compounds. Under conditions that stabilize the keto form, its reactivity is comparable to that of other ketones. Of particular interest are hydrazines and other difunctional compounds, which react with acetoacetic esters to yield heterocyclic compounds (e.g., 3-methyl-5-pyrazolones):

$$CH_3-\overset{\overset{O}{\|}}{C}-CH_2-COOR + H_2N-NHR^1 \longrightarrow$$

$$CH_3-\underset{\underset{N-NHR^1}{\|}}{C}-CH_2-COOR + H_2O$$

$$\downarrow H^+ \text{ or } OH^-$$

$$\underset{\underset{R^1}{N-N}}{H_3C-C\diagup\overset{CH_2}{\diagdown}C=O} + ROH$$

Reaction with ureas leads to uracil derivatives and with hydroxylamines leads to isoxazoles. The reactivity of the *ester group* is comparable to that of other carboxylic esters. Acetoacetic esters are used as starting materials in the synthesis of acetoacetamides.

## 3.2. Production

Diketene is used as starting material in the production of acetoacetic esters and amides. The use of diketene makes strict safety regulations necessary. Enrichment of diketene in the reaction mixture must be avoided, therefore reaction temperature, flow, and pressure must be carefully controlled. A temporary decrease in catalytic activity (temperature drop, dosage inaccuracies) or inaccuracies in the addition of reactants can result in the strong rapid exothermic decomposition of enriched diketene. The reaction mixture or vapor must also be kept from contact with stored diketene.

$$\underset{H_2C-C=O}{H_2C=C-O} + ROH \xrightarrow{\text{Catalyst}} CH_3-\overset{\overset{O}{\|}}{C}-CH_2-COOR$$

**Production of Acetoacetic Esters.** The methyl and ethyl esters are the preferred acetoacetic esters for industrial use. They are produced in continuous processes, usually from *crude diketene* and methanol or ethanol. Basic catalysts such as amines or their acetates are commonly used, although strong acids also have catalytic action. The exothermic reaction is carried out at 100–140 °C and 1.5–2 bar. The residence time in the reactor must be short (0.5–2 h) to avoid formation of condensation products of acetoacetic esters (dipyrones).

Since crude diketene always contains a certain amount of acetic anhydride, acetic acid and acetic esters are formed as byproducts. Together with excess alcohol and acetone (the latter formed in side reactions from diketene) acetic acid and esters constitute a low-boiling fraction in the crude acetoacetic ester, which must be removed by distillation. Crude diketene contains oligomers and polymers of ketene. These substances partially react with the alcohol and form the high-boiling fraction in the crude product. After purification of the ester by distillation, they remain as a partially solid residue, which is combusted.

**Production of Acetoacetamides.** *Distilled diketene* is used for the production of acetoacetamides. The reaction is carried out in aqueous solution or suspension or in dilute acetic acid. Because of the basic reactant amine, no catalyst is needed.

$$H_2C=C-O \quad \quad \quad R \quad \quad \quad \quad \quad \quad O \quad \quad \quad R$$
$$\quad \quad | \quad | \quad + HN \quad \quad \longrightarrow \quad CH_3-\overset{\|}{C}-CH_2-CON$$
$$H_2C-C=O \quad \quad R^1 \quad \quad \quad \quad \quad \quad \quad \quad \quad \quad R^1$$

Acetoacetamide is produced by reacting a dilute aqueous ammonia solution with diketene and is used in further reactions (e.g., to pyrazolones) as an aqueous solution. Aromatic amides of acetoacetic acid, the so-called arylides, crystallize from the cooled reaction mixture, and are separated and dried. When no diketene is available, acetoacetic esters are used as starting materials for the production of acetoacetamides. The alcohol released from the amidation reaction must be recycled.

## 3.3. Uses and Economic Aspects

*Acetoacetic esters* are important starting materials in the synthesis of intermediates for drugs, dyes, and agrochemicals. The worldwide installed production capacity of acetoacetic esters is more than 80 000 t/a. More than 80% of all esters produced are methyl and ethyl acetoacetates (**1** and **2**, respectively, in Table 1).

The main application of *acetoacetamides* is in the production of dyes and organic pigments. The most important amides in pigment production are acetoacetanilide (acetoacetaminobenzene, **3**) and acetoacet-*m*-xylidide (4-acetoacetamino-1,3-dimethylbenzene, **4**), followed by acetoacet-*o*-anisidide (2-acetoacetamino-1-methoxybenzene, **5**) and acetoacet-*o*-chloroanilide (2-chloro-1-acetoacetaminobenzene, **6**). The so-called cresidine (3-acetoacetamino-4-methoxytoluene-6-sulfonic acid, ammonium salt, **7**) is an intermediate in the production of reactive dyes.

*Producers.* The main producers of acetoacetic compounds are BASF (Germany), BP Chemicals (UK), Colour Chem. (India), DAICEL (Japan), Eastman Kodak (United

States), Hoechst (Germany), Lonza (United States, Switzerland, Germany), Nippon Gosai (Japan), and Wacker Chemie (Germany).

## 3.4. Toxicology and Environmental Protection

*Acetoacetic esters* have a low acute toxicity. The $LD_{50}$ values are 3230 mg/kg (rat) [31] for methylacetoacetate and 3980 mg/kg (rat) [32] for ethylacetoacetate. Chronic diseases caused by acetoacetic esters have not been reported. The methyl ester causes reversible irritation of skin and mucous membranes [33]. Acetoacetic esters are completely degradable in biological wastewater treatment and have low toxicity for fish and bacteria.

The toxicity and degradability of *acetoacetamides* are similar to those of the esters.

To avoid negative environmental effects, the effluents of acetoacetic compound production must be purified in a biological wastewater treatment plant. In handling and shipping acetoacetic esters the rules for products with low toxicity must be obeyed:

| | |
|---|---|
| RID/ADR | 3/32 c |
| ADNR | 3/4/K 3 |
| ICAO/IATA | not classified |

## 4. Levulinic Acid

Levulinic acid [*123-76-2*], β-acetylpropionic acid, γ-ketovaleric acid, 4-oxopentanoic acid, $H_3C-CO-CH_2-CH_2-COOH$, $M_r$ 116.11, is the simplest and most important γ-oxocarboxylic acid. It was first described and named levulinic acid by GROTE and TOLLENS in 1875 [34]. Only since 1960 has levulinic acid gained commercial importance [35], [36].

**Physical Properties.** Levulinic acid forms colorless crystals that are readily soluble in water, ethanol, and diethyl ether. Important physical properties are given below:

| | |
|---|---|
| mp | 37 °C |
| bp | |
| (101.31 kPa) | 245–246 °C |
| (1.33 kPa) | 137–139 °C |
| Specific density (20 °C) | 1.14 |
| Refractive index (25 °C) | 1.4396 |
| Flash point | 138 °C |

**Chemical Properties.** Levulinic acid reacts both as a ketone and as a carboxylic acid. Angelica lactone [*591-12-8*] is formed by elimination of water on prolonged heating [37], [38]. γ-Valerolactone [*108-29-2*] is formed by catalytic hydrogenation of levulinic acid [39]. Condensation reactions with aldehydes can occur at the α-, β-, or δ-carbon atoms, depending on conditions [35]. Levulinic acid is the starting compound in the synthesis of many heterocycles. 5-Methyl-2-pyrrolidone [108-27-0] is formed by reductive amination [40], and 6-keto-3-methyl-1,4,5,6-tetrahydropyridazone by reaction of levulinic acid with hydrazine [41].

**Production.** Levulinic acid is produced industrially from sugar polymers such as cellulose or starch via the monomeric hexoses [36] according to many patented processes. The reaction is usually acid catalyzed: D-glucose [50-99-7] is formed first and is then isomerized enzymatically to D-fructose [57-48-7]. D-Fructose is subsequently converted to hydroxymethylfurfural [67-47-0], an intermediate that reacts further to form levulinic acid. The classical levulinic acid synthesis, the treatment of D-fructose with hydrochloric acid, is also being used [36].

A newer patented process is based on the use of petrochemical raw materials. Levulinic acid is obtained by ozonolysis of unsaturated hydrocarbons in a relatively sophisticated process [42].

**Uses.** Levulinic acid is used as a raw material for the synthesis of pharmaceuticals [43], [44], for the production of plasticizers [45], and as an auxiliary in electroplating [45].

The calcium salt is used in calcium therapy [46], and levulinic acid esters are used in the production of cosmetics [47].

**Storage and Quality Specifications.** Levulinic acid is stored and transported in tightly closed polyethylene containers, protected from light. It is stable for long periods at room temperature.

**Toxicology.** The $LD_{50}$ (oral, rat) is 1850 mg/kg. Levulinic acid can cause acid burns, and concentrated solutions can cause irritation of the skin and mucous membranes [36].

# 5. Acetonedicarboxylic Acid

Acetonedicarboxylic acid [542-05-2], 3-oxoglutaric acid, β-ketoglutaric acid, 3-oxopentane dicarboxylic acid, HOOC–CH$_2$–CO–CH$_2$–COOH, $M_r$ 146.10, was first isolated and described by v. Pechmann [48], [49].

**Physical Properties.** Acetonedicarboxylic acid forms colorless crystals that are readily soluble in water and ethanol, and sparingly soluble in trichloromethane and diethyl ether. It melts with decomposition at 138 °C. The dissociation constants (in 0.01 M solution) are $K_1^{25} = 4.68 \times 10^{-4}$ and $K_2^{25} = 5.37 \times 10^{-5}$ [50].

**Chemical Properties.** On heating above its melting point, acetonedicarboxylic acid decomposes into carbon dioxide and acetone [49]. When acetonedicarboxylic acid is warmed in water, a two-step decomposition occurs with acetoacetic acid [541-50-4] as an intermediate that decomposes further to acetone and carbon dioxide [50]. The decomposition is catalyzed by metal ions and protons [51]. Phenols react with acetonedicarboxylic acid in the presence of dehydrating agents to form coumarin derivatives [52]–[54].

**Production.** Acetonedicarboxylic acid is produced from citric acid [77-92-9] by many industrial processes that differ only slightly [55], [56]. Citric acid is treated with oleum and reacts to yield acetonedicarboxylic acid via decarbonylation and dehydration.

Other possible production methods include reaction of acetone with carbon dioxide [57], [58], oxidation of citric acid with chlorosulfuric acid [59], or reaction of ketene with phosgene [60].

**Uses.** Acetonedicarboxylic acid is an important starting material in the production of pharmaceutically active alkaloids [61]–[63]. It is also suitable as a stabilizer for natural fats and oils [64], [65].

**Storage and Quality Specifications.** Large quantities of acetonedicarboxylic acid can be stored for short periods in tightly closed polyethylene containers in refrigerated areas. The acid must not be stored in metal containers. Laboratory-scale quantities of the pure substance can be kept for longer periods over phosphorus pentoxide in a desiccator [66]. The commercial product has a concentration of ca. 98%; the purity is determined by titration.

**Toxicology.** Acetonedicarboxylic acid can cause irritation and acid burns to the eyes, respiratory passages, and skin.

# 6. Other Oxocarboxylic Acids

*Oxaloacetic acid* [328-42-7] (2-oxosuccinic acid), an important intermediate in the tricarboxylic acid cycle, is known only in the form of its cis and trans enol tautomers, hydroxymaleic acid and hydroxyfumaric acid, respectively [67], [68]. Its diethyl ester, obtained by Claisen condensation of diethyl oxalate [95-92-1] and ethyl acetate [69], is used in the production of higher α-oxocarboxylic acids (by decarboxylation of the condensation product) and higher carboxylic acids (retro-Claisen condensation).

*α-Oxoglutaric acid* [328-50-7], which plays an important role in amino acid metabolism [70], is obtained by condensation of diethyl succinate [123-25-1] with diethyl oxalate [95-92-1] [71], [72]. It was used as the starting material for the production of L-glutamic acid [56-86-0] in a microbial process [73].

# 7. References

[1] A. J. L. Cooper, J. Z. Ginos, A. Meister, *Chem. Rev.* **83** (1983) 321.
[2] A. E. Martell, *Acc. Chem. Res.* **22** (1989) 115.
[3] M. Walser, J. R. Williamson: *Metabolism and Clinical Implications of Branched Chain Amino- and Ketoacids,* Elsevier, Amsterdam 1981.
[4] F. Weygand, W. Steglich, H. Tanner, *Justus Liebigs Ann. Chem.* **658** (1962) 128.
[5] B. Bell, H. Chrobaczek, W. Kohl, *Chem. Ztg.* **109** (1985) 167.
[6] R. Robinson, *J. Chem. Soc.* 1930, 745; 1934, 1543.
[7] H. Stetter, *Angew. Chem.* **88** (1976) 695.
[8] K. Schreiber: *Die Brenztraubensäure und ihr Stoffwechsel,* Editio Cantor, Aulendorf 1956.
[9] J. J. Berzelius, *Ann. Phys.* **36** (1835) 1.
[10] A. Schellenberger, K. Winter, *Chem. Ber.* **92** (1959) 793.
[11] L. Wolff, *Justus Liebigs Ann. Chem.* **317** (1901) 1.
[12] H. Goldfine, *Biochim. Biophys. Acta* **40** (1960) 557.
[13] A. Schellenberger, E. Podany, *Chem. Ber.* **91** (1958) 1781.
[14] A. Bistrzycki, B. v. Siemiradzki, *Ber. Dtsch. Chem. Ges.* **39** (1906) 58.
[15] S. Patai, S. Dayagi, *J. Chem. Soc.* 1958, 3058.
[16] R. M. Herbst, L. L. Engel, *J. Biol. Chem.* **107** (1934) 505.
[17] G. Hahn, A. Hansel, *Ber. Dtsch. Chem. Ges.* **71** (1938) 2163.
[18] O. Hinsberg, *Justus Liebigs Ann. Chem.* **237** (1887) 327.
[19] G. B. Elion, G. H. Hitchings, P. B. Russel, *J. Am. Chem. Soc.* **72** (1950) 78.
[20] Bayer, DE 2 830 953, 1978 (W. Meyer, H. Rudolf, E. Cleur, E. Schoenhals).
[21] J. W. Howard, W. A. Fraser, *Org. Synth. Coll. Vol.* **1** (1932) 475.
[22] J. D. Riedel, DE 281 902, 1913.
[23] C. H. Boehringer Sohn, DE 523 190, 1931 (F. Zumstein).
[24] BASF, EP-A 0 313 850, 1988 (B. Cooper).
[25] Y. Izumi, Y. Matsumura, Y. Tani, H. Yamada, *Agric. Biol. Chem.* **46** (1982) 2673.
[26] Dow Chemical, US 3 524 880, 1966 (L. H. Lee, D. E. Ranck).
[27] BASF, DE-OS 3 219 355, 1982 (U. R. Samel, L. Hupfer).
[28] *Sichere Chemiearbeit* **29** (1977) 87.
[29] S. Gould: *Mechanisms and Structure in Organic Chemistry,* 2nd ed., Henry Holt and Company, New York 1959.
[30] L. F. Fieser, M. Fieser: *Organische Chemie,* 2nd ed., Verlag Chemie, Weinheim 1972, p. 535.
[31] H. F. Smyth et al.: *J. Ind. Hyg. Toxicol.* **30** (1948) 63.
[32] H. F. Smyth, C. P. Carpenter, C. S. Weil, *J. Ind. Hyg. Toxicol.* **31** (1949) 60.
[33] *Food Cosmet. Toxicol.* **16** (1978) 815.
[34] A. v. Grote, B. Tollens, *Justus Liebigs Ann. Chem.* **175** (1875) 181.
[35] M. Kitano, F. Tanimoto, M. Okabayashi, *CEER Chem. Econ. Eng. Rev.* **7** (1975) 25.
[36] V. Sunjic, J. Horvat, B. Klaic, S. Horvat, *Kem. Ind.* **33** (1984) 593.
[37] J. Thiele, R. Tischbein, E. Lossow, *Justus Liebigs Ann. Chem.* **319** (1901) 180.
[38] Newport Ind., US 2 761 869, 1956; US 2 809 203, 1959 (R. H. Leonard).
[39] Quaker Oats, US 2 786 852, 1957 (A. P. Dunlop, J. W. Maden).
[40] L. Frank, W. R. Schmitz, B. Zeidman, *Org. Synth. Coll. Vol.* **3** (1955) 328.
[41] W. G. Overend, L. F. Wiggins, *J. Chem. Soc.* 1947, 239.
[42] BP Chem., GB 1 596 651, 1981 (J. Russel, C. G. Gasser).

[43] Edelfettwerke W. Schlueter, Glyco Iberica, DE-OS 2 360 795, 1975.
[44] Henkel, DE-OS 2 533 101, 1975 (H. Moeller, H. Schnegelberger, R. Osberghaas).
[45] Leningrad Technological Institute, SU 722 943, 1980 (Y. M. Postolov et al.).
[46] A. R. Carnes, D. D. Mann, DE-OS 2 830 636, 1979.
[47] Y. Hikotaro, JP 73 43 178, 1973.
[48] H. v. Pechmann, *Ber. Dtsch. Chem. Ges.* **17** (1884) 929, 2542.
[49] H. v. Pechmann, *Justus Liebigs Ann.* **261** (1891) 155.
[50] R. W. Hay, K. N. Leong, *J. Chem. Soc. A* 1971, 3639.
[51] J. E. Prue, *J. Chem. Soc.* 1952, 2331.
[52] H. P. Kansara, N. H. Shah, *J. Univ. Bombay Sci.* **17 A** (1948) 47.
[53] K. A. Thakar, *J. Indian Chem. Soc.* **40** (1963) 397.
[54] V. M. Dixit, A. N. Kanakudati, *J. Indian Chem. Soc.* **28** (1951) 323.
[55] *Beilstein*,**3**, 789; **3 (1)**, 275; **3 (2)**, 482; **3 (3)**, 1369.
[56] R. Adams, H. M. Chiles, C. F. Rassweiler, *Org.Synth. Coll. Vol.* **1** (1932) 9.
[57] Mitsui Toatsu Chem Inc., JP-Kokai 75 71 625, 1975 (M. Kawamata, H. Tanabe).
[58] Montedison, DE-OS 2 429 627, 1974 (E. Alneri, G. Bottaccio, V. Carletti, G. Lana).
[59] C. H. Boehringer Sohn, DE 1 160 841, 1961 (F. Gerner).
[60] Akzo, DE-OS 2 409 342, 1974 (N. Heyboer).
[61] R. C. Menzies, R. Robinson, *J. Chem. Soc.* **125** (1924) 2163.
[62] Sadolin + Halmblad, GB 791 770, 1958 (N. Elming, P. Nedenskov).
[63] C. Schöpf, G. Lehmann, *Justus Liebigs Ann. Chem.* **518** (1935) 1.
[64] Secretary of Agriculture, US 2 605 186, 1951 (A. W. Schwab, H. A. Moser, C. D. Evans).
[65] A. Mosca, IT 494 153, 1951.
[66] E. O. Wilg, *J. Phys. Chem.* **32** (1928) 961.
[67] Merck Index 10th ed., 6782, Rahway, N.J. 1983.
[68] *Beilstein*,**3**, 777; **3 (1)**, 272; **3 (2)**, 478; **3 (3)**, 1359; **3 (4)**, 1808.
[69] Höchster Farbenwerke, DE 43 847, 1887.
[70] K. Heyns, *Angew. Chem.* **61** (1949) 474.
[71] L. Friedman, E. Kosower, *Org. Synth. Coll. Vol.* **3** (1955) 510.
[72] E. M. Bottorff, L. L. Moore, *Org. Synth. Coll. Vol.* **5** (1973) 687.
[73] Int. Minerals and Chem. Corp., US 2 933 434, 1960 (R. C. Good).

# Pentanols

PETER LAPPE, Hoechst Aktiengesellschaft, Werk Ruhrchemie, Oberhausen, Federal Republic of Germany (Chaps. 2–7)

THOMAS HOFMANN, Hoechst Aktiengesellschaft, Frankfurt, Federal Republic of Germany (Chap. 8)

| | | | | | |
|---|---|---|---|---|---|
| 1. | Introduction | 3611 | 6. | Quality Specifications and Chemical Analysis | 3620 |
| 2. | Physical Properties | 3611 | 7. | Storage and Transportation | 3620 |
| 3. | Chemical Properties | 3613 | 8. | Toxicology and Occupational Health | 3621 |
| 4. | Production | 3615 | | | |
| 5. | Uses | 3618 | 9. | References | 3622 |

## 1. Introduction

Pentanols (amyl alcohols), $C_5H_{11}OH$, $M_r$ 88.15, are saturated aliphatic $C_5$-alcohols. The eight structural pentanol isomers are often named by common names in addition to the systematic IUPAC nomenclature (see Table 1). Four of the pentanol isomers are primary alcohols, three are secondary alcohols, and 2-methyl-2-butanol is a tertiary alcohol.

Pentanols are found mainly in fruit, honey, tea, cheese, and alcoholic beverages. They were first discovered in fusel oils and were side products in the fermentation of starch-containing materials.

Three of the $C_5$-alcohol isomers—2-pentanol, 2-methyl-1-butanol, and 3-methyl-2-butanol—are optically active.

## 2. Physical Properties

The most important physical properties of the eight pentanol isomers are listed in Table 2. The pentanols are colorless liquids of characteristic odor, with the exception of 2,2-dimethyl-1-propanol, which is solid at room temperature. The pentanols are only sparingly miscible with water at 20 °C. However, they are almost completely miscible with most organic solvents. Binary liquid–liquid equilibrium constants for pentanols in water at 0–90 °C are given in [1].

2-Pentanol, 2-methyl-1-butanol, and 3-methyl-2-butanol can be separated into their optically active isomers by suitable separation processes. Some specific rotations are compiled in Table 3. Many binary and tertiary azeotropic mixtures of various pentanol

**Table 1.** Pentanol isomers

| Formula | Systematic name | Common name | CAS registry no. |
|---|---|---|---|
| $CH_3CH_2CH_2CH_2CH_2OH$ | 1-pentanol | n-amyl alcohol | [71-41-0] |
| $CH_3CH_2CH_2CH(OH)CH_3$ | 2-pentanol | sec-amyl alcohol | [6032-29-7] |
| $CH_3CH_2CH(OH)CH_2CH_3$ | 3-pentanol | | [584-02-1] |
| $CH_3CH_2CH(CH_3)CH_2OH$ | 2-methyl-1-butanol | active primary amyl alcohol | [137-32-6] |
| $(CH_3)_2CHCH_2CH_2OH$ | 3-methyl-1-butanol | isoamyl alcohol | [123-51-3] |
| $CH_3CH_2C(CH_3)(OH)CH_3$ | 2-methyl-2-butanol | tert-amyl alcohol | [75-85-4] |
| $(CH_3)_2CHCH(OH)CH_3$ | 3-methyl-2-butanol | sec-isoamyl alcohol | [598-75-4] |
| $(CH_3)_3CCH_2OH$ | 2,2-dimethyl-1-propanol | neopentyl alcohol | [75-84-3] |

**Table 2.** Physical properties of pentanols

| | 1-Pentanol | 2-Pentanol | 3-Pentanol | 2-Methyl-1-butanol | 3-Methyl-1-butanol | 2-Methyl-2-butanol | 3-Methyl-2-butanol | 2,2-Dimethyl-1-propanol |
|---|---|---|---|---|---|---|---|---|
| mp, °C | −78.5 | <−60 | <−75 | <−70 | −117.2 | −11.9 | | 53 |
| bp, °C | 137.8 | 119.3 | 115.6 | 128.0 | 130.5 | 101.8 | 112.0 | 113–114 |
| Density $\varrho$, g/cm$^3$ | 0.8144 | 0.8090 | 0.8218 | 0.816 | 0.812 | 0.809 | 0.819 | |
| Refractive index $n_D^{20}$ | 1.4099 | 1.4053 | 1.4098 | 1.4097 | 1.4078 | 1.4052 | 1.4095 | |
| Flash point, °C | 50 | 33 | 40 | 43 | 44 | 21 | 26 | 36 |
| Ignition temperature, °C | 310 | 340 | 360 | 400 | 365 | 435 | | 430 |
| Viscosity, mPa·s | 3.31 (25 °C) | 0.53 (20 °C) | 4.12 (25 °C) | 5.09 (20 °C) | 3.86 (23.8 °C) | 3.79 (25 °C) | | |
| Specific heat at 20 °C, J g$^{-1}$ K$^{-1}$ | 2.98 | 3.40 | 2.85 | 2.50 | 2.28 | 3.15 | 2.63 | 2.69 |
| Heat of vaporization, J/g | 505 | | 405 | 419 | 441 | 445 | | |
| Vapor pressure at 20 °C, kPa | 0.14 | 0.53 | 0.93 | 0.40 | 0.27 | 1.60 | 1.07 | |
| Solubility in H$_2$O at 20 °C, wt% | 2.6 | 4.7 | | 2.2 | 2.5 | | | 3.0 |
| Expansion coefficient | 0.00092 | 0.00095 | 0.00149 | 0.00078 | 0.00090 | 0.00133 | | |

isomers with other liquids are known. The boiling points and compositions of binary systems with water are listed in Table 4.

**Table 3.** Specific rotations []$_D$ of optically active pentanols

| Compound | Specific rotation | |
|---|---|---|
| (R)-2-Methyl-1-butanol | + 3.1° | (20 °C) |
| (S)-2-Methyl-1-butanol | − 5.88° | (18 °C) |
| (S)-3-Methyl-2-butanol | + 5.34° | (20 °C) |
| (R)-2-Pentanol | −16.8° | (20 °C) |
| (S)-2-Pentanol | +16.6° | (20 °C) |

**Table 4.** Binary azeotropic mixtures with water

| Compound | bp of compound, °C | bp of azeotropic mixture, °C | Water in azeotrope, wt% |
|---|---|---|---|
| 1-Pentanol | 137.8 | 95.8 | 54.4 |
| 2-Pentanol | 119.3 | 91.7 | 36.5 |
| 3-Pentanol | 115.4 | 91.7 | 36.0 |
| 3-Methyl-1-butanol | 132.05 | 95.15 | 49.6 |
| 3-Methyl-2-butanol | 112.9 | 91.0 | 33.0 |
| 2-Methyl-2-butanol | 102.85 | 87.35 | 27.5 |

# 3. Chemical Properties

The chemical behavior of pentanols is determined primarily by the hydroxyl group. The most important reactions are dehydration, oxidation, and esterification.

**Dehydration.** Catalytic dehydration of alcohols for the production of olefins is a well-established reaction [4]. Isomeric mixtures of the $C_5$-olefins usually are obtained during dehydration of pentanols. *2-Pentanol* yields mainly 2-pentene [5] with ZnO – $Al_2O_3$ as catalyst at 270 – 370 °C; 1-pentene is produced through the use of thorium oxide-containing catalysts [6]. Many other studies of the influence of transition-metal oxides on the mechanism and selectivity of the catalytic dehydration of 2-pentanol have been carried out by B. H. DAVIS [7].

2-Methyl-1-butene is obtained from *2-methyl-1-butanol* on special aluminum oxide contacts with a selectivity of 86 – 90 % [8].

Dehydration of *3-methyl-1-butanol* yields mainly 3-methyl-1-butene [9] or 2-methyl-2-butene [10], depending on reaction conditions. A mixture of these two olefins is produced by the dehydration of 2-methyl-2-butanol [11].

Production of dipentyl ethers during the dehydration of pentanols is a commonly observed side reaction. Dipentyl ethers are normally obtained from pentanols under the influence of catalysts such as synthetic bentonites [12], modified montmorillonites [13], or solid superacids (e.g., Nafion-H) [14].

Dehydration of 2,2-dimethyl-1-propanol in the presence of nickel- or palladium- containing catalysts yields the corresponding dipentyl ether [15]. Syntheses of other dipentyl ethers are described in [16] – [18].

**Oxidation.** Primary and secondary pentanols are transformed to aldehydes, carboxylic acids or ketones under usual conditions; catalytic oxidation dehydrogenation is preferred industrially. Thus, 2-methylbutanal is obtained from 2-methyl-1-butanol by using MnO–NiO–MgO as catalyst [19]. Other suitable oxidation dehydrogenation catalysts for the synthesis of $C_5$-aldehydes from pentanols include lead(II) acetate [20], MgO–B, Al, Si, P, Sb [21], and copper(I) chloride–phenanthroline [22]. Further suitable oxidants are osmium tetroxide [23], dimethylsulf-oxide–trifluoroacetic anhydride [24], $CrO_2Cl_2$ [25], and $(Ph_3BiCl)_2O$ [26].

Recommended catalysts for the conversion of secondary pentanols to the corresponding ketones include modified bentonites [27], $(Ph_3P)_2Ru(O_2CCF_3)CO$ [28], $ZnO–Cr_2O_3–CuO–CdO$ [29], and phosphine-containing ruthenium(IV)–oxobipyridine complexes [30]. Other suitable oxidants are ozone [31], potassium permanganate [32], silver peroxy disulfate [33], nitrogen dioxide [34], and *tert*-butyl hypochlorite–pyridine [35].

The $C_5$-carboxylic acids can be obtained from primary pentanols under appropriate reaction conditions. 2,2-Dimethylpropionic acid (pivalic acid) and 2-methylbutanoic acid are obtained by oxidizing the corresponding alcohols with $KMnO_4–CuSO_4$ in the presence of potassium hydroxide [36]. Other suitable oxidizing agents are ozone [37] and nickel oxide–hydroxide [38].

**Esterification.** Pentanols can react with carboxylic acids or carboxylic anhydrides to yield esters by various methods. Examples are continuous reaction of 2-pentanol with acrylic acid under the catalytic action of acid ion exchangers in the liquid phase [39]; azeotropic esterification of formic acid with 1-pentanol with diisopropyl ether as entraining agent [40]; and reaction of bicyclo[2.2.0]oct-2-en-5,6-dicarboxylic anhydride with 1-pentanol [41].

Another frequently used method for the production of esters is transesterification. Methyl acetate reacts with 1-pentanol in the presence of catalysts such as the oxides of aluminum, titanium, tin, or zirconium at 20 °C to give pentyl acetate in high yield [42]. Diesters such as methyl succinate react with 2-methyl-2-butanol in the presence of lithium to give 2-methylbutyl succinate [43].

The conversion of pentanols to pentyl valerates (part of the pentanol is oxidized to the corresponding aldehyde which reacts with unconverted pentanol via the hemiacetal to the valerate) is catalyzed particularly by ruthenium-containing compounds. Thus, 2-methylbutyl 2-methylbutyrate is obtained from 2-methyl-1-butanol in the presence of $RuH_2(PPh_3)_4$ [44]; 2,2-dimethylpropyl 2,2-dimethylpropionate is formed from 2,2-dimethyl-1-propanol by using $Ru_3(CO)_{12}$ as catalyst [45]; and pentyl valerate is obtained from 1-pentanol and calcium hypochlorite with a yield of 83% [46].

Pentanols can react with isocyanates and aminonitriles in a similar manner. Phenylisocyanate reacts with 2,2-dimethyl-1-propanol in the presence of lithium alkoxides or di-*n*-butyltin diacetate to yield the corresponding *N*-phenylcarbamate [47]. 3-Methylbutyl 2-aminopropionate is obtained by reaction of 2-aminopropionitrile with 3-methyl-1-butanol [48].

**Other Reactions.** Pentanols are often used as building blocks in many chemical syntheses. For example, 2-pentanol reacts with methyl acrylate to yield 5-methyl-5-propyltetrahydrofuran-2-one, and 1-pentanol reacts with methyl allylacetate (methyl 4-pentenoate) to give 6-pentyltetrahydropyran-2-one [49]. N-Heterocyclic compounds can be produced by reacting pentanols with iminodiacetonitrile [50] or acetic anhydride and ammonia [51]. Substituted thiophenes can be synthesized by gas-phase reaction of 2-pentanol with carbon disulfide under the catalytic action of $Cr_2O_3 - Al_2O_3 - K_2CO_3$ [52].

Cyclization of pentanols such as 2-methyl-2-butanol and 2-methyl-1-butanol in the presence of $V_2O_5 - MoO_3$ at 360–410 °C yields citraconic anhydride (3-methyl-2,5-furandione) [53].

Other industrially important reaction products of $C_5$-alcohols include pinacolone produced from 2-methyl-2-butanol [54]; neopentylamine from 2,2-dimethyl-1-propanol [55]; xanthates from 1-pentanol and other $C_5$-alcohol isomers (especially 2-methyl-1-butanol), carbon disulfide, and sodium hydroxide [56]; and dipentyl carbonate from 1-pentanol and carbon dioxide [57].

Pentanols with a methyl group in the 2-position as well as the corresponding $C_{10}$-alcohols can be produced from pentanols under the catalytic action of nickel, palladium, or rhodium and a sodium alkoxide base (Guerbet reaction) [58]. Various ethers and glycols can be obtained from pentanols and epichlorohydrin [59] and from pentanols and formaldehyde [60]. Finally $C_5$-alcohols can be transformed to pentyl formates by rhodium-catalyzed formylation reaction [61] and to dipentyl tricyclo[5.2.1.0$^{2.6}$]dec-3-ene-8,9-dicarboxylate by reaction with dicyclopentadiene and carbon monoxide in the presence of $PdCl_2 - CuCl_2$ catalyst [62].

# 4. Production

Only a few industrial processes are known for the production of pentanols, namely, hydroformylation of butenes, hydrolysis of chloropentanes, and hydration of pentenes. Production of $C_5$-alcohols (mainly 3-methyl-1-butanol and 2-methyl-1-butanol) from fusel oils by fermentation has been replaced mainly by the oxo synthesis (discovered by O. ROELEN) and currently has little industrial importance. Production of pentanols from pentane by the Sharples process (chlorination of pentanes and subsequent hydrolysis) has also been largely replaced by oxo synthesis.

Hydroformylation of butenes (1-butene, 2-butene, and isobutene) and subsequent hydrogenation yield 1-pentanol, 2-methyl-1-butanol, and 3-methyl-1-butanol almost exclusively. The remaining $C_5$-alcohol isomers are obtained chiefly by hydration of $C_5$-olefins, hydrogenation of corresponding carbonyl compounds, and other special techniques.

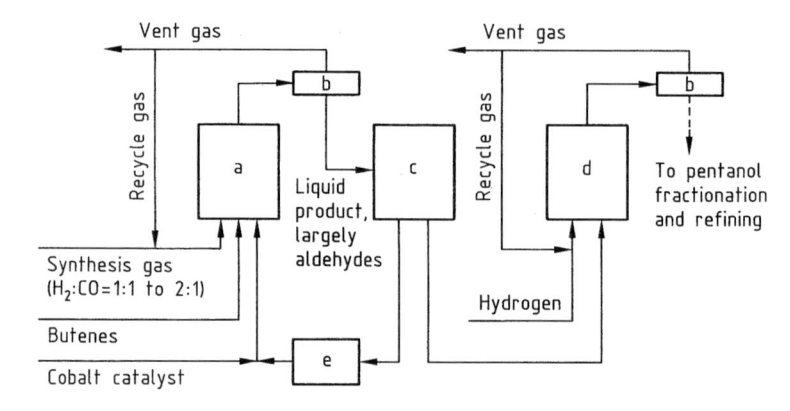

**Figure 1.** Flow diagram for the production of pentanols [63] a) Oxo reactor (100–200 °C, 10–45 MPa); b) Separator; c) Catalyst removal (decobalting); d) Hydrogenation unit (50–250 °C, 0.3–24 MPa); e) Catalyst recovery

**Hydroformylation of Butenes.** The most important process for the production of pentanols is hydroformylation of butenes. The $C_4$-olefins are reacted with a carbon monoxide–hydrogen mixture under the catalytic action of cobalt- or rhodium- containing compounds to yield $C_5$-aldehydes. Hydrogenation of the latter gives the corresponding $C_5$-alcohols. The flow sheet for this process is given in Figure 1.

Raffinate I from various cracking processes (after the separation of 1,3-butadiene) is used as raw material and contains ca. 44–49 vol% isobutene, 24–28 vol% 1-butene, 19–21 vol% 2-butene, and 8–11 vol% $C_4$-hydrocarbons [64]. Further processing of this $C_4$-raffinate converts isobutene by reaction with methanol to methyl *tert*-butyl ether (→ Methyl Tert-Butyl Ether). The remaining $C_4$-olefin mixture (raffinate II) is the most important raw material in the industrial production of 1-pentanol and 2-methyl-1-butanol (→ Butenes).

Cobalt-catalyzed hydroformylation of *n-butenes* and subsequent hydrogenation yield a mixture containing ca. 70% 1-pentanol and 30% 2-methyl-1-butanol [65]. If rhodium–triphenylphosphane is used as catalyst, a mixture containing up to 90% 1-pentanol and 10% 2-methyl-1-butanol is obtained.

*2-Butene* can be converted to mixtures of $C_5$-aldehydes of varying composition, depending on the rhodium-containing catalyst system and reaction conditions. Thus, a ratio of 1.13:1 of 1-pentanal to 2-methylbutanal is obtained with a rhodium–phosphite catalyst at 130 °C [66]; a ratio of 0.2:1 is obtained by hydroformylation with a rhodium–tricyclohexylphosphane catalyst at 110 °C [67]. The dependence of the $n:i$ ratio on catalyst species and temperature during conversion of 1-butene and 2-butene with a rhodium–triarylphosphane catalyst system is given in [68].

Hydroformylation of *isobutene* with rhodium or cobalt catalysts yields mainly 3-methyl-1-butanol [69]; small amounts of 2,2-dimethyl-1-propanal occur chiefly in the cobalt-catalyzed oxo synthesis [65].

Apart from raffinate II and pure olefins, other $C_4$-olefin mixtures can be used as raw materials for hydroformylation. Thus, hydroformylation of an olefin mixture containing

**Table 5.** Hydration of pentenes

| Olefin | Alcohol | Process catalyst | Reference |
|---|---|---|---|
| 1-Pentene | 2-pentanol | zeolites, aluminosilicates | [75] |
| 2-Methyl-2-butene | 2-methyl-2-butanol | acid ion exchangers | [76] |
| 2-Methyl-1-butene/2-Methyl-2-butene | 2-methyl-2-butanol | acid ion exchangers | [77] |
| 2-Methyl-2-butene | 2-methyl-2-butanol | zeolites | [78] |
| 2-Methyl-2-butene | 2-methyl-2-butanol | layer silicates | [79] |
| 2-Methyl-2-butene | 2-methyl-2-butanol | formic acid | [80] |
| 2-Methyl-2-butene/2-Methyl-1-butene | 2-methyl-2-butanol | HCl, $SbCl_3$, $ZnCl_2$ | [81] |
| 2-Methyl-2-butene/2-Methyl-1-butene | 2-methyl-2-butanol | acetic acid | [82] |

1-butene, 2-butene, and isobutene (ratio 1:1:2) with $Co(CO)_4$ as catalyst at 100 °C yields a mixture of 1-pentanol, 2-methyl-1-butanol, and 3-methyl-1-butanol in a ratio of 15:5.2:11 [70]. More detailed information on the hydroformylation of butenes is given in [65].

Catalysts based on CuO–ZnO [71], nickel [72], zirconia [73], and Cu–$TiO_2$ [74] are used industrially for the subsequent hydrogenation of $C_5$-aldehydes.

**Hydration of Pentenes.** 2-Pentanol, 3-pentanol, 2-methyl-2-butanol, and 3-methyl-2-butanol are produced mainly by hydration of pentenes (see Table 5 for details). Apart from classical hydration processes, many modifications of the hydroboration reaction are described in the literature, which are applied especially to stereoselective reactions [83]. The reaction of the monobromoborane–dimethyl sulfide complex with 2-methyl-2-butene yields 3-methyl-2-butanol with high selectivity [84]. 2-Pentene reacts with dimesitylborane at 65 °C to give a mixture containing 76% 2-pentanol and 24% 3-pentanol [85]. Several other hydroboration reactions are described in [86].

**Other Production Processes.** An important process for production of the three secondary pentanols—2-pentanol, 3-pentanol, and 3-methyl-2-butanol—is hydrogenation of the corresponding ketones. Catalysts for these reactions include not only commercially available nickel, copper, and palladium catalysts [87] but also a range of specialized catalyst systems such as cationic rhodium complexes [88], ruthenium–polystyrene [89], and $ClMgAlH_4$ [90].

Various acid-stable catalysts were developed particularly for the production of 2,2-dimethyl-1-propanol from pivalic acid. Pivalic acid is converted to 2,2-dimethyl-1-propanol with a yield of 94.3% in the presence of rhenium(VII) oxide ($Re_2O_7$) or osmium tetroxide ($OsO_4$) in dioxane at 90 °C and 100 bar [91]. Other active catalyst systems include CuO–ZnO–$Al_2O_3$ [92], zirconia [93], and $HTcO_4$ or $KReO_4$–$HClO_4$ [94].

In addition to pivalic acid its esters react with complex hydrides [95] or Raney Ni–Sn [96]; pivalic anhydride, (reaction with organoaluminum hydrides) [97] and pivalonitrile [in the presence of $ZrO(OH)_2$] can give $C_5$-alcohols in high yield [98].

Use of carbon monoxide and hydrogen to produce higher alcohols was investigated intensely in the 1980s. However, despite the development of various catalyst systems [99], this synthesis yields a mixture of many different alcohols and hence has not been used for industrial production of pentanols. The latter also applies to the synthesis of pentanols from the homologization (i.e., chain elongation by one or more methylene groups) of lower alcohols such as methanol and ethanol [100].

An interesting process for the production of 3-methyl-1-butanol is based on the reaction of isobutene with formaldehyde (Prins reaction). The intermediates formed—3-methyl-3-buten-1-ol or 4,4-dimethyl-1,3-dioxane (depending on conditions)—can be hydrogenated to 3-methyl-1-butanol by a catalyst system such as $L_2Cu(OAc)_2/Cr_2O_3 - ZnO - BaO$ (L=quinoline) [101]. Other suitable catalysts for this reaction are given in [102]. This process is used by several companies for the production of isoprene. Cleavage of the dioxane derivative occurs at 240–400 °C on carbon or $H_3PO_4 - Ca_3(PO_4)_2$ catalyst [64].

3-Methyl-2-butanol not only is formed by hydration of 2-methyl-2-butene, but also occurs as a byproduct during the production of isoprene via acetylene–acetone [103].

Finally, different methods for production of $C_5$-alcohols by oxidation of $C_5$-hydrocarbons exist [104]: oxidation of isopentane with Co–Mo stearate as catalyst at 110–140 °C yields 2-methyl-2-butanol [105].

# 5. Uses

The range of applications of 1-pentanol, 2-methyl-1-butanol, and 3-methyl-1-butanol (obtainable by the oxo synthesis) are discussed separately because of their individual significance.

**1-Pentanol.** 1-Pentanol is particularly useful as a solvent, extracting agent [106], and starting material for lubricant additives (zinc dipentyl dithiophosphates) [107] and for auxiliaries in flotation (xanthates). It is used as solvent in the reaction of substituted diamines with diisocyanates [108] and for the production of sulfurized olefins [109], polyacrylates [110], and pyridine-2,3-dicarboxylic acid derivatives [111]. 1-Pentanol is employed as an extracting agent in the purification of phosphoric acid [112] and the separation of strontium chloride from aqueous metal chloride solutions [113].

Other industrially important application of 1-pentanol are the synthesis of starting materials for the production of liquid crystals [114], dyes [115], flavorings [116], and special catalysts [117].

1-Pentanol is used as starting material for several syntheses in the pharmaceutical industry. Examples include the production of isoindolinone derivatives [118], 2-alkylbenzimidazoles [119], and condensed pyridine–pyrazine derivatives [120]. In the plant protection sector it is used as starting material in the synthesis of cyclopropanecarboxylates [121] and phenylurea derivatives [122].

1-Pentanol is employed in the production of many esters that are applied in different fields.

**2-Methyl-1-butanol,** like 1-pentanol, is used mainly as a solvent and an extracting agent and for the production of various esters; examples are given in [123]. Other important applications of 2-methyl-1-butanol include its use in the production of flotation reagents [124], liquid crystals [125], plant protection agents [126], fragrances [127], and corrosion inhibitors [128]. 2-Methyl-1-butanol is also part of the catalyst system employed in the polymerization of caprolactam [129]; it is used as a fuel additive [130] and in the extraction of crude oil [131]. A mixture of 1-pentanol and 2-methyl-1-butanol is generally used in practical applications instead of either of the pure alcohols.

**3-Methyl-1-Butanol.** 3-Methyl-1-butanol is used mainly as a solvent and extracting agent in several practical applications [132], [133]. An important use is in the production of di(3-methylbutyl) ether; the latter is a component of a catalyst system used for the industrial production of polypropylene [134]. Other uses of 3-methyl-1-butanol in catalysis are discussed in [135].

3-Methyl-1-butanol is used in the production of herbicides based on 1,2,4-triazolecarboxamides [136], $N$-methyl-4-pyridone derivatives [137], and substituted phenylhydrazones [138]. It is also employed for the synthesis of fragrances and flavors: 3-methyl-1-butanol reacts e.g., with ethylidenenorbornene to give the corresponding norbornyl ether [139], with bicyclopentadiene to yield tricyclo[5.2.1.0$^{2.6}$]dec-3-en-8(9)-yl-3-methylbutyl ether [140], and with vanillin alcohol to give the corresponding vanillin ether derivative [141].

3-Methyl-1-butanol is also used for the production of various additives [142], corrosion inhibitors [143], and azo dyes [144]. It is a component of surfactants [145].

**Other Pentanols.** *2-Methyl-2-butanol* is employed as solvent for coating materials based on epoxy resins and polyurethanes [146], in the production of aryl pyruvic acids [147], and in the oxidation of olefins to the corresponding carbonyl compounds [148]. It is also used as a stabilizing agent for 1,1,1-trichloroethane, chlorinated $C_3 - C_4$-hydrocarbons, and $N,N,$-dialkylallyl amines [149] – [153]. The reaction of 2,2,6,6-tetramethyl-4-cyano-4-hydroxypiperidine with 2-methyl-2-butanol yields highly active light stabilizers for polymers [154].

*2-Pentanol* is used as a solvent for primers containing phosphoric acid [150] and for the extraction of 2-hydroxy-4-methylthiobutyric acid [151].

*3-Methyl-2-butanol* is employed as a stabilizing agent for tetrachloroethane [155].

*2-Pentanol, 3-pentanol,* and *2-methyl-2-butanol* are used in the production of catalyst systems for dimerization and polymerization, and for the synthesis of methanol from carbon monoxide and hydrogen [156].

Other applications of $C_5$-alcohols that are inaccessible by the oxo synthesis are in the production of lubricant additives [157], cleaning agents [158], defrosting agents [159], various herbicides [160], and many fragrances [161].

## 6. Quality Specifications and Chemical Analysis

The eight $C_5$-alcohol isomers are available commercially as pure substances or as mixtures of isomers. Generally, high purity is required because of the variety of applications.

The quality of $C_5$-alcohols is controlled by determination of various characteristics such as density, refractive index, color, water content, boiling point, and hydroxyl value. These properties are usually determined according to ASTM, DIN, or BS methods ($\rightarrow$ Alcohols, Aliphatic). Isomeric purity and traces of byproducts are best determined by gas chromatography. Polar or nonpolar packed columns, as well as appropriate capillary columns, can be employed for separation. A thermal conductivity detector system is used for general purity determination; a flame ionization detector is employed for trace analysis (of byproducts); suitable carrier gases are hydrogen, nitrogen, and helium.

The commercially available $C_5$-alcohol mixture obtained from hydroformylation of 1-butene with subsequent hydrogenation contains ca. 70% 1-pentanol and 30% 2-methyl-1-butanol. This product is marketed under the name amyl alcohol. Typical specifications for amyl alcohol may be summarized as follows:

| | |
|---|---|
| Purity (gas chromatography) | min. 98 wt% |
| Boiling range at 101.3 kPa, min. 95 vol% (DIN 53171) | 133–139 °C |
| Density at 20 °C (DIN 51757) | 0.813–0.817 g/cm$^3$ |
| Refractive index $n_D$ at 20 °C (DIN 51423, part 2) | 1.410–1.411 |
| Hazen color number (DIN 53409) | max. 10 APHA |
| Water content (DIN 51777, modif.) | max. 0.2 wt% |
| $fp$ | < −60 °C |
| Viscosity at 20 °C (DIN 51562) | ca. 4.3 mPa · s |
| Solubility in water at 20 °C | 2.5 wt% |
| Water absorption at 20 °C | 10 wt% |
| Vapor pressure at 20 °C | 0.3 kPa |

## 7. Storage and Transportation

Pentanols can be stored or dispatched in normal or enameled steel containers, provided that ingress of moisture is prevented. Stainless steel containers can also be used. National and international safety regulations relating to air, sea, inland water, rail and road transport and current safety data for pentanols must be observed.

**Table 6.** Safety data for pentanols

| Compound | LD$_{50}$,[a] mg/kg | Flash point,[b] °C | Ignition temperature,[c] °C | Temperature class[d] | Group and hazard class[e] | Class and number of GGVE/GGVS/RID/ADR/ADNR | IMDG code | UN no. |
|---|---|---|---|---|---|---|---|---|
| 1-Pentanol | 3030 | 50 | 310 | T2 | AII | 3/3 | 3.3 | 1105 |
| 2-Methyl-1-butanol | 4920 | 43 | 400 | T2 | AII | 3/3 | 3.3 | 1105 |
| 3-Methyl-1-butanol | 1300 | 44 | 365 | T2 | AII | 3/3 | 3.3 | 1105 |
| Amyl alcohol | 2690 | 43 | 365 | T2 | AII | 3/3 | 3.3 | 1105 |

[a] Oral, rat.
[b] DIN 51 755/51 758.
[c] DIN 51 794.
[d] DIN 57 165/VDE 0165.
[e] According to VbF.

Flash points of the eight $C_5$-alcohol isomers lie between 30 and 50 °C; they belong to safety class A II (VbF). Some safety specifications for the pentanols available through the oxo synthesis are compiled in Table 6.

# 8. Toxicology and Occupational Health

**Kinetics and Metabolism.** Pentanols are absorbed quickly after oral, dermal, and inhalative administration. *Primary pentanols* are very quickly oxidized (especially in the liver) to form the corresponding aldehydes. The latter quickly oxidize further to acids. Oxidation of *secondary pentanols* to the corresponding ketones proceeds more slowly and takes place only partly in the liver. *Tertiary pentanol* is mainly excreted unchanged or as glucuronide. Elimination takes 3.5 – 9 h for primary pentanols, 13 – 16 h for secondary pentanols, and 50 h for tertiary pentanol [162]. The metabolite valeric acid is eliminated very slowly and is probably responsible for the hangover effects observed after intake of fusel oils [163].

**Acute Toxicity.** Pentanols have stronger narcotic and toxic effects than the lower alcohols. The LD$_{50}$ values after oral administration are in the range 1.0 – 5.77 g/kg body weight; primary pentanols have the lowest toxicity and tertiary alcohol has the highest [164]. The main symptoms are CNS depression and liver or kidney damage [165], [166]. Pentanols have an irritating effect on the skin and mucous membranes. Humans who inhale pentanol vapors exhibit irritation of the eyes and trachea, headache, dizziness, nausea, vomiting, diarrhea, breathing problems, coughing, and central nervous symptoms [164]. 3-Methyl-1-butanol is not a sensitizer in humans [167]. 2-Methyl-2-butanol was previously used as a sedative.

**Subchronic and Chronic Toxicity.** Daily oral administration of *1-pentanol* to rats for 13 weeks (dose 1000 mg/kg body weight) did not produce any toxic effects [168]. Daily oral administration of *3-methyl-1-butanol* to rats for 17 weeks (dose 1000 mg/kg body weight) had a slightly adverse effect on body weight gain and caused a slight reduction in food intake [169]. Rats that received 3-methyl-1-butanol for their lifetime, either as an oral dose of 81 mg/kg body weight twice a week or as a subcutaneous dose of 32.5 mg/kg body weight per week, showed higher tumor incidence. Furthermore, haematotoxic and hepatotoxic effects were observed [170]. Repeated dermal application of *tertiary pentanol* to rabbits (dose 3440 mg/kg body weight per day) resulted in serious systemic effects and death. Animals that received a dose of 344 mg/kg body weight per day under similar conditions showed slight irritation at the site of application. Thirteen-week inhalative exposure of 3600 mg/m$^3$ of 2-methyl-2-butanol to rats, mice, and dogs resulted in impairment of motor coordination, lacrimation, and increased liver weight in rats and dogs. Inhalative exposure of 810 mg/m$^3$ of 2-methyl-2-butanol caused slight irritation of mucous membranes; exposure to 180 mg/m$^3$ of *tert*-pentanol had no adverse effects [164].

**Mutagenicity.** 2-Methyl-2-butanol did not exhibit mutagenic activity in the Ames test [164]. In contrast, higher reversion rates were observed with pentanol on *Escherichia coli* [171]. After incubation of V 79 cells (fibroblasts) with 1-pentanol, an increase in the number of cells with more than 22 chromosomes was reported [172].

**Teratogenicity.** Exposure of pregnant rats to 1-pentanol at a concentration of 14 000 mg/m$^3$ of air resulted in decreased food intake in dams. The juvenile exhibited retarded bone development; no malformations were found [173].

**Occupational Health.** A TLV – TWA value of 360 mg/m$^3$ and an MAK value of 360 mg/m$^3$ have been set for 3-methyl-1-butanol [174], [175].

# 9. References

[1] R. Stephenson, J. Stuart, M. Tabak, *J. Chem. Eng. Data* **29** (1984) 287 – 290. Z. Maczynska, *Solubility Data Ser.* **15** (1984) 159 – 161. A. Maczyski, *Solubility Data Ser.* **15** (1984) 203 – 209.

[2] L. Adembri, *Ann. Chim. (Rome)* **46** (1956) 62 – 68. M. Windholz et al. in: *The Merck Index*, 9th ed., Merck, Rahway, NJ, 1976, no. 5905. *Handbook of Chemistry and Physics*, 55th ed., CRC Press, Cleveland, Ohio, 1974/75. Beilstein **E IV 1**, 1655 – 1656, 1666.

[3] L. H. Horsley: "Azeotropic Data III," *Adv. Chem. Ser.* **116** (1973).

[4] J. Falbe (ed.): "Grundgerüste," *Methodicum Chimicum*, vol. **4**, Thieme Verlag, Stuttgart 1980.

[5] Phillips Petroleum Co., US 4 260 845, 1981 (T. K. Shioyama).

[6] Institut Français du Pétrole, FR 1 603 748, 1971 (J. M. Bonnier, J. P. Damon, B. Delmon).

[7] B. H. Davis, *J. Catal.* **79** (1983) 58 – 69. A. H. Dabbagh, B. H. Davis, *J. Catal.* **110** (1988) 416 – 418. A. H. Dabbagh, B. H. Davis, *J. Mol. Catal.* **47** (1988) 123 – 127.

[8] Phillips Petroleum Co., US 4 529 827, 1985 (C. A. Drake); US 4 602 119, 1986 (C. A. Drake).
[9] Phillips Petroleum Co., US 4 234 752, 1980 (Y. Wu, S. J. Marwil).
[10] V. P. Kasumova, A. M. Ragimova, *Model. Optim. Khim. Protsessov* **3** (1977) 128–131; *Chem. Abstr.* **89** (1978)163 154.
[11] Hindustan Lever Ltd., IN 161 104, 1987 (S. U. Kulkarni).
[12] British Petroleum Co., GB 2 164 636, 1986 (P. A. Diddams, W. Jones, J. M. Thomas).
[13] British Petroleum Co., GB 2 179 563, 1987 (J. A. Ballantine et al.).
[14] M. He, X. Fu, G. Zeng, *Xinan Shifan Daxue Xuebao, Ziran Kexueban* **2** (1987) 114–116; *Chem. Abstr.* **109** (1988) 189 789.
[15] H. Pines, J. Hensel, J. Simonik, *J. Catal.* **24** (1972) 206–210. E. Licht, Y. Schaechter, H. Pines, *J. Catal.* **55** (1978) 191–197.
[16] J. A. Ballantine et al., *J. Chem. Soc., Chem. Commun.* 1981, no. 9, 427–428.
[17] J. Simonik, H. Pines, *J. Catal.* **24** (1972) 197–205.
[18] E. Licht, Y. Schaechter, H. Pines, *J. Catal.* **31** (1973) 110–118.
[19] Phillips Petroleum Co., US 4 304 943, 1981 (G. Bjornson).
[20] B. M. Choudary, P. N. Reddy, L. M. Kantam, Z. Jamil, *Tetrahedron Lett.* **26** (1985) 6257–6258.
[21] Nippon Shokubai Kagaku Kogyo Co., EP-A 183 225, 1986 (Y. Shimasaki, Y. Hino, M. Ueshima).
[22] C. Jallabert, H. Riviere, *Tetrahedron* **36** (1980) 1191–1194.
[23] V. P. Somaiah, B. K. Reddy, B. Sethuram, N. T. Rao, *Indian J. Chem. Sect. A* **27A** (1988) no. 10, 876–879; *Chem. Abstr.* **111** (1989) 6716.
[24] S. L. Huang, K. Omura, D. Swern, *J. Org. Chem.* **41** (1976) 3329–3331.
[25] J. San Filippo, C. I. Chern, *J. Org. Chem.* **42** (1977) 2182–2183.
[26] Agence Nationale de Valorisation de la Recherche, FR 2 441 602, 1980 (D. Barton, W. Motherwell).
[27] Mitsui Petrochemical Industries, JP-Kokai 62 142 134, 1987 (T. Fujita, K. Mizuno, K. Saeki).
[28] A. Dobson, S. D. Robinson, *Inorg. Chem.* **16** (1977) 137–142.
[29] Wacker-Chemie, DE-OS 2 358 254, 1975 (O. Sommer, F. Knoerr).
[30] M. E. Marmion, K. J. Takeuchi, *J. Chem. Soc. Dalton Trans.* 1988, no. 9, 2385–2391.
[31] W. L. Waters et al., *J. Org. Chem.* **41** (1976) 889–891.
[32] S. N. Murthy, E. V. Sundaram, *Indian J. Chem. Sect. A* **16 A** (1978) no. 9, 806–808; *Chem. Abstr.* **90** (1979) 38 354.
[33] C. Walling, D. M. Camaioni, *J. Org. Chem.* **43** (1978) 3266–3271.
[34] Kogai Boshi Chosa Kenkyusho, DE-OS 2 826 065, 1979 (W. Ando, I. Nakaoka).
[35] J. N. Milovanovic, M. Vasojevic, S. Gojkovic, *J. Chem. Soc. Perkin Trans. 2* 1988, no. 4, 533–535.
[36] C. W. Jefford, Y. Wang, *J. Chem. Soc. Chem. Commun.* 1988, no. 10, 634–635.
[37] S. Miyazaki, Y. Suhara, *J. Am. Oil Chem. Soc.* **55** (1978) 536–538.
[38] J. Kaulen, H. J. Schaefer, *Tetrahedron* **38** (1982) 3299–3308.
[39] Knapsack, FR 2 186 457, 1974.
[40] W. Werner, *J. Chem. Res. Synop.* 1980, no. 6, 196.
[41] A. S. Kyazimov et al., *Dokl. Akad. Nauk. Az. SSR* **39** (1983) 39–42; *Chem. Abstr.* **102** (1985) 24 153.
[42] Japan Tobacco Inc., EP-A 230 286, 1987 (H. Matsushita, M. Shibagaki, K. Takahashi).
[43] Ciba-Geigy, EP-A 278 914, 1988 (U. Frei, R. Kirchmayr).
[44] S. Murahashi et al., *J. Org. Chem.* **52** (1987) 4319–4327. S. Murahashi et al., *Tetrahedron Lett.* **22** (1981) 5327–5330.
[45] Y. Blum, Y. Shvo, *J. Organomet. Chem.* **263** (1984) 93–107.
[46] S. O. Nwaukwa, P. M. Keehn, *Tetrahedron Lett.* **23** (1982) 35–38.

[47] W. J. Bailey, J. R. Griffith, *J. Org. Chem.* **43** (1978) 2690–2692.
[48] BASF AG, DE-OS 3 145 736, 1983 (W. Boell, M. Kroener, K. H. Beyer, D. Hertel).
[49] Ube Industries, JP 53 036 466, 1978 (S. Yamazaki et al.). Asahi Chemical Ind., JP-Kokai 57 156 481, 1982; JP-Kokai 57 167 980, 1982.
[50] BASF AG, DE-OS 3 242 195, 1984 (F. Brunnmueller, M. Kroener).
[51] Eastman Kodak Co., EP-A 96 641, 1983 (H. G. Rajoharison, C. M. A. Roussel).
[52] Miland-Yorkshire Tar Distillers, DE-OS 2 225 443, 1973 (N. R. Clark, W. E. Webster).
[53] Pfizer Inc., DE-OS 2 522 568, 1975 (R. G. Berg, B. E. Tate).
[54] Kuraray Co., JP-Kokai 54 148 711, 1979 (S. Kyo, H. Tsuchiya, H. Tanako); JP-Kokai 54 066 613, 1979 (H. Tanaka, H. Tsuchiya, S. Kyo); DE-OS 2 918 521, 1979 (S. Kyo, H. Tsuchiya, H. Tanaka).
[55] Bayer AG, EP-A 22 532, 1981 (F. Werner et al.).
[56] CIL Inc., ZA 80 4227, 1981 (D. J. Gannon, D. T. F. Fung). J. Sejbl, *Rudy* **31** (1983) 163–165;*Chem. Abstr.* **99** (1983) 175 192.
[57] Sun Abbott Co., JP-Kokai 57 058 645, 1982.
[58] J. Sabadie, G. Descotes, *Bull. Soc. Chim. Fr.* 1983, no. 9–10, Part 2, 253–256. P. L. Burk, R. L. Pruett, K. S. Campo, *J. Mol. Catal.* **33** (1985) 1–21.
[59] Kanebo, JP-Kokai 54 115 307, 1979 (K. Honda, Y. Nishijima).
[60] British Petroleum Co., EP-A 147 983, 1985 (C. G. Griggs); EP-A 147 133, 1985 (C. G. Griggs, D. A. Pippard).
[61] G. Jenner, G. Bitsi, *J. Mol. Catal.* **45** (1988) 235–246.
[62] Nissan Chemical Ind., JP-Kokai 59 139 376, 1984; JP-Kokai 60 104 080, 1985.
[63] H. Bieber, *Encycl. Chem. Process. Des.* **3** (1977) 278–289.
[64] K. Weissermel, H. J. Arpe: *Industrielle Organische Chemie,* 3rd ed., VCH Verlagsgesellschaft, Weinheim 1988.
[65] W. J. Scheidmeir, *Chem. Ztg.* **96** (1972) 383–387.
[66] Union Carbide Corp., US 4 717 775, 1988 (E. Billig, et al.).
[67] Exxon EP-A 195 656, 1986 (D. A. Young).
[68] Exxon Res. and Eng., WO 84 3697, 1984 (A. A. Oswald, J. S. Merola, J. C. Reisch, R. V. Kastrup).
[69] UOP Inc., US 4 438 287, 1984 (T. Imai). Inst. de Cercetari Ing. Tehnologica, RO 83 795, 1984 (I. Muja, D. Goidea).
[70] Monsanto Co., EP-A 94 456, 1983 (G. E. Barker, D. Forster).
[71] Union Carbide Corp., DE-OS 3 737 277, 1988 (J. E. Logsdon, R. H. Loke, J. S. Merriam, R. W. Voight); EP-A 8767, 1980 (C. C. Pai).
[72] J. M. Bonnier, J. P. Damon, J. Masson, *Appl. Catal.* **30** (1987) 181–184.
[73] M. Shibagaki, K. Takahashi, H. Matsushita, *Bull. Chem. Soc. Jpn.* **61** (1988) 3283–3288. Japan Tobacco Inc., JP-Kokai 86 204 143, 1986 (S. Mizusaki et al.).
[74] F. S. Delk, A. Vavere, *J. Catal.* **85** (1984) 380–388.
[75] Toa Nenryo Kogyo, EP-A 127 486, 1984 (Y. Okumura, S. Kamiyama, H. Furukawa, K. Kaneko).
[76] EC Erdölchemie, EP-A 325 144, 1989 (R. Malessa, B. Schleppinghoff).B. Schleppinghoff, C. Gabel, H. L. Niederberger, *Erdöl Erdgas* **104** (1988) 173–177. Dow Chemical Co., US 4 182 920, 1980 (J. H. Giles, J. H. Stultz, S. W. Jones).
[77] Z. Prokop, K. Setinek, *Collect. Czech. Chem. Commun.* **52** (1987) 1272–1279.
[78] Centre National de la Recherche Scientifique, Institut Français du Pétrole, BE 904 797, 1986 (F. F. Francois et al.).
[79] J. A. Ballantine et al., *J. Mol. Catal.* **26** (1984) 57–77.

[80]  A. A. Makhnin, A. F. Frolov, K. A. Aronovich, *Kinet. Katal.* **22** (1981) 344–348; *Chem. Abstr.* **95** (1981) 79557. A. A. Makhnin et al., *Khim Tekhnol. Topl. Masel.* 1979, 34–35; *Chem. Abstr.* **91** (1979) 6161.

[81]  Japan Oil Co., JP-Kokai 72 39007, 1972 (A. Nambu); JP-Kokai 72 39013, 1972 (A. Nambu).

[82]  Gulf Res. and Development Co., US 3285977, 1966 (A. M. Henke, R. C. Odioso, B. K. Schmid).

[83]  Y. Aoyama et al., *J. Org. Chem.* **52** (1987) 2555–2559. Aldrich-Boranes Inc., EP-A 188235, 1986 (H. C. Brown). H. C. Brown, U. S. Racherla, *J. Org. Chem.* **51** (1986) 895–897. H. C. Brown, N. Ravindran, *J. Org. Chem.* **42** (1977) 2533–2534.

[84]  H. C. Brown, N. Ravindran, *Synthesis* 1977, 695–697.

[85]  A. Pelter, S. Singaram, H. C. Brown, *Tetrahedron Lett.* **24** (1983) 1433–1436.

[86]  H. C. Brown, M. C. Desai, P. K. Jadhav, *J. Org. Chem.* **47** (1982) 5065–5069. H. C. Brown, P. K. Jadhav, A. K. Mandal, *J. Org. Chem.* **47** (1982) 5074–5083. H. C. Brown, J. A. Sikorski, *Organometallics* **1** (1982) 28–37. H. C. Brown, B. Singaram, C. P. Mathew, *J. Org. Chem.* **46** (1981) 2712–2717. M. Anez, G. Uribe, L. Mendoza, R. Contreras, *Synthesis* 1981, 214–216.

[87]  Hoechst AG, Katalysatoren Hoechst, Frankfurt 1979.

[88]  H. Fujitsu et al., *J. Chem. Soc. Perkin Trans. 1* 1981, 2650–2655.

[89]  F. Ciardelli, P. Pertici, *Z. Naturforsch. B Anorg. Chem. Org. Chem.* **40B** (1985) 133–140.

[90]  Studiengesellschaft Kohle, DE-OS 3536797, 1987 (B. Bogdanovic, M. Schwickardi).

[91]  Kao Corp., JP-Kokai 87210056, 1987 (Y. Kajiwara, Y. Inamoto, J. Tsuji).

[92]  Air Products and Chemicals Inc., EP-A 180210, 1986 (S. A. Butter, I. Stoll).

[93]  Japan Tobacco Inc., EP-A 285786, 1988 (H. Matsuhita, M. Shibagaki, K. Takahashi).Mitsubishi Chemical Industries Co., JP-Kokai 87108832, 1987 (T. Maki, M. Nakajima, T. Yokoyama, T. Setoyama).

[94]  Akademie der Wissenschaften, DD 142050, 1980 (M. Wahren et al.).

[95]  H. C. Brown, S. Narasimhan, *J. Org. Chem.* **49** (1984) 3891–3898. K. Soai, H. Oyamada, M. Takase, A. Ookawa, *Bull. Chem. Soc. Jpn.* **57** (1984) 1948–1953.

[96]  Institut Français du Pétrole, EP-A 172091, 1986 (J. P. Bournonville, J. P. Candy, G. Mabilor).

[97]  Anic S.p.A., DE-OS 2945348, 1980 (G. Dozzi, S. Cucinella); DE-OS 2849767, 1979 (S. Cucinella).

[98]  Japan Tobacco Inc., EP-A 271092, 1988 (H. Matsushita, M. Shibagaki, K. Takahashi).

[99]  Standard Oil Co., US 4677091, 1987 (T. J. Mazanec, J. G. Frye). Eastman Kodak Co., US 4748144, 1988 (J. R. Monnier, G. R. Apai, M. J. Hanrahan). Belge Secretaire General des Services de Programmation, EP-A 283635, 1988 (A. Noels et al.). W. X. Pau, R. Cao, G. L. Griffin, *J. Catal.* **114** (1988) 447–456. J. G. Nunan et al., *J. Catal.* **116** (1989) 195–221.E. Tronconi et al., *Appl. Catal.* **47** (1989) 317–333. Metallgesellschaft, DE-OS 3403492, 1985 (G. Cornelius et al.). Phillips Petroleum Co., US 4537909, 1985 (F. N. Lin, F. Pennella).

[100]  Metallgesellschaft, DE-OS 3641774, 1988 (W. Hilsebein, E. Supp, F. Möller, P. König). Dow Chemical Co., EP-A 180719, 1986 (G. J. Quarderer et al.). Union Oil Co. of California, US 4540836, 1985 (D. M. Fenton).

[101]  A., V. Devekki et al., *Zh. Obshch. Khim.* **53** (1983) 2652–2653; *Chem. Abstr.* **100** (1984) 156303.

[102]  N. N. Gonchakova et al., *Neftekhimiya* **21** (1981) 567–575; *Chem. Abstr.* **95** (1981) 188395. Kuraray Co., JP-Kokai 74 11806, 1974 (S. Kyo, T. Yasui). Mitsubishi Gas Chemical Co., JP 73 43084, 1973 (A. Miyamoto et al.).

[103]  C. Moberg, L. Rakos, *J. Organomet. Chem.* **335** (1987) 125–131. G. D. Zakumbaeva, L. A. Beketaeva, S. Z. Aitmagambetova, *Izv. Akad. Nauk. Kaz. SSR Ser. Khim.* 1979, 17–21; *Chem.*

[104] *Abstr.* **92** (1980) 11 749. N. A. Zakarina, V. P. Kuidina, A. M. Khisametdinov, *Izv. Akad. Nauk. Kaz. SSR Ser. Khim.* 1982, 30–33; *Chem. Abstr.* **96** (1982) 217 175. D. V. Sokolskii et al., *Zh. Khim.* **23** (1987) 1430–1432; *Chem. Abstr.* **108** (1988) 166 935. Institut Français du Pétrole, FR 2 175 689, 1973 (H. Mimoun, C. Bocard, I. Seree de Roch). K. S. Suslick, B. R. Cook, M. Fox, *J. Chem. Soc. Chem. Commun.* 1985, 580–582. B. R. Cook, T. J. Reinert, K. S. Suslick, *J. Am. Chem. Soc.* **108** (1986) 7281–7286. C. A. Dewar, C. J. Suckling, R. Higgins, *J. Chem. Res. Synop.* 1979, 335. R. V. Nazarova et al., *Issled. Obl. Khim. Vysokomol. Soedin. Neftekhim.* 1977, 23; *Chem. Abstr.* **93** (1980) 45 891.

[105] S. A. Yurzhenko, V. A. Lobov, *Neftepererab. Neftekhim. (Kiev)* **15** (1977) 92–95; *Chem. Abstr.* **88** (1978) 89 034.

[106] Hoechst AG, Alkohole Hoechst, Frankfurt 1984.

[107] Standard Oil Co., BE 896 226, 1983.

[108] Texaco Inc., US 4 761 465, 1988 (G. P. Speranza, J. J. Lin, M. Cuscurida).

[109] Lubrizol Corp., US 4 764 297, 1988 (R. L. Sowerby, S. A. Di Biase).

[110] Rohm and Haas Co., US 4 594 363, 1986 (R. M. Blankenship, A. Kowalski).

[111] Sugai Chemical Industry Co., EP-A 274 379, 1988 (T. Yamashita, M. Kodama, S. Shimada).

[112] K. P. Ehlers, *Ber. Bunsen-Ges. Phys. Chem.* **83** (1979) 1113–1116.

[113] VEB Erdoel-Erdgas Gommern, DD 243 687, 1987 (H. Richter et al.).

[114] Kanto Chemical Co., JP-Kokai 01 51538, 1989 (K. Suzuki, A. Sugiura, T. Fujii). Mitsui Toatsu Chemicals, EP-A 305 938, 1989 (M. Nakatsuka et al.). Alps Electric Co., JP-Kokai 62 294 663, 1987 (M. Kano). ICI, EP-A 259 995, 1988 (D. A. Jackson, P. A. Gemmell).

[115] Hoechst AG, DE-OS 3 644 661, 1988 (K. Guenther, G. Mau). Hitachi Chemical Co., EP-A 284 370, 1988 (S. Tai et al.).

[116] International Flavors and Fragrances Inc., US 4 435 428, 1984 (R. M. Boden, M. R. Hanna, T. J. Tyszkiewicz). BASF AG, DE-OS 3 116 913, 1982 (F. Thoemel, W. Hoffmann).

[117] L. Cerveny et al., *React. Kinet. Catal. Lett.* **33** (1987) 471–476. Nisso Petrochemical Industries Co., DE-OS 3 025 434, 1981 (K. Kogoma et al.).

[118] Rhône-Poulenc, FR 2 607 506, 1988 (J. D. Bourzat et al).

[119] Lonza, CH 653 022, 1985 (R. Bergamin, B. Bourgeois, J. Jovanovic).

[120] Shionogi and Co., EP-A 157 260, 1985 (I. Adachi, T. Yamamori, M. Ueda). Lion Corp., DE-OS 3 432 983, 1985 (S. Suzuki).

[121] Roussel-Uclaf, CA 1 248 455, 1989 (J. Martel, J. Tessier, A. Teche).

[122] Takeda Chemical Industries, JP-Kokai 62 205 053, 1987 (J. Yamada, H. Miki, T. Uekado).

[123] Celanese Corp., US 4 470 881, 1984 (L. Berg). BASF Wyandotte Corp., US 4 426 291, 1984 (S. G. Sharangpani, F. G. Spence). S. C. Johnson and Son Inc., EP-A 71 116, 1983 (G. O. Schulz, D. M. Wilson).

[124] Lubrizol Corp., WO 89 00 457, 1989 (A. C. Clark, E. P. Richards, D. R. Shaw).

[125] Centre National de la Recherche Scientifique, FR 2 612 182, 1988 (N. Huu Tinh, C. Salleneuve, C. J. P. Destrade). Idemitsu Kosan Co., EP-A 258 898, 1988 (K. Morita, S. Uchida, S. Hachiya).Merck, WO 87 05 018, 1987 (R. Eidenschink et al.). Canon K. K., DE-OS 3 629 446, 1987 (K. Yoshinaga, K. Katagiri, T. Kobayashi, K. Shinjo).

[126] Monsanto Co., FR 2 479 200, 1981 (G. H. Alt); FR-A 2 479 202, 1981 (G. H. Alt); US 4 284 564, 1981 (G. H. Alt, J. P. Chupp).

[127] Toray Industries Inc., WO 84 00 161, 1984 (M. Naruto, H. Kondo, G. Hata).

[128] Bayer AG, DE-OS 2 535 685, 1977 (H. D. Block).

[129] Stamicarbon, EP-A 238 143, 1987 (J. J. M. Bongers, A. A. Van Geenen).

[130] Union Carbide Corp., BR 8 306 714, 1984 (E. J. Derderian). Texaco Inc., US 4 539 014, 1985 (W. M. Sweeney); US 4 713 087, 1987 (R. H. Jenkins et al.).
[131] M. A. Dahami, W. D. Constant, J. M. Wolcott, *Fuel* **67** (1988) 1242–1248. Marathon Oil Co., US 3 613 786, 1971 (S. C. Jones, W. O. Roszelle, M. A. Svaldi).
[132] Mitsubishi Gas Chemical Co., EP-A 127 888, 1984 (T. Isshiki, T. Yui, H. Uno, M. Abe). Mobil Oil Corp., US 4 532 084, 1985 (A. O. M. Okorodudu). Soken Chemical Eng. Co., JP 62 045 580, 1987 (S. Morikawa). V. Macho et al., CS 254 225, 1988.
[133] Haarmann und Reimer, DE-OS 2 729 075, 1979 (K. Bauer et al.). Negev Phosphates, DE-OS 3 438 655, 1986 (A. Iosef, M. Bar-On, J. Oren, D. Kellerman). M. Muhammed, Y. Zhang, *Hydrometallurgy* **21** (1989) 277–292.
[134] Solvay, EP-A 288 109, 1988 (P. Fiasse); EP-A 261 727, 1988 (P. Fiasse, A. Bernard). Chisso Corp., EP-A 284 287, 1988 (J. Masuda, T. Shiraishi, N. Komori, T. Uwai). BP Chemicals, EP-A 238 345, 1987 (J. C. A. Bailly, S. Sandis).
[135] Nippon Oil Co., EP-A 143 334, 1985 (M. Matsuno, M. Kudoh, H. Imai). A. Sarpudeen, I. Tajuddin, Q. Anwaruddin, *Eur. Polym. J.* **20** (1984) 1019–1020. K. Kishore et al., *J. Appl. Polym. Sci.* **31** (1986) 2829–2837.
[136] Kureha Chemical Industry Co., JP 63 230 676, 1988 (T. Shida et al.). Kureha Chemical Industry Co., JP 63 230 664, 1988 (T. Shida, Y. Kubota, I. Ichinose).
[137] Bayer AG, DE-OS 3 430 232, 1986 (H. J. Knops et al.).
[138] Kureha Chemical Industry, EP-A 285 294, 1988 (T. Shida et al.).
[139] International Flavors and Fragrances, US 4 379 060, 1983 (M. A. Sprecker); US 4 311 861, 1982 (M. A. Sprecker); US 4 369 133, 1983 (M. A. Sprecker).
[140] BASF AG, DE-OS 2 642 519, 1978 (W. Hoffmann, L. Schuster).
[141] Takasago Perfumery Co., JP 57 009 729, 1982.
[142] Standard Oil Co., US 4 155 958, 1979 (E. K. Fields). VEB Petrolchemisches Kombinat Schwedt, DD 201 146, 1983 (H. Brendel et al.). Lubrizol Corp., WO 88 04 685, 1988 (T. F. Steckel).
[143] L. Komora et al., CS 229 760, 1986.
[144] Ciba-Geigy, DE-OS 2 922 227, 1979 (R. Portmann).
[145] Phillips Petroleum, US 4 504 399, 1985 (P. R. Stapp). UOP Inc., US 4 501 675, 1985 (T. P. Malloy, R. J. Swedo).
[146] Lechler Chemie, DE-OS 3 033 613, 1982 (W. Oechsner, O. Vogt). Dow Chemical Co., US 4 778 863, 1988 (C. S. Wang, Z. K. Liao).
[147] Sagami Chemical Research Center, JP 60 061 550, 1985; JP 62 116 541, 1987 (K. Hirai, T. Fuchigami).
[148] Shell Internationale Research Maatschappij, GB 2 167 406, 1986 (B. L. Feringa).
[149] Dow Chemical Co., US 4 808 440, 1989 (E. L. Tasset).
[150] Monsanto Co., EP-A 143 100, 1985 (D. A. Ruest, M. Takano, L. R. Wolf); EP-A 142 488, 1985 (D. A. Ruest, M. Takano, L. R. Wolf).
[151] Rhône-Poulenc, GB 2 024 242, 1980 (Y. Correia, J. Lesparre). Toyo Soda Mfg. Co., JP 61 189 237, 1986 (Y. Oda, J. Mizoguchi, K. Yabuta); JP 61 191 629, 1986 (Y. Oda, J. Mizoguchi, K. Yabuta). Dow Chemical Co., US 4 309 301, 1982 (N. Ishibe, T. G. Metcalf).
[152] Deutsche Solvay Werke, EP-A 309 958, 1989 (W. Dilla, G. Jakobson).
[153] Wako Pure Chemical Industries, JP 49 125 309, 1974 (K. Yamanishi, Y. Takezoe).
[154] Hoechst AG, DE-OS 2 742 582, 1979 (H. Wiezer, N. Mayer, G. Pfahler).
[155] Asahi Chemical Ind., JP 69 20 965, 1969 (Y. Hashiguchi, T. Owaki).

[156] Lithium Corp. of America, US 4 555 498, 1985 (C. W. Kamienski). E. Alder, H. Fuellbier, W. Gaube, DD 152 778, 1981;DD 152 779, 1981.Hercules Inc., EP-A 41 362, 1981 (W. P. Long, A. S. Matlack). Shell Internationale Research Maatschappij, EP-A 306 114, 1989 (S. T. Sie, E. Drent, W. W. Jager); EP-A 289 067, 1988 (S. T. Sie, E. Drent, W. W. Jager).
[157] Asahi Denka Kogyo, JP 86 87 690, 1986 (S. Katsumata et al.).
[158] Asahi Glass Co., JP 72 20 232, 1972 (H. Ukibashi, K. Kirimoto, A. Hara).
[159] Hoechst AG, GB 2 027 046, 1980 (H. Peters, E. Surma).
[160] American Cyanamid Co., US 4 188 487, 1980 (M. Los). Ciba-Geigy, DD 206 525, 1984 (W. Kunz, W. Eckhardt). Bayer, DE-OS 3 643 851, 1988 (K. Sasse et al.).
[161] Mitsui Petrochemical Industries, DE-OS 2 427 465, 1974 (H. Miki, H. Hasui).International Flavors and Fragrances Inc., DE-OS 2 330 648, 1974 (J. B. Hall, M. G. Beets, L. E. Lala, W. I. Taylor). Mitsui Petrochemical Ind., DE-OS 2 427 465, 1979 (H. Miki, H. Hasui). Lucta S.A., ES 4 993 976, 1981 (A. Gavalda Sola).
[162] E. Browning: *Toxicity and Metabolism of Industrial Solvents,* Elsevier Publ., Amsterdam 1965.
[163] E. Rüdell et al., *Blutalkohol* **18** (1981) no. 5,315–325.
[164] V. K. Rowe, S. B. McCollister in G. D. Clayton, F. E. Clayton (eds.): *Patty's Industrial Hygiene and Toxicology,* 3rd ed., vol. **2 C,** Wiley-Interscience, New York 1982, p. 4527.
[165] R. A. Scala, E. G. Burtis, *Am. Ind. Hyg. Assoc. J.* **34** (1973) 493–499.
[166] I. F. H. Purchase, *S. Afr. Med. J.* **43** (1969) 795–798.
[167] D. L. J. Opdyke, *Food Cosmet. Toxicol.* **16** (1978) Suppl. 1, 785–787.
[168] K. R. Butterworth et al., *Food Cosmet. Toxicol.* **16** (1978) 203–207.
[169] F. M. B. Carpanini et al., *Food Cosmet. Toxicol.* **11** (1973) 713–724.
[170] W. Gibel, K. Lohs, G. P. Wildner, *Arch. Geschwulstforsch.* **45** (1975) no. 1, 19–24.
[171] H. Hilscher, E. Geissler, H. Lohs, W. Gibel, *Acta Biol. Med. Ger.* **23** (1969) 843–852.
[172] A. önfelt, *Mutat. Res.* **182** (1987) 135–154.
[173] B. K. Nelson et al., *J. Am. Coll. Toxicol.* **8** (1989) no. 2, 405–410.
[174] ACGIH, TLV's—Threshold Limit Values and Biological Indices for 1989–1990, Cincinatti, Ohio, 1989.
[175] DFG: *Maximale Arbeitsplatzkonzentrationen und biologische Arbeitsstofftoleranzwerte 1989,* VCH Verlagsgesellschaft, Weinheim 1989.

# Peroxy Compounds, Organic

HERBERT KLENK, Degussa AG, Hanau, Federal Republic of Germany (Chaps. 1–3, 5, 7, 8)
PETER H. GÖTZ, Degussa AG, Hanau, Federal Republic of Germany (Chaps. 1–3, 5, 7, 8)
RAINER SIEGMEIER, Degussa AG, Hanau, Federal Republic of Germany (Chaps. 4, 6, 9)
WILFRIED MAYR, Degussa AG, Hanau, Federal Republic of Germany (Chap. 10)

| | | |
|---|---|---|
| 1. | Introduction | 3630 |
| 2. | Alkyl Hydroperoxides | 3632 |
| 2.1. | Properties | 3633 |
| 2.2. | Production | 3635 |
| 2.3. | Uses | 3637 |
| 3. | Dialkyl Peroxides | 3638 |
| 3.1. | Properties | 3638 |
| 3.2. | Production | 3639 |
| 3.3. | Uses | 3641 |
| 4. | Peroxycarboxylic Acids | 3641 |
| 4.1. | Properties | 3642 |
| 4.2. | Production | 3644 |
| 4.2.1. | Peroxycarboxylic Acids from Carboxylic Acids and Hydrogen Peroxide | 3644 |
| 4.2.2. | Peroxycarboxylic Acids from Carboxylic Anhydrides and Hydrogen Peroxide | 3647 |
| 4.2.3. | Peroxycarboxylic Acids from Carboxylic Acid Chlorides and Hydrogen Peroxide | 3648 |
| 4.2.4. | Peroxycarboxylic Acids by Autoxidation of Aldehydes | 3648 |
| 4.3. | Uses | 3649 |
| 5. | Diacyl Peroxides | 3649 |
| 5.1. | Properties | 3650 |
| 5.2. | Production | 3651 |
| 5.3. | Uses | 3653 |
| 6. | Peroxycarboxylic Esters | 3653 |
| 6.1. | Properties | 3654 |
| 6.2. | Production | 3657 |
| 6.3. | Uses | 3662 |
| 7. | α-Oxyperoxides | 3662 |
| 7.1. | Properties | 3663 |
| 7.2. | Production | 3666 |
| 7.3. | Uses | 3669 |
| 8. | α-Aminoperoxides | 3669 |
| 8.1. | Properties | 3669 |
| 8.2. | Production | 3671 |
| 8.3. | Uses | 3672 |
| 9. | Safety Measures | 3672 |
| 9.1. | Hazards | 3672 |
| 9.2. | Transportation and Storage | 3673 |
| 9.3. | Handling | 3675 |
| 10. | Toxicology | 3675 |
| 11. | References | 3678 |

# 1. Introduction

Peroxides are characterized by the presence of the peroxo group –O–O–. In organic peroxides this group is bound to at least one carbon atom, or is bound to a carbon atom via another atom [1]–[14].

The first synthesis of an organic peroxide was that of dibenzoyl peroxide in 1858 [15]. Since then, a large number of organic peroxides have been synthesized, mainly on a laboratory scale. Interest in organic peroxides arose mainly in connection with biological processes [16] and the aging of organic compounds such as rubber, fats [17], oils, paints, and solvents, in which peroxides are often formed as reactive intermediates. Since the mid-1970s a surprisingly large number, with regard to the relative instability of the peroxo group, of naturally occurring peroxo compounds have been isolated and characterized.

Industrial interest in organic peroxides began in the early 1900s with the introduction of benzoyl peroxide as a bleaching agent. The consumption of synthetic plastics has increased since the 1940s, resulting in the increasing importance of peroxides as initiators for the radical polymerization of vinyl monomers. Additional applications include their use as cross-linking agents for polyolefins, as hardeners for elastomers, as adhesion promoters for metal–plastic composites (e.g., polyethylene coatings), and in pharmaceuticals. Various organic peroxides have become industrially important as selective oxidation and epoxidation reagents. Peroxides are frequently generated in situ because of handling problems.

More than 65 organic peroxides in more than 100 formulations (liquid, solid, paste, powder, solutions, dispersions) are produced in industrial quantities today. An overview of the consumption of organic peroxides in the United States is given in Figures 1 [18] and Fig. 2.

Formally, organic peroxides can be regarded as derivatives of hydrogen peroxide in which the hydrogen atoms are substituted by various organic groups. The following classes can be distinguished:

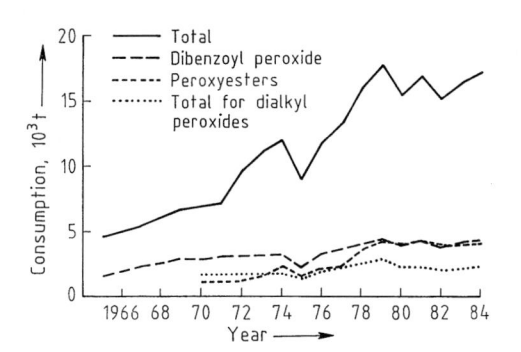

**Figure 1.** Consumption trend for organic peroxides in the United States

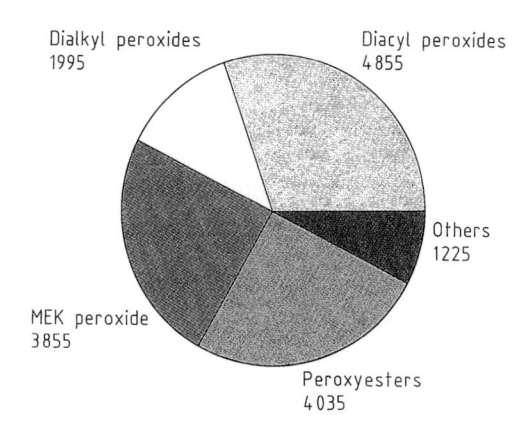

**Figure 2.** Consumption of organic peroxides (in tonnes) in the United States in 1983
MEK peroxide = methyl ethyl ketone peroxide

| | |
|---|---|
| Alkyl hydroperoxides | $R-O-O-H$ |
| Dialkyl peroxides | $R^1-O-O-R^2$ |
| Peroxycarboxylic acids | $R-\overset{\overset{O}{\|\|}}{C}-O-O-H$ |
| Diacyl peroxides | $R^1-\overset{\overset{O}{\|\|}}{C}-O-O-\overset{\overset{O}{\|\|}}{C}-R^2$ |
| | $R^1-\overset{\overset{O}{\|\|}}{C}-O-O-SO_2-R^2$ |
| Peroxycarboxylic esters | $R^1-\overset{\overset{O}{\|\|}}{C}-O-O-R^2$ |
| Peroxycarbonates | $R^1-O-\overset{\overset{O}{\|\|}}{C}-O-O-R^2$ |
| α-Oxy-, α-amino peroxides | $R^1-\overset{\overset{R^2}{\|}}{\underset{X}{C}}-O-O-R^3(H)$ |

$X = -OH, -OR^4, -O-O-, -NH_2, -NHR^5, -NR^5R^6$

Compounds in which the peroxy group is bonded to one or two atoms other than carbon are also regarded as organic peroxides, as long as additional organic substituents are present. Peroxides with the general structures

$$-\overset{|}{\underset{|}{C}}-O-O-M \quad \text{and} \quad -\overset{|}{\underset{|}{C}}-M-O-O-M-\overset{|}{\underset{|}{C}}-$$

where M is an element of groups 1, 2, 4, 5, 6, 8 or 12 – 16 of the periodic table have been synthesized in large numbers [19]. Furthermore, representatives of other substance classes, in which the peroxy group is connected to a heteroatom, are known. With few exceptions, compounds of this type are only of academic interest.

Homolytic cleavage of the O–O bond, readily induced by light or heat, is a typical reaction of organic peroxides. The reactivity depends strongly on the structure of the organic peroxy compound. In addition, heterolytic and hydrolytic cleavage, and rearrangements are also known. The oxidation potential decreases in the following order: peroxycarboxylic acids > hydroperoxides > diacyl peroxides > peroxycarboxylic esters > dialkyl peroxides.

Within a given class of compounds the stability increases with increasing carbon content. The thermal stability is often characterized by the half-life or the 10-h half-life temperature, whereby the radical initially formed upon decomposition is very short-lived, i.e., $t_{1/2} < 10^{-3}$ s [20]. Typical 10-h half-life temperatures of commercial peroxides lie between 25 and 170 °C. Decomposition is accelerated by various transition metals (promoters).

There are two main methods for the production of organic peroxides: (1) from hydrogen peroxide and its derivatives and (2) by oxidation with oxygen.

Analytical methods for peroxides include the usual spectroscopic [21] and chromatographic procedures [22], as well as various methods based on oxidative titrations (iodometry) [23] and electrochemical techniques [24]. Particular attention must be paid to safety factors when dealing with peroxides. Many peroxy compounds are not only heat sensitive, but are also sensitive to shock, impact, and light. Peroxides tend to decompose explosively. These compounds are frequently handled in desensitized form.

## 2. Alkyl Hydroperoxides

The formal substitution of a hydrogen atom in hydrogen peroxide by an alkyl group gives a hydroperoxide with the general formula R–OOH. Primary, secondary, and tertiary hydroperoxides are known; vinyl hydroperoxides have not yet been synthesized.

Hydroperoxides occur widespread in nature. They occur mainly as intermediates during decomposition, and by aging of organic compounds under the influence of atmospheric oxygen (autoxidation):

$$-\overset{|}{\underset{|}{C}}-H \xrightarrow{O_2} -\overset{|}{\underset{|}{C}}-OOH$$

These compounds are oxidized further to frequently strong-smelling aldehydes, ketones, and carboxylic acids; an example is the smell of rancid fat [25]. Autoxidation is used industrially in the drying of paints.

## 2.1. Properties

**Physical Properties.** Hydroperoxides are generally sufficiently stable to be melted or distilled at a temperature below 80 °C. Decomposition increases markedly with increasing temperature; under certain circumstances explosions may occur. The thermal stability is mainly determined by the dissociation energy of the O–O bond; for homolytic cleavage a typical value is 180–184 kJ/mol [26], [27]. A summary of the physical properties of some important hydroperoxides is shown in Table 1.

The acidity of hydroperoxides is greater than that of the corresponding alcohols, so that in most cases salts can be prepared. This property is often exploited for isolation and purification [28], [29]. The acid strengths ($pK_a$ at 20 °C) of some hydroperoxides are as follows [30]:

| | |
|---|---|
| Methyl hydroperoxide | 11.5 |
| Ethyl hydroperoxide | 11.8 |
| Isopropyl hydroperoxide | 12.1 |
| Cumene hydroperoxide | 12.6 |
| tert-Butyl hydroperoxide | 12.8 |
| Triphenylmethyl hydroperoxide | 13.07 |

When dissolved in noncoordinating solvents hydroperoxides form mainly dimers and trimers with six-membered ring structures.

**Chemical Properties.** Alkyl hydroperoxides can react with retention or cleavage of the O–O bond. In the first instance, substitution of the hydrogen atom on the peroxy oxygen atom leads to a whole series of products:

$$R-OOH + R'-OH \longrightarrow R-OO-R'$$

$$R-OOH + -\underset{X}{\underset{|}{C}}=\underset{|}{C}- \longrightarrow -\underset{ROO}{\underset{|}{C}}-\underset{X}{\overset{H}{\underset{|}{C}}}-$$

$$R-OOH + R'-\overset{O}{\underset{\|}{C}}-Cl \longrightarrow ROO-\overset{O}{\underset{\|}{C}}-R'$$

$$R-OOH + \underset{|}{\overset{|}{C}}=O \longrightarrow ROO-\underset{|}{\overset{|}{C}}-OH \longrightarrow ROO-\underset{|}{\overset{|}{C}}-OOR$$

Reactions in which the O–O bond is cleaved are important for the quantitative analysis of hydroperoxides (e.g., by iodometry). Under basic conditions some hydroperoxides decompose to give carbonyl compounds, for example:

$$H_3C-\underset{OOH}{\underset{|}{CH}}-CH_3 \xrightarrow{OH^-} H_3C-\underset{\underset{O}{\|}}{C}-CH_3 + H_2O$$

**Table 1.** Physical properties of some alkyl hydroperoxides

| Name | CAS registry no. | Formula | $M_r$ | bp, °C (kPa) | mp, °C | $n_D$, 20 °C | $d$ (20 °C) | Active oxygen content, % |
|---|---|---|---|---|---|---|---|---|
| Methyl hydroperoxide | [3031-73-0] | H$_3$C–OOH | 48.04 | 538–40 (8.6) | | 1.3641 | 0.9967[a] | 33.3 |
| Ethyl hydroperoxide | [3031-74-1] | CH$_3$CH$_2$–OOH | | 43–44 (4) | | | | |
| n-Propyl hydroperoxide | [6068-96-8] | CH$_3$CH$_2$CH$_2$–OOH | 76.09 | 35 (2.7) | | 1.3890[b] | 0.9040[b] | 21.0 |
| Isopropyl hydroperoxide | [3031-75-2] | (H$_3$C)$_2$CH–OOH | 76.09 | 107–109 (101) | | 1.8861[c] | 0.8927[c] | 21.0 |
| tert-Butyl hydroperoxide | [75-91-2] | (H$_3$C)$_3$C–OOH | 90.12 | 35 (2.3) | 4.0 | 1.4013 | 0.8960 | 17.7 |
| Cyclohexyl hydroperoxide | [766-07-4] | C$_6$H$_{11}$–OOH | 116.16 | 57 (0.16) | 20 | 1.4622 | | |
| Cumene hydroperoxide | [80-15-9] | Ph–C(CH$_3$)$_2$–OOH | 152.19 | 53 (0.013) | | 1.5242 | 1.0619 | 10.5 |
| Octadecyl hydroperoxide | [56537-19-0] | n-C$_{18}$H$_{37}$–OOH | 286.50 | | 49–50 | | | |

[a] At 15 °C.
[b] At 25 °C.
[c] At 23 °C.

Tertiary hydroperoxides decompose under basic conditions to form the corresponding alcohol and oxygen:

$$(H_3C)_3C-OOH \xrightarrow{OH^-} (H_3C)_3C-OH + \tfrac{1}{2}O_2$$

Under acidic conditions hydroperoxides decompose with rearrangement; a classic example is the industrial synthesis of acetone and phenol from cumene hydroperoxide by the Hock process. In the presence of a polymer with acidic groups, hydrogen peroxide and isobutene can be obtained from *tert*-butyl hydroperoxide [31]. The thermal decomposition of primary alkyl hydroperoxides takes place with homolytic cleavage, whereas for secondary and tertiary hydroperoxides homolytic cleavage can be observed for both thermal and photolytic decomposition [32]. The radicals formed in these processes are reactive species that can react, for example, with compounds containing C=C double bonds, either by addition or by substitution of a hydrogen atom. Peroxy radicals react with alkyl groups in the presence of metal ions (M) to give dialkyl peroxides and the corresponding alcohol in yields of up to 90%:

$$2R-OOH + R'-H \xrightarrow{M^+} R-OOR' + R-OH + H_2O$$

In the absence of a suitable substrate, symmetrical dialkyl peroxides are formed with elimination of oxygen:

$$3\,R-OOH \xrightarrow{M^+} R-OH + R-OO-R + H_2O + O_2$$

## 2.2. Production

Several methods are known for the production of alkyl hydroperoxides. Essentially three major production routes exist, which can be differentiated according to the origin of the peroxy group:

1) From hydrogen peroxide
2) From oxygen (atmospheric oxygen, singlet or triplet $O_2$)
3) From another peroxy compound

Only a few processes can be carried out on a large scale and are of industrial importance.

**Alkylation of Hydrogen Peroxide.** Hydrogen peroxide can be alkylated under acidic or basic conditions by alkyl halides, alkyl sulfonates [33], carboxylic and sulfonic acid esters, ethers, tosylhydrazine, and cyclopropanes [34]–[36]. Dialkyl peroxides, which are often difficult to remove, are formed in addition to the monoalkylation product. A further limitation of this method is the safety measures required, as the lower molecular mass hydroperoxides in particular tend to undergo explosive decomposition. In another method, primary, secondary, and tertiary hydroperoxides are formed by reaction of an alkyl halide with hydrogen peroxide in the presence of silver trifluoroacetate [37]. The relatively stable tertiary hydroperoxides [38] (e.g., *tert*-butyl hydroperoxide) are of particular industrial importance.

$$(CH_3)_3C-OH \xrightarrow{H_2O_2/H_2SO_4} (CH_3)_3C-OOH$$

The remarkably stable acetylene mono- and dihydroperoxides can also be prepared in this way. These compounds have found some industrial interest as starting materials for the corresponding dialkyl peroxides, which are used as high-temperature polymerization initiators [39]:

$$\underset{\underset{OH}{|}}{\overset{\overset{CH_3}{|}}{H_3C-C}}-C\equiv C-\underset{\underset{OH}{|}}{\overset{\overset{CH_3}{|}}{C}}-CH_3 \xrightarrow{H_2O_2 \atop H_2SO_4} \underset{\underset{OOH}{|}}{\overset{\overset{CH_3}{|}}{H_3C-C}}-C\equiv C-\underset{\underset{OOH}{|}}{\overset{\overset{CH_3}{|}}{C}}-CH_3$$

Primary and secondary hydroperoxides can only be obtained in moderate yields from hydrogen peroxide and an alkylating agent [40].

Halogen-substituted hydroperoxides can be obtained in good yields by the reaction of electron-rich olefins with hydrogen peroxide and a halogen [41]:

$$\underset{H_3C-\underset{|}{\overset{|}{C}}=CH_2}{\overset{H_3C}{|}} \xrightarrow{H_2O_2/Br_2} \underset{H_3C-\underset{OOH}{\overset{|}{C}}-CH_2Br}{\overset{H_3C}{|}}$$

Halogen-substituted hydroperoxides can also be formed by peroxymercuration with subsequent substitution of the acetoxymercury(II) group with halogens [42].

$$Ph-CH=CH_2 \xrightarrow{H_2O_2/Hg(OAc)_2} Ph-\underset{HOO}{\overset{|}{CH}}-\underset{Hg(OAc)}{\overset{|}{CH_2}}$$
$$\xrightarrow{I_2} Ph-\underset{OOH}{\overset{|}{CH}}-CH_2I$$

**Oxidation with Oxygen.** The insertion of oxygen into an alkane C–H bond leads initially to hydroperoxides. When this reaction is carried out with atmospheric oxygen it is known as autoxidation. It frequently leads to a multitude of products because of further spontaneous reactions. Therefore, this reaction is of limited synthetic use. Moreover, these reactions are generally very slow and require high temperature (usually > 100 °C).

The process involves a radical mechanism, which can be started by radical initiators (peroxides [43], azo compounds [44], halogens, or hydrogen halides [45]), photochemically, thermally [46], and by various transition metal ions.

Initiation: $RH \longrightarrow R\cdot + H\cdot$
Chain propagation: $R\cdot + O_2 \longrightarrow ROO\cdot$
$ROO\cdot + RH \longrightarrow ROOH + R\cdot$
Chain termination: $2\ R\cdot \longrightarrow R\text{–}R$
$ROO\cdot + R\cdot \longrightarrow ROOR$
$2\ ROO\cdot \longrightarrow O_2 + ROOR$

Tertiary carbon atoms are preferentially attacked because of the lower dissociation energy of the C–H bond compared with those of primary and secondary centers. Selective preparation of hydroperoxides can be achieved mainly by reaction of activated hydrogen atoms (benzylic hydrogen atoms, hydrogen atoms in the α-position to carbonyl groups). This method is of industrial importance for the production of cumene hydroperoxide, *tert*-butyl hydroperoxide, 1,4-diisopropylbenzene hydroperoxide, pinane hydroperoxide, and phenethyl hydroperoxide. In contrast to the oxidation of saturated hydrocarbons, autoxidation takes place even at room temperature.

*tert-Butyl Hydroperoxide* [75-91-2]. The oxidation of isobutane with oxygen in the gas phase at 160 °C and a residence time of ca. 3 min leads to a 70% yield of *tert*-butyl hydroperoxide with a conversion of 80%. Hydrogen bromide is used as the initiator [47]. Byproducts include di-*tert*-butyl peroxide, *tert*-butanol, and various alkyl bromides. Another process operates in the liquid phase at 130 °C and 3.5 MPa without a catalyst. The hydroperoxide is obtained in a yield of ca. 60% with an isobutane conversion of < 25% [48].

*Cumene Hydroperoxide* [80-15-9]. Cumene is oxidized with atmospheric oxygen at 120 °C. The selectivity is 95 % at a conversion of 15 % [49] – [53].

A method for the preparation of hydroperoxides that are not easily accessible by other routes is the reaction of organometallic compounds (Grignard reagents, zinc and cadmium compounds) with oxygen at 0 °C. This method leads to tertiary hydroperoxides in good yields, and primary and secondary hydroperoxides in moderate yields [54].

$$R-MgX \xrightarrow{O_2} R-O-O-MgX \xrightarrow{H^+} R-O-OH$$

This method is particularly suitable for the production of low molecular mass hydroperoxides. Alternative synthetic routes via boranes [55] and ozonides are also known.

Peroxygenation of alkenes with triplet oxygen proceeds via the intermediate allyl radical; numerous allyl hydroperoxides can be formed due to resonance of the allyl radical:

$$RCH=CH-CH_2R' \xrightarrow{-H\cdot} [R\overset{\cdot}{C}H-CH=\overset{}{C}HR' \longleftrightarrow RCH-\overset{}{C}H=\overset{\cdot}{C}HR']$$
$$\longrightarrow RCH=CH-\underset{OOH}{\overset{|}{C}HR'} + RCH-CH=CHR' \atop \underset{OOH}{|}$$

Various alkynes can be converted to propargyl hydroperoxides (2-alkynyl hydroperoxides) by treatment with triplet oxygen.

Alkenes that contain at least one allylic hydrogen atom undergo an allylic shift on reaction with singlet oxygen to give 2-alkenyl hydroperoxides [56] – [60]:

$$\underset{}{\overset{R}{\underset{|}{C}}}H=CH-CH_2-H + {}^1O_2 \longrightarrow \underset{OOH}{\overset{R}{\underset{|}{C}}H-CH=CH_2}$$

## 2.3. Uses

*tert*-Butyl hydroperoxide is used as an oxidant in the Halcon process for the production of propylene oxide. The standard commercial product (80 %, stabilized with water and phosphoric acid) is suitable for curing polyester resins and for the emulsion polymerization of styrene –butadiene rubbers.

In addition to the synthesis of phenol, cumene hydroperoxide is also used for the polymerization of dienes and for the cross-linking of sulfide rubbers. Furthermore, it is the starting material for dicumyl peroxide. The usual commercial product is an 80 % solution in cumene. Various heterocyclic hydroperoxides are employed as cross-linking agents for olefin polymers [61].

# 3. Dialkyl Peroxides

Dialkyl peroxides have the general formula $R-O-O-R'$, where $R$ and $R'$ represent alkyl-, alkenyl-, or aryl groups. Both symmetrical ($R = R'$) and mixed ($R \neq R'$) peroxides are known as well as numerous cyclic peroxides. Only the relatively stable tertiary dialkyl peroxides have so far gained any industrial importance.

## 3.1. Properties

**Physical Properties.** Several dialkyl peroxides can be distilled at atmospheric pressure or under vacuum without decomposition. Their dissociation energy is about 170–180 kJ/mol [2], [3]. Primary dialkyl peroxides such as dimethyl peroxide are shock sensitive and explosive. The physical properties of some dialkyl peroxides are listed in Table 2.

**Chemical Properties.** Dialkyl peroxides can be cleaved homolytically by the action of heat or light to give alkoxy radicals:

$$R-O-O-R' \xrightarrow[\text{or heat}]{h\nu} R-O\cdot + \cdot O-R'$$

The subsequent reactions of these radicals depend on their structure, on the conditions, and on the temperature [62], [63]. In the absence of a substrate the main products result from disproportionation:

$$H_3C-O-O-CH_3 \longrightarrow HCHO + CH_3OH$$

The reaction of the intermediate methoxy radical with formaldehyde then leads to methanol and carbon monoxide:

$$H_3C-O\cdot + HCHO \longrightarrow CH_3OH + \cdot CHO$$
$$\cdot CHO + H_3C-O\cdot \longrightarrow CH_3OH + CO$$

Secondary dialkyl peroxides lead analogously to the corresponding ketones and alcohols. Tertiary dialkyl peroxides undergo rearrangement to form ketones, alcohols, and epoxides, in addition to lower hydrocarbons. The main products of the decomposition of di-*tert*-butyl peroxide are acetone and ethane:

$$3(CH_3)_3C-O-O-C(CH_3)_3$$
$$\longrightarrow H_3C-\overset{O}{\underset{\|}{C}}-CH_3 + CH_4 + C_2H_6 +$$
$$(CH_3)_3C-OH + CH_3-\overset{O}{\underset{\|}{C}}-CH_2-CH_3 + (CH_3)_2C\overset{O}{\underset{\diagdown}{\diagup}}CH_2$$

Numerous synthetic reactions are possible in the presence of reactive substrates [64], [65], of which the most important is the production of tertiary alkyl esters [66].

## 3.2. Production

Dialkyl peroxides can be obtained from numerous starting materials by reaction with oxygen, hydrogen peroxide, alkyl hydroperoxides, and in some cases potassium peroxide.

**Oxidation with Oxygen.** For the production of acyclic dialkyl peroxides, the autoxidation of hydrocarbons with molecular oxygen is only suitable in certain cases. This method can only be used when the intermediate radical formed is stabilized, e.g., by alkyl or aryl substituents, or by allylic conjugation. In these reactions, either a C–H bond is cleaved initially [67], (e.g., in the case of 2-methylpropane) or formal insertion of oxygen can take place following cleavage of a C–C or a C–X bond (X = halogen):

$$Ph_3C-CPh_3 \longrightarrow 2\,Ph_3C^{\cdot} \xrightarrow{2\,O_2} 2\,Ph_3C-O-O^{\cdot}$$
$$Ph_3C-O-O^{\cdot} + Ph_3C^{\cdot} \longrightarrow Ph_3C-O-O-CPh_3$$

With various alkenes, such as styrene, acrylates, acrylonitrile, and vinyl chloride, autoxidation leads to polymeric dialkyl peroxides with the general structure:

$$n\,HC{=}CH_2 \atop \phantom{n\,H}X \quad \xrightarrow{n\,O_2} \quad {+}CH_2{-}CH{-}O{-}O{+}_n \atop \phantom{{+}CH_2{-}C}X$$

X = C₆H₅, COOR, CN, halogen

The oxidation of conjugated dienes (photo-oxygenation [68]) with oxygen leads either to polymers or to cyclo adducts (hetero-Diels–Alder adducts), i.e., the 1,4-epiperoxides, which are rarely accessible by other routes. The autoxidation of α-terpinene to give ascaridole is an example of the synthesis of an epiperoxide [69]:

The photosensitized autoxidation of polynuclear aromatic hydrocarbons has been known for a long time [70].

The [2+2] cycloaddition of singlet oxygen to alkenes offers a simple route to 1,2-dioxetanes [71]–[76], which are of particular interest in connection with chemi- and bioluminescence.

**Table 2.** Physical properties of some dialkyl peroxides

| Name | CAS registry no. | Formula | $M_r$ | mp, °C | bp, °C (kPa) | $n_D$ | $d_4^{20}$ | Active oxygen content, % |
|---|---|---|---|---|---|---|---|---|
| Dimethyl peroxide | [690-02-8] | $CH_3OOCH_3$ | 62.07 | | 13.5 (101) | $1.3503^a$ | 0.8680 | 25.8 |
| Ethyl methyl peroxide | [70299-48-8] | $CH_3CH_2OOCH_3$ | 76.09 | | 40 (98.5) | $1.3590^b$ | $0.8337^b$ | 21.1 |
| Diethyl peroxide | [628-37-5] | $CH_3CH_2OOCH_2CH_3$ | 90.12 | | 64 (98.5) | $1.3701^b$ | $0.8266^b$ | 17.8 |
| tert-Butyl methyl peroxide | [51392-67-7] | $(CH_3)_3COOCH_3$ | 104.15 | −102 | 23 (15.8) | | | 15.4 |
| Di-tert-butyl peroxide | [110-05-4] | $(CH_3)_3COOC(CH_3)_3$ | 146.23 | −40 | 111 (101) | $1.3860^c$ | $0.791^c$ | 10.9 |
| Dicumyl peroxide | [80-43-3] | $PhC(CH_3)_2OOC(CH_3)_2Ph$ | 270.37 | 39 | | $1.5630^d$ | | 5.9 |

[a] At 0 °C. [b] At 17 °C. [c] At 25 °C. [d] Supercooled melt at 21 °C.

**Alkylation of Hydrogen Peroxide.** The dialkylation of hydrogen peroxide using alkyl halides, alcohols, alkenes, or alkyl sulfates gives symmetrical dialkyl peroxides and takes place under conditions similar to those described for monoalkylation in Section 2.2.

*Dicumyl peroxide* [80-43-3]. The reaction of an $H_2O_2$–urea adduct with 2-phenyl-2-propanol at 35 °C in the presence of a mineral acid catalyst gives dicumyl peroxide in good yields [77].

*Di-tert–amyl peroxide* [10508-09-5] is formed in 68 % yield by slow addition of *tert*-amyl alcohol to a mixture of $H_2O_2$ and $H_2SO_4$ at 20 °C [78].

**Alkylation of Hydroperoxides.** The best method for the production of both symmetrical and unsymmetrical dialkyl peroxides is the alkylation of hydroperoxides with alcohols, olefins, esters, ethers, halides, or epoxides [79].

*Dicumyl Peroxide.* The reaction of cumene hydroperoxide with 2-phenyl-2-propanol in the presence of a mineral acid catalyst gives dicumyl peroxide in almost quantitative yield [49]–[53], [80]–[82].

*Di-tert-butyl peroxide* [110-05-4] is synthesized from *tert*-butyl hydroperoxide and *tert*-butanol [49]–[53], [80]–[83]. This reaction can also be carried out with an acidic ion-exchange resin as catalyst, the water produced in the reaction being removed by azeotropic distillation [84].

*Bis(tert-amylperoxy)-2,5-dimethylhexane* [78-63-7] is formed in 79 % yield from 2,5-dihydroperoxy-2,5-dimethylhexane and *tert*-amyl alcohol at 50 °C in the presence of sulfuric acid [85].

Phenols can be converted to dialkyl peroxides (quinone peroxides) by dehydrogenation [86]–[88]. These compounds have been investigated primarily in connection with the mechanism of action of phenolic antioxidants [89]–[91].

## 3.3. Uses

Due to their relatively high thermal stability, dialkyl peroxides are used mainly as catalysts for suspension and bulk polymerization [92], [93]. They also find use as hardeners for unsaturated polyester resins and for the cross-linking of polymers [94]. In addition to di-*tert*-butyl peroxide [95] and dicumyl peroxide, bis(*tert*-alkylperoxy) alkanes are employed. The dialkyl peroxides are commercially available as technical-grade material and as 40 – 50 % solutions.

# 4. Peroxycarboxylic Acids

Peroxycarboxylic acids have a wide range of applications in preparative and industrial chemistry on account of their high oxidation potential. Due to their instability some peroxycarboxylic acids are either generated in situ or used in desensitized form.

## 4.1. Properties

**Physical Properties.** Short-chain aliphatic peroxycarboxylic acids are infinitely miscible with water and possess an unpleasant, pungent odor. They are explosive when pure or highly concentrated. The tendency to explode decreases with increasing chain length. Long-chain peroxycarboxylic acids (> $C_6$) are increasingly water-insoluble solids. Aromatic peroxycarboxylic acids are only slightly soluble in water; both types are, however, soluble in various organic solvents.

In the crystalline state peroxycarboxylic acids associate to form dimers (**2**), whereas in the gas phase, in the molten state, and in solution in noncoordinating solvents monomers (**1**) with intramolecular hydrogen bonds are present [96].

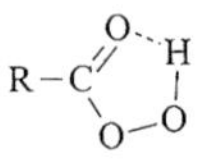

In the presence of donor compounds (Do) such as tetrahydrofuran, 1,4-dioxane, or N,N-dimethylacetamide, peroxycarboxylic acids form complexes with cleavage of the intramolecular hydrogen bonds [97], [98]:

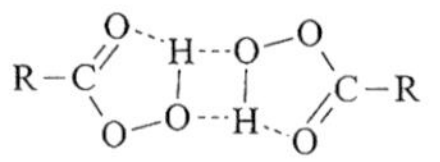

Peroxycarboxylic acids exhibit $pK_a$ values that are 3–4 units lower than those of the corresponding carboxylic acids [99] due to the absence of resonance stabilization in the peroxycarboxylate anions. As a result of the weak substituent effect, the $pK_a$ values all lie in a small range. Table 3 lists physical properties for some peroxycarboxylic acids.

Under certain circumstances peroxycarboxylic acids can decompose explosively on heating or under impact. The dissociation energy of the O–O bond (80–90 kJ/mol) is considerably lower than that of the hydroperoxides or the dialkyl peroxides [100]–[102].

**Chemical Properties.** Peroxycarboxylic acids have the highest oxidation potential of all organic peroxides, whereby the corresponding carboxylic acid is formed with release of active oxygen. The stability of aqueous solutions of peroxycarboxylic acids depends on the pH. Bases cause decomposition with loss of active oxygen, whereas in acidic media an equilibrium is established.

$$R-CO_3H + H_2O \underset{}{\overset{H^+}{\rightleftharpoons}} R-CO_2H + H_2O_2$$

With the exception of performic acid, neutral solutions are largely hydrolytically stable. Peroxycarboxylic acids decompose thermally by two mechanisms [103]: radical and concerted:

Table 3. Physical properties of some peroxycarboxylic acids-,4

| Name | CAS registry no. | Formula | $M_r$ | mp, °C | bp, °C (kPa) | $pK_a$ | Active oxygen content, % | Other |
|---|---|---|---|---|---|---|---|---|
| Peroxyformic acid (performic acid) | [107-32-4] | $HCO_3H$ | 62.0 | $-18^a$ | $50 (13.3)^a$ | 7.1 | 25.8 | |
| Peroxyacetic acid (peracetic acid) | [79-21-0] | $H_3C-CO_3H$ | 76.1 | 0 | 25 (1.6) | 8.2 | 21.1 | $n_D^{20}$ 1.3974 $d_4^{20}$ 1.0375 |
| Peroxypropanoic acid | [4212-43-5] | $H_3C-CH_2-CO_3H$ | 90.1 | $-13$ | 25 (2.7) | 8.1 | 17.8 | $n_D^{20}$ 1.4041 |
| Peroxybutanoic acid | [13122-71-9] | $H_3C-(CH_2)_2-CO_3H$ | 104.1 | $-10$ | 26–29 (1.6) | 8.2 | 15.4 | $n_D^{20}$ 1.4125 $d_4^{21}$ 1.0745 |
| Peroxynonanoic acid | [3058-35-3] | $H_3C-(CH_2)_7-CO_3H$ | 174.2 | 35 | | ca. 8.5 | 9.2 | |
| Peroxydodecanoic acid | [2388-12-7] | $H_3C-(CH_2)_{10}-CO_3H$ | 216.3 | 50 | | 10.8 | 7.4 | |
| Diperoxyglutaric acid | [28317-46-6] | $HO_3C-(CH_2)_3-CO_3H$ | 164.1 | 80–100 | | ca. 7–8.5 | 19.5 | |
| Diperoxyadipic acid | [5824-51-1] | $HO_3C-(CH_2)_4-CO_3H$ | 178.1 | 116–117 | | 7.6 | 18.0 | |
| Diperoxyoctanedioic acid | [28317-47-7] | $HO_3C-(CH_2)_6-CO_3H$ | 206.2 | 89 | | ca. 7.5–8 | 15.5 | |
| Diperoxynonanedioic acid | [1941-79-3] | $HO_3C-(CH_2)_7-CO_3H$ | 220.2 | 90 | | ca. 8.0 | 14.5 | |
| Diperoxydecanedioic acid | [5796-85-0] | $HO_3C-(CH_2)_8-CO_3H$ | 234.2 | 98 | | ca. 8–8.5 | 13.7 | |
| Diperoxydodecanedioic acid | [66280-55-5] | $HO_3C-(CH_2)_{10}-CO_3H$ | 262.3 | >100 (decomp.)$^b$ | | $9.7^{b,c}$ | 12.2 | |
| Peroxybenzoic acid (perbenzoic acid) | [93-59-4] | $C_6H_5-CO_3H$ | 138.1 | 41–42 | | 7.8 | 11.6 | |
| 3-Chloroperoxybenzoic acid | [937-14-4] | $m\text{-}Cl\text{-}C_6H_4\text{-}CO_3H$ | 172.6 | 88 | | 7.7 | 9.3 | |
| 4-Nitroperoxy-benzoic acid | [943-39-5] | $p\text{-}NO_2\text{-}C_6H_4\text{-}CO_3H$ | 183.1 | 138 (decomp.) | | 7.4 | 8.7 | |
| Monoperoxy-phthalic acid | [2311-91-3] | $o\text{-}HO_2C\text{-}C_6H_4\text{-}CO_3H$ | 182.1 | ca. 110 (decomp.)$^a$ | | ca. 6–8 | 8.8 | |

$^a$ 90 % pure.
$^b$ 95 % pure.
$^c$ In $H_2O/CH_3CN$.

$$R-CO_3H \begin{array}{c} \nearrow \\ \searrow \end{array} \begin{array}{l} R-OH + CO_2 \quad (1) \\ R-CO_2H + 1/2 O_2 \quad (2) \end{array}$$

In the absence of oxygen, the radical decomposition (Eq. 1), in which the first step is homolytic cleavage of the O−O bond, dominates. Heavy metal ions accelerate the radical decomposition even at low temperatures. In contrast, in the presence of radical inhibitors the concerted reaction (Eq. 2) takes place. It is facilitated by the stereochemistry of the peroxycarboxyl group, which is able to form a five-membered cyclic transition state [103], [104].

In the presence of urea, starch, cholic acid, or zeolites, straight-chain peroxycarboxylic acids form inclusion compounds [105], [106]. Isolable 1:1 complexes are obtained with triphenylphosphine oxide, triphenylarsine oxide, and heteroaromatic N-oxides [107].

Qualitative detection of peroxycarboxylic acids is facilitated with reagents that give colored products on oxidation, e.g., alkali-metal iodides, titanium sulfate, or phenolphthalein. For quantitative analysis potentiometry [3], polarography [5], and iodometric titration are suitable. IR and NMR spectroscopy are also useful for determination of peroxycarboxylic acids [108].

## 4.2. Production

Peroxycarboxylic acids can be produced by numerous methods, some of which are shown in Figure 3. Of these methods, however, only the first five are synthetically and industrially important. Production from diacyl peroxides is only of interest for perbenzoic acid, as the starting material, dibenzoyl peroxide, is commercially available.

### 4.2.1. Peroxycarboxylic Acids from Carboxylic Acids and Hydrogen Peroxide

The classical synthesis of peroxycarboxylic acids is the direct, usually acid-catalyzed, equilibrium reaction between carboxylic acids and hydrogen peroxide. With lower aliphatic carboxylic acids, aqueous solutions of peroxycarboxylic acids are obtained, which also contain carboxylic acid and hydrogen peroxide.

The equilibrium mixture is produced by simply mixing the components; the position of equilibrium depends on the molar ratio of the starting materials and on the concentration of the hydrogen peroxide. The equilibrium composition of the system acetic acid − hydrogen peroxide − paracetic acid − water is shown in Table 4 [109].

Equilibrium peracetic acid of this type, with a variable peracetic acid content of up to 40 %, is commercially available.

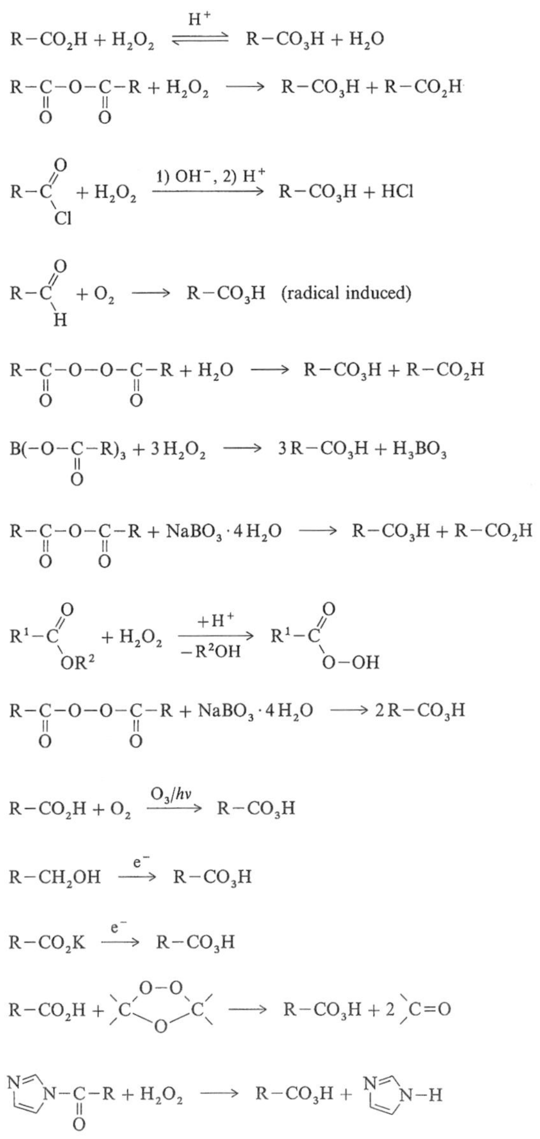

**Figure 3.** Reactions used for the synthesis of peroxyacids

$$R-CO_2H + H_2O_2 \underset{}{\overset{H^+}{\rightleftharpoons}} R-CO_3H + H_2O$$

$$R-\underset{\underset{O}{\|}}{C}-O-\underset{\underset{O}{\|}}{C}-R + H_2O_2 \longrightarrow R-CO_3H + R-CO_2H$$

$$R-\underset{Cl}{\overset{O}{C}} + H_2O_2 \xrightarrow{1)\ OH^-,\ 2)\ H^+} R-CO_3H + HCl$$

$$R-\underset{H}{\overset{O}{C}} + O_2 \longrightarrow R-CO_3H \quad \text{(radical induced)}$$

$$R-\underset{\underset{O}{\|}}{C}-O-O-\underset{\underset{O}{\|}}{C}-R + H_2O \longrightarrow R-CO_3H + R-CO_2H$$

$$B(-O-\underset{\underset{O}{\|}}{C}-R)_3 + 3\,H_2O_2 \longrightarrow 3\,R-CO_3H + H_3BO_3$$

$$R-\underset{\underset{O}{\|}}{C}-O-\underset{\underset{O}{\|}}{C}-R + NaBO_3 \cdot 4\,H_2O \longrightarrow R-CO_3H + R-CO_2H$$

$$R^1-\underset{OR^2}{\overset{O}{C}} + H_2O_2 \xrightarrow[-R^2OH]{+H^+} R^1-\underset{O-OH}{\overset{O}{C}}$$

$$R-\underset{\underset{O}{\|}}{C}-O-O-\underset{\underset{O}{\|}}{C}-R + NaBO_3 \cdot 4\,H_2O \longrightarrow 2\,R-CO_3H$$

$$R-CO_2H + O_2 \xrightarrow{O_3/h\nu} R-CO_3H$$

$$R-CH_2OH \xrightarrow{e^-} R-CO_3H$$

$$R-CO_2K \xrightarrow{e^-} R-CO_3H$$

$$R-CO_2H + \underset{}{\overset{O-O}{C\diagdown O\diagup C}} \longrightarrow R-CO_3H + 2\,C{=}O$$

$$\underset{}{\overset{N{\diagup\hspace{-6pt}\diagdown}}{N}}{-}\underset{\underset{O}{\|}}{C}{-}R + H_2O_2 \longrightarrow R-CO_3H + \overset{N{\diagup\hspace{-6pt}\diagdown}}{N}{-}H$$

The reaction can also be carried out in the presence of organic solvents [110], [111]. Catalyst-free equilibrium mixtures can be obtained by the use of acidic, preferably perfluorinated, ionexchange resins [112]. Strong aliphatic carboxylic acids such as formic acid or trifluoroacetic acid do not require a catalyst, particularly when used in a large excess relative to hydrogen peroxide.

The reaction between a carboxylic acid and hydrogen peroxide can also be carried out in the gas phase over a boric acid–phosphoric acid catalyst [113].

**Table 4.** Equilibrium composition and $H_2O_2$ conversion in the system acetic acid (AA) – hydrogen peroxide – peracetic acid (PAA) – water at 25 °C

| [$H_2O_2$] (aqueous solution) | Molar ratio AA : $H_2O_2$ | Composition, wt% (mol%) | | | | $H_2O_2$ conversion, % |
|---|---|---|---|---|---|---|
| | | AA | PAA | $H_2O$ | $H_2O_2$ | |
| 50 | 1:1    | 22.0 (9.7)  | 29.1 (16.2) | 31.5 (58.4) | 16.0 (15.7) | 37 |
|    | 2:1    | 22.8 (11.6) | 45.2 (29.2) | 23.6 (50.9) | 7.2 (8.2)   | 70 |
|    | 4:1    | 18.6 (11.1) | 62.4 (47.2) | 15.3 (38.5) | 2.4 (3.2)   | 76 |
|    | 6.66:1 | 13.8 (9.0)  | 73.3 (60.8) | 10.4 (28.8) | 9.2 (1.3)   | 86 |
|    | 10:1   | 10.1 (7.0)  | 81.0 (70.8) | 7.4 (21.6)  | 0.5 (0.7)   | 90 |
| 90 | 1.5:1  | 45.0 (27.8) | 35.0 (27.4) | 14.0 (36.5) | 6.0 (8.3)   | 77 |

**Table 5.** Azeotropic mixtures of peroxycarboxylic acids and water

| Compound | Peroxycarboxylic acid, wt% | Water, wt% |
|---|---|---|
| Peracetic acid | 61.0 | 39.0 |
| Peroxypropanoic acid | 51.5 | 48.5 |
| Peroxybutanoic acid | 42.0 | 58.0 |
| Peroxyisobutanoic acid | 55.5 | 44.5 |

Pure aqueous solutions of peroxycarboxylic acids are obtained from the reaction mixtures by distilling off the peroxycarboxylic acid and water under vacuum [114]. The compositions of azeotropes of some lower peroxycarboxylic acids with water are listed in Table 5.

The production of anhydrous peroxycarboxylic acids can be achieved by three main routes:

1) By extraction of the equilibrium mixture or the pure aqueous solution of peroxycarboxylic acid with organic solvents such as phosphate esters [115], carboxylic esters [116], or chlorinated or aromatic hydrocarbons [117] – [122]. Residual water in the extracts can be removed by azeotropic distillation.
2) Removal of the water by azeotropic distillation with an organic solvent under vacuum [123] – [125]. For example, peroxypropionic acid [4212-43-5] is obtained in 89% yield by heating a mixture of hydrogen peroxide, propionic acid, boric acid, phosphoric acid, and dichloroethane at 70 °C for 2 h and then distilling off the dichloroethane – water azeotrope under vacuum [126].
3) By selective absorption of the peroxycarboxylic acid and the corresponding carboxylic acid from the vapor in the distillative production of aqueous solutions of peroxycarboxylic acids in a phosphate ester, followed by desorption with a carboxylic ester [127].

It is not necessary for longer chain, water-insoluble mono- and dicarboxylic acids to be in solution to undergo reaction. In the presence of sulfuric acid the corresponding peroxycarboxylic acids can be obtained in good yields from aqueous suspensions

[128]–[131]. The additional use of a solvent, from which the peroxycarboxylic acid formed will be isolated, is also possible [132]. The isolated compounds are frequently difficult to handle; in these cases they may be stabilized by addition of inorganic salts during the reaction (in situ desensitizing) [133], [134].

**Diperoxydodecanedioic Acid** [66280-55-5], desensitized. Dodecanedioic acid is added to a mixture of 50% hydrogen peroxide, 96% sulfuric acid, and trioctylphosphine oxide. After 2–3 h at 60 °C the mixture is allowed to cool, diluted with sodium sulfate solution, the pH adjusted to 3.5 with sodium hydroxide, and, depending on the desired peracid concentration, conditioned with sodium sulfate. Diperoxydodecanedioic acid is obtained in 94% yield.

Aromatic carboxylic acids can be converted to the corresponding peroxycarboxylic acids with methanesulfonic acid as catalyst [135]–[137]. The use of concentrated sulfuric acid as catalyst has also been described.

**Diperoxyisophthalic acid** [1786-87-4] [138] is obtained in 77% yield as an 88% solution from a mixture of 85% hydrogen peroxide, 96% sulfuric acid, and isophthalic acid.

## 4.2.2. Peroxycarboxylic Acids from Carboxylic Anhydrides and Hydrogen Peroxide

The reaction of lower aliphatic carboxylic anhydrides with hydrogen peroxide gives anhydrous solutions of peroxycarboxylic acids in the corresponding carboxylic acid, which is formed by reaction of the water introduced with the hydrogen peroxide with excess anhydride. The reaction is particularly suitable for the synthesis of monoperoxydicarboxylic acids, such as mono-peroxyphthalic acid or monoperoxysuccinic acid, from cyclic anhydrides [139], [140]. Inorganic bases, ammonia, or metal salts can be used as catalysts [141]–[143].

**Monoperoxyphthalic Acid** [2311-91-3]. The reaction of phthalic anhydride with hydrogen peroxide in 1,2-dichloroethane at 30 °C in the presence of a catalytic amount of ammonia gives, after acidification of the reaction mixture, 90% monoperoxyphthalic acid in >96% yield.

Furthermore, the diperoxycarboxylic acids can be obtained under acidic conditions by using a large excess of hydrogen peroxide [144]–[146].

**Decylbutanediperoxycarboxylic Acid** [87458-73-9]. The reaction of 1 mol of decylsuccinic anhydride in dichloromethane with 3.7 mol of 50% hydrogen peroxide in the presence of 6.8 mol of 95% sulfuric acid for 30 min gives decylbutanediperoxycarboxylic acid in 97% yield with a purity of 98%.

Explosive diacyl peroxides can be formed as byproducts of the acylation of hydrogen peroxide with lower aliphatic carboxylic acid anhydrides, the amount formed depending on the reaction conditions.

### 4.2.3. Peroxycarboxylic Acids from Carboxylic Acid Chlorides and Hydrogen Peroxide

The base-catalyzed reaction of carboxylic acid chlorides with hydrogen peroxide is the most important method for the production of aromatic peroxycarboxylic acids, which are difficult to synthesize by other routes [147]–[150]. Water [151], or water mixed with organic solvents [152], [153], can be used as solvent.

**3-Chloroperbenzoic Acid** [*937-14-4*] [154]. The reaction of 3-chlorobenzoyl chloride with hydrogen peroxide in alkaline, aqueous dioxane in the presence of $MgSO_4 \cdot 7\,H_2O$ gives, after acidification, 80–85% 3-chloroperbenzoic acid.

The reaction of carboxylic acid chlorides with 85% hydrogen peroxide in organic solvents such as dichloromethane, diethyl ether, or tetrahydrofuran in the presence of pyridine has also been described. However, the pyridine should not be added in equimolar amounts because of the formation of potentially explosive diacyl peroxides [155], [156].

**Peroxydodecanoic Acid** [*2388-12-7*]. The reaction of dodecanoyl chloride in tetrahydrofuran with 85% hydrogen peroxide in the presence of pyridine gives 70% peroxydodecanoic acid after 20 min reaction in an ice bath.

This method is also suitable for the production of sterically crowded peroxycarboxylic acids [157], [158].

### 4.2.4. Peroxycarboxylic Acids by Autoxidation of Aldehydes

Aldehydes react with oxygen in a radical chain process to give carboxylic acids. The intermediate peroxycarboxylic acids can be obtained as the main product by appropriate control of the reaction conditions. The reaction can be initiated photochemically or by the use of ozone and it can be catalyzed by metal salts, predominantly cobalt and iron compounds, or by 2,2′-bipyridyl [159]–[163]. The reaction can be carried out in the vapor phase at standard pressure [164], [165] or in the liquid phase [166], [167].

**C$_7$-Peroxycarboxylic Acid Mixtures.** The oxidation of isomeric C$_7$ aldehydes with oxygen in the presence of 2,2′-bipyridyl at 60 °C gives 62% of the corresponding peroxycarboxylic acid mixture [168].

The method is suitable for the production of aromatic peroxycarboxylic acids when the autoxidation is carried out at room temperature in carboxylic esters as solvent. Substituted benzaldehydes, however, show considerable differences in reactivity [169].

## 4.3. Uses

The largest use of peroxycarboxylic acids is in the production of epoxides. Compounds with an isolated double bond that cannot be oxidized with hydroperoxides or hydrogen peroxide in the absence of a catalyst are epoxidized smoothly and stereospecifically by peroxycarboxylic acids. Cyclic ketones, such as cyclohexanone, are converted to lactones; the production of ε-caprolactone, an important intermediate for the production of polyesters and polyurethanes, is carried out on a large scale. Phenols can be hydroxylated with peroxycarboxylic acids to give dihydroxybenzenes [170]–[172].

Amine $N$-oxides and -sulfoxides, obtained by reaction of peroxycarboxylic acids with amines and sulfides are used as detergents and in the pharmaceutical industry. The use of peroxycarboxylic acids for disinfection [173]–[177], in environmental protection, and for bleaching is increasing due to their high oxidizing power and ecological safety. Low-temperature bleaches based on aromatic monoperoxycarboxylic acids and long chain aliphatic diperoxycarboxylic acids are of interest for laundry detergents [178]–[182].

# 5. Diacyl Peroxides

Diacyl peroxides possess the following structural formula:

$$R-\overset{O}{\underset{\|}{C}}-O-O-\overset{O}{\underset{\|}{C}}-R'$$

The best known compound of this type is dibenzoyl peroxide. Both symmetrical (R = R′) and unsymmetrical compounds are known. Furthermore, peroxydicarbonates of type (**3**) and sulfur-containing unsymmetrical diacyl peroxides of type (**4**), where R is an alkyl or cycloalkyl group, are also of interest.

$$R-O-\overset{O}{\underset{\|}{C}}-O-O-\overset{O}{\underset{\|}{C}}-O-R' \qquad R-SO_2-O-O-\overset{O}{\underset{\|}{C}}-R'$$
$$\qquad\qquad 3 \qquad\qquad\qquad\qquad 4$$

**Table 6.** Physical properties of some diacyl peroxides

| Name | CAS registry no. | Formula | $M_r$ | mp, °C | Decomposition temperature, °C | Active oxygen content, % |
|---|---|---|---|---|---|---|
| Diacetyl peroxide | [110-22-5] | $H_3C-\overset{O}{\underset{\|}{C}}-O-O-\overset{O}{\underset{\|}{C}}-CH_3$ | 118.09 | 30 (decomp.) | ca. 30 | 13.5 |
| Acetyl benzoyl peroxide | [644-31-5] | $H_3C-\overset{O}{\underset{\|}{C}}-O-O-\overset{O}{\underset{\|}{C}}-Ph$ | 180.16 | 37–39 | ca. 40 | 8.9 |
| Dibenzoyl peroxide | [94-36-0] | $Ph-\overset{O}{\underset{\|}{C}}-O-O-\overset{O}{\underset{\|}{C}}-Ph$ | 242.23 | 106–7 | ca. 110 | 6.6 |
| Diisopropyl peroxydicarbonate | [105-64-6] | $(H_3C)_2CHO-\overset{O}{\underset{\|}{C}}-O-O-\overset{O}{\underset{\|}{C}}-OCH(CH_3)_2$ | 206.19 | 8–10 | 35–38 | 7.8 |
| Acetyl cyclohexylsulfonyl peroxide | [3179-56-4] | $H_3C-\overset{O}{\underset{\|}{C}}-O-O-\overset{O}{\underset{\|}{\underset{O}{S}}}-C_6H_{11}$ | 222.26 | 35–36 | 36 | 7.2 |

## 5.1. Properties

**Physical Properties.** As a result of the low dissociation energy of the O–O bond (ca. 120–130 kJ/mol [6], [7], [183]), diacyl peroxides are among the best radical-forming compounds. Most diacyl peroxides are low melting solids, which frequently decompose explosively even at low temperature. A summary of physical properties of some important diacyl peroxides is given in Table 6.

Extreme care must be taken when handling diacyl peroxides that have not ben desensitized as they are very sensitive toward shock, impact, and friction (see Chap. 9).

**Chemical Properties.** The products formed by thermal or photochemical decomposition [184] and the resulting homolytic cleavage of the O–O bond are shown below [185]:

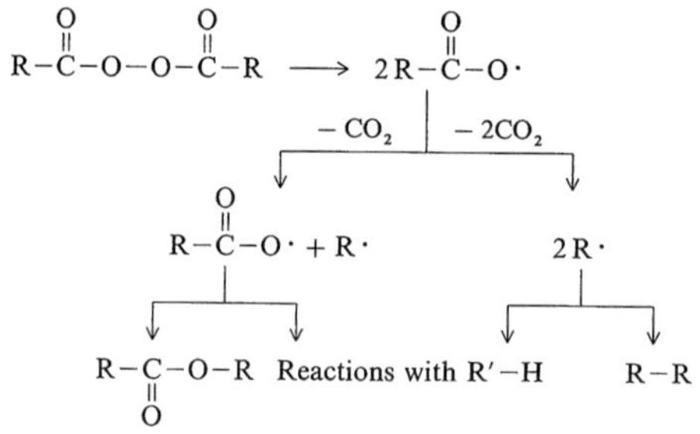

$$R-\overset{O}{\underset{\|}{C}}-O-O-\overset{O}{\underset{\|}{C}}-R \longrightarrow 2\,R-\overset{O}{\underset{\|}{C}}-O\cdot$$

$$\underset{-CO_2}{\downarrow} \quad \underset{-2CO_2}{\downarrow}$$

$$R-\overset{O}{\underset{\|}{C}}-O\cdot + R\cdot \qquad 2\,R\cdot$$

$$\downarrow \qquad \downarrow \qquad \downarrow$$

$$R-\underset{\underset{O}{\|}}{C}-O-R \quad \text{Reactions with } R'-H \quad R-R$$

The radicals formed induce further homolytic cleavage, so that the decomposition is autocatalytically accelerated in a chain reaction.

The heterolytic decomposition of diacyl peroxides is catalyzed by strong Lewis acids and highly polar solvents, and leads to rearrangement products [186]:

$$\underset{R-C-O-O-C-R}{\overset{O\quad\ O}{\|\quad\ \|}} \longrightarrow \underset{RO-C-O-C-R}{\overset{O\quad\ O}{\|\quad\ \|}} \longrightarrow$$

Further products

Acid hydrolysis of a diacyl peroxide gives a mixture of the acid and the peroxyacid [187]:

$$\underset{R-C-O-O-C-R}{\overset{O\quad\ O}{\|\quad\ \|}} \xrightarrow{H_2O/H^+}$$

$$\underset{R-C-OH}{\overset{O}{\|}} + \underset{R-C-O-O-H}{\overset{O}{\|}}$$

The reduction of diacyl peroxides leads to various products, depending on the reducing agent. Reaction with trialkylphosphines gives the corresponding carboxylic anhydride [188]; the carboxylic acid is formed with iodide solution, ammonium borohydrides, and with lithium aluminum hydride [189]. The reaction with chromium(II) perchlorate in aqueous ethanol gives a quantitative yield of carbon dioxide and the hydrocarbon [190].

Dialkyl peroxydicarbonates are heat, shock, and impact sensitive. They decompose at low temperature like the diacyl peroxides and at room temperature explosive decomposition may take place, particularly with the lower peroxycarbonates:

$$\underset{RO-C-O-O-C-OR}{\overset{O\quad\ O}{\|\quad\ \|}} \longrightarrow 2\,RO\cdot + 2\,CO_2$$

$$2\,RO\cdot + R'-H \longrightarrow R-O-R' + R-OH$$

The risk of decomposition can be minimized by the use of additives such as iodine, oxygen, or substituted aromatics [191], [192].

## 5.2. Production

**By Autoxidation.** Only a few, mainly unsymmetrical, diacyl peroxides are produced by autoxidation. The sulfoxidation of hydrocarbons, leading to acetyl alkyl sulfonyl peroxides, is industrially important [193]–[195]. The reaction is induced by UV light [196]–[198] and by lithium and magnesium salts of carboxylic acids [199].

**From Peroxycarboxylic Acids.** The reaction of peroxycarboxylic acids with acid chlorides is used most frequently for the production of unsymmetrical diacyl peroxides [200]–[202].

*3-Chlorobenzoyl 2-Chlorobutyryl Peroxide* [23128-41-8]. Treatment of 85 % 3-chloroperbenzoic acid with 2-chlorobutyryl chloride in chloroform at −2 °C in the presence of aqueous sodium carbonate solution leads to the diacyl peroxide in almost quantitative yield with a purity of 90 %.

Chloroformate esters can be converted to acyl alkyl peroxycarbonates by reaction with peroxy acids in the presence of a base [203]:

$$R-\underset{\underset{O}{\|}}{C}-O-OH + R'-O-\underset{\underset{O}{\|}}{C}-Cl \longrightarrow R-\underset{\underset{O}{\|}}{C}-O-O-\underset{\underset{O}{\|}}{C}-OR'$$

An analogous reaction with phosgene leads to diacyl diperoxycarbonates [204]:

$$2\,R-\underset{\underset{O}{\|}}{C}-O-OH + COCl_2 \longrightarrow$$

$$R-\underset{\underset{O}{\|}}{C}-O-O-\underset{\underset{O}{\|}}{C}-O-O-\underset{\underset{O}{\|}}{C}-R$$

Diperoxydicarboxylic bis(carboxylic) dianhydrides can be obtained either by reaction of dicarboxylic acid dichlorides with peroxyacids, or by reaction of diperoxydicarboxylic acids with acid chlorides [205]:

$$Cl-\underset{\underset{O}{\|}}{C}-(CH_2)_n-\underset{\underset{O}{\|}}{C}-Cl + 2\,R-\underset{\underset{O}{\|}}{C}-O-OH$$
$$\downarrow$$
$$R-\underset{\underset{O}{\|}}{C}-O-O-\underset{\underset{O}{\|}}{C}-(CH_2)_n-\underset{\underset{O}{\|}}{C}-O-O-\underset{\underset{O}{\|}}{C}-R$$
$$\uparrow$$
$$HO-O-\underset{\underset{O}{\|}}{C}-(CH_2)_n-\underset{\underset{O}{\|}}{C}-O-OH + 2\,R-\underset{\underset{O}{\|}}{C}-Cl$$

**Acylation of Hydrogen Peroxide or Sodium Peroxide.** Industrially, the most important method for the preparation of symmetrical diacyl peroxides is the reaction of carboxylic acid chlorides or anhydrides with hydrogen peroxide, sodium peroxide, or alkali-metal peroxoborates [206]–[214].

*Diphthaloyl Peroxide* [37051-42-6]. After 20 h reaction at 25 °C, 78 % diphthaloyl peroxide is obtained from phthalic anhydride and hydrogen peroxide. About 14 % monoperoxyphthalic acid is formed as byproduct [215].

*4,4'-Di-tert-butylbenzoyl Peroxide* [1712-79-4]. Very pure crystalline product is obtained in quantitative yield by treatment of 4-*tert*-butylbenzoyl chloride with alkaline hydrogen peroxide (without solvent at 30–35 °C) in the presence of alkyl sulfonates to reduce the agglomeration of the peroxide particles [216].

For the preparation of hydrolytically sensitive diacyl peroxides, adducts of hydrogen peroxide (e.g., with urea) are used instead of the parent compound [217]. Dialkyl peroxydicarbonates are explosive when concentrated; they can be produced from chloroformate esters and alkaline hydrogen peroxide [218], [219].

*Dicetyl peroxydicarbonate* [26322-14-5] is obtained by treating 35% alkaline hydrogen peroxide with cetyl chloroformate (hexadecyl chloroformate) at 30 °C, in 98% yield after recrystallization from carbon tetrachloride [220].

## 5.3. Uses

Because of their facile thermal decomposition, diacyl peroxides are used, mainly as radical initiators for polymerization (especially diisopropyl peroxydicarbonate), and for cross-linking and hardening of polyester resins. Dialkyl peroxydicarbonates are gaining importance as low-temperature initiators for vinyl chloride polymerization.

The oxidizing power of diacyl peroxides is employed in organic synthesis and in bleaching additives for laundry detergents.

Diacyl peroxides are available commercially only in desensitized form (in solution or as a paste) because of their explosive nature. They should be stored and transported at low temperature.

# 6. Peroxycarboxylic Esters

The alkyl esters of peroxycarboxylic acids have been known since the beginning of the twentieth century. They can be separated into the following groups, based on the various possible structures:

| | |
|---|---|
| Esters of peroxy-carboxylic acids | $R-\overset{\overset{O}{\|}}{C}-O-O-R'$ |
| Esters of monoperoxy-dicarboxylic acids | $HO-\overset{\overset{O}{\|}}{C}-(CH_2)_n-\overset{\overset{O}{\|}}{C}-O-O-R'$ |
| Esters of diperoxydi-carboxylic acids | $R-O-O-\overset{\overset{O}{\|}}{C}-(CH_2)_n-\overset{\overset{O}{\|}}{C}-O-O-R'$ |
| Esters of monoperoxy-carbonic acid | $R-O-\overset{\overset{O}{\|}}{C}-O-O-R'$ |

| | |
|---|---|
| Esters of diperoxy-carbonic acid | R—O—O—C(=O)—O—O—R' |
| Alkylene bis(peroxy esters) of carboxylic acids | R—C(=O)—O—O—$(CH_2)_n$—O—O—C(=O)—R' |
| Esters of peroxy-carbamic acid | >N—C(=O)—O—O—R |
| Esters of peroxy-sulfonic acids | R—S(=O)(=O)—O—O—R' |

Monoperoxyesters of dibasic acids are best described by the *OO* or the *O* nomenclature, in order to indicate which alkyl group is attached to the peroxy group; hence the compound

$$(CH_3)_2CH-O-\underset{\underset{O}{\|}}{C}-O-O-C(CH_3)_3$$

is named *OO-tert*-butyl-*O*-isopropyl monoperoxycarbonate.

## 6.1. Properties

**Physical Properties.** Peroxycarboxylic esters are thermally unstable because of the low dissociation energy of the O–O bond (ca. 147–164 kJ/mol). Therefore, they should be distilled only in exceptional cases and under very mild conditions. Purification by recrystallization or by chromatography is preferable. A summary of the important physical properties of some peroxyesters is given in Table 7.

Peroxyesters can be characterized by means of their refractive index and their density. Qualitative and quantitative determination is possible by polarography, gas chromatography, NMR and, in particular, IR spectroscopy, with characteristic bands in the carbonyl region at 1750–1820 cm$^{-1}$ and in the fingerprint region between 800 and 900 cm$^{-1}$ [221], [222]. They can also be used for quantitative determination of concentration for kinetic measurements.

**Chemical Properties.** Esters of peroxycarboxylic acids decompose thermally by radical, ionic, and concerted mechanisms. The most important process in their applications is the thermally or photochemically induced homolytic decomposition of *tert*-butyl peroxycarboxylates to give free radicals [223]–[225]. This process has been thoroughly investigated and can take place by way of two mechanisms. The actual mode of decomposition is determined by the polar and steric effects of the group R attached to the acyl group [226]:

Table 7. Physical properties of some peroxyesters

| Name | CAS registry no. | Formula | $M_r$ | mp, °C | bp, °C (kPa) | $n_D^{20}$ | $d_4^{20}$ | Active oxygen content, % |
|---|---|---|---|---|---|---|---|---|
| Performic acid, tert-butyl ester | [819-50-1] | HC—O—O—C(CH$_3$)$_3$<br>‖<br>O | 118.1 | | 28 (1.3) | 1.3965[a] | | 13.5 |
| Peracetic acid, tert-butyl ester | [107-71-1] | H$_3$C—C—O—O—C(CH$_3$)$_3$<br>‖<br>O | 132.2 | | 50–51 (2.0) | | 0.888 | 12.1 |
| Peroxyisobutyric acid, tert-butyl ester | [109-13-7] | (CH$_3$)$_2$CH—C—O—O—C(CH$_3$)$_3$<br>‖<br>O | 160.2 | 45 | | | | 10.0 |
| Perbenzoic acid, tert-butyl ester | [614-45-9] | C$_6$H$_5$—C—O—O—C(CH$_3$)$_3$<br>‖<br>O | 194.2 | 8 | 75–77 (0.27) | 1.5007 | | 8.2 |
| Perbenzoic acid, cumyl ester | [7074-00-2] | C$_6$H$_5$—C—O—O—C(CH$_3$)$_2$—C$_6$H$_5$<br>‖<br>O | 256.3 | 45 | | | 1.043 | 6.2 |
| Peroxynonanoic acid, tert-butyl ester | [22913-02-6] | n-C$_8$H$_{17}$—C—O—O—C(CH$_3$)$_3$<br>‖<br>O | 230.3 | | 52–53 (0.001) | 1.4271[b] | 0.8868[c] | 6.9 |
| Monoperoxymaleic acid, tert-butyl ester | [1931-62-0] | HO$_2$C—CH=CH—C—O—O—C(CH$_3$)$_3$<br>‖<br>O | 188.2 | 114–116 (decomp.) | | | | 8.5 |
| Monoperoxyphthalic acid, tert-butyl ester | [15042-77-0] | HO$_2$C—C$_6$H$_4$—C—O—O—C(CH$_3$)$_3$<br>‖<br>O | 238.2 | 104 | | | | 6.7 |
| Diperoxyadipic acid, di-tert-butyl ester | [22158-52-7] | (H$_3$C)$_3$C—O—O—C—(CH$_2$)$_4$—C—O—O—C(CH$_3$)$_3$<br>‖          ‖<br>O           O | 290.4 | 42–43 | | | | 5.5 |
| OO-tert-Butyl-O-isopropyl monoperoxycarbonate | [2372-21-6] | (CH$_3$)$_2$CH—O—C—O—O—C(CH$_3$)$_3$<br>‖<br>O | 176.2 | 58–59 | | | | 9.1 |
| Diperoxycarbonic acid, di-tert-butyl ester | [3236-56-4] | (H$_3$C)$_3$C—O—O—C—O—O—C(CH$_3$)$_3$<br>‖<br>O | 206.2 | 54–55 (0.07) | | 1.4106 | | 15.5 |
| 2,5-Dimethyl-2,5-bis(benzoylperoxy)hexane | [2618-77-1] | (H$_3$C)$_2$C—(CH$_2$)$_2$—C(CH$_3$)$_2$<br>\|                              \|<br>O                             O<br>\|                              \|<br>O                             O<br>\|                              \|<br>C=O                       C=O<br>\|                              \|<br>C$_6$H$_5$                   C$_6$H$_5$ | 386.4 | 118 | | | | 8.3 |
| Peroxycarbamic acid, tert-butyl ester | [18389-96-3] | H$_2$N—C—O—O—C(CH$_3$)$_3$<br>‖<br>O | 133.1 | 51 | | | | 12.0 |

[a] At 25 °C. [b] At 30 °C. [c] $d_5^{30}$.

1) Cleavage of the O–O bond and subsequent reactions

$$(CH_3)_3C-O-O-\underset{\underset{O}{\|}}{C}-R \longrightarrow (CH_3)_3C-O^{\bullet} + {}^{\bullet}O-\underset{\underset{O}{\|}}{C}-R$$

$$R-\underset{\underset{O}{\|}}{C}-O^{\bullet} \quad \begin{array}{c} \longrightarrow R^{\bullet} + CO_2 \\ \overline{+ HX} \\ \longrightarrow R-COOH + X^{\bullet} \end{array}$$

$$(CH_3)_3C-O^{\bullet} \quad \begin{array}{c} \longrightarrow (CH_3)_2CO + H_3C^{\bullet} \\ \overline{+ HX} \\ \longrightarrow (CH_3)_3C-OH + X^{\bullet} \end{array}$$

2) Simultaneous cleavage of the O–O and the C–R bonds

$$(CH_3)_3C-O-O-\underset{\underset{O}{\|}}{C}-R \longrightarrow (CH_3)_3C-O^{\bullet} + CO_2 + R^{\bullet}$$

The kinetics of thermal decomposition of tertiary peroxyesters is influenced by the structure of the peroxyester [227] and by the polarity of the solvent [228], [229]. The stability decreases in the following order: *tert*-butyl > *tert*-amyl > tert-octyl > cumyl. In addition, the stability depends on the degree of α-substitution of the carboxylic acid moiety [230], [231]. Primary and secondary peroxycarboxylic esters decompose rapidly by a concerted mechanism to give a carboxylic acid and a carbonyl compound [232], [233]:

$$R^1-\underset{\underset{O-O}{\diagdown}}{\overset{\overset{O--H}{\diagup}}{C}}\underset{R_3}{\overset{R^2}{\diagup}}C \xrightarrow{\Delta} R^1-COOH + O=C\underset{R_3}{\overset{R^2}{\diagup}}$$

The radical-induced decomposition, in which a radical adds to the O–O bond resulting in cleavage, is usually undesired because fewer radicals are formed than expected for complete decomposition of a peroxycarboxylate. From one radical and one peroxyester molecule, which should normally give two radicals, only one radical is formed together with an ester molecule [234]:

$$R''^{\bullet} + R_3C-O-O-\underset{\underset{O}{\|}}{C}-R' \longrightarrow R_3C-O^{\bullet} + R''-O-\underset{\underset{O}{\|}}{C}-R'$$

Radical-induced decompositions are therefore not as effective as thermal homolysis, but can serve to extend the temperature range for the use of peroxycarboxylic esters.

Tertiary peroxyesters are more stable towards radical-induced decomposition than diacyl peroxides. This stability is more pronounced for peroxyesters that have an α-hydrogen atom or an unsaturated alkyl group [235].

Metal ions, especially copper(I) ions, catalyze the decomposition of peroxyesters [236], [237]:

$$R'-\underset{\underset{O}{\|}}{C}-O-O-R + Cu^+ \longrightarrow R'-\underset{\underset{O}{\|}}{C}-O^- + R-O^{\bullet} + Cu^{2+}$$

This reaction is of preparative use in the acyloxylation of compounds with reactive C–H bonds.

The ionic decomposition of peroxycarboxylic acids occurs with rearrangement [238], [239]:

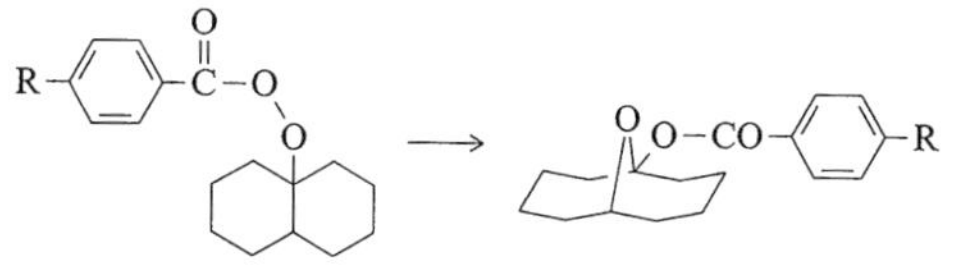

Peroxycarboxylic esters undergo hydrolysis much more readily than the corresponding carboxylic esters, with formation of carboxylic acids and hydroperoxides:

$$R_3C-O-O-C(=O)-R' \xrightarrow{H_2O} R_3C-O-OH + R'-C(=O)-OH$$

For the determination of peroxycarboxylic acids, conventional iodometric titration [240] is not usually applicable because it is always accompanied by an acid-catalyzed decomposition that does not consume iodine. The reaction of the peroxycarboxylic ester with sodium ethoxide or methanolic potassium hydroxide, followed by either iodometric titration of the resulting hydroperoxide [241], or metal-catalyzed iodometric titration [242] is more suitable.

## 6.2. Production

Peroxycarboxylic esters cannot be produced by direct esterification of peroxycarboxylic acids. All of the literature methods start from the corresponding hydroperoxide. The most stable and, therefore, most readily accessible peroxyesters are the tertiary peroxycarboxylic esters. The most important production processes include the reaction of hydroperoxides with carboxylic acids, carboxylic acid halides, anhydrides, and N-acylimidazoles (Fig. 4).

**From Hydroperoxides and Carboxylic Acids.** Hydroperoxides can be esterified with carboxylic acids to give peroxyesters. The reaction is generally catalyzed with p-toluenesulfonic acid and pyridine [243], dicyclohexyldiimide [244], [245], or phosphate esters [246]. The reaction probably takes place via an intermediate activated carboxylic acid derivative. Mineral acids or boron trioxide are also suitable as catalysts for the production of peroxyesters of $C_2-C_{10}$ aliphatic carboxylic acids and peroxyesters of aromatic carboxylic acids [247]. The water of reaction can be removed by azeotropic distillation with toluene, benzene, or cyclohexane.

**From Hydroperoxides and Carboxylic Acid Halides.** The reaction of carboxylic acid halides, in particular the chlorides, with hydroperoxides is the most important and

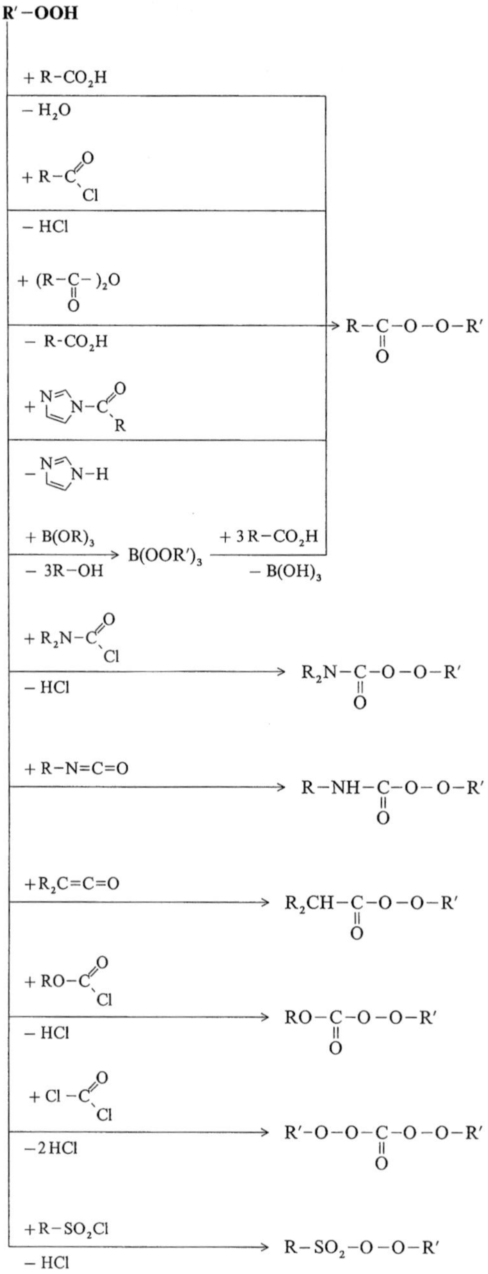

**Figure 4.** Reactions used for the synthesis of peroxyesters from hydroperoxides R′–OOH

industrially most frequently employed method for the synthesis of peroxycarboxylic esters. The process is usually carried out with high selectivity under Schotten–Baumann conditions in aqueous [248]–[251] or aqueous–organic media. Ether or a chlorinated hydrocarbon is used as the organic solvent [252]–[255]. In addition to the use of uniform starting materials, mixtures of carboxylic acid chlorides and/or hydroperoxides are also used, since the resulting mixtures of peroxyesters possess special properties for their application as polymerization initiators [256], [257].

*tert-Butyl Peroxyoctadecanoate* [2123-89-9]. Octadecanoyl chloride (0.165 mol) is added at 10 °C to a mixture of 0.2 mol *tert*-butyl hydroperoxide in petroleum naphtha. Subsequently, 0.2 mol of pyridine is added to this mixture in such a way that the temperature does not exceed 15 °C. After standing for 40 min at room temperature, the mixture is shaken with 10 % hydrochloric acid. The product crystallizes from the organic phase, is washed with water and dried; yield 96%.

The reaction takes place smoothly at low temperature in the presence of organic bases, preferably pyridine [258]–[265], but triethylamine [266]–[268], 2,6-dimethylpyridine [269] or 1,4-diazabicyclo[2.2.2]octane [270] have also been used. This method permits the use of an anhydrous system and is therefore suitable for acyl chlorides that are poorly soluble in aqueous media. In this way peroxyesters of carbamic acids, carbonic acid [271], and acetylene carboxylic acids can be prepared.

$$R_2N-C(=O)-Cl + R'-OOH \xrightarrow[-HCl]{Base} R_2N-C(=O)-O-O-R'$$

$$Cl-C(=O)-Cl + 2R-OOH \xrightarrow[-2HCl]{Base} R-O-O-C(=O)-O-O-R$$

$$Ph-C\equiv C-C(=O)Cl + (H_3C)_3C-OOH \xrightarrow[-HCl]{Base} Ph-C\equiv C-C(=O)-O-O-C(CH_3)_3$$

Di- or tetraperoxyesters can be produced both from the corresponding dicarboxylic acid dichlorides or tetracarboxylic acid tetrachlorides and hydroperoxides [272]–[274], as well as from bishydroperoxides and carboxylic acid chlorides [275]:

$$\underset{Me_3Si}{\overset{Me_3Si}{\triangle}}\!\!\!\begin{matrix}COCl\\COCl\end{matrix} \xrightarrow[pyridine]{(CH_3)_3COOH} \underset{Me_3Si}{\overset{Me_3Si}{\triangle}}\!\!\!\begin{matrix}CO_3CMe_3\\CO_3CMe_3\end{matrix}$$

Primary and secondary alkyl peroxycarboxylic esters are best prepared from the alkali metal salts of the corresponding hydroperoxide [276]–[280], since they are not readily accessible by other methods.

**From Hydroperoxides and Carboxylic Anhydrides.** Hydroperoxides, alkyl metal peroxides, or alkyl nonmetal peroxides can also be acylated by carboxylic anhydrides:

$$(R^1-CO)_2O + R^2-O-OH \xrightarrow{-R^1COOH} R^1-C(=O)-O-O-R^2$$

The reaction is carried out in the presence of alkali-metal hydroxide solution, organic bases such as pyridine, or alkali-metal acetates [281], [282]. This method is particularly suitable for the synthesis of dicarboxylic acid mono-*tert*-alkyl peroxyesters [283], [284]. The use of mineral acids as catalysts is only possible for acid-stable hydroperoxides.

*tert-Butyl Monoperoxysuccinate* [28884-42-6]. *tert*-Butyl hydroperoxide (0.13 mol) and 0.06 mol pyridine are added in succession to 0.12 mol succinic anhydride at 45–55 °C. After 15 min the mixture is cooled, extracted with chloroform, and the extract washed with hydrochloric acid and water. The product is isolated in 83% yield by distillation from the dried organic phase.

Alkylperoxy organogermanes, -stannanes, -silanes, and -stibines are cleaved heterolytically at the O–M bond by carboxylic anhydrides [285]–[288]:

$$(H_3C-CO)_2O + (H_3C)_3C-O-O-MR_3 \xrightarrow{-H_3C-CO-OMR_3} H_3C-C(=O)-O-O-C(CH_3)_3$$

$$X\begin{pmatrix}C(=O)\\O\\C(=O)\end{pmatrix}O + (H_3C)_3C-O-O-MR_3 \longrightarrow X\begin{pmatrix}C(=O)-O-MR_3 \; (-ER_n)\\C(=O)-O-O-C(CH_3)_3\end{pmatrix}$$
(or $-ER_n$)

$X = -(CH_2)_2-$, $C_6H_4$, $-CH=CH-$

$M = Ge^{III}, Sn^{IV}$    $ER_n = Si(CH_3)_3, Sb(C_6H_5)_4$
$R = C_2H_5, C_6H_5$

The corresponding dicarboxylic mono-*tert*-butyl peroxyesters can be obtained by treatment with 0.5 N sulfuric acid or by introducing dry hydrogen chloride [289].

**From Hydroperoxides and N-Acylimidazoles.** The acylation of alkylhydroperoxides with the very reactive *N*-acylimidazoles, available from carboxylic acids and *N,N'*-carbonylimidazoles, is particularly useful when the corresponding acyl chloride cannot readily be obtained pure [290]–[292]. Isolation and purification of the *N*-acylimidazole intermediate is not necessary. The reaction can be carried out from imidazole in a one-pot process [293]:

$$4 \underset{N}{\overset{\diagup\!\!\!\diagdown}{\underset{\diagdown\!\!\!\diagup}{\phantom{N}}}}NH \xrightarrow[-2\left[\underset{N}{\overset{\diagup\!\!\!\diagdown}{\underset{\diagdown\!\!\!\diagup}{\phantom{N}}}}NH_2^+\right]Cl^-]{+\,COCl_2} \underset{N}{\overset{\diagup\!\!\!\diagdown}{\underset{\diagdown\!\!\!\diagup}{\phantom{N}}}}N-\overset{O}{\underset{\|}{C}}-N\underset{N}{\overset{\diagdown\!\!\!\diagup}{\underset{\diagup\!\!\!\diagdown}{\phantom{N}}}}$$

$$\xrightarrow[-CO_2]{+\,2R^1-COOH} 2\,\underset{N}{\overset{\diagup\!\!\!\diagdown}{\underset{\diagdown\!\!\!\diagup}{\phantom{N}}}}N-\overset{O}{\underset{R^1}{C}}$$

$$\xrightarrow[-2\,\underset{N}{\overset{H}{\underset{\diagdown\!\!\!\diagup}{N}}}]{+\,2R^2-O-OH} 2R^1-\overset{O}{\underset{O-O-R^2}{C}}$$

This method is of limited use for the production of peroxycarboxylic esters derived from carboxylic acids containing a C=C double bond [294].

**Other Methods.** The acid-catalyzed reaction of hydroxycarboxylic acids with hydrogen peroxide in ether leads to cyclic peroxycarboxylic esters [295]:

$$\underset{CH_3}{\overset{CH_2-COOH}{R-\underset{|}{\overset{|}{C}}-OH}} \xrightarrow[H^+]{H_2O_2} \underset{H_3C}{\overset{R}{\underset{\diagdown}{C}}}\underset{O-O}{\overset{CH_2}{\underset{\diagup}{\phantom{C}}}}C=O$$

Esters of peroxychloroacetic acid are obtained by acidolysis of dialkyl peroxides with chloroacetic acid [296]:

$$Cl-CH_2-\overset{O}{\underset{OH}{C}} + (H_3C)_3C-O-O-C(CH_3)_3 \xrightarrow[-(H_3C)_3C-OH]{+H^+}$$

$$Cl-CH_2-\overset{O}{\underset{O-O-C(CH_3)_3}{C}}$$

Ketenes and isocyanates add hydroperoxides under neutral or weakly acidic conditions to give the peroxyesters of carboxylic or carbamic acids [297]–[299]:

$$H_2C=C=O + R'-OOH \longrightarrow H_3C-\underset{\underset{O}{\|}}{C}-O-O-R'$$

$$R-N=C=O + R'-OOH \longrightarrow R-NH-\underset{\underset{O}{\|}}{C}-O-OR'$$

This method is particularly suitable for the production of base-sensitive peroxycarboxylic esters. The reaction of triperoxyborates, produced from borate esters and hydroperoxides, with carboxylic acids leads similarly to the corresponding peroxycarboxylic esters [300]:

$$B(OR)_3 \xrightarrow[+ 3\,R'-OOH]{} B(O-O-R')_3 + 3\,R'OH$$

$$B(O-O-R')_3 \xrightarrow[+ 3\,R-\underset{\underset{O}{\|}}{C}-OH]{} 3\,R-\underset{\underset{O}{\|}}{C}-O-O-R' + B(OH)_3$$

The acid-catalyzed reaction is carried out in aromatic or cycloaliphatic solvents such as benzene, toluene, or cyclohexane. However, the reaction is not suitable for primary hydroperoxides.

## 6.3. Uses

The most important industrial application of peroxyesters is in the initiation of polymerization reactions, and as hardeners for resins. The more reactive types (e.g., *tert*-butyl or cumyl peroxypivalate) are used mainly as initiators for the polymerization of ethylene or vinyl chloride. *tert*-Butyl perbenzoate is used predominantly as a hardener for polyester resins.

The usual commercial products are of technical grade or are product mixtures that have been desensitized by the addition of an organic solvent (aliphatic hydrocarbons, carboxylic esters). Frozen peroxide emulsions can be stored safely and exhibit very good storage stability [301].

Peroxycarboxylates play an important role in preparative organic chemistry in, for example, inter- and intramolecular epoxidation of olefins in the presence of copper salts [302]. The peroxyesters react with strong nucleophiles, such as Grignard reagents or phosphines, with transfer of the alkoxy group. The copper-catalyzed acyloxylation of C–H bonds occurs by a radical mechanism. In this way, α-acyloxy derivatives of olefins, ethers, and thioethers can be obtained [303]. Thermal homolysis of peroxyesters in protic solvents can be used for the decarboxylation of carboxylic acids.

# 7. α-Oxyperoxides

α-Oxyperoxides possess the structural element

$$X-O-O-\underset{|}{\overset{|}{C}}-O-Y$$

where X = –H, –R, and Y = –H, –R, –OH, –OR. The substituents X and Y may also be part of a cyclic group. A selection of possible structures is given in Table 8.

Cyclic alkoxyperoxides of the type

$$-HC\underset{O}{\overset{O-O}{\diagdown\diagup}}CH-$$

**Table 8.** Structure and nomenclature of α-oxyperoxides

| Substituents | | |
|---|---|---|
| X | Y | Name |
| H | H | α-hydroxyhydroperoxide |
| H | R | α-alkoxyhydroperoxide |
| H | OH | geminal bishydroperoxide |
| H | OR | α-hydroperoxyperoxide |
| R | H | α-hydroxyperoxide |
| R | R | α-alkoxyperoxide |
| R | OR | geminal diperoxide, peroxyketal |

are known as ozonides. α-Oxyperoxides are unstable and highly reactive and therefore capable of numerous reactions.

## 7.1. Properties

**Physical Properties.** The stability of α-oxyperoxides varies widely depending on the structure, and range from explosive to being stable up to the melting or boiling point. Physical properties of some selected compounds are listed in Table 9.

**Chemical Properties.** As the chemical behavior of α-oxyperoxides depends markedly on the structure, only a few typical reactions are described here.

The thermolysis of α-oxyperoxides leads to a complex mixture of products [304]. Dicyclohexylidene diperoxide and tricyclohexylidene diperoxide give $C_{12}$ or $C_{15}$ cycloalkanes which are not readily accessible by other routes, as well as lactones with 12- or 15-membered rings [305], [306].

The decomposition of dimeric cyclopentanone and cyclohexanone peroxides (dicyclopentylidene diperoxide, dicyclohexylidene diperoxide) in the presence of a Cu(I) catalyst in aqueous hydrochloric acid leads to 5-chlorovaleric acid and 6-chlorocaproic acid, respectively, in high yield [307], [308].

Cycloaliphatic α-oxyperoxides decompose to form open-chain dicarboxylic acids in the presence of iron or copper salts as catalysts [305], [309], whereas dehydration of these compounds leads to cyclic geminal alkylidene triperoxides (1, 2, 4, 5, 7, 8-hexaoxycyclononanes).

Heating α-oxyperoxides in water causes elimination of hydrogen peroxide to give the dimeric α,α'-dihydroxyperoxides which can be dehydrated to give ozonides [310]:

**Table 9.** Physical properties of some -oxyperoxides

| Name | CAS registry no. | Formula | $M_r$ | mp, °C | bp, °C (kPa) | Active oxygen content, % |
|---|---|---|---|---|---|---|
| 2-Chloro-1-hydroperoxy-cyclohexanol | [15250-08-5] | | 166.71 | 76 | | 9.6 |
| α,α′-Dihydroxydicyclo-hexyl peroxide | [2407-94-5] | | 230.30 | 69–71 | | 6.9 |
| α-Hydroperoxy-α′-hydroxy-dicyclohexyl peroxide | [78-18-2] | | 246.30 | 78 | | 13.0 |
| α,α′-Bishydroperoxydicyclo-hexyl peroxide | [2699-12-9] | | 262.30 | 82–83 | | 18.3 |
| Dicyclohexylidene diperoxide | [183-84-6] | | 228.29 | 127–129 | | 14.0 |
| Tricyclohexylidene triperoxide | [182-01-4] | | 342.43 | 93 | | 14.0 |
| Dicyclopentylidene diperoxide | [311-38-6] | | 200.23 | 95–98 | | 16.0 |
| Diisopropylidene diperoxide | [1073-91-2] | | 148.16 | 132 | | 21.6 |
| 2,2-Bis(*tert*-butylperoxy)-propane | [4262-61-7] | | 220.31 | | 69–70 (2.0) | 14.5 |
| 2,2-Bis(*tert*-butylperoxy)-butane | [2167-23-9] | | 234.34 | | 50 (0.27) | 13.7 |
| 1,1-Bis(*tert*-butylperoxy)-cyclohexane | [3006-86-8] | | 260.37 | | 52–54 (0.02) | 12.3 |
| Ethylene ozonide (1,2,4,-trioxolane) | [289-14-5] | | 76.05 | | 18 (2.1) | 21.0 |
| Isobutene ozonide (3,3-dimethyl-1,2,4-trioxolane) | [22409-33-2] | | 104.10 | | 42–42.5 (18.0) | 15.4 |
| Tetramethylethylene ozonide (3,3,5,5-tetramethyl-1,2,4-trioxolane) | [15592-50-4] | | 132.16 | 133–134 | | 12.1 |

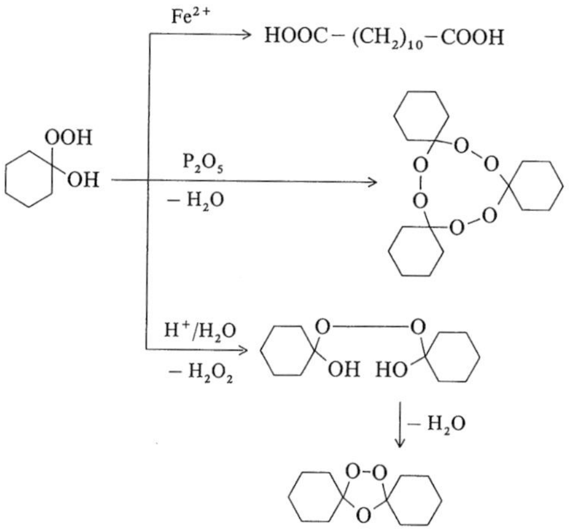

Ozonides (cyclic α-alkoxyperoxides) give a multitude of products upon thermal or radical induced decomposition [311]. The reactive species is the carbonyl oxide (**6**), formed from the primary ozonide (or 1,2,3-trioxolane) (**5**):

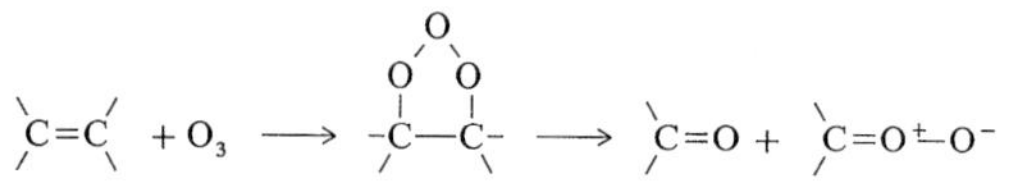

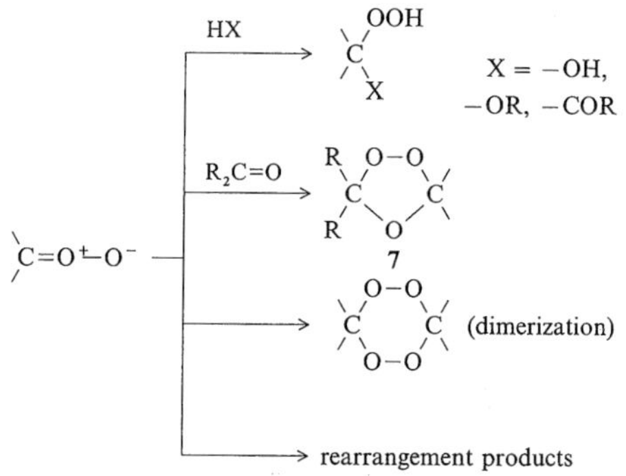

The primary ozonide (**5**) is unstable and generally not isolable [312], [313]. In contrast, ozonides of structure (**7**), with the exception of the simplest representatives such as ethylene ozonide which is explosive, are stable compounds that can be isolated. They only tend to decompose spontaneously under the catalytic influence of heavy metals [313].

Ozonides can be reduced to aldehydes and oxidized to carboxylic acids [304].

## 7.2. Production

Three main routes exist for the production of α-oxyperoxides:

1) Autoxidation of ethers, acetals, aldehydes, or alcohols
2) Addition of hydrogen peroxide, alkyl hydroperoxides, or peroxycarboxylic acids to carbonyl compounds
3) Ozonolysis of unsaturated compounds

**α-Oxyperoxides by Autoxidation.** Autoxidation is not suitable for the preparation of α-oxyperoxides because the reaction usually leads to mixtures of products. Ethers are readily oxidized by atmospheric oxygen to yield ether hydroperoxides, from which the explosive polymeric ether peroxides are formed [186], [314]:

$$C_2H_5-O-C_2H_5 \xrightarrow{O_2} \underset{OC_2H_5}{\overset{CH_3}{\underset{|}{CH}-OOH}} \xrightarrow{-C_2H_5OH} -(\underset{|}{\overset{CH_3}{CH}}-O-O)_n-$$

Acetals are easier to autoxidize than ethers. Dimers of structure **10** are formed from 1,3-dioxolanes (**8**) via the hydroperoxide (**9**), when R = H or CH$_3$ [315]. Bulky groups impede the dimeriza-tion. Autoxidation of vinyl-1,3-dioxolanes (R = –CH = CH$_2$) leads to heterocyclic α-oxyperoxides (**12**), via formation of an unstable allyl peroxide (**11**) intermediate [316].

*2-Hydroperoxy-2-isopropyl-1,3-dioxolane* [34594-19-9] (R = –CH(CH$_3$)$_2$) is obtained in good yield by the reaction of 2-isopropyl-1,3-dioxolane with atmospheric oxygen at 75 °C [317].

Autoxidation of alcohols gives mixtures of α-hydroxyhydroperoxides and α,α'-bishydroperoxyperoxides [318] – [320]:

$$\underset{OH}{\overset{H}{\underset{|}{R_2C}}} \longrightarrow \underset{OH}{\overset{OOH}{\underset{|}{R_2C}}} + \underset{HOO}{\overset{O-O}{\underset{|}{R_2C}}}\underset{OOH}{\overset{}{\underset{|}{CR_2}}}$$

2-Hydroperoxy-2-propanol, the primary oxidation product of 2-propanol, decomposes

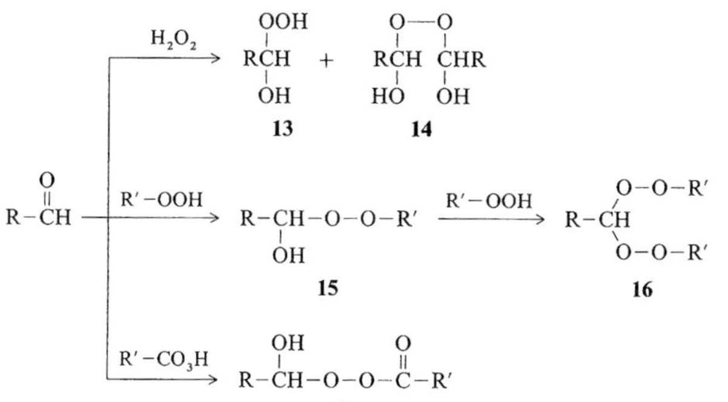

$$\text{R-CH} \begin{array}{c} \xrightarrow{H_2O_2} \\ \\ \xrightarrow{R'-OOH} \\ \\ \xrightarrow{R'-CO_3H} \end{array}$$

Figure 5. Reactions of aldehydes with hydrogen peroxide, alkyl hydroperoxides, and peroxycarboxylic acids

(Figure 5 shows structures 13, 14, 15, 16, 17 as described in the text.)

in the presence of water to give hydrogen peroxide and acetone. This process is employed industrially for the production of hydrogen peroxide [321], [322].

The autoxidation of aldehydes proceeds via intermediate α-hydroxyalkyl peroxycarboxylates to give the corresponding peroxycarboxylic acids:

$$2\,R-\overset{O}{\underset{H}{C}} \xrightarrow{O_2} R-\overset{O}{C}-\overset{OH}{\underset{O-O}{CH-R}} \longrightarrow R-\overset{O}{C}\diagdown_{OOH} + R-\overset{O}{C}\diagdown_{H}$$

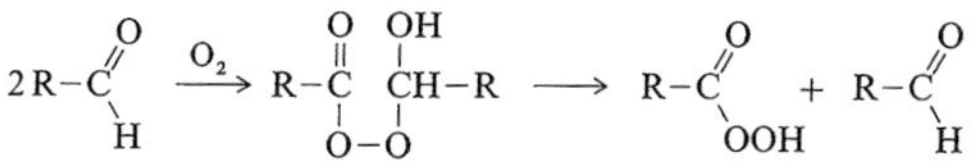

This reaction is used industrially for the production of peroxyacetic acid from acetaldehyde.

**α-Oxyperoxides by Addition of Hydrogen Peroxide, Alkyl Hydroperoxides, or Peroxycarboxylic Acids to Carbonyl Compounds.** Aldehydes react with hydrogen peroxide to give mixtures of α-hydroxyhydroperoxides (**13**) and α,α'-dihydroxyperoxides (**14**). Aldehydes react with alkylhydroperoxides under acid catalysis to form α-hydroxyperoxides (**15**), which react with excess hydroperoxide to form peroxyacetals (**16**) [323], [324]. With peroxycarboxylic acids, aldehydes and ketones form α-hydroxyalkyl peroxycarboxylic esters (**17**), which may also be obtained by acylation of α-hydroxyhydroperoxides [325]. These reactions are summarized in Figure 5.

In the presence of concentrated sulfuric acid, aldehydes react with hydrogen peroxide to form cyclic geminal alkylidenedi- and triperoxides (1,2,4,5-tetraoxanes and 1,2,4,5,7,8-hexaoxacyclononanes) [326].

Treatment of ketones with hydrogen peroxide gives a multitude of products, most of which arise from rearrangements. For example, the following compounds are formed from cyclohexanone:

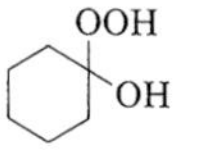

α-Hydroperoxycyclohexanol

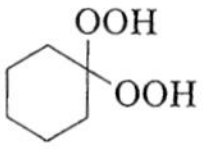

α,α-Dihydroperoxycyclohexane

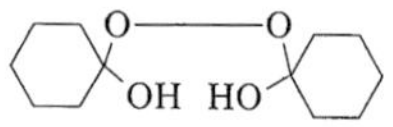

α,α'-Dihydroxydicyclohexyl peroxide

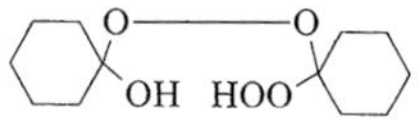

α-Hydroperoxy-α'-hydroxydicyclohexyl peroxide

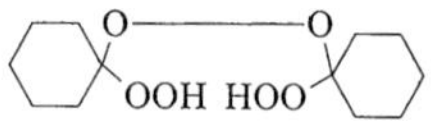

α,α'-Dihydroperoxydicyclohexyl peroxide

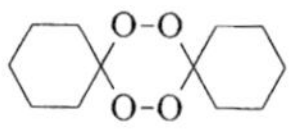

Dicyclohexylidene diperoxide (3,6-dicyclohexylidene-1,2,4,5-tetroxane)

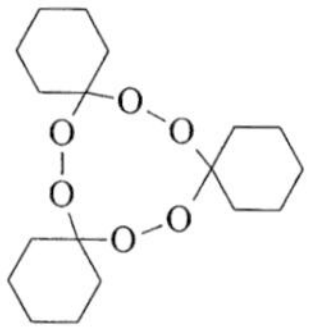

Tricyclohexylidene triperoxide (3, 6, 9-tricyclohexylidene-1, 2, 4, 5, 7, 8,-hexaoxacyclononane)

α,α'-Dihydroperoxydicyclohexyl Peroxide [2699-12-9]. Treatment of cyclohexanone with hydrogen peroxide in acetonitrile in the presence of perchloric acid results in formation of α,α'-dihydroperoxydicyclohexyl peroxide in good yield. The compound can be recrystallized from hexane.

The α-oxyperoxides of acetone [327], cyclohexanone [305], [326], [328], [329], methyl ethyl ketone [327], and acetylacetone [330] have been studied in detail. Ketones react with alkyl hydroperoxides to give peroxyketals, whereby the peroxy hemiketals normally cannot be isolated [32], [327], [331], [332]:

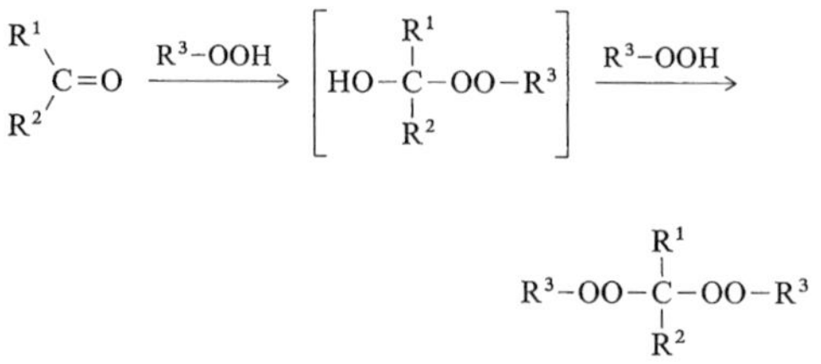

$$R^1R^2C=O \xrightarrow{R^3-OOH} \left[ HO-\underset{R^2}{\underset{|}{\overset{R^1}{\overset{|}{C}}}}-OO-R^3 \right] \xrightarrow{R^3-OOH} R^3-OO-\underset{R^2}{\underset{|}{\overset{R^1}{\overset{|}{C}}}}-OO-R^3$$

Reaction of α,α'-dihydroperoxyperoxides with ketones leads to formation of cyclic geminal alkylidenedi- and triperoxides, which are used as starting materials for the synthesis of macrocyclic dienes [305], [306], [333]–[335].

**α-Oxyperoxides by Ozonolysis.** In the ozonolysis of olefins, other α-oxyperoxides in addition to the ozonides are formed [304], [336]. The reaction is of little preparative use as it is generally not very selective for α-oxyperoxides, and gives defined compounds only for tetrasubstituted olefins [311].

## 7.3. Uses

α-Oxyperoxides are mainly used as vulcanization accelerators, cross-linking agents [337], [338] and hardeners, in particular methyl ethyl ketone peroxide and methyl isobutyl ketone peroxide, as well as cyclohexanone peroxide and various peroxyketals. In combination with transition metal salts a wide temperature range can be covered (e.g., thermal hardening of polyesters), so that suitable catalyst systems are available for various materials [339]–[342].

Ozonides are used as disinfectants [343]. Ozonolysis is applied industrially using fatty acids, e.g., in the production of nonanedioic acid and nonanoic acid from oleic acid [344], [345], or of decanedioic acid from cyclodecene [346]. α-Oxyperoxides are commercially available only in desensitized form because of their explosiveness. Phosphates, dibutyl phthalate, and aliphatic hydrocarbons are typical additives.

# 8. α-Aminoperoxides

In addition to the hydroperoxy or peroxy group, α-aminoperoxides bear a nitrogen atom on the same carbon atom; both functional groups may be part of a ring.

## 8.1. Properties

**Physical Properties.** The physical properties of some α-aminoperoxides are listed in Table 10. Their stability varies widely, ranging from high shock and impact sensitivity with the lower homologues to a thermal stability up to the melting point.

**Chemical Properties.** Some typical reactions of α-aminoperoxides are as follows [347]:

**Table 10.** Physical properties of some -aminoperoxides

| Name | CAS registry no. | Formula | $M_r$ | mp, °C | bp, °C (kPa) | Active oxygen content, % |
|---|---|---|---|---|---|---|
| 1-Amino-1-hydroperoxycyclohexane | [24075-24-9] | | 131.18 | 57–58 (decomp.) | | 12.2 |
| 1,1'-Peroxydicyclohexylamine | [21842-28-4] | | 211.30 | 40–42 | 94–97 (0.05) | 7.6 |
| 1,1'-Peroxydicyclopentylamine | [21842-27-3] | | 183.25 | 22–23 | | 8.7 |
| 2,2'-Peroxydiisopropylamine | [24075-03-4] | | 131.18 | | 50 (1.6) | 12.2 |
| 2,2'-Peroxydiisobutylamine | [24075-04-5] | | 159.23 | | 66–68 (1.6) | 10.1 |
| 1-Amino-1-*tert*-butylperoxy-cyclohexane | [24829-66-1] | | 187.28 | | 36–38 (0.03) | 8.5 |
| 1-Cyclohexylamino-1-hydroperoxy-cyclohexane | [2808-61-9] | | 212.31 | 88–89 | | 7.5 |
| 1-Hydroperoxy-1-azocyclohexane | [32140-97-9] | | 225.31 | 73–75 | | 7.1 |

The unstable 1-amino-1-hydroperoxycyclohexane reacts with ketones to form cyclic α-aminoperoxides. With aqueous ammonia in alcohol and catalytic amounts of sodium

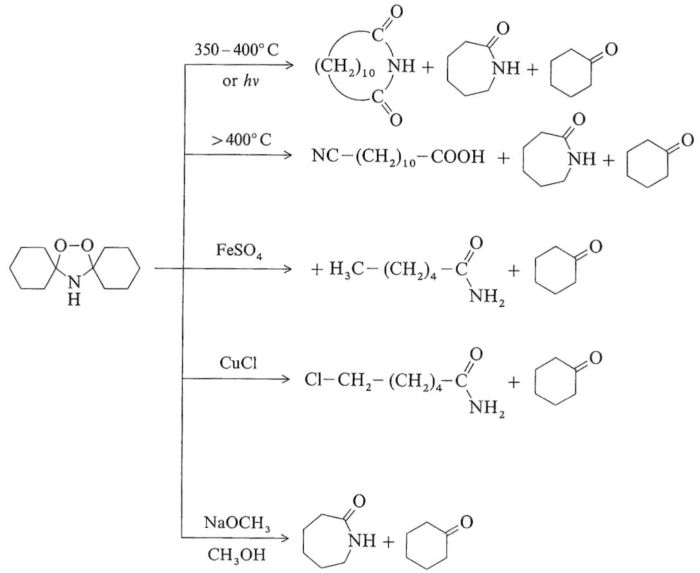

**Figure 6.** Reactions of 1,1′-peroxydicyclohexylamine

tungstate, cyclohexanone oxime is formed. Acylation gives a dimeric product with elimination of hydrogen peroxide. Catalytic decomposition in the presence of iron leads to cyclohexanone together with small amounts of caproic acid and its amide. Cyclohexanone, together with ε-caprolactam, is also the pyrolysis product of 1-amino-1-hydroperoxycyclohexane.

1,1′-Peroxydicyclohexylamine rearranges, both thermally and photochemically, to decane-1,10-dicarboxylic acid imide; byproducts include ε-caprolactam and cyclohexanone. These reactions are shown in Figure 6.

Thermolysis at >400 °C leads mainly to 11-cyanoundecanoic acid.

Radical decomposition with iron or copper salts gives caproic acid amide and cyclohexanone.

ε-Caprolactam is formed from 1,1′-peroxydicyclohexylamine in good yield by treatment with sodium methoxide in methanol.

## 8.2. Production

α-Aminoperoxides can be produced by the same methods as α-oxyperoxides. Autoxidation is important only for Schiff's bases [348]. High yields of α-aminoperoxides are obtained by the Mannich reaction of amines with formaldehyde and hydroperoxides:

$$R_2NH + CH_2O + R'-OOH \xrightarrow{-H_2O} R_2N-CH_2-O-O-R'$$

The most useful synthetic method is the reaction of carbonyl compounds with hydrogen peroxide in the presence of ammonia or primary amines [349] – [358]:

$$R_2C=O \xrightarrow[NH_3]{H_2O_2} R_2C\!\!\begin{array}{c}NH_2\\OOH\end{array} \xrightarrow{R_2C=O} R_2C\!\!\begin{array}{c}H\\N\\O-O\end{array}\!\!CR_2$$

The initially formed α-aminoperoxides are frequently not isolable and react further with the carbonyl compound to give 1,2,4-dioxazolidines.

Cyclopentanone forms 1,1'-peroxydicyclopentylamine, whereas 1-aminohydroperoxides are formed from large-ring ketones.

*1-Amino-1-hydroperoxycyclohexane* [24075-24-9]. A solution of 30% $H_2O_2$ is added to a mixture of methanol, ammonia, and cyclohexanone with stirring at 0 °C. The precipitated solid is removed by filtration and the crude product obtained in 88% yield [359]. The use of alkyl hydroperoxides instead of hydrogen peroxide gives 1-aminoperoxides.

## 8.3. Uses

α-Aminoperoxides are important intermediates in the synthesis of ε-caprolactam [360] – [363], of cyclohexanone oxime [347], of derivatives of terminal dicarboxylic acids [347], [349], and in the production of nylon-12 from cyclohexanone [364]. More recently, they are also being considered for use as initiators in polymerization of styrene [365], as disinfectants [366], and as antimalarial agents [367].

# 9. Safety Measures

## 9.1. Hazards

Organic peroxides are frequently very unstable compounds which can decompose spontaneously and explosively under thermal or mechanical stress. Such decomposition can be caused by shock, impact, friction, or by the catalytic effect of impurities. Organic peroxides are therefore subject to the regulations for explosives [368].

Specific safety and protective measures must be observed when working with organic peroxides [369], [370]. To reduce the hazards involved, commercially available peroxides are usually desensitized by addition of inert solids or liquids. Typical desensitizing agents are water, inert solvents (aliphatic and halogenated hydrocarbons, phthalate and phosphate esters, acetals), and inert inorganic solids (sulfates, phosphates, borates,

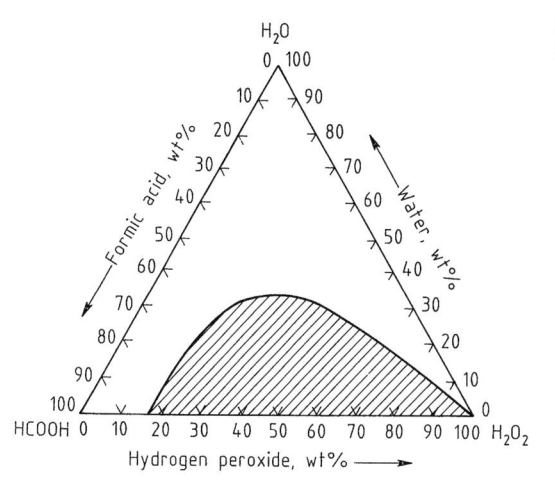

**Figure 7.** Explosive range (hatched area) for $H_2O_2$–$HCOOH$–$H_2O$

silicates, carbonates). Therefore peroxides are available in solid, liquid, or paste form. Investigation of the hazards associated with organic peroxides involves a wide range of test methods which are described in the relevant legal regulations. In Germany BAM is responsible for conducting and evaluating such tests. Similar institutions exist in other countries [371]. In the most important tests the following properties of peroxides are investigated:

1) Detonation or deflagration ability (BAM 2″ steel tube test)
2) Shock sensitivity (BAM falling hammer test)
3) Friction sensitivity
4) Heat sensitivity under defined confinement in a steel tube (Koenen test)
5) Explosive energy in a lead block (Trauzl method)

The explosive range of peroxide-containing mixtures can be displayed graphically, as shown in Figure 7 for the $H_2O_2$–$HCOOH$–$H_2O$ system [372], [373]. For industrial applications, knowledge of explosive ranges is essential, in particular for reactions in which performic or peracetic acid is generated in situ. In these cases the aqueous phase that is usually present is extremely explosive.

The explosive range can be investigated, for example, by 2″ steel tube tests. Passage through this range must be avoided during a reaction by appropriate process control.

## 9.2. Transportation and Storage

Organic peroxides are subject to the regulations for dangerous materials [374]. The establishment of the maximum permitted ambient temperature for storage and transportation is based on the SADT (self-accelerating decomposition temperature), above which peroxide decomposition is self-accelerating. The SADTs of some commercial

**Table 11.** Self-accelerating decomposition temperature (SADT) of commercial peroxides

| Peroxide | CAS registry no. | Supplied form | SADT, °C |
|---|---|---|---|
| tert-Butyl hydroperoxide | [75-91-2] | 70 % in water | 88 |
| | | solution in phosphate | ca. 65 |
| Cumene hydroperoxide | [80-15-9] | 80 % water | 80 |
| Di-tert-butyl peroxide | [110-05-4] | technical grade (liquid) | 79–80 |
| 2,5-Dimethyl-2,5-bis(tert-butylperoxy)-hexane | [2618-77-1] | liquid | 82 |
| Dicumyl peroxide | [80-43-3] | technical grade (powder) | 88 |
| | | 40 % (powder with chalk) | > 70 |
| Peracetic acid | [79-21-0] | 40 % in acetic acid/$H_2O$ | 55 |
| | | 15 % in acetic acid/$H_2O$/$H_2O_2$ | 60 |
| Dibenzoyl peroxide | [94-36-0] | aqueous, technical grade | 68 |
| | | technical grade (flakes) | ca. 50 |
| | | 50 % in plasticizer | 52 |
| | | 50 % in silicone oil | 63 |
| Diacetyl peroxide | [110-22-5] | 25 % solution in phthalate | ca. 35 |
| tert-Butyl perbenzoate | [614-45-9] | technical grade | ca. 60 |
| Diisopropyl peroxydicarbonate | [105-64-6] | 40 % solution in phthalate | ca. 5 |
| Acetyl cyclohexanesulfonyl peroxide | [3179-56-4] | water-damp granules | ca. 15 |
| | | 28 % solution in phthalate | ca. 5 |
| Methyl ethyl ketone peroxide | [1338-23-4] | 40 % solution in phthalate | 63 |
| | | 50 % solution in phosphate | ca. 65 |
| Methyl isobutyl ketone peroxide | [37206-20-5] | 60 % solution in phthalate | 64 |
| Cyclohexanone peroxide | [12262-58-7] | moistened with water | ca. 50 |
| | | 50 % solution in phosphate | ca. 50 |
| 2,2-Bis(tert-butylperoxy)butane | [2167-23-9] | 50 % solution in phthalate | 63 |

peroxides are listed in Table 11. Organic peroxides and their formulations must be transported and stored at a temperature significantly below the SADT. For thermally very unstable peroxides maximum transportation temperatures are specified.

In Germany, the "Safety Precautions for Organic Peroxides" are applied for storage [370]. National transportation regulations are based on the recommendations for the transportation of dangerous goods by the United Nations. The UN has reexamined the classification of organic peroxides and new regulations have been introduced [375]. Organic peroxides are assigned to the hazard groups A to G by means of a flow diagram, which also indicates the permitted amount for a given container and container type. Following the new regulations, some peroxides with a low active oxygen content may be exempted. Until then, the following dangerous goods regulations are valid for the Federal Republic of Germany, within which the organic peroxides are assigned mainly to Class 5.2:

| | |
|---|---|
| Road transportation: | GGVS with Appendices A and B, ADR |
| Transportation by rail: | GGVE/RID |
| Maritime: | GGVSee with Appendix, IMDG |
| Air Freight: | ICAO Technical Instructions, IATA-DGR |

## 9.3. Handling

Safety glasses and gloves must be worn when handling organic peroxides. Protective clothing is required when working with larger amounts of peroxides [369]. In the event that skin or eye contact with peroxides occurs, the afflicted area should be immediately rinsed with water for a sufficiently long period (at least 15 min). Contaminated clothing should be removed. A physician or eye specialist should be consulted without delay.

Cleanliness at the workplace is an essential prerequisite for working safely with peroxides because even small amounts of contaminants such as ash, dust, or traces of heavy metals could lead to decomposition. The containers and equipment employed must be stable against peroxides and must not promote decomposition. Suitable materials are glass, porcelain, earthenware, stainless steel, and certain plastics such as teflon or polyethylene. When setting up apparatus suitable precautions must be taken as decomposition can lead to a considerable pressure increase. Often it is best to work with open containers. Peroxide samples removed from their original containers should not be returned because of the risk of contamination. Preliminary experiments should be carried out with small quantities of material. Furthermore, peroxides should only be added in small amounts, in order to avoid high local concentrations. Transferring and weighing of peroxides must be carried out carefully, as even slight friction or impact with a spatula or a scoop could cause an explosion with the more sensitive compounds.

Spills of water-soluble peroxides can be cleared up without risk by the use of large amounts of water. Compounds that are insoluble in water must be diluted and desensitized with sufficiently large amounts of appropriate solvents. After being absorbed in an inert material they should be destroyed immediately. The hazard associated with moist peroxides increases considerably with increasing dryness. Before absorbing these materials they should therefore be dampened sufficiently with water. Small quantities of peroxides can be disposed of most simply by incineration. For larger quantities, information can often be obtained from the producer.

## 10. Toxicology

Organic peroxides generally possess a pronounced effect on body tissue, and marked damage must be expected at the site of direct contact. Nevertheless, there are many peroxide compounds whose effect on the organism is less pronounced or that lead to no noticeable effect on health. The chemical structure and physical properties such as water solubility, fat solubility, and vapor pressure, as well as the type and composition of the formulation (concentration, solvent, additives), can be important for penetration of human tissue and in reactions with cell compartments and macromolecules. In contrast, the influence of UV and ionizing radiation is not so predominant in the effect

**Table 12.** Acute toxicity data for organic peroxides

| Compound | CAS registry no. | Irritation* Skin | Irritation* Eye | Acute Toxicity $LD_{50}$, mg/kg (rat, oral) | Acute Toxicity $LC_{50}$, (4 h, inhalative) |
|---|---|---|---|---|---|
| tert-Butyl hydroperoxide | [75-91-2] | +++ | +++ | 406, 790** | 500 ppm |
| Cumene hydroperoxide | [80-15-9] | +++ | +++ | 382 | 220 ppm |
| Diisopropylbenzene peroxide | [26762-93-6] | +++ | | | |
| p-Menthane hydroperoxide | [37837-09-5] | +++ | | | |
| Di-tert-butyl peroxide | [110-05-4] | (+) | (+) | 6750 | > 4000 ppm |
| Dicumyl peroxide | [80-43-3] | (+) | (+) | 4100 | 0.09 mg/L |
| Peracetic acid | [79-21-0] | +++ | +++ | 17.5, 1410** | 0.45 mg/L |
| Diperoxydodecanedioic acid | [66280-55-5] | − | ++ | > 5000 | |
| Diacetyl peroxide | [110-22-5] | +++ | | | |
| Dicyclohexyl peroxide | [1758-61-8] | +++ | | | |
| Dibenzoyl peroxide | [94-36-0] | (+) | ++ | > 5000 | 24 mg/L |
| Dilauroyl peroxide | [105-74-8] | − | + | > 500 | 100 mg L$^{-1}$ d$^{-1}$ |
| Diisopropyl peroxydicarbonate | [105-64-6] | (+) | (+) | 2140 | |
| Acetyl cyclohexyl-sulfonyl peroxide | [3179-56-4] | +++ | + | (moderate) | |
| tert-Butyl 2,2-dimethylperoxypropionate | [927-07-1] | − | − | 4300 | |
| tert-Butyl peracetate | [107-71-1] | ++ | ++ | | |
| tert-Butyl 2-ethylperoxyhexanoate | [3006-82-4] | − | | 675 | > 8.2 mg/L |
| tert-Butyl perbenzoate | [614-45-9] | (+) | +++ | 10000 | > 200 mg/L |
| Methyl ethyl ketone peroxide | [1338-23-4] | +++ | +++ | 1012, 3800** | > 200 mg/L |
| Cyclohexanone peroxide | [12262-58-7] | +++ | +++ | 484 | 200 ppm |

\* Degree of irritation: − no effect, (+) weak, + slight, ++ moderate, +++ severe (caustic). \*\* $LD_{50}$ value after dermal application (rabbit).

of organic peroxides on the mammalian organism, and is not comparable with the formation and effect of endogeneous radicals produced by radiation.

The main effects associated with contact with peroxides are local irritation and burns of the skin, the mucous membranes of the eyes and the respiratory tract [376] (see Table 12).

Local tissue reactions may appear immediately following exposure or may only become apparent after a few days. Healing of the damaged tissue is frequently very slow and irreversible effects in the tissue are posssible. In the skin, irritation appears in the form of reddening (erythema) and swelling (edema), as well as burns (necrosis, corrosion), which are characterized by severe degenerative changes and scarring. In direct contact with the eye, even very small amounts of peroxides can cause inflammation of the conjuctiva, nacreous opacity of the cornea, or loss of the eye [377]. This can often be avoided by immediate and intensive rinsing with water. Nevertheless, this procedure must be carried out for a sufficiently long period, especially with contamination by solid or powdery material, as even strongly diluted solutions can cause irritation. For example, dry benzoyl peroxide possesses a smaller irritant effect than a 10% aqueous solution.

Inhalation of the vapor of volatile peroxides also leads to irritation or burns of the mucous membranes of the eye or the upper respiratory tract (nose, throat, bronchia). Inflammation of the respiratory tract or even lung edema can result from prolonged exposure [378].

Sensitization of the skin by peroxides has been observed mainly with dibenzoyl peroxide [379], [380]. In animal experiments methyl ethyl ketone peroxide and cyclohexanone peroxide exhibited ambiguous positive results [381], [382], whereas diperoxydodecandioic acid was clearly negative [383]. Acute poisoning may result from the inhalation and ingestion of organic peroxides, but in practice the hazard of resorptive effects (systemic toxicity) is low. After swallowing, peroxides are moderately toxic (see Table 12). In animal experiments, the acute toxicity for substances administered orally, dermally and by inhalation was characterized by tremors, decrease in muscle tone, clonic convulsions, hyperactivity, and dyspnea. Furthermore, irritation, hemorrhages, and inflammation were detected in the mucous membranes of the small intestine or respiratory tract [384], [385]. After chronic exposure to peroxy compounds, effects on the general condition, on the red blood cells, and impairment of the immune system in the blood (SH groups, peroxidase activity) towards oxidative reagents (oxidative stress) may appear [386]. Data exist only for a few organic peroxide compounds with respect to their carcinogenic, mutagenic, or reproductive toxicological potential. These data are not always sufficiently unambiguous to allow estimation of the risk associated with chemicals of this group. Observations on humans are not available. In long-term studies with rats and mice no substance-related tumors appeared after oral or subcutaneous application of benzoyl peroxide for up to two years [387].

After repeated application of organic peroxides to the skin of mice (skin painting studies) none of the tested compounds were found to be an absolute carcinogen. In two-stage experiments on the chemical carcinogenicity (initiating or promoting properties in the formation of tumors), none of the compounds examined showed a complete carcinogenic or tumor-initiating activity. Nevertheless, a number of compounds (benzoyl peroxide [388]–[390], *tert*-butyl hydroperoxide [391], lauroyl peroxide [391], peracetic acid [392], perbenzoic acid [393], chloroperbenzoic acid [394], and methyl ethyl ketone peroxide [381]) were found to promote skin tumors in mice and hamsters. For example, with dimethylbenzanthracene as tumor initator, subsequent treatment of the skin of mice with benzoyl peroxide led to skin carcinomas, whereas with benzoyl peroxide alone only hyperpigmentation, formation of skin scales, and a low level of hyperplasia of the skin was found [388]–[390]. When UV radiation was used as a tumor initiator, weak promoting activity was exhibited by methyl ethyl ketone peroxide [394], but not by benzoyl peroxide [395], [396]. Investigations of the mechanism of action of organic peroxides in vivo and in vitro showed an influence on the activity of the calcium-dependent protein kinase [397] and enzymes of the cytochrome P 450 peroxidase system [398], [399], as well as the participation of glutathione (SH groups) in the reduction [400] of the exogenously administered peroxides. In more recent investigations, the activation of oncogenes has been reported [401].

Testing of organic peroxides for mutagenicity by means of experiments with bacterial assays (*Salmonella typhimurium, Escherichia coli*) showed essentially no increased rate of mutation (dilauroyl peroxide [393], benzoyl peroxide [393], dibenzoyl peroxide [402], di-*tert*-butyl peroxide [393], peracetic acid [393]). However, both positive and negative results have been reported for *tert*-butyl hydroperoxide [393], [403], dicumyl peroxide [393], [404], monoperoxysuccinic anhydride [405], and disuccinoyl peroxide [404]. Direct interaction with DNA in the form of single strand breaks [406], [407] or as sister chromatid exchanges (SCE) [408] could be demonstrated by in vitro experiments. DNA transformation with *Bacillus subtilis* bacteria was observed with disuccinoyl peroxide [409], but not with di-*tert*-butyl peroxide [410]. In the mouse lymphoma forward test *tert*-butyl perbenzoate acted as a weak mutagen [408].

Reproductive toxicology experiments have been carried out with three-day-old chicken embryos. Cyclohexanone peroxide, cumene hydroperoxide, methyl ethyl ketone peroxide, benzoyl peroxide, acetylacetone peroxide, *tert* -butyl perbenzoate, dicumyl peroxide, and lauroyl peroxide led to embryotoxic effects. Malformation was only observed to a small extent [411].

# 11. References

General References

[1] S. Patai: *The Chemistry of Peroxides*, John Wiley & Sons, Chichester 1983.
[2] A. G. Davies: *Organic Peroxides*, Butterworth, London 1961.
[3] E. G. E. Hawkins: *Organic Peroxides*, E. and F. F. Spon, London 1961.
[4] J. O. Edwards: *Peroxide Reaction Mechanisms*, Interscience Publ., New York 1961.
[5] D. Swern: *Organic Peroxides*, Wiley-Interscience, New York vol. **I,** 1970, vol. **II,** 1971, vol. **III,** 1972.
[6] *Methodicum Chimicum*, vol. **5,** Thieme Verlag, Stuttgart 1975, p. 723.
[7] *Ullmann,* 4th ed., **17,** 661.
[8] D. Swern in D. Barton, D. W. Ollis (eds.): *Comprehensive Organic Chemistry,* vol. **1,** Pergamon Press, Oxford 1979, p. 909.
[9] *Kirk-Othmer,* 3rd ed., **17,** 27.
[10] W. M. Weigert, P. Kleinschmidt, "Organische Peroxide, Synthesen und Anwendungen," *Chem. Ztg.* **98** (1974) 583.
[11] J. C. Mitchell, *Chem. Soc. Rev.* **14** (1985) 339.
[12] W. M. Weigert et al., *Chem. Ztg.* **99** (1975) 106.
[13] *Houben-Weyl,* 4th ed., **E 13.**
[14] W. Weigert: *Wasserstoffperoxid und seine Derivate,* Hüthig Verlag, Heidelberg 1978.

Specific References

[15] B. C. Brodie, *Justus Liebigs Ann. Chem.* **108** (1858) 79.
[16] W. Adam, A. J. Bloodworth, *Annu. Rep. Prog. Chem. Sect. B* **75** (1978) 342.
[17] E. N. Frankel, W. E. Neff, E. Selke, D. D. Brooks, *Lipids* **23** (1988) 295.
[18] Modern Plastics International, Sept.–Oct. 1964–1984.
[19] *Houben-Weyl,* 4th ed., **E 13/1,** 176–236.
[20] D. Griller, K. U. Ingold, *Acc. Chem. Res.* **9** (1976) 13.
[21] W. H. T. Davison, *J. Chem. Soc.* 1951 2456; S. S. Ivanchev, A. I. Yurzhenko, Y. N. Anisimov, *Zh. Fiz. Khim.* **39** (1965) 1900; G. D. Mendenhall, *Angew. Chem.* **89** (1977) 220; K. D. Gundermann, *Chem. Ztg.* **99** (1975) 279.
[22] A. Rieche, M. Schulz, *Angew. Chem.* **70** (1958) 694; G. Dobson, A. Huges, *J. Chromatogr.* **16** (1964) 416; S. W. Bukata, L. L. Zabrocki, M. F. McLaughlin, *Anal. Chem.* **35** (1963) 885.
[23] R. D. Mair, A. J. Graupner, *Anal. Chem.* **36** (1964) 194.

[24]  M. L. Whisman, B. H. Eccleston, *Anal. Chem.* **30** (1958) 1638.
[25]  E. N. Frankel, W. E. Neff, E. Selke, D. D. Brooks, *Lipids* **23** (1988) 295.
[26]  K. S. Pitzer, *J. Am. Chem. Soc.* **70** (1948) 2140.
[27]  S. W. Benson, *J. Chem. Phys.* **40** (1964) 1007.
[28]  J. D'Ans, H. Gold, *Chem. Ber.* **92** (1959) 2559.
[29]  N. A. Sokolov, Yu. A. Aleksandrov, *Russ. Chem. Rev. (Engl. Transl.)* **47** (1978) 172.
[30]  A. J. Everett, G. J. Minkoff, *Trans. Faraday Soc.* **49** (1953) 410; I. M. Kolthoff, A. I. Medalia, *J. Am. Chem. Soc.* **71** (1949) 3789; D. Barnard, K. R. Hargrave, G. M. C. Higgins, *J. Chem. Soc.* 1956, 2845.
[31]  Atochem, EP 237 402, 1987.
[32]  L. J. Durham, C. F. Wurster, Jr., H. S. Moscher, *J. Am. Chem. Soc.* **80** (1958) 327, 332; M. S. Kharasch, A. Fono, W. Nudenberg, *J. Org. Chem.* **16** (1951) 113.
[33]  H. Kropf, F. Woehrle, *J. Chem. Res. Synop.* 1987, 387.
[34]  G. Rücker, K. Frey, *Liebigs Ann. Chem.* 1987, 389.
[35]  W. Adam et al., *J. Org. Chem.* **43** (1978) 1154.
[36]  N. A. Porter, C. B. Ziegler, F. F. Khouri, D. H. Roberts, *J. Org. Chem.* **50** (1985) 2252.
[37]  P. G. Cookson, A. G. Davies, B. P. Roberts, *J. Chem. Soc. Chem. Commun.* 1976, 1022.
[38]  N. A. Milas, D. M. Surgenor, *J. Am. Chem. Soc.* **68** (1946) 205, 643; N. A. Milas, L. H. Perry, *J. Am. Chem. Soc.* **68** (1946) 1938.
[39]  N. A. Milas, O. L. Mageli, *J. Am. Chem. Soc.* **74** (1952) 1471.
[40]  H. R. Williams, H. S. Mosher, *J. Am. Chem. Soc.* **76** (1954) 2984, 2987.
[41]  M. Schulz, A. Rieche, K. Kirschke, *Chem. Ber.* **100** (1967) 370.
[42]  E. Schmitz, A. Rieche, O. Brede, *J. Prakt. Chem.* **312** (1970) 30.
[43]  E. J. Lorand, E. I. Edwards, *J. Am. Chem. Soc.* **77** (1955) 4035.
[44]  H. Hock, H. Kropf, *Angew. Chem.* **69** (1957) 313.
[45]  E. R. Bell et al., *Ind. Eng. Chem.* **41** (1949) 2597.
[46]  G. A. Russell, *J. Am. Chem. Soc.* **78** (1956) 1047.
[47]  Shell, US 2 403 772, 1946.
[48]  Shell, US 2 845 461, 1958.
[49]  Halcon, DE-OS 1 518 996, 1965.
[50]  Halcon, DE-OS 1 568 774, 1966.
[51]  Halcon, FR 1 484 567, 1966.
[52]  Halcon, DE-OS 1 668 200, 1967.
[53]  L. F. Martin: *Organic Peroxide Technology*, Noyes Data Corp., London 1973.
[54]  C. Walling, S. A. Buckler, *J. Am. Chem. Soc.* **77** (1955) 6032; C. F. Cullis, E. Fersht, *Combust. Flame* **7** (1963) 185.
[55]  G. Wilke, P. Heimbach, *Justus Liebigs Ann. Chem.* **652** (1962) 7.
[56]  R. W. Denny, A. Nickon, *Org. React. (N.Y.)* **20** (1973) 133.
[57]  A. P. Schaap: *Singlet Molecular Oxygen*, Dowden, Hutchinson & Ross, Stroudsburg, Pa., 1976.
[58]  K. Gollnick, H. J. Kuhn in H. H. Wasserman, R. W. Muray (eds.): *Singlet Oxygen*, Academic Press, New York 1979, p. 287.
[59]  I. Saito, T. Matsuura, *Tetrahedron* **41** (1985) 2237.
[60]  A. A. Frimer, L. M. Stephenson in A. A. Frimer (ed.): *Singlet $O_2$*, vol. **II**, CRC Press, Boca Raton, Fla., 1985, p. 67.
[61]  Akzo, EP 250 024, 1987.
[62]  J. H. Raley, F. F. Rust, W. E. Vaughan, *J. Am. Chem. Soc.* **70** (1948) 88, 95, 1338; P. L. Hanst, J. G. Calvert, *J. Phys. Chem.* **63** (1959) 104; L. J. Durham, H. S. Moscher, *J. Am. Chem. Soc.* **82**

(1960) 4537; **84** (1962) 2811; E. S. Huyser et al., *J. Am. Chem. Soc.* **86** (1964) 2401, 4148; R. Hiatt, D. J. LaBlanc, C. Thankachan, *Can. J. Chem.* **52** (1974) 4090; E. S. Huyser, F. Tang, *J. Org. Chem.* **43** (1978) 1016.

[63] E. G. E. Hawkins, *J. Chem. Soc.* 1959, 2169, 2801.
[64] G. Baker et al., *J. Chem. Soc.* 1965, 6965, 6970.
[65] Akzo, EP 273 990 A 1, 1988.
[66] Texaco Dev. Corp., CA 743 559, 1966.
[67] E. R. Bell et al., *Ind. Eng. Chem.* **41** (1949) 2597.
[68] G. O. Schenk, *Angew. Chem.* **64** (1952) 12; *Nachr. Chem. Tech.* **19** (1971) 446; W. Adam, *Angew. Chem.* **86** (1974) 683; W. Adam, *Chem. Ztg.* **99** (1975) 142.
[69] G. O. Schenck, K. Ziegler, *Naturwissenschaften* **32** (1944) 125.
[70] C. Moureu, C. Dufraisse, P. M. Dean, *C. R. Hebd. Séances Acad. Sci.* **182** (1926) 1440; C. Dufraisse, E. Etienne, J. Rigaudy, *C. R. Hebd. Séances Acad. Sci.* **226** (1948) 1773.
[71] A. A. Frimer, *Chem. Rev.* **79** (1979) 359.
[72] A. P. Schaap, K. A. Zaklika in H. H. Wassermann, R. W. Murray (eds.): *Singlet Oxygen*, Academic Press, New York 1979, p. 174.
[73] P. D. Bartlett, M. E. Landis in H. H. Wassermann, R. W. Murray: *Singlet Oxygen*, Academic Press, New York 1979, p. 244.
[74] W. Adam, G. Cilento, *Angew. Chem.* **95** (1983) 525.
[75] W. Adam, F. Yany in A. Hassner (ed.): *The Chemistry of Heterocyclic Compounds, Small Ring Heterocycles*, vol. **42**, part 3, John Wiley & Sons, Chichester 1983, p. 351.
[76] A. L. Baumstark in A. A. Frimer (ed.): *Singlet $O_2$*, vol. **II**, part 1, CRC Press, Boca Raton, Fla., 1985, p. 1.
[77] M. Schulz, O. Johansson, DL 63 491, 1967.
[78] Union Rhein, Braunkohlen Kraftstoff, DE-OS 1 911 176, 1969.
[79] JP-Kokai 01 61453 A 2 and 89 61453, 1989.
[80] Bergwerksgesellschaft Hibernia, US 3 254 130, 1966.
[81] Noury & van der Lande, US 3 337 639, 1967.
[82] Pennwalt Corp., DE 3 034 235, 1989.
[83] Massachusetts Institute of Technology, US 3 505 363, 1970.
[84] Wallace & Tiernan, US 3 308 163, 1967.
[85] FMC Corp., US 3 135 805, 1964.
[86] E. R. Altwicker, *Chem. Rev.* **67** (1967) 511.
[87] K. U. Ingold, *Chem. Rev.* **61** (1961) 575.
[88] M. S. Kharasch, P. Pauson, W. Nudenberg, *J. Org. Chem.* **18** (1953) 322.
[89] L. R. Mahoney, *Angew. Chem.* **81** (1969) 555.
[90] L. Reich, S. S. Stivala: *Autoxidation of Hydrocarbons and Polyolefins*, Marcel Dekker, New York 1969, p. 139.
[91] J. Pospísil, *Adv. Polym. Sci.* **36** (1980) 69.
[92] T. Endo, K. Suga, Y. Orikasa, S. Kojima, *J. Appl. Polym. Sci.* **37** (1989) 1815.
[93] Nippon Oils and Fats, JP 63 254 110 A 2 and 88 254 110, 1988.
[94] Toray Silicone, EP 304 676, 1989.
[95] Johnson, EP 303 057 A 2, 1989.
[96] W. H. Richardson in S. Patai (ed.): *The Chemistry of Peroxides*, Wiley-Interscience, New York 1983, p. 129.
[97] J. Skerjanc, A. Regent, B. Plesnicar, *J. Chem. Soc. Chem. Commun.* 1980, 1007.
[98] R. Kavcic, B. Plesnicar, *J. Org. Chem.* **35** (1970) 2033.

[99] W. E. Parker, L. P. Witnauer, D. Swern, *J. Am. Chem. Soc.* **79** (1957) 1929.
[100] S. W. Benson, *J. Chem. Phys.* **40** (1964) 1007.
[101] K. S. Pitzer, *J. Am. Chem. Soc.* **70** (1948) 2140.
[102] A. H. Sehon et al., *Can. J. Chem.* **41** (1963) 1819, 1826.
[103] D. Lefort et al., *Bull. Soc. Chim. Fr.* 1959, 1385; 1961, 2219.
[104] W. E. Parker, L. P. Witnauer, D. Swern, *J. Am. Chem. Soc.* **80** (1958) 323.
[105] L. Heslinga, W. Schwaiger, *Rec. Trav. Chim. Pays-Bas* **85** (1966) 75.
[106] W. Schlenk, Jr., *Fortschr. Chem. Forsch.* **2** (1951) 92; F. Cramer, *Angew. Chem.* **68** (1956) 115; F. Schwochow, L. Puppe, *Angew. Chem.* **87** (1975) 659.
[107] B. Plesnicar, R. Kavcic, D. Hadzi, *J. Mol. Struct.* **20** (1974) 457.
[108] *Houben-Weyl*, 4th ed., **E 13,** pp. 1418, 1438.
[109] *Kirk-Othmer*, 1st. suppl. vol., 1957, p. 662.
[110] Propylox, EP 4407, 1978.
[111] Produits Chimiques Ugine Kuhlmann, EP 223 96, 1979.
[112] Degussa, DE 2 018 713, 1970.
[113] Degussa, DE-OS 2 125 160, 1971.
[114] Degussa, DE 1 170 960, 1962.
[115] Degussa, DE-OS 2 145 157, 1971.
[116] Degussa, DE-OS 2 141 155, 1971.
[117] K. Fancovic et al., CS 198 936, 1978.
[118] Bayer, Degussa, DE-OS 2 519 287, 1975.
[119] Bayer, Degussa, DE 2 519 293, 1975.
[120] Bayer, Degussa, DE-AS 2 519 295, 1975.
[121] Bayer, Degussa, DE 2 856 665, 1978.
[122] Bayer, Degussa, DE-OS 2 519 295, 1975.
[123] Bayer, DE-OS 2 038 318, 1970.
[124] Mitsui Toatsu Chem., JP 63 159 365, 1988.
[125] Produits Chimiques Ugine Kuhlmann, EP 83 894, 1982.
[126] Mitsui Toatsu Chem., JP 63 115 857, 1988.
[127] Degussa, DE-OS 3 720 562, 1987.
[128] Procter & Gamble, US 4 233 235, 1979.
[129] Procter & Gamble, US 4 244 884, 1979.
[130] Degussa, Henkel, DE 3 418 450, 1984.
[131] Degussa, Henkel, DE-OS 3 539 036, 1987.
[132] Food Machinery, US 4 172 086, 1977.
[133] Degussa, Henkel, DE 2 930 546, 1978.
[134] Degussa, Henkel, DE 3 320 497, 1983.
[135] L. S. Silbert, E. Siegel, D. Swern, *J. Org. Chem.* **27** (1962) 1336; *Org. Synth.* **43** (1963) 93.
[136] Pittsburg Plate Glass, US 3 655 738, 1969.
[137] Pittsburg Plate Glass, US 3 880 914, 1975.
[138] Degussa, Henkel, DE 2 929 839, 1979.
[139] Degussa, DE 3 426 792, 1984.
[140] Interox Chem., GB 7 936 178, 1978.
[141] G. B. Payne, *J. Org. Chem.* **24** (1959) 1354.
[142] L'Air Liquide, FR 1 452 737, 1965.
[143] Kharkov Polytechnic Inst., SU 615 066, 1976.
[144] Degussa, Henkel, DE 3 438 529, 1984.

[145] Degussa, Henkel, DE 3 539 036, 1985.
[146] Monsanto Co., US 4 680 145, 1987.
[147] Food Machinery, US 3 231 605, 1966.
[148] Food Machinery, US 3 247 244, 1966.
[149] Food Machinery, US 3 462 480, 1969.
[150] B. A. Palei et al., *Zh. Prikl. Khim. (Leningrad)* **61** (1988) 123.
[151] P. R. H. Speakman, *Chem. Ind. (London)* 1978, 579.
[152] Y. Ogata, Y. Sawaki, *Tetrahedron* **23** (1967) 3327.
[153] Food Machinery, US 3 485 869, 1967.
[154] R. N. McDonald, R. N. Steppel, J. E. Dorsey, *Org. Synth.* **50** (1970) 15.
[155] Agence Nationale de Valorisation, FR 2 324 626, 1975.
[156] J. Y. Nedelec, J. Sorba, D. Lefort, *Synthesis* 1976, 821.
[157] J. Rebek, L. Marshall, R. Wolak, J. McManis, *J. Am. Chem. Soc.* **106** (1984) 1170.
[158] J. Rebek, L. Marshall, J. McManis, R. Wolak, *J. Org. Chem.* **51** (1986) 1649.
[159] Mitsui Toatsu Chem., JP 73 13 527, 1973.
[160] Daicel Co., DE 2 165 168, 1970.
[161] M. R. Yazdanbakhch, Dissertation, Universität Hamburg 1977.
[162] Mitsubishi, JP 55 27 904, 1976.
[163] S. Levush, I. Garbuzyuk, Z. Prisyazhnyuk, V. Bryukhovetskii, SU 979 336, 1981.
[164] V. A. Bryukhovetskii, S. S. Levush, F. B. Moin, V. U. Shevchuk, *Kinet. Katal.* **17** (1976) 1130.
[165] D. Paronikyan, I. Vardanyan, A. Nalbandyan, *Arm. Khim. Zh.* **36** (1983) 20.
[166] Mitsubishi, DE 2 515 033, 1975.
[167] E. Sandler et al., *Neftekhimiya* **23** (1983) 674.
[168] Mitsubishi, JP 55 43 058, 1978.
[169] C. R. Dick, R. F. Hanna, *J. Org. Chem.* **29** (1964) 1218.
[170] Bayer, DE 2 658 943, 1976.
[171] Bayer, EP 46 534, 1980.
[172] Oxiran Chem. Corp., DE 2 364 181, 1972.
[173] Food Machinery, US 4 026 798, 1977.
[174] J. Gutknecht, *Pharm. Technol.* **9** (1988) 46.
[175] Peroxid-Chemie, EP 302298, 1989.
[176] Interox, EP 233 731, 1987.
[177] J. A. L. Fraser, *Spec. Chem.* **7** (1987) 178, 180, 182, 184, 186.
[178] H. Krüger, P. Kuzel, H. Schwab, *Chem. Ztg.* **99** (1975) 132.
[179] Degussa, DE-OS 3 709 347, 1988.
[180] Degussa, DE-OS 3 709 348, 1988.
[181] Kao Corp., JP 63 112 698, 1988.
[182] T. Bolsman, R. Kok, A. D. Vreugdenhil, *J. Am. Oil Chem. Soc.* **65** (1988) 1211.
[183] I. Jaffe, E. J. Prosen, M. Szwarc, *J. Chem. Phys.* **27** (1957) 416.
[184] E. D. Skakovskii et al., *Vestsi Akad. Navuk BSSR, Ser. Khim. Navuk* 1987, 26.
[185] C. E. Boozer et al., *Chem. Abstr.* **59** (1963) 2609; J. E. Guillet, US 3 232 921, 1966; J. E. Guillet et al., *Ind. Eng. Chem. Prod. Res. Dev.* **3** (1964) 257.
[186] A. Rieche, *Angew. Chem.* **70** (1958) 251; R. C. Lamb, J. G. Pacifici, P. W. Ayers, *J. Am. Chem. Soc.* **87** (1965) 3928; R. C. Lamb et al., *J. Org. Chem.* **31** (1966) 147.
[187] A. M. Clover, G. F. Richmond, *J. Am. Chem. Soc.* **29** (1903) 179.
[188] L. Horner, B. Anders, *Chem. Ber.* **95** (1965) 2470.
[189] E. A. Sullivan, A. A. Hinckley, *J. Org. Chem.* **27** (1962) 3731.

[190] J. K. Kochi, P. E. Mocadlo, *J. Org. Chem.* **30** (1965) 1134.
[191] S. G. Cohen, D. B. Sparrow, *J. Am. Chem. Soc.* **72** (1950) 611.
[192] H. C. McBay, O. Tucker, *J. Org. Chem.* **19** (1954) 869.
[193] Pennwalt Corp., DE-OS 1 954 965, 1970.
[194] Pennwalt Corp., DE-OS 2 053 463, 1971.
[195] Pennwalt Corp., DE 2 631 910, 1975.
[196] Pennwalt Corp., US 3 580 955, 1971.
[197] Pennwalt Corp., US 3 706 783, 1972.
[198] Pfizer+Co., US 3 438 882, 1969.
[199] Koppers Comp., US 3 397 245, 1968.
[200] G. Schroeder, R. Lombard, *Bull. Soc. Chim. Fr.* 1964, 542.
[201] Argus Chem. Corp., US 3 502 701, 1970.
[202] Argus Chem. Corp., US 4 032 605, 1976.
[203] Canadian Industries, US 3 108 093, 1963.
[204] Eastman Kodak Co., CA 725 695, 1966.
[205] Eastman Kodak Co., CA 725 694, 1966.
[206] Montecatini, DE 1 192 181, 1965.
[207] Elektrochem. Werke, NL 6 517 064, 1966.
[208] Rhône-Progil, DE-OS 2 256 255, 1973.
[209] Teknar Apex Co., US 3 728 402, 1973.
[210] H. Spaeth, R. Kneissl, DE 1 668 355, 1974.
[211] Akzo, DE-OS 2 628 272, 1976.
[212] Peroxid-Chemie, DE 1 768 199, 1974.
[213] Sanken Kako Co., JP 74 40 220, 1974.
[214] Dow Chem. Co., US 3 674 858, 1972.
[215] Interox, BE 848 246, 1975.
[216] Société Chimiques de Carbonnages, DE-AS 2 334 269, 1974.
[217] D. F. DeTar, L. A. Carpino, *J. Am. Chem. Soc.* **77** (1955) 6370.
[218] Akzo, DE-OS 2 428 184, 1975.
[219] Goodyear, DE 1 099 738, 1970.
[220] Liljeholmens Stearinfabriks, DE-AS 1 947 434, 1969.
[221] P. D. Bartlett, R. R. Hiatt, *J. Am. Chem. Soc.* **80** (1958) 1398.
[222] Y. N. Anisimov, S. S. Ivanchev, *Zh. Anal. Khim.* **21** (1966) 113.
[223] C. Rüchardt, V. Golzke, G. Range, *Chem. Ber.* **114** (1981) 2769.
[224] J. J. Davies, J. G. Miller, *J. Am. Chem. Soc.* **99** (1977) 245.
[225] L. Thijs, S. N. Gupta, D. C. Neckers, *J. Org. Chem.* **44** (1979) 4123.
[226] S. O. Lawesson, G. Schroll in S. Patai (ed.): *The Chemistry of Carboxylic Acids and Esters*, Interscience, New York 1969, p. 670.
[227] T. Komai, K. Matsuyama, M. Matsushima, *Bull. Chem. Soc. Jpn.* **61** (1988) 1641.
[228] Y. Nakashio, T. Yamamoto, M. Hirota, *Nippon Kagaku Kaishi* 1988, 69.
[229] E. M. Gavryliv, R. G. Makitra, Ya. N. Pirig, *Org. React. (Tartu)* **24** (1987) 3.
[230] C. S. Sheppard, V. R. Kamath, *Polym. Eng. Sci.* **19** (1979) 597.
[231] W. Duismann, C. Rüchardt, *Justus Liebigs Ann. Chem.* 1976, 1834.
[232] B. G. Dixon, G. B. Schuster, *J. Am. Chem. Soc.* **101** (1979) 3116.
[233] B. G. Dixon, G. B. Schuster, *J. Am. Chem. Soc.* **103** (1981) 3068.
[234] C. Rüchardt, *Fortschr. Chem. Forsch.* **6** (1966) 251.
[235] T. Ochiai, *Bull. Chem. Soc. Jpn.* **49** (1976) 2522, 2641.

[236] S. Oliveri-Vigh, F. T. Hainsworth, *J. Pharm. Sci.* **67** (1978) 1035.
[237] D. J. Rawlinson, M. Konieczny, G. Sosnovsky, *Z. Naturforsch. B Anorg. Chem. Org. Chem.* **34 B** (1979) 76, 80.
[238] I. S. Voloshanovskii, S. S. Ivanchev, *Zh. Obshch. Khim.* **44** (1974) 892.
[239] R. Criegee, *Ber. Dtsch. Chem. Ges. B* **77** (1944) 22.
[240] R. D. Mair, A. J. Graupner, *Anal. Chem.* **36** (1964) 194.
[241] P. D. Bartlett, J. L. Kice, *J. Am. Chem. Soc.* **75** (1953) 5591; Technisches Merkblatt "Peroxyester", Luperox GmbH, Günsburg 1976.
[242] L. S. Silbert, D. Swern, *Anal. Chem.* **30** (1958) 385.
[243] V. V. Shibanov, V. A. Fedorova, T. I. Yurzhenko, *Visn. L'viv. Politekh. Inst.* 197424.
[244] W. Adam, A. Alzerreca, J. C. Liu, F. Yani, *J. Am. Chem. Soc.* **99** (1977) 5768.
[245] U. Aeberhard et al., *Helv. Chim. Acta* **66** (1983) 2740.
[246] Y. Hamada, A. Mizuno, T. Ohno, T. Shioiri, *Chem. Pharm. Bull.* **32** (1984) 3683.
[247] Halcon Int., DE-OS 2 013 469, 1970.
[248] Akzo, NL 852 261, 1985.
[249] P. M. Kohn, *Chem. Eng.* **85** (1978) 16, 88.
[250] Pennwalt Corp., DE 3 004 330, 1979.
[251] Argus Chem. Corp., EP 20 846, 1979.
[252] Nippon Oils & Fats, JP 51 076 213, 1975.
[253] Pennwalt Corp., US 4 129 586, 1977.
[254] Nippon Oils & Fats, JP 52 156 828, 1976.
[255] Kenobel, EP 271 462, 1988.
[256] Pennwalt Corp., DE 2 757 440, 1977.
[257] Nippon Oils & Fats, JP 82 92 005, 1980.
[258] P. D. Bartlett, R. R. Hiatt, *J. Am. Chem. Soc.* **80** (1958) 1398.
[259] P. Lorenz, C. Rüchardt, E. Schacht, *Chem. Ber.* **104** (1971) 3429.
[260] B. Maillard, A. Kharrat, C. Gardrat, *Tetrahedron* **40** (1984) 3531.
[261] M. Schulz, J. Römbach, *Z. Chem.* **21** (1981) 404.
[262] C. Rüchardt, V. Golzke, G. Range, *Chem. Ber.* **114** (1981) 2769.
[263] P. E. Eaton, Y. S. Or, S. J. Branca, B. K. R. Shankar, *Tetrahedron* **42** (1986) 1621.
[264] Standard Oil Co., US 4 220 746, 1978.
[265] M. Shiozaki, N. Ishida, H. Maruyama, T. Hiraoka, *Heterocycles* **20** (1983) 279.
[266] L. Thijs, S. N. Gupta, D. C. Neckers, *J. Org. Chem.* **44** (1979) 4123.
[267] P. Gottschalk, D. C. Neckers, *J. Org. Chem.* **50** (1985) 3498.
[268] D. C. Neckers, US 4 416 826, 1980.
[269] C. Rüchardt, G. Hamprecht, *Chem. Ber.* **101** (1968) 3957.
[270] G. A. Olah, D. G. Parker, N. Yoneda, *J. Org. Chem.* **42** (1977) 32.
[271] Nippon Oils & Fats, EP 266 966, 1988.
[272] Pennwalt Corp., US 4 075 236, 1978.
[273] G. Maier et al., *J. Am. Chem. Soc.* **109** (1987) 5183.
[274] V. A. Federova, V. A. Donchak, V. A. Puchin, Y. Y. Musabekov, *Khim. Tekhnol.* 1980, 23.
[275] Pennwalt Corp., US 4 097 408, 1978.
[276] H. Hock, H. Kropf, *Chem. Ber.* **88** (1955) 1544.
[277] W. Duismann, C. Rüchardt, *Justus Liebigs Ann. Chem.* 1976, 1834.
[278] M. R. Vilenskaya, G. A. Petrovskaya, Y. V. Panchenko, V. A. Puchin, *Zh. Org. Khim.* **20** (1984) 2112.

[279] G. A. Razuvaev, T. N. Brevnova, N. F. Cherepennikova, S. K. Ratushnaya, *Zh. Obshch. Khim.* **46** (1976) 857.
[280] D. E. Falvey, G. B. Schuster, *J. Am. Chem. Soc.* **108** (1986) 7419.
[281] W. E. Cass, A. K. Bahl, *J. Org. Chem.* **39** (1974) 3602.
[282] Nippon Oils & Fats, JP 75 007 055, 1966.
[283] I. I. Artym, S. S. Ivanchev, M. M. Soltys, M. A. Kovbuz, *Zh. Prikl. Khim. (Leningrad)* **49** (1976) 2056.
[284] G. S. Bylina, L. R. Uvarova, *Zh. Org. Khim.* **11** (1975) 2503.
[285] V. A. Dodonov, V. V. Chesnokov, E. B. Eresov, *Zh. Obshch. Khim.* **45** (1975) 1524.
[286] V. A. Dodonov, V. V. Chesnokov, T. A. Urchenko, *Zh. Obshch. Khim.* **46** (1976) 1293.
[287] V. A. Dodonov, T. I. Starostina, *Zh. Obshch. Khim.* **47** (1977) 843.
[288] G. A. Razuvaev et al., *Izv. Akad. Nauk SSSR Ser. Khim.* 1980, 821.
[289] V. A. Dodonov, T. I. Starostina, *Zh. Obshch. Khim.* **47** (1977) 843.
[290] S. Sustmann, C. Rüchardt, *Chem. Ber.* **108** (1975) 3043.
[291] A. Kharrat, C. Gardrat, B. Maillard, *Bull. Soc. Chim. Belg.* **95** (1986) 535.
[292] A. L. J. Beckwith, V. W. Bowry, G. Moad, *J. Org. Chem.* **53** (1988) 1632.
[293] R. Hecht, C. Rüchardt, *Chem. Ber.* **96** (1963) 1281.
[294] E. Montaudon, M. Campagnole, M. J. Bourgeois, B. Maillard, *Bull. Soc. Chim. Belg.* **91** (1982) 725.
[295] F. D. Greene, W. Adam, G. A. Knudsen, *J. Org. Chem.* **31** (1966) 2087.
[296] A. M. Kukova, L. V. Zhitina, V. L. Antonovskii, G. S. Kirichenko, *Khim. Promst. (Moscow)* 1980, 464.
[297] Shell, US 2 608 570, 1952.
[298] E. J. Pedersen, *J. Org. Chem.* **23** (1958) 252.
[299] B. G. Dixon, G. B. Hunter, *J. Am. Chem. Soc.* **101** (1979) 3116.
[300] Halcon Int., DE-OS 1 814 344, 1967.
[301] Pittsburg Plate Glass, US 3 988 261, 1975.
[302] I. Saito, T. Mano, R. Nagata, T. Matsuura, *Tetrahedron Lett.* **28** (1987) 1909.
[303] G. Bouillon, C. Lick, K. Schank in S. Patai: *The Chemistry of Peroxides*, John Wiley & Sons, New York 1983.
[304] P. S. Bailey, F. J. Garcia-Sharp, *J. Org. Chem.* **22** (1957) 1008; A. D. Jenkins, D. W. G. Style, *J. Chem. Soc.* 1953, 2357; A. I. Kirillov, *J. Org. Chem. USSR (Engl. Transl.)* **1** (1965) 1226, 1411; C. W. Cooper, *J. Chem. Soc.* 1951, 1340; C. S. Marvel, V. E. Nichols, *J. Org. Chem.* **6** (1941) 269; P. S. Bailey, *Chem. Rev.* **58** (1958) 925.
[305] P. R. Story et al., *J. Am. Chem. Soc.* **90** (1968) 817; P. R. Story et al., *J. Org. Chem.* **35** (1970) 3059; P. Busch, P. R. Story, *Synthesis* 1970, 181; J. R. Sanderson, P. R. Story, *J. Org. Chem.* **39** (1974) 3463.
[306] Research Corp., DE 1 807 337, 1968.
[307] Montecatini, DE 1 178 055, 1958.
[308] Montecatini, DE 1 216 282, 1959.
[309] Distillers Comp., GB 740 747, 1965.
[310] G. O. Schenck, H. D. Becker, *Angew. Chem.* **70** (1958) 504; A. Rieche, K. Meister, *Chem. Ber.* **72** (1939) 1938.
[311] R. Criegee, G. Werner, *Justus Liebigs Ann. Chem.* **564** (1949) 9.
[312] R. Criegee, G. Schröder, *Chem. Ber.* **93** (1960) 689.
[313] R. Criegee, *Angew. Chem.* **87** (1975) 765; R. Criegee, *Chem. Unserer Zeit* **7** (1973) 75.
[314] H. Rein, R. Criegee, *Angew. Chem.* **62** (1950) 120.

[315] A. Rieche, E. Schmitz, E. Beyer, *Chem. Ber.* **91** (1958) 1935; A. Rieche, H. E. Seyfarth, A. Hesse, *Angew. Chem.* **78** (1966) 269; H. E. Seyfarth, A. Rieche, A. Hesse, *Chem. Ber.* **100** (1967) 624.
[316] C. K. Ikeda, R. A. Braun, B. E. Sorenson, *J. Org. Chem.* **29** (1964) 286.
[317] J. Wolpers, W. Ziegenbein, *Tetrahedron Lett.* 1971, 3889.
[318] Distillers Comp., GB 740 747, 1965.
[319] N. Brown, A. W. Anderson, C. E. Schweitzer, *J. Am. Chem. Soc.* **77** (1955) 1760; C. Dufraisse, S. Ecary, *C. R. Hebd. Seances Acad. Sci.* **223** (1946) 735.
[320] G. O. Schenk, H. D. Becker, K.-H. Schulte-Elte, C. H. Krauch, *Chem. Ber.* **96** (1963) 509.
[321] Shell, US 2 871 104 (1955).
[322] *Eur. Chem. News*, Jan. 1969, 24.
[323] F. H. Dickey, F. F. Rust, W. E. Vaughan, *J. Am. Chem. Soc.* **71** (1949) 1432; F. H. Dickey et al., *Ind. Eng. Chem.* **41** (1949) 1673.
[324] Montecatini, DE 1 259 894, 1965; DE-AS 2 025 974, 1970.
[325] B. Phillips, F. C. Frostick, P. S. Starcher, *J. Am. Chem. Soc.* **79** (1957) 5982; J. d'Ans, K. Dossow, J. Mattner, *Angew. Chem.* **66** (1954) 633.
[326] M. S. Kharasch, G. Sosnovsky, *J. Org. Chem.* **23** (1958) 1322.
[327] N. A. Milas, A. Golubovic, *J. Am. Chem. Soc.* **81** (1959) 3361, 5824, 6461.
[328] T. Ledaal, *Acta Chem. Scand.* **21** (1967) 1656.
[329] R. Criegee, W. Schorrenberg, J. Becke, *Justus Liebigs Ann. Chem.* **565** (1949) 7.
[330] A. Rieche, C. Bischoff, *Chem. Ber.* **95** (1962) 77.
[331] Pennwalt Corp., DE-OS 1 923 085, 1969.
[332] Montecatini, DE-AS 1 293 771, 1966.
[333] Research Corp., DE-OS 2 034 736, 1970.
[334] Story Chem. Corp., DE-OS 213 316, 1971.
[335] K. Wulz, H. A. Brune, *Tetrahedron* **27** (1971) 3669.
[336] K. Griesbaum, M. Meister, *Chem. Ber.* **120** (1987) 1573.
[337] Showa Denko K. K., JP 62 121747 A 2 and 87 121747, 1987.
[338] JP-Kokai 63 22816 A 2 and 88 22816, 1988.
[339] Pennwalt Corp., EP 273 090, 1988.
[340] Q. T. Wiles et al., *Ind. Eng. Chem.* **41** (1949) 1679; Pennwalt Lucidol, product bulletin "Keton Peroxides" Philadephia, Pa.
[341] Montecatini, GB 1 034 266, 1966.
[342] Wallace & Tiernan, GB 1 047 830, 1966.
[343] K. Gäbelein, *Umsch. Wiss. Tech.* **75** (1975) 478.
[344] Emery, US 2 813 113, 1953.
[345] A. Rieche, DE 565 158, 1931.
[346] S. D. Rasumowskij et al., DE-OS 2 217 661, 1972.
[347] E. G. E. Hawkins, *J. Chem. Soc. C* 1969, 2663, 2671, 2678, 2682, 2686.
[348] R. Criegee, G. Lohaus, *Chem. Ber.* **84** (1951) 219.
[349] BP Chemicals, DE-OS 1 643 640, 1967.
[350] BP Chemicals, DE-OS 1 670 246, 1967.
[351] BP Chemicals, DE-OS 1 768 387, 1968.
[352] Degussa, DE-OS 2 003 269, 1970.
[353] Degussa, DE-OS 2 004 440, 1970.
[354] BP Chemicals, DE-OS 2 049 560, 1970.
[355] BP Chemicals, DE-OS 2 109 923, 1971.
[356] BP Chemicals, DE-AS 1 695 503, 1967.

[357] BP Chemicals, DE-AS 2 116 500, 1971.
[358] BP Chemicals, US 3 576 817, 1971.
[359] E. G. E. Hawkins, *J. Chem. Soc.* 1969, 2663.
[360] BP Chemicals, DE-OS 1 803 872, 1968.
[361] BP Chemicals, DE-OS 1 814 288, 1968.
[362] BP Chemicals, DE-OS 1 814 289, 1968.
[363] Rhône-Poulenc, DE-OS 2 048 575, 1970.
[364] *Jpn. Chem. Week,* June 1978, 1; *Chem. Age,* June 1978, 14; Ube, DE-AS 2 504 323, 1975.
[365] M. V. Pokhmurskaya, V. N. Nosan, N. I. Klim, *Vysokomol. Soedin. Ser. B* **30** (1988) 781.
[366] Atochem, EP 241 390 A 2, 1987.
[367] J. L. Vennerstrom, *J. Med. Chem.* **32** (1989) 64.
[368] Gesetz über explosionsgefährliche Stoffe (Sprengstoffgesetz) from Sept. 13, 1976; Bundesgesetzblatt I (1976), pp. 2737–2787; 1st–5th Verordnung zum Sprengstoffgesetz.
[369] Organische Peroxide, Merkblatt M 001 4/80 der Berufsgenossenschaft der chemischen Industrie, Jedermann-Verlag, Heidelberg 1989.
[370] Unfallverhütungsvorschrift "Organische Peroxide" der Berufsgenossenschaft der chemischen Industrie, Heidelberg (Entwurf 1989).
[371] OECD-IGUS, Organisation for Economic Cooperation and Development, International Group of Experts on the Explosion Risks of Unstable Substances.
[372] R. Siegmeier, unpublished results.
[373] K. Eckwert, E. Peukert, L. Jeromin, *Fat Sci. Technol.* **90** (1988) 511.
[374] Gefahrstoffverordnung of Aug. 1986, Bundesgesetzblatt I, p. 1470.
[375] 8th Recommendations on the Transport of Dangerous Goods, 6th rev. ed., United Nations, New York 1988.
[376] Berufsgenossenschaft Chemie, Merkblatt "Organische Peroxide", Verlag Chemie, Weinheim 1980.
[377] H. J. Küchle, *Zentralbl. Arbeitsmed. Arbeitsschutz* **8** (1958) 25.
[378] Register of Safety Information of Chemical Products, National Board of Labour Protection, Tampere, Finnland, Nov. 1981.
[379] F. A. Bahmer, A. Schulze-Dirks, H. Zaun, *Dermatosen Beruf Umwelt* **32** (1984) 21.
[380] H. J. Bandmann, M. Agathos, *Hautarzt* **36** (1985) 670.
[381] L. Aringer: *Arbetarskyddsstyrelsen,* Publikationssevice 17 184, Solna, Sweden 1985.
[382] R. E. Gosselin, H. C. Hodge, R. P. Smith, M. N. Gleason: *Clinical Toxicology of Commercial Products,* The Williams & Wilkins Co., Baltimore 1976.
[383] Degussa, unpublished results.
[384] E. P. Floyd, M. S. Stokinger, H. E. Stockinger, *Ann. Ind. Hygiene Assoc. J.* **19** (1958) 205.
[385] Degussa, unpublished results.
[386] F. I. Orlova, *Sb. Nauchn. Tr. Kuibyshev. Nauchno Issled. Inst. Gig* **6** (1971) 101.
[387] M. Sharratt, A. C. Frazer, O. C. Forbes, *Food Cosmet. Toxicol.* **2** (1964) 527.
[388] C. J. Molloy, M. A. Gallo, J. D. Laskin, *J. Soc. Cosmet. Chem.* **35** (1984) 197.
[389] J. Schweizer, H. Loehrke, L. Edler, K. Goerttler, *Carcinogenesis (London)* **8** (1987) 479.
[390] L. W. Updyke, A. Chuthaputti, R. W. Pheifer, G. K W. Yim, *Carcinogenesis (London)* **9** (1988) 1943.
[391] A. J. P. Klein-Szanto, T. J. Slaga, *J. Invest. Dermatol.* **79** (1982) 30.
[392] F. G. Bock, H. K. Myers, H. W. Fox, *J. Nat. Cancer Inst. (U.S.)* **55** (1975) 1359.
[393] T. Yamaguchi, Y. Yamashita, *Agric. Biol. Chem.* **44** (1980) 1675.
[394] M. K. Logani, C. P. Sambuco, P. D. Forbes, R. E. Davies, *Food Chem. Toxic.* **22** (1984) 879.

[395] O. H. Iversen, *J. Invest. Dermatol.* **86** (1986) 442.
[396] J. H. Epstein, *J. Invest. Dermatol.* **91** (1988) 114.
[397] T. E. Donnelly, Jr., J. C. Pelling, C. L. Anderson, D. Dalbey, *Carcinogenesis (London)* **8** (1987) 1871.
[398] M. Ando, A. L. Tappel, *Toxicol. Appl. Pharmacol.* **81** (1985) 517.
[399] M. Ando, *Nippon Eiseigaku Zasshi* **41** (1986) 587.
[400] D. Galaris, E. Cadenas, P. Hochstein, *Free Radicals Biol. Med.* **6** (1989) 473.
[401] J. C. Pelling et al., *Carcinogenesis (London)* **8** (1987) 1871.
[402] M. Ishadate, Jr., T. Sofuni, K. Yoshikawa, *Hen'Igen to Dokusei* **3** (1980) 80.
[403] *Bibra P. Watts Bull.* **20** (1981) 514.
[404] M. Chevallier, D. Luzzatti, *C. R. Hebd. Séances Acad. Sci.* **250** (1960) 1572.
[405] F. H. Sobels, *Dros Info Serv.* **32** (1958) 159.
[406] H. L. Gensler, G. T. Bowden, *Carcinogenesis (London)* **4** (1983) 1507.
[407] H. C. Birnboim, *Carcinogenesis (London)* **7** (1986) 1511.
[408] Phillips Petroleum Co., TSCA-Mitteilung, Sect. 8e (1985).
[409] G. B. Freese et al., *Mutat. Res.* **4** (1967) 517.
[410] K. Heindorff, A. Michaelis, O. Aurich, R. Rieger, *Mutat. Res.* **142** (1985) 23.
[411] A. Korhonen, K. Hemminki, H. Vainio, *Environ. Res.* **33** (1984) 54.

# Phenol

WILFRIED JORDAN, Phenolchemie GmbH, Gladbeck, Federal Republic of Germany
HEINRICH VAN BARNEVELD, Phenolchemie GmbH, Gladbeck, Federal Republic of Germany
OTTO GERLICH, Phenolchemie GmbH, Gladbeck, Federal Republic of Germany
MICHAEL KLEINE-BOYMANN, Phenolchemie GmbH, Gladbeck, Federal Republic of Germany
JOCHEN ULLRICH, Phenolchemie GmbH, Gladbeck, Federal Republic of Germany

| | | |
|---|---|---|
| 1. | Introduction | 3689 |
| 2. | Physical Properties | 3690 |
| 3. | Chemical Properties | 3691 |
| 4. | Production | 3693 |
| 4.1. | Cumene Oxidation (Hock Process) | 3694 |
| 4.2. | Toluene Oxidation | 3697 |
| 4.3. | Other Processes | 3701 |
| 4.3.1. | Sulfonation of Benzene | 3701 |
| 4.3.2. | Chlorination of Benzene | 3701 |
| 4.3.2.1. | Alkaline Hydrolysis | 3701 |
| 4.3.2.1. | Steam Hydrolysis | 3702 |
| 4.3.3. | Dehydrogenation of Cyclohexanol – Cyclohexanone Mixtures | 3703 |
| 4.3.4. | Acetoxylation | 3704 |
| 5. | Environmental Protection | 3704 |
| 6. | Quality Specifications | 3705 |
| 7. | Storage and Transportation | 3705 |
| 8. | Economic Aspects | 3706 |
| 9. | Toxicology and Occupational Health | 3709 |
| 10. | References | 3710 |

# 1. Introduction

Phenol, hydroxybenzene, carbolic acid, $C_6H_5OH$ [108-95-2], discovered in 1834 by F. RUNGE, is the parent substance of a homologous series of compounds with the hydroxy group bonded directly to the aromatic ring. Phenol occurs as a free component or as an addition product in natural products and organisms. For example, it is a component of lignin, from which it can be liberated by hydrolysis. As a metabolic product it is normally excreted in quantities of up to 40 mg/L in human urine [1]. Higher quantities are formed in coking or low-temperature carbonization of wood, brown coal, or hard coal and in oil cracking. Initially phenol was extracted exclusively from hard coal tar, and only after consumption had risen significantly was it also produced synthetically. The earlier methods of synthesis (via benzenesulfonic acid and chlorobenzene) have been replaced by modern processes, mainly by the Hock process starting from cumene. Phenol has achieved considerable importance as the starting material for numerous intermediates and finished products.

## 2. Physical Properties

Phenol has a melting point of 40.9 °C, crystallizes in colorless prisms, and has a characteristic, slightly pungent odor. In the molten state it is a clear, colorless, mobile liquid. In the temperature range up to 68.4 °C its miscibility with water is limited (see Fig. 1), above this temperature it is completely miscible. The melting and solidification points of phenol are lowered quite considerably by water (see Fig. 2). A mixture of phenol and ca. 10 wt% water is called phenolum liquefactum, because it is liquid at room temperature (see Fig. 2). Phenol is readily soluble in most organic solvents (aromatic hydrocarbons, alcohols, ketones, ethers, acids, halogenated hydrocarbons, etc.) and somewhat less soluble in aliphatic hydrocarbons. Phenol forms azeotropic mixtures with water and other substances (see Table 1). Other physical data of phenol follow [3], [4]:

| | |
|---|---|
| $M_r$ | 94.11 |
| bp (101.3 kPa) | 181.75 °C |
| mp | 40.9 °C |
| Relative density | |
| 0 °C | 1.092 |
| 20 °C | 1.071 |
| 50 °C | 1.050 |
| Vapor density (air = 1) | 3.24 |
| Refractive index | |
| $n_D^{41}$ | 1.5418 |
| $n_D^{50}$ | 1.5372 |
| Dynamic viscosity | |
| 20 °C | 11.41 mPa · s |
| 50 °C | 3.421 mPa · s |
| 100 °C | 1.5 mPa · s |
| 150 °C | 0.67 mPa · s |
| Cubic expansion coefficient (50 °C) | $8.8 \times 10^{-7}$ L g$^{-1}$ K$^{-1}$ |
| Dissociation constant in water (20 °C) | $1.28 \times 10^{-10}$ |
| Heat of fusion (41 °C) | 120.6 kJ/kg |
| Heat of vaporization (182 °C) | 511 kJ/kg |
| Heat of formation (20 °C) | − 160 kJ/kg |
| Heat of combustion | − 32 590 kJ/kg |
| Specific heat | |
| 0 °C | 1.256 kJ kg$^{-1}$ K$^{-1}$ |
| 20 °C | 1.394 kJ kg$^{-1}$ K$^{1}$ |
| 50 °C | 2.244 kJ kg$^{-1}$ K$^{-1}$ |
| 100 °C | 2.382 kJ kg$^{-1}$ K$^{-1}$ |
| Specific heat (vapor state) | |
| 27 °C | 1.105 kJ kg$^{-1}$ K$^{-1}$ |
| 527 °C | 2.26 kJ kg$^{-1}$ K$^{-1}$ |
| Flash point (DIN 51 758) | 81 °C |

| | |
|---|---|
| Lower explosion limit in air at 101.3 kPa (corresponds to 73 °C product temperature i.e., vapor–liquid equilibrium) | 1.3 vol% (50 g/m$^3$) |
| Saturation concentration in air (20 °C) | 0.77 g/m$^3$ |
| Ignition temperature (DIN 51794) | 595 °C |
| Temperature class (VDE, EN) | T 1 (> 450 °C) |
| Autoignition temperature [5] | 715 °C |
| No allocation to a hazard class according to VbF because of a solidification point > 35 °C | |
| Flash point of mixtures with water | |
| 1–4% water | ca. > 82 °C |
| > 4.5% water | > 100 °C |
| Specific resistance (45 °C) | > 1.8 × 10$^7$ Ω · cm |
| Mixture with 5% water | 4.9 × 10$^6$ Ω · cm |
| Vapor pressure | |
| 20 °C | 0.02 kPa |
| 36.1 °C | 0.1 kPa |
| 48.5 °C | 0.25 kPa |
| 58.3 °C | 0.5 kPa |
| 69 °C | 1 kPa |
| 84.8 °C | 2.5 kPa |
| 98.7 °C | 5 kPa |
| 114.5 °C | 10 kPa |
| 137.3 °C | 25 kPa |
| 158 °C | 50 kPa |
| 181.5 °C | 100 kPa |
| 207.5 °C | 200 kPa |
| 247.7 °C | 500 kPa |
| 283 °C | 1000 kPa |

# 3. Chemical Properties

Phenol can be considered as the enol of cyclohexadienone. Whereas the tautomeric keto–enol equilibrium lies far to the ketone side in the case of aliphatic ketones, for phenol it is shifted almost completely to the enol side.

The reason for this stabilization is the formation of the aromatic system. The resonance stabilization is very high because of the contribution of the *o*- and *p*-quinoid resonance structures. In the formation of the phenolate anion, the contribution of quinoid resonance structures can stabilize the negative charge.

In contrast to aliphatic alcohols, phenol therefore forms salts with aqueous alkali

**Table 1.** Azeotropic mixtures formed by phenol

| Second component | Pressure, kPa | bp, °C | Proportion of phenol, % |
|---|---|---|---|
| H$_2$O | 101.3 | 94.5 | 9.21 |
| | 70.8 | 90.0 | 8.29 |
| | 39.2 | 75.0 | 7.2 |
| | 16.9 | 56.3 | 5.5 |
| Isopropylbenzene | 101.3 | 149 | 2 |
| n-Propylbenzene | 101.3 | 158.5 | 14 |
| α-Methylstyrene | 101.3 | 162 | 7 |

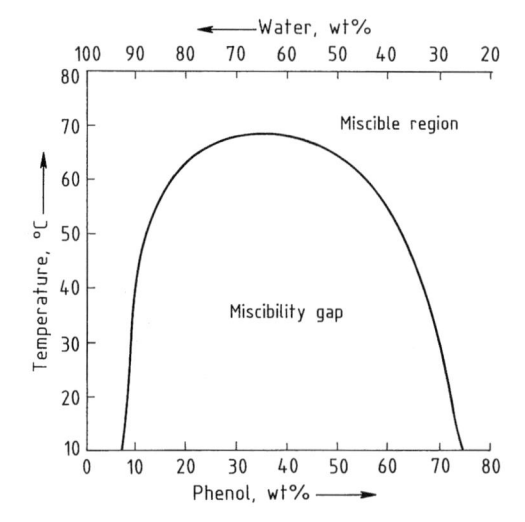

**Figure 1.** Miscibility of phenol and water

hydroxide solutions. At room temperature phenol can be liberated from the salts even with carbon dioxide. At temperatures near the boiling point of phenol, phenol can displace carboxylic acids, e.g., acetic acid, from their salts and phenolates are formed.

The contribution of o- and p-quinoid resonance structures allows electrophilic substitution reactions such as chlorination, sulfonation, nitration, nitrosation, and mercuration. The introduction of two or three nitro groups into the benzene ring can only be achieved indirectly because of the sensitivity of phenol towards oxidation. Examples are (1) nitration of chlorobenzene followed by ipso substitution of the chlorine by a hydroxy group or (2) nitration of phenoldisulfonic acid. Nitrosation in the para position can be carried out even at ice bath temperature. Phenol reacts readily with carbonyl compounds in the presence of acid or basic catalysts. Formaldehyde reacts with phenol to give hydroxybenzyl alcohols, and synthetic resins on further reaction. Reaction of acetone with phenol yields bisphenol A.

The reaction in the presence of acid catalysts is used to remove impurities from synthetic phenol. Olefinic impurities or carbonyl compounds, e.g., mesityl oxide, can be polymerized into higher molecular mass compounds by catalytic quantities of sulfuric

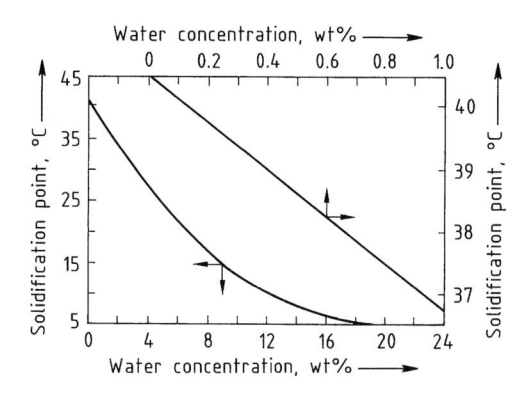

**Figure 2.** Effect of water on the solidification point of phenol

acid [6] or acidic ion exchangers [7] and can thus be separated easily from phenol by distillation.

Phenol readily couples with diazonium salts to yield colored compounds, which can be used for the photometric detection of phenol as in the case of diazotized 4-nitroaniline [8].

Salicylic acid (2-hydroxybenzoic acid) can be produced by the Kolbe–Schmitt reaction from sodium phenolate and carbon dioxide, whereas potassium phenolate gives the para compound. Alkylation and acylation of phenol can be carried out with aluminum chloride as catalyst, methyl groups can also be introduced by the Mannich reaction.

Diaryl ethers can only be produced under extreme conditions.

$$C_6H_5-ONa + Br-C_6H_5 \xrightarrow{200\,°C/Cu} C_6H_5-O-C_6H_5 + NaBr$$

With oxidizing agents phenol readily forms a free radical, which can dimerize to form diphenols or can be further oxidized to form dihydroxybenzenes and quinones. Since phenol radicals are relatively stable, phenol is a suitable radical scavenger and can also be used as an oxidation inhibitor. This latter property can also be undesirable, e.g., the autoxidation of cumene can be inhibited by small quantities of phenol.

# 4. Production

Small quantities of phenol (cresylic acids) are isolated from tars and coking plant water produced in the coking of hard coal and the lowtemperature carbonization of brown coal as well as from the wastewater from cracking plants. By far the greatest proportion is obtained by oxidation of benzene or toluene. Although direct oxidation of benzene is in principle possible, the phenol formed is immediately oxidized further. Therefore, alternative routes must be chosen, e.g., via halogen compounds which are subsequently hydrolyzed or via cumene hydroperoxide which is then cleaved catalytically. The following processes were developed as industrial syntheses for the production of phenol:

1) Sulfonation of benzene and production of phenol by heating the benzene sulfonate in molten alkali hydroxide
2) Chlorination of benzene and alkaline hydrolysis of the chlorobenzene
3) Chlorination of benzene and steam hydrolysis of the chlorobenzene (Raschig process, Raschig–Hooker, Gulf oxychlorination)
4) Alkylation of benzene with propene to isopropylbenzene (cumene), oxidation of cumene to the corresponding *tert*-hydroperoxide, and cleavage to phenol and acetone (Hock process)
5) Toluene oxidation to benzoic acid and subsequent oxidizing decarboxylation to phenol (Dow process)
6) Dehydrogenation of cyclohexanol–cyclohexanone mixtures (Scientific Design)

Of the processes named, only the Hock process (cumene oxidation) and the toluene oxidation are important industrially. The other processes were given up for economic reasons. In the Hock process acetone is formed as a byproduct.

This has not, however, hindered the expansion of this process, because there is a market for acetone. New plants are now run predominantly on the cumene process.

## 4.1. Cumene Oxidation (Hock Process)

The cumene–phenol process is based on the discovery of cumene hydroperoxide [*80-15-9*] and its cleavage to phenol [*108-95-2*] and acetone [*67-64-1*] published in 1944 by H. Hock and S. Lang [9]. This reaction was developed into an industrial process shortly after World War II by the Distillers Co. in the United Kingdom and the Hercules Powder Co. in the United States. The first plant was put into operation in 1952 by Shawinigan in Canada and had an initial capacity of 8000 t/a of phenol.

Today phenol is predominantly produced by this process in plants in the United States, Canada, France, Italy, Japan, Spain, Finland, Korea, India, Mexico, Brazil, Eastern Europe, and Germany with an overall capacity of $5 \times 10^6$ t/a. In addition to the economically favorable feedstock position (due to the progress in petrochemistry since the 1960s), the fact that virtually no corrosion problems occur and that all reaction stages work under moderate conditions with good yields was also decisive for the rapid development of the process.

To produce cumene [*98-82-8*], benzene [*71-43-2*] is alkylated with propene [*115-07-1*] using phosphoric acid (UOP process) or aluminum chloride as catalyst.

Two reaction steps form the basis of the production of phenol from cumene:

1) Oxidation of cumene with oxygen to cumene hydroperoxide:

$$\underset{\text{CH}}{\text{H}_3\text{C}\diagdown\diagup\text{CH}_3}\text{–C}_6\text{H}_5 + \text{O}_2 \longrightarrow \underset{\text{C–OOH}}{\text{H}_3\text{C}\diagdown\diagup\text{CH}_3}\text{–C}_6\text{H}_5$$

$$\left( + \underset{\text{C–OH}}{\text{H}_3\text{C}\diagdown\diagup\text{CH}_3}\text{–C}_6\text{H}_5 + \underset{\text{C=O}}{\text{CH}_3}\text{–C}_6\text{H}_5 \right)$$

2) Cleavage of cumene hydroperoxide in an acidic medium to phenol and acetone:

$$\underset{\text{C–OOH}}{\text{H}_3\text{C}\diagdown\diagup\text{CH}_3}\text{–C}_6\text{H}_5 \xrightarrow{\text{Acid}} \text{C}_6\text{H}_5\text{–OH} + \text{CH}_3\text{–CO–CH}_3$$

**Oxidation.** The oxidation of cumene with air or oxygen-enriched air is carried out in steel or stainless steel reactors, which work on the bubble column principle (see Fig. 3). The reactors (a), which can be more than 20 m high, are usually connected in cascade series to achieve an optimal residence time distribution. Series of two to four reactors are common in industry. The oxidation is performed at 90–120 °C and 0.5–0.7 MPa. It is carried out both in an alkali-stabilized system, (through the addition of soda) at pH 7–8 [10], [11] and in an unstabilized system at pH 3–6 [12].

The waste gas from the reactors is purified by a two-stage condensation of organic impurities, which consist predominantly of cumene. Water is used as coolant in the first stage and refrigerant in the second stage. In an activated carbon adsorption plant located downstream, the residual impurities are removed to concentrations below the limiting value of the TA Luft (first general government regulation in the German Immission Protection Law, 27 Feb. 1986).

The oxidation is autocatalytic, i.e., the reaction rate increases with increasing hydroperoxide concentration. The rather slow reaction can be accelerated using catalysts such as metal phthalocyanines [13]. Since these compounds simultaneously catalyze the decomposition of the hydroperoxide formed, however, they have not been used industrially.

The reaction is exothermic; ca. 800 kJ are released per kilogram of cumene hydroperoxide. The heat of reaction is removed by cooling.

The main byproducts of the autoxidation, which takes place by a free-radical chain mechanism, are $\alpha,\alpha$-dimethyl benzyl alcohol and acetophenone. Both products are also formed in the thermal decomposition of cumene hydroperoxide, which occurs to a large extent at temperatures exceeding 130 °C. Sulfur-containing compounds (e.g., sulfides, disulfides, thiols, sulfoxides, and thiophenes) and phenols inhibit autoxidation and must be removed from the cumene.

The oxygen content of the waste gas from the reactors is 1–6 vol %. The critical oxygen concentration for the *ignition* of a cumene–nitrogen–oxygen mixture is ca.

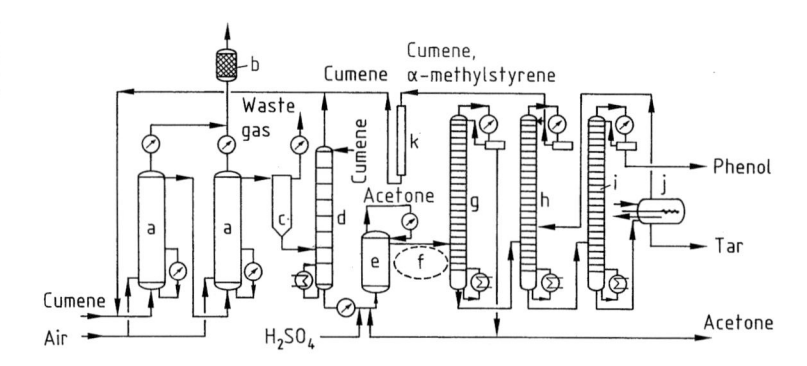

**Figure 3.** Production of phenol from cumene
a) Cumene oxidation reactor; b) Waste gas purification; c) Gas separator; d) Concentration; e) Cleavage reactor; f) Catalyst separation; g) Acetone column; h) Cumene column; i) Phenol column; j) Cracking; k) Hydrogenation

8.5 vol% at the reaction pressure of 0.5–0.6 MPa. Water vapor in the gas volume renders the system slightly more inert; the critical oxygen concentration is about 1 vol% higher in this case. The *explosion limits* of cumene vapor–air mixtures are 0.8–8.8 vol% cumene at normal pressure. At 0.5 MPa the upper explosion limit is shifted to 10.3 vol% cumene [14]. The oxidation mixture contains ca. 20–30% cumene hydroperoxide. If the oxidation is carried out in alkali-stabilized systems, the oxidation mixture is washed with water to remove inorganic salts before unreacted cumene is distilled off in vacuum from the oxidation mixture to achieve a hydroperoxide concentration of 65–90% before the cleavage.

**Cleavage.** The acid-catalyzed cleavage of cumene hydroperoxide to phenol and acetone follows an ionic mechanism [15]. Sulfuric acid is used almost exclusively as the catalyst in industry. Other acids such as perchloric acid, phosphoric acid, *p*-toluenesulfonic acid, sulfur dioxide, and acidic ion exchangers are also effective catalysts.

The acid-catalyzed cleavage is carried out in two different ways. In the *homogeneous phase*, an excess of acetone is charged to the cleavage reactor (e) and 0.1–2% sulfuric acid is sufficient for cleavage. The reaction temperature is the boiling point of the cumene hydroperoxide–acetone mixture which is determined by its acetone content. The heat removal from the strongly exothermic cleavage reaction (ca. 1680 kJ per kilogram of cumene hydroperoxide) is achieved by means of evaporation of acetone from the reaction system [16], [17].

In the *heterogeneous phase reaction*, cumene hydroperoxide is cleaved with 40–45% sulfuric acid at a concentrate–acid ratio of 1:5. Mixing is performed in a centrifugal pump; the actual reaction occurs in a subsequent cooling system in which a stable temperature range of 50–60 °C is maintained by appropriate sizing of the individual coolers. To limit the formation of byproducts, short residence times of about 45–60 s are used [18]. To avoid corrosion, a special chromium–nickel steel alloyed with copper is used as material in the cooling system.

The cumene hydroperoxide is split down to a residual concentration of < 0.1%. Most of the α,α-dimethyl benzyl alcohol [*617-94-7*] formed as a byproduct in the oxidation step is dehydrated to α-methylstyrene [*98-83-9*]. Typical byproducts under the acidic cleavage conditions are 4-cumylphenol [*599-64-4*] and mesityl oxide [*141-79-7*].

**Workup.** Prior to distillation of the reaction mixture, the sulfuric acid must be neutralized or removed to reduce corrosion damage to the columns and to avoid acid-catalyzed condensation reactions in the cleavage mixture. The neutralization can be carried out with aqueous sodium hydroxide or aqueous phenolate solution [19]; the aqueous, salt-containing phase (pH 5–6) is then separated from the reaction mixture. A further process for acid separation is the extraction of the cleavage mixture with aqueous solutions of neutral salts such as sodium sulfate [20] or water [21].

In the distillative workup the *crude acetone* is first distilled (g). It is purified in an alkaline scrubber and subsequently distilled again to give the pure product. Cumene and α-methylstyrene are then distilled from the bottoms product of the crude acetone column, partly by azeotropic distillation with water (h) [22]. The α-methylstyrene contained in cumene is either hydrogenated on a nickel contact to cumene, which is then recycled to the process or obtained in high purity by fractional distillation and marketed. The *crude phenol* distilled off in the crude phenol column (i) is purified by extractive distillation with water [23] or by passing over an acidic ion-exchange resin [7]. In the latter process, impurities which are difficult to separate by distillation, such as mesityl oxide, are condensed to high-boiling compounds. Phenol is obtained with a purity of > 99.9% by subsequent distillation.

Another distillative workup process differs from the process just described in that after separation of the crude acetone by distillation, the hydrocarbons contained in the crude phenol (e.g., cumene and α-methylstyrene) and other impurities are distilled off overhead with an auxiliary extraction liquid, such as diethylene glycol which dissolves the phenol. The phenol is then distilled off as a pure product from the higher-boiling auxiliary liquid in a second distillation step [24].

## 4.2. Toluene Oxidation

This phenol process developed originally by Dow (United States) [25]–[30] has been carried out up until now in three industrial plants—in the United States, Canada, and the Netherlands. After initial problems the plant in the Netherlands was taken over completely by the DSM and brought up to industrial standard [31]. Snia Viscosa (Italy) uses the toluene oxidation only for the production of benzoic acid as an intermediate in the production of caprolactam [32]–[34].

The process runs in two stages. In the first stage toluene [*108-88-3*] is oxidized with atmospheric oxygen in the presence of a catalyst to benzoic acid [*65-85-0*] in the liquid phase. In the second stage the benzoic acid isolated is decarboxylated catalytically in the presence of atmospheric oxygen to produce phenol.

$$\text{C}_6\text{H}_5\text{-CH}_3 \xrightarrow[-H_2O]{O_2(\text{air}), \text{cat.}} \text{C}_6\text{H}_5\text{-COOH}$$

$$\xrightarrow[-CO_2]{O_2(\text{air}), \text{cat.}} \text{C}_6\text{H}_5\text{-OH}$$

**Reaction Details.** The *oxidation of toluene* is a radical chain reaction involving peroxy radicals [32]:

$$\text{Ar-CH}_3 \longrightarrow \text{Ar-CH}_2\cdot \longrightarrow \text{Ar-CH}_2\text{-OO}\cdot \longrightarrow \text{Ar-CH}_2\text{-OOH} \longrightarrow \text{Ar-CH}_2\text{-O}\cdot \longrightarrow \text{Ar-CHO}$$
$$\text{Ar-CHO} \longrightarrow \text{ArCO}\cdot \longrightarrow \text{ArCO-OO}\cdot \longrightarrow \text{ArCO-OOH} \longrightarrow \text{ArCO-O}\cdot \longrightarrow \text{ArCOOH}$$

The activation energy of the exothermic oxidation of toluene to benzoic acid is 136 kJ/mol [32]. The large number of radicals produced during the reaction leads to the formation of several byproducts, e.g., benzyl alcohol (1%), benzaldehyde (1–2%), benzyl benzoate (1–2%), biphenyl (0.1–0.2%), and methylbiphenyls (0.7–1%). In addition, formic acid, acetic acid, and carbon monoxide are formed [26]. The oxidation occurs in the liquid phase at 120–150 °C and ca. 0.5 MPa in the presence of a cobalt catalyst. 30% of the toluene is converted into benzoic acid. The selectivity is 90%.

The *oxidizing decarboxylation* of benzoic acid proceeds via Cu(II) benzoate, which reacts further to give benzoylsalicylic acid. The latter then either hydrolyzes to salicylic acid and benzoic acid or decarboxylates to phenyl benzoate. Phenyl benzoate hydrolyzes in the presence of water to phenol and benzoic acid. During the formation of benzoylsalicylic acid, Cu(II) is reduced to Cu(I) which is present as Cu(I) benzoate. Copper(I) is reoxidized to Cu(II) with atmospheric oxygen at a very high reaction rate.

[Reaction scheme showing: 2 benzoic acid + Cu²⁺ → copper(II) benzoate → (−CO₂) phenyl benzoate → (H₂O) phenol + benzoic acid; and benzoylsalicylic acid → (H₂O) salicylic acid + benzoic acid → (−CO₂) phenol]

The reaction proceeds at 220–250 °C under normal pressure or a small excess pressure (up to 0.25 MPa). The selectivity is ca. 90%. Byproducts are benzene and high-boilers which form tarry residues after workup. The reaction is exothermic, 1425 kJ per kilogram phenol are produced [28]. In total 1.45 t toluene, 0.3 kg cobalt, and 3 kg copper are consumed per tonne of phenol obtained.

*Catalysts.* For the *toluene oxidation* a soluble cobalt catalyst is required, e.g., cobalt naphthenate or benzoate in quantities of 100 to 300 ppm.

For the *decarboxylation* step soluble copper (II) salts must be added. In the reaction mixture ca. 1–5% Cu is present. The selectivity is improved by the addition of metal salts, mainly magnesium salts which act as promoters. According to later developments molybdenum (III) compounds bring about higher selectivity than the Cu (II) salts [35].

A new variant of the process developed by Lummus uses a copper-containing catalyst, over which the benzoic acid can be oxidized in the vapor phase [36]. As a result of the very high space velocity, a 50% conversion, and yields of ca. 90%, quantities of 1 kg phenol per liter reaction volume can be achieved. It is reported that tarry residues are not formed. The catalyst must, however, be regenerated more frequently than in the original process.

**Process Description** (Fig. 4). *Toluene Oxidation.* Fresh toluene is fed into the oxidation reactor (a) together with recycled toluene and the corresponding quantity of catalyst. Air is blown into the liquid toluene using air spargers. During the reaction, oxygen is consumed to a residual concentration of < 4 vol%. The waste gas containing toluene and the water formed in the reaction are cooled and then led into a waste-gas purification system (b) where most of the toluene is removed. The condensate obtained on cooling consists of toluene and water loaded with formic and acetic acids. Toluene is recycled into the process and the water phase is disposed of or subjected to suitable treatment. Since oxidation produces excess heat, the reactor must be equipped with a cooling device. The reaction product, benzoic acid, is separated from unreacted toluene in the toluene stripping column (c). The

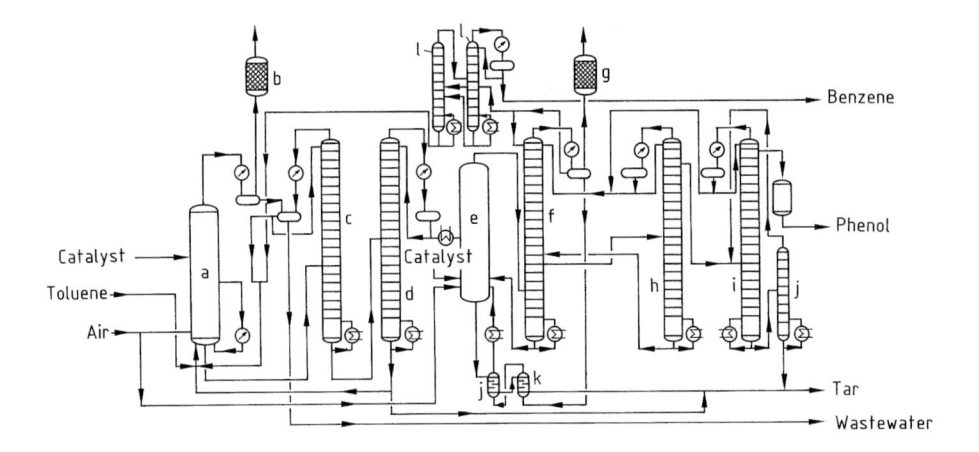

**Figure 4.** Production of phenol from toluene
a) Toluene oxidation reactor; b) Waste gas purification; c) Toluene stripping column; d) Benzoic acid column; e) Benzoic acid oxidation reactor; f) Water–hydrocarbon stripping column; g) Waste gas purification; h) Crude phenol column; i) Phenol column; j) Residue extraction; k) Benzene column; l) Phenol residue column

toluene is recycled to the oxidation reactor. The bottom product of the stripping column, which consists of benzoic acid and higher-boiling byproducts, reaches the benzoic acid column (d) where pure benzoic acid is distilled off overhead. The bottom product from this column, consisting of byproducts, is disposed of or can be led back into the oxidation reactor.

*Oxidizing Decarboxylation of Benzoic Acid.* The purified benzoic acid is led to the decarboxylation reactor (e) and at the same time the necessary quantity of catalyst is added. This stage also takes place in the liquid phase. Air is introduced through a sparger and water vapor is led in. The phenol formed is removed from the reactor as a vapor and separated from the inert gases, which contain some toluene and benzene, in the water–hydrocarbon stripping column (f). The stream of inert gases is freed from residual toluene and benzene in the waste gas purification (g).

Phenol is drawn off from the side of the stripping column (f) and distilled in the crude phenol column (h). The higher-boiling fractions from the bottom of the phenol column (i) are distilled in column (l). The bottoms product of column (f) which still contains benzoic acid, is recycled to the reactor (e). Tar, which is formed in the reactor, must be removed continually. The benzoic acid contents of the tar is extracted with water. The water phase is recycled to the process and the tar is incinerated in an incineration plant with heat recovery. In the incineration complete removal of the catalyst components from the flue gas must be ensured.

For heating the benzoic acid column a thermal oil is generally used.

Benzene is separated by distillation from recycle toluene in the benzene columns (k) and is obtained as a pure, marketable product.

*Materials.* Because of the aggressive nature of the product the plant must be constructed from stainless CrNiMo steel (material no. 4571). For the benzoic acid area, Hastelloy is used.

## 4.3. Other Processes

### 4.3.1. Sulfonation of Benzene

The sulfonation of benzene was developed by A. WURTZ and A. KEKULÉ in 1867 and is described by the following steps [37]:

1) Formation of benzenesulfonic acid by reaction of benzene with concentrated sulfuric acid.
2) Reaction of the benzenesulfonic acid with sodium sulfite or sodium carbonate to form sodium benzenesulfonate.
3) Heating of benzenesulfonate in molten sodium hydroxide to yield sodium phenolate.
4) Reaction of sodium phenolate with sulfur dioxide or carbon dioxide to give the free phenol.

This multistep process was the first industrial phenol synthesis. The process was continually improved to lower the quantity of process materials required and the amount of byproducts. The sulfur dioxide or carbon dioxide liberated in the reaction of benzenesulfonic acid with sodium sulfite or carbonate, respectively, is used in the last step for the liberation of the phenol. The reaction of the benzenesulfonate with sodium hydroxide occurs at 320–340 °C. Although the reaction could have been developed as a continuous process [38], [39], the costs for the process compared with the cumene process were higher, despite the yield of over 90% based on benzene. This is due to high energy requirements, consumption of chemicals, and environmental contamination by the large amount of sodium sulfate produced. Currently, no plants are known which run this process.

### 4.3.2. Chlorination of Benzene

Chlorobenzene is a readily accessible substance which forms phenol when hydrolyzed. Processes for the production of phenol based on this compound were therefore developed and run. Although they have the advantage that no acetone is coproduced as in the synthesis of phenol by the cumene process (see p. 302), they no longer have any economic importance.

#### 4.3.2.1. Alkaline Hydrolysis

The fact that phenol can be obtained by the alkaline hydrolysis of chlorobenzene was discovered in 1872 by L. DUSART and CH. BARDY and the reaction mechanism was elucidated by K. H. MEYER, F. HUISGEN, and J. SAUER [40]. The byproducts of the hydrolysis were investigated by A. LÜTTRINGHAUS and D. AMBROS [41]. The equation for the reaction can be subdivided into the following steps:

Electrolysis:

$$2\,NaCl + 2\,H_2O \longrightarrow Cl_2 + 2\,NaOH + H_2$$

Benzene chlorination:

$$C_6H_6 + Cl_2 \longrightarrow C_6H_5Cl + HCl$$

Hydrolysis:

$$C_6H_5Cl + 2\,NaOH \longrightarrow C_6H_5ONa + NaCl + H_2O$$
$$\underline{C_6H_5ONa + HCl \longrightarrow C_6H_5OH + NaCl}$$
$$C_6H_6 + H_2O \longrightarrow C_6H_5OH + H_2O$$

### 4.3.2.1. Steam Hydrolysis

**Raschig–Hooker Process.** The Raschig–Hooker process, which was developed by Raschig in Ludwigshafen [42], [43] and further developed by Hooker, proceeds according to:

$$\underline{\begin{array}{l}2\,C_6H_6 + 2\,HCl + O_2 \longrightarrow 2\,C_6H_5Cl + 2\,H_2O \\ 2\,C_6H_5Cl + 2\,H_2O \longrightarrow 2\,C_6H_5OH + 2\,HCl\end{array}}$$
$$2\,C_6H_6 + O_2 \longrightarrow 2\,C_6H_5OH$$

In the first step benzene is oxychlorinated in the vapor phase on an iron–copper catalyst. Subsequent hydrolysis in the gas phase at 450 °C on a tricalcium phosphate catalyst gives an aqueous solution containing hydrochloric acid and phenol, which must then be worked up.

The low benzene conversion (8–15 %) to avoid byproducts, high energy consumption, and high plant investment costs because of the corrosive action of the HCl–phenol solution, make this process economically unattractive compared to the cumene process, despite a maximal content of 95 % phenol in the reaction products.

**Gulf Oxychlorination.** At the Gulf Research and Development Co., Pittsburgh, a variant of the oxychlorination was developed [44]–[47]. The process, which has not yet been run in an industrial plant, is shown schematically in Figure 5. In this process the oxychlorination of benzene is performed in the liquid phase at 50–170 °C under the equilibrium pressure and with catalytic quantities of nitric acid.

At 80 % conversion the selectivity based on monochlorobenzene is 95 %, whereas in the Raschig process at 10 % conversion a typical monochlorobenzene selectivity of only ca. 90 % is achieved.

In the Gulf process, hydrolysis occurs on a copper-doped lanthanum phosphate catalyst. A phenol solution containing hydrochloric acid must also be worked up in this process, so that special construction materials are required, as in the Raschig process.

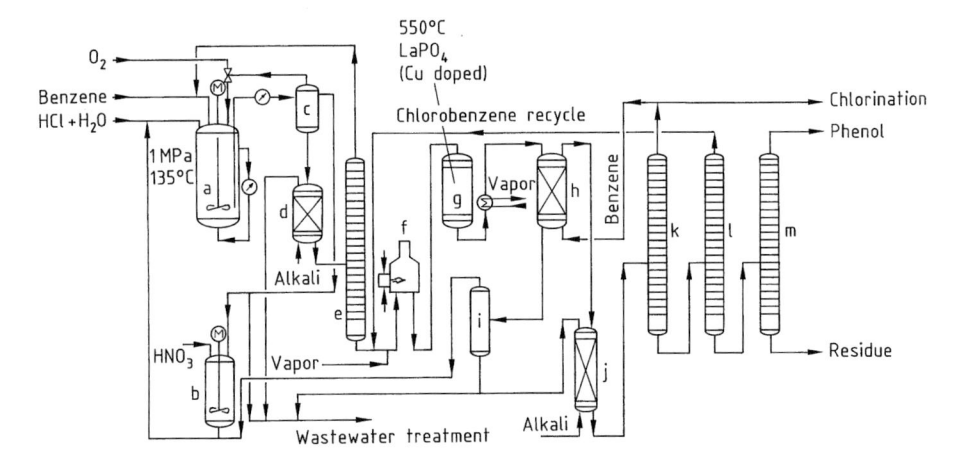

**Figure 5.** Production of phenol by the oxychlorination of benzene (Gulf process)
a) Chlorination; b) Catalyst charging vessel; c) Separation vessel; d) Alkali wash; e) Benzene distillation; f) Heating; g) Hydrolysis; h) Extractor; i) HCl stripper; j) Alkali wash; k) Benzene column; l) Chlorbenzene column; m) Phenol column

Whether the hydrolysis of chlorobenzene will ever be used industrially is in great doubt because of the environmental problems of the chlorination reaction.

## 4.3.3. Dehydrogenation of Cyclohexanol – Cyclohexanone Mixtures

Phenol can be obtained by dehydrogenation of the mixture of cyclohexanone and cyclohexanol produced in the oxidation of cyclohexane.

$$n\, C_6H_{11}OH + m\, C_6H_{10}O \longrightarrow (n+m)\, C_6H_5OH + (3\,n + 2m)\, H_2$$

The process developed by Scientific Design works with a solid bed catalyst consisting of ca. 3% platinum on activated carbon. Other dehydrogenation catalysts e.g., those containing nickel or cobalt, can also be used [48]–[52].

Using a cyclohexanol–cyclohexanone mixture in a ratio of 10:1, 1.805 t of the mixture are required to produce 1 t of phenol (corresponding to a yield of 97.8% in the dehydrogenation step). Since phenol forms azeotropic mixtures with cyclohexanone and cyclohexanol, an extractive distillation should be carried out.

The process is not economic if phenol is produced as the sole product starting from cyclohexane. A smaller plant which was erected in Australia and started up in 1965, was replaced three years later for economic reasons by one which worked according to the cumene process, [51]. A phenol production step can only be of interest in combination with a large-scale plant for cyclohexane oxidation, such as those constructed for caprolactam production.

## 4.3.4. Acetoxylation

In a process developed by Hoechst, benzene [71-43-2] is used as the starting material [53], [54]. Benzene is oxidized in the presence of acetic acid [64-19-7] (acetoxylation) with palladium as catalyst. The acetoxybenzene (phenyl acetate) obtained is hydrolyzed to phenol and acetic acid. The process has not yet been developed on an industrial scale.

# 5. Environmental Protection

**Wastewater Purification.** Phenol-containing wastewater may not be conducted into open water without treatment because of the toxicity of phenol. Phenol-containing wastewater from the production of phenol by the cumene process—which can contain 1–3% phenol—is generally freed by extraction from most of its phenol content. Cumene [98-82-8] and acetophenone [98-86-2] serve as extracting agents, both being produced by the process. Other solvents used industrially for the extraction of phenol from wastewater are butyl acetate (Phenosolvan process), benzene, diisopropyl ether, and methyl isobutyl ketone. In this way phenol is removed, depending on the process, down to a residual concentration of 20 to 500 mg/L. The remaining phenol is removed in the biological purification stage in a sewage treatment plant.

The removal of phenol can also be carried out (1) by steam distillation processes based on the steam volatility of phenol; (2) by adsorption on surface-active materials, such as activated carbon and ion exchanger resins; and (3) by decomposition of the phenol with oxidizing agents such as hydrogen peroxide.

**Waste Gas Purification.** Phenol vapor from the waste gas stream is best removed by *absorption* in absorption towers, which contain a packing layer of 1 to 2 m in height to improve mass transfer. Water or 3–20% sodium hydroxide is used as the scrubbing agent. *Adsorption* on activated carbon provides another possibility for the purification of waste gases. The limiting concentration of phenol (class 1) of 20 mg/m$^3$ (STP) (quantity emitted from 0.1 kg/h) given in the TA Luft (clean air regulation in the Federal Republic of Germany) can be met by adsorption or absorption processes. Also the maximum concentrations of other organic coproducts such as acetone, cumene, α-methylstyrene, and methanol mandated by the TA Luft can be adhered to without any problem using adsorption or absorption.

Table 2. Quality specifications for phenol

| Pharmacopeia | Water content, % | Color | Content, % | Nonvolatile impurities, % | Boiling range, °C | Solidification point, °C |
|---|---|---|---|---|---|---|
| DAB 9 | 0.5 | colorless, faintly pink, or faintly yellow | 98.5 – 100.5 | 0.05 | 180 – 182 | |
| U.S.P. XXI | 0.5 | colorless, faintly pink | 99.0 – 100.5 | 0.05 | | $\geq 39.0$ |
| B.P. 88 | | colorless, faintly pink | 99.0 – 100.5 | 0.05 | | 40 – 41 |
| E.P. 89* | | colorless, faintly pink, or faintly yellow | 99.0 – 100.5 | 0.05 | | $\geq 39.5$ |

* E.P. = European Pharmacopeia.

# 6. Quality Specifications

The quality specifications of the various pharmacopeias are generally met by synthetically produced phenol. These specifications are limited to identity tests, assays, determinations of acidically-reacting impurities, nonvolatile impurities, color, water solubility, solidification point, water content, and boiling range (Table 2).

The quality specifications of the various users of phenol go beyond these criteria to some extent. Impurities such as hydroxyacetone, mesityl oxide, 2-methylbenzofuran, or 3-methylcyclopentanone can result in coloration of downstream products in further processing. The concentration of lower carboxylic acids can increase the extent of undesired side reactions. Metal ions can damage catalysts or lead to coloration of reaction products. The concentrations of these components in phenol must therefore be limited.

Synthetic phenol is normally colorless. *Coloration* is determined by comparison with calibrated color scales.

The *solidification point* of phenol is a sensitive criterion for impurities. It is, however, strongly affected by the water content (see Fig. 2). The effect of water on this method of testing can be excluded by the addition of gypsum or dry molecular sieve 4 A.

ASTM methods (D 1078-78; D 1493-67; D 1631-61 etc.) are used as the methods of investigation. Factory specifications of the producer and manufacturers are also used.

# 7. Storage and Transportation

Large quantities of phenol are transported as liquids in road or rail tank cars equipped with heating coils. Steel is generally used as the container material. If the phenol is to remain colorless for as long as possible, the walls of the containers must be of stainless steel. Pure phenol is therefore stored predominantly in stainless steel tanks, which are insulated to avoid heat loss. On storing liquid phenol it must be ensured that the temperature remains below 70 °C. The lower explosion limit for phenol – air mixtures is reached at a saturation temperature of 73 °C, corresponding to a phenol vapor fraction in the mixture of 1.36 vol %. At storage temperatures > 70 °C the tank gas volume is blanketed with nitrogen.

Smaller quantities of phenol are transported in steel drums. If the phenol contains water, steel drums are not suitable because of the coloration of the phenol which occurs. In this case drums of galvanized sheet metal are necessary.

For phenol the following transport regulations are in force:

| | |
|---|---|
| GGVE/RID | Class 6.1, no. 13 b |
| | RN 601 |
| | Hazard no. 68 |
| GGVS/ADR | Class 6.1, no. 13 b |
| | RN 2601 |
| | Hazard no. 68 |
| ADNR | Class 6.1 (IVa), no. 13 c |
| | RN 6401 |
| IMDG Code | Class 6.1 |
| UN no. | 2312 |

# 8. Economic Aspects

Most of the phenol produced is processed further to give phenol–formaldehyde resins. The quantities of phenol used in the production of caprolactam via cyclohexanol–cyclohexanone (→ Cyclohexanol and Cyclohexanone) have decreased because phenol has been replaced by cyclohexane as the starting material for caprolactam. The production route starting from phenol is less hampered by safety problems than that starting from benzene, which proceeds via cyclohexane oxidation.

Bisphenol A [2,2-bis(4-hydroxyphenyl)propane], which is obtained from phenol and acetone (→ Phenol Derivatives), has become increasingly important as the starting material for polycarbonates and epoxy resins. Aniline can be obtained from phenol by the Halcon process, where the hydroxy group of phenol is substituted by means of ammonolysis. Adipic acid is obtained from phenol by oxidative cleavage of the aromatic ring. Alkylphenols, such as cresols, xylenols, 4-*tert*-butylphenol, octylphenols, and nonylphenols are produced by alkylation of phenol with methanol or the corresponding olefins. Salicylic acid is synthesized by addition of $CO_2$ to phenol (Kolbe synthesis). Chlorophenols are also directly obtained from phenol.

All these products have considerable economic importance because they are used for the production of a wide range of consumer goods and process materials. Examples are: preforms, thermosets, insulating foams, binders (e.g., for mineral wool and molding sand), adhesives, laminates, impregnating resins, raw materials for varnishes, emulsifiers and detergents, plasticizers, herbicides, insecticides, dyes, flavors, and rubber chemicals.

**Table 3.** Percentage distribution of phenol consumption among different products

|  | Western Europe, % | | United States, % | Japan, % | World, % |
|---|---|---|---|---|---|
|  | 1973 | 1984 | 1974 | 1973 | 1989 |
| Phenolic resins | 37 | 33 | 48 | 60 | 41 |
| Caprolactam | 22 | 25 | 15 |  | 21 |
| Adipic acid | 12 |  | 3 |  |  |
| Bisphenol A | 11 | 22 | 13 | 18 | 15 |
| Aniline |  | 2 |  | 9 | 3 |
| Other products | 18 | 18 | 21 | 14 | 20 |
| Total consumption, $10^6$ t | 0.954 |  | 1.040 | 0.211 | 4.450 |

**Table 4.** Worldwide breakdown of phenol consumption, $10^6$ t

| United States | | Western Europe | | Eastern Europe | | Asia | | Others | | World | |
|---|---|---|---|---|---|---|---|---|---|---|---|
| 1985 | 1989 | 1985 | 1989 | 1985 | 1989 | 1985 | 1989 | 1985 | 1989 | 1985 | 1989 |
| 1.20 | 1.634 | 1.04 | 1.225 | 0.810 | 0.820 | 0.25 | 0.545 | 0.135 | 0.230 | 3.435 | 4.450 |

Table 3 shows the percentage distribution of phenol consumption among the most important downstream products. A breakdown of phenol consumption worldwide is given in Table 4.

In Table 5 the available production capacity for phenol is summarized and presented as a percentage for the individual production processes. Besides the availability and price of the raw materials, the saleability of possible coproducts, such as acetone from the cumene process, is very important for the economy of the process. In the past the proportion of the production processes has moved in favor of the cumene process, particularly in the United States. In Japan a new plant based on the toluene process is being built (capacity 120 000 t, predicted start-up in 1991). In Western Europe only expansion of the capacity of existing plants have been undertaken. However, at present the capacities available in Western Europe are no longer sufficient for covering the requirements at the current rate of increase, and additional capacity is thus created by building a new plant.

The *cumene process* has been able to maintain its leading role because a market exists for acetone (0.6 parts acetone are produced to 1 part phenol). To carry out the *toluene process* economically a low toluene price relative to that of benzene is a prerequisite. The classical processes of *benzene sulfonation* and the *alkaline chlorobenzene hydrolysis* could not hold their own ground because of the considerable quantities of sodium sulfate and sodium chloride formed as byproducts. The high running costs of the *Raschig–Hooker process* led to the abandonment of the process. The similar *Gulf oxychlorination process* developed a few years ago has not yet come into industrial use. The *Scientific Design process* starting from cyclohexane was abandoned.

Table 5. Production capacity for phenol

| | Western Europe | | | | United States | | | | Japan | Asia | | Eastern Europe | | Others | World |
|---|---|---|---|---|---|---|---|---|---|---|---|---|---|---|---|
| | 1973 | 1978 | 1985 | 1989 | 1974 | 1977 | 1985 | 1989 | 1977 | 1985 | 1989 | 1985 | 1989 | 1989 | 1989 |
| Capacity, $10^6$ t/a | 1.14 | 1.48 | 1.30 | 1.45 | 1.25 | 1.5 | 1.50 | 1.90 | 0.28 | 0.40 | 0.60 | 0.89 | 0.90 | 0.14 | 5.0 |
| Cumene process, % | 87 | 88 | 92 | 93 | 87 | 95 | 98 | 98 | 100 | 99 | 99 | | | | 96 |
| Toluene process, % | 8 | 10 | 8 | 7 | 2 | 2 | 2 | 2 | | | | | | | 3 |
| Other processes, % | 2 | | | | 9 | 2 | | | | | | | | | |
| Phenol from tar, % | 3 | 2 | | | 2 | 1 | | | | | | | | | |

# 9. Toxicology and Occupational Health

**Toxicology.** Because of its germicidal effect phenol was used for a long time (in the last century) as a very effective disinfectant (carbolic acid, Lister spray). However, because of its protein-degenerating effect, it often had a severely corrosive effect on the skin and mucous membranes. Phenol only has limited use in pharmaceuticals today because of its toxicity. Phenol occurs in normal metabolism and is harmless in small quantities according to present knowledge.

Phenol is toxic in high concentrations. It can be absorbed through the skin, by inhalation, and swallowing. The typical main absorption route is the *skin*, through which phenol is resorbed relatively quickly, simultaneously causing caustic burns on the area of skin affected. Besides the corrosive effect, phenol can also cause sensitization of the skin in some cases. Resorptive poisoning by larger quantities of phenol—which is possible even over small affected areas of skin—rapidly leads to paralysis of the central nervous system with collapse and a severe drop in body temperature. If the skin is wetted with phenol or phenolic solutions, decontamination of the skin must therefore be carried out immediately. After removal of contaminated clothing, polyglycols (e.g., lutrol) are particularly suitable for washing the skin.

On skin contamination local anesthesia sets in after an initial painful irritation of the area of skin affected. Hereby the danger exists that possible resorptive poisoning is underestimated. If phenol penetrates deep into the tissue, this can lead to phenol gangrene through damage to vessels. The effect of phenol on the *central nervous system*—sudden collapse and loss of consciousness—is the same for humans and animals. In animals a state of cramp precedes these symptoms because of the effect phenol has on the motor activity controlled by the central nervous system.

Caustic burns on the cornea heal with scarred defects. Possible results of *inhalation* of phenol vapor or mist are dyspnea, coughing, cyanosis, and lung edema. *Swallowing phenol* can lead to caustic burns on the mouth and esophagus and stomach pains. Severe, though not fatal, phenol poisoning can damage inner organs, namely kidneys, liver, spleen, lungs, and heart. In addition, neuropsychiatric disturbances have been described after survival of acute phenol poisoning. Most of the phenol absorbed by the body is excreted in urine as phenol and/or its metabolites. Only smaller quantities are excreted with the faeces or exhaled [55]–[59].

**Acute Toxicity.** In animals the $LD_{50}$ was determined for different species and forms of administration. For oral administration the following values are reported: mouse 500 mg/kg; dog 500 mg/kg; rabbit 420 mg/kg [60].

Concentrations of 200 ppm and 20 ppm phenol in aqueous solution were administered to rats. The $LD_{50}$ values determined were 340 mg/kg and 530 mg/kg respectively [61].

In the summary and evaluation of data on the lethal dose for humans, a value of 140 mg/kg was reported as the minimum lethal oral dose [62].

**Chronic Toxicity.** Drinking water investigations showed that the growth, fertility, and general health of rats was normal up to 5000 ppm (no observed adverse effect level, NOAEL) [63]. In investigations by the National Cancer Institute a dose-dependent decrease in weight gain was observed [64]. Loss of appetite, marasmus, headache, rapid fatigue, and severe chronic insomnia are given as symptoms of chronic phenol intoxication in humans after long-term intake of excessive phenol concentrations [55].

**Carcinogenicity.** The published reports on the carcinogenic and cocarcinogenic investigations give ambiguous indications [64]–[69]. The cocarcinogenic experiments by VAN DUUREN et al. gave the result that the tumor-producing effect of benzo[*a*]pyrene was slightly weakened by phenol [67].

Phenol should be classified in group D according to the EPA evaluation criteria for cancer data (1987). This means that the existing basis for an evaluation of the carcinogenic potential is insufficient.

**Occupational Health.** The following limiting values were laid down for work protection:

| | |
|---|---|
| United States: | 19 mg/m$^3$ or 5 ppm |
| Federal Republic of Germany: | 19 mg/m$^3$ or 5 ppm |
| Soviet Union: | 5 mg/m$^3$ or 1.5 ppm |

Healthy working people can be exposed to this concentration in air over 8 h per day (MAK value). In Germany allowed peak concentrations are also laid down:

| | |
|---|---|
| Short period value: | 2×MAK value; 38 mg/m$^3$ |
| Length of short period: | 5 min, momentary value |
| Frequency per shift: | 8× |

Phenol exposure can be determined from the phenol concentration in urine. The permitted BAT value is 300 mg/L biological tolerance value of a working material issued in 1989.

# 10. References

[1] R. Hoschek, W. Fritz: *Taschenbuch für den medizinischen Arbeitsschutz.* 4th ed., Enke Verlag, Stuttgart 1978, p. 315.
[2] L. H. Horsley: *Azeotropic Data I and II,* Amer. Chem. Soc., Washington, D.C., 1952 and 1962.
[3] VDI: *VDI-Wärmeatlas,* 4th ed., VDI-Verlag, Düsseldorf 1984.
[4] K. Nabert, G. Schön: *Sicherheitstechnische Kennzahlen brennbarer Gase u. Dämpfe,* 2nd ed., Dtsch. Eichverlag, Berlin 1963. 6th supplement 1.7.1990.

[5] C. I. Hilado, S. W. Clark, *Chem. Eng. (N. Y.)* **75** (1972) Sept. 4, 75.
[6] Phenolchemie, DE 1 126 887, 1959 (H. Sodomann, B. Hauschulz, H.-W. Schwermer).
[7] Phenolchemie, DE 1 668 952, 1968 (H. Sodomann, B. Hauschulz, G. Rasner, J. Mertmann).
[8] Fachgruppe Wasserchemie in der GDCh: *Deutsche Einheitsverfahren zur Wasser-, Abwasser- und Schlamm-Untersuchung,* 3rd. ed., Verlag Chemie, Weinheim 1960.
[9] H. Hock, S. Lang, *Ber. Dtsch. Chem. Ges. B* **77** (1944) 257.
[10] Distillers Co., GB 653 761, 1948 (G. P. Armstrong, T. Bewley, K. H. W. Turck).
[11] Distillers Co., DE 819 092, 1949 (G. P. Armstrong, T. Bewley, K. H. W. Turck).
[12] Phenolchemie, DE 1 131 674, 1960 (H. Sodomann, B. Hauschulz).
[13] H. Kropf, *Justus Liebigs Ann. Chem.* **637** (1960) 73.
[14] J. C. Butler, W. P. Webb, *Ind. Eng. Chem.* **2** (1957) no. 1, 42.
[15] M. S. Kharasch, A. Fono, W. Nudenberg, *J. Org. Chem.* **15** (1950) 748.
[16] Hercules Powder Co., US 2 663 735, 1949 (L. J. Filar, M. A. Taves).
[17] Distillers Co., DE 861 251, 1950 (L. J. Filar, M.A. Taves).
[18] Phenolchemie, DE 1 112 527, 1959 (H. Sodomann, B. Hauschulz, G. Berg).
[19] Phenolchemie, DE 1 128 859, 1957 (H. Sodomann, R. Tomaschek, M. Hanke).
[20] Distillers Co., GB 805 048, 1956 (M. D. Cooke).
[21] G. Böhme et al., DE 91 643, 1971.
[22] Rütgers, DE 976 125, 1963 (K. F. Lang, H. Schildwächter).
[23] Distillers Co., DE 1 167 353, 1961 (T. Bewley et al.).
[24] Progil Electrochimie, DE 1 146 067, 1958.
[25] W. W. Kaeding, R. O. Lindblom, R. G. Temple, *Ind. Eng. Chem.* **53** (1961) no. 10, 805.
[26] W. W. Kaeding, *Hydrocarbon Process.* **43** (1964) no. 11, 173.
[27] W. W. Kaeding, R. O. Lindblom, R. G. Temple, H. I. Mahon, *Ind. Eng. Chem. Process Des. Dev.* **4** (1965) 1, 97.
[28] Dow Chemical, US 2 727 926, 1954 (W. W. Kaeding, R. O. Lindblom, R. G. Temple).
[29] Dow Chemical, US 2 954 407, 1957 (W. H. Taplin).
[30] Dow Chemical, US 3 235 588, 1962 (C. W. Weaver).
[31] L. van Dierendonck et al., *Adv. Chem. Ser.* **133** (1974) 432; *Eur. Chem. News* (1971) 5. Feb., 28.
[32] M. Taverna, *Riv. Combust.* **22** (1968) no. 4, 203.
[33] I. Donati, G. S. Sioli, M. Taverna, *Chim. Ind. (Milan)* **50** (1968) no. 9, 997.
[34] Snia Viscosa, DE 1 189 977, 1962 (L. Notarbartolo et al.).
[35] G. Navazio, A. Scipioni, *Chim. Ind. (Milan)* **50** (1968) no. 10, 1086.
[36] A. P. Gelbein, A. S. Nislick, *Hydrocarbon Process.* **57** (1978) no. 11, 125.
[37] *Ullmann,* 3rd ed., **13,** 427.
[38] Stamicarbon, GB 753 922, 1954.
[39] Monsanto, US 2 632 028, 1949 (J. F. Adams, R. L. Bauer, G. E. Taylor).
[40] R. Huisgen, J. Sauer, *Angew. Chem.* **72** (1960) 97.
[41] A. Lüttringhaus, D. A. Ambros, *Chem. Ber.* **89** (1956) 463–474.
[42] W. Prahl, W. Mathes, BIOS no. 507, 22.
[43] W. Prahl, W. Mathes, *Angew. Chem.* **52** (1939) 591–592.
[44] Gulf Research & Development Co., US 3 591 644, 1971 (V. A. Notaro, C. M. Selwitz).
[45] Gulf Research & Development Co., US 3 591 645, 1971 (C. M. Selwitz).
[46] C. M. Selwitz, V. A. Notaro: "Symposium on Advances in Petrochemical Technology," *Am. Chem. Society Meeting,* New York, Aug. 1972.
[47] R. L. Rennard Jr., W. L. Kehl: "Symposium on Advances in Petrochemical Technology," *Am. Chem. Society Meeting,* New York, Aug. 1972.

[48] Scientific Design, GB 929 680, 1961.
[49] Scientific Design, GB 939 613, 1961.
[50] Scientific Design, GB 1 002 083, 1962 (C. N. Winnick).
[51] *Chem. Ind. (Düsseldorf)* **19** (1967) June, 383.
[52] *Hydrocarbon Process.* **44** (1965) no. 11, 255.
[53] H. J. Arpe, L. Hörning, *Erdöl Kohle Erdgas Petrochem. Brennst. Chem.* **28** (1970) 2, 79.
[54] Hoechst, DE-OS 1 643 355, 1967 (M. Buldt, H. J. Arpe).
[55] W. B. Deichmann, M. L. Keplinger in G. D. Clayton, F. E. Clayton (eds.): *Patty's Industrial Hygiene and Toxicology*, 3rd ed., vol **2 A,** Wiley-Interscience, New York 1981, pp. 2569–2570.
[56] W. Wirth, G. Hecht, Chr. Gloxhuber: *Toxikologie Fibel*, Thieme Verlag, Stuttgart 1971.
[57] R. Ludewig, K. Lohs: *Akute Vergiftungen*, G. Fischer Verlag, Stuttgart 1974.
[58] B. Neundörfer, E. Wolpert, *M. M. W. Münch. Med. Wochenschr.* **118** (1976) 1177.
[59] B. Baranowska-Dutkiewicz, *Int. Arch. Occup. Environ. Health* **49** (1981) 99–104.
[60] Datenbank für wassergefährdende Stoffe, version June 22, 1988.
[61] W. B. Deichmann, S. Witherup, *J. Pharmacol. Exp. Ther.* **80** (1944) 233–400.
[62] R. M. Bruce, J. Santodonato, M. W. Neal, *Toxicol. Ind. Health* **3** (1987) 535–568.
[63] V. G. Heller, L. Purzell, *J. Pharmacol. Exp. Ther.* **63** (1938) 99–107.
[64] National Cancer Institute USA: "Bioassay of Phenol for Possible Carcinogenicity," *Carcinog. Tech. Rep. Ser. U.S. Natl. Cancer Inst.* 1980, no. 203.
[65] R. K. Boutwell, D. K. Bosch, *Cancer Res.* **19** (1959) 413–424.
[66] M. H. Salaman, O. M. Glendenning, *Br. J. Cancer* **11** (1957) 434–444.
[67] B. L. Van Duuren, B. M. Goldschmidt, *JNCI J. Natl. Cancer Inst.* **46** (1971) 1039–1044.
[68] B. L. Van Duuren, C. Katz, B. M. Goldschmidt, *JNCI J. Natl. Cancer Inst.* **51** (1973) 703–705.
[69] B. L. Van Duuren, B. M. Goldschmidt, *JNCI J. Natl. Cancer Inst.* **56** (1976) 1235–1242.

# Phenol Derivatives

*Individual keywords:* → Aminophenols; → Chlorophenols; → Cresols and Xylenols; → Hydroquinone; → Resorcinol. Chlorohydroxybenzenesulfonic Acids and Aminohydroxybenzoic Acids → Benzenesulfonic Acids and Their Derivatives; Nitrophenols → Nitro Compounds, Aromatic

HELMUT FIEGE, Bayer AG, Leverkusen, Federal Republic of Germany (Sections 1.1–1.5.2)

HEINZ-WERNER VOGES, Hüls Aktiengesellschaft, Marl, Federal Republic of Germany (Sections 1.5.3–1.5.11)

TOSHIKAZU HAMAMOTO, Ube Industries, Ltd., Ube, Japan (Chap. 2)

SUMIO UMEMURA, Ube Industries, Ltd., Ube, Japan (Chap. 2)

TADAO IWATA, Mitsui Petrochemical Industries, Ltd., Yamaguchi, Japan (Chap. 3)

HISAYA MIKI, Mitsui Petrochemical Industries, Ltd., Yamaguchi, Japan (Chap. 3)

YASUHIRO FUJITA, Mitsui Petrochemical Industries, Ltd., Yamaguchi, Japan (Chap. 3)

HANS-JOSEF BUYSCH, Bayer AG, Leverkusen, Federal Republic of Germany (Chaps. 4 and 5)

DOROTHEA GARBE, Haarmann & Reimer GmbH, Holzminden, Federal Republic of Germany (Chap. 6)

WILFRIED PAULUS, Bayer AG, Uerdingen, Federal Republic of Germany (Chap. 7)

| | | | | | |
|---|---|---|---|---|---|
| 1. | Alkylphenols | 3714 | 2. | Catechol | 3758 |
| 1.1. | Physical Properties | 3714 | 2.1. | Physical Properties | 3758 |
| 1.2. | Chemical Properties | 3715 | 2.2. | Chemical Properties | 3759 |
| 1.3. | Occurrence, Formation, Isolation | 3718 | 2.3. | Production | 3760 |
| | | | 2.4. | Uses | 3761 |
| 1.4. | Production, General | 3721 | 2.5. | Economic Aspects | 3762 |
| 1.4.1. | Alkylation of Phenols | 3721 | 2.6. | Toxicology | 3762 |
| 1.4.2. | Other General Processes | 3726 | 3. | Trihydroxybenzenes | 3763 |
| 1.5. | Industrially Important Alkylphenols | 3727 | 3.1. | Pyrogallol | 3763 |
| 1.5.1. | Polymethylphenols | 3727 | 3.2. | Hydroxyhydroquinone | 3765 |
| 1.5.2. | Ethylphenols | 3730 | 3.3. | Phloroglucinol | 3766 |
| 1.5.3. | Isopropylphenols | 3732 | 4. | Bisphenols (Bishydroxyarylalkanes) | 3768 |
| 1.5.4. | sec-Butylphenols | 3735 | | | |
| 1.5.5. | tert-Butylphenols | 3736 | | | |
| 1.5.6. | tert-Pentylphenols | 3745 | 4.1. | Physical Properties | 3768 |
| 1.5.7. | Higher Alkylphenols | 3747 | 4.2. | Chemical Properties | 3770 |
| 1.5.8. | Cycloalkylphenols | 3752 | 4.3. | Production | 3771 |
| 1.5.9. | Aralkylphenols | 3754 | 4.4. | Analysis, Testing, Storage | 3772 |
| 1.5.10. | Alkenylphenols | 3756 | 4.5. | Uses, Economic Aspects | 3773 |
| 1.5.11. | Indanols | 3757 | | | |

| | | | |
|---|---|---|---|
| 4.6. | Toxicology ............... 3774 | 6. | Phenol Ethers ............ 3778 |
| 5. | Hydroxybiphenyls ........ 3774 | 6.1. | Properties ............... 3778 |
| 5.1. | Physical Properties ........ 3774 | 6.2. | Production .............. 3778 |
| 5.2. | Chemical Properties. ....... 3775 | 6.3. | Representative Phenol Ethers. 3779 |
| 5.3. | Production .............. 3775 | 7. | Halogen Derivatives of Phenolic Compounds ....... 3782 |
| 5.4. | Analysis, Quality Specifications, Storage ...... 3776 | 7.1. | Introduction ............ 3782 |
| 5.5. | Uses .................... 3777 | 7.2. | Representative Compounds .. 3785 |
| 5.6. | Toxicology ............... 3777 | 8. | References. .............. 3793 |

# 1. Alkylphenols

In this article alkylphenols are defined as monohydric phenols in which one or more ring hydrogen atoms have been replaced by the same or a different alkyl, alkenyl, cycloalkyl, cycloalkenyl, or aralkyl group, but which have no other substituents. Indanols are treated as disubstituted alkylphenols. For cresols and xylenols, see → Cresols and Xylenols.

In the nomenclature, alkylphenols are usually considered as phenol derivatives; the alkyl groups are ordered alphabetically and numbered starting from the hydroxyl group. In industry, products from cresol or xylenol are often described as derivatives of the corresponding cresols or xylenols, e.g., 6-*tert*-butyl-2,4-dimethylphenol is called 6-*tert*-butyl-2,4-xylenol. A very complicated system which is, e.g., found in handbooks considers alkylphenols as benzene derivatives. Here, the substituents are ordered alphabetically, and numbered such that the sum of all numbers gives the smallest possible value. 6-*tert*-Butyl-2,4-dimethylphenol is thus referred to as 1-*tert*-butyl-2-hydroxy-3,5-dimethylbenzene. According to this system cresol derivatives are also occasionally described as toluene derivatives, e.g., 2,6-di-*tert*-butyl-4-methylphenol is called 3,5-di-*tert*-butyl-4-hydroxytoluene (BHT). Some alkylphenols have common names.

## 1.1. Physical Properties

Pure alkylphenols are colorless liquids or form low-melting crystals, which frequently darken with time. They show a tendency to supercooling. Within groups of substitutional isomers the para isomers have the highest melting points. In some cases two melting points are observed (occurrence of a metastable phase).

The boiling points are > 200 °C at normal pressure and increase with molecular mass. Whereas the boiling point difference between meta und para isomers is very small, the para compounds usually boil 10 – 20 °C higher than the isomeric ortho compounds. The sharper rise in boiling point on para substitution remains if more alkyl groups are present. A particularly sharp rise in boiling point results if a 3- or 5-alkylphenol is *tert*-butylated in the 4-position [1]. This effect is used industrially for the separation of alkylphenols which cannot be separated by distillation.

The relatively small rise in boiling point on ortho substitution is caused by steric hindrance of hydrogen bond formation. This steric hindrance is reflected in many other physicochemical properties of alkylphenols (e.g., spectra, dissociation energies, acidity of the O–H bond, association, dipole moments [2], [3]).

Physical properties of individual alkylphenols are listed in the corresponding sections. Further data for ethylphenols can be found in [4], for isopropylphenols in [5]–[8] tert-butyl- and other phenols in [9]–[11]. Vapor pressure curves for many alkylphenols are given in [12] and thermochemical data under normal conditions in [13]. The relationship between vapor pressure and association of alkylphenols is given in [14].

Alkylphenols are readily soluble in polar organic solvents (lower alcohols, ketones, esters, carboxylic acids); their solubility in hydrocarbons increases with the number and size of the alkyl substituents on the ring. The opposite trend occurs with regard to solubility in water, higher alkylphenols are virtually insoluble and lower alkylphenols are soluble in water only at increased temperatures to a noticeable extent. Their water-solubility shows a temperature-dependence similar to that of phenol, but, with considerably higher critical solution temperatures (i.e., the temperature where unlimited miscibility with water occurs) [15], [16]. Within a group of substitutional isomers the water-solubility decreases with increasing proximity of the alkyl substituents to the hydroxyl group. The presence of other water-soluble organic compounds increases the solubility of alkylphenols and lowers the critical solution temperature [17].

Water can have considerable solubility in alkylphenols [15], [16] and hydrates may be formed, e.g., 6-*tert*-butyl-3-methylphenol, $C_{11}H_{16}O \cdot 1/4\ H_2O$, *mp* 37 °C [10].

## 1.2. Chemical Properties

The chemical behavior of alkylphenols deviates from the reactions typical of phenol with increasing number, size, and degree of branching of the alkyl groups. This change in properties is particularly pronounced in 2,6-di-*tert*-alkylphenols [18]. The alkyl groups themselves can also show typical reactions.

**Acidity.** Alkylphenols are less acidic than phenol. The acidity decreases in the direction of meta > para > ortho substitution. The decrease in acidity due to the presence of ortho alkyl groups becomes more pronounced the more these groups are branched in their α-positions. This is mainly because of steric hindrance of solvation in the phenolate anion formed [2].

These effects are reflected in the solubility of alkylphenols in alkali: 4-*tert*-butylphenol, 2-isopropylphenol, 2,6-diethylphenol, 4-*tert*-butyl-2,5-dimethylphenol, and 4-*tert*-

butyl-2,6-dimethylphenol are soluble in dilute aqueous sodium hydroxide solution, 2-*tert*-butyl-4-methylphenol, 6-*tert*-butyl-2,4-dimethylphenol, 4,6-di-*tert*-butyl-2-methylphenol, 2,6-di-*sec*-butyl-4-methylphenol, and 2,4,6-triisopropylphenol are insoluble in alkali. The latter are, however, soluble in the so-called Claisen solution (350 g potassium hydroxide in 250 g water made up to 1 L with methanol). In contrast to these partially sterically hindered alkylphenols, phenols with two bulky alkyl groups (e.g., *tert*-butyl, *tert*-pentyl) in the ortho and one alkyl group in the para position are also insoluble in the Claisen solution [19]. Examples of such sterically hindered (shielded) phenols are 2,6-di-*tert*-butyl-4-methylphenol, 2,6-di-*tert*-pentyl-4-methylphenol, and 2,4,6-tri-*tert*-butylphenol. The phenolates of these compounds can only be produced under special conditions [20], e.g., by treatment with solid potassium hydroxide and azeotropic removal of the water by an inert organic entrainer [21]. o-*tert*-Butylphenols without para alkyl groups behave differently; for example 2-*tert*-butylphenol and 6-*tert*-butyl-3-methylphenol are soluble in aqueous sodium hydroxide, and 2,6-di-*tert*-butylphenol is soluble to a certain extent in the Claisen solution. This is presumably because alkali salts of the corresponding tautomeric cyclohexadienones can be formed relatively easily. The insolubility of pentamethyl-, 4-*tert*-octyl, 4-nonyl,- or 4-dodecylphenol in cold aqueous alkalis is apparently due to the increase in the hydrophobic nature of the molecule because of the increasing number or length of the alkyl groups. These phenols are readily soluble in the Claisen solution. Alkylphenols which are insoluble in aqueous alkali but readily soluble in the Claisen solution are called cryptophenols.

The different solubility of the alkylphenols in alkalis can be used for their separation. As a result of the low acidity of the alkylphenols, they can even be liberated from their alkali phenolate solutions by carbon dioxide (p$K$ 6.4).

**Ether Formation.** The hydroxyl group can be etherified, for example, with alkyl halides, halogenated carboxylic acids, or dialkyl sulfates in aqueous alkaline media. Bulky ortho alkyl groups hinder the reaction. In these cases the etherification can be forced by, for example, treating the anhydrous alkali phenolate with dimethyl sulfate in a solvent. If phase-transfer catalysts are used, it is possible to work in the presence of water. An example is the etherification of 2,4,6-tri-*tert*-butylphenol with dimethyl sulfate in aqueous sodium hydroxide–dichloromethane in 93% yield in the presence of phase-transfer catalysts [22]. Alkylene oxides add to the hydroxyl group in the presence of basic catalysts to form alkylphenolmono- and polyglycol ethers (alkoxylation). The hydroxyl group can be etherified with alkenes in the presence of Friedel–Crafts catalysts under mild conditions.

**Esterification.** The hydroxyl group can generally be esterified successfully with acid anhydrides or acid chlorides. In the case of sterically hindered phenols, conditions have to be modified as with etherification [19]. With the exception of 2,6-di-*tert*-alkylphenols the hydroxyl groups of the alkylphenols react with isocyanates to form urethanes, which can be used for identification of the alkylphenols [23].

**Substitution of the Hydroxyl Group.** At 250 °C and in the presence of palladium – charcoal catalysts the OH group can be exchanged to yield the corresponding dialkylaniline [24]. This reaction occurs even with dialkylphenols such as 6-*tert*-butyl-2-methylphenol.

**Hydrogenation.** The hydrogenation of alkylphenols with hydrogen in the presence of Raney nickel or noble-metal catalysts leads to alkylcyclohexanols and/or alkylcyclohexanones. The latter are formed particularly in the case of sterically hindered alkylphenols [25], [26]. At high temperatures (250 – 350 °C) and in the presence of transition-metal – aluminum oxide catalysts, hydrogenolytic elimination of the alkyl groups and the hydroxyl group can also occur [27].

**Oxidation.** The coloration of alkylphenols in the presence of air indicates that they are sensitive to oxidation. Oxidation generally begins with the formation of phenoxy radicals. These radicals are relatively stable in the case of sterically hindered alkylphenols and are the reason why these phenols are particularly strong autoxidation inhibitors. In the further course of the oxidation, many different compounds, some intensely colored, can be formed: hydroquinones, quinols, quinones, ketones, aldehydes, diphenoquinones, di-, tri-, and polyphenols, di-, tri-, and polyethers, and other C – C and/or C – O-linked products. Depending on the reaction conditions, oxidizing agent, catalyst, and nature and position of the alkyl group(s), certain products are formed preferentially [28] – [30]. Thus the formation of diphenoquinones is observed particularly in the case of 2,6-dialkylphenols with bulky alkyl groups. The 4-methyl group of 2,6-dialkyl-4-methylphenols can be often oxidized selectively to an aldehyde group (e.g., with pentyl nitrite, *tert*-butyl hydroperoxide – bromine, or oxygen) [28], [31], [32]. Strong oxidizing agents degrade the phenyl ring, e.g., the oxidation of alkylated phenols with potassium permanganate gives fatty acids.

**Electrophilic Aromatic Substitution.** Alkylphenols readily undergo electrophilic substitution. Substitution generally occurs in the positions ortho or para to the hydroxyl group, provided that they are not already occupied by other substituents. In particular when several or bulky alkyl groups are present, substitution (displacement) of alkyl groups by the electrophile can occur in addition to the normal substitution of hydrogen, or the reaction may stop at the stage of the cyclohexadienone, formed initially by addition of the electrophile. These types of anomalies are observed, for example, in nitrations [33], chlorinations [34], and in the reaction with trichloromethane – alkali (Reimer – Tiemann aldehyde synthesis) [35]. The nitration occurs even with dilute nitric acid and requires inert dilution agents and the careful maintenance of low temperatures to avoid undesired secondary reactions or a dangerous reaction process [33]. Nitrosation [36], chlorination [34], alkylation [37], and sulfonation, for example, take place with similar ease. The nitroso group preferably enters the ring para to the OH group. The nitrosophenols can also be present in the tautomeric quinone oxime form.

On heating alkylphenols, particularly in the presence of Friedel–Crafts catalysts, isomerization, transalkylation, and dealkylation can occur [38]. The tendency towards these reactions increases sharply in the order primary < secondary < tertiary alkyl groups. The isomerizations proceed until an equilibrium state is reached in which the thermodynamically more stable 3- and 3,5- (i.e., meta) isomers prevail. Groups in the para and ortho positions are cleaved particularly easily, those in the meta position less easily. Dealkylation can also occur during distillation of alkylphenols, if acids or potential acid-forming substances in the starting material are not completely removed or rendered ineffective (by addition of alkali). 1,1,3,3-Tetramethyl-5-indanol—which can be considered as ring-closed 3,4-di-*tert*-butylphenol and generally behaves as a typical phenol—however, shows remarkable stability towards dealkylation even in the presence of acids. The thermal stability of other indanols is also much higher than that of alkylphenols.

Heating alkali phenolates of alkylphenols with free ortho or para positions in the presence of carbon dioxide in polar, aprotic solvents such as dimethylformamide leads to hydroxyalkylbenzoic acids (Kolbe–Schmidt synthesis) [39]. Aldehyde groups can be very selectively introduced into the position para to the hydroxy group with zinc cyanide–hydrochloric acid. If this position is occupied, introduction into the ortho position occurs (Gattermann–Adams aldehyde synthesis) [40]. In the presence of alkali, formaldehyde reacts at room temperature with free ortho and para positions of alkylphenols to form hydroxymethyl-alkylphenols. Under acidic conditions or at higher temperatures these hydroxymethyl compounds condense with alkylphenols to form bis(hydroxyalkylphenyl)-methanes, and undergo autocondensation to yield high molecular mass alkylphenol–formaldehyde resins [41], [42]. The condensation of alkylphenols with formaldehyde in the presence of primary or secondary amines leads to Mannich bases [42]. The acidic condensation of alkylphenols with other aldehydes or ketones gives bis(hydroxyalkylphenyl)alkanes (see Section 4.3). Thiobisphenols are readily obtained with sulfur dichloride [43] and with disulfur dichloride, dithiobisphenols [44].

Alkenylphenols show a tendency towards dimerization and polymerization.

## 1.3. Occurrence, Formation, Isolation

Some alkylphenols such as 4-ethylphenol, 4-isopropylphenol (*p*-cumenol), 6-isopropyl-3–methylphenol (thymol), 5-isopropyl-2-methylphenol (carvacrol), 2,6-diisopropyl-3-methylphenol, 2,4-diisopropyl-5-methylphenol, 2,5-diisopropyl-4-methylphenol, and 4-allylphenol (chavicol) are components of natural essential oils, some of which have been used for centuries for the production of perfumes. 2,3,4-Trimethylphenol, 5-ethyl-2-methylphenol, and 2,3-xylenol were detected in the defensive secretions of opilionides [45], and 2,6-di-*tert*-butyl-4-methylphenol is contained in Australian brown coal [46]. 3-(Pentadecen-8-yl)phenol, 3-(pentadecadien-8,11-yl)phenol, and 3-(pentadecatrien-

**Table 1.** Alkylphenol content in bituminous coal tars

| Alkylphenols | Low-temperature carbonization, wt% | Coking, wt% |
| --- | --- | --- |
| Ethylphenols | 42 | 47 |
| Trimethylphenols | 24 | 13 |
| Ethylmethylphenols | 13 | 15 |
| Propylphenols | 10 | 15 |
| Methylpropylphenols | 0.5 | 5 |
| Tetramethylphenols | 1 | 0.5 |
| Diethylphenols |  | <0.5 |
| Butylphenols |  | <0.5 |
| Indanols | 9 | 4 |

8,11,14-yl)phenol have been identified as the main components of cardanol extracted from cashew nutshell oil on an industrial scale [47]. 3-Pentadecylphenol (hydrocardanol) is obtained from it by hydrogenation [48].

Alkylphenols are obtained in great number in the *purely thermal cracking* (low-temperature carbonization, coking), in the *oxidizing thermal cracking* (gasification), and the *hydrogenating thermal cracking* of natural products such as bituminous coal, brown coal, peat, wood, lignin, and in the *coking thermal cracking* of crude oil fractions (→ Cresols and Xylenols). Thus many different alkylphenols were detected in the various coal tars, some of them with several methyl, ethyl, *n*-propyl, isopropyl, *n*-butyl, isobutyl, *sec*-butyl, allyl, 1-cyclopentenyl, cyclohexyl, and/or indanoyl groups and various indanols and indenols. The unsaturated compounds are found in small quantities, mainly in low-temperature tars. Without taking the cresols and xylenols into account, about a hundred different monohydric phenols in total have been found in the various coal tars [49]. For the history of tar phenols see [50], [51].

The product distribution of the various cracking processes differs in the nature, but more in the ratio and particularly in the quantity in which the alkylphenols are produced, which depends on the starting material and the operating conditions. The mass ratio range of alkylphenols isolated from bituminous coal tars is given in Table 1.

More details concerning product composition in bituminous coal coking are given in [52], and in various bituminous coal lowtemperature carbonization processes in [53], [54].

Brown coal low-temperature tar contains a relatively high proportion of ethylphenols [54], [55]. Alkylphenols isolated from crude oil have a higher content of phenols with longer side chains, e.g., propyl- and butylphenols [56], [57]. In the hydrogenation of coal many indanols are formed [58], [59].

The strong influence of the cracking process on the total amount of monohydric phenols produced and on the proportion of alkylphenols among them is shown in Table 2.

In the distillative workup of the crude phenols obtained in the different cracking processes (→ Cresols and Xylenols), 2-ethylphenol (*bp* 204.3 °C) is distilled over with the *m*-/*p*-cresol fraction. The residue, which has a very complex composition, is usually worked up by batch distillation. Since the boiling points of the alkylphenols lie very

**Table 2.** Phenol yields in the hydrogenation, low-temperature carbonization, and coking of bituminous coal, and in the cracking of gas oils

| Reforming process | kg phenols per t $H_2O$-free raw material | |
|---|---|---|
| | Total * | Alkylphenols ** |
| Hydrogenation | 75 | 25 |
| Low-temperature carbonization | 15 | 3 |
| Coking | 0.7 | 0.04 |
| Cracking | 1 | 0.04 |

\* Phenol + cresols + xylenols + alkylphenols (without phenol ethers etc.).
\*\* Excluding cresols and xylenols.

close together and overlap to some extent with those of the xylenols and other compounds which may be present (e.g., phenol ethers), the production of certain alkylphenols in pure form is not possible by distillation. The products in the distillation range of ca. 205–285 °C are therefore separated only into particular fractions which depend on the application purposes and are characterized by their boiling ranges.

The distillate boiling between ca. 205–225 °C is generally called the xylenol fraction. If it originates from high-temperature bituminous coal tar, it can contain up to 30% alkylphenols, predominantly 3- and 4-ethylphenol and methylethylphenols [52]. In xylenol fractions from East German brown coal low-temperature tars the xylenol content can be below 30% and the 3-/4-ethylphenol content 40% [60]. A 216–218 °C xylenol fraction obtained from brown coal low-temperature tar phenols for the varnish industry contains predominantly 3- and 4-ethylphenol.

For the separation of 3- and 4-*ethylphenol* the same processes as for the separation of *m*- and *p*-cresol must be used (→ Cresols and Xylenols). Similar to *p*-cresol, 4-ethylphenol can be separated by esterification with oxalic acid [61]. The aforementioned 216–218 °C brown coal xylenol fraction, for example, is a suitable starting material for isolation of 3- and 4-ethylphenol.

*2-Ethylphenol* can be separated from the fraction boiling between 203 and 207 °C (in which it is enriched) as the quinaldine addition compound [62]. Other alkylphenols which can be enriched in narrow-boiling fractions and isolated pure by special methods, include 2,3,5-trimethylphenol (isopseudocumenol) [63], 2,3,6-trimethylphenol [64], 2,4,6-trimethylphenol (mesitol), 5-ethyl-3-methylphenol, 4-indanol, and 5-indanol [65].

Most of the alkylphenols with a boiling point > 230 °C are used as a mixture. The composition of an alkylphenol fraction boiling between 230 and 285 °C obtained from vertical retort bituminous coal tar is given in [66].

## 1.4. Production, General

### 1.4.1. Alkylation of Phenols

Alkylphenols are produced industrially mainly by catalytic alkylation of phenol, cresols, or xylenols with olefins which are readily accessible petrochemicals. Alcohols or halogenated hydrocarbons are only used as the alkylating agent if they are more readily available (e.g., cyclohexanol) or if the corresponding olefin does not exist (e.g., methanol, benzyl chloride). Other production processes for alkylphenols are less important than alkylation.

**Ortho/Para Alkylation and Meta Isomerization in the Presence of Friedel–Crafts Catalysts** [67]–[72]. The following Friedel–Crafts catalysts are mainly used in industry: sulfuric acid, phosphoric acid, *p*-toluenesulfonic acid, strongly acidic cation-exchange resins (e.g., Amberlite JR-112, Lewatit S 100), acid-activated silicates of the Montmorillonite type (bleaching clays), synthetic aluminum silicates of the crack catalyst type, boron trifluoride, aluminum chloride, iron(III) chloride, and zinc chloride.

The catalysts react with the alkylating agent to form an alkyl cation which is as branched (stable) as possible. The alkyl cation enters the nucleus preferentially in the positions ortho and para to the OH group according to the rules of electrophilic substitution. In addition, particularly under mild conditions, etherification of the OH group occurs. For the mechanism see [67]–[69].

The ratio of isomers formed initially is kinetically controlled and determined (1) by the position and inductive effect of the methyl (alkyl) group(s) already on the ring; (2) the reactivity of the alkyl cation; (3) the steric situation on the ring; and (4) the incoming alkyl group. Since the monoalkylation products are generally more reactive than the starting phenol, they are further alkylated, provided that there is no steric hindrance. The ratio of mono- to di- and trialkylphenols is proportional to the ratio of phenol to alkylating agent.

During the further course of the reaction, rearrangements (isomerizations, transalkylations, disproportionations) occur and the product composition shifts in the direction of a thermodynamically more stable equilibrium state, in which phenols with meta alkyl groups finally prevail. During this process phenol ethers first rearrange to alkylphenols, the amount of di- and trialkylphenols is lowered, and the proportion of o-alkylphenols is shifted in favor of the p-alkylphenols, provided that the para position is free and is not sterically hindered.

Possible side reactions are (1) dealkylations; (2) di-and oligomerizations of the olefin used as alkylating agent or formed by dealkylation; (3) isomerization and cleavage in the alkyl groups (particularly in the case of larger and branched alkyl groups or olefins); (4) readdition of the olefins formed by cleavage, di- or oligomerization, (5) dehydrogenation of the alkyl to alkenyl groups and their reaction with olefins to form indanols; and (6) reactions of the catalyst with reaction components, e.g., the formation of phenolsulfonic acids and alkylsulfonates if sulfuric acid is used as catalyst. Ortho and para alkyl groups are much more prone to rearrangement and cleavage reaction than meta alkyl groups. The tendency towards rearrangement and cleavage reactions of ortho and para alkyl groups increases in the order of primary < secondary < tertiary alkyl groups. Methyl groups can only be made to migrate with great difficulty.

How fast a reaction mixture reaches thermodynamic equilibrium and the amount of side reactions that occur, depends on the nature of the phenol and that of the alkylating agent and on the severity of the reaction conditions, in particular on the activity (type) and quantity of the catalyst, on the temperature, the reaction time, and occasionally also on the water content of the reaction mixture. The reaction conditions must be carefully adjusted to achieve optimal product compositions.

High yields of a particular alkylphenol are often possible by making use of the rearrangement equilibria, so that the undesired ortho-/para-mono-, di-, and/or trialkylphenols formed as byproducts are separated from the desired product and recycled to the alkylation reaction. This is, however, not the case for 2-alkylphenols and 2,6-dialkylphenols: these compounds can only be obtained with the coproduction of considerable quantities of the corresponding phenols alkylated in the 4-position even under carefully controlled alkylation conditions and skillfully exploited rearrangement conditions. This is because, as previously indicated, ortho and para alkylations always occur simultaneously in the presence of Friedel–Crafts catalysts and the ortho isomer readily rearranges to the para isomer which is often thermodynamically more stable. This behavior has been intensively investigated in the *tert*-butylation of phenol with isobutene [73]–[78]. Good yields of 2-alkyl- or 2,6-dialkylphenols are possible if the para position is blocked by an alkyl group which will not migrate under the reaction conditions, e.g., a methyl group.

**Ortho-Alkylation.** In the presence of *aluminum phenolates*, phenols can be selectively alkylated with olefins in the ortho position. The catalyst is produced in the alkylation reactor before the addition of the olefin by heating the dry phenol with 1–2 wt% aluminum sand, powder, or turnings to 120–180 °C (with hydrogen formation). Depending on the olefin–phenol ratio, 2-alkyl- or 2,6-dialkyl-phenols are preferentially formed in the alkylation; at very high ratios 2,4,6-trialkylphenols are also formed. Currently 2,6-dialkylphenols, in particular, are produced using this process, which was developed independently by both Bayer [79] and Ethyl Corp. [80] in 1954. An overview of the reaction is given in the publications [81]–[84].

It is assumed that ortho selectivity is caused by the formation of a strongly acidic complex (**2**) from the aluminum phenolate (**1**) and the olefin, which directs the electrophilic attack of the olefin to the sterically more favorable position ortho to the OH group. The reaction ends with the reversal of the aluminum phenolate formation by proton exchange with excess phenol [85]:

al = Al/3

Apart from the special ortho-directing effect and a somewhat lower activity of the aluminum phenolate, which can be compensated for by using higher temperatures, the process is very similar to the alkylation in the presence of Friedel–Crafts catalysts. Thus the reactivity of the olefins decreases in the order $R_2C=CH_2 > R-CH=CH_2 \approx R-CH=CH-R > CH_2=CH_2$, and the alkyl group with maximum possible branching is formed. The ortho alkylation is also kinetically controlled and at the beginning of the reaction is accompanied by reversible ether formation which increases with increasing temperature. At too high a temperature and with longer reaction times dealkylation and the formation of para isomers also occur, especially if the alkylation has taken place more readily. Therefore, if a high yield of the ortho isomer is desired, it is necessary to work at the lowest possible temperature (which must nevertheless be sufficiently high for rapid ether cleavage) and at a high olefin concentration, and to stop the reaction at the optimum point in time. If lower olefins are used as alkylating agents, it is necessary to work under pressure in order to achieve a high olefin concentration [81], [82]. To avoid dealkylation and isomerization in the subsequent distillative workup, the aluminum phenolate catalyst is deactivated by, for example, addition of water or sodium hydroxide to the reaction mixture after the end of the reaction [86]. For ethyl- and sec-alkylphenols it may be possible to do without the catalyst removal, if distillation is carried out very carefully.

The ortho-selective effect of the aluminum phenolate is also seen in transalkylations. Thus under kinetically controlled conditions 2,4-di-*tert*-butylphenol and phenol react to form predominantly 2-*tert*-butylphenol [87]. In the case of dealkylation in the presence of aluminum phenolate, ortho-*tert*-alkyl groups are eliminated more rapidly than those in the para position [82], [88].

The aluminum phenolate catalyst can be obtained by treating phenol with aluminum or with reactive aluminum compounds such as aluminum alkoxides, aluminum trialkyls [89], dialkylaluminum chloride [90], or aluminum chloride [91]. To achieve smooth dissolution of aluminum, it can be activated by addition of mercury(II) chloride [81]. If toluenesulfonic acid is added during the aluminum phenolate, a hydrolysis-resistant mixed catalyst of similar ortho-selectivity is obtained [92]. Aluminum salts of the ortho phenolsulfonic acids are also suitable catalysts. Since they are not as readily soluble in the reaction mixture as the aluminum phenolates, they can be filtered off after the reaction has finished and reused [93]. Also compounds such as aluminum thiophenolate [94], aluminum diphenyldithiophosphate [95], polyphenoxy aluminum oxanes [96], and aluminum phenolate bonded to phenol–formaldehyde resin [97] have been proposed as ortho-alkylation catalysts.

In addition to aluminum phenolate which is used exclusively in industry, the phenolates of, e.g., magnesium, zinc, iron [79], [80], zirconium, hafnium, niobium, tantalum [98], gallium [99], and boron (aryl borates) [100], also have a pronounced ortho-selective effect. Titanium only shows this effect as the catecholate [101], or as the titanium phenolate – arylsulfonate mixed salt [102]. Zirconium, hafnium, niobium, and tantalum phenolates split off the 6-*tert*-butyl group from 3-alkyl-4,6-di-*tert*-butylphenols very selectively, whereas titanium phenolates are unsuitable for this partial dealkylation [103], [104].

*γ-Aluminum oxide* [105] and other modifications of aluminum oxide (apart from α-Al$_2$O$_3$) [106] also direct the alkyl group very selectively into the ortho position, if olefins are reacted with phenols in the liquid phase at 200 – 350 °C and 2 – 20 MPa. In industry these types of aluminum oxide are used particularly for the production of ortho-monoalkylphenols, because they lead to a clearly higher ortho-mono/ortho-dialkylphenol ratio than aluminum phenolate if the same olefin/phenol ratio is used (e.g., 1:1). Moreover, they have the advantage over aluminum phenolate—unless it is fixed in a resin [97]—that they can be readily separated after the reaction by, for example, filtration, and reused or used in a fixed-bed form in continuous processes. For the effect of the double bond position in the olefin on the yield of ortho-alkylphenols on γ-Al$_2$O$_3$, see [107].

High selectivities were observed, for example, with oxides of the elements of groups 15 and 17 of the Periodic Table [108], and with high-purity silicon dioxide [109]. The acitivity of the metal compounds can be increased by applying them to a carrier such as Al$_2$O$_3$ or silicates [108], [110].

Relatively high ortho selectivities are also observed if olefins are heated with phenols *in the absence of catalysts* in the liquid phase to 300 – 400 °C under pressure [111]. This so-called *thermal alkylation* has been known for a long time [112], but has so far remained unimportant industrially. The conversions are moderate and the yields based on olefin are poor. Long-chain linear α-olefins become bonded to the phenol preferentially at their C-2 atoms to give ortho-monoalkylphenols. Unlike the situation with Friedel – Crafts catalysis, bonding at C-1 has been observed, but hardly any at the middle C-atoms of the olefin chain, and no para substitution [113]. It is assumed that the thermal ortho-alkylation takes place via a 6-membered, cyclic transition state, into which the phenolic OH group is incorporated [111], [113].

**Alkylation Rules.** The position that an alkyl group can occupy in the phenyl ring depends on the size of the entering group and on the alkyl group already present. For cresols and xylenols the following rules are generally valid:

1) *Secondary alkyl groups* such as isopropyl, *sec*-butyl, cyclopentyl, and cyclohexyl can enter the ring between the hydroxyl and a meta-methyl group, but not between two methyl groups meta to each other. For example, propylation of 3,5-xylenol can yield 2,6-diisopropyl-3,5-dimethylphenol, but not 4-isopropyl-3,5-dimethylphenol.

2) *Tertiary alkyl groups* such as *tert*-butyl, *tert*-pentyl, and *tert*-octyl are unable to enter the ring between the hydroxyl and a methyl group meta to it. This means that, for example, 3,5-xylenol cannot be *tert*-butylated.
3) Unlike *sec*-alkyl groups, *tert*-alkyl groups do not migrate to positions next to a para methyl group. 4-Methyl-3-*tert*-butylphenol therefore cannot be synthesized by isomerization of 4-methyl-2-*tert*-butylphenol in the presence of alkylation catalysts.

For the polyalkylation of phenols with various alkyl groups, the most sensitive (tertiary) group must be introduced last. For example, for the production of 4-*tert*-butyl-2-cyclopentylphenol the *sec*-cyclopentyl group is first introduced (with aluminum phenolate) and then the *tert*-butyl group (with a Friedel–Crafts catalyst).

**General Information on Alkylation of Phenol with Olefins.** Alkylation is carried out predominantly in the liquid phase without a solvent. In the presence of Friedel–Crafts catalysts, normal or slightly increased pressure and a temperature range of ca. 30–150 °C are generally used. Higher pressures and/or temperatures can be necessary for the meta isomerization or for alkylations in the presence of ortho-selective catalysts (see below). The plant is generally inerted with nitrogen before alkylation.

To be able to adapt to the different starting materials and changing market demands with regard to the nature and quality of the alkylphenol, most alkylphenols are produced batchwise in multipurpose plants. For compounds produced in large quantities such as thymol, 2-*sec*-butylphenol, 4-*tert*-butylphenol, 2,6-di-*tert*-butylphenol, 2,6-di-*tert*-butyl-4-methylphenol, octylphenol, and nonylphenol there is a trend towards specialized continuous or semibatch plants.

The choice of catalyst depends, e.g., on the reactivity of the phenol and of the alkylating agent, the desired product composition (isomer distribution, byproducts), the cost, or the available possibilities with regard to the corrosion resistance of the plant, catalyst removal from the reaction mixture, disposal of the catalyst and/or its decomposition products, workup of the product mixture, the quantity of the product, and its value.

*Protonic acids* such as $H_2SO_4$ are readily available, cheap, and so active that small quantities (1–2 wt%) are sufficient to obtain high space–time yields at relatively low temperatures. However, they are corrosive and must be removed or rendered ineffective in order to obtain a stable, distillable product. This requires careful scrubbing of the alkylphenol with alkali hydroxides and water, thus producing wastewater which must subsequently be treated. A similar situation exists if metal halides and metal phenolates are used.

*Solid inorganic catalysts* (acid-activated bleaching clays, oxides of aluminum and silicon etc.) must be added in finely divided form at higher temperatures and in larger quantities (2–5 wt%, for isomerizations even up to 10 wt%) to compensate for their lower activity. They are, however, less corrosive than protonic acids and have the advantage in process technology that they can be separated simply by filtration. However, filtration must be carried out carefully because catalyst particles which have passed through the filter can lead to dealkylation at high distillation temperatures. The disposal of the depleted catalyst loaded with phenol can be problematic.

*Cation exchangers* [71] are reusable (long service life), highly selective for monoalkylations, noncorrosive, can be readily separated from the alkylated phenol, and do not produce any wastewater. These advantages are best realized if they are used in continuous alkylations as bead polymers arranged in a fixed bed. When used in suspended form they can lead to formation of abrasion particles, which must be completely separated or neutralized by addition of alkali to the alkylated phenol, so that dealkylation does not occur in the subsequent distillations. The upper operating temperature of the cation exchanger is ca. 140 °C; at this temperature, however, clear loss of activity must be taken into account in continuous operation. Powder cation exchangers have a higher activity and are more effective for dialkylation than the bead exchangers [69].

After separation of the catalyst the reaction mixture is worked up by fractional distillation in vacuum. Unreacted starting material, undesired ortho and para isomers, and high-boilers are recycled to the reaction as far as possible. Isolation of a particular isomer is often laborious and frequently unnecessary: for many industrial applications alkylphenols are only separated from starting material and high-boilers by distillation, and are used as a mixture of isomers. Occasionally, the separation of high-boiling substances is not required, and stripping of the alkylation product is sufficient as the workup step. In special cases even the catalyst separation and distillation are not performed and the crude alkylation product is led directly into the subsequent reaction [114].

## 1.4.2. Other General Processes

Alkylphenols which are not readily accessible or totally inaccessible by alkylation, e.g., those with long primary alkyl groups ($\geq 3$), can be obtained as follows:

*Reduction of acylphenols* (phenolketones) which are obtained by Friedel–Crafts acylation of phenols or by the Fries rearrangement of phenol esters [115], [116]:

$$R-CO-C_6H_4-OH \xrightarrow{4[H]} R-CH_2-C_6H_4-OH + H_2O$$

*Alkali fusion of alkylbenzenesulfonates*, which are obtained by sulfonation of alkylbenzenes [117] ($\rightarrow$ Cresols and Xylenols):

$$R-C_6H_4-SO_3Na + 2\,NaOH$$
$$\longrightarrow R-C_6H_4-ONa + Na_2SO_3 + H_2O$$

*Cleavage of bis(4-hydroxyphenyl)alkanes*, either alkali-catalyzed [118], [119] with hydrogen in the presence of a hydrogenation catalyst [120], [121], or with a hydrogen donor such as cyclohexanol:

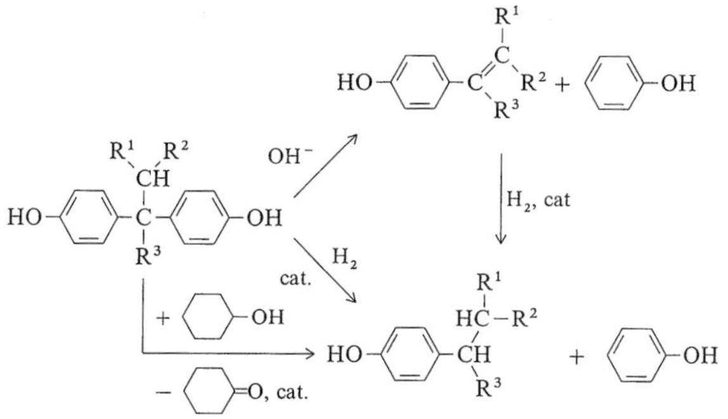

*Claisen rearrangement of allyl aryl ethers* into allylphenols [122], if necessary with subsequent rearrangement into vinylphenols [123] or hydrogenation to alkylphenols:

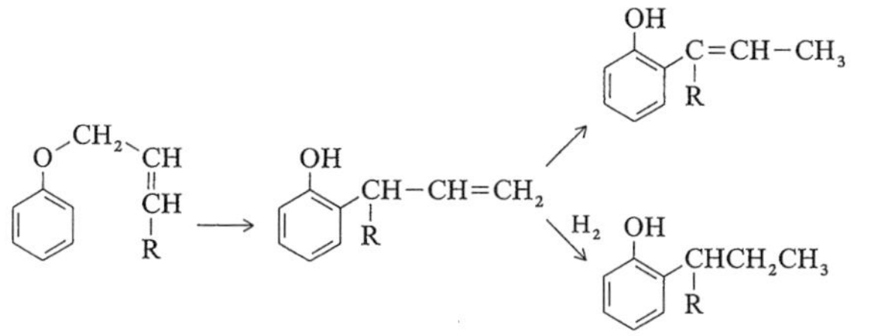

The Claisen rearrangement is particularly suitable for the introduction of alkyl groups into the position ortho to an OH group, the other methods are preferentially used for the production of para-alkylphenols.

Other possibilities can be found in Houben–Weyl (see general references). Alternatives to phenol alkylation are also discussed in the following section.

# 1.5. Industrially Important Alkylphenols

## 1.5.1. Polymethylphenols

2,3,6-Trimethylphenol, 2,4,6-trimethylphenol, and other polymethylphenols are formed in small quantities as byproducts in the production of cresol and particularly xylenol by methylation of phenol (→ Cresols and Xylenols, → Cresols and Xylenols).

**2,3,6-Trimethylphenol** [2416-94-6] (physical properties see Table 3) is produced selectively by gas phase methylation of *m*-cresol with methanol at 300–460 °C under normal pressure on ortho-selective metal oxide catalysts, as used for the selective methylation of phenol (→ Cresols and Xylenols). The reaction occurs in multitube reactors with a fixed catalyst. The reaction temperature to be maintained depends on the catalyst used in each case. Iron oxide catalysts modified with oxides of other metals (e.g., Zn [124], Cr and Sn [125], or Mg and Si [126]) are particularly suitable for the

# Phenol Derivatives

**Table 3.** Physical properties of some polymethylphenols and ethylphenols

| Name (common name) | Formula | CAS registry number | $M_r$ | mp, °C | bp at 101.3 kPa, °C | bp at 2.66 kPa, °C | $\Delta H_V$ at bp 101.3 kPa, kJ/mol | $d_4$, g/cm³ (°C) | $n_d$ (°C) | pK at 20 °C |
|---|---|---|---|---|---|---|---|---|---|---|
| **Polymethylphenols** | | | | | | | | | | |
| 2,3,4-Trimethylphenol | $C_9H_{12}O$ | [526-85-2] | 136.19 | 81 | 235–237 | 125.6 | | | | 10.72 |
| 2,3,5-Trimethylphenol (isopseudocumenol) | $C_9H_{12}O$ | [697-82-5] | 136.19 | 93.7 | 235.3 | 158 | 50.0 | 0.963 (80) | | |
| 2,3,6-Trimethylphenol | $C_9H_{12}O$ | [2416-94-6] | 136.19 | 62 | 226 | $^a$(13.3) | | | | |
| 2,4,5-Trimethylphenol (pseudocumenol) | $C_9H_{12}O$ | [496-78-6] | 136.19 | 72 | 232 | 124 | | | 1.5070 | 10.72 |
| 2,4,6-Trimethylphenol (mesitol) | $C_9H_{12}O$ | [527-60-6] | 136.19 | 69 | 220.4 | 110 | | | 1.5105 (65) | |
| 3,4,5-Trimethylphenol | $C_9H_{12}O$ | [527-54-8] | 136.19 | 109 | 248 | 141.5 | | | | 10.51 |
| 2,3,4,5-Tetramethylphenol (prehnitenol) | $C_{10}H_{14}O$ | [488-70-0] | 150.21 | 82–87 | 256–260 | | | | | |
| 2,3,4,6-Tetramethylphenol (isodurenol) | $C_{10}H_{14}O$ | [3238-38-8] | 150.21 | 79–81 | 250 | | | | | |
| 2,3,5,6-Tetramethylphenol (durenol) | $C_{10}H_{14}O$ | [527-35-5] | 150.21 | 119 | 248.6 | 126.6 | | | | |
| Pentamethylphenol | $C_{11}H_{16}O$ | [2819-86-5] | 164.24 | 128 | 267 | | | | | |
| **Ethylphenols** | | | | | | | | | | |
| 2-Ethylphenol | $C_8H_{10}O$ | [90-00-6] | 122.16 | -3.3; -28$^b$ | 204.5 | 100.6 | 48.2 | 1.0146 (25) | 1.5335 (20) | 10.47 |
| 3-Ethylphenol | $C_8H_{10}O$ | [620-17-7] | 122.16 | -4.5 | 218.4 | 114.4 | 50.9 | 1.0076 (25) | | 10.17 |
| 4-Ethylphenol | $C_8H_{10}O$ | [123-07-9] | 122.16 | 45.1 | 218 | 114.3 | 50.9 | 1.054 (25)$^c$ | 1.5239 (25)$^d$ | 10.38 |
| 2,4-Diethylphenol | $C_{10}H_{14}O$ | [936-89-0] | 150.21 | | 229 | | | 0.9794 (25) | 1.5218 (25) | |
| 2,6-Diethylphenol | $C_{10}H_{14}O$ | [1006-59-3] | 150.21 | 38–39 | 218 | 112 | | | 1.5105 (50) | |
| 3,5-Diethylphenol | $C_{10}H_{14}O$ | [1197-34-8] | 150.21 | 77 | 248 | 131 | | | 1.5033 (75) | 10.31 |
| 3-Ethyl-2-methylphenol | $C_9H_{12}O$ | [1123-73-5] | 136.19 | 71 | 227 | | | | | |
| 4-Ethyl-2-methylphenol | $C_9H_{12}O$ | [2219-73-0] | 136.19 | | 224.5 | | | | | |
| 6-Ethyl-2-methylphenol | $C_9H_{12}O$ | [1687-64-5] | 136.19 | 147 | 213 | | | | | |
| 4-Ethyl-3-methylphenol | $C_9H_{12}O$ | [1123-94-0] | 136.19 | 26 | 228–230 | | | | | |
| 5-Ethyl-3-methylphenol | $C_9H_{12}O$ | [698-71-5] | 136.19 | 51.6 | 235.7 | 129 | 51.3 | | 1.5293 (20)$^d$ | |
| 6-Ethyl-3-methylphenol | $C_9H_{12}O$ | [1687-61-2] | 136.19 | 43.4 | 224.2 | | | | | |
| 2-Ethyl-4-methylphenol | $C_9H_{12}O$ | [3855-26-3] | 136.19 | 25.7 | 222.3 | | | | | |
| 3-Ethyl-4-methylphenol | $C_9H_{12}O$ | [6161-67-7] | 136.19 | 33 | 234–235 | | | | | |

$^a$ At 13.3 kPa.
$^b$ Metastable.
$^c$ Solid.

process. At temperatures of ca. 350 °C and a liquid hourly space velocity (LHSV) of ca. 1 h$^{-1}$, an initial mixture of *m*-cresol, methanol, and steam in a molar ratio of 1:6:1 gives 2,3,6-trimethylphenol in yields of 90–95% (relative to *m*-cresol) at virtually complete *m*-cresol conversion. Small quantities of 2,5-xylenol and 2,4,6-trimethylphenol are the main byproducts. About half the methanol decomposes to $H_2$, $CH_4$, and $CO_2$. The reaction mixture is worked up by distillation. For the workup by recrystallization from aliphatic hydrocarbons, see [127]. The commercial product is ≥ 99% pure.

2,3,6-Trimethylphenol can also be obtained by methylation of 2,6-xylenol on γ-aluminum oxide, in particular if the reaction is carried out in a trickle bed [128]. Under optimum conditions (a molar ratio 2,6-xylenol–methanol of 2:1, 355 °C, 3.2 MPa, LHSV 4 h$^{-1}$, multitube reactor), however, a selectivity of only 50% is achieved at 32% conversion [129]. From the xylenol–trimethylphenol fractions thus obtained the 2,3,6-trimethylphenol must be isolated in pure form [121] by relatively sophisticated procedures, i.e., by alkaline countercurrent extraction [130] or, after spreading out the boiling points by *tert*-butylation of the mixture, by fractional distillation, debutylation, and a second fractionation [131].

Another industrial route to 2,3,6-trimethylphenol developed by BASF involves the condensation of diethyl ketone in the presence of base (NaOH) with methyl vinyl ketone [132] or crotonaldehyde [133] to give 2,3,6- and 2,5,6-trimethyl-2-cyclohexen-1-ones, respectively. These compounds are subsequently dehydrogenated in the gas phase at 250–300 °C on noble metal carrier catalysts [134]:

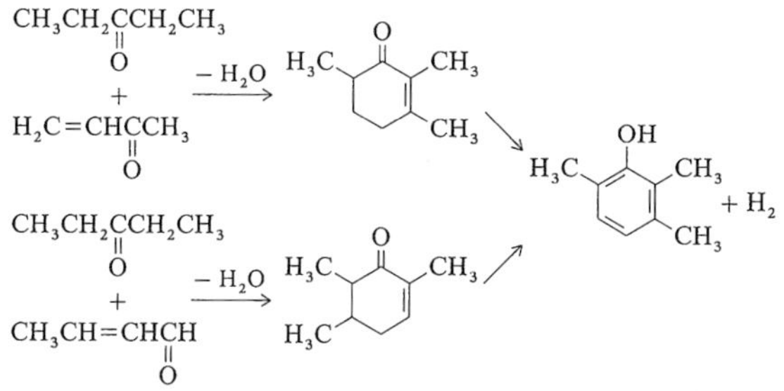

The condensation to give the trimethylcyclohexenones takes place in excess boiling diethyl ketone in yields of ca. 80% [132], [133]. The selectivity of the dehydrogenation is 97% on palladium–spinel catalysts, if the trimethylcyclohexanone that is formed as a byproduct in up to 10–15% yield by hydrogenation is recycled into the dehydrogenation [134]. The dehydrogenation can also be performed at 220–225 °C on 1% palladium–charcoal catalysts [135].

**2,4,6-Trimethylphenol** [527-60-6] (physical properties see Table 3) is the main product (>80%) of the gas phase alkylation of phenol with excess methanol (molar ratio 1:5) at 450 °C on magnesium oxide catalysts, which are doped with alkali or other metal oxides (e.g., $V_2O_5$, SrO). 2,6-Xylenol and 2,3,4,6-tetramethylphenol are the main byproducts [136], [137].

**2,3,5-Trimethylphenol** [697-82-5] and *2,3,5,6-tetramethylphenol* [527-35-5] (physical properties of both compounds are given in Table 3) are obtained by methylation of 3,5-xylenol in the presence of ortho-selective catalysts. A mixture of 3,5-xylenol – methanol vapors (molar ratio 1:5) is continuously reacted in a tubular reactor on a mixed MgO – UO$_3$ catalyst at 400 °C and an LHSV of 1 h$^{-1}$. The conversion of 3,5-xylenol is 75 %; a 2,3,5-trimethylphenol selectivity of ca. 75 % and a 2,3,5,6-tetramethylphenol selectivity of ca. 25 % are obtained [138].

In the same way *p*-cresol and 2,4-xylenol can be converted into *2,4,6-trimethylphenol*; 2,3- and 2,5-xylenol into *2,3,6-trimethylphenol*; and 3,4-xylenol into a mixture of *2,3,4-trimethylphenol, 2,4,5-trimethylphenol* [496-78-6] (physical properties see Table 3) and *2,3,4,6-tetramethylphenol* [3238-38-8] (physical properties see Table 3) on ortho-selective methylation catalysts.

2,4,6-Trimethylphenol can be isomerized in the presence of AlCl$_3$ – HCl at 85 °C to a mixture of 62 % 2,3,6-, 26 % 2,3,5-, and 12 % 2,4,6-trimethylphenols [139]. Under more stringent conditions all trimethylphenols rearrange in the presence of AlCl$_3$ – HCl to form the thermodynamically stable 2,3,5-trimethylphenol in good yields [140].

**Producers.** Producers of 2,3,6-trimethylphenol are, for example, Synthetic Chemicals in the United Kingdom, BASF and Lowi in Germany, and Honshu Chem. Ind. in Japan. 2,4,6-Trimethylphenol is produced by Lowi in Germany, General Electric in the United States, and Konan Chemicals, Mitsubishi Gas, and Nippon Kagaku in Japan. 2,3,5-Trimethylphenol is produced by Shell Chemicals in the United Kingdom.

**Uses.** *2,3,6-Trimethylphenol* is an important starting material for vitamin E synthesis; it also serves as a comonomer for the modification of polyphenyleneoxide resins (PPO). *2,3,5-Trimethylphenol* and *2,4,6-trimethylphenol* [137] are also starting materials for vitamin E. A mixture of the *N*-methylcarbamates of 2,3,5-trimethylphenol and 3,4,5-trimethylphenol is used as an insecticide.

The *tar alkylphenols* with a boiling point > 230 °C (to which the trimethylphenols and methylethylphenols belong) show bactericidal, fungicidal, and insecticidal action. Therefore these alkylphenols are contained in formulations of impregnating oils used for wood protection.

## 1.5.2. Ethylphenols

*2-Ethylphenol* [90-00-6] and *2,6-diethylphenol* [1006-59-3] (for physical properties of both compounds see Table 3) are produced industrially by ortho-alkylation of phenol with ethylene in high pressure autoclaves at 320 – 340 °C and 20 MPa in the presence of 1 – 2 % aluminum phenolate. If a molar ratio of phenol – ethylene of ca. 1:2 is used, the yield (related to converted phenol) is ca. 32 % 2-ethylphenol and ca. 39 % 2,6-diethylphenol after 6 h reaction time. About 17 % 6-*sec*-butyl-2-ethylphenol is formed as a byproduct [81].

Cresols and xylenols with free ortho positions can be ethylated similarly. The ethylation of 3,5-xylenol leads very smoothly to *2-ethyl-3,5-dimethylphenol* and *2,6-diethyl-3,5-dimethylphenol* [81].

Phenol can also be ethylated with ethanol, [141]–[144]. The conditions are similar to those for the methylation of phenol with methanol. A yield of 60% 2-ethylphenol and 22% 2,6-diethylphenol is achieved using an $Fe_2O_3$ catalyst doped with silicon and magnesium oxides in the gas phase at 360 °C (molar ratio phenol–ethanol–$H_2O$ 1:6:1, LHSV of 0.6 $h^{-1}$) [126]. 2-Ethyl- and 2,6-diethylphenol are also formed preferentially on $Al_2O_3$ in the gas phase at ca. 300 °C [142]. With increasing temperature and residence time the ortho-selectivity decreases and the yield of 3- and 4-ethylphenol increases [143]. At 400–450 C° mixtures of isomers with a relatively high proportion of meta product can be obtained, in particular after increasing the isomerization and transalkylation activity of the catalyst by combination of the catalyst with other compounds [144]. If phenol is reacted with ethylene instead of ethanol, ca. 77% conversion of phenol is achieved. Phenol and ethylene react, for example, in the gas phase at 420 °C on a phosphoric acid–$SiO_2$ catalyst to form an ethylphenol mixture, of which ca. 44% is 3-ethylphenol [145]. The processes give product mixtures from which the pure meta or para isomers can only be isolated at relatively high cost. An especially high 4-ethylphenol selectivity (ca. 96%) is possible according to [146] with ethylene–phenol–$H_2O$ vapor mixtures (molar ratio 1:0.86:1.3) at 400 °C and low phenol conversions (ca. 7%) on H-ZSM-5-zeolite modified with tetramethoxysilane. As far as is known these processes have not been exploited industrially.

*4-Ethylphenol* [123-07-9] (for physical properties see Table 3) is produced industrially by sulfonation of ethylbenzene under mild (kinetically controlled) conditions [147] and subsequent alkali fusion of the 4-ethylbenzenesulfonic acid obtained [148], [149]. The purity of the commercial product is 98%.

*3-Ethylphenol* [620-17-7] (for physical properties see Table 3) is also produced by sulfonation of ethylbenzene and subsequent alkali melt of the 3-ethylbenzenesulfonic acid obtained. By sulfonation under severe (thermodynamically controlled) conditions (200 °C) the 4-ethylbenzenesulfonic acid formed initially isomerizes to form a mixture of 2-, 3-, and 4-ethylbenzenesulfonic acids in a ratio of ca. 2:58:40. The 2- and 4-ethylbenzenesulfonic acids are selectively hydrolyzed back to ethylbenzene and $H_2SO_4$ by blowing steam through the mixture at 150–170 °C. 3-Ethylphenol is then obtained by alkali melt of the remaining 3-ethylbenzenesulfonic acid at 330–340 °C in up to 98% purity [150]–[152].

*Producers.* Producers of 2-ethylphenol are, for example, EMS-Dottikon in Switzerland and Coalite in the United Kingdom. 4-Ethylphenol is produced, e.g., by Synthetic Chemicals and Coalite in the United Kingdom and Maruzen Oil Co. and Konan Chemicals in Japan. Producers of 3-ethylphenol are, e.g., Coalite and Synthetic Chemicals in the United Kingdom and Taoka Chem. Co. in Japan.

**Uses.** 3-Ethylphenol and 4-ethylphenol are contained in the xylenol fraction boiling between ca. 205 and 225 °C obtained from tar phenols, which is used, for example, for the production of phenolic resins. A xylenol fraction boiling between 216 and 218 °C

**Table 4.** Physical properties of isopropylphenols

| Name (common name) | CAS registry no. | mp, °C | bp, °C |
|---|---|---|---|
| 2-Isopropylphenol | [88-69-7] | 16.8 | 215.5 |
| 3-Isopropylphenol | [618-45-1] | 25.7 | 223.5 |
| 4-Isopropylphenol | [99-89-8] | 63 | 223 |
| 2,4-Diisopropylphenol | [2934-05-6] | 21.6 | 258 |
| 2,6-Diisopropylphenol | [2078-54-8] | 19.6 | 236.5 |
| 3,5-Diisopropylphenol | [26886-05-5] | 53.1 | 257.5 |
| 2,4,6-Triisopropylphenol | [2934-07-8] | 28.7 | 255 |
| 5-Isopropyl-2-methyl-phenol (carvacrol) | [499-75-2] | 2.5 | 237.4 |
| 6-Isopropyl-2-methyl-phenol | [3228-04-4] | −16.5 | 225 |
| 2-Isopropyl-3-methyl-phenol | [3228-01-1] | 70–70.5 | 229 |
| 4-Isopropyl-3-methyl-phenol | [3228-02-2] | 114 | 245 |
| 5-Isopropyl-3-methyl-phenol (isothymol) | [3228-03-3] | 51 | 241 |
| 6-Isopropyl-3-methyl-phenol (thymol) | [89-83-8] | 51.5 | 232.4 |
| 2-Isopropyl-4-methyl-phenol | [4427-56-9] | 34 | 234 |

containing predominantly 3- and 4-ethylphenol is obtained from brown coal low-temperature tar and used in the varnish industry.

Pure 2-ethylphenol and 3-ethylphenol are starting materials for photochemicals. Pure 4-ethylphenol is the starting material for the production of 4-vinylphenol and of various antioxidants [e.g., 2,6-di-*tert*-butyl-4-ethylphenol and 2,2′-methylenebis(6-*tert*-butyl-4-ethylphenol)], which are used in rubber and polymers. 4-Ethylphenol is an intermediate for pharmaceuticals and dyes.

## 1.5.3. Isopropylphenols

**Properties.** Some physical properties of isopropylphenols mentioned in this article are listed in Table 4.

**Production.** Isopropylphenols are formed by reaction of phenol or alkylphenols with propene or isopropanol in the presence of acid catalysts of the Lewis or Brønsted type or of aluminum phenolate. An extensive study of the alkylation of phenol, both with propanol and isopropanol in the presence of alumina has been published [153].

Industrially, *2-isopropylphenol* is produced from phenol and propene using $\gamma$-$Al_2O_3$ as the catalyst. The alkylation is carried out continuously in a steel tube filled with $\gamma$-$Al_2O_3$ pressed into tablets, with graphite as a tabletting aid. The reaction temperature is 250–300 °C and the pressure is about 10 MPa, to keep both the phenol and the propene safely in the liquid phase. If a phenol:propene molar ratio of 1.3:1 is used, 2-isopropylphenol is obtained with a selectivity of about 85% at complete propene conversion [154].

*2,6-Diisopropylphenol* is formed as a byproduct (10–12%) in this process along with smaller amounts of *4-isopropylphenol* (1–2%) and *isopropyl phenyl ether* [2741-16-4] (*bp* 176.8 °C). The proportion of ether formed increases considerably at temperatures below

250 °C. Isopropyl phenyl ether, once formed, remains unchanged over the alumina catalyst even at 250–300 °C. Isopropyl phenyl ether can be transformed mainly to 2-isopropylphenol by treatment at 85 °C with $BF_3$ [155] or with strongly acidic ion-exchange resins.

Monoisopropylphenol mixtures enriched in the 4-isomer can be synthesized from phenol and propene or isopropanol in the presence of zeolite catalysts [156], [157]. When a 1:1 molar mixture of phenol and isopropanol is passed over a ZSM-5-catalyst at 250 °C and LHSV = 1 $h^{-1}$, a phenol conversion of 20% and a selectivity for 4-isopropylphenol of 63% is achieved [156]. 1:1 Mixtures of 2- and 4-isopropylphenols are obtained by transalkylation of phenols with cumene in a $HF-BF_3$ medium at 30 °C [158].

Isomerization and transalkylation occur if (mixtures of) 2- and/or 4-isopropylphenol(s) are heated to 110 °C for 6 h in $BF_3$-containing hydrogen fluoride. The main reaction product is *3-isopropylphenol* which is obtained in 61% yield [159].

3- and 4-Isopropylphenol are minor byproducts of commercialized resorcinol and hydroquinone syntheses (→ Resorcinol; → Hydroquinone) that use 1,3- and 1,4-diisopropylbenzene, respectively as starting materials. Production of 3- or 4-isopropylphenol via this autoxidation route is, however, not economical.

*2,6-Diisopropylphenol* is produced from phenol and propene (molar ratio 1:2.1) with homogeneously dissolved aluminum phenolate as catalyst, which is formed by dissolution of aluminum granules (1 wt%) in the educt phenol. The alkylation is carried out in an autoclave or pressure tube and takes place at 220 °C within 3 h. The workup procedure involves: (1) treatment of the reaction product with aqueous mineral acid to destroy the phenolate catalyst; (2) phase separation, (3) neutralization of the acid entrained in the organic phase with a base; and (4) a water wash. The product is subsequently distilled to yield 77% 2,6-diisopropylphenol and 10% 2-isopropylphenol [160], [161]. The wastewater is contaminated with phenolic compounds; therefore the procedure needs to be improved because of ecological considerations.

*2,4,6-Triisopropylphenol* is obtained in 70% yield by reacting 1 mol phenol with 4 mol propene at 250 °C under pressure in the presence of fluorided alumina [162]. Diisopropylphenols are formed as byproducts. Mixtures of mono- di-, and triisopropylphenols can be produced from phenol and propene with aluminum chloride (at 50–150 °C) [163], [164] or with activated clays (at 100–150 °C) as catalysts [165]. These mixtures find various applications in the plastics industry. If desired, the proportion of 3- or 3,5-isomers can be increased by (1) raising the temperature (up to 200 °C), (2) increasing the amount of catalyst (up to 20%) and/or (3) increasing the reaction time (>20 h) [166]. The reaction mixtures are fractionated and fractions with desired composition are recycled to the alkylation reactor. If $AlCl_3$ is used as the catalyst, the mixture can be reacted with phosphorus oxychloride to form phosphoric acid tris(isopropyl phenyl ester) [163], [167]. Phenol alkylation and subsequent esterification of the alkylphenol can thus be carried out in one reaction vessel without intermediate separation of the catalyst and fractionation of the reaction mixture.

Several isopropylation products of the methylphenols (cresols) are used industrially. The most important is the *m*-cresol-derived *6-isopropyl-3-methylphenol* (*thymol*), a precursor of l-menthol. *m*-Cresol with a purity $\geq 98.5\%$ must be used as the starting material for thymol production, because 2-isopropyl-4-methylphenol, derived from *p*-cresol (the main contaminant of commercial *m*-cresol), cannot be easily separated from thymol.

In the industrial process employed by Bayer [168], *m*-cresol and propene, both in the liquid state, are pumped through a pressure tube filled with activated alumina. The process is performed at a molar ratio of *m*-cresol–propene of 1:0.7, at 350–365 °C, ca. 5 MPa, and an LHSV of 0.25 h$^{-1}$. The reaction product consists of ca. 25% *m*-cresol, 60% thymol, and 15% other products. Thymol of $\geq 99.5\%$ purity is obtained by rectification of the crude product. Undesired alkylates of *m*-cresol (2-, 4-, and 5-isopropyl-*m*-cresol; 2,6- and 4,6-diisopropyl-*m*-cresol) are recycled to the reaction zone to achieve a high overall yield of the desired thymol.

The influence of the reaction conditions on the formation of thymol from *m*-cresol and propene in the liquid phase in the presence of $\gamma$-Al$_2$O$_3$ has been described [170]. Impregnation with metal sulfates, e.g., FeSO$_4$, increases the activity of $\gamma$-Al$_2$O$_3$ [171], [172]. Aluminum phenolate [173] and aqueous zinc halide–hydrogen halide solution [174] have also been used as catalysts. With these systems selectivities for thymol formation are about 75–80%.

In 1988, a gas phase process for thymol manufacture was developed using medium-pore sized zeolites (e.g., erionite, mordenite, or ZSM-23) as heterogeneous catalyst; 80–100% of the exchangeable cations in the zeolite are present as protons. Reaction temperatures are lower (230–270 °C) than in the older liquid-phase process and the reaction pressure is normal or slightly elevated [169].

*5-Isopropyl-3-methylphenol* (*isothymol*) can be produced by employing catalysts that possess high isomerization and transalkylation activities. Examples include finely ground synthetic silica–alumina [175] and synthetic zeolites with faujasite structure, especially the wide-pored Y-type [176]. Acid-activated clays are less suitable as catalysts because they can lead to corrosion at the relatively high temperatures that are required for isomerization. Usually, the isothymol synthesis is conducted in two steps. First, 1 mol *m*-cresol is reacted with < 1 mol propene at 100–200 °C in the presence of one of the aforementioned catalysts (10 wt% silica–alumina or 5 wt% zeolite Y). The alkylate is then kept in an autoclave at 300 °C until the thermodynamic equilibrium is established (ca. 5–10 h). The equilibrated mixture contains 5-, 6- and 4-isopropyl-*m*-cresols in a 75:20:5 ratio together with unreacted *m*-cresol and higher alkylates. The higher the excess of *m*-cresol used initially, the lower is the proportion of higher alkylates obtained. Using an efficient fractionating column, isothymol is distilled from the reaction mixture after removal of the catalyst by filtration. Undesired isomers of isothymol and higher alkylates can be recycled to the isomerization–transalkylation step for optimization of the yield. Isothymol can also be separated from its isomers, by alkylation (with isobutene) [177] or sulfonation (with sulfuric acid) of the isomeric mixture [178]. In contrast to its isomers, isothymol does not react with both compounds and thus can be separated from its alkylated/sulfonated isomers by a less efficient separation column. Isopropylation of *o-* and *p*-cresol yields rather complex product mixtures, depending on reaction conditions.

*o-Cresol* is isopropylated under kinetic control and with aluminum phenolate as catalyst to form *6-isopropyl-2-methylphenol* in good yield [173], which in turn can easily be isomerized to *4-isopropyl-2-methylphenol* by treatment with catalytic amounts of sulfuric acid at 60 °C [179]. Under thermodynamic control, at 300 °C, and with an activated clay as catalyst, the main product is *5-isopropyl-2-methylphenol (carvacrol)* [180].

*p-Cresol* reacts with propene in the presence of aluminum phenolate [173] or $BF_3$–phosphoric acid [181] as catalysts to give 2-isopropyl-4-methylphenol and 2,6-diisopropyl-4-methylphenol under kinetic control. Conditions which establish the thermodynamic equilibrium lead to the additional formation of *3-isopropyl-4-methylphenol* and *2,5-* and *3,5-diisopropyl-4-methylphenols*. Product distributions resulting from the reaction of cresols with isopropanol in the presence of $AlCl_3$ and phosphoric acid have been studied [182]–[184].

If *xylenols* are to be isopropylated, the substitution rules given on page p. 3724 can be applied to predict the composition of the resulting product mixtures.

**Uses.** Isopropylphenols have found relatively few commercial applications. As nontoxic substitutes the phosphate esters of isopropylphenol mixtures are used for cresylphosphates as secondary plasticizers and flame retardants in vinyl polymers, mainly poly(vinyl chloride) [162], [163]. 2-Isopropylphenol is an efficient antiskinning aid in air-drying coatings. 6-Isopropyl-3-methylphenol (thymol) is the starting material for an industrial l-menthol synthesis [185]. Thymol and carvacrol (5-isopropyl-2-methylphenol) are used as antiseptics. The *N*-methylcarbamate derived from 5-isopropyl-3-methylphenol is a very efficient insecticide [176]. 2,4,6-Triisopropylphenol can be used as an antioxidant in motor fuels and lubricants [162]; its reaction products with alkylene oxides are reported to be useful nonionic surfactants with good surface activity and low foaming properties [186].

## 1.5.4. *sec-*Butylphenols

**2-*sec-*Butylphenol** [*89-72-5*], *mp* 12 °C, *bp* 226–228 °C, is obtained with high selectivity by reaction of phenol with *n*-butenes at 250–300 °C using $\gamma$-$Al_2O_3$ as catalyst and at a pressure of 3.5–8.0 MPa which keeps both reactants in the liquid phase. If the reaction is performed batchwise, the catalyst is finely ground and kept in suspension by stirring. In a continuous operation the catalyst is pressed into tablets with 10–20 % graphite as a tabletting aid, and positioned in a pressure tube. Typically, phenol and *n*-butene(s) in a 4:1 molar ratio are pumped over the catalyst at 250 °C, 3.5 MPa, with a residence time of 1 h. Conversions are 22 % (phenol) and 90–96 % (*n*-butenes); the selectivities for 2-*sec*-butylphenol and 4-*sec*-butylphenol [*99-71-8*] (*mp* 56–61 °C, *bp*

135–136 °C at 3.33 kPa) are 95% and 2.5%, respectively. *sec* -Butyl phenyl ether [*10574-17-1*] (*bp* 101–103 °C at 5.8 kPa) is formed as a byproduct [187]. A gas-phase alkylation employing a multitubular reactor has been described [188].

Aluminum phenolate (prepared in situ by dissolving about 3 wt% of aluminum granules or chips in the educt phenol) can also be used as the alkylating catalyst. Alkylation takes place with *n*-butene(s) at temperatures of 150–180 °C and by increasing pressure up to ca. 5 MPa. Depending on the amount of butene(s) introduced, either 2-*sec*-butylphenol or 2,6-di-*sec*-butylphenol [*5510-99-6*] (*mp* 42 °C, *bp* 225–260 °C) is obtained as the principal product. The usual workup (treatment with aqueous mineral acid to destroy the catalyst) can be circumvented by flash distillation of the crude reaction product at temperatures < 225 °C (vacuum). Under these conditions dealkylation is prevented and the catalyst-containing distillation residue can be recycled back to the alkylation step [189]. Alternatively, the distillation residue can be subjected to a transalkylation reaction after addition of phenol and heating to 250–350 °C [190].

*Producers.* 2-*sec*-Butylphenol is commercially produced by Hüls, Ethyl, Pearson, Shell, and Schweizerische Teerindustrie AG (STIAG). On request 2,6-di-*sec*-butylphenol will also be provided. Product purities are about 98%. The main use of 2-*sec*-butylphenol is in the synthesis of insecticides, acaricides, and herbicides.

## 1.5.5. *tert*-Butylphenols

**Properties.** Boiling and melting points of selected *tert*-butylated phenols as well as a qualitative note on their solubility in alkali are listed in Table 5. Solubility in 2.5 N NaOH can be useful for separating substances with similar boiling points.

**Production.** *2-tert-Butylphenol* is obtained as the main alkylation product if phenol is reacted with isobutene in the presence of sulfonated polystyrene–polydivinylbenzene ion-exchange resins. The acidity of the ion-exchange resins is depressed to an exchange capacity of about 2.5 meq/g (dry weight) by having been used for related reactions, e.g., for the production of higher alkylphenols [191]. A preconditioned catalyst of this kind that is placed in a tower reactor on a supporting sieve, permits production of 2-*tert*-butylphenol on an industrial scale. Reaction of phenol and isobutene in a 1:0.42 molar ratio at 95 °C gives a product mixture which has the following composition after removal of unconverted phenol:

| | |
|---|---|
| 2-*tert*-butylphenol | 39.7% |
| 4-*tert*-butylphenol | 38.5% |
| 2,4-di-*tert*-butylphenol | 21.4% |
| highboilers | 0.4% |

2-*tert*-Butylphenol is obtained in a purity of 99% by distillation; 4-*tert*-butylphenol and 2,4-di-*tert*-butylphenol are marketable byproducts.

Table 5. Physical properties of -butylphenols

| Name | CAS registry no. | mp, °C | bp, °C 101.3 kPa | bp, °C (kPa) | Solubility in 2.5 N NaOH* |
|---|---|---|---|---|---|
| 2-tert-Butylphenol | [88-18-6] | −7 | 224 | 99 (1.33) | s |
| 3-tert-Butylphenol | [585-34-2] | 43 | 240 | 129.5 (2.66) | s |
| 4-tert-Butylphenol | [98-54-4] | 99.5 | 239.8 | 114 (1.33) | s |
| 2,4-Di-tert-butylphenol | [96-76-4] | 56.5 | 264 | 146 (2.66) | i |
| 2,6-Di-tert-butylphenol | [128-39-2] | 39 | 253 | 133 (2.66) | i |
| 3,5-Di-tert-butylphenol | [1138-52-9] | 87–89 | | | |
| 2,4,6-Tri-tert-butylphenol | [732-26-3] | 131 | 278 | 158 (2.66) | i |
| 4-tert-Butyl-2-methylphenol | [98-27-1] | 28 | 247 | 174 (13.3) | s |
| 6-tert-Butyl-2-methylphenol | [2219-82-1] | 30–32 | 230 | 159 (13.3) | i |
| 4-tert-Butyl-3-methylphenol | [2219-72-9] | 72 | | 152 (2.66) | s |
| 5-tert-Butyl-3-methylphenol | [4892-31-3] | 50 | | 142 (2.66) | s |
| 2-tert-Butyl-5-methylphenol | [88-60-8] | 23 | | 129.5 (2.66) | i |
| 2-tert-Butyl-4-methylphenol | [2409-55-4] | 51–52 | 232.7 | 127 (2.66) | s |
| 4,6-Di-tert-butyl-2-methylphenol | [616-55-7] | 52 | 269 | 149.5 (2.66) | i |
| 4,6-Di-tert-butyl-3-methylphenol | [497-39-2] | 62 | 282 | 167 (2.66) | i |
| 2,6-Di-tert-butyl-4-methylphenol | [128-37-0] | 70–71 | 266 | 147 (2.66) | i |
| 6-tert-Butyl-2,4-dimethylphenol | [1879-09-0] | 22–23 | 249 | 131 (2.66) | i |
| 4-tert-Butyl-2,5-dimethylphenol | [17696-37-6] | 71 | 264 | 193 (13.3) | s |
| 4-tert-Butyl-2,6-dimethylphenol | [879-97-0] | 82–83 | 248 | 176 (13.3) | s |
| 2-tert-Butyl-4,5-dimethylphenol | [1445-23-4] | 46 | | 187 (13.3) | s |

* s = soluble; i = insoluble.

2-tert-Butylphenol is also formed with relatively high selectivity by reaction of phenol and isobutene over a γ-Al$_2$O$_3$-catalyst that has been tabletted with 10% graphite and preconditioned by heating to 650 °C in an air stream. Phenol and isobutene conversions are 52 and 56% at 150–160 °C, a molar ratio phenol:isobutene of 1:1, a pressure of 3.5 MPa, and a residence time of 1 h. The selectivity for 2-tert-butylphenol is 88% based on phenol conversion; 4-tert-butylphenol (2.5%) and 2,4-di-tert-butylphenol (5%) are formed as byproducts. tert-Butyl phenyl ether [6669-13-2], bp 185 °C, is isolated as a byproduct (or intermediate). The amount of tert-butyl phenyl ether formed decreases with increasing residence times and reaction temperatures, and increases with age of catalyst. The ether can be recycled to the reaction zone.

4-tert-Butylphenol is obtained by reaction of phenol with isobutene at normal or slightly elevated pressure and ca. 80–140 °C in the presence of strongly acidic catalysts such as sulfuric [192] or phosphoric acids [193], boron trifluoride [194], activated clays [195]–[197], [204], zeolites [202], [203] or strongly acidic ion-exchange resins [198]–[201]. 4-tert-Butylphenol is obtained in ca. 70% yield if tert-butyl phenyl ether is heated to 100 °C in the presence of a clay catalyst that has been activated with sulfuric acid [204]. Silica–alumina (75% SiO$_2$) which has been sintered at 600 °C is said to catalyze isomerizations – trans-alkylations. Mixtures of phenol with 2-tert-butylphenol and 2,4-di-tert-butylphenol give 4-tert-butylphenol at 165 °C [205]. In the industrial process (Fig. 1), dehydrated strongly acidic sulfonated polystyrene – polydivinylbenzene ion-exchange resins of the macroporous form are employed which are placed on supporting sieves in column-type reactors. For maximum yields two reactors in series,

**Table 6.** Composition (wt%) of a phenol–isobutene alkylation product. Catalyst: cation-exchange resin; molar ratio 1.4:1 (phenol–isobutene)

| Component | Alkylation | Isomerization–transalkylation |
|---|---|---|
| Isobutene | 0.19 | 0.07 |
| Phenol | 30.17 | 22.16 |
| 4-*tert*-Butylphenol | 33.92 | 75.08 |
| 2-*tert*-Butylphenol | 10.19 | 1.62 |
| 2,4-Di-*tert*-butylphenol | 23.71 | 0.82 |
| 2,4,6-Tri-*tert*-butylphenol + 4-*tert*-octylphenol | 0.41 | 0.22 |

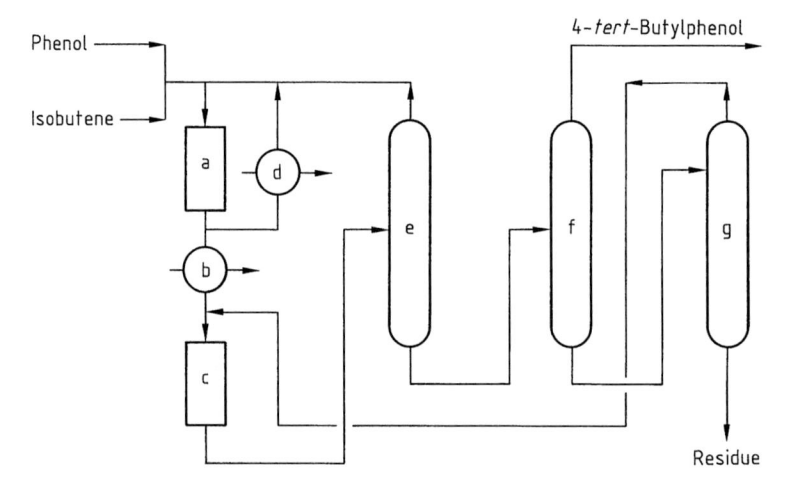

**Figure 1.** 4-*tert*-Butylphenol plant
a) Alkylator; b) Cooler; c) Isomerizer-transalkylator; d) Cooler; e) Distillation (phenol, 2-*tert*-butylphenol); f) Distillation; g) Vaporizer (2,4-di-*tert*-butylphenol recovery)

held at different temperatures, must be used. The first reactor, where the exothermic alkylation takes place (−90 kJ per mole isobutene reacted), is held at 90 – 100 °C; the second reactor (isomerization – transalkylation reactor) is held at 120 °C. (Isomerization and transalkylation reactions are thermoneutral to slightly endothermic.) The reactants are introduced to the top of the first reactor in a phenol – isobutene molar ratio of 1.4:1. Composition of the reaction products leaving the alkylation and the isomerization – transalkylation zones are given in Table 6. The reaction product leaving the isomerizer is worked up by continuous distillation under reduced pressure (4 – 10 kPa). A flow sheet of the industrial process [198] is depicted in Figure 1. Product yield is ≥ 95% and product purity is > 98%. 2,4-Di-, 2,4,6-tri-*tert*-butyl-, and 4-*tert*-octylphenol are formed as byproducts in the 4-*tert*-butylphenol production process. Their proportion increases with decreasing phenol – isobutene molar ratio and/or less stringent reaction conditions. 2,4-Di-*tert*-butylphenol can be made from phenol [206] or 4-*tert*-butylphenol [207] by alkylation with isobutene under mild reaction conditions and in the presence of small amounts of sulfuric acid at temperatures below 80 °C.

2,4,6-Tri-*tert*-butylphenol can be obtained from either of the three aforementioned phenols under these reaction conditions [208].

*3-tert-Butylphenol* is obtained from (1) 4- or 2-*tert*-butylphenol [209], [210], or (2) mixtures of 2,4- and 2,6-di-, or 2,4,6-tri-*tert*-butylphenol with phenol by heating to 170–200 °C in the presence of acid-activated clays [211], (synthetic) silica–alumina, modified zeolite of the Y-type [212], [213], or trifluoromethanesulfonic acid [214]. In the equilibrium state the monoalkylated fraction contains ≥ 70% 3-*tert*-butylphenol and ca. 20% of the 4-isomer; the dialkylated fraction contains ca. 80% 3,5-di-*tert*-butylphenol [212]. A selective debutylation of the 4-isomer can be achieved. Thus, heating a 70:30 mixture of 3- and 4-*tert*-butylphenols to 160–170 °C for 5 h, after addition of 1–2 wt% $H_2SO_4$, followed by neutralization of the mineral acid and washing, yields 3-*tert*-butylphenol in > 90% purity [215].

3-*tert*-Butylphenol can also be synthesized by heating 2-chloro-*tert*-butylbenzene with potassium-*tert*-butylate in *tert*-butanol to 240 °C for 4 h in an autoclave [216].

*3,5-Di-tert-butylphenol* is formed by heating isomeric mixtures of di-*tert*-butylphenols with acid-activated clays to 175–190 °C [217]. 3-*tert*-Butylphenol and/or 3,5-di-*tert*-butylphenol (depending on the chosen degree of alkylation) are obtained by reaction of phenol with isobutene at –78 to +25 °C in hydrofluoric acid [218]. Rearrangements of mono-*tert*-butylphenols in hydrofluoric acid as the solvent have been studied [219].

*2,6-Di-tert-butylphenol* (and 2-*tert*-butylphenol) is produced from phenol and isobutene in the presence of aluminum phenolate which is homogeneously dissolved in the phenol. The process, which has considerable industrial importance, was originally developed by Bayer [220].

Batchwise processes are preferred for industrial production using a stainless steel stirred-tank reactor provided with an efficient cooling device, e.g., an internal cooling coil. The reactor is charged with molten phenol, and flushed with pure nitrogen to remove all atmospheric oxygen. Metallic aluminum (1 wt%) in the form of granules or powder is added, and the reactor is heated to 120–180 °C for 0.5–1 h until the metal has dissolved to form aluminum phenolate under evolution of hydrogen which is vented. After cooling to 100 °C, liquid isobutene (1.8–2 mol per mole of phenol) is introduced into the reactor with stirring. During the reaction time of about 2 h, the pressure rises from 0.1–0.4 MPa to ca. 1.5 MPa. After stirring for a further 30 min, the pressure decreases to about 0.5 MPa. The reactor is depressurized and the reaction mixture is carefully washed with dilute sodium hydroxide and water. Rectification in vacuum yields about 75% 2,6-di-*tert*-butylphenol, 10% 2-*tert*-butylphenol, and a distillation residue consisting mainly of 2,4,6-tri-*tert*-butylphenol. 2-*tert*-Butylphenol can be recycled to the reactor. If desired, a higher amount of the monoalkylated phenol can be produced as a marketable product by using less than 1.8 mol isobutene per mole of phenol. The composition of the reaction products is dependent on the amount of isobutene used (see Table 7).

Exact control of temperature is important during the alkylation reaction, because at temperatures much lower than 100 °C *tert*-butyl phenyl ether formation prevails, whereas at higher temperatures ortho selectivity is progressively lost.

Industrially, the classical process provides an overall selectivity of about 85%. Improvements have been proposed. According to an invention of Ethyl [221], selectivity

**Table 7.** Compositions (wt%) of phenol alkylation products using aluminum triphenolate as the catalyst (1 wt% Al dissolved in the educt phenol)

| Component | Molar ratio isobutene : phenol | |
|---|---|---|
| | 1.8 | 2.1 |
| Phenol | 4.4 | 0.6 |
| 2-*tert*-Butylphenol | 10.3 | 1.7 |
| 4-*tert*-Butylphenol | | 0.2 |
| 2,6-Di-*tert*-butylphenol | 73.5 | 75.9 |
| 2,4-Di-*tert*-butylphenol | 2.2 | 0.9 |
| 2,4,6-Tri-*tert*-butylphenol | 9.6 | 19.4 |

can be as high as 95% if 2-*tert*-butylphenol is used as starting material instead of phenol and if aluminum tris(2-*tert*-butylphenolate) is used as catalyst which is prepared in situ by reaction of 2-*tert*-butylphenol with triethylaluminum (ethane is evolved and vented). The reaction temperature should be preferably 0–10 °C. *tert*-Butyl phenyl ether is not formed because the catalyst is not acidic. The process is advantageous for producers that have an independent source of 2-*tert*-butylphenol. A simpler nonaqueous workup procedure has also been described [222]. Addition of small amounts of glycols and glycol ethers to the reaction mixture safely deactivates the aluminum phenolate catalyst and allows for direct recovery of products by distillation. Disposal of alkaline wash waters that are contaminated by phenolic bodies is therefore unnecessary.

Aluminum phenolate catalysis can be applied to produce 2-methyl-6-*tert*-butylphenol from *o*-cresol and isobutene.

Mono- and dibutylations of mixed *m*- and *p*-cresols, and of mixed 2,4- and 2,5-xylenols with gaseous or liquefied isobutene are carried out industrially in the presence of concentrated sulfuric acid (0.5–2 wt%). In the first step the corresponding phenolsulfonic acids are formed as the actual alkylation catalysts. A water content of more than 0.1% in the educt phenols can be compensated for by using a higher proportion of sulfuric acid or by using oleum. Reactions occur at 50–80 °C and atmospheric to slightly elevated pressures.

For monobutylation of cresols an isobutene–cresol molar ratio of 0.7–1 is applied; the maximum yield of 80% is obtained at a molar ratio of 1 [223], [224]. *p*-Cresol is converted to 2-*tert*-butyl-4-methylphenol (with some 2,6-di-*tert*-butyl-4-methylphenol), whereas *m*-cresol forms 2-*tert*-butyl-5-methylphenol (and some 4,6-di-*tert*-butyl-3-methylphenol). 4-*tert*-Butyl-3-methylphenol is formed from *m*-cresol under very mild, kinetically controlled conditions (very little $H_2SO_4$, < 40 °C) [225]. *o*-Cresol gives a mixture of 6-*tert*-butyl-2-methyl- and 4-*tert*-butyl-2-methylphenols, along with some 4,6-di-*tert*-butyl-2-methylphenol. The butylation products are usually worked up by neutralization of the catalyst with a slight excess of concentrated sodium hydroxide solution and subsequent distillation. Unconverted cresols and dibutylated compounds are recycled. The yield of monobutylated products is ≥ 90%.

6-*tert*-Butyl-3-methylphenol can also be made by transbutylation of *m*-cresol with 4,6-di-*tert*-butyl-3-methylphenol, which is an intermediate of the *m*/*p*-cresol separation process (see below). The transbutylation is achieved by heating the mixture to 80–100 °C in the presence of acid-activated clays [226].

4-*tert*-Butyl-3-methylphenol is obtained only with difficulty. It can be formed by controlled debutylation of 4,6-di-*tert*-butyl-3-methylphenol by heating with the phenolates of aluminum [227] or (better) of zirconium [228], [229] under reflux conditions. Trans- and debutylation mixtures are worked up by fractionation, after deactivation of the catalyst by addition of alkali. 4-*tert*-Butyl-3-methylphenol can then be separated from 2-*tert*-butyl-5-methyl- and 2,6-di-*tert*-butyl-3-methylphenols because of its solubility in alkali [230].

5-*tert*-Butyl-3-methylphenol can be prepared by isomerization of 2-*tert*-butyl-5-methylphenol using substantial quantities of aluminum chloride at 30 °C [231].

2,6-Di-*tert*-butyl-4-methylphenol, frequently abbreviated as BHT (butylated hydroxytoluene), is a widely used antioxidant. Production of BHT involves reaction of pure and anhydrous ($< 0.1$ wt% $H_2O$) *p*-cresol with two or slightly more than two moles of isobutene in the presence of 2 wt% concentrated sufuric acid at 70 °C and at an isobutene partial pressure of ca. 0.1 MPa. The reaction time is about 5 h. The reaction mixture contains about 95% of BHT; some monobutylated phenols are formed as byproducts [232]. The usual workup procedure includes removal of acid by several washings with water at 70 °C, crystallization of the crude product, and a recrystallization from ethanol–water. The mother and washing liquids are first distilled at normal pressure for recovery of the ethanol. The remaining still bottoms separate into (1) an organic phase which is distilled under vacuum to recover 2-*tert*-butyl-4-methylphenol (to be recycled); and (2) an aqueous phase which is subjected to heat treatment (200 °C) to hydrolyze alkylphenolsulfonic acids into alkylphenols (which are recycled) and dilute sulfuric acid (which is neutralized and discarded) [233].

Alternatively, *m*-/*p*-cresol mixtures which cannot be separated by distillation, are first dibutylated and, after neutralization of the acid catalyst, are fractionated under vacuum to yield 2,6-di-*tert*-butyl-4-methylphenol (*bp* 147 °C at 2.66 kPa) and 4,6-di-*tert*-butyl-3-methylphenol (*bp* 167 °C at 2.66 kPa). The latter is subsequently reconverted to *m*-cresol by heating with 1 wt% oleum at 160–200 °C. The butyl group in the 4-position is eliminated first to yield 2-*tert*-butyl-5-methylphenol as intermediate. If required, this compound can be recovered by distillation of the partially debutylated product after neutralization of the acidic catalyst.

Several syntheses for 2,6-di-*tert*-butyl-4-methylphenol have been developed which are not based on *p*-cresol. At Shell, 2,6-di-*tert*-butylphenol was condensed with formaldehyde to yield 4,4′-methylenebis(2,6-di-*tert*-butylphenol) [118-82-1] which was subsequently heated to 200 °C with NaOH–methanol to give BHT in 80% yield [234]. This synthesis can be conducted in one step by heating 2,6-di-*tert*-butylphenol with formaldehyde in the presence of NaOH–methanol [235]. In the Soviet Union, 2,6-di-*tert*-butyl-4-dimethylaminomethylphenol [88-27-7] (a Mannich base) is produced from 2,6-di-*tert*-butylphenol, formaldehyde, and dimethylamine, which in turn can be hydrogenated (Raney nickel, 120 °C) to give BHT in excellent yields [236], [237].

According to an invention of Ethyl, primary straight-chain alkyl groups can be introduced to the (ortho- or) para-position of phenols by heating them with *n*-alcohols that are presaturated with potassium hydroxide. However, this process does not take place with methanol. 2,6-Di-*tert*-butyl-4-ethylphenol [*4130-42-1*] and 4-butyl-2,6-di-*tert*-butylphenol [*5530-30-3*] became available by this "alkaline alkylation method" [238].

Mixtures of 2,4- and 2,5-xylenol can be alkylated with isobutene in the presence of concentrated sulfuric acid at 40–60 °C [239], [240]. 2,4-Xylenol reacts faster than 2,5-xylenol. Usually the reaction is terminated when all the 2,4-xylenol is converted. Fractionation of the neutralized and washed product mixture yields unconverted 2,5-xylenol, 6-*tert*-butyl-2,4-dimethylphenol, and a certain amount of the higher boiling 4-*tert*-butyl-2,5-dimethylphenol. Alternatively, the reaction mixture is first washed with dilute alkali to remove unchanged 2,5-xylenol (and, if present, 2,4-xylenol) and its mono-*tert*-butyl derivative. The insoluble residue is then distilled to give 6-*tert*-butyl-2,4-dimethylphenol [240].

**Uses.** 2-*tert*-Butylphenol is a starting material for the synthesis of antioxidants and agrochemicals. Fragrance compounds are made from *cis*-2-*tert*-butylcyclohexanol [*7214-18-8*] which is obtained by hydrogenation of 2-*tert*-butylphenol in the presence of Pd–Al$_2$O$_3$ or Ru–Al$_2$O$_3$ catalysts [241]. Thermodynamic aspects of the stereoselective hydrogenation have been studied [242]. The acetate of the cis isomer [*20298-69-5*] has an orris-like fragrance.

3-*tert*-Butylphenol has found little if any commercial application. 4-*tert*-Butylphenol is used on a large scale for the production of a variety of phenol–formaldehyde resins. If used as the single phenolic component, oil-soluble resins result; if used in admixture with phenol and *m*-cresol, resins with controllable cross-linking are obtained. These resins, which are applied as binders in surface coatings, lacquers, and varnishes, have good air-drying properties and durability against the deteriorating influences of light, weather, and chemicals. 4-*tert*-Butylphenol-derived novolac resins with low viscosity are used as tackifiers for oil-extended SBR–natural rubber blends [243] and as plasticizers for cellulose acetate. Under slightly modified conditions, formaldehyde and 4-*tert*-butylphenol (as well as other 4-alkylated phenols) condense to form cyclic condensation products called calixarenes because of their beaker- or cup-like shape [244], [245]. These compounds possess clathrating properties for cations.

R = *tert*-butyl, higher alkyl; $n = 4, 6, 7, 8$

An oil-soluble calixarene (R=*n*-hexyl, *n*=6) [*117397-61-2*] is a very efficient extraction agent for uranyl ions from aqueous solutions [246].

4-*tert*-Butylphenol has antioxidant properties and is used as a stabilizer in rubber and chlorinated hydrocarbons, and, formerly, in soaps. The esters of 4-*tert*-butylphenol with phosphorous and phosphoric acids are UV stabilizers for plastics and lacquers. Esters of 5-*tert*-butylsalicylic acid [*16094-31-8*], which is obtained from 4-*tert*-butylphenol by Kolbe synthesis, are used for the same purpose. 4-*tert*-Butylphenol serves as a chain-length regulator in the production of polycarbonate resins.

The alkoxylates (i.e., reaction products with alkylene oxides) of 4-*tert*-butylphenol are emulsifiers and wetting agents which can also act as demulsifiers, e.g., in crude oil. Hydrogenation of 4-*tert*-butylphenol yields *cis*- and *trans*-4-*tert*-butylcyclohexanols [*98-52-2*], whose esters with acetic acid are widely used as a substitute for lemon grass oil, especially in soap perfumes. 4-*tert*-Butylcyclohexanone [*98-53-3*] has a strong camphor-like odor and is used in scenting detergents and soaps.

The main use of *2,4-di-tert-butylphenols* is in the manufacture of its triphosphite [*31570-04-4*] that is employed as a (co-)stabilizer for poly(vinyl chloride), and of its benzotriazole derivatives that are used as UV absorbers in polyolefins.

*2,6-Di-tert-butylphenol* is an indispensable building block in the synthesis of higher molecular mass antioxidants and light-protection agents for plastics, especially polyolefins. Frequently used intermediates are 3,5-di-*tert*-butyl-4-hydroxybenzyl alcohol [*88-26-6*] (**3**), 3,5-di-*tert*-butyl-4-hydroxybenzyl chloride [*955-01-1*] (**4**), methyl-3-(3,5-di-*tert*-butyl-4-hydroxyphenyl)- propionate [*6386-38-5*] (**5**), and 3,5-di-*tert*-butyl-4-hydroxybenzoic acid [*1421-49-4*] (**6**). These compounds are prepared from 2,6-di-*tert*-butylphenol by mild hydroxymethylation (**3**) [247],

$$(CH_3)_3C\text{-}C_6H_2(OH)(R)\text{-}C(CH_3)_3$$

R = (3): $-CH_2OH$; (4): $-CH_2Cl$
(5): $-CH_2CH_2COOCH_3$;
(6): $-COOH$

chloromethylation (**4**) [248], conjugate addition to methyl acrylate (**5**) [249], and carboxylation with carbon dioxide (**6**) [250]. 2,6-Di-*tert*-butylphenol reacts with formaldehyde to give 4,4'-methylenebis(2,6-di-*tert*-butylphenol) [*118-82-1*], an effective antioxidant for lubricating oils.

In the 1980s much research was focussed on the oxidative coupling of 2 molecules of 2,6-di-*tert*-butylphenol to form 3,3',5,5'-tetra-*tert*-butyldiphenoquinone which is then reduced to the corresponding 3,3',5,5'-tetra-*tert*-butyl-4,4'-diphenol [*2455-14-3*]. After removal of the butyl groups this compound yields 4,4'-dihydroxydiphenyl [*92-88-6*] which is a building block for thermoresistant polymers.

2,4,6-*Tri-tert-butylphenol* is the starting material for the synthesis of 2,6-di-*tert*-butyl-4-methoxyphenol [*489-01-0*] which is a powerful antioxidant. The first step involves oxidation of 2,4,6-tri-*tert*-butylphenol with $Cl_2$ in methanol in the presence of $Na_2CO_3$ followed by debutylation with sulfuric acid in methanol.

*Mono-tert-butyl derivatives of* **m**- *and p-cresol* are precursors for a number of commercially important antioxidants and light protection agents of the bisphenol and thiobisphenol type. Examples are 2,2′-methylenebis(6-*tert*-butyl-4-methylphenol) [*119-47-1*], 4,4′-methylenebis(6-*tert*-butyl-3-methylphenol) [*2872-08-4*], as well as the corresponding 2,2′-thiobis(6-*tert*-butyl-4-methylphenol) [*90-66-4*] and 4,4′-thiobis(2-*tert*-butyl-5-methylphenol) [*96-69-5*].

An important UV absorber, 2-(3-*tert*-butyl-2-hydroxyphenyl-5-methyl)-2*H*-5-chlorobenzotriazole [*3896-11-5*], is produced from *2-tert-butyl-5-methylphenol*. 2-*tert*-Butyl-5-methylphenol is also used for the preparation of musk ambrette (6-*tert*-butyl-3-methyl-2,4-dinitroanisole) [*83-66-9*] which is a perfume fixative. The *N*-methylcarbamate of 5-*tert*-butyl-3-methylphenol is an insecticide. *2,6-Di-tert-butyl-4-methylphenol* (BHT) has found broad application as an antioxidant in polymers, in technical oils, as well as in edible oils and fat-containing foods where it retards the development of rancidity.

*2,6-Di-tert-butyl-4-ethylphenol* and *4-butyl-2,6-di-tert-butylphenol* are also used as antioxidants. *2-tert-Butyl-4-ethylphenol* is used as a precursor for the synthesis of other antioxidants.

**Economic Aspects.** The worldwide annual production capacity for *tert*-butylphenols is about 128 000 t, which can be broken down as follows:

| | |
|---|---|
| Western Europe | 53 000 t |
| United States | 60 000 t |
| Japan | 12 000 t |

**Table 8.** Toxicological data of *tert*-butylphenols

|  | LD$_{50}$ (oral, rat), mg/kg | Effect on | |
|---|---|---|---|
|  |  | Skin | Eyes |
| 2-*tert*-Butylphenol | 868 | severe irritation | cauterization |
| 4-*tert*-Butylphenol | 4000 | moderate irritation | severe irritation |
| 2,4-Di-*tert*-butylphenol | 3250 | moderate irritation | moderate irritation |
| 2,6-Di-*tert*-butylphenol | 9200 | irritation not reported | irritation not reported |

About 25 000 t/a of 2,6-di-*tert*-butylphenol and about 15 000 t/a of 2,4-di-*tert*-butylphenol are produced worldwide. The consumption of BHT as a stabilizer of polymers is about 15 000 t/a; the total consumption of BHT is higher by a factor of 2–3.

**Toxicology and Occupational Health.** Toxicological data for butylphenols are listed in Table 8. 4-*tert*-Butylphenol was found to cause leucoderma, a depigmentation of the skin (vitiligo), in a few sensitive individuals [251], [252]. The condition is likely to be long-lasting and cannot be treated. The appearance of vitiligo is greatly reduced if skin contact is avoided and there is good ventilation. The MAK value for 4-*tert*-butylphenol is 0.08 mL/m$^3$ (0.08 ppm), equivalent to 0.5 mg/m$^3$.

**Storage, Packaging, Transportation.** In the presence of atmospheric oxygen, *tert*-butylphenols, like other alkylphenols, are prone to discolorations which are enhanced under the influence of light, elevated temperatures, traces of heavy metals (copper, iron), dissolved alkali, and moisture. During storage and transportation these influences must be kept to a minimum. *tert*-Butylphenols are therefore stored under an atmosphere of pure nitrogen.

In the liquid (molten) state, *tert*-butylphenols are stored and transported in stainless steel containers under a blanket of nitrogen. 2-*tert*-Butylphenol is transported in road or rail tank cars and in 200 L drums. 4-*tert*-Butylphenol is transported in the liquid state in thermostated road or rail containers at a temperature of about 110 °C. In the solid state it is stored under nitrogen in polyethylene bags of about 20 kg weight each that are palletted to a maximum weight of 1 t. As a solid 4-*tert*-butylphenol must be stored at < 40 °C. 2,4-Di-*tert*-butylphenol is transported in thermostated road tank cars at a temperature of 70° in the liquid state and iron drums (200 L) are used for the solid material. Transport classifications for butylphenols are given in Table 9.

## 1.5.6. *tert*-Pentylphenols

**Properties.** Only a few physical constants of the *tert*-pentylphenols are available (see Table 10). Synonyms for the *tert*-pentyl group are 2-methyl-2-butyl or *tert*-amyl.

**Production.** *tert*-Pentylphenols are obtained from the reaction of phenol with 2-methyl-1-butene, 2-methyl-2-butene, and/or 3-methyl-1-butene in the presence of acidic

**Table 9.** Classification of butylphenols in land-, sea-, and air-transportation

| Name | State* | mp, °C | Flash point, °C | Land GGVS/GGVE, ADR/RID | | Sea GGVSee, IMDG code (Amdt 25-89) | | | Air ICAO code, IATA-DGR | | |
|---|---|---|---|---|---|---|---|---|---|---|---|
| | | | | Class | No. | Class | UN no. | Page | PG | Class | UN no. | PG |
| 2-sec-Butylphenol | l | ca. 14 | ca. 101 | 8 | 66 b | 8 | 1760 | 8147 | II | 8 | 1760 | II |
| 2-tert-Butylphenol | l | ca. −38 | ca. 80 | 6.1 | 14 c | 6.1 | 2228 | 6091 | III | 6.1 | 2228 | III |
| 4-tert-Butylphenol | s | ca. 100 | ca. 115 | 6.1 | 14 c | 6.1 | 2229 | 6091 | III | 6.1 | 2229 | III |
| 2,4-Di-tert-butylphenol | s | ca. 53 | ca. 129 | | | | | | | | | |
| 2,6-Di-tert-butylphenol | l/s | ca. 39 | ca. 118 | | | | | | | | | |

* l = liquid; s = solid.

**Table 10.** Physical properties of -pentylphenols

| Name | CAS registry no. | mp, °C | bp, °C | |
|---|---|---|---|---|
| | | | 101.3 kPa | 2.66 kPa |
| 2-*tert*-Pentyl-phenol | [3279-27-4] | | | 124 |
| 4-*tert*-Pentyl-phenol | [80-46-6] | 95 | 266 | 142 |
| 3-*tert*-Pentyl-phenol | [27336-20-5] | | | 143* |
| 2,4-Di-*tert*-pentylphenol | [120-95-6] | 27 | | 166 |
| 2,6-Di-*tert*-pentylphenol | [3279-20-7] | | 283 | 165 |

* At 3.33 kPa.

catalysts, since methylbutenes isomerize to form the most stable cation which is the tertiary pentyl. When phenol and olefin are employed in equimolar amounts, and boron trifluoride [253], phosphoric acid [254], or strongly acidic ion-exchange resins are used as catalyst [255], yields as high as 90 – 95 % 4-*tert*-pentylphenol are obtained at 100 – 120 °C. With a molar excess of olefin, 2,4-di-*tert*-pentylphenol is also formed, especially if acid-activated clays are used as catalysts [256], [257]. The residual $C_5$-fraction of an isoprene extraction plant can be used as a source of methylbutenes, provided that the cyclopentene fraction has been selectively hydrogenated on a Pd/$Al_2O_3$ catalyst prior to the alkylation [258].

**Uses.** Condensates of 4-*tert*-pentylphenol with formaldehyde are compounded with a drying oil and used as oil-soluble resins for application in surface coatings. Ethoxylation of 4-*tert*-pentylphenol forms surface-active substances which can be used as demulsifiers in crude oils. 2,4-Di-*tert*-pentylphenol is used for the preparation of photographic color developers, textile processing aids, and light stabilizers.

## 1.5.7. Higher Alkylphenols

**Properties.** This group of compounds includes alkylphenols containing at least one alkyl group with six or more carbon atoms. Physical data and specifications of higher alkylphenols of commercial importance are given in Table 11. The data represent standards set by producers with modern plants, e.g., Hüls. Higher alkylphenols are generally mixtures of isomers, due to the varying composition and chemical structure of the higher olefins employed for their production. The alkyl groups ($C_8$, $C_9$, $C_{12}$) are bonded via a tertiary (or, less frequently, secondary) carbon atom to the aromatic ring. In addition, the alkyl chains are branched.

**Production.** Higher alkylphenols are obtained by alkylation of phenol with "higher olefins". Linear or branched α-olefins as well as linear or branched olefins with internal double bonds can be used as alkylating agents. Industrially, the most important higher olefins are

# Phenol Derivatives

Table 11. Physical properties and specifications of commercial higher alkylphenols

| Alkylating olefin | Product name | CAS registry no. of product | EINECS/ TSCA[h] | Molecular formula | mp/setting point, °C | bp (2.0 kPa), °C | Flash point, °C | Density (20 °C), g/mL | Viscosity, mPa·s | | Color (state) |
|---|---|---|---|---|---|---|---|---|---|---|---|
| | | | | | | | | | 20 °C | 80 °C | |
| Diisobutene (DIB) | 4-(1,1,3,3-tetramethyl-butyl)phenol (4-tert-octylphenol) | [140-66-9] | +/+ | $C_{14}H_{22}O$ | 84–85 | 158 | | 0.89 (90 °C) | | | white (s) |
| Diisobutene (DIB) | tert-octylphenol[a] | [27193-28-8] | +/+ | $C_{14}H_{22}O$ | 80 (min) | 162 | ca. 147 | 0.901 (100 °C) | | 7 (100 °C) | colorless (l), white (s) |
| Di-n-butene (DNB) | isooctylphenol[b] | [11081-15-5] | +/− | $C_{14}H_{22}O$ | ca. −10 | 163 | ca. 152 | 0.966 | 1180 | 9 | colorless (l) |
| Propene trimer (PTri) | nonylphenol[c] | [25154-52-3] | +/+ | $C_{15}H_{24}O$ | ca. −8 | 173 | ca. 155 | 0.949 | 2500 | 12 | colorless (l) |
| Propene trimer (PTri) | 4-nonylphenol, branched[d] | [84852-15-3] | +/+ | $C_{15}H_{24}O$ | ca. −14 | 174 | ca. 166 | 0.952 | 3160 | 14 | colorless (l) |
| Propene trimer (PTri) | 2,4-dinonylphenol[e] | [137-99-5] | +/+ | $C_{24}H_{42}O$ | ca. −13 | 224 | ca. 180 | 0.912 | 3300 | | colorless–pale yellow (l) |
| Propene tetramer (PTet) | dodecylphenol, branched[f] | [121158-58-5] | +/− | $C_{18}H_{30}O$ | ca. −7 | 187 | ca. 165 | 0.944 | 12000 | 25 | colorless–pale yellow (l) |
| Tri-n-butene (TNB) | 4-isododecylphenol[g] | [27459-10-5] | −/+ | $C_{18}H_{30}O$ | ca. −7 | 195 | ca. 180 | 0.940 | 7000 | 22 | colorless–pale yellow (l) |

[a] With >90 % 4-tert-octylphenol and ≤2 % 2-tert-octylphenol. [b] With ≥90 % 4- and ≤10 % 2-isomer. [c] With ca. 90 % 4- and ca. 10 % 2-isomer. [d] With 98 % 4-nonylphenol. [e] With 95 % (min) dinonylphenol. [f] Physical data for TNB product, with <10 % 2-isomer. [g] Physical data for PTet product, with ca. 5 % 2-isomer. [h] EINECS = European Inventory of Existing Chemical Substances; + = registered, − = not registered.

1) diisobutene, a 4:1 mixture of 2,4,4-trimethyl-1-pentene [*107-39-1*], bp 101.4 °C, and 2,4,4-trimethyl-2-pentene [*107-40-4*], bp 104.5 °C;
2) di-*n*-butene [*25377-83-7*] (mixture of octenes) with boiling range 116 – 126 °C;
3) propene trimer [*27215-95-8*] (mixture of nonenes) with boiling range 134 – 143 °C;
4) propene tetramer [*9003-07-0*] (mixture of dodecenes) with usual boiling range 185 – 205 °C; and
5) tri-*n*-butene [*72317-18-1*] (mixture of dodecenes) with boiling range 197 – 203 °C

A review of phenol alkylations with higher olefins has been published [259]. The preferred catalysts are ion-exchange resins, acid-activated clays, and synthetic aluminosilicates [260]. Occasionally boron trifluoride, *p*-toluenesulfonic acid, or phosphoric acid [193] are employed.

*5-tert-Octylphenol*, produced from phenol and diisobutene, consists mainly of 4-(1,1,3,3-tetramethylbutyl)phenol; commercial products always contain small amounts of the ortho isomer. The ortho – para isomer ratio depends on the alkylation conditions.

Boron trifluoride or its complexes with phenol or ethers are recommended as catalysts for the alkylation of phenol (and the cresols) with diisobutene and with other higher olefins. A reaction temperature of 50 – 85 °C and 1 – 2 wt % of $BF_3$ with respect to the phenol is required. A phenol – diisobutene molar ratio of 1.2 – 1.4 : 1 yields 90 – 95 % octylphenol in less than 30 min at 50 °C [261]. Catalysis with $BF_3$ is suitable for batchwise production. Processes in continuously operated stirred-tank and tube reactors have been also described [262], [263]. The $BF_3$ catalyst can be removed from the reaction product by extraction with hot water [262] or by precipitation with milk of lime or concentrated ammonia [264]. Various procedures for a (partial) recovery of $BF_3$ have been described [263], [265], [266]. If further reactions are to be carried out with the primary alkylate, the removal of $BF_3$ can sometimes be postponed to a later stage, for example, if the alkylate is reacted with sulfur dichloride to give thiobisphenols [267].

Sulfonated polystyrene – polydivinylbenzene ion-exchange resins of the macroporous type are currently the catalysts of choice. They are applied in the form of beads of 0.3 – 1.5 mm in diameter. Column-type reactors filled with the catalyst resin are employed. The reaction of higher olefins with phenol is slower than that of isobutene, permitting easier control of the heat of reaction and the temperature gradient along the catalyst bed. Sometimes the catalyst activity is deliberately reduced by special measures. Nonetheless, high space velocities can be maintained because isomerization reactions are less important with higher olefins than with isobutene. Alkylations with higher olefins can be carried out at or close to normal pressure because of their high boiling points. Plants are similar to those described for the production of 4-*tert*-butylphenol (two column-type reactors in series, connected by a heat exchanger allowing for an intermediate cooling of the primary reaction product leaving the first reactor). Depending on the olefin, the alkylations are run adiabatically or close to isothermal conditions.

In a process specially designed for the production of 4-octylphenol [268], phenol and diisobutene (1.5:1 molar ratio, LHSV = 1 $h^{-1}$) are passed over an ion-exchange resin that is placed in a single column reactor. The reaction temperature is maintained at

100–105 °C by internal cooling coils. The resin, with an exchange capacity of ca. 400 meq H$^+$/100 g, is kept in the hydrated form (water content 10–15%) by using phenol with a water content of 1–2 wt%. The water decreases the catalyst activity to a desired level, thus spreading the evolution of reaction heat and effectively suppressing the depolymerization of diisobutene, and, consequently, the formation of 4-*tert*-butylphenol. The reaction is carried out to a diisobutene conversion of about 95%. The reaction product is distilled to remove water, diisobutene, and phenol and has the following composition: 0.5–1% 4-*tert*-butylphenol, 2–3% 2-*tert*-octylphenol, 93–96% 4-*tert*-octylphenol, and 2–3% dialkylphenols. The product can be purified further by fractionation under vacuum. 4-*tert*-Octylphenol is marketed in the liquid state in thermostated containers or in the form of pastilles or flakes.

Hüls, a major producer of higher alkylphenols in Western Europe, operates plants with a total annual capacity of 50 000 t that produce *nonylphenol* (the major product; propene trimer as alkylating olefin), *octylphenols* (di-*n*-butene and diisobutene as alkylating olefins), and *dodecylphenol* (propene tetramer and tri-*n*-butene as alkylating olefins). The process uses two equally-sized reactors in series which are both operated adiabatically. The heat of reaction that is generated in the first reactor is partially removed by a heat exchanger connecting the reactors. Macroporous, anhydrous ion-exchange resins of different acidity are employed as catalysts. The exchange capacities are about 80 meq H$^+$/100 mL in the first, and about 140 meq H$^+$/100 mL in the second reactor. The nonylphenol process is carried out as follows [269]:

*Process Description.* A mixture of phenol and propene trimer (molar ratio 1.7:1) is preheated to 70 °C and pumped to the top of the first reactor at an hourly rate of about 30 mass units per 1 mass unit of catalyst. The product leaving the first reactor with a temperature of 120 °C is cooled to 100 °C and led to the second reactor. The temperature of the reaction product leaving the second reactor is about 125 °C. Product compositions after the first and second step are (in wt%): propene trimer 28.9 and 3.8, phenol 46.1 and 27.5, nonylphenol 22.4 and 65.8, and dinonylphenol 2.6 and 2.9. The product of the second reaction step is worked up by distillation. Recovered phenol and propene trimer, as well as dinonylphenol, are recycled. Nonylphenol thus can be obtained in >95% yield. In general, the product purity is >98% and the ratio of 4–2-nonylphenol is ca. 90:10. A nonylphenol with a purity of 99.7% and a 4–2-nonylphenol ratio of at least 95:5, obtained by careful distillation, is also being commercialized by Hüls. Dinonylphenol is formed as a byproduct and is normally recycled to the reaction zone. It can be isolated from the nonylphenol bottoms by further distillation under vacuum in 50–98% purity, depending on market requirements.

Other nonylphenol processes that are performed in stirred-tank reactors, with a suspension of a finely ground ion-exchange resin as the catalyst, are becoming obsolete.

The production of *dodecylphenol* from phenol and propene tetramer is carried out with a comparatively large excess of phenol; molar ratios of 3:1 (phenol–propene tetramer) are common. In this case the alkylation can be conducted adiabatically at moderate temperatures, rising from about 50 °C to about 100 °C along the catalyst bed. An ion-exchange resin with an exchange capacity of 200 meq H$^+$/100 g serves as catalyst. Distillation of the reactor effluent and recyclization of unconverted phenol gives product yields of 90–95%, depending on the quality of the propene tetramer.

Comparable yields of dodecylphenol are obtained if activated clays are used as catalysts. The reaction times, however, are considerably longer (several hours) with clays than with ion-exchange resins or boron trifluoride. Batch [270] and continuous [271] processes have been described. The batch process, working at 80 °C, uses a finely ground catalyst, that is maintained in suspension by stirring; the continuous process, which is operated at about 150 °C, uses a granulated catalyst in a column-type reactor.

**Uses.** Higher alkylphenols, mainly 4-octyl-, 4-nonyl-, and 4-dodecylphenols are predominantly reacted with ethylene oxide to form the corresponding ethoxylates that find many applications as technical nonionic surfactants. Extensive use in household detergents is prohibited because of insufficient biodegradability. Ethoxylates of nonylphenol with 4–5 ethylene oxide units are oil-soluble, those with 8–9 ethylene oxide units are used as wetting agents, e.g., in textile scouring. The main application of ethoxylates of dodecylphenol, dioctyl-, and dinonylphenols is the field of technical emulsifiers. Sulfonated dodecylphenol ethoxylates have been suggested as foam-stabilizing surfactants in miscible-gas enhanced oil recovery. The alkali salts of the esters of sulfuric and phosphoric acid with nonylphenol ethoxylates are anionic surfactants, which are used in metalworking and as emulsifiers for insecticides and herbicides.

Condensates of higher alkylphenols with formaldehyde are the basis for oil-soluble resins and lacquers.

Higher alkylphenols are employed as starting materials for the synthesis of antioxidants; an important example is tris(4-nonylphenyl)phosphite [*31631-13-7*]. Derivatives of 4-octylphenol are used as light stabilizers. 4-Octyl-, 4-nonyl-, and 4-dodecylphenol, as well as their mixtures with the corresponding di- and trialkylphenols are used as age-protection additives. Dodecylphenol is the starting material for lubricating oil additives: dodecylphenol is first sulfurized to give a bisphenol disulfide, and subsequent carboxylation (Kolbe synthesis) yields the corresponding salicylic acids. The calcium and magnesium salts of these acids are widely used as dispersants and antioxidants in lubricating oils.

**Economic Aspects.** The annual worldwide capacity for higher alkylphenols is about 540 000 t (Western Europe 166 000 t, Eastern Europe 110 000 t, North America 233 000 t, South America 7000 t, Japan 24 000 t).

**Toxicology and Occupational Health.** Octyl-, nonyl-, and dodecylphenols are relatively nontoxic; $LD_{50}$ values (oral, rat) are 2000 mg/kg or higher. Contact with skin and eyes must be avoided because irritation or even burns can occur.

**Storage, Packaging, and Transportation.** Octyl-, nonyl-, and dodecylphenols will discolor if oxygen is not excluded. Storage and transportation facilities are therefore blanketed with nitrogen. The phenols are stored at 50–70 °C to permit sufficient pumpability.

The products are shipped in stainless steel or rail tank cars, or in 200 L drums. Transport classifications for higher alkylphenols are given in Table 12.

## 1.5.8. Cycloalkylphenols

**Properties.** Physical properties of cyclopentyl- and cyclohexylphenols are listed in Table 13.

**Production.** *Cyclopentylphenols* are produced by reaction of phenols with cyclopentene. Monoalkylation in the position ortho to the phenolic hydroxyl group is achieved with $\gamma$-$Al_2O_3$ as the catalyst at about 300 °C [249], whereas aluminum phenolate at 200 °C leads to 2,6-disubstitution [192], [272]. Disubstitution at the 2,4-positions occurs with boron trifluoride at 80 °C, with *p*-toluenesulfonic acid at 100–200 °C, or with active clays at about 150 °C [272]. Extended reaction times lead to isomerizations. Thus, dialkylation of *p*-cresol will give a mixture of 4-methyl-2,5-, 4-methyl-2,6-, and 4-methyl-3,5-dicyclopentylphenols after 15 h on active clay at 150 °C [272] or after 20 h on ion-exchange resins at 80 °C [273].

*Cyclohexylphenols* can be obtained from cyclohexene and phenols, under the conditions described for cyclopentylphenols. In addition to the aforementioned catalysts, phosphoric acid can also be used [232]. However, for reasons of price and availability cyclohexanol rather than cyclohexane is predominantly employed as the alkylating reagent. With aluminum silicates as catalysts, *p*-cresol reacts with cyclohexanol to give 2-cyclohexyl-4-methylphenol, whereas with phenol a mixture of 2- and 4-cyclohexylphenols is formed, which can be separated by distillation. Likewise, 2-methylcyclohexanol forms 4-methyl-2-(1-methylcyclohexyl)phenol with *p*-cresol and a mixture of 2- and 4-(1-methylcyclohexyl)phenol, with phenol. Alkylation of 2,4-xylenol with 2-methylcyclohexanol yields 2,4-dimethyl-6-(1-methylcyclohexyl)phenol. Industrially, cycloalkylphenols are produced by Bayer and ICI.

**Uses.** *2-Cyclopentylphenol* is a precursor for toner components in xerography. The *N*-methylcarbamate derived from 3-cyclopentylphenol is used as an insecticide. 2,6-Dicyclopentyl-4-methylphenol is reported to be an efficient antioxidant [274].

*2- and 4-Cyclohexylphenols* are intermediates in the synthesis of pharmaceuticals and agrochemicals. The most important application of cyclohexylphenols is in the field of rubber chemicals. Thus, 2,4-dimethyl-6-(1-methylcyclohexyl)phenol is an excellent antioxidant that is especially suitable for white or light-colored rubber products. The condensates of 2-cyclohexyl-4-methylphenol and of 4-methyl-2-(1-methylcyclohexyl)phenol with formaldehyde are well-known nonstaining antioxidants used in natural and synthetic rubbers and in thermoplastics such as polyolefins, ABS resins, and high-impact polystyrene.

**Table 12.** Classification of higher alkylphenols in land-, sea-, and air-transportation

| Name | State | Land | | Sea | | | | Air | | |
|---|---|---|---|---|---|---|---|---|---|---|
| | | GGVS/GGVE, ADR/RID | | GGVSee, IMDG code | | | | ICAO code, IATA-DGR | | |
| | | Class | No. | Class | UN no. | Page | Packaging group | Class | UN no. | Packaging group |
| 4-*tert*-Octylphenol [mainly 4-(1,1,3,3-tetramethylbutyl)phenol] | s | 8 | 65 c | 8 | 1759 | 8151 | III | 8 | 1759 | III |
| Octyl-, nonyl-, dodecyl-phenols, and dinonylphenol (lower grade)[a] | l | 6 | 66 b | 8 | 1760 | 8147 | II | 8 | 1760 | II |
| Dinonylphenol, conc.[b] | l | | | | | | | | | |

[a] ca. 1:1 mixtures of nonyl- and dinonylphenols.
[b] Dinonylphenol concentration 95% or higher.

**Table 13.** Physical properties of cycloalkylphenols

| Name | CAS registry no. | mp, °C | bp, (kPa) °C |
|---|---|---|---|
| 2-Cyclopentylphenol | [1518-84-9] | 40 | 152 (2.66) |
| 4-Cyclopentylphenol | [1518-83-8] | 64 | 166 (2.66) |
| 2,6-Dicyclopentyl-4-methylphenol | [41505-40-2] | 36–37 | 102–103 (0.003) |
| 2-Cyclohexylphenol | [119-42-6] | 55.7 | 169 (3.3) |
| | | | 238 (101.3) |
| 4-Cyclohexylphenol | [1131-60-8] | 128 | 181 (3.3) |
| | | | 295 (101.3) |
| 2-Cyclohexyl-4-methylphenol | [1596-09-4] | 56 | 172 (2.27) |
| | | | 293 (101.3) |
| 2-Cyclohexyl-6-methylphenol | [4855-68-9] | 70.5 | 192 (6.67) |
| 6-*tert*-Butyl-2-cyclohexyl-4-methylphenol | [51806-69-0] | | 183 (1.86) |
| | | | 320 (101.3) |
| 4-Methyl-2-(1-methylcyclohexyl)phenol | [16152-65-1] | 60–61 | 155 (1.33) |
| 2,4-Dimethyl-6-(1-methylcyclohexyl)phenol | [77-61-2] | 28–29 | 179 (2.93) |

**Table 14.** Physical data of aralkylphenols

| Name | CAS registry no. | mp, °C | bp, °C |
|---|---|---|---|
| 2-Benzylphenol | [28944-41-4] | 53–54.5 | 312 |
| 4-Benzylphenol | [101-53-1] | 83–85 | 198–200 (1.33 kPa) |
| *p*-Cumylphenol | [599-64-4] | 70–73 | 335 |

## 1.5.9. Aralkylphenols

**Properties.** Physical properties of selected aralkylphenols, described in this article, are listed in Table 14.

**Production.** Mixtures of *2-* and *4-benzylphenols* (ratio about 3:2) are obtained by heating benzyl chloride in an excess of phenol to 150 °C. The product mixture is obtained in 50% yield if a molar ratio phenol–benzyl chloride of 2:1 is used; with a ratio of 10:1 the yield is 90% of the theoretical yield [275]. Dibenzylphenols are the main byproducts. In the presence of Friedel–Crafts catalysts, preferably zinc chloride, the reaction rate is enhanced and the temperature can be reduced to below 100 °C [276]. The reaction product can be distilled. Generally, a separation of single isomers is not carried out. Cresols and xylenols can be benzylated in the same way. It was found, that the 3- and 4-benzylated derivatives of 2,6-xylenol are formed in parallel, not consecutive reactions [277].

Benzylations of phenolic compounds can also be achieved with benzyl alcohol under acid catalysis, preferably with *p*-toluenesulfonic acid. This has been used for ortho benzylation of bisphenols, whose para-positions are substituted [278].

(α-*Methylbenzyl*)*phenols* (styrenated phenols) [61788-44-1] are produced on a relatively large scale by reaction of phenol with styrene in the presence of sulfuric or *p*-toluenesulfonic acid or activated clays [279]. Phosphoric acid [280] and oxalic acid are also suitable catalysts. Depending on the amount of styrene applied, mixtures of mono-, di-, and tristyrenated phenol are obtained. As an example, when 2.1 mol styrene are slowly added, with stirring, to 1 mol phenol in the presence of 0.002 mol $p$-toluenesulfonic acid and at 110 – 120 °C, a product is obtained with less than 1 % of unreacted phenol, and with mono-, di-, and tristyrenated phenol concentrations of ca. 12, 46, and 41 %, respectively. Styrenated phenol is produced by many manufacturers and traded as a colorless to pale yellow viscous oil. Styrenations, as described, can also be applied to the common alkylphenols, such as the cresols and xylenols.

The styrenated phenols are predominantly used as such. Sometimes the reaction mixtures are used as intermediates. For example, the reaction product from $p$-cresol and styrene is first distilled under vacuum to remove unreacted starting materials, and the residue, mainly 4-methyl-2-($\alpha$-methylbenzyl)phenol, is condensed with aldehydes to give the corresponding bisphenols. The alkylation of partially styrenated phenol with isobutene or diisobutene is another example. After an appropriate workup, which includes neutralization or filtration of the catalyst and distillation of low-boilers, the product can be used as an antioxidant.

If $\alpha$-methylstyrene is used instead of styrene, $\alpha,\alpha$-dimethylbenzyl groups (cumyl groups) are introduced into the phenol moiety [281], [282]. The catalyst of choice is an acidic ion-exchange resin arranged as a fixed bed in a column reactor as described for the production of the higher alkylphenols. Very pure $p$-cumylphenol can be obtained by this method, which is required for certain applications in the polymer field.

The alkylation of phenols with styrene, $\alpha$-methylstyrene, and indene can also be performed with aluminum phenolate as catalyst [173], [283]. Substitution of phenol or $o$-cresol with styrene takes place predominantly in the ortho-position, whereas with $\alpha$-methylstyrene, mainly para substitution is found.

$p$-*Cumylphenol* is formed to a certain extent during acid-catalyzed cleavage of cumene hydroperoxide ($\rightarrow$ Phenol). The high boiling still bottoms of phenol – acetone plants are a potential source for this aralkylphenol, which is, however, not exploited.

**Uses.** *2-Benzylphenol* and the mixtures of *2-* and *4-benzylphenol* have a certain commercial significance as disinfectants.

(α-*Methylbenzyl)phenols* (styrenated phenols) find extensive use as protective agents against thermal and oxidative aging of natural rubber and synthetic styrene – butadiene rubbers and latices, especially for application in white and light-colored articles.

*4-(α,α-Dimethylbenzyl)phenol* ($p$-cumylphenol) can be condensed with formaldehyde to give acid- and base-resistant resins. $p$-Cumylphenol of high purity, as supplied, for instance, by Hüls, serves as a molecular-mass controlling component in the production of polycarbonates.

## 1.5.10. Alkenylphenols

**Production.** 4-Vinylphenol [2628-17-3] (mp 73.5 °C, sublimation range 70–80 °C at 0.004 kPa) can be obtained by gas-phase dehydrogenation of 4-ethylphenol at 550–600 °C and 0.1 MPa on an iron oxide catalyst and in the presence of steam and an aromatic hydrocarbon as diluents [284]. The alkenylphenol, which polymerizes readily, can be separated from the less acidic unconverted 4-ethylphenol by extracting the condensed reactor effluent with alkali at 30 °C [285]. The Maruzen Oil Co. is a supplier of 4-vinylphenol.

4-Isopropenylphenol [4286-23-1] and other 4-(alken-2-yl)phenols are obtained in yields up to 95 % of the theoretical by alkali-catalyzed cleavage of the corresponding bis(4-hydroxyphenyl)-alkanes (bisphenols) [286], [287]. As an example, bisphenol A is cleaved by heating to 190–240 °C at a pressure of ≤ 5 kPa in the presence of 0.1 wt % sodium hydroxide. This reaction is carried out on a small technical scale by Mitsui Toatsu. The resulting phenol–4-isopropenylphenol mixture is removed from the resinous material by distillation, followed by separation of phenol by rectification. The residual material is a fairly stable mixture of oligomers of 4-isopropenylphenol [64054-77-9], consisting of ≥ 80 % of the linear dimers (**7**) and (**8**), with dimer (**7**) prevailing [288].

The addition of radical scavengers does not prevent the formation of dimers and oligomers, which show the reactive behavior of the underlying monomer in many consecutive reactions. When the oligomers are heated to 150–200 °C under slightly reduced pressure and the evolving vapors are absorbed in high-boiling alcohols (n-octanol or ethylene glycol), a solution of monomeric 4-isopropenylphenol with remarkable stability is obtained [289].

Alkylations of phenol with conjugated dienes take place primarily as 1,4-additions. However, these reactions are rarely carried out on an industrial scale. At low temperatures (20–25 °C), and with phosphoric or p-toluenesulfonic acid as catalyst, 1,3-butadiene and isoprene react to give the expected alk-3-en-2-ylphenols. These tend to undergo cyclization reactions: ortho-alkenylphenols cyclize to form chromans and para-alkenylphenols form indanols. A concise review on phenol alkylation with conjugated dienes has been published [290]. A modified aluminum phenolate catalyst has been described for the reaction of various phenols with 1,3-pentadiene to form the expected pent-3-en-2-ylphenols [291].

"*Cardanol*" [*37330-39-5*], also known as "anarcardol", is a mixture of meta-$C_{15}$-alkenylphenols, and is obtained by decarboxylation of anarcardic acid, which is the main constituent of cashew nut-shell oil. The alkenylphenols have been identified as 3-(penta-dec-8-en-1-yl)-, 3-(pentadeca-8,11-dien-1-yl)-, and 3-(pentadeca-8,11,14-trien-1-yl)phenols [292].

**Uses.** Mixtures of *4-isopropenylphenol oligomers* with phenol(s) and self-drying oils are used in the lamination of paper [288]. A hydroquinone synthesis has been suggested involving the addition of hydrogen peroxide to the double bond of 4-isopropenylphenol, followed by acid-catalyzed cleavage of the 4-hydroxycumene hydroperoxide.

*4-Isopropenylphenol* serves as the starting material for the synthesis of indanols (see below). *Cardanol* is a component of special phenol–formaldehyde resins, which have self-drying properties because of the unsaturated character of the $C_{15}$-substituent.

## 1.5.11. Indanols

*5-Indanol* [*1470-94-6*], mp 54–55 °C, bp 255 °C, is produced from indan [*496-11-7*] by a sulfonation–alkali-fusion process [293]. A synthesis for 1,1,3,3-tetraalkylindanols was developed by Bayer [294]–[296], which involves the reaction of isoolefins with secondary alkenylphenols (or compounds generating them under the reaction conditions) at 100–250 °C and in the presence of acidic catalysts. An example is the reaction of isobutene with bisphenol A in the presence of acid-activated clays which yields 1,1,3,3-tetramethyl-5-indanol [*53718-26-6*], mp 115 °C, bp (2.66 kPa) 156 °C, in a yield of 80% of the theoretical, and 4-*tert*-butylphenol as a coproduct.

**Uses.** 5-Indanol is a precursor for the synthesis of drugs. 1,1,3,3-Tetramethyl-5-indanol and other alkylindanols are used as photochemicals [297].

# 2. Catechol

Catechol [120-80-9], (pyrocatechol, 1,2-dihydroxybenzene, 1,2-benzenediol), $C_6H_4(OH)_2$, $M_r$ 110.11 is a crystalline compound with a phenolic odor. It was first obtained in 1839 by REINSCH by dry distillation of catechin. Currently, it is produced industrially from phenol. It is mainly used as raw material for the synthesis of polymerization inhibitors, perfumes, drugs, pesticides, dyes, in fur dyeing and leather tanning, as well as in photographic developers, deoxygenating agents, and analytical reagents.

**Occurrence.** Catechol has been found in, e.g., crude beet sugar, crude wood tar, some species of eucalyptus, onion, and coal. Catechol and many of its derivatives are obtained by dry distillation and other processes from, e.g., tannin, lignin, woods, and bituminous coal. It also occurs in tobacco smoke and is present in human and horse urine in the form of its sulfuric ester.

## 2.1. Physical Properties

Catechol forms colorless monoclinic crystals. It sublimes, is steam volatile, and discolors in air or light. It readily dissolves in cold and warm water and in cold and warm hydrophilic organic solvents (e.g., ethanol and acetone). Catechol is readily soluble in warm, but only sparingly soluble in cold hydrophobic solvents (e.g., benzene and chloroform). Physical properties of catechol may be summarized as follows [298]:

| | |
|---|---|
| mp | 104 – 105 °C |
| bp (at 0.1 MPa) | 246 °C |
| $d_4^{15}$ | 1.37 |
| $d_4^{121}$ (fusion) | 1.15 |
| Solubility in water | |
| at 15 °C | 25 g/100 g solution |
| at 20 °C | 31.2 g/100 g solution |
| at 100 °C | 98.2 g/100 g solution |
| Specific heat | |
| at 25 °C | 0.14 kJ mol$^{-1}$ K$^{-1}$ |
| at 104.3 °C (fusion) | 0.24 kJ mol$^{-1}$ K$^{-1}$ |
| Heat of fusion | 22.78 kJ/mol |
| Heat of sublimation at 36 °C | 80.81 kJ/mol |
| Heat of vaporization | 103.6 kJ/mol |
| Heat of formation | 1.2 kJ/mol |
| Flash point (closed cup) | 127 °C |

Dissociation constant
 $K_1$(18 °C)      $3.3 \times 10^{-10}$
 $K_1$ (30 °C)      $7.5 \times 10^{-10}$
 $K_2$ (30 °C)      $8.4 \times 10^{-13}$
Vapor pressure
 168 °C      10 kPa
 188 °C      20 kPa
 211 °C      40 kPa
 235 °C      80 kPa

## 2.2. Chemical Properties

Catechol gives a green coloration with iron-(III) chloride and turns red upon addition of a small amount of sodium hydroxide or ammonia. These colorations are specific for catechol and can be used for its detection and identification. Catechol can form stable coordination compounds with almost all metals and is therefore used as an analytical reagent for metals. Catechol is a weak acid and forms mono- and disalts with alkali hydroxides or carbonates. Many heavy metal salts of catechol, in particular its lead salts, are virtually insoluble in water. Thus, the reaction of catechol with lead acetate is used for quantitative analysis and separation of catechol from its isomers hydroquinone and resorcinol. Catechol is the strongest reducing agent of the three benzenediol isomers and can react with a solution of heavy metal salts to form fine precipitates of the elemental metals. 1,2-Benzoquinone is formed by careful oxidation of catechol with silver oxide or silver carbonate on celite. *cis,cis*-Muconic acid monomethyl ester [*61186-96-7*] can be obtained by oxidative cleavage of catechol with oxygen in the presence of copper(I) chloride and methanol [299].

Catechol undergoes many typical reactions of phenol. It reacts with acyl halides to form the corresponding mono- and diesters, which are converted into phenolic ketones by Fries rearrangement with aluminum chloride as catalyst. An aldehyde group can be introduced into the aromatic nucleus by Reimer–Tiemann reaction with chloroform and alkali, or by the addition of glyoxylic acid and subsequent oxidative decarboxylation.

Mono- and diethers of catechol can be prepared by the usual methods. Catechol can undergo cyclization reactions because of its two adjacent hydroxyl groups. It reacts with dichloromethane to form methylenedioxybenzene and with bis(2-chloroethyl)ether to yield dibenzo-18-crown-6-polyether [*14605-55-1*] [300]. Ammonolysis of catechol gives 2-aminophenol. Catechol couples with aryldiazonium salts to form azo compounds, which can be reduced to 4-aminocatechol.

Catechol forms bis(dihydroxyphenyl)methane by condensation reaction with formaldehyde, and polycyclic compounds (e.g., hystazarin and alizarin) by condensation with phthalic anhydride.

Ring mono-substitutions occur at the 3- and 4-positions. Catechol can thus be alkylated, halogenated, nitrated, carboxylated, and sulfonated.

**Analysis.** Catechol can be determined directly in mixtures containing phenol and other benzenediols by gas chromatography without prior separation [301].

## 2.3. Production

Catechol was formerly produced by low-temperature carbonization of coal, but this process is currently carried out only in rare cases.

**Hydrolysis of 2-Chlorophenol.** Industrially, catechol can be produced by hydrolysis of 2-chlorophenol with an aqueous solution of barium hydroxide and sodium hydroxide. Barium is recovered as carbonate and recycled to the process by converting the carbonate into hydroxide, which complicates the process. Therefore, improved processes in which only caustic alkali is used have been developed. For example, 1 mol 2-chlorophenol is reacted with 2.3 mol 4–8% sodium hydroxide solution in the presence of a copper(II) sulfate or copper(I) oxide at 190 °C for 3 h in a copper autoclave. The conversion of 2-chlorophenol is 96–99%, the selectivity for catechol is 81–86% [302]. The reaction mixture is then neutralized with sulfuric acid and the crude catechol formed is extracted, the solvent is recovered, and catechol is separated and purified by distillation. The industrial production of catechol was carried out by this method up to 1973.

**Hydroxylation of Phenol.** Catechol is currently manufactured together with hydroquinone by direct hydroxylation of phenol with peroxides. At present, three plants using this process are in operation worldwide; each plant uses a somewhat different type of peroxide or catalyst. As the reaction is exothermic and the benzenediols formed are more easily oxidized than phenol, the reaction is carried out in a large excess of phenol.

In France, *Rhône-Poulenc* react phenol with 70% hydrogen peroxide (molar ratio 20:1) in the presence of phosphoric acid and catalytic amounts of perchloric acid at 90 °C; catechol and hydroquinone are obtained in a ratio of ca. 3:2 [303]. This reaction proceeds electrophilically and phosphoric acid serves as a masking reagent preventing side reactions (formation of resorcinol by a radical reaction which gives a lower yield) caused by trace amounts of metallic ions. After the reaction, phosphoric and perchloric acids are removed by washing with water, then the reaction mixture is simultaneously extracted by diisopropyl ether, distilled, and continuously separated.

*Brichima SpA* (now Enichem) in Italy uses heavy metal compounds (e.g., small quantities of ferrocene and/or cobalt salts) as catalyst and reacts phenol with 60% aqueous hydrogen peroxide at 40 °C. Catechol and hydroquinone are produced in the ratio of 1.5–4.1 [304]. This reaction occurs via a free-radical chain mechanism and is thus very fast.

In Japan, *Ube Industries* produces catechol together with hydroquinone by hydroxylation of phenol with ketone peroxides formed in situ from a ketone and hydrogen peroxide in the presence of an acid catalyst [305][306]. The process is carried out by adding a trace amount of acid (e.g., sulfuric or sulfonic acid), a small volume of ketone, and 60% aqueous hydrogen peroxide to phenol at 70 °C. The ketone peroxide that is formed in situ reacts rapidly and electrophilically with phenol, and catechol and hydroquinone are obtained in a molar ratio of about 3:2 in more than 90% yield (based on phenol reacted). When a solid acid such as clay is used as a catalyst, the molar ratio of catechol and hydroquinone is about 1:1 [307]. As only a small amount of catalyst is used, no corrosion occurs and the reaction mixture can be distilled without removing the catalyst after the reaction. The added ketone can be recycled to the process by recovering it by distillation.

The *separation and purification methods* used in the three aforementioned processes are basically the same. That is, the reaction mixture is separated by distillation in different distillation columns. Water is removed, low-boiling fractions (solvents, ketone etc.) and unreacted phenol are recovered and recycled, the catechol fraction is made into a flaked product, and hydroquinone is purified by recrystallization of the corresponding fraction from water.

Processes, in which hydroxylation of phenol is carried out in the presence of catalysts such as strong acids or sulfur dioxide with nonaqueous hydrogen peroxide solution, have been patented [308], [309].

Peracids formed in situ from carboxylic acids and hydrogen peroxide or synthesized by air oxidation of aldehydes are utilized as hydroxylating agents [310]. In the case of hydroxylation using peracetic acid (synthesized through autoxidation of acetaldehyde), chelating reagents and phosphates are used as additives which improve the yield of benzenediols. Nevertheless, because of problems such as complicated separation of products and corrosion of apparatus, the peracid processes have not been realized industrially.

**Dehydrogenation of 1,2-Cyclohexanediol.** Recently, a patent has been published claiming that catechol can be obtained in 90% yield by dehydrogenation of 1,2-cyclohexanediol with a Pd/Te catalyst system at 300 °C [311]. Catechol is reportedly formed as the single product.

In addition, catechol may be obtained via the following synthetic routes: (1) alkali fusion of 2-phenolsulfonic acid and phenol-2,4-disulfonic acid; (2) oxidation of salicylaldehyde with hydrogen peroxide in aqueous alkaline solution; (3) demethylation of guaiacol with hydrobromic acid or aluminum chloride; and (4) hydrolysis of 2-aminophenol using a hydrogen halide.

## 2.4. Uses

Catechol itself is used as a photographic developer, analytical reagent, and oxygen scavenger (antioxidant); most catechol is used in the form of its derivatives: guaiacol (2-methoxyphenol) [*90-05-1*] and veratrole (1,2-dimethoxybenzene) [*91-16-7*] can be synthesized by *O*-methylation of catechol. Vanillin (4-hydroxy-3-methoxybenzaldehyde) *121-33-5*], derived from guaiacol is used as a flavoring agent. Ethylvanillin(3-ethoxy-4-hydroxybenzaldehyde) [*121-32-4*] derived from guethol (2-ethoxyphenol, a homologue of guaiacol) is not found in nature, and its flavor is 3–4 times stronger than that of vanillin. It is thus a valuable aroma compound. Eugenol (2-methoxy-4-allylphenol) [97-53-0], safrole (5-allyl-1,3-benzodioxole) [94-59-7], and piperonal [3,4-(methylenedioxy)benzaldehyde] [120-57-0], which can be prepared from the latter, are useful fragrances in perfumery.

In medicine, potassium guaiacol sulfonate (4-hydroxy-3-methoxy-benzenesulfonic acid monopotassium salt) [16241-25-1] and guaiacol glyceryl ether [3-(2-methoxyphenoxy)-1,2-propanediol] [93-14-1] are used as expectorants. L-α-Methyldopa (3-hydroxy-α-methyl-L-tyrosine) [555-30-6] and L-dopa (3-hydroxy-L-tyrosine) [59-92-7] derived from vanillin are used as an antihypertensive and as an antiparkinsonism drug, respectively. In addition, trimethoprim [738-70-5] derived from vanillin is used as an antiinfective; carbazochrome [69-81-8] and carbazochrome sodium sulfonate [51460-26-5] are used as hemostatics; papaverine [58-74-2] is used as an antispasmodic, vasodilator drug and smooth muscle relaxant.

Two important carbamate insecticides that are produced from catechol are used as agricultural chemicals — carbofuran [1563-66-2] (2,3-dihydro-2,2-dimethyl-7-benzofuranyl methylcarbamate, trade name Furadan), and propoxur [114-26-1] (2-isopropoxyphenyl-N-methylcarbamate, trade name Baygon) developed by Bayer.

4-tert-Butylcatechol [98-29-3], which is prepared by ring alkylation of catechol, is applied as a polymerization inhibitor during manufacturing and storage of monomers such as styrene and butadiene.

## 2.5. Economic Aspects

The consumption of catechol in the world is estimated to be about 20 000 t/a in 1990. It is assumed that 50% of that is used as starting material for insecticides, 35–40% for perfumery and drugs, and 10–15% for polymerization inhibitors and others. As the vanillin process is changing from production using lignin in pulp waste liquor to the synthesis from catechol, the demand for catechol used in the production of vanillin is expected to increase in the future [312]. The market price of catechol in May 1990 was 4.2–5.7 $/kg [313].

## 2.6. Toxicology

Catechol is readily absorbed from the gastrointestinal tract and through the skin and readily metabolized and excreted into urine. General symptoms of catechol intoxication are similar to those of phenol intoxication. Administration of a large dose of catechol induces strong suppression of the central nervous system and long-term vascular hypertension. The prolonged contact of catechol with skin leads to eczematous dermatitis or ulcers. In addition, it causes severe eye irritation.

Comprehensive references exist on catechol toxicity [314]. Some characteristic values of catechol toxicity are as follows [315]:

| | |
|---|---|
| $LD_{50}$ (oral, rat) | 3890 mg/kg |
| LDLo (subcutaneous, rat) | 200 mg/kg |
| $LD_{50}$ (subcutaneous, mouse) | 247 mg/kg |
| Threshold limit value | 5 ppm |

# 3. Trihydroxybenzenes

## 3.1. Pyrogallol

Pyrogallol [*87-66-1*], $C_6H_3(OH)_3$, $M_r$ 126.11 is generally called 1,2,3-trihydroxybenzene, 1,2,3-benzenetriol (IUPAC name), or pyrogallic acid. Pyrogallol was first isolated by SCHEELE by dry distillation of gallic acid (3,4,5-trihydroxybenzoic acid) in 1786. Pyrogallol derivatives occur in many natural products, such as tannin, anthocyanin, and alkaloids [316].

**Physical Properties.** Pyrogallol forms colorless needles or leaflets, the color gradually changes to dark gray on contact with air or light. It is soluble in water and polar solvents, and slightly soluble in chloroform and carbon disulfide. It sublimes without decomposition upon gradual heating. Important physical data are as follows:

| | |
|---|---|
| *mp* | 133 – 134 °C |
| *bp* (101.3 kPa) | 309 °C (partial decomp.) |
| (17 kPa) | 171 °C |
| $d_4^4$ | 1.453 |
| Dissociation constant (20 °C) | |
| $K_1$ | $1.5 \times 10^{-10}$ |
| $K_2$ | $4.5 \times 10^{-12}$ |
| Solubility (25 °C) | |
| water | 38 wt% |
| ethanol | 50 wt% |
| pyridine | soluble |
| ether | 45 wt% |
| benzene | slightly soluble |
| chloroform | slightly soluble |
| carbon disulfide | slightly soluble |

**Chemical Properties.** Pyrogallol is the strongest reducing agent among the polyhydroxybenzenes. The basic aqueous solution quickly absorbs gaseous oxygen to form a dark brown precipitate. Because of this property, pyrogallol is often used for quantitative analysis of oxygen.

Reactions of pyrogallol are quite similar to those of phenols. The hydroxy groups are etherified and esterified by the conventional methods to give mono-, di-, and trisubstituted products. Heating of pyrogallol in aqueous potassium bicarbonate solution mainly gives pyrogallol-4-carboxylic acid (2,3,4-trihydroxybenzoic acid); gallic acid is formed as a byproduct. Formylation, acylation, and the Mannich reaction produce 4-substituted cyclohexenones as a major product. Bromination with bromine eventually yields 1,2,6,6-tetrabromocyclohexene-3,4,5-trione. Reaction with phosgene gives pyrogallol carbonate, and with thionylbromide 4,5,6-tribromopyrogallol. The highly sensitive color reactions of aqueous pyrogallol solution with heavy metal ions are useful for the quantitative analysis of metals. Precious metal ions (e.g., $Ag^+$, $Au^{2+}$, $Hg^{2+}$) are reduced to elemental metals.

A complex prepared from diethylzinc and pyrogallol is used for a catalyst for alternating polymerization of carbon dioxide and propylene oxide [317].

**Production.** Pyrogallol is commercially produced by decarboxylation of gallic acid, which is prepared by hydrolysis of tannin.

The reaction is carried out batchwise. A 50% aqueous solution of gallic acid is heated to 175–200 °C in an autoclave. During the reaction, the pressure increases to 1.2 MPa. After the evolution of carbon dioxide is completed, the solution is cooled. The reaction proceeds quantitatively. Decolorization of the reaction mixture with charcoal followed by distillation gives a crude product which is purified by sublimation or vacuum distillation [318].

Pyrogallol is expensive to obtain because the starting materialgallic acid is prepared from a natural product of limited availability. Therefore, new methods for producing gallic acid have been sought. Base-catalyzed condensation of trimethyl propane-1,2,3-tricarboxylate with the dimethyl ketal of mesoxalic acid, followed by hydrolysis and decarboxylation gives gallic acid in 74% yield [319]. The reaction of the ketal and glutaric ester gives pyrogallol without generation of gallic acid [320].

Other methods for producing pyrogallol include oxidation of resorcinol with hydrogen peroxide [321], hydrolysis of 2,6-diamino-4-butylphenol [322], demethylation of 4-substituted 2,6-dimethoxyphenols [323], oxidation of 2,6-dimethylphenol [324], hydrolysis of 2,2,6,6-tetrachlorocyclohexanone [325], deoximation of 1,2,3-cyclohexanetrion-1,3-dioxime [326], and dehydrogenation of 1,2,3-trihydroxycyclohexane [327].

**Uses.** Pyrogallol is used in photography, lithography, and in hair dyes. It is also used as an antioxidant and stabilizer. The use of pyrogallol in the field of cosmetics and medicines is currently declining because of its pronounced toxicity.

**Toxicology.** Ingestion of pyrogallol may cause severe irritation. Toxicity data are as follows [328]:

| | |
|---|---|
| $LD_{50}$ (oral, mouse) | 300 mg/kg |
| $LD_{50}$ (subcutaneous, mouse) | 566 mg/kg |

## 3.2. Hydroxyhydroquinone

Hydroxyhydroquinone [533-73-3], $C_6H_3(OH)_3$, $M_r$ 126.11 is usually called 1,2,4-trihydroxybenzene, 1,2,4-benzenetriol (IUPAC name), or hydroxyquinol. Its derivatives are found as ethers or quinoids in plants.

**Physical Properties.** Hydroxyhydroquinone forms colorless platelets or prisms that quickly colorize in air. It is soluble in water and polar solvents, and slightly soluble in chloroform and carbon disulfide. Important physical data are as follows:

| | |
|---|---|
| mp | 140.5 – 141 °C |
| Dissociation constant (20 °C) | |
| $K_1$ | $8.3 \times 10^{-10}$ |
| $K_2$ | $1.5 \times 10^{-12}$ |
| Solubility | |
| water | soluble |
| ethanol | soluble |
| pyridine | soluble |
| ether | soluble |
| benzene | slightly soluble |
| chloroform | slightly soluble |
| carbon disulfide | slightly soluble |

**Chemical Properties.** Hydroxyhydroquinone is, like other phenols, a strong reducing agent. Its basic aqueous solution absorbs gaseous oxygen to produce black precipitates of the humic acid type. Hydroxyhydroquinone shows typical reactions of phenols. Derivatives of the tautomeric keto form are also known: halogenation of hydroxyhydroquinone, for example, gives 1,2,5,5-tetrahalocyclohexene-3,4,6-trione. Condensation with ethylacetoacetate produces dihydroxycoumarin; reaction with phthalic anhydride yields hydroxyhydroquinonephthalein. Monosubstitution of a hydroxy group by an amino group occurs easily at room temperature and gives 2,4-dihydroxyaniline [329].

**Production.** Hydroxyhydroquinone is synthesized by hydrolysis of 1,2,4-triacetoxybenzene, which is prepared by the acid-catalyzed reaction of *p*-benzoquinone with acetic anhydride [330].

Other production methods for hydroxyhydroquinone are oxidation of resorcinol with hydrogen peroxide [331] and Dakin oxidation of 2,4- or 3,4-dihydroxybenzaldehydes or 2,4- or 3,4-dihydroxyacetophenones with alkaline hydrogen peroxide solution.

**Uses.** Hydroxyhydroquinone is used as a stabilizer, antioxidant, polymerization inhibitor and in hair dyes.

**Toxicology.** Toxicity data are as follows [328]:

LDLo (intraperitoneal, mouse) 125 mg/kg
$LD_{50}$ (subcutaneous, mouse) 500 mg/kg

## 3.3. Phloroglucinol

Phloroglucinol [*108-73-6*], $C_6H_3(OH)_3$, $M_r$ 126.11 is also called phloroglucine, 1,3,5-trihydroxybenzene, 1,3,5-benzenetriol (IUPAC name), or cyclohexane-1,3,5-trione. In 1855, Hlasiwetz found phloroglucinol in the hydrolysis products of phloretin, which was obtained from the bark of fruit trees. The name phloroglucinol originates from the Greek and means "sweet bark". Its derivatives (e.g., flavones, anthocyanides, and xanthins) are widely distributed in the plant kingdom.

**Physical Properties.** Phloroglucinol crystallizes in the anhydrous form or as the dihydrate. Recrystallization from water gives the dihydrate, which is colorless, oderless, sweet in taste, and forms rhombic crystals. The dihydrate is converted into the anhydrous form by heating at 110 °C. Phloroglucinol sublimes at higher temperatures with partial decomposition. Important physical data are as follows:

| | |
|---|---|
| mp (anhydrous form) | 218–220 °C |
| mp (dihydrate) | 116–117 °C |
| Dissociation constant (30 °C) | |
| $K_1$ | $3.56 \times 10^{-9}$ |
| $K_2$ | $1.32 \times 10^{-9}$ |
| Solubility | |
| water | 1 wt% |
| ethanol | 9 wt% |
| pyridine | 75 wt% |
| diethyl ether | soluble |
| benzene | slightly soluble |

In conductometric titration, the behavior of phloroglucinol resembles that of benzenediols. The third proton is not dissociated. According to $^1H$ NMR, the dianion of phloroglucinol forms stable keto-type tautomers in solution. Phloroglucinol itself, however, is present in the enolic form, which has been confirmed by UV, IR, NMR, and X-ray spectra [332].

**Chemical Properties.** Because of the keto–enol tautomerism, phloroglucinol reacts to form triphenol derivatives and triketone derivatives.

Phloroglucinol behaves as a triketone, e.g., towards hydroxylamine to form the trioxime, and adds sodium bisulfite to yield the corresponding mono-, di-, and tri-substituted products. Triphenol behavior is shown in etherification and esterification that form mono-, di-, and trisubstituted phenols. Friedel–Crafts reaction and nitration also take place.

Phloroglucinol is also a reducing agent. The alkaline aqueous solution absorbs gaseous oxygen. The absorption of oxygen by phloroglucinol proceeds more slowly than by pyrogallol. Phloroglucinol reduces Fehling's solution and precious metal ions (e.g., $Au^{2+}$, $Ag^+$, and $Pt^{2+}$). Reaction with methyl iodide causes ring methylation under basic conditions. Reaction with aqueous ammonia produces 5-aminoresorcinol or 3,5-diaminophenol. Phloroglucinol forms phloroglucide upon heating. Hydrogenation gives 1,3,5-trihydroxycyclohexane. Coupling with diazonium salts takes place easily.

**Production.** Phloroglucinol can be prepared by various processes. 2,4,6-trinitrotoluene is oxidized with sodium dichromate to give 2,4,6-trinitrobenzoic acid, which is subsequently decarboxylated. The nitro groups are then reduced with iron–hydrochloric acid to form 1,3,5-triaminobenzene. Subsequent nucleophilic substitution of the amino groups by hydroxyl groups produces phloroglucinol. However, waste-disposal of the acidic filtrates which contain chromium and iron poses a problem [333].

In the 1980s, a process was developed to produce phloroglucinol from 1,3,5-triisopropylbenzene via the corresponding trihydroperoxide [334]. The yield is increased by addition of hydrogen peroxide to the reaction mixture in the acidolysis step [335]. Another method for producing phloroglucinol is the Beckmann rearrangement of triacetylbenzene trioxime followed by hydrolysis [317].

Other methods include Hofmann rearrangement of benzene-1,3,5-tricarboxylic acid triamide [337]; reaction of 1,3,5-tribromobenzene [338] or hexachlorobenzene [339] with alkoxide followed by acidolysis; and hydrolysis of 4-chlororesorcinol with potassium hydroxide [340].

**Uses.** Phloroglucinol is used as a coupler in diazotyping. Phloroglucinol and the diazo compound react with each other to form a large molecule which has a black color. In the drying process, the diazo and hydroxy compounds are both contained in the coating. Phloroglucinol is also used for photocopying [333].

**Toxicology.** Toxicity data are as follows [328]:

| | |
|---|---|
| $LD_{50}$ (intraperitoneal, mouse) | 4050 mg/kg |
| $LD_{50}$ (oral, mouse) | 4550 mg/kg |
| $LD_{50}$ (subcutaneous, mouse) | 991 mg/kg |

# 4. Bisphenols (Bishydroxyarylalkanes)

In this review bisphenols are defined as compounds consisting of two phenol nuclei linked by a hydrocarbon bridge.

**History.** The importance of bisphenols has increased continually with the growth of the plastics industry from about the end of the 1930s. They are used as intermediates for high-grade thermosets, thermoplastics, and raw materials for varnishes; and as antioxidants for rubber, plastics, oils, and fats. 2,2-Bis(4-hydroxyphenol)propane (bisphenol A), first described in 1891 by A. DIANIN, is by far the most important compound in this series. It has been produced industrially from phenol and acetone to be used as a starting material for epoxy resins since about 1945. The development of polycarbonates, the most important thermoplastics based on bisphenol A, began in the 1950s.

## 4.1. Physical Properties

Bisphenols are colorless, odorless substances, and most of them are solid at room temperature. The melting points of the most important bisphenols are between 100 and 200 °C [341]. Bisphenols are virtually insoluble in water. Their solubility in organic solvents is determined by their substituents. Whereas bisphenol A is only readily soluble in polar media such as ethers and alcohols, bisphenols with large aliphatic groups in the molecule are soluble in araliphatic and aliphatic hydrocarbons. The alkali

**Table 15.** Physical properties of some bisphenols

|  | CAS registry no. | Structural formula | Molecular formula | $M_r$ | mp, °C | bp, °C |
|---|---|---|---|---|---|---|
| Bis-(4-hydroxyphenyl)methane | [620-92-8] | 9 | $C_{13}H_{12}O_2$ | 200.2 | 158 | |
| 2,2-Bis(4-hydroxyphenyl)propane (bisphenol A) | [80-05-7] | 10 | $C_{15}H_{16}O_2$ | 228.3 | 156–157 | 190 (130 Pa) |
| 4,4-Bis(4-hydroxyphenyl)valeric acid | [126-00-1] | 11 | $C_{17}H_{18}O_4$ | 286.3 | 173 | |
| 2,2-Bis(4-hydroxy-3-methylphenyl)propane | [79-97-0] | 12 | $C_{17}H_{20}O_2$ | 256.3 | 136 | |
| 2,2-Bis(4-hydroxyphenyl)cyclohexane (bisphenol Z) | [843-55-0] | 13 | $C_{18}H_{20}O_2$ | 268.4 | 188 | |
| 2,2'-Methylene bis(6-*tert*-butyl-3-methylphenol) | [119-47-1] | 14 | $C_{23}H_{32}O_2$ | 340.5 | 131 | |
| α,α'-Bis(4-hydroxyphenyl)-*p*-diisopropylbenzene | [2167-51-3] | 15 | $C_{24}H_{26}O_2$ | 346.3 | 196 | 230 (13 Pa) |
| Bis(2-hydroxy-3-cyclohexyl-5-methylphenyl)methane | | 16 | $C_{27}H_{36}O_2$ | 392.6 | 128 | |
| Bis(4-hydroxy-3,5-di-*tert*-butylphenyl)methane | [118-82-1] | 17 | $C_{29}H_{44}O_2$ | 424.7 | 154 | |

salts of bisphenols are water-soluble. However, their solubility decreases drastically with increasing substitution. The boiling points of bisphenols are very high because of the size of the molecule and its polarity. For this reason and because of the decomposition frequently observed during boiling, bisphenols are rarely distilled. Some important physical properties of bisphenols are summarized in Table 15. Physical properties for the industrially important bisphenol A may be summarized as follows:

| | |
|---|---|
| Density at 20 °C | 1.04 g/cm³ |
| at 160 °C | 1.065 g/m³ |
| Bulk density | 0.492 g/cm³ |
| Heat of fusion | 147 J/g |
| *bp* at 101.3 kPa | 360 °C |
| at 1.4 kPa | 240 °C |
| at 0.4 kPa | 222 °C |
| Heat of varporization at 101.3 kPa | 404 J/g |
| Flash point | 227 °C |
| Ignition temperature | 510 °C |
| Solubility in H₂O at 83 °C | 0.344 wt % |
| Solubility in acetone, alcohol | good |
| Solubility in methylene chloride | ca. 1 wt % |

## 4.2. Chemical Properties

The chemical properties of the bisphenols are determined by the phenolic OH groups, the aromatic rings, and the alkyl bridge. They therefore undergo the same reactions as the corresponding substituted monophenols (esterification and etherification at the OH group, substitution and hydrogenation on the aromatic ring, see Section 1.2). They are also suitable as building blocks for high molecular mass linear polyesters and polyethers because of their bifunctionality [341]–[343].

Bisphenols which are alkylated ortho to the OH group readily trap radicals and are therefore suitable as stabilizers.

Bisphenols fragment purely thermally, or better catalytically, according to:

$$HO-C_6H_4-C(CH_3)_2-C_6H_4-OH \xrightarrow{H_2, cat.} HO-C_6H_4-CH(CH_3)_2 + C_6H_5-OH$$

$$HO-C_6H_4-C(CH_3)_2-C_6H_4-OH \xrightarrow{OH^-} HO-C_6H_4-C(=CH_2)(CH_3) + C_6H_5-OH$$

Under hydrogenation conditions bisphenol A is cleaved to give 4-isopropylphenol [344]; alkali-catalyzed cleavage gives 4-isopropenylphenol [345] in good yields. Both compounds are difficult to obtain by other methods. The alkali-catalyzed cleavage of various bisphenols has been investigated [346]. The cleavage can also be catalyzed by acids to form indans and spirobisindans [347]:

$$HO-C_6H_4-C(CH_3)_2-C_6H_4-OH \xrightarrow{H} [H_2C=C(CH_3)-C_6H_4-OH] \longrightarrow$$

spirobisindan (with H₃C, CH₃ and HO, OH groups) + indan (HO-C₆H₃(CH₃)(OH) with H₃C, CH₃ substituents)

The purely thermal cleavage is generally less straightforward [345].

## 4.3. Production

Bisphenols can be produced (1) by condensation of ketones or aldehydes with phenols, (2) by reaction of, e.g., diols or dihalides with phenols, and (3) by condensation of hydroxymethylphenols with phenols.

*Condensation of Ketones or Aldehydes with Phenols.* Acid- or alkali-catalyzed condensation of 1 mol ketone or aldehyde with 2 mol phenol yields bisphenols, whose aromatic nuclei are separated by one carbon [348]. Acetone and cyclohexanone give particularly smooth para addition and high yields of 4,4′-dihydroxydiphenylalkanes [345]:

Aldehydes generally react with un- and monosubstituted phenols to give mixtures of isomers and polynuclear phenols of the novolac type. 2,4-, 2,5-, 2,6-, 2,3,6-, and 2,3,5,6-substituted phenols, however, react readily and often in excellent yields with aldehydes (particularly formaldehyde) to form the corresponding bisphenols [349] – [353]. These reactions are frequently alkali-catalyzed. In other cases, however, alkali gives less pure products and poorer yields [345], [353]. Highly alkylated bisphenols are also obtained by subsequent alkylation of preformed bisphenols in the presence of acid catalysts [354]. Suitable acid catalysts are hydrogen chloride, highly concentrated hydrochloric acid, sulfuric acid (up to 75%), boron trifluoride, acid-activated clays and sulfonated cross-linked polystyrenes [344], [345], [355] – [359].

Cocatalysts containing SH groups e.g., thiols and mercaptoacids, are used to increase the reaction rate [344], [345], [357], [360], [361].

Organic solvents such as aromatic and chlorinated hydrocarbons or glacial acetic acid serve as the reaction medium. Aqueous emulsions may also be used. To prevent side reactions, the phenol which is being reacted must frequently be added in excess and thus functions simultaneously as the reaction medium.

*Reaction of Diols, Diolefins etc. with Phenols.* Bisphenols can also be produced by the acid-catalyzed reaction of diols and dihalides [359], [362], [363] or alkynes or allenes [364] with phenols. Divinylbenzenes react with phenols at elevated temperature without catalysts [365].

*Condensation of Hydroxymethylphenols with Phenols.* Unsymmetrical dihydroxydiphenylmethanes are formed by condensation of hydroxymethylphenols (from phenol and formaldehyde) with phenols with a different substitution pattern [352]. The synthesis of unsymmetrical bisphenols substituted at the central carbon atom can be achieved by alkylation of phenols with vinylphenols [341], [366].

The workup of the bisphenol reaction mixture depends very much on the structure of the products and the purity requirements. Basic purification is not necessary for bisphenols used as stabilizers, whereas bisphenol A for polycondensations must be of high purity. Possible workup procedures are, for example, the simple distillation of volatile components from the reaction mixture, washing out starting materials, or recrystallization from suitable solvents. Distillations of bisphenols are generally not carried out because of their thermal lability.

The mechanism for the formation of bisphenols, in particular that for the formation of bisphenol A and its byproducts, is essentially understood [345].

**Bisphenol A.** There are many processes for the industrial synthesis of this most important bisphenol [357], [367]–[371]. Currently, hydrogen chloride or sulfonated cross-linked polystyrenes [367], [372]–[374] are used as the catalyst which are usually arranged as a fixed bed over which the reaction mixture is passed (Fig. 2). The reaction of phenol with acetone takes place at 50–90 °C, the molar ratio phenol–acetone is up to 15:1. Bisphenol A crystallizes as an adduct with 1 mol phenol, after separation of the hydrogen chloride by distillation or neutralization. The use of ion exchangers is preferred to that of hydrogen chloride because they are less corrosive. The yield is normally 80–95 %.

To isolate the bisphenol, the whole reaction mixture can be fractionally distilled, whereby bisphenol A itself is distilled over particularly carefully under high vacuum, separated from resinous byproducts, and subsequently recrystallized under pressure at elevated temperature [375]. Crude bisphenol A can also be purified by extracting into hot heptane or mixtures of aromatics [376] or by recrystallization from aromatics [377]. A very high purity product with polyester quality is obtained if the bisphenol A–phenol 1:1 adduct is separated, recrystallized from phenol, and the phenol removed by distillation [370], [378]. These processes are mostly carried out continuously [367], [369], [371], [375], [378]. The acetone–phenol mixture produced in the Hock phenol synthesis from cumene hydroperoxide can be used as starting material for bisphenol A synthesis [379].

## 4.4. Analysis, Testing, Storage

Many bisphenols can be analyzed by gas chromatography. The quality of bisphenols can also be tested by measurement of the solidification point and the color index of solutions (bisphenol A in 30 % methanolic solution: APHA color index $\leq 50$).

Bisphenols are usually sold as solids, rarely as liquids; bisphenol A is supplied as white flakes which sometimes have a yellow-brown coloration. It can be stored in this form, but not in the molten stage for an unlimited time in stainless steel containers. The pneumatic conveying of bisphenol A must be performed under nitrogen because of the danger of dust explosions.

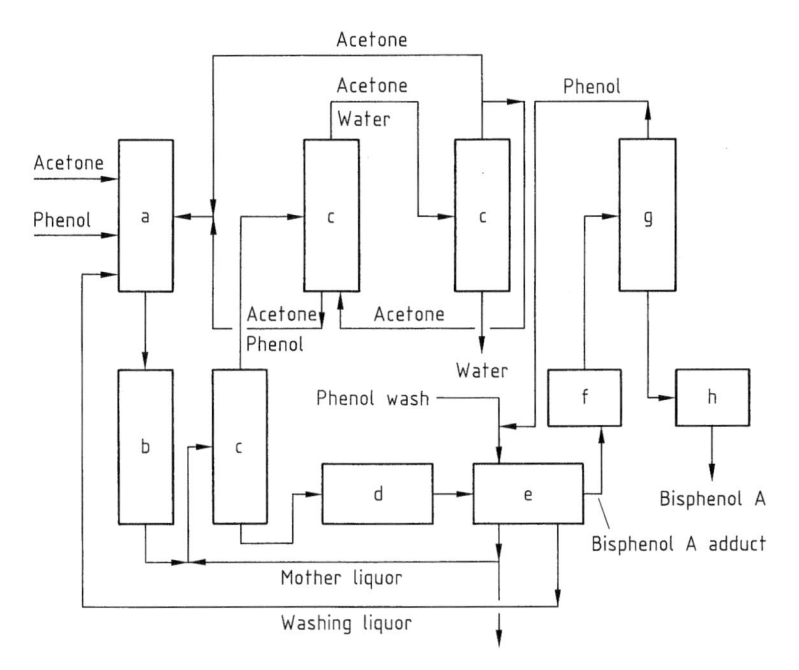

**Figure 2.** Flowchart of a bisphenol A plant according to [378]
a) Storage tank; b) Reactor; c) Column; d) Crystallizer; e) Separator; f) Melt; g) Desorber; h) Flaking off

## 4.5. Uses, Economic Aspects

Bisphenol A is the most important bisphenol which is used industrially. Since it is mainly used for the production of high-grade plastics, the demand has increased sharply with industrial development.

In 1976, production of bisphenol A in the United States (producers: Shell, General Electric, Dow, Aristech, Union Carbide) amounted to 168 000 t, and in Western Europe (producers: Bayer, Shell, Rhône-Poulenc, General Electric, Dow) 136 000 t were consumed. In line with the rapid industrial growth of the 1980s, world bisphenol A production capacity rose to $10^6$ t, a figure which will presumably be exceeded in 1990. About 50 % of this capacity is located in the United States, over 30 % in Western Europe, and the remainder mostly in Japan. However, due to the current economic decline, the new overcapacity is only gradually being utilized.

Whereas in the mid-1970s ca. 50 % of the bisphenol A was converted to epoxy resins and only 40 % to polycarbonates [341], [380], in 1989 over 50 % was used for polycarbonates and ca. 40 % for epoxy resins. The remainder is used for a wide range of products such as aromatic polyesters with phthalic acid; polysulfones; polyetherketones; unsaturated, hydrolysis-resistant polyesters [381]; antioxidants; phenoxy resins from bisphenol A and epichlorohydrin [382]; and phenolic resins elastified by incorporation of bisphenol A [383]. However, the majority is used for flame retardants, especially tetrabromobisphenol A [384].

Polyester resins can be linked with phenolic resins via 4,4 bis(4-hydroxyphenyl)pentanoic acid (see Table 15, [382]). Bisphenol Z is used in the production of polycarbonate films [385].

Bisphenol Z and bisphenol A substituted with various ortho-alkyl groups counteract coccidiosis in poultry [386] and bisphenols containing nitrogen heterocycles are said to act as laxatives [387].

## 4.6. Toxicology

Bisphenols are far less toxic than phenol itself. Therefore some of them can even be used as antioxidants in cosmetics and foods.

Bisphenol A has an $LD_{50}$ (mouse) of 4 g/kg, limits the weight gain in rats in feeding experiments, and has a clearly estrogenic effect. The permeation of bisphenol A through the skin is practically negligible compared with phenol. It is, however, reported that bisphenol A leads to minor skin irritation.

# 5. Hydroxybiphenyls

## 5.1. Physical Properties

Hydroxybiphenyls are colorless solids with a weak odor (if any). They are only sparingly soluble in water, but most of them dissolve readily in sodium hydroxide to form salts. Polar organic solvents such as alcohols, ethers, esters, and ketones have an outstanding solvent power for hydroxybiphenyls, whereas that of nonpolar solvents such as aromatic and aliphatic hydrocarbons is lower. The para-hydroxybiphenyls are generally less soluble than the *o*-hydroxybiphenyls. The solubilities can be affected considerably by substitution. Thus, e.g., (**22**) is soluble in various organic solvents but insoluble in water and 10% sodium hydroxide.

Some physical properties are given in Table 16. 2-Hydroxybiphenyl (**18**) has the greatest industrial importance; further physical properties for 2-hydroxybiphenyl are summarized as follows:

| Solubility (g/100 g solvent at 25 °C) in | |
|---|---|
| methanol | 800 |
| ethanol | 590 |
| acetone | 660 |
| white spirit | 20 |
| water | 0.07 |
| at 100 °C | 0.13 |
| 10% NaOH | 50 |
| Heat of fusion | 91 J/g |

| | | 138 °C |
|---|---|---|
| Flash point | | 138 °C |
| Ignition temperature | | 520 °C |
| Vapor pressure (100 °C) | | 70 Pa |

**Table 16.** Physical properties of some hydroxybiphenyls

| Name | CAS registry no. | Structural formula | Molecular formula | $M_r$ | mp, °C | bp, °C | $d_4^{20}$ |
|---|---|---|---|---|---|---|---|
| 2-Hydroxybiphenyl | [90-43-7] | 18 | $C_{12}H_{10}O$ | 170.1 | 57 | 286 | 1.219 |
| 4-Hydroxybiphenyl | [92-69-3] | 19 | $C_{12}H_{10}O$ | 170.1 | 165 | 312 | |
| 2,2'-Dihydroxybiphenyl | [1806-29-7] | 20 | $C_{12}H_{10}O_2$ | 186.2 | 109 | | |
| 4,4'-Dihydroxybiphenyl | [92-88-6] | 21 | $C_{12}H_{10}O_2$ | 186.2 | 274–275 | | |
| 3,5,3',5'-Tetra-*tert*-butyl-4,4'-dihydroxybiphenyl | [2455-14-3] | 22 | $C_{28}H_{42}O_2$ | 410.6 | 185 | | |

## 5.2. Chemical Properties

The chemical properties of hydroxybiphenyls are very similar to those of phenol itself. Phenyl substitution has a small but insignificant deactivating effect on the other phenyl nucleus; in the ortho-position it also causes steric hindrance, which is, however, still lower than that of the *tert*-butyl group.

Accordingly, hydroxybiphenyls can undergo reactions at the OH group involving etherification, esterification, and salt formation in the usual way. Because of their bifunctionality, dihydroxybiphenyls, especially 4,4-dihydroxybiphenyl, can be converted into polycondensation products which have achieved a limited industrial importance as liquid crystal polymers.

Substitution at the ring (halogenation, sulfonation, nitration, and alkylation) also takes place as expected. In addition, 4,4'-dihydroxybiphenyls can be oxidized to yield diphenoquinones [389] and 2,2'-dihydroxybiphenyls can undergo ring-closure reactions to form dibenzofurans, carbazoles, and dibenzothiophenes [390].

## 5.3. Production

**2-Hydroxybiphenyl (2-Phenylphenol).** The starting material for the synthesis of 2-hydroxybiphenyl (**18**) is cyclohexanone, which is first autocondensed to form 2-cyclohexenylcyclohexanone. This reaction is catalyzed by acidic ion exchangers or by metal salts of higher aliphatic carboxylic acids [391]. 2-Cyclohexenylcyclohexanone is subsequently dehydrogenated to (**16**) in the gas phase with noble metal catalysts on an inert carrier material [391]–[393].

2-Hydroxybiphenyl is formed as a byproduct in the pressure phenol synthesis from chlorobenzene (→ Phenol).

**4-Hydroxybiphenyl** (**19**) is formed alongside (**18**) as a byproduct in the alkaline pressure hydrolysis of chlorobenzene to phenol in 5–6% yield. The mixture consists of 2/3 ortho and 1/3 para isomer which can be separated by fractional distillation.

4-Hydroxybiphenyl can also be produced by sulfonation of biphenyl and alkaline fusion of the *para*-biphenylsulfonic acid [394].

**2,2'-Dihydroxybiphenyl** (**20**) is obtained by alkaline fusion of biphenylene oxide at ca. 300 °C (sometimes catalyzed by fluorene or carbazole [395], [396]) or by debutylation of 3,3',5,5'-tetra-*tert*-butyl-2,2'-dihydroxybiphenyl, which itself is produced by the oxidative coupling of 2,4-di-*tert*-butylphenol [397]. 2,2'-Dihydroxy-5,5'-dimethylbiphenyl can be synthesized accordingly.

**4,4'-Dihydroxybiphenyl** (**21**) can also be synthesized in excellent yields by debutylation of 3,3',5,5'-tetra-*tert*-butyl-4,4'-dihydroxybiphenyl [398]. Other production processes are the alkaline fusion of 4,4'-biphenyldisulfonic acid, the boil down of benzidinebisdiazonium salts [399], and the decarboxylation of 4,4'-dihydroxybiphenyl-2-carboxylic acid [400]. Symmetrically alkylated 4,4'-dihydroxybiphenyls such as (**22**) or 4,4'-dihydroxy-2,2',3,3',5,5',6,6'-hexamethylbiphenyl are preferably obtained by the oxidative coupling of the corresponding substituted monophenols [398], [401]. Here, diphenoquinones are generally formed in the first step which must be reduced. Phenol itself always gives isomeric mixtures of dihydroxybiphenyls on coupling with various oxidants [402]. These mixtures are also formed by alkaline fusion of 4,4'-dihydroxybiphenylsulfone [403].

## 5.4. Analysis, Quality Specifications, Storage

The purity of 2-hydroxybiphenyl can be determined by gas chromatography or by titration [404]. The melting point should be 56–58 °C.

Commercially 2-hydroxybiphenyl is supplied as virtually colorless and odorless flakes or in the form of the sodium salt (NaOH content < 1.0%). In the absence of light and moisture it can be stored indefinitely.

## 5.5. Uses

**2-Hydroxybiphenyl** shows a broad effectiveness against bacteria, yeasts, and mold, but low toxicity towards warm-blooded organisms [405] and very good biodegradability [406]. Therefore it is increasingly used as a preservative and fungicide and for the production of disinfectants in the leather, paper, glue, textile, plastics, and cosmetics industries. 2-Hydroxybiphenyl and its sodium salt are used for the preservation of citrus fruits in their skins (approval according to EC guideline no. E 321).

2-Hydroxybiphenyl is also used as a dyeing accelerant for dyeing synthetic fibers, e.g., polyethylene terephthalate and cellulose triacetate (Levegal OPS, Bayer).

Sulfonation of 2-hydroxybiphenyl and condensation with formaldehyde yields a dispersion agent that is used in the formulation of plant protection agents in aqueous solution [407].

*Trade names* for 2-hydroxybiphenyl are:

Preventol O extra (Bayer, content > 99.5%)
Preventol ON extra (Bayer, sodium salt)
Dowicide 1 (DOW), Dowicide A (DOW, sodium salt)
Cotane (Coalite)

**4-Hydroxybiphenyl** is an intermediate for the production of varnish resins and nonionogenic emulsifiers (trade names: Emulgator W and Emulvin W, Bayer) which are used in the plant protection, polyurethane, and dyeing sectors.

**Other Hydroxybiphenyls.** A range of possible applications have been proposed for **21**, **21**, and **22**: e.g., as a fungicide (**20**), an antiseptic for feeds (**20**), (**21**), a toner for xerography (**20**), starting materials for polyesters, polycarbonates, and powder coatings (**21**). Mixtures of isomers of chlorinated and brominated dihydroxybiphenyls (e.g., chlorinated **20**) are used for flameproofing plastics and have been proposed for combatting algae and bacteria. However, dihydroxybiphenyls are actually only used as stabilizers and antioxidants, principally **22**, in oils, rubber, and polyolefins.

## 5.6. Toxicology

The toxicity of 2-hydroxybiphenyl towards warm-blooded organisms is usually low for a phenolic compound. The $LD_{50}$ for rats is 2700–3000 mg per kilogram body weight. Accumulation in rat tissue was not established even after feeding over a period of 2 years with feeds containing 2% 2-hydroxybiphenyl. Daily doses of 200 mg/kg did not produce any harmful effects even after administration for several weeks. Human skin is neither irritated nor sensitized by application of a 5% solution in oil [404],

[405]. 2-Hydroxybiphenyl is toxic to fish in concentrations > 2 mg/L. However, since it is rapidly and completely degraded in biological sewage treatment plants, fishing grounds are not endangered [406].

# 6. Phenol Ethers

Phenol ethers are organic compounds of the general formula Ar – O – R, in which Ar represents a substituted or unsubstituted phenyl ring and R is a substituted or unsubstituted aliphatic or aromatic hydrocarbon group.

## 6.1. Properties

**Physical Properties.** Most phenol ethers have a pleasant, characteristic smell. Some phenol ethers are important in the flavoring and fragrance industry. The ethers described in this article are those, whose importance lies mainly in other areas. The lower phenol ethers are colorless liquids that are insoluble in water and soluble in organic solvents.

**Chemical Properties.** The ether linkage in phenol ethers is very stable. The carbon – oxygen bond in completely aromatic phenol ethers such as diphenyl ether is cleaved only under drastic conditions, e.g., with alkalis at high temperature and pressure; in contrast, the aliphatic carbon – oxygen bond in anisole, for example, is split by treatment with strong acids such as hydrogen iodide to form phenol and an alkyl iodide. Sterically hindered methoxy groups can be cleaved selectively with $BCl_3$. The phenyl ring is susceptible to substitution reactions, including hydroxylation, nitration, halogenation, and sulfonation. Phenol ethers can be alkylated or acylated by Friedel – Crafts reactions. Aromatic – aliphatic ethers, however, can also be cleaved in the presence of Friedel – Crafts catalysts at high temperatures.

## 6.2. Production

Nearly all phenol ethers can be prepared by the reaction of phenols with alkyl or aryl halides (preferably chlorides) in weakly basic aqueous media:

$C_6H_5ONa + RCl \rightarrow C_6H_5OR + NaCl$

For the preparation of phenol ethers with short-chain aliphatic groups, dialkyl sufates are often employed as alkylating agents:

$$C_6H_5OH + (RO)_2SO_2 \xrightarrow{NaOH} C_6H_5OR + (RO)SO_3Na + H_2O$$

The second alkyl group present in the monoalkyl sulfate can be reacted using higher temperatures. Another method for the preparation of aromatic–aliphatic ethers is the etherification of phenols with aliphatic alcohols in the presence of an acidic ion exchanger [408]. This reaction is performed using excess phenol to suppress formation of the dialkyl ether [409].

## 6.3. Representative Phenol Ethers

**Phenyl methyl ether** [*100-66-3*] (**23**), anisole, $C_7H_8O$, $M_r$ 108.15, mp −37.5 °C, bp (101.3 kPa) 155 °C, $d_4^{20}$ 0.9954, is a colorless liquid with a spicy-sweet smell.

OCH₃ on benzene ring
**23**

Anisole can be produced by all the aforementioned methods. It is an important intermediate in the synthesis of organic compounds, for example fragrances and pharmaceuticals. The compound is also used as a solvent and as a heat transfer medium.

**Phenyl ethyl ether** [*103-73-1*] (**24**), phenetole, $C_8H_{10}O$, $M_r$ 122.17, mp −29.5 °C, bp (101.3 kPa) 170 °C, $d_4^{20}$ 0.966, is a colorless liquid which is prepared by etherification of phenol with ethyl chloride or diethyl sulfate. Phenyl ethyl ether is an intermediate in numerous organic syntheses.

OC₂H₅ on benzene ring
**24**

**Diphenyl ether** [*101-84-8*] (**25**), $C_{12}H_{10}O$, $M_r$ 170.21, mp 26.8 °C, bp (101.3 kPa) 257.9 °C, bp (1.34 kPa) 121 °C, $d_4^{20}$ 1.0748, occurs, e.g., in lemon balm oil.

Ph–O–Ph
**25**

The compound forms colorless crystals that possess a geranium-like aroma. Diphenyl ether is formed as a byproduct in the high-pressure hydrolysis of chlorobenzene. Because of its high thermal stability, a eutectic mixture of diphenyl ether and biphenyl is employed as a heat transfer medium. Diphenyl ether is the starting material for organic intermediates and is used as a fragrance component. The compound is also employed as a processing aid in the production of polyesters [410]; several polybromine derivatives of diphenyl ether are used as fire retardants.

**2-Phenoxyethanol** [*122-99-6*] (**26**), $C_8H_{10}O_2$, $M_r$ 138.17, mp −2 °C, bp (101.3 kPa) 237 °C, bp (2.4 kPa) 134–135 °C, $d^{20}$ 1.1020, is a colorless liquid with a faint rose-like aroma.

C₆H₅—OCH₂CH₂OH

**26**

2-Phenoxyethanol is produced by the hydroxyethylation of phenol, for example, in the presence of alkali-metal hydroxides or alkali-metal borohydrides [411]. The compound is known under the trade name Cellosolve as a solvent for cellulose acetate. 2-Phenoxyethanol is a synthetic intermediate in the production of plasticizers, pharmaceuticals, and fragrances. In the perfume industry the compound finds use as a fixing agent.

**Phenoxyacetic acid** [*122-59-8*] (**27**), $C_8H_8O_3$, $M_r$ 152.16, mp 98–99 °C, bp (101.3 kPa) 285 °C (decomp.), forms colorless crystals that possess a faint honey-like odor and are soluble in the common organic solvents.

C₆H₅—OCH₂COOH

**27**

Phenoxyacetic acid can be prepared by reaction of phenol with chloroacetic acid in dilute aqueous sodium hydroxide, or by hydrolysis of its methyl ester. The methyl ester is obtained by reaction of phenol and methyl chloroacetate in the gas phase with zinc oxide–aluminum oxide as catalyst [412]. Phenoxyacetic acid is an intermediate for fragrances and dyes and is a starting material for the synthesis of penicillin V.

**4-Methylanisole** [*104-93-8*] (**28**), $C_8H_{10}O$, $M_r$ 122.17, bp (101.3 kPa) 176.5 °C, $d_{25}^{25}$ 0.9698, is a colorless liquid with a pungent odor, which in dilute solution has a flower-like fragrance. 4-Methylanisole is a starting material for the preparation of *p*-anisaldehyde.

CH₃—C₆H₄—OCH₃

**28**

**2-Methoxyphenol** [*90-05-1*] (**29**), guaiacol, $C_7H_8O_2$, $M_r$ 124.15, bp (101.3 kPa) 205 °C, bp (3.2 kPa) 106.5 °C, $d_4^{21}$ 1.128, forms colorless prisms that have a sweet, faint phenol odor. Guaiacol is present in many essential oils and is also found as a flavor compound in many foods. Guaiacol is prepared by monomethylation of catechol using methyl halides or dimethyl sulfate. Monomethylation of catechol can also be accomplished with methanol in the presence of phosphoric acid, phosphates, or an ion exchanger [413]. Guaiacol is an important synthetic intermediate in the flavoring and fragrance industries (e.g., for the synthesis of vanillin), as well as in the pharmaceutical industry.

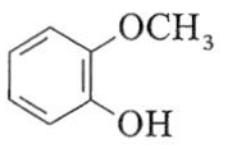

OCH₃, OH
**29**

**2-Ethoxyphenol** [94-71-3] (**30**), $C_8H_{10}O_2$, $M_r$ 138.17, mp 29 °C, bp (101.3 kPa) 217 °C, bp (5.3 kPa) 68 °C, is prepared analogously to guaiacol by monoetherification of catechol. The compound is an intermediate in the synthesis of ethyl vanillin.

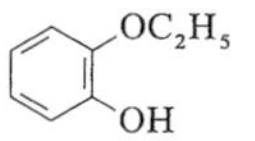

OC₂H₅, OH
**30**

**2-Isopropoxyphenol** [4812-20-8] (**31**), $C_9H_{12}O_2$, $M_r$ 152.20, bp (101.3 kPa) 217 °C, bp (14.7 kPa) 100 – 102 °C, is prepared by monoetherification of catechol with isopropyl halides in the presence of alkali-metal salts and either a phase transfer catalyst [414] or a polyhydroxyalkyl monoether in an autoclave [415]. This compound is the starting material for the synthesis of the pesticide Propoxur (2-isopropoxyphenyl methylcarbamate).

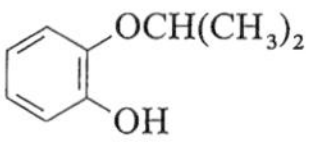

OCH(CH₃)₂, OH
**31**

**Veratrole** [91-16-7] (**32**), $C_8H_{10}O_2$, $M_r$ 138.17, mp 22.5 °C, bp (101.3 kPa) 206 °C, bp (1.3 kPa) 90 °C, $d_{25}^{25}$ 1.084, is prepared by methylation of catechol or guaiacol. This compound is an intermediate in the synthesis of alkaloids and pharmaceuticals.

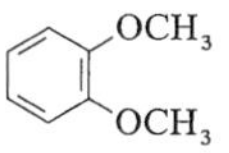

OCH₃, OCH₃
**32**

**1,3-Benzodioxole** [274-09-9] (**33**), $C_7H_6O_2$, $M_r$ 122.13, bp (101.3 kPa) 172 °C, is a colorless liquid. This compound is prepared by the reaction of catechol and dihalomethanes, for example, by use of concentrated aqueous alkaline solution in the presence of tetraalkylammonium or phosphonium salts [416], sometimes in the presence of alkyl iodides [417]. 1,3-Benzodioxole is an important intermediate in organic synthesis, especially for the preparation of alkaloids.

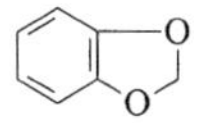

**33**

# 7. Halogen Derivatives of Phenolic Compounds

The phenolic derivatives described in this chapter include halogenated alkylphenols, cycloalkylphenols, arylphenols, alkylarylphenols, phenoxyphenols, dihydroxydiphenyl sulfides (thiobisphenols) and halogenated salicylanilides. They all possess antimicrobial properties and are therefore widely used as microbicides. Depending on their chemical and physical properties, the compounds can be used (1) as preservatives for functional fluids, (2) to control slime in process water systems, (3) to protect various materials e.g., textiles, paper, and wood against microbial decay, and (4) as active ingredients in disinfectants (for halogenated phenol derivatives used in disinfectants. Pentachlorophenol (PCP) is not discussed in this chapter even though it can be widely used as a biocide. The use of PCP as a biocide is rapidly declining because of its ecotoxicity and general toxicity. The production, use, and marketing of PCP and PCP-containing products has been banned in various countries, e.g., in the Federal Republic of Germany from January 1, 1990.

Similarly, 2,4,5- und 2,4,6-trichlorophenol are not included in this chapter although they possess strong microbicidal properties. There is pressure to find suitable substitutes for these compounds because they are the starting materials in the formation of the highly toxic tetrachlorodibenzo-*p*-dioxins.

## 7.1. Introduction

Phenol (formerly known as carbolic acid) was discovered in 1834. In 1860, the antiseptic properties of phenol were recognized and used for antibacterial treatment of surgical instruments and bandaging material and for treatment of wounds. This was the beginning of the use of chemical methods to kill microbes or to inhibit their growth. At the same time, the toxic and skin-cauterizing properties of phenol itself led to the isolation and syntheses of hundreds of phenol derivatives with the aim of finding less toxic phenol derivatives and more effective compounds which required lower concentrations for use. With the discovery that chemicals could be used to protect materials against microbial attack and decay, phenol derivatives with suitable chemical and physical properties were investigated for such uses. This development, which can now be considered complete, also led to discovery of the correlation between structure and effectiveness and the mechanism of action of phenol derivatives [426].

Phenol derivatives are so-called membrane-active microbicides. Initially, they are reversibly adsorbed on the surface of microbe cells and inhibit metabolic processes. If applied in higher concentrations, the phenol derivatives dissolve fairly quickly and readily in lipids according to their chemical and physical properties and penetrate into the microbe cell. They react with the protoplasma and cell proteins and inhibit enzymes, which leads to destruction of the microbial cell. The applied concentration determines whether the phenol derivatives act microbistatically or microbicidally.

Phenol derivatives show optimal antimicrobial effect in neutral and acid media, i.e., in the undissociated state, since only in this state are they membrane-active and exhibit affinity for the negatively charged surface of the microbial cell wall; the negatively charged phenolate anions are repelled. Therefore, if optimal advantage is to be taken of the antimicrobial power of phenol deravatives during application, their $pK_a$ values must be taken into account. The $pK_a$ values for some important phenolic microbicides are listed in the following:

| | |
|---|---|
| 2-Hydroxybiphenyl | 11.6 |
| 6-Isopropyl-3-methylphenol | 10.6 |
| 4-Chloro-3,5-dimethylphenol | 9.7 |
| 4-Chloro-3-methylphenol | 9.6 |
| 2,4,6-Trichlorophenol | 8.5 |
| 4-Chloro-2-(2,4-dichloro)phenoxyphenol | 7.9 |
| 2,2′-Methylenebis(4-chlorophenol) | |
| $\quad pK_a$ (1) | 8.7 |
| $\quad pK_a$ (2) | 12.6 |

Halogenation of phenol derivatives produces active substances with substantially increased antimicrobial effect. The acidity of the compounds is also more pronounced; the degree of dissociation of halogenated phenol derivatives increases with the number of halogen atoms (see above). The combination of alkylation and halogenation — the latter preferably in the 4-position — produces phenolic microbicides which have great practical importance as active ingredients in disinfectants and for the preservation of materials. These include 2-benzyl-4-chlorophenol, 4-chloro-3-methylphenol, and 4-chloro-3,5-dimethylphenol. The minimum inhibitory concentrations listed in Table 17 give an indication of the range of effect of halogenated phenol derivatives which have particular practical importance as microbicides.

It is often claimed that phenols, especially chlorinated phenols, show high oral toxicity, are toxic percutaneously, and are in general difficult or impossible to degrade biologically. This disqualification of phenols is a very unreliable generalization and is not verified by the facts. The whole class of phenolic microbicides must not be looked at uniformly and unjustly disqualified, just because some representatives of this class of compounds indeed possess properties such as high ecotoxicity, are persistent, and can contain dioxins. Even some of the halogenated phenol derivatives have toxicological properties and ecological effects which satisfy the standards for microbicides used for preservation of materials and for disinfectants that are set by a civilization with an increased sense of responsibility for the environment. As shown by the curves in Figure 3, there are phenol derivatives including halogenated phenols which can be biodegraded quickly and effectively, and others, e.g., pentachlorophenol which are only biodegraded slowly and only if present in very low concentrations [418]. The biodegradation of phenols does not only occur in biological sewage plants; surface waters

**Table 17.** Minimum inhibitory concentration (MIC) of halogenated phenol derivatives in nutrient agar

| Test organisms | MIC, mg/L | | | | |
|---|---|---|---|---|---|
| | PCMC [a] | PCMX [b] | Chlorophene [c] | Dichlorophene [d] | Fenticlor [e] |
| *Bacillus subtilis* | 150 | 75 | 10 | 100 | |
| *Bacillus punctatum* | 200 | 100 | 10 | 50 | |
| *Escherichia coli* | 250 | 200 | 3500 | 100 | 75 |
| *Leuconostoc mesenteroides* | 200 | 100 | 10 | 5 | 3500 |
| *Proteus vulgaris* | 200 | 200 | 100 | 50 | |
| *Pseudomonas aeruginosa* | 800 | 1000 | 5000 | >5000 | |
| *Staphylococcus aureus* | 200 | 100 | 20 | 5 | 35 |
| *Desulfovibrio desulfuricans* | 35 | 50 | 50 | 20 | |
| *Candida albicans* | 200 | 75 | 50 | 50 | 50 |
| *Torula rubra* | 50 | 100 | 50 | 50 | 50 |
| *Aspergillus niger* | 100 | 100 | 100 | 100 | 50 |
| *Aureobasidium pullulans* | 30 | 50 | 20 | 35 | <20 |
| *Chaetomium globosum* | 80 | 50 | 20 | 20 | 75 |
| *Cladosporium herbarum* | 200 | 100 | 100 | 200 | 35 |
| *Coniophora puteana* | 100 | 35 | 5 | 2 | <20 |
| *Penicillium citrinum* | 100 | 50 | 75 | 50 | 20 |
| *Penicillium glaucum* | 100 | 35 | 50 | 50 | 10 |
| *Polyporus versicolor* | 5000 | 75 | 50 | 50 | 35 |
| *Trichoderma viride* | 140 | 100 | 100 | 50 | 50 |
| *Trichophyton pedis* | 100 | 50 | 10 | 10 | 10 |

[a] PCMC = 4-chloro-3-methylphenol.
[b] PCMX = 4-chloro-3,5-dimethylphenol.
[c] Chlorophene = 2-benzyl-4-chlorophenol.
[d] Dichlorophene = 2,2'-methylenebis(4-chlorophenol).
[e] Fenticlor = 2,2'-thiobis(4-chlorophenol).

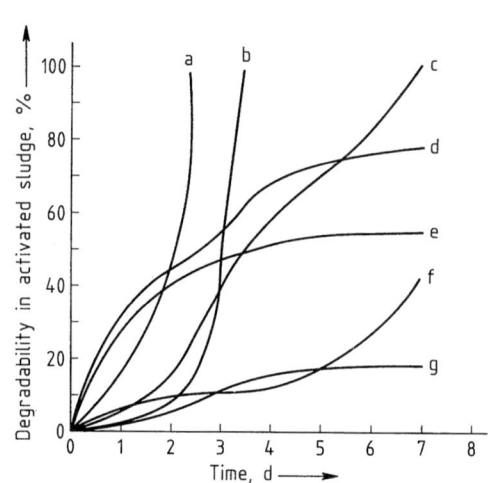

**Figure 3.** Biodegradability of various phenol derivatives
a) 2-Hydroxybiphenyl, 100 ppm; b) 4-Chloro-3-methylphenol, 50 ppm; c) Benzylphenol, 20 ppm; d) 2-Benzyl-4-chlorophenol, 50 ppm; e) Dichlorophene, 20 ppm; f) 4-Chloro-3,5-dimethylphenol, 10 ppm; g) Pentachlorophenol, 10 ppm

also possess a considerable potential to biodegrade phenol derivatives including pentachlorophenol [419].

## 7.2. Representative Compounds

**4-Chloro-3-methylphenol** [*59-50-7*] (**34**), *p*-chloro-*m*-cresol, PCMC, $C_7H_7ClO$, $M_r$ 142.59, has a density $\varrho$ 1.37 g/cm$^3$ at 20 °C, a vapor pressure of 8 Pa at 20 °C, of 0.7 kPa at 100 °C, a *mp* of 63–65 °C, a *bp* of ca. 239 °C, the flash point is ca. 118 °C, and the ignition temperature ca. 590 °C. Its solubility in water is 4 g/L. 4-Chloro-3-methylphenol is readily soluble in alkali solutions and organic solvents, and has a p$K_a$ of 9.6.

*Toxicological Data* [426], [427]. LD$_{50}$ (oral, rat) 5129 mg/kg, LD$_{50}$ (dermal, rat) > 500 mg/kg (exposure 7 d). PCMC irritates the skin and mucous membranes but is not significantly sensitizing and not mutagenic. LCLo for fish 0.5 mg/L (96 h exposure).

*Preparation and Uses.* 4-Chloro-3-methylphenol is prepared by chlorination of 3-methylphenol with, e.g., SO$_2$Cl$_2$. PCMC is of major importance as a preservative for aqueous functional fluids and as a raw material for disinfectants. This is due to its strong and wide-ranging antimicrobial effect combined with favorable chemical and physical properties and good biodegradability [426].

**2-Benzyl-4-chlorophenol** [*120-32-1*] (**35**), Chlorophene, $C_{13}H_{11}ClO$, $M_r$ 218.69, has a density $\varrho$ 1.22 g/cm$^3$ at 20 °C, a vapor pressure of 10 Pa at 100 °C, a *mp* of 48–49 °C, and a *bp* (1.5 kPa) of 190–196 °C. Its flash point is ca. 188 °C, and the ignition temperature ca. 490 °C. 2-Benzyl-4-chlorophenol has a solubility in water of 0.4 g/L at 20 °C. It is readily soluble in alkali solutions and organic solvents and has a p$K_a$ of 11.0.

*Toxicological Data.* LD$_{50}$ (oral, rat) > 5000 mg/kg, LD$_{50}$ (dermal, rat) > 2500 mg/kg, LC$_{50}$ (inhaled for 4 h, rat) 2.5 mg/L. Chlorophene strongly irritates the skin and mucous membranes but is not toxic percutaneously and is neither mutagenic nor teratogenic. Its LCLo for *Leuciscus idus* is 0.5 mg/L (48 h exposure). Organisms in activated sludge tolerate 10–50 mg/L Chlorophene and are able to degrade it [426], [427].

*Preparation and Uses.* Chlorophene can be prepared by reaction of 2-benzylphenol with SO$_2$Cl$_2$ at 100 °C. It is used as an active agent in disinfectants.

**4-Chloro-2-cyclopentylphenol** [*13347-42-7*] (**36**), $C_{11}H_{13}ClO$, $M_r$ 196.68, *mp* 23.7 °C, *bp* 303 °C, has a solubility in water of 0.24 g/L at 20 °C, and is readily soluble in organic solvents.

*Toxicological Data.* LD$_{50}$ (oral, rat) 2460 mg/kg, LD$_{50}$ (oral, rabbit) 420 mg/kg, LD$_{50}$ (dermal, rabbit) 850 mg/kg. 4-Chloro-2-cyclopentylphenol strongly irritates the skin and mucous membranes but is not sensitizing.

*Preparation.* 4-Chloro-2-cyclopentylphenol is prepared by shaking 4-chlorophenol and cyclopentene in concentrated sulfuric acid; 2,6-dicyclopentyl-4-chlorophenol (*bp* 235–240 °C at 1.8 kPa) is produced as a byproduct. 4-Chloro-2-cyclopentylphenol is no longer used as a microbicide in disinfectants because of its toxicity.

**5-Chloro-2-hydroxybiphenyl** [*1331-46-0*] (**37**), monochloro-*o*-phenylphenol, MCOPP, C$_{12}$H$_9$ClO, $M_r$ 204.66, *mp* 36–37 °C, *bp* (1.5 kPa) 178 °C, is very sparingly soluble in water and readily soluble in organic solvents.

*Toxicological Data.* LD$_{50}$ (oral, rat) 3500 mg/kg. 5-Chloro-2-hydroxybiphenyl is toxic percutaneously, irritates the skin and mucous membranes, and can cause photoallergic contact dermatitis. 5-Chloro-2-hydroxybiphenyl is no longer used as a microbicide in disinfectants because of these properties.

*Preparation.* 5-Chloro-2-hydroxybiphenyl is obtained along with 2-chloro-6-hydroxybiphenyl when chlorine is introduced into 2-hydroxybiphenyl.

**4-Chloro-3,5-dimethylphenol** [*88-04-0*] (**38**), *p*-chloro-*m*-xylenol, PCMX, C$_8$H$_9$ClO, $M_r$ 156.61, *mp* 114–116 °C, *bp* 246 °C, has a vapor pressure of 0.3 kPa at 100 °C, its solubility in water is 0.33 g/L at 20 °C. 4-Chloro-3,5-dimethylphenol is soluble in alkali solutions and organic solvents, and has a p$K_a$ of 9.7.

*Toxicological Data.* LD$_{50}$ (oral, rat) 3830 mg/kg. PCMX mildly irritates the skin and mucous membranes, has a low sensitization potential, is not mutagenic in the Ames test, and is not teratogenic in rats; the LCLo for *Leuciscus idus* is 1 mg/L. Organisms in activated sludge tolerate 15 mg/L PCMX and slowly degrade the substance [426], [427].

*Preparation and Uses.* 4-Chloro-3,5-dimethylphenol is prepared by reaction of 3,5-dimethylphenol with chlorine at 25–30 °C or with SO$_2$Cl$_2$ starting at 30–40 °C [420]. It is used as a raw material for disinfectants, as an active component in deodorants, soaps, skin preparations for dermatological disorders, and antiseptics, and as a preservative for aqueous functional fluids.

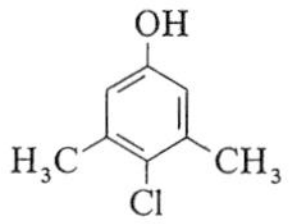

**38**

**4-Chloro-6-isopropyl-3-methylphenol** [*89-68-9*] (**39**), 4-chlorothymol, $C_{10}H_{13}ClO$, $M_r$ 184.67, *mp* 64 °C, *bp* 259–263 °C, has a solubility in water of 0.3 g/L at 20 °C, is soluble in alkali solutions and readily soluble in organic solvents.

*Toxicological Data.* $LD_{50}$ (oral, mouse) 2460 mg/kg.

*Preparation and Uses.* 4-Chloro-6-isopropyl-3-methylphenol is prepared by reaction of equimolar amounts of propene and 4-chloro-3-methylphenol in the presence of bleaching clay under pressure at 140–160 °C or by chlorination of thymol with $SO_2Cl_2$ in chloroform. On account of its microbicidal properties [421], chlorothymol is of importance as a raw material for disinfectants [422].

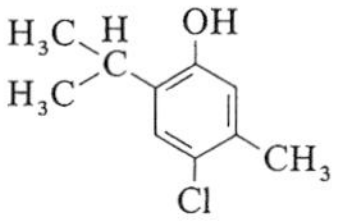

**39**

**4-Benzyl-2-chloro-6-methylphenol** [*65053-91-0*] (**40**), $C_{14}H_{13}ClO$, $M_r$ 232.72, has a *bp* (27 Pa) of 128–129 °C, and an $n_D^{20}$ of 1.5944.

*Toxicological Data.* $LD_{50}$ (oral, mouse) 1700 mg/kg, $LD_{50}$ (subcutaneous, mouse) 1000 mg/kg, $LD_{50}$ (intravenous, mouse) 96 mg/kg [423].

*Preparation.* 4-Benzyl-2-chloro-6-methylphenol is prepared from 2-chloro-6-methylphenol (or the forerunnings formed in the production of 4-chloro-2-methylphenol by chlorination of 2-methylphenol) by reaction with benzyl chloride in the presence of anhydrous zinc chloride starting at 100 °C and reaching completion after 5 h at 150–155 °C [423].

*Uses.* The substance shows a strong antimicrobial effect especially towards bacteria and fungi, and in particular dermatophytes. The microbicide is used mostly for epidermal applications, e.g., in pharmaceuticals, cosmetics, and soaps because it is tolerated well by skin and has a low skin-sensitization potential.

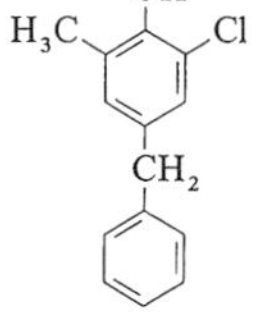

**40**

**4-Bromo-2,6-dimethylphenol** [*2374-05-2*] (**41**), 4-bromo-2,6-xylenol, $C_8H_9BrO$, $M_r$ 201.08, has a *mp* of 80–81 °C.

*Toxicological Data.* $LD_{50}$ (i.p., mouse) 650 mg/kg.

*Preparation and Uses.* 4-Bromo-2,6-dimethylphenol is prepared by bromination of 2,6-dimethylphenol in glacial acetic acid at 15 °C and used as a microbicide in disinfectants.

OH
H₃C      CH₃

Br
**41**

**2,4-Dichloro-3,5-dimethylphenol** [*133-53-9*] (**42**), dichloro-*m*-xylenol, DCMX, $C_8H_8Cl_2O$, $M_r$ 191.06, mp 95–96 °C (sublimation), *bp* 250 °C, flash point 134 °C, has a solubility in water of 0.2 g/L at 20 °C, is soluble in alkali solutions, and readily soluble in organic solvents.

*Toxicological Data.* $LD_{50}$ (oral, rat) is 2810–4120 mg/kg.

*Preparation and Uses.* DCMX is obtained by reaction of 4-chloro-3,5-dimethylphenol with *N*-chloroacetamide in glacial acetic acid with concentrated HCl. It is used as a preservative in aqueous functional fluids and as a microbicide in disinfectants, but in comparison with PCMX (**38**) is of lesser importance because it has a more intense smell and is less soluble in water.

OH
       Cl
H₃C      CH₃
    Cl
**42**

**Tetrabromo-2-methylphenol** [*576-55-6*] (**43**), 3,4,5,6-tetrabromo-*o*-cresol, $C_7H_4Br_4O$, $M_r$ 423.75, mp 205–208 °C, is practically insoluble in water. The solubility (g/L) in various solvents is as follows: ethanol 85, isopropanol 80, ethylene glycol 9, propylene glycol 8, vaseline oil 4. Tetrabromo-2-methylphenol discolors in light and in contact with traces of iron.

*Toxicological Data.* Tetrabromo-2-methylphenol has an $LD_{50}$ (oral, rat) > 6400 mg/kg and is well tolerated by skin and mucous membranes.

*Preparation and Uses.* Tetrabromo-2-methylphenol is obtained by reaction of *o*-cresol with bromine in tetrachloromethane in the presence of aluminum and iron powder. It is used as a preservative for cosmetics, as an active component in deodorants and disinfectants, and as an active agent for antimycotic finishing of textile materials; it is, however, of minor importance [427].

OH
Br      CH₃
Br   Br
   Br
**43**

**5-Chloro-2-(2,4-dichlorophenoxy)phenol** [*3380-34-5*] (**44**), 2,4,4′-trichloro-2′-hydroxydiphenylether, Triclosan, $C_{12}H_7Cl_3O_2$, $M_r$ 289.55, mp 60–61 °C, decomposes at

280–290 °C. The vapor pressure is 2.6 Pa at 100 °C. The solubility (g/L) in various solvents is as follows: water 0.01 at 20 °C, 0.039 at 50 °C, in 0.1 mol/L NaOH 23; it is readily soluble in organic solvents. Triclosan has a p$K_a$ of 7.9, is unstable in light, and discolors in contact with traces of heavy metals. Under certain conditions such as higher alkalinity and heat, conversion to chlorinated dibenzo-*p*-dioxines can occur.

*Toxicological Data* [427]. LD$_{50}$ (oral, rat or dog) > 5000 mg/kg, LD$_{50}$ (intravenous, rat) 29 mg/kg, LD$_{50}$ (intraperitoneal, rat) 198 mg/kg. Triclosan is neither mutagenic nor teratogenic, is tolerated well by the skin, and has shown no evidence of sensitization potential, LCLo for fish is ca. 0.6 mg/L (exposure 48 h). Activated sludge organisms tolerate up to 10 mg/L Triclosan and degrade the substance slowly.

*Preparation and Uses.* Triclosan is produced by treatment of 2,4,4′-trichloro-2′-methoxydiphenyl ether with AlCl$_3$ in benzene under reflux. It is highly active against staphylococci and is used as an active agent in deodorants and antiseptic soaps.

**44**

**2,2′-Methylenebis(4-chlorophenol)** [*97-23-4*] (**45**), Dichlorophene, C$_{13}$H$_{10}$Cl$_2$O$_2$, $M_r$ 269.14, *mp* 176 °C, has a density $\varrho$ 1.5 g/cm$^3$ at 20 °C, a vapor pressure < 1 Pa at 100 °C, and is not steam-volatile. The solubility (g/L) in different solvents is as follows: water 0.07 at 20 °C, 0.2 at 50 °C; 10% NaOH 500, butanol 400; toluene 12. Dichlorophene has a p$K_a$ (1) of 8.7 and p$K_a$ (2) of 12.6.

*Toxicological Data* [426], [427]. LD$_{50}$ (oral, rat) 3300 mg/kg. Dichlorophene does not irritate the skin, mildly irritates the mucous membranes, and skin sensitization is not significant. Dichlorophene is neither mutagenic nor teratogenic. LCLo for *Leuciscus idus* 0.5 mg/L (72 h exposure); data on biodegradability are given in Figure 3.

*Preparation and Uses.* Dichlorophene is produced by reaction of 4-chlorophenol with 0.5 M aqueous formaldehyde solution with the addition of sulfuric acid at 50–65 °C. On account of its bactericidal and fungicidal properties, the pure nearly odorless compound can be used in deodorants and antiseptic soaps, for antimicrobial finishing of textiles and paper, and as a preservative for metalworking lubricants.

**45**

**4,4′-Methylenebis(2,6-dichlorophenol)** [*3933-88-8*] (**46**), C$_{13}$H$_8$Cl$_4$O$_2$, $M_r$ 338.04, *mp* 184–185 °C, is produced by introducing chlorine into a solution of 4,4′-methylenebis(2-chlorophenol) in glacial acetic acid at room temperature or by treating 2,6-dichlorophenol with formalin in concentrated sulfuric acid at −10 °C. It has antimicrobial properties, but because of its high toxicity it no longer has practical importance as a microbicide.

**2,2'-Methylenebis(3,4,6-trichlorophenol)** [70-30-4] (**47**), Hexachlorophene, $C_{13}H_6Cl_6O_2$, $M_r$ 406.92, has an *mp* of 164–165 °C; its solubility (g/L) in various solvents is as follows: acetone 1010, ethanol 500, toluene 56. Hexachlorophene is soluble in alkalis but practically insoluble in water. Its $pK_a$ values are 5.4 and 10.9.

*Toxicological Data* [426], [427]. $LD_{50}$ (oral, rat) 59 mg/kg, $LD_{50}$ (oral, mouse) 80 mg/kg. Hexachlorophene is neither mutagenic nor teratogenic; it is however, neurotoxic, irritates the skin and mucous membranes, and is absorbed through the skin.

*Preparation and Uses.* Hexachlorophene is formed by reaction of 2,4,5-trichlorophenol with paraformaldehyde in fuming sulfuric acid (20% $SO_3$) at 65–135 °C. Hexachlorophene is a strong bacteriostat which also has a microbicidal effect at high concentrations. The use of Hexachlorophene as an antimicrobial agent in cosmetics, medicinal soaps, detergent solutions, and textiles has, however, been discontinued because of its toxicity. For details on the correlation between chemical structure and biological activity of chlorinated methylenebisphenols and their physical, chemical, and pharmacological properties, see [424].

**2,2'-Methylenebis(6-bromo-4-chlorophenol)** [15435-29-7] (**48**), bromochlorophene, $C_{13}H_8Br_2Cl_2O_2$, $M_r$ 426.95, *mp* 188–191 °C, its solubility (g/L) in different solvents is as follows: water < 0.01, 95% ethanol 0.5, propanol 0.7, isopropanol 0.4, 1.2-propylene glycol 2.5, paraffin oil 0.05.

*Toxicological Data* [427]. $LD_{50}$ (oral, rat) 3700 mg/kg, $LD_{50}$ (oral, mouse) 1550 mg/kg, $LD_{50}$ (dermal, rat) > 10 000 mg/kg. Bromochlorophene does not irritate the skin and there is no evidence of sensitization.

*Preparation and Uses.* Bromochlorophene is produced by bromination of 2,2'-methylenebis(4-chlorophenol) in glacial acetic acid in the cold. It is used as a microbicide in cosmetics, e.g., deodorants, mouthwashes, and toothpastes.

**2,2′-Thiobis(4-chlorophenol)** [97-24-5] (**49**), Fenticlor, $C_{12}H_8Cl_2O_2S$, $M_r$ 287.18, *mp* 174 °C, has a solubility in water of 30 mg/L and is readily soluble in alkalis and organic solvents.

*Toxicological Data.* $LD_{50}$ (oral, rat) 3250 mg/kg; photosensitization of the skin is possible.

*Preparation and Uses.* Fenticlor is produced (1) by the reaction of 4-chlorophenol with sulfur dichloride in the presence of $AlCl_3$ or (2) by chlorination of 2,2′-thiobisphenol with chlorine in glacial acetic acid or with $SO_2Cl_2$ in dichlorobenzene. It shows a tendency to discoloration. Fenticlor has a broad range of antimicrobial action; it is, e.g., effective against bacteria, fungi, yeasts, and algae. Fenticlor is therefore used as a preservative for aqueous functional fluids and to control algae and slime in process water circulation systems [426].

**49**

**2,2′-Thiobis(4,6-dichlorophenol)** [97-18-7] (**50**), Bithionol, $C_{12}H_6Cl_4O_2S$, $M_r$ 356.07, *mp* 188 °C, is virtually insoluble in water, readily soluble in ethanol, diethyl ether, acetone, glacial acetic acid, and dilute alkalis.

*Toxicological Data.* $LD_{50}$ (oral, rat) 1430 mg/kg, $LD_{50}$ (oral, mouse) 2100 mg/kg.

*Preparation and Uses.* Bithionol is prepared by reaction of 2,4-dichlorophenol with sulfur dichloride and $AlCl_3$ in carbon disulfide or tetrachloromethane. It is used as a microbicide to protect aqueous functional fluids from biodeterioration and to control algae and slime formation in process water systems [426].

**50**

**2,2′-Thiobis(3,4,6-trichlorophenol)** [3161-14-6] (**51**), $C_{12}H_4Cl_6O_2S$, $M_r$ 424.97, *mp* 157 – 159 °C, is practically insoluble in water, readily soluble in ethanol, diethyl ether, acetone, and glacial acetic acid. It is prepared by reaction of 2,4,5-trichlorophenol with $SCl_2$ and $AlCl_3$ in 1,2-dichloroethane at 45 °C. Its use as a microbicide is declining because of its toxicity.

**51**

**3,3′-Thiobis(2,4,6-trichlorophenol)** [104294-11-3] (**52**), $C_{12}H_4Cl_6O_2S$, $M_r$ 424.97, *mp* 285 °C, is prepared by reaction of 2,4,6-trichlorophenol with $SCl_2$ in the presence of

AlCl$_3$ in carbon disulfide or tetrachloromethane. Its use as a microbicide is declining.

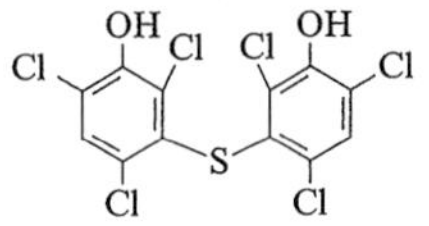

**52**

**5,2′-Dichlorosalicylanilide** [*6626-92-2*] (**53**), C$_{13}$H$_9$Cl$_2$NO$_2$, $M_r$ 282.14, mp 188 °C, is produced by heating equimolar quantities of 3-chlorosalicylic acid and 2-chloroaniline in the presence of PCl$_3$ in chlorobenzene or toluene.

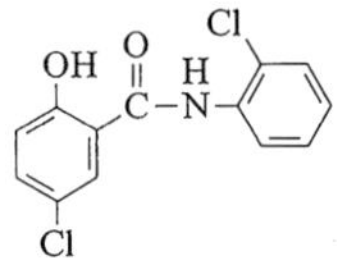

**53**

**5,3′,4′-Trichlorosalicylanilide** [*642-84-2*] (**54**), Anobial, C$_{13}$H$_8$Cl$_3$NO$_2$, $M_r$ 316.59, mp 244–246 °C, is virtually insoluble in water, soluble in acetone and methanol. It is produced by reaction of 3-chlorosalicylic acid and 3,4-dichloroaniline with PCl$_3$ in boiling toluene. It is more effective as a microbicide than DBS (**55**), but its use as a microbicide [425] is declining.

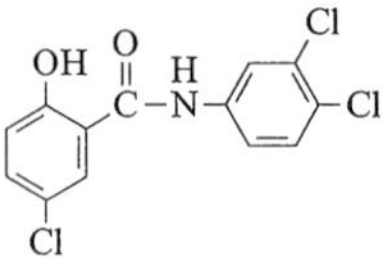

**54**

**4,3′,4′-Trichlorosalicylanilide** [*58622-66-5*] (**54a**) C$_{13}$H$_8$Cl$_3$NO$_2$, $M_r$ 316.59, mp 221–222 °C, is soluble in acetone and methanol and is light-sensitive. It is produced by reaction of 4-chlorosalicylic acid and 3,4-dichloroaniline with AlCl$_3$ and PCl$_3$ in nitrobenzene. It was formerly used a microbistat in deodorants and soaps [425].

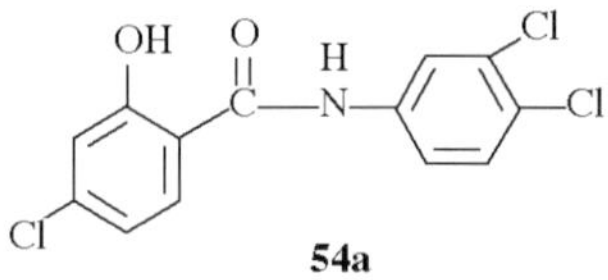

**54a**

**5,4′-Dibromosalicylanilide** [*87-12-7*] (**55**), DBS, Disanyl, C$_{13}$H$_9$Br$_2$NO$_2$, $M_r$ 371.21, mp 238–245 °C, is very sparingly soluble in water, sparingly soluble in ethanol, and soluble in acetone and dimethylformamide. Its LD$_{50}$ (oral, mouse) is 1700 mg/kg; photosensitization of the skin is possible. 5,4′-Dibromosalicyclanilide is produced from salicylanilide and bromine in acetic acid. DBS has broad antimicrobial activity [425] but its use as a microbicide in, e.g., paper and plastics is declining.

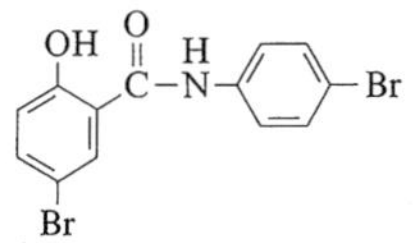

**3,5,4'-Tribromosalicylanilide** [87-10-5] (**56**), TBS, Trisanyl, $C_{13}H_8Br_3NO_2$, $M_r$ 449.96, mp 227–228 °C, is practically insoluble in water, soluble in hot acetone, and readily soluble in dimethylformamide. Its $LD_{50}$ (oral, rat) is 570 mg/kg; photosensitization of the skin is possible. 3,5,4'-Tribromosalicylanilide is produced by direct bromination of salicylanilide in aqueous acetic acid at 50–55 °C or in water containing an emulsifier at 50–65 °C. Its use as a microbicide for the protection of materials is declining [425], [427].

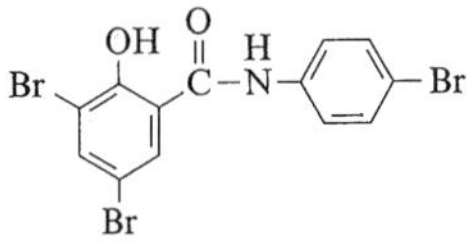

# 8. References

General References

*Beilstein*, **6 H**, 470 ff; **6 I**, 234 ff; **6 II**, 442 ff, **6 III**, 1655 ff, **6 IV**, 3011 ff.
*Houben-Weyl*, 6/1 c, parts 1 and 2.
*Rodd's Chemistry of Carbon Compounds*, 2nd. ed., Elsevier, Amsterdam, vol. **III**, A, pp. 289 ff., suppl. III A, pp. 161 ff.
A. Dierichs, R. Kubika: *Phenole und Basen*, Akademie-Verlag, Berlin 1958.
*Ullmann*, 4th ed., **18**, 191–215.

Specific References

[1] R. S. Bownman, D. R. Stevens, *J. Org. Chem.* **15** (1950) 1172–1176.
[2] G. A. Nikiforov, V. V. Ershov, *Russ. Chem. Rev. (Engl. Transl.)* **39** (1970) 644–654.
[3] P. Demerseman et al., *Bull. Soc. Chim. Fr.* 1971, 201–210.
[4] D. P. Biddiscombe et al., *J. Chem. Soc.* 1963, 5764–5768.
[5] G. Bertholon, *Bull. Soc. Chim. Fr.* 1967, 2977–2982.
[6] R. Larmatine, R. Perrin, *Bull. Soc. Chim. Fr.* 1969, 443–445.
[7] J. Bassus et al., *Bull. Soc. Chim. Fr.* 1974, 3031–3038.
[8] G. Bertholon, R. Perrin, *Bull. Soc. Chim. Fr.* 1975, 1537–1544.
[9] W. A. Pardee, W. Weinrich, *Ind. Eng. Chem.* **36** (1944) 595–603.
[10] R. Hanley et al., *J. Chem. Soc.* 1964, 4404–4406.
[11] H. Stage, E. Müller, P. Faldix, *Erdöl Kohle* **6** (1953) 375–380.
[12] S. Ohe: *Computer Aided Data Book of Vapour Pressure*, Data Book Publishing Co., Tokyo 1976.
[13] G. Bertholon et al., *Bull. Soc. Chim. Fr.* 1971, 3180–3187.
[14] A. Ksiazczak, J. J. Kosinski, *Fluid Phase Equilib.* **55** (1990) 17–37.

[15] E. Terres et al., *Brennst. Chem.* **36** (1955) 289–301.
[16] L. U. Erichsen, E. Dobbert, *Brennst. Chem.* **36** (1955) 338–345.
[17] E. Terres et al., *Brennst. Chem.* **36** (1955) 359–372.
[18] T. H. Coffield et al., *J. Am. Chem. Soc.* **79** (1957) 5019–5023.
[19] G. H. Stillson et al., *J. Am. Chem. Soc.* **67** (1945) 303–307; **68** (1946) 722.
[20] C. A. Brown, *Synthesis* 1974, 427–428.
[21] *Houben-Weyl,* **6/2,** 35–41.
[22] A. McKillop, J.-C. Fiaud, R. P. Hug, *Tetrahedron* **30** (1974) 1379–1382.
[23] J. B. McKinley et al., *Ind. Eng. Chem. Anal. Ed.* **16** (1944) 304.
[24] Ethyl Corp., DE 2 208 827, 1972 (J. C. Wollensak).
[25] *Houben-Weyl,* **7/2 a,** 849–853.
[26] O. M. Kut, U. R. Dätwyler, G. Gut, *Ind. Eng. Chem. Prod. Res. Dev.* **27** (1988) 215–225.
[27] V. Pencev et al., *Erdöl Kohle Erdgas Petrochem. Brennst. Chem.* **23** (1970) 571–574.
[28] *Houben-Weyl,* **6/1 c,** 1121–1129.
[29] D. C. Nonhebel et al., *J. Chem. Res. Synop.* 1977, 12–16, *J. Chem. Res. Miniprint,* 0201–0283.
[30] F. R. Hegwill, G. B. Howie, *Aust. J. Chem.* **31** (1978) 907–917.
[31] *Houben-Weyl,* **7/1,** 152, 158.
[32] Bayer AG, EP 330 036, 1989 (A. Schnatterer, H. Fiege).
[33] *Houben-Weyl,* **10/1,** 576–584; 825–828.
[34] *Houben-Weyl,* **5/3,** 679–688.
[35] *Houben-Weyl,* **7/2 b,** 1413–1422.
[36] *Houben-Weyl,* **10/1,** 1028–1037.
[37] *Houben-Weyl,* **6/1 c,** 925–1019.
[38] *Houben-Weyl,* **6/1 c,** 1073–1090.
[39] *Houben-Weyl,* **6/1 c,** 1086–1090.
[40] *Houben-Weyl,* **7/1,** 20–29.
[41] *Houben-Weyl,* **14/2,** 193–230.
[42] *Houben-Weyl,* **6/1 c,** 1021–1060.
[43] M. Weinberg, *Rev. Chim. (Bucharest)* **21** (1967) 669–671.
[44] V. A. Parfenova, V. I. Isagulyants, *Khim. Tekhnol. Topl. Masel* **1** (1974) 7–8.
[45] B. Roach, T. Eisner, J. Meinwald, *J. Chem. Ecol.* **6** (1980) 511–516; *Chem. Abstr.* **93,** 12 90 24 k.
[46] P. D. Swann et al., *Fuel* **52** (1973) 154–155.
[47] J. H. P. Tyman, *J. Chromatogr.* **111** (1975) 277–284, 285–292.
[48] V. S. Pansare, A. B. Kulkarni, *J. Indian Chem. Soc.* **41** (1964) 251.
[49] J. Maak, P. Buryan, *Chem. Listy* **69** (1975) 457–518.
[50] M. Sy, G. Lejeune, *Chim. Ind. (Paris)* **78** (1957) 619–628.
[51] K. F. Lang, I. Eigen, *Fortschr. Chem. Forsch.* **8** (1967) 93–130.
[52] P. Buryan, J. Maak, V. M. Nabivach, *J. Chromatogr.* **148** (1978) 203–210.
[53] H. Pichler, G. Schwarz, *Brennst. Chem.* **50** (1969) 72–78. H. Pichler, P. Hennenberger, *Brennst. Chem.* **50** (1969) 341–346.
[54] A. Siller, *J. Prakt. Chem. Ser. 4* **1** (1955) 209–224.
[55] V. Kusy, *Erdöl Kohle Erdgas Petrochem. Brennst. Chem.* **23** (1970) 575–580.
[56] R. S. Aries, S. A. Savitt, *Chem. Eng. News* **28** (1950) 316–321.
[57] R. E. Maple: *Symposium on Refining Petroleum for Chemicals,* Am. Chem. Soc. New York City Meeting, 7–12, Sept. 1969, pp. D 105–113.
[58] E. O. Woolfolk, M. Orchin, *Ind. Eng. Chem.* **42** (1950) 552–556.
[59] Y. Sugimoto, Y. Miki, S. Yamada, M. Oba, *Nippon Kagaku Kaishi* 1984, 755–763.

[60] S. Preis, *J. Prakt. Chem. Ser. 4* **1** (1955) 157–171, 172–176, 177–186.
[61] VEB-Leuna-Werke, DE 1 003 222, 1955 (S. Preis).
[62] Yorkshire Tar Distillers, GB 736 604, 1952 (D. W. Milner, R. Flathers).
[63] Gesellschaft für Teerverwertung, DE 1 127 908, 1959 (H. Wille, L. Rappen). Shell, US 2 370 554, 1941 (D. B. Luten, S. B. Thomas).
[64] Shell, US 2 336 720, 1941 (A. De Benedictis, D. B. Luten).
[65] J. G. M. Thorne, *Chem. Process. (Chicago)* (1970) March, 29–31.
[66] R. E. Dean, E. N. White, D. Mc Neil, *J. Appl. Chem. (London)* **9** (1959) 629–641.
[67] G. A. Olah (ed.): *Friedel–Crafts and Related Reactions*, vol. II/1, Interscience Publ., New York 1964.
[68] *Houben-Weyl*, **6/1 c**, 925–1019.
[69] O. N. Tsvetkov et al., *Int. Chem. Eng.* **7** (1967) 104–113, 113–121.
[70] P. S. Belov, K. D. Korenev, A. Y. Estigneev, *Khim. Tekhnol. Topl. Masel* **4** (1981) 58–61.
[71] J. Klein, H. Widdecke, *Chem. Ing. Tech.* **51** (1979) 568; **53** (1981) 954–957.
[72] M. M. Sharma et al., *Ind. Eng. Chem. Prod. Res. Dev.* **29** (1990) 29–34, 1025–1031.
[73] Shell, US 2 923 745, 1957 (V. W. Buls, R. S. Miller).
[74] Shell, US 3 177 259, 1963; US 3 116 336, 1960 (J. L. Van Winkle).
[75] J. I. De Jong, *Recl. Trav. Chim. Pays-Bas* **83** (1964) 469–476.
[76] Consolidation Coal, US 3 461 175, 1965 (M. D. Kulik, R. J. Laufer).
[77] Consolidation Coal, US 3 408 410, 1965 (R. J. Laufer, M. D. Kulik).
[78] Consolidation Coal, US 3 418 380, 1965 (R. J. Laufer, M. D. Kulik).
[79] Bayer, DT 944 014, 1954 (R. Stroh, R. Seydel).
[80] Ethyl Corp., US 2 831 898, 1954 (G. G. Ecke, A. J. Kolka).
[81] R. Stroh, R. Seydel, W. Hahn in W. Foerst (ed.): *Neuere Methoden der präparativen organischen Chemie*, vol. **II**, Verlag Chemie, Weinheim 1960, pp. 231–246.
[82] A. J. Kolka et al., *J. Org. Chem.* **22** (1957) 642–646.
[83] G. G. Knapp et al.: *Proc. 7th World Petroleum Congress*, vol. **5,** Elsevier, London 1967, pp. 403–413.
[84] K. N. Kulieva, *Khim. Tekhnol. Topl. Masel* **10** (1989) 38–40.
[85] R. Gompper, *Angew. Chem.* **76** (1964) 412–423.
[86] Hoechst, DE-OS 2 039 062, 1970 (E. Reindl).
[87] F. R. J. Willemse et al., *Recl. Trav. Chim. Pays-Bas* **90** (1971) 5–13.
[88] Koppers Comp. Inc. US 3 091 646, 1960 (G. Leston).
[89] Ethyl Corp., DE 1 493 622, 1964 (T. H. Coffield, G. G. Knapp, J. P. Napolitano).
[90] Shell, US 3 268 595, 1962 (K. L. Mai).
[91] Shell, DE 1 154 484, 1960 (V. W. Buls).
[92] Koppers Comp. Inc., US 3 267 154, 1963 (T. Hokama).
[93] Koppers Comp. Inc., US 3 267 153, 1963 (G. Leston).
[94] Consolidation Coal, US 3 032 595, 1958 (N. B. Neuworth, R. J. Laufer, E. P. Previc).
[95] Shell, FR 1 398 153, 1964 (J. C. E. Bolle, J. M. Tomaszewski).
[96] M. V. Kurashev, N. N. Korneev, *Neftekhimiya* **28** (1988) 176–182.
[97] Ethyl Corp., EP 206 085 (D. E. Goins, S. W. Holmes, E. A. Burt).
[98] Koppers Comp. Inc., US 3 331 879, 1965 (G. Leston).
[99] M.-F. Berny, *Bull. Soc. Chim. Fr.* 1969, 973–976.
[100] Monsanto, GB 1 008 592, 1963 (B. B. Millward).
[101] Koppers Comp. Inc., US 3 267 155, 1963 (G. Leston).
[102] Koppers Comp. Inc., US 3 267 152, 1963 (T. Hokama).

[103] Koppers Comp. Inc., US 3 346 649, 1966 (G. Leston).
[104] Koppers Comp. Inc., DE-AS 1 230 432, 1964 (G. Leston).
[105] Bayer, DE 1 142 873, 1961 (W. Hahn).
[106] Ethyl Corp., US 3 367 981, 1963 (J. P. Napolitano).
[107] I. T. Golubtschenko et al., *Dokl. Akad. Nauk SSSR* **261** (1981) 891–894.
[108] Bayer, DE 1 159 960, 1960 (W. Hahn).
[109] UK Wesseling, DE-OS 2 040 228, 1970 (C. Kalav).
[110] T. Yamanaka, *Bull. Chem. Soc. Jpn.* **49** (1976) 2669–2673.
[111] E. A. Goldsmith et al., *J. Org. Chem.* **23** (1958) 1871–1876.
[112] S. Skraup, W. Beifuss, *Ber. Dtsch. Chem. Ges.* **60** (1972) 1070.
[113] R. G. Anderson, S. H. Sharman, *J. Am. Oil Chem. Soc.* **48** (1971) 107–112.
[114] Geigy (UK) Ltd., GB 1 146 173, 1966 (W. Pickels, D. R. Randell).
[115] *Houben-Weyl,* **6/1 c,** 1060–1072.
[116] J. Becht, W. Gerhardt, Akademie Wiss. DDR, Int. Tagung Grenzflächenaktive Stoffe, Berlin 1977, part 1, pp. 143–149.
[117] *Houben-Weyl,* **6/1 c,** 202–208.
[118] *Houben-Weyl,* **6/1 c,** 1081–1083.
[119] Bayer AG, EP 297 391, 1988 (R. Dujardin, W. Ebert).
[120] Schering-Kahlbaum, DT 467 640, 1928; DT 479 352, 1929 (H. Jordan).
[121] Bayer, DT 1 105 428, 1958 (G. Schuckmann, H. Schnell).
[122] *Houben-Weyl,* **6/1 c,** 502–550.
[123] *Houben-Weyl,* **6/1 c,** 1081.
[124] Kuraray, JP 50 76 033, 1973.
[125] Mitsui Toatsu Chemicals, JP 55 07 731, 1978; *Chem. Abstr.* **93 (15):** 14 99 78c.
[126] Bayer AG, EP 102 493, 1983 (J. Käsbauer, K. F. Wedemeyer).
[127] Kuraray, JP 62 255 445, 1986 (H. Tamai, T. Kunitomi); *Chem. Abstr.* **109 (19):** 17 00 21a.
[128] B. E. Leach, *J. Org. Chem.* **43** (1978) 1794–1797. Continental Oil Co., US 3 979 464, 1975.
[129] Continental Oil Co., DE 2 811 471, 1978 (R. Poe, J. F. Scamehorn, C. R. Schupbach).
[130] Continental Oil Co., US 3 985 812, 1974 (E. Del Bel, D. C. Jones, M. B. Neuworth).
[131] Continental Oil Co., US 3 862 248, 1968 (D. C. Jones, M. B. Neuworth).
[132] BASF, DE 1 668 874, 1968 (L. Arnold, H. Pasedach, H. Pommer).
[133] BASF, DE 1 793 037, 1968 (L. Arnold, H. Pasedach, H. Pommer).
[134] BASF, EP 0 123 233, 1984 (N. Goetz, H. Laas, P. Tavs, L. Hupfer, K. Baer).
[135] N. G. Baranova et al., *Khim.-Farm. Zh.* **22** (1968) 736–739; *Chem. Abstr.* **110** (13): 114 380u.
[136] Teijin Ltd., US 3 968 172, 1973/76 (Y. Ichikawa, Y. Yamanaka, T. Naruchi, O. Kobayashi, K. Sakota).
[137] Y. Ichikawa et al., *Ind. Eng. Chem. Prod. Res. Dev.* **18** (1979) 373–375.
[138] Rütgerswerke AG, DE-AS 1 254 155, 1965 (M. Froitzheim, K. F. Lang, L. Rappen, E. Schweym, J. Turowski).
[139] General Electric Co., US 4 503 270, 1983 (J. J. Talley).
[140] D. E. Pearson, C. A. Buehler, *Synthesis* 1971, 471–472.
[141] T. Kotinagawa, K. Shimokawa, *Bull. Chem. Soc. Jpn.* **47** (1974) 1535–1536.
[142] Rütgerswerke AG, DE 2 346 498, 1973 (H.-D. Hausigk, G. Löhnert).
[143] K. Aomura et al., *Kogakubu Kenkyu Hokoku (Hokkaido Diagaku)* **76** (1975) 147–153; *Chem. Abstr.* 84, 16 688r.
[144] M. Inoue, S. Emoto, *Chem. Pharm. Bull.* **20** (1972) 232–237.
[145] M. Inoue et al., *Yaki Gosei Kagaku Kyokai Shi* **28** (1970) 1127–1132.

[146] Maruzen Petrochemical Co., EP 320 936, 1988 (T. Yamagishi, T. Idai, E. Takahasi).
[147] Maruzen Oil Co., JP 53 063 346, 1976 (F. Suganuma, Y. Hirose, M. Hayano, T. Nakamura).
[148] Maruzen Oil Co., JP 52 106 825, 1976 (M. Sato et al.).
[149] Maruzen Oil Co., JP 53 119 839, 1977 (F. Suganuma et al.).
[150] Taoka Chemical Co., EP 80 880, 1982 (M. Wada, S. Maki).
[151] Taoka Chemical Co., JP 59 108 730, 1982.
[152] Taoka Chemical Co., JP 59 076 033, 1982.
[153] L. H. Klemm, D. R. Taylor, *J. Org. Chem.* **45** (1980) 4320–4329.
[154] Bayer DE 1 142 873, 1961 (W. Hahn).
[155] F. J. Sowa, H. D. Hinton, J. A. Nieuwland, *J. Am. Chem. Soc.* **54** (1932) 2019.
[156] Mobil Oil, US 4 391 998, 1981 (M. M. Wu).
[157] Sh. G. Sadykhov, Ch. K. Rasulov, Yu. K. Dzhaforov, Ch. K. Salamanova, *Azerb. Khim. Zh.* 1981, no. 5, 53–55; *Chem. Abstr.* **96** (1982) 162 239u.
[158] N. Yoneda, Y. Takahashi, C. Tajiri, A. Suzuki, *Nippon Kagaku Kaishi* 1977 no. 6, 831–836; *Chem. Abstr.* **87** (1977) 134 237n.
[159] PCUK Ugine Kuhlmann, US 4 423 254, 1982 (G. A. Olah).
[160] R. Stroh, R. Seydel, W. Hahn, *Angew. Chem.* **69** (1957) 699.
[161] Ethyl Corp., US 3 766 276, 1979 (L. E. Goddard).
[162] UOP, US 4 275 248, 1980 (B. E. Firth).
[163] Geigy (UK), GB 1 146 173, 1966 (W. Pickles, D. R. Randell).
[164] FMC Corp., US 3 936 410, 1973 (F. L. Terhune, G. A. Rampy).
[165] FMC Corp., US 3 859 395, 1973 (F. L. Terhune, G. A. Rampy).
[166] Bayer, DE 1 280 255, 1965 (W. Schulte-Huermann, A. Kersting).
[167] FMC Corp., BE 830 497, 1975 (R. A. Aal, N. Chen, J. K. Chapman).
[168] Bayer, DE 2 528 303, 1975 (W. Biedermann, H. Köller, K. Wedemeyer).
[169] Bayer, DE 3 824 284, 1988 (P. Wimmer, H. J. Buysch, L. Puppe).
[170] M. Nitta, *Bull. Chem. Soc. Jpn.* **47** (1974) 2360–2364.
[171] M. Nitta, *Bull. Chem. Soc. Jpn.* **47** (1974) 2897–2898.
[172] T. Yamanaka, *Bull. Chem. Soc. Jpn.* **49** (1976) 2669–2673.
[173] R. Stroh, R. Seydel, W. Hahn in W. Foerst (ed.): *Newer Methods of Preparative Organic Chemistry,* vol. **2,** Academic Press, New York 1963, p. 337 ff.
[174] Union Rheinische Kraftstoff, DE-OS 2 139 622, 1971 (E. Biller, D. Kühne).
[175] Bayer, DE-OS 2 242 628, 1972 (A. Klein, K. Wedemeyer).
[176] Koppers Co., US 3 992 455, 1976 (G. Leston).
[177] Sumitomo Chem. Co., DE-OS 2 340 218, 1973 (H. Suda, N. Kotera, S. Hasegawa).
[178] Koppers Co., US 4 046 818, 1975 (G. Leston).
[179] H. Hart, E. A. Haglund, *J. Org. Chem.* **15** (1950) 396.
[180] Rhein. Kampfer Fabrik, GB 325 855, 1928.
[181] L. S. Charcenko, V. Zavgorodnij, *Ukr. Khim. Zh. (Russ. Ed.)* **30** (1964) 187–190.
[182] J. Bassus, R. Perrin, *C. R. Hebd. Seances Acad. Sci. Ser. C,* **264** (1967) 1444–1446.
[183] J. Bassus, *Bull. Soc. Chim. Fr.* 1974, 3031–3038.
[184] G. Bertholon, C. Decoret, *Bull. Soc. Chim. Fr.* 1975, 1530–1536.
[185] *Chem. Eng. (N.Y.)* **85** (1978), no. 12, 62–63.
[186] Ajinomoto KK, JP 57 046 931, 1980; *Chem. Abstr.* **97** (1982) 164 951d.
[187] Bayer DE 1 142 873, 1961 (W. Hahn).
[188] Combinatul Petrochimic, Ploiesti, RO 66 499, 1978 (J. Hersccvici et al.); *Chem. Abstr.* **98** (1983) 109 290u.

[189] Ethyl Corp. US 3 766 276, 1970 (L. E. Goddard).
[190] Ethyl Corp. US 3 933 927, 1973 (L. E. Goddard).
[191] Hüls AG, DE 3 443 736, 1984 (H. Alfs).
[192] Bayer, DE 944 014, 1954 (R. Stroh, R. Seydel).
[193] Nalco Chem. Co., US 4 092 367, 1976 (B. W. Bridwell, C. E. Johnson).
[194] Hoechst, DE-AS 1 813 840, 1968 (J. Bohunek).
[195] Progil, FR 1 336 080, FR 1 336 081, 1981 (M. E. DeGeorges, J. Berthoux).
[196] Progil, DE-OS 2 034 369, 1971 (M. Berthoux, M. Jean, G. Schwachhofer).
[197] Union Rheinische Kraftstoff, DE-OS 2 021 525, 1970 (C. Kalav).
[198] Union Carbide, DE-AS 1 443 346, 1961 (S. Kaufmann, R. E. Nicolson).
[199] B. Loev, J. T. Massengale, *J. Org. Chem.* **22** (1957), 988–989.
[200] BASF, DE-OS 2 526 644, 1975 (F. Merger et al.).
[201] Chem. Werke Hüls AG, DE-OS 2 745 589, 1977 (H. Alfs, G. Boehm, H. Steiner).
[202] Sh. G. Sadykhov et al., *Azerb. Khim. Zh.* 1982 no. 1, 26–28; *Chem. Abstr.* **97** (1982) 162 490q.
[203] A. Corma, H. Garcia, J. Primo, *J. Chem. Res, Synop.* 1988 no. 1, 40–41.
[204] I. I. de Jong, *Rec. Trav. Chim. Pays-Bas* **83** (1964) 472.
[205] Hitachi Chemical Co., JP 7 400 823, 1970; *Chem. Abstr.* **81** (1974) 25 364g.
[206] ICI, GB 701 264, 1950 (W. Kunz).
[207] ICI, GB 1 068 693, 1965 (A. S. Briggs).
[208] Union Rheinische Kraftstoff, DE-OS 2 608 241, 1976 (B. Haas, E. Meisenburg).
[209] Coalite & Chem. Prod., DE 1 186 873, 1961 (H. F. Bondy, F. R. Moore).
[210] Pennsalt Chem. Corp., US 3 014 079, 1958 (J. F. Olin).
[211] Chevron Res. Co., DE-OS 1 965 165, 1969 (G. K. Kohn).
[212] A. P. Bolton et al., *J. Org. Chem.* **33** (1968) 3415–3418.
[213] Goi Kasei KK, JP-Kokai 6 019 739, 1983; *Chem. Abstr.* **102** (1985) 220 560x.
[214] Stauffer Chem. Co., US 4 103 096, 1977 (S. L. Giolito, S. B. Mirviss).
[215] Hodogaya Chem. Co., DE-OS 1 960 747, 1969 (N. Onodera, H. Mitsuta, M. Wataya).
[216] Ethyl Corp., US 4 774 368, 1987 (D. R. Brackenridge).
[217] Coalite & Chem. Prod., BE 661 950, 1965; *Chem. Abstr.* **64** (1966) 6 561g.
[218] Phillips Petr. Co., US 3 878 255, 1972 (J. R. Norell).
[219] J. R. Norell, *J. Org. Chem.* **38** (1973) 1929.
[220] R. Stroh, R. Seydel, W. Hahn, *Angew. Chem.* **69** (1957) 699.
[221] Ethyl Corp., US 3 355 504, 1967 (T. H. Coffield, G. G. Knapp, J. P. Napolitano).
[222] Maruzen Petrochemical Co., JP-Kokai 6 150 935, 1984; *Chem. Abstr.* **105** (1986) 114 725g.
[223] D. R. Stevens, *Ind. Eng. Chem.* **35** (1943) 655–660.
[224] J. K. Gehlawat, M. M. Sharma, *J. Appl. Chem.* **20** (1970) 93–98.
[225] Gulf Res. & Dev., US 2 560 666, 1948 (D. R. Stevens, R. S. Borrman).
[226] ICI, DE-OS 2 423 356, 1973 (J. Atkinson, D. Ball).
[227] Koppers Co., US 3 091 646, 1960 (G. Leston).
[228] Koppers Co., US 3 346 649, 1963 (G. Leston).
[229] Koppers Co., DE-AS 1 230 432, 1963 (G. Leston).
[230] *Houben-Weyl*, E 4, vol. **6/1 c**, 959.
[231] Sumitomo Chem. Co., GB 1 296 179, 1972 (K. Kamoshita, S. Nakai).
[232] C. L. Zundel, L. Choron, DE-OS 1 518 460, 1963.
[233] Bayer, DE-OS 2 602 149, 1976 (H. Fiege et al.).
[234] Shell, US 2 841 622, 1957 (D. G. Norton, F. C. Davis).
[235] Shell, US 2 841 623, 1957 (D. G. Norton, R. C. Morris).

[236] Sterlitamak Petrochem., GB 1 512 941, 1977 (N. V. Zakharova et al.).
[237] Novokuib. Giprokauchuka, DE-OS 2 749 278, 1977 (N. V. Zakharova et al.).
[238] Ethyl Corp., US 3 919 333, 1974 (J. C. Wollensak).
[239] Shell, GB 557 519, 1942 (D. B. Luten, A. DeBenedictis).
[240] ICI, GB 582 057, 1945 (H. A. Basterfield).
[241] Chem. Werke Hüls AG, DE 3 401 343, 1983 (W. Otte, R. Nehring, M. ZurHausen).
[242] O. M. Kut, U. R. Daetwyler, G. Gut, *Ind. Eng. Chem. Res.* **27** (1988) no. 2, 215–225.
[243] F. F. Wolny, J. J. Lamb, *Elastomerics* **116** (1984) no. 4, 40–43.
[244] C. D. Gutsche, B. Dhawan, K. H. No, R. Muthukrishan, *J. Am. Chem. Soc.* **103** (1981) no. 13, 3782–3792.
[245] B. Dhawan, S. Chen, C. D. Gutsche, *Makromol. Chem.* **188** (1987) no. 5, 921–950.
[246] Konebo Ltd., JP-Kokai 6 399 031, 1986 (Y. Kondo et al.); *Chem. Abstr.* **109** (1988) 210 747u.
[247] Shell, US 3 052 728, 1962 (R. C. Morris, A. L. Rochlin).
[248] N. P. Neureiter, *J. Org. Chem.* **28** (1963) no. 12, 3486.
[249] J. R. Geigy AG, FR 1 343 301, 1963 (E. A. Meier, M. Dexter).
[250] Ciba Geigy AG, EP 186 629, 1986 (S. D. Pastor, J. D. Spivack, P. Odorisio).
[251] H. J. Florian, H. M. Schian, R. E. Tiller in: *Tagungsbericht Verband deutscher Betriebs- und Werksärzte e.V.*, Gentner-Verlag, Stuttgart 1989, pp. 267–274.
[252] C. D. Calnan, *Proc. R. Soc. Med.* **66** (1973) no. 3, 258–260.
[253] A. V. Topchiev, J. M. Paushkin, M. V. Kurosev, *Dokl. Akad. Nauk SSSR* **130** (1960) 559; *Chem. Abstr.* **54** (1960) 10 921d.
[254] Ethyl Corp., US 3 933 927, 1973 (L. E. Goddard).
[255] V. I. Isagulyants, N. A. Slavskaja, *Z. Prikl. Khim. (Leningrad)* **33** (1960) 953; *Chem. Abstr.* **54** (1960) 16 414a.
[256] I. F. Radzevenchuk, *Zh. Org. Khim.* **1** (1965), 1017–1020; *Chem. Abstr.* **63** (1965) 11 407d.
[257] I. F. Radzevenchuk, E. Koltsova, *Zh. Obshch. Khim.* **38** (1968) 204; *Chem. Abstr.* **69** (1968) 16 019y.
[258] Mitsubishi Petrochemical, JP-Kokai 62 153 235, 1985 (M. Imanari, H. Iwane, T. Sugawara); *Chem. Abstr.* **108** (1988) 21 490y.
[259] O. N. Tsevtkov et al., *Int. Chem. Eng.* **7** (1967) 104–121.
[260] V. A. Soldatova et al., *Azerb. Khim. Zh.* 1985 no. 4, 40; *Chem. Abstr.* **105** (1986) 190 578b.
[261] G. G. Trigo, *Chim. Ind. (Paris)* **86** (1961) 549–556.
[262] Centre de Technologie Chimique, FR 1 209 863, 1956.
[263] BASF, FR 1 419 289, 1964 (E. Rotter, H. Buelow).
[264] Nopco Chem. Co., US 3 168 577, 1961 (R. Weinstein, I. M. Rose, W. R. Christian).
[265] Jefferson Chem. Co., US 3 000 964, 1953 (J. G. Milligan).
[266] Thompson-Hayward Chem. Co., GB 1 142 233, 1966.
[267] Standard Oil, US 2 655 544, 1953 (G. M. McNulty, T. Cross).
[268] Soc. des Produits Chim. du Sidorbe-Sinova, FR 2 228 749, 1973.
[269] Chemische Werke Hüls AG, DE-AS 2 346 273, 1973 (H. Alfs et al.).
[270] California Research Corp., US 2 732 408, 1953 (J. K. Foote).
[271] Calumet Petrochem., US 4 055 605, 1977 (C. L. Jarreau).
[272] Agfa-Gevaert, DE-OS 2 440 678, 1974 (A. Klein et al.).
[273] Nippon Zeon Co., DE-OS 2 527 402, 1975 (K. Goto, T. Natsumme, H. Asai).
[274] *Chem. Mark. Rep.* (1978, 17.04.) 7, 36.
[275] L. McMaster, W. M. Bruner, *Ind. Eng. Chem.* **28** (1936) 505–506.
[276] W. F. Short, M. L. Stewart, *J. Chem. Soc.* 1929, 553.

[277] P. McLaughlin, V. M. Greedom, B. Miller, *Tetrahedron Lett.* 1978, 3537–3540.
[278] Bayer, DE 2 804 215, 1978 (H. J. Buysch, M. Matner, H. Freese).
[279] N. N. Alekseev et al., *Zh. Prikl. Khim. (Leningrad)* **46** (1973) 218–220; *Chem. Abstr.* **78** (1973) 110 716.
[280] Ya. M. Paushkin et al., *Neftekhimiya* **9** (1969) 842–847; *Chem. Abstr.* **72** (1970) 78 572.
[281] Naucno-Issledovatelskij Institut, DE-OS 1 543 791, 1966 (A. A. Grinberg et al.).
[282] Goodyear, DE-OS 2 600 204, 1975 (W. S. Hollingshead),
[283] J. A. M. Laan, J. P. Ward, *Chem. Ind. (London)* 1987 no. 1, 34–35.
[284] Maruzen Oil Co., DE-OS 2 608 407, 1976 (K. Kanezaki).
[285] Maruzen Oil Co., DE-OS 2 637 923, 1976 (M. Sato et al.).
[286] Bayer, DE 1 235 894, 1958 (H. Krimm, H. Schnell).
[287] J. Kahovec et al., *Collect. Czech. Chem. Commun.* **36** (1971) 1896–1994.
[288] H. Schnell, H. Krimm, *Angew. Chem.* **75** (1963) 662.
[289] Mitsui Toatsu Chem. Inc., DE-OS 2 438 432, 1973 (K. Mimaki et al.).
[290] *Houben-Weyl*, E 4, vol. **6/1 c,** 986–990.
[291] F. B. Gershanov et al., SU 1 191 445, 1983; *Chem. Abstr.* **104** (1986) 224 590d.
[292] J. H. P. Tyman, *J. Chromatogr.* **111** (1975) 277–292.
[293] Rütgerswerke AG, DE-OS 2 208 253, 1972 (E. Pastorek, H. Miele, W. Orth).
[294] Bayer, DE-OS 2 304 588, 1973 (A. Klein, K. Wedemeyer, J. Thies).
[295] Bayer, DE-OS 2 319 079, 1973 (A. Klein, K. Wedemeyer, J. Thies).
[296] Bayer, DE-OS 2 356 813, 1973 (A. Klein, K. Wedemeyer).
[297] *Chem. Eng. (N.Y.)* **85** (1978) 99.
[298] *Ullmann*, 4th ed., **18**, 219–222.
[299] J. Tsuji, H. Takayanagi, *J. Am. Chem. Soc.* **96** (1974) 7349–7350.
[300] C. J. Pedersen, *Science (Washington, D.C.)* **241** (1988) 536–540.
[301] P. Buryan, J. Macak, *J. Chromatogr.* **150** (1978) 246–249.
[302] Ube Industries, JP-Kokai 47 39 039, 1971 (T. Nagaoka et al.).
[303] J. Varagnat, *Ind. Eng. Chem. Prod. Res. Dev.* **15** (1976) 212–215.
[304] P. Maggioni, F. Minisci, *Chim. Ind. (Milan)* **59** (1977) 239–242.
[305] Ube Industries, US 4 078 006, 1976, JP-Kokai 50 130 727, 1974 (S. Umemura, N. Takamitu, T. Hamamoto, N. Kuroda).
[306] T. Hamamoto, N. Kuroda, N. Takamitu, S. Umemura, *Nippon Kagaku Kaishi* 1980, 1850–1854.
[307] Ube Industries, JP-Kokai 52 118 436, 1976 (S. Umemura, N. Takamitu, T. Hamamoto, N. Kuroda).
[308] Bayer AG., US 4 053 523, 1975 (H. Seifert, H. Waldmann, W. Schwerdtel, W. Swodenk).
[309] Degussa AG, US 4 618 730, 1984 (K. Drauz, A. Kleemann).
[310] Agency of Industrial Science and Technology,Oxirane Chemical Co. Ltd., DE 2 364 181, 1973 (J. Imamura, M. Ando, K. Sasaki, T. Iio).
[311] Mitsubishi Kasei Co., Ltd., JP-Kokai 58 55 439, 1981 (T. Maki, K. Murayama).
[312] *Chem. Mark. Rep.* **235** (1989) Jan. 23, 4.
[313] *Chem. Mark. Rep.* (1990) May 7, 33.
[314] M. A. Liebert, *J. Am. Coll. Toxicol.* **5** (1986) 123–164.
[315] N. I. Sax: *Dangerous Properties of Industrial Materials*, 6th ed., Van Nostrand Reinhold Co., New York, 1984, p. 2339.
[316] A. Critechlow, R. D. Haworth, P. L. Pauson, *J. Chem. Soc.* 1951, 1318.
[317] P. Gorecki, W. Kuran, *J. Polym. Sci. Polym. Lett. Ed.* **23** (1985) 299.

[318]   Przedsiebiorstwo przemyslowo-Handlowe "Polskie Odczynniki Chemiczne", PL 83 989, 1979; PL 87 534, 1979 (C. Osnowski).
[319]   M. T. Shipchandler, C. A. Peters, C. D. Hurd, *J. Chem. Soc. Perkin Trans. 1* 1975, 1400.
[320]   IMC Chemical Group, Inc., US 4 046 817, 1977 (M. T. Shipchandler).
[321]   Degussa AG, DE 3 607 924, 1987 (G. Prescher, G. Ritter, H. Sauenstein).
[322]   Mitsubishi Chem. Ind. Ltd., DE 2 445 336, 1975 (H. Obara, J. Onodera, A. Matsukuma, K. Yoshida).
[323]   Fisons Ltd., DE 2 627 874, 1979 (D. Baldwin, P. S. Gates).
[324]   Fisons Ltd., EP 25 659, 1981 (J. F. Harris).
[325]   Fisons Ltd., US 4 268 694, 1981 (J. F. Harris, B. J. Magill).
[326]   Fisons Ltd., EP 13 085, 1980 (J. F. Harris).
[327]   Mitsubishi Chem. Ind. Ltd., EP 31 530, 1981 (T. Maki, K. Murayama).
[328]   NIOSH: "Registry of Toxic Effects of Chemical Substances", Cincinnat, Ott, 1985 – 1986.
[329]   R. L. Lautz, E. Michel, *Bull. Soc. Chim. Fr.* 1961, 2402.
[330]   E. B. Vliet, *Org. Synth. Coll. Vol.* **1** (1964) 317.
[331]   Degussa AG, DE 3 607 924, 1986 (G. Prescher, G. Ritter, H. Sauerstein).
[332]   R. J. Hight, T. J. Batterham, *J. Org. Chem.* **29** (1964) 475.
[333]   M. L. Kastens, J. F. Kaplan, *Ind. Eng. Chem.* **42** (1950) 402.
[334]   Mitsui Petrochem. Ind. Ltd., JP-Kokai 83 150 529, 1983.
[335]   Mitsui Petrochem. Ind. Ltd., EP 88-302 689, 1988.
[336]   AKZO GmbH, DE 2 621 431, 1976 (H. Zengel, M. Bergfeld).
[337]   AKZO GmbH, DE 2 502 429, 1975 (H. G. Zengel, M. Bergfeld).
[338]   Kalle AG, DE 1 195 327, 1965 (S. Pietsch).
[339]   AKZO GmbH, DE 2 840 597, 1980 (R. Zielke, H. Maegerlein).
[340]   Andeno N. V., DE 2 231 005, 1973 (A. J. J. Hendrickx, N. A. De Heij).
[341]   H. Schnell: *Chemistry and Physics of Polycarbonates Polymer Reviews.*, vol. **9,** Wiley Interscience, New York 1964, pp. 77 – 98.
[342]   G. Bier, *Polymer* **15** (1974) 527 – 535. K. Hazama, *Jpn. Plast. Age* **14** (1976) 39 – 44.
[343]   H. Kittel: *Kunststoffjahrbuch*, 10th ed., Pansegrau Verlag, Berlin 1968, p. 10.
[344]   Bayer, DE-AS 1 105 428, 1958.
[345]   H. Schnell, H. Krimm, *Angew. Chem.* **75** (1963) 662 – 668.
[346]   A. Burawoy, J. T. Chamberlain, *J. Chem. Soc. (London)* **1949**, 626. Bayer, DE-AS 1 768 749, 1968. *Houben-Weyl,* vol. **I/1 c**, p. 1081 ff.
[347]   Bayer, DE-OS 2 645 020, 1976.
[348]   *Houben-Weyl,* vol. **VI/1 c**, p. 995, 1021 – 1060; vol. **XIV/2,** p. 195 ff.
[349]   A. Burawoy, J. T. Chamberlain, *J. Chem. Soc. (London)* 1949, 624 – 626.
[350]   Consolidation Coal, US 3 027 412, 1959.
[351]   Ethyl Corp., US 2 807 653, 1955.
[352]   Shell, Int. Res., BE 593 606, 1961.
[353]   A. H. Filbey, T. H. Coffield, *J. Org. Chem.* **22** (1957) 1435.
[354]   Goodyear Tire & Rubber Co., DE-AS 1 495 985, 1964.
[355]   Union Carbide Corp., US 2 858 342, 1955.
[356]   Esso Res. Eng. Co., DE-AS 1 068 270, 1957.
[357]   Rhône-Poulenc, DE-AS 1 071 713, 1958.
[358]   Bayer, DE-AS 1 051 864, 1953.
[359]   Bayer, US 3 689 572, 1966.
[360]   Goodrich Co., US 2 468 982, 1946.

[361] Dow, DE 905 977, 1950.
[362] Bayer, DE-OS 2 204 380, 1972.
[363] Union Carbide Corp., US 3 419 624, 1964. BASF, DE-OS 2 050 800, 1970.
[364] Union Carbide Corp., DE 1 056 620, 1956.
[365] Bayer, US 3 808 279, 1971.
[366] Bayer, BE 611 184, 1961.
[367] Dow, DE-AS 1 244 796, 1962.
[368] Dow, US 2 191 831, 1938.
[369] Hooker Chem. Corp., DE-AS 1 238 038, 1959.
[370] Dow, US 2 623 908, 1951.
[371] Bataafsche, DE-AS 1 030 836, 1953.
[372] Hercules Powder, US 3 172 916, 1960.
[373] Union Carbide Corp., FR 1 237 656, 1959; DE-AS 1 242 237, 1958; R. A. Reinicker, B. C. Gates, *AIChE J.* **20** (1974) no. 5, 933–940.
[374] Dow, DE-OS 2 164 339, 1971. *Eur. Chem. News* 1965, July 16, 38, 40.
[375] Hooker Chem. Corp., DE-AS 1 254 637, 1963.
[376] Gulf Oil Canada Ltd., US 3 493 622, 1967. Shell Dev. Co., US 2 845 464, 1954.
[377] Koningklijke Zwavelzuurfabrieken, DE-AS 1 272 302, 1961.
[378] Union Carbide Corp., DE-AS 1 244 796, 1962.
[379] Hibernia, DE-AS 1 025 418, 1953.
[380] J. Bussink, *Kunststoffe* **66** (1976) 600.
[381] *Kunststoffhandbuch*, vol. **8**, Hanser Verlag, München 1973, pp. 258, 261, 283.
[382] H. Kittel: *Lehrbuch der Lacke und Beschichtungen*, vol. **I/1**, Verlag Heenemann, Berlin 1971, pp. 355–357.
[383] *Kunststoffhandbuch*, vol. **10**, Hanser Verlag, München 1973, pp. 53, 109, 124.
[384] General Electric, DE-OS 2 243 226, 1972. Bayer, DE-OS 2 148 598, 1971.
[385] Bayer, DE-OS 2 354 533, 1973.
[386] Dow, US 2 538 725, 1949; US 2 535 015, 1948.
[387] Dr. H. Thomae, US 2 753 351, 1953.
[388] G. Bornmann, A. Loeser, *Arzneim. Forsch.* **9** (1959) 9.
[389] R. Willstätter, L. Kalb, *Chem. Ber.* **38** (1905) 1235. P. P. T. Sah, *Re. Trav. Chim. Pays-Bas* **59** (1940) 454. N. M. Cullinane, C. G. Davies, G. J. Davies, *J. Chem. Soc.* 1936, 1435.
[390] N. M. Cullinane, C. G. Davies, *Re. Trav. Chim. Pays-Bas* **55** (1936) 881.
[391] ICI, DE-AS 1 913 182, 1969.
[392] Bayer, DE-AS 2 049 809, 1970.
[393] Agency of Industrial Science and Technology, Tokio, DE-AS 2 211 721, 1972.
[394] Monsanto Chem. Co., US 2 368 361, 1942.
[395] ICI, GB 922 679, 1960.
[396] Rütgerswerke, GB 529 936, 1939.
[397] M. Tashiro, H. Watanabe, O. Tsuge, *Org. Prep. Proced. Int.* **6** (1974) no. 3, 117–122.
[398] General Electric, US 3 631 208, 1971. A. S. Hay, *J. Org. Chem.* **34** (1969) 1160–1161.
[399] Hoechst, DE-AS 1 148 326, 1968. H. Musso, W. Steckelberg, *Justus Liebigs Ann. Chem.* **693** (1966) 187–196. J. Bourdon, M. Calvin, *J. Org. Chem.* **22** (1957) 101–116.
[400] *Chem. Abstr.* **52**, 10 991g (1958).
[401] Sun Oil, US 3 281 435, 1962. C. R. H. J. De Jonge, H. M. van Dort, L. Vollbracht, *Tetrahedron Lett. 1970*, 1881–1884.
[402] UCC, US 3 322 838, 1963. General Electric, NL 6 410 238, 1963.

[403] UBE Industries Ltd., JP 73 36 152, 73 34 153, 1973.
[404] *WHO Tech. Rep. Ser.* 1965, 309.
[405] H. C. Hodge et al., *J. Pharmacol. Exp. Ther.* **104** (1952) 202. F. C. Mac Intosh, *Analyst (London)* **70** (1945) 334.
[406] O. Pauli, G. Franke: *Biodeterior. Mater. Proc. Int. Biodeterior. Symp. 2nd* (1972) 52–60.
[407] Bayer, DE-OS 1 719 417, 1965.
[408] G. D. Kharlampovich, V. N. Vinogradova, SU 197 613, 1967.
[409] Bayer AG, DE-OS 2 655 826, 1976 (R. Neumann, H.-H. Schwarz, K.-H. Arnold).
[410] Union Carbide, EP 73 492, 1983 (L. M. Maresca, M. Matzner).
[411] National Distillers and Chemical Corp., DE-OS 3 312 684, 1982 (E. G. Harris).
[412] Dow Chemical Co., US 4 613 682, 1986 (K. A. Eickholt).
[413] Brichima SpA., BE 890 589, 1980.
[414] Brichima SpA., EP-A 151 392, 1985 (P. Maggione, F. Minisci, M. Correale).
[415] Bayer AG, EP-A 52 314, 1982 (G. Buettner, A. Jufat, U. Allenbach, M. Lenthe).
[416] Brichima SpA., DE-OS 2 703 640, 1976 (P. Maggione).
[417] Takasago Perfumery, JP 84 46 949, 1976.
[418] W. Paulus, H. Genth in T. A. Oxley, S. Barry (eds.): *Biodeterioration 5*, John Wiley & Sons, New York 1983, pp. 701–712.
[419] D. Liu, *Toxicity Assessment* **4** (1989) 115–127.
[420] W. W. Cocker, US 2 350 677.
[421] E. Klarmann et al., *J. Am. Soc.* **55** (1939) 2576–2583.
[422] R. S. Law, *J. Soc. Chem. Ind. London Trans. Commun.* **60** (1941) 66.
[423] VEB Jenapharm, DE-OS 2 706 747, 1977.
[424] G. F. Reddish: *Antiseptics, Disinfectants, Fungicides and Sterilization*, Lea & Febiger, Philadelphia 1954, p. 250.
[425] K. H. Wallhäusser: *Praxis der Sterilisation – Desinfektion – Konservierung*, Georg Thieme, Stuttgart 1984, pp. 416–420.
[426] W. Paulus in *Microbicides for the Protection of Materials – a Handbook*, Chapman and Hall, London 1993, pp. 141–198.
[427] H. P. Fiedler in *Lexikon der Hilfsstoffe für Pharmazie, Kosmetik und angrenzende Gebiete*, Editio Cantor, Aulendorf 1989.

# Phenylacetic Acid

DOROTHEA GARBE, Haarmann & Reimer GmbH, Holzminden, Federal Republic of Germany

| | | | | |
|---|---|---|---|---|
| 1. | Physical Properties ....... 3805 | 5. | Quality Specifications...... 3806 |
| 2. | Chemical Properties........ 3805 | 6. | Economic Aspects ......... 3806 |
| 3. | Production .............. 3806 | 7. | Toxicology............... 3807 |
| 4. | Uses ................... 3806 | 8. | References............... 3807 |

Phenylacetic acid [103-82-2], α-toluic acid, $C_8H_8O_2$, $M_r$ 136.15, was initially synthesized by CANNIZZARO in 1845.

$$\text{C}_6\text{H}_5\text{-CH}_2\text{COOH}$$

Phenylacetic acid occurs in Japanese peppermint oil, in neroli oil, and in traces in rose oil. Phenylacetic acid and its esters are volatile constituents of many fruits and foods [1].

## 1. Physical Properties

Phenylacetic acid forms colorless crystals that are sparingly soluble in cold water but readily soluble in ethanol and diethyl ether. Important physical properties are: $mp$ 78 °C, $bp$ (101.3 kPa) 265.5 °C, $d_4^{79.8}$ 1.0809, $K$ (25 °C) $5.56 \times 10^{-5}$. It has a sweet, honeylike odor when highly diluted; the concentrated solution smells suffocating and unpleasant.

## 2. Chemical Properties

Phenylacetic acid forms salts and esters; it reacts with thionyl chloride to yield phenylacetyl chloride. The methylene group can undergo condensation and substitution reactions. Substitution on the phenyl ring occurs primarily at the *para*-position.

## 3. Production

Two routes exist for the synthesis of phenylacetic acid; both methods employ benzyl chloride as starting material.

*Hydrolysis of Benzyl Cyanide.* Benzyl chloride reacts with sodium cyanide to yield benzyl cyanide; the latter is then hydrolyzed to phenylacetic acid:

$$C_6H_5CH_2Cl \xrightarrow{NaCN} C_6H_5CH_2CN \xrightarrow{Hydrolysis} C_6H_5CH_2COOH$$

*Hydrolysis of Phenyl Acetates.* Benzyl chloride reacts with carbon monoxide and alcohols, in the presence of a suitable catalyst, to form phenyl acetates [2], [3]; the latter are hydrolyzed to phenylacetic acid:

$$C_6H_5CH_2Cl \xrightarrow[\text{Catalyst}]{CO/ROH} C_6H_5CH_2COOR \xrightarrow{Hydrolysis} C_6H_5CH_2COOH$$

## 4. Uses

Phenylacetic acid is used mostly in the form of its esters; it is added in small amounts to perfumes and flavor compositions. Most of the phenylacetic acid produced is used in the synthesis of Penicillin G. Addition of phenylacetic acid during fermentation increases the yield of Penicillin G.

## 5. Quality Specifications

Commercial phenylacetic acid should have an acid number $\geq 98$, a solidification point $\geq 76\,°C$, and a flash point $> 100\,°C$.

## 6. Economic Aspects

The annual worldwide production of phenylacetic acid is several hundred tonnes. The most important producers are Calaire (France), Prom (Denmark), and Orbis Products and Givaudan (United States).

# 7. Toxicology

The acute oral $LD_{50}$ for phenylacetic acid in rats and rabbits exceeds 5 g/kg. A 2% solution in petrolatum produced no irritation after a 48-hour closed-patch test on humans and no sensitization in a maximization test on 25 volunteers [4].

# 8. References

[1] H. Maarse, E. A. Visscher: *Volatile Compounds in Foods, Qualitative Data,* TNO-Division for Nutrition and Food Research, TNO-CIVO Food Analysis Institute, Zeist, Netherlands, 1988.
[2] Dynamit Nobel, DE – OS 2 240 398, 1972 (M. El Chahawi, H. Richtzenhain).
[3] Montedison SpA, FR 2 486 070, 1982 (G. Cainelli, M. Foa, A. U. Ronchi, A. Gardano).
[4] Research Institute for Fragrance Materials, *Food Cosmet. Toxicol.* **14** (1975) Suppl. 901 – 902.

# Phenylene- and Toluenediamines

ROBERT A. SMILEY, Wilmington, Delaware 19880-0336, United States

| | | | | |
|---|---|---|---|---|
| 1. | Introduction ............ 3809 | 6. | Quality Specifications....... 3813 |
| 2. | Physical Properties ........ 3810 | 7. | Storage and Transportation .. 3813 |
| 3. | Chemical Properties........ 3810 | 8. | Uses ................... 3814 |
| 4. | Production .............. 3811 | 9. | Toxicology and Occupational Health.................. 3816 |
| 5. | Environmental Protection ... 3813 | 10. | References............... 3817 |

## 1. Introduction

*Phenylenediamine* is the common name for diaminobenzene, which exists as three isomers: 1,2-, 1,3-, and 1,4-diaminobenzene. The 1,2-isomer is commonly called *o*-phenylenediamine; the 1,3-isomer is *m*-phenylenediamine; and the 1,4-isomer is *p*-phenylenediamine.

The *toluenediamines*, with six possible isomers, are commonly named by using numbers to designate the position of the amino groups with respect to the toluene methyl group, i.e., 2,3-diaminotoluene or 2,3-toluenediamine (3-methyl-1,2-phenylenediamine) [2687-25-4]; 2,4-diaminotoluene or 2,4-toluenediamine (4-methyl-1,3-phenylenediamine) [95-80-7]; 2,5-diaminotoluene or 2,5-toluenediamine(2-methyl-1,4-phenylenediamine) [95-70-5]; 2,6-diaminotoluene or 2,6-toluenediamine (2-methyl-1,3-phenylenediamine) [823-40-5]; 3,4-toluenediamine (4-methyl-1,2-phenylenediamine) [496-72-0]; and 3,5-toluenediamine (5-methyl-1,3-phenylenediamine) [108-71-4]. An 80:20 mixture of the 2,4- and 2,6-isomers is used and sold as *m*-toluenediamine. Commercially available *o*-toluenediamine is a 60:40 mixture of the 3,4- and 2,3-isomers.

Like most simple aromatic amines, the phenylene- and toluenediamines can be prepared by electrophilic aromatic nitration, followed by reduction of the introduced nitro group (usually catalytic hydrogenation) to an amino group. This chemistry was developed in the early 1800s. Until the mid-1900s, phenylene- and toluenediamines were used principally for the preparation of dyes, but they are currently more important in the production of high-performance textile fibers, agricultural chemicals, and diisocyanates for the preparation of a variety of polyurethane products.

Table 1. Melting and boiling points of phenylene- and toluenediamines

| Compound | CAS registry no. | mp, °C | bp, °C |
|---|---|---|---|
| o-Phenylenediamine | [95-54-5] | 102–103 | 256–258 |
| m-Phenylenediamine | [108-45-2] | 62–63 | 284–287 |
| p-Phenylenediamine | [106-50-3] | 145–147 | 267 |
| 2,3-Toluenediamine | [2687-25-4] | 63–64 | 255 |
| 2,4-Toluenediamine | [95-80-7] | 97–99 | 283–285 |
| 2,5-Toluenediamine | [95-70-5] | 64 | 273–274 |
| 2,6-Toluenediamine | [823-40-5] | 104–106 | |
| 3,4-Toluenediamine | [496-72-0] | 91–93 | 265 (sub.)* |
| 3,5-Toluenediamine | [108-71-4] | >0 | 283–284 |

* sub.=sublimation.

## 2. Physical Properties

The melting and boiling points of the phenylene- and toluenediamines are listed in Table 1. All of these compounds are white crystalline solids when pure, except 3,5-toluenediamine which is a liquid at room temperature. They are all readily soluble in hot water and common organic solvents such as alcohols, ketones, and ethers. Because of their limited solubility in cold water, o- and p-phenylenediamines can be recrystallized from water.

## 3. Chemical Properties

The chemical properties of the phenylene- and toluenediamines are typical of aromatic amines; they react as bases in aqueous solution and form stable salts with acids. Their most notable chemical characteristic is their ease of oxidation to complex colored compounds, which is the basis for their use in dyes and hair colorants. Oxidation also occurs if they are stored in the presence of air. The various isomers differ greatly in chemical reactivity due to the relative positions of the two amino groups. Phenylene- and tolueneamines with ortho amino groups, for example, readily form heterocyclic compounds, whereas the m- and p-isomers do not.

The aromatic rings of phenylenediamines can be catalytically hydrogenated to the corresponding diaminocyclohexanes; for example, hydrogenation of p-phenylenediamine over a ruthenium on alumina catalyst produces cis- and trans-1,4-cyclohexanediamine [15827-56-2] (cis), [2615-25-0] (trans)] in a ratio of 1:3 [1].

**o-Phenylenediamine.** Benzimidazole [51-17-2] is formed by a ring-closure reaction of o-phenylenediamine with formic acid. Alkyl-substituted benzimidazoles are obtained by reaction of o-phenylenediamine with nitriles or aldehydes in the presence of copper(II) acetate. When o-phenylenediamine is reacted with vicinal dicarbonyl compounds, quinoxalines result; reaction with glyoxal [107-22-2], for example, gives quinoxaline [91-19-0]. Benzotriazole [95-14-7] is obtained by the reaction of o-phenylenediamine with nitrous acid in dilute sulfuric acid [2].

The oxidation of o-phenylenediamine with aqueous iron(III) chloride produces 2,3-diaminophenazine [655-86-7]; oxidation with silver oxide in diethyl ether gives o-benzoquinonediimine [4710-40-1]; and oxidation with air in a solution of copper(I) chloride in pyridine is reported to give 1,4-dicyano-1,3-butadiene [821-60-3] [3].

*m*-**Phenylenediamine** can be readily acylated with acid chlorides to form substituted amides. With diacid chlorides it forms polyamides; for example, with isophthaloyl chloride a poly(*m*-phenyleneisophthalamide) [24938-60-1] is produced, which is the starting material for textile fibers. When *m*-phenylenediamine is reacted with excess nitrous acid, both amino groups are diazotized and couple with unreacted *m*-phenylenediamine to give 5,5′-[(4-methyl-*m*-phenylene)bisazo]bis(toluene-2,4-diamine), Bismarck Brown, one of the first azo dyes.

*p*-**Phenylenediamine** is more readily oxidized than the *o*- or *m*-isomer. Acidic manganese dioxide or potassium dichromate produces 1,4-benzoquinone [106-51-4]. When *p*-phenylenediamine is oxidized with hydrogen peroxide, for example in a hair dye, the first product is *p*-benzoquinonediimine [4377-73-5], which can react with unreacted *p*-phenylenediamine to form a brown dye known as Bandrowsky Base [20048-27-5]. The diimine will also react rapidly with other aromatic amines and phenols to form a variety of colorants.

The amino groups of *p*-phenylenediamine undergo reactions typical of aromatic amines, (e.g., diazotization, acylation, alkylation, and phosgenation).

The chemical properties of the toluenediamines are identical to those of the phenylenediamines, i.e., the same type of compounds can be made from them. The methyl group may slow down the reaction rate of an adjacent amine group due to steric hindrance, but this effect is minor in most cases.

# 4. Production

Preparation of phenylene- and toluenediamines almost always starts with aromatic nitration followed by reduction, although some exceptions exist. The initial starting material in most cases is either benzene or toluene.

*o-Phenylenediamine.* The principal route to *o*-phenylenediamine is by the reduction of 2-nitroaniline [88-74-4], which is produced by the amination of 2-chloronitrobenzene [88-73-3] with ammonia [4]. Hydrogenation can be achieved with a variety of reducing agents, such as iron powder [5], hydrazine [6], or hydrogen sulfide [7], but commercially it is performed catalytically in the liquid phase over a palladium catalyst [8], [9]. Other routes to *o*-phenylenediamine include the amination of 1,2-dichlorobenzene [95-

*50-1*] with ammonia [10], [11] and the direct amination of aniline [12]. The yield in the latter process is, however, low. A method starting from cyclohexane, sulfur, and ammonia has also been described [13].

*m-Phenylenediamine.* The *m*-isomer is synthesized by catalytic hydrogenation of 1,3-dinitrobenzene [*99-65-0*], which is obtained as the principal product in the dinitration of benzene with a mixture of sulfuric and nitric acids [14] – [16]. A certain amount of *o*- and *p*-phenylenediamines (about 10%) are also produced as byproducts, but these can be removed by chemical treatment of the crude nitration product either prior to hydrogenation [17] or during isolation of the *m*-diamine by distillation. Hydrogenation is usually performed in water or methanol over a supported palladium or Raney nickel catalyst [18], [19].

*p-Phenylenediamine.* Routes to *p*-phenylenediamine include reaction of 1,4-dichlorobenzene [*106-46-7*] [10], [11], hydroquinone [*123-31-9*], [20], [21], or aniline [*62-53-3*] [22] with ammonia; hydrogenation of 4-nitroaniline [*100-01-6*] [23]; Hoffmann degradation of terephthalamide [*3010-82-0*] [24]; and diazotization of aniline to a triazine followed by rearrangement [25], [26].

The most important commercial routes are the catalytic hydrogenation of 4-nitroaniline and aniline diazotization. The *4-nitroaniline route* is identical to the preparative method for the *o*-isomer; 4-chloronitrobenzene is aminated with ammonia to the nitroaniline, followed by hydrogenation over a palladium catalyst. A disadvantage of this process is the 65:35 ortho–para ratio obtained during nitration of chlorobenzene because of the difficulty of matching the demand for the fixed ratio of isomers. The diazotization process used by Du Pont in the United States, overcomes this problem because it is based only on aniline.

In the *Du Pont process*, aniline is diazotized using nitrogen oxides [25]. The diazo compound reacts with excess aniline to produce 1,3-diphenyltriazine [*136-35-6*], which rearranges under acidic conditions to 4-aminoazobenzene [*60-09-3*]. 4-Aminoazobenzene is catalytically cleaved and hydrogenated to give *p*-phenylenediamine and aniline. The aniline is recovered and recycled.

One route to *p*-phenylenediamine that does not involve any nitration and does not start with benzene is the *Akzo process*, which can start with dimethyl terephthalate [*120-61-6*] or waste poly(ethylene terephthalate). Either ester can be reacted with ammonia to produce terephthalamide. The diamide is chlorinated to give a chloroamide that is then reacted with a base such as sodium hydroxide to produce *p*-phenylenediamine (Hoffmann degradation) [24].

*Toluenediamines.* From a commercial standpoint, the most important toluenediamine isomers are the meta, or 2,4- and 2,6-isomers, which are obtained as a mixture by hydrogenation of the corresponding dinitrotoluenes. The processes involved are similar to those used for the manufacture of *m*-phenylenediamine, but using toluene as the starting material instead of benzene. About 4% of the 2,3- and 3,4-isomers are formed as byproducts during toluene dinitration, and the same isomeric diamines subsequently appear as impurities in the mixture of *m*-diamines. They can be removed by distillation and are sold as commercial *o*-toluenediamine.

**Table 2.** Phenylenediamine product specifications

| | Isomer | | |
|---|---|---|---|
| | Ortho | Meta | Para |
| Water, % (min.) | 0.2 | 0.1 | 0.1 |
| fp, °C | 2.7 | | |
| Purity, % | 99.5 | 99.8 | 99.5 |
| Content ortho isomer, ppm | | 200 | 1000 |
| Meta isomer, ppm | 1* | | 1000 |
| Para isomer, ppm | 1* | 100 | |
| Insolubles, % (in water) | 0.1 | | |
| Nitrobodies as dinitrobenzene, % | | 0.1 | |
| Aniline, ppm | | | 500 |

* Combined meta and para content=0.25%.

# 5. Environmental Protection

**Aquatic Toxicity.** The $LC_{50}$ of *o*-phenylenediamine determined with fathead minnows is 44 mg/L (96 h exposure). For the *p*-isomer it is only 0.028 mg/L, which indicates that *p*-phenylenediamine is very toxic to aquatic life. The $LC_{50}$ of the *m*-isomer is greater than 1 mg/L but less than 50 mg/L [27]–[29]. The $LC_{50}$ for toluenediamine is 430 mg/L for golden shiners and 1000 mg/L for minnows [30].

# 6. Quality Specifications

All three *phenylenediamine isomers* are available commercially as flaked solids; the *m*-isomer is also available as cast solid and molten liquid. The commmercial products are light tan to brown in color. They tend to discolor during storage, and the presence of air, moisture, or elevated temperature hastens color deterioration. However, color alone is not a good criterion of purity because highly colored material may still be above the minimum purity specifications. Typical sales specifications are shown in Table 2.

*Toluenediamines* are available as cast solids in drums and molten liquids in tank cars. The solid *m*-isomer mixture is yellow to tan, whereas the *o*-mixture is light gray to purple. Like the phenylenediamines, these compounds tend to darken on storage. Product specifications are shown in Table 3.

# 7. Storage and Transportation

All aromatic diamines should be stored in tightly closed, upright containers in a cool, well-ventilated area away from heat and sources of ignition.

**Table 3.** Toluenediamine product specifications

| | Isomer | |
| --- | --- | --- |
| | Ortho | Meta |
| Purity, % (min.) | 97.0 | 99.0 |
| Normalized isomer ratio, wt% | | |
| 2,4- | | 79–81 |
| 2,6- | | 19–21 |
| o- | 99.5 (min.) | |
| Other | 0.5 (max.) | |
| Ortho isomers (2,3- and 3,4-) | | 0.3 (max.) |
| Para and meta isomers (2,5- and 3,5-) | | 1.0 (max.) |
| Nitrobodies as dinitrotoluene, % | | 0.03 (max.) |
| Moisture, % | 1.0 (max.) | 0.1 (max.) |
| Toluidines, % | | 0.05 (max.) |

In the United States, the shipping of o-phenlyenediamine other than by air (DOT) is not regulated. m-Phenylenediamine and p-phenylenediamine are shipped under hazard class ORM-A and UN no. 1673. For international shipping (IMO) by water or air, all three isomers are shipped under the name phenylenediamine and hazard class Poison β, 6.1.

# 8. Uses

For about a hundred years, phenylene- and toluenediamines were used mainly in the manufacture of dyes and dye intermediates. However, in the mid-1900s these compounds became very important starting materials for a variety of polymers, which now accounts for most of their consumption.

*o-Phenylenediamine.* A significant proportion of o-phenylenediamine is used for the preparation of benzimidazole-derived agricultural fungicides such as benomyl [17804-35-2]. This is produced in the United States by Du Pont and sold under the trade name Benlate, and in the Soviet Union under the name Uzgen. Other fungicides made from o-phenylenediamine include thiophanate methyl (Topsin), a turf fungicide made by Nippon Soda for sale by others, and fuberidazole produced by Bayer.

A small but high-value use of o-phenylenediamine is in the preparation of a variety of substituted benzimidazoles used as veterinarian anthelmintics. These compounds are available under trade names such as Equizole (thiabendazole) (Merck, Sharpe & Dohme) and Panacur (fenbendazole) (Hoechst), [31].

Benzotriazole and tolyltriazole are prepared, respectively, from o-phenylenediamine and an isomeric mixture of o-toluenediamines, by reaction with nitrous acid. These compounds are widely used as corrosion inhibitors in automobile antifreezes and other aqueous solutions because of their specific inhibition of copper and copper alloys.

Mercaptobenzimidazole [583-39-1] obtained from o-phenylenediamine and potassium ethyl xanthate is used as an antioxidant for rubber (Antioxidant MB, Mobay). A substituted benzotriazole derived from *o*-phenylenediamine is sold by Ciba–Geigy under the trade name Tinuvin P as an ultraviolet light absorber for stabilizing plastics and other organic materials against discoloration and deterioration [32].

*m-Phenylenediamine.* The fastest growing use for *m*-phenylenediamine is in the production of the aramid fiber poly(*m*-phenyleneisophthalamide), which is obtained by reaction of *m*-phenylenediamine with isophthaloyl chloride [*99-63-8*] [33]. This material is used in areas where high-temperature resistance or fire retardancy is required. Examples are protective clothing and aircraft interiors. The polymer is made in the United States by Du Pont (Nomex), in Japan by Teijin, and in the Soviet Union.

*m*-Phenylenediamine is also used for curing epoxy resins to impart high-temperature strength and good resistance to chemical solvents. Such cured resins are used in filament-wound casings and adhesives.

*m*-Phenylenediamine is still used as the starting material for many dyes. Examples include Basic Brown 1 (Bismarck Brown), Basic Orange 2, Direct Black 38, and Developed Black BH.

*p-Phenylenediamine.* Aramid textile fibers are obtained from poly(*p*-phenylene terephthalamide) [*24938-64-5*], which is produced by the reaction of *p*-phenylenediamine with terephthaloyl chloride [*100-20-9*] [33]–[35]; the two suppliers are Du Pont (Kevlar) and Akzo (Twaron). These fibers are used for reinforcing tires and V-belts, as a substitute for asbestos in brake linings, in reinforcing plastics for aerospace and sporting applications, and for a variety of other uses where high strength and tensile modulus are required [36].

Reaction of *p*-phenylenediamine with phosgene produces *p*-phenylene diisocyanate (PPDI), a symmetrical diisocyanate that gives highly crystalline polyurethanes reported to have good high temperature properties as both thermoplastics and cast elastomers [37].

*p*-Phenylenediamine is still employed in hair dyes and is the main component of oxidation hair dyes. *N,N'*-disubstituted *p*-phenylenediamines are used extensively as photographic chemicals, as fuel additives, and in the compounding of rubber. However, most of these compounds are made by the reductive alkylation of *p*-nitroaniline rather than from the parent diamine.

*Toluenediamines.* The largest use for toluenediamines is in the production of diisocyanates for the polyurethane industry. These diisocyanates are synthesized by reaction of *m*-toluenediamine (80:20 mixture of the 2,4- and 2,6-isomers) with phosgene. The resulting diisocyanates (trade name TDI) are used principally in polyurethanes as flexible foams for upholstery and other types of padding.

# 9. Toxicology and Occupational Health

All aromatic diamines are toxic when ingested and may cause skin and eye irritation. In some individuals, inhalation can lead to respiratory problems and asthma. Thus, as a class these diamines should be treated as very hazardous materials, and all contact with the body should be avoided by use of appropriate protective equipment.

**o-Phenylenediamine.** In addition to causing skin rashes, evidence exists that o-phenylenediamine can permeate the skin in toxic amounts. A single application to the skin of rabbits produced weight loss, loss of coordination, weakness, darkening of the urine, and death. Currently, this compound is not considered a carcinogen, but it produces noninheritable genetic damage in bacterial and mammalian cell cultures [27].

**m-Phenylenediamine.** The toxic effects observed in animals from short exposure to m-phenylenediamine by inhalation, ingestion, or skin contact include liver and kidney damage. Since no carcinogenic acitivity has been noted in animal tests, m-phenylenediamine is not considered a carcinogen, although it is mutagenic in bacterial cultures and may have embryotoxic activity [28].

**p-Phenylenediamine.** Toxic effects observed in animals from single exposures to p-phenylenediamine by inhalation include tremors, cyanosis, and severe weight loss. Toxic effects by ingestion include weight loss and lethargy. This compound does not have carcinogenic effects on animals. Tests for reproductive activity have not yet been performed [29].

p-Phenylenediamine is such a strong dye-former in the presence of air and moisture that everything in contact with it, whether skin, clothes, or walls, is almost immediately dyed a dark blueish black. Thus, exposure to p-phenylenediamine in even small amounts or lack of good industrial hygiene in operations that involve its use can be readily detected.

**o-Toluenediamines.** The toxic inhalation and ingestion effects of the commercial mixture of o-toluenediamines can be assumed to be similar to those of phenylenediamines, but they do not appear to be primary skin irritants. The Ames mutagenic assay of o-toluenediamines is positive. Because the mixture can contain up to 0.5% of the meta isomers, the toxic effects of these compounds must be considered when working with commercial o-toluenediamine.

**m-Toluenediamine.** Application of m-toluenediamine to the skin in hair dyes does not appear to produce toxic or carcinogenic effects [38], but it does appear to be a carcinogen when ingested by rats [39]. It is an active mutagen in the Ames test but has not yet been tested for embryotoxicity.

# 10. References

[1] Du Pont, US 3 636 108, 1977 (L. Brake).
[2] J. B. Wright, *J. Am. Chem. Soc.* **71** (1949) no. 5, 203.
[3] D. Barton, W. D. Ollis in I. O. Sutherland (ed.): *Comprehensive Organic Chemistry*, vol. **2**, Pergamon Press, Oxford 1979,p. 177.
[4] Du Pont, US 3 929 889, 1975 (E. N. Squire).
[5] GAF, US 2 956 082, 1960 (L. M. Schenck).
[6] C. Budeanu, RU 60 417, 1976.
[7] S. Bogdal, M. Milewska, *Przem. Chem.* **58** (1979) no. 9, 481.
[8] Hoechst, BR 1 344 796, 1974 (R. Mees, J. Ribka).
[9] UOP, US 3 230 259, 1966 (B. Levy).
[10] Toyo Soda, JP 5 3 77 023, 1978 (M. Ohshio et al.).
[11] Japan Synthetic Rubber, US 4 521 622, 1985 (N. Andoh, S. Fujiwara).
[12] Du Pont, BR 1 327 494, 1973 (E. N. Squire).
[13] F. J. Weigert, *J. Org. Chem.* **46** (1981) no. 9, 1936.
[14] Atlantic Refining, US 2 951 571, 1960 (G. A. Bonetti).
[15] A. D. Mesurc & J. G. Tillett, *J. Chem. Soc. Phys. Org.* **7** (1966) 669.
[16] A. Ujhidy, M. Magyar, R. Berkes, *Acta Chem. (Budapest)* **69** (1971) no. 1, 107.
[17] Du Pont, US 3 329 719, 1967 (D. J. Crowley).
[18] Du Pont, US 3 328 465, 1967 (L. Spiegler).
[19] Teijin, JP 45 490, 1970 (T. Yamanaka).
[20] Olin Mathieson, BR 1 019 750, 1966.
[21] Du Pont, US 4 031 106, 1977 (T. W. Del Pesco).
[22] Asahi, JP 7 7 42 829, 1977 (S. Enomoto, T. Kamiyama, Y. Takahashi, E. Fujimoto).
[23] Sumitomo, DE 2 331 900, 1974.
[24] Akzo, BR 1 364 229, 1974 (H. Zengel, M. Bergfeld); DE 2 216 117, 1973 (H. Zengel, M. Bergfeld).
[25] Du Pont, US 4 020 052, 1977 (J. K. Detrick).
[26] Du Pont, US 4 279 815, 1981 (F. F. Herkes).
[27] Du Pont, *o*-phenylenediamine, Material Safety Sheet, E. I. du Pont de Nemours & Co. Inc. (Environmental Affairs), Wilmington, Del., 1987.
[28] Du Pont, *m*-Phenylenediamine, Material Safety Data Sheet, E. I. du Pont de Nemours & Co. Inc. (Chemicals & Pigments Dept.), Wilmington, Del., 1987.
[29] Du Pont, *p*-phenylenediamine, Material Safety Sheet, E. I. du Pont de Nemours & Co. Inc. (Environmental Affairs), Wilmington, Del., 1987.
[30] Air Products, Nitro/Amino Aromatic Intermediates, Product Bulletin, Air Products and Chemicals, Inc., Allentown, Pa., 1983.
[31] *Veterinary Pharmaceuticals and Biologicals,* Harwal Publishing Co., Media, Pa., 1980–1981.
[32] *Merck Index,* 11th ed., Merck & Co, Rahway, N. J., 1989.
[33] Du Pont, US 375 361, 1973 (J. H. Jensen).
[34] Teijin, JP 4 8 23 198, 1973 (S. Ozawa).
[35] Du Pont, US 3 869 430, 1975 (H. Blades).
[36] Akzo, BR 1 547 802, 1979.
[37] Du Pont, Paraphenylenediisocyanate, Product Bulletin, E. I. du Pont de Nemours & Co., Inc. (Petrochemicals Dept.), Wilmington, Del., 1990.

[38]   C. Burnett et al., *Food Cosmet. Toxicol.* **13** (1975) 353.
[39]   R. H. Cardy, *JNCI, J. Natl. Cancer Inst.* **26** (1979) no. 4, 1107.

# Phosphorus Compounds, Organic

Jürgen Svara, Hoechst AG, Werk Knapsack, Hürth-Knapsack, Federal Republic of Germany (Chaps. 1–12)

Norbert Weferling, Hoechst AG, Werk Knapsack, Hürth-Knapsack, Federal Republic of Germany (Chaps. 1–12)

Thomas Hofmann, Hoechst Marion Roussel, Hattersheim, Federal Republic of Germany (Chap. 13)

| | | |
|---|---|---|
| 1. | Introduction | 3820 |
| 2. | Phosphines | 3820 |
| 2.1. | Properties | 3820 |
| 2.2. | Production | 3821 |
| 2.2.1. | Primary and Secondary Phosphines | 3821 |
| 2.2.2. | Tertiary Phosphines | 3823 |
| 2.3. | Uses | 3824 |
| 3. | Halophosphines | 3825 |
| 3.1. | Properties | 3826 |
| 3.2. | Production | 3826 |
| 3.3. | Uses | 3827 |
| 4. | Phosphonium Salts | 3827 |
| 4.1. | Properties | 3828 |
| 4.2. | Production | 3828 |
| 4.3. | Uses | 3830 |
| 5. | Phosphine Oxides and Sulfides | 3830 |
| 5.1. | Properties | 3831 |
| 5.2. | Production | 3831 |
| 5.3. | Uses | 3833 |
| 6. | Phosphonous Acid Derivatives | 3834 |
| 6.1. | Production | 3834 |
| 6.2. | Properties and Uses | 3835 |
| 7. | Phosphinic Acids and their Derivatives | 3835 |
| 7.1. | Properties | 3835 |
| 7.2. | Production | 3836 |
| 7.3. | Phosphinic Acids and Phosphinate Esters | 3837 |
| 7.4. | Thiophosphinates | 3838 |
| 7.5. | Other Phosphinic Acid Derivatives | 3838 |
| 8. | Phosphites and Hydrogenphosphonates | 3838 |
| 8.1. | Properties | 3838 |
| 8.2. | Production | 3839 |
| 8.3. | Trialkyl Phosphites | 3839 |
| 8.4. | Triaryl Phosphites and Alkyl Aryl Phosphites | 3840 |
| 8.5. | Dialkyl and Diaryl Phosphonates | 3841 |
| 8.6. | Alkyl Phosphonates | 3841 |
| 9. | Phosphonic Acids and their Derivatives | 3842 |
| 9.1. | Properties | 3842 |
| 9.2. | Production | 3842 |
| 9.3. | Phosphonic Acids and Phosphonocarboxylic Acids | 3844 |
| 9.4. | Esters of Phosphonic Acid | 3845 |
| 9.5. | Other Derivatives of Phosphonic Acid | 3847 |
| 10. | Esters of Phosphoric Acid | 3847 |
| 10.1. | Properties | 3847 |
| 10.2. | Production | 3848 |
| 10.3. | Trialkyl Phosphates | 3849 |

| | | | | |
|---|---|---|---|---|
| 10.4. | Triaryl and Alkyl Aryl Phosphates .............. 3849 | 11.3. | Monothiophosphates ....... 3852 |
| 10.5. | Mono- and Dialkyl Phosphates 3850 | 11.4. | Dithiophosphates ......... 3853 |
| 11. | Esters of Thiophosphoric Acid 3851 | 12. | Economic Aspects ......... 3855 |
| 11.1. | Properties ................ 3851 | 13. | Toxicology ............... 3856 |
| 11.2. | Production ............... 3851 | 14. | References ............... 3860 |

# 1. Introduction

For the purpose of this article, organophosphorus compounds include not only compounds containing a phosphorus – carbon bond, but all compounds of phosphorus with an organic group in the molecule.

Organophosphorus compounds are characterized by an unusually large variety of structures and a wide range of uses. The reason is that phosphorus is able to form stable compounds with a coordination number of 1 to 6, and an oxidation state of −III to +V. Of the wide range of possible compounds, only those of industrial importance are treated here.

The organic chemistry of phosphorus has been described in many reviews. In particular, the chemistry of phosphorus with a low coordination number has been intensively investigated in recent years [1]–[9].

# 2. Phosphines [10]

Phosphines are compounds in which phosphorus has a coordination number of 3 and an oxidation state of −3. They are derived from phosphine [7803-51-2], $PH_3$. Successive formal substitution of 1, 2, or 3 hydrogen atoms of $PH_3$ by alkyl or aryl groups gives primary, secondary, and tertiary phosphines.

## 2.1. Properties

Under normal conditions, $PH_3$, methylphosphine $CH_3PH_2$, trifluoromethylphosphine [420-52-0], $CF_3PH_2$, and bis(trifluoromethyl)phosphine [460-96-8], $(CF_3)_2PH$, are gases. Most of the known phosphines are colorless liquids at 0.1 MPa and 20 °C, usually immiscible with water, and have a density of 0.8–0.85 g/cm$^3$. *tert*-Alkylphosphines with a chain length >$C_{12}$ tend to crystallize at room temperature. Phosphines that contain a rigid organic group, e.g., 9H-9-phosphabicyclononane [13887-02-0], [13396-80-0] are crystalline, as are compounds in which the van der Waals forces

**Table 1.** Physical properties of primary phosphines

| Formula | CAS no. | $M_r$ | Density, g/cm$^3$ | $^{31}$P NMR, ppm | bp at 103 kPa, °C | $n_D^{20}$ | Flash point, °C |
|---|---|---|---|---|---|---|---|
| CH$_3$PH$_2$ | [593-54-4] | 48.02 | | −160.2 | −14 | | a |
| C$_2$H$_5$PH$_2$ | [593-68-0] | 62.05 | | −128 | 25 | | a |
| n-C$_4$H$_9$PH$_2$ | [1732-74-7] | 90.10 | 0.769 | −140 | 86.7 | 1.4477 | a |
| c-C$_6$H$_{11}$PH$_2$ | [822-68-4] | 116.13 | 0.875 | −110 | 145 | 1.4860 | a |
| C$_6$H$_5$PH$_2$ | [638-21-1] | 110.09 | 1.001 | −118.7 | 160 | 1.5736 | a |

$^a$ Spontaneously flammable.

resulting from the functional groups leads to formation of crystalline phases [e.g., tris(2-cyanoethyl)phosphine].

Phosphines have an intense and penetrating odor, often described as resembling garlic, even in the ppb range. For phosphine itself, the odor threshold is extremely dependent on the purity of the gas. High-purity PH$_3$ (5 N), as used in the semiconductor industry, is detectable by smell only when the concentration is at a dangerously toxic level (> 1000 ppm), but when PH$_3$ is produced by the hydrolysis of AlP or Mg$_3$P$_2$, it can be detected even at ca. 20 ppb. The MAK value for PH$_3$ is 0.1 ppm and the TLV value, 0.3 ppm. There are no MAK or TLV limits for any other phosphines, with the exception of phenylphosphine (see Chap. 13).

Phosphine, PH$_3$, like some other low molecular mass phosphines (e.g., phenylphosphine, methylphosphine, dimethylphosphine, and trimethylphosphine), is very toxic. The LC$_{50}$ and LD$_{50}$ values show that the acute toxicity decreases with increasing molecular mass.

The presence of an electron lone pair gives rise to the three principal properties of the phosphines:

1) Oxidizability
2) Ability to function as a ligand in complexes
3) Nucleophilicity

The physical properties of some primary, secondary, and tertiary phosphines are given in Tables 1, 2, 3.

## 2.2. Production

### 2.2.1. Primary and Secondary Phosphines

The radical-induced addition of alkenes to PH$_3$ produces primary and secondary phosphines with good selectivity, provided that linear α-alkenes are not used. Commercially, 9H-9-phosphabicyclononane (American Cyanamid, Hoechst) and diisobutylphosphine [1732-72-5] (American Cyanamid) are produced by this method [11]. Azobisisobutyronitrile (AIBN) is used as the initiator.

**Table 2.** Physical properties of secondary phosphines

| Formula | CAS no. | $M_r$ | Density, g/cm$^3$ | $^{31}$P NMR, ppm | bp (p), °C (Pa) | mp, °C | $n_D^{20}$ | Flash point, °C |
|---|---|---|---|---|---|---|---|---|
| (CH$_3$)$_2$PH | [676-59-5] | 62.05 | | −99.5 | 21.1 (1.03 × 10$^5$) | < −64 | 1.4960 | a |
| (C$_2$H$_5$)$_2$PH | [627-49-6] | 90.10 | 0.7862 | −55.5 | 85 (1.03 × 10$^5$) | −51 | 1.4470 | a |
| (n-C$_4$H$_9$)$_2$PH | [1732-72-5] | 146.20 | 0.808 | −69.5 | 183 (1.03 × 10$^5$) | | 1.456 | a |
| (c-C$_6$H$_{11}$)$_2$PH | [829-84-5] | 198.27 | 0.98 | −28.3 | 281 (1.03 × 10$^5$) | | 1.514 | a |
| (C$_6$H$_5$)$_2$PH | [829-85-6] | 186.17 | 1.07 | −41 | 116 (346) | | 1.6270 | a |
| C$_8$H$_{14}$PH$^b$ | asym. [13396-80-0] sym. [13887-02-0] | 142.17 | 1.06$^c$ | −48.9 −55.1 | 220 (1.03 × 10$^5$)$^c$ | 50$^c$ | | 50$^c$ |

$^a$ Spontaneously flammable. $^b$ asym. (P); sym. (P). $^c$ Mixture of isomers.

**Table 3.** Physical properties of tertiary phosphines

| Formula | CAS no. | $M_r$ | Density, g/cm$^3$ | $^{31}$P NMR, ppm | bp at 103 kPa, °C | mp, °C | $n_D^{20}$ | Flash point, °C |
|---|---|---|---|---|---|---|---|---|
| (CH$_3$)$_3$P | [594-09-2] | 76.07 | 0.748 | −62 | 40 | −85 | | a |
| (C$_2$H$_5$)$_3$P | [554-70-1] | 118.15 | 0.801 | −20.4 | 127 | | 1.456 | a |
| (n-C$_4$H$_9$)$_3$P | [998-40-3] | 202.31 | 0.817 | −32.2 | 240 | | 1.4635 | 37.2 |
| (c-C$_6$H$_{11}$)$_3$P | [2622-14-2] | 280.41 | | +7 | | 76 | | |
| (C$_6$H$_5$)$_3$P | [603-35-0] | 262.27 | 1.2 | −5.4 | 360 | 80 | | 182 |
| (NCCH$_2$CH$_2$)$_3$P | [4023-53-4] | 193.19 | | −23 | | 98−99 | | 278 |

$^a$ Spontaneously flammable.

PH$_3$ + (cyclooctene) $\xrightarrow[\text{80−100 °C}]{\text{AIBN, 80−150 bar}}$ (cyclooctyl-PH) + (cyclooctyl-PH)

Primary phosphines are produced selectively, though only in moderate yield, by the acid-catalyzed reaction of PH$_3$ with alkenes [12]. In contrast to the radical-catalyzed reaction, this reaction yields Markownikoff products. *tert*-Butylphosphine can be obtained from PH$_3$ and isobutene in this way. This substance has been proposed as a substitute for high-purity PH$_3$ for use in the semiconductor industry (American Cyanamid) [13].

PH$_3$ + H$_2$C=C(CH$_3$)$_2$ $\xrightarrow{\text{CH}_3\text{SO}_3\text{H}}$ H$_2$P−C(CH$_3$)$_3$

## 2.2.2. Tertiary Phosphines

**Reaction of PCl$_3$ with Organometallic Compounds.** The Grignard reaction with phosphorus trichloride offers an industrial route to tertiary phosphines (M & T or Atochem, Hokko) [14], [15].

$$PCl_3 + 3\,RMgCl \xrightarrow{-MgCl_2} PR_3$$

R = phenyl, cyclohexyl, butyl, octyl, tolyl

Triphenylphosphine, which is important as a catalyst ligand and as an intermediate in the synthesis of Wittig reagents, is also produced by BASF in a reaction analogous to the Wurtz reaction [16].

$$PCl_3 + 3\,C_6H_5Cl + 6\,Na \longrightarrow P(C_6H_5)_3 + 6\,NaCl$$

**Addition of Unsaturated Compounds to PH$_3$.** Trialkylphosphines are obtained by the radical-induced addition of alkenes to PH$_3$ (American Cyanamid, Hoechst, Nippon Chemical Industries) [17] – [19].

$$PH_3 + 3\,H_2C=CHR \xrightarrow{I} P(CH_2CH_2R)_3$$

R = H, C$_n$H$_{2n+1}$ ($n$ = 1–18)

I = Initiator (e.g., azobisisobutyronitrile)

Both the Grignard process and the PH$_3$ route have advantages and disadvantages. For example, trimethylphosphine and triarylphosphines cannot be obtained by the PH$_3$ process. Only linear α-alkenes react smoothly to give tertiary phosphines, and phosphines with sterically hindered groups are generally not obtainable in this way. On the other hand, the Grignard process has a potential pollution problem due to the byproduct magnesium salts, meaning that extra cost is incurred.

Tris(2-cyanoethyl)phosphine, which is of interest as an intermediate, can be obtained either directly by radical- or base-catalyzed reaction of PH$_3$ with acrylonitrile [20], or by reacting tris(hydroxymethyl)phosphine [2767-80-8] with acrylonitrile [21]. Tris(hydroxymethyl)phosphine can be obtained by reacting PH$_3$ with formaldehyde in the presence of metal catalysts (e.g., Cd$^{2+}$), but is more conveniently produced from the commercially more readily available tetrakis(hydroxymethyl)phosphonium chloride or sulfate (see Chap. 4) by reaction with aqueous NaOH with loss of one equivalent of formaldehyde.

PH$_3$ →
- CH$_2$=CH–CN
- 1) H$_2$CO/Cd$^{2+}$
- 2) CH$_2$=CH–CN
- 1) H$_2$CO/H$_2$O/HCl
- 2) NaOH/–NaCl/–H$_2$CO
- 3) CH$_2$=CH–CN

→ P(CH$_2$CH$_2$CN)$_3$

The industrial production of tertiary phosphines with differing groups on the phosphorus atom is carried out by a number of methods.

Primary or secondary phosphines can be further alkylated with alkenes (sometimes bearing functional groups) under free-radical catalysis.

$$\text{CyclohexylPH} \xrightarrow[\text{Initiator}]{\text{Eicosene}} \text{CyclohexylP(C}_{20}\text{H}_{41}\text{)} \quad (1)$$

$$\text{H}_3\text{C-CH(CH}_3\text{)-PH}_2 \xrightarrow[\text{Initiator}]{\text{Allyl alcohol}} sec\text{-BuP(CH}_2\text{CH}_2\text{CH}_2\text{OH)}_2$$

One cyanoethyl group of tris(2-cyanoethyl)phosphine can be cleaved under basic conditions if the phosphine compound is first converted to a quaternary compound by treatment with an alkyl halide:

$$\text{P(CH}_2\text{CH}_2\text{CN)}_3 + \text{BrCH}_2\text{CH}_2\text{SO}_3^-\text{Na}^+ \longrightarrow$$

$$\text{(NCCH}_2\text{CH}_2\text{)}_3\text{P}^+\text{CH}_2\text{CH}_2\text{SO}_3^- \xrightarrow{\text{CH}_3\text{ONa}} \quad (2)$$

$$\text{Na}^{+\,-}\text{O}_3\text{SCH}_2\text{CH}_2\text{P(CH}_2\text{CH}_2\text{CN)}_2$$

This method can be used to produce chiral phosphines with the phosphorus atom as chirality center.

**Synthesis from Arylphosphines**. Tertiary phosphines with one or two phenyl groups can be obtained either from the corresponding chloroarylphosphines or from triphenylphosphine [22].

$$(C_6H_5)_2PCl \xrightarrow[\text{2) }o\text{-Chlorobenzoic acid}]{\text{1) 2 Na}}$$

$$(C_6H_5)_3P \xrightarrow[\substack{\text{2) NH}_4\text{Cl} \\ \text{3) }o\text{-Chlorobenzoic acid}}]{\text{1) 2 Na}} \text{(C}_6\text{H}_5\text{)}_2\text{P-C}_6\text{H}_4\text{-COOH} \quad (3)$$

## 2.3. Uses

Primary and secondary phosphines are intermediates, not usually isolated, in the production of tertiary phosphines and their derivatives. Also, phosphinic and dithiophosphinic acids (see Chap. 7), obtained by oxidation of secondary phosphines, have been used to a limited extent for solvent extraction and flotation processes (Cyanex 272, Aerofine 3418 A, American Cyanamid). High-purity *tert*-butylphosphine [*2501-94-2*] (Cypure, American Cyanamid) can partially replace the more toxic $PH_3$ in the production and doping of semiconductors in the electronics industry [13].

**Table 4.** Physical properties of halophosphines

| Formula | CAS no. | $M_r$ | Density, g/cm$^3$ | bp (p), °C (Pa) | mp, °C | $n_D^{20}$ | Flash point, °C |
|---|---|---|---|---|---|---|---|
| CH$_3$PCl$_2$ | [676-83-5] | 116.92 | 1.32 | 81 (1.03×10$^5$) | <−64 | 1.4960 | 23 [a] |
| C$_6$H$_5$PCl$_2$ | [644-97-3] | 178.99 | 1.318 | 224 (1.03×10$^5$) | −51 | 1.5958 | 102 |
| Cl$_2$PC$_6$H$_4$C$_6$H$_4$PCl$_2$ | [35346-33-9] | 355.96 | | 175 (665) | 63 | | |
| (C$_6$H$_5$)$_2$PCl | [1079-66-9] | 220.64 | 1.19 | 320 (1.03×10$^5$) | 15 | 1.634 | 138 [b] |

[a] Pensky–Martens closed cup.
[b] Cleveland open cup.

Triphenylphosphine (TPP) has the greatest industrial importance of all the tertiary phosphines as a ligand in homogeneous catalysis (hydroformylation, hydrogenation, and oligomerization), and as a starting material in the preparation of Wittig reagents, especially for the synthesis of vitamin A and in the production of β-carotene [23]. TPP can be sulfonated by a process developed by Rhône-Poulenc and Hoechst. The rhodium complexes of sulfonated TPP used in hydroformylation are water soluble and can therefore easily be separated from the organic phase [24], [25].

The phosphinoethanesulfonate salt, obtainable by Equation (2), is also water-soluble. It is used as an antioxidant and complexing agent for silver in the production of positive transparencies [26].

The tertiary bicyclic phosphines, produced according to Equation (1), enable long chain α-alkenes to be hydroformylated and hydrogenated in high yield to give alcohols for detergents [27].

The ligand formed in Equation (3) is of major industrial importance. It is used as a nickel complex to catalyze the oligomerization of ethylene to form linear α-alkenes in the Shell Higher Olefin Process (SHOP) [28].

Trialkylphosphines are used as starting materials in the production of phosphonium salts (see Chap. 4) as well as tertiary phosphine oxides and sulfides (see Chap. 5). They are also used directly as catalysts for the trimerization of diisocyanates [29].

Triethylphosphine functions as a ligand in a gold(I) complex for oral treatment of arthritis (Auranofin, Smith-Kline & French) [30].

# 3. Halophosphines

The halophosphines RPX$_2$ (phosphonous acid dihalides) and R$_2$PX (phosphinous acid halides) are important starting materials in the production of numerous organophosphorus compounds.

## 3.1. Properties

Halophosphines are generally colorless liquids of high density. Due to their high reactivity and water and air sensitivity, they are used only as intermediates in the production of plant protection agents, stabilizers for plastics, and catalysts. Some physical properties of the industrially important compounds are listed in Table 4.

## 3.2. Production

**From Hydrocarbons and PCl$_3$.** Alkyl- and aryldichlorophosphines are obtained by the free radical reaction of PCl$_3$ with hydrocarbons in the gas phase at high temperature. Halogen compounds are used as catalysts [31]–[35]. In the case of aromatic compounds, yellow phosphorus may be added as an additional component [36].

$$CH_4 + PCl_3 \xrightarrow[CCl_4]{550-650\,°C} CH_3PCl_2$$

$$C_6H_6 + PCl_3 \xrightarrow[C_6H_5Cl]{450-600\,°C} C_6H_5PCl_2$$

**By the Friedel–Crafts Reaction.** The reaction of phosphorus trichloride with aromatic hydrocarbons in the Friedel–Crafts reaction takes place under considerably milder conditions than the gas-phase reaction. However, complexes of halophosphine and AlCl$_3$ are produced, which must be decomposed by adding POCl$_3$, ketones, or esters, forming more stable aluminum chloride complexes [37], [38].

$$C_6H_6 + PCl_3 + AlCl_3 \longrightarrow C_6H_5PCl_2 \cdot AlCl_3 + HCl$$

$$\text{biphenyl} + 2\,PCl_3 \xrightarrow[\text{2) POCl}_3]{\text{1) AlCl}_3} Cl_2P\text{-C}_6H_4\text{-C}_6H_4\text{-}PCl_2$$

**By the Kinnear–Perren Reaction.** Alkyltrichlorophosphonium tetrachloroaluminates can be obtained from phosphorus trichloride, alkyl chlorides, and aluminum trichloride. Alkyldichlorophosphines can be obtained from the phosphonium salts by reduction and complexing the AlCl$_3$.

$$RCl + PCl_3 + AlCl_3 \longrightarrow RPCl_3^+ AlCl_4^-$$

$$3\,RPCl_3^+AlCl_4^- \begin{cases} \xrightarrow{+2P+3KCl} 3\,RPCl_2 + 2\,PCl_3 + 3\,KAlCl_4 \\ \xrightarrow{+2Al+5KCl} 3\,RPCl_2 + 5\,KAlCl_4 \end{cases}$$

**By Disproportionation or Comproportionation.** Chlorodiphenylphosphine and PCl$_3$ are obtained by the disproportionation of dichlorophenylphosphine in the gas phase at elevated temperature [39]. Similarly, triphenylphosphine or triphenylphosphine oxide can be comproportionated with PCl$_3$ or dichlorophenylphosphine [40]–[43].

$$2\,C_6H_5PCl_2 \longrightarrow (C_6H_5)_2PCl + PCl_3$$
$$(C_6H_5)_3P + 2\,PCl_3 \longrightarrow 3\,C_6H_5PCl_2$$
$$(C_6H_5)_3PO + 2\,PCl_3 \longrightarrow 2\,C_6H_5PCl_2 + C_6H_5P(O)Cl_2$$

**From Elemental Phosphorus.** The reaction of red phosphorus with perfluoroiodoalkanes to give a mixture of perfluoroalkyldiiodophosphine and bis(perfluoroalkyl)iodophosphine is of industrial importance [44].

$$2\,P_{red} + 3\,RI \longrightarrow RPI_2 + R_2PI$$
$$R = C_nF_{2n+1},\ n = 6\text{–}12$$

## 3.3. Uses

Dichloromethylphosphine (Hoechst) [45] is used in the production of methylphosphinic acid [*4206-94-4*], CH$_3$PH(O)OH, which is used in the production of the total herbicide Gluphosinate. Another application is in the production of a cyclic phosphinocarboxylic anhydride, which is used in the flameproofing sector (see Sections 7.3 and 7.5). Dichlorophenylphosphine (Akzo, Ferro Corp.) is converted to phenylphosphinic acid [*1779-48-2*] by hydrolysis (see Section 7.3), or disproportionated to give chlorodiphenylphosphine. It has largely lost its importance as a precursor for the insecticide EPN (see Section 8.5). The bifunctional biphenyl derivative Cl$_2$P–C$_6$H$_4$–C$_6$H$_4$–PCl$_2$ [*35346-33-9*] (Sandoz) is used in the production of the plasticizer Sandostab P-EPQ (see Section 6.2).

Chlorodiphenylphosphine (Akzo) is an important starting material for the production of UV-hardening paint systems (see Section 5.3). Mixtures of perfluoroalkyliodophosphines are converted by hydrolysis to perfluoroalkylphosphonic and perfluoroalkylphosphinic acids, which are used for cleaning and degreasing (Fluowet PP, Hoechst).

## 4. Phosphonium Salts [46]

Phosphonium salts are four-coordinate phosphorus compounds of the type [R$_n$PH$_{4-n}$]$^+$ X$^-$, where the phosphorus atom carries a positive charge. In the industrially important phosphonium compounds the group R is alkyl, benzyl, or aryl. The counterion X$^-$ can be organic or inorganic. If both charges are present on the same molecule, the compound is referred to as a phosphabetaine. Phosphonium salts cannot in all cases be easily distinguished from phosphoranes with a five-coordinate phosphorus

Table 5. Physical properties of phosphonium salts

| Formula | CAS no. | $M_r$ | Density, g/cm$^3$ | $^{31}$P NMR, ppm | bp at 103 kPa, °C | mp, °C | Flash point, °C |
|---|---|---|---|---|---|---|---|
| [(CH$_3$)$_4$P]Cl | [1941-19-1] | 126.57 | | +24.7 | 400 | | > 300 |
| [(CH$_2$OH)$_4$P]Cl | [124-64-1] | 190.56 | 1.33 | +25.2 | | 150 | |
| [(C$_4$H$_9$)$_4$P]Br | [3115-68-2] | 339.34 | | +34 | | 95 | 290 |
| [(C$_4$H$_9$)$_3$PC$_{16}$H$_{33}$]Br | [14937-45-2] | 507.66 | | +32.6 | | 55 | |
| [(C$_4$H$_9$)$_3$PCH$_3$]I | [1702-42-7] | 344.26 | | +32.0 | | 145 | 285 |
| [(C$_6$H$_5$)$_4$P]Br | [2751-90-8] | 419.31 | | +20.8 | | 295 | 295 [a] |
| [(C$_6$H$_5$)$_3$PCH$_2$OCH$_3$]Cl | [4009-98-7] | 342.80 | | | | 180 | |

[a] Thermal decomposition.

atom, because they can exist in either form, depending on the solvent and/or the state of aggregation.

## 4.1. Properties

Phosphonium salts, with few exceptions, are colorless, crystalline solids with good to very good solubility in water. Compounds with short-chain (C$_1$–C$_4$) and long-chain (> C$_8$) alkyl groups in the same molecule exhibit surfactant properties. Like the corresponding ammonium compounds, they are biocides (see Section 4.3), though in general their bactericidal properties are weaker. Their toxicity to fish is lower than that of the corresponding ammonium salts. Many phosphonium salts have an extremely irritant effect on skin and mucous membranes and moderate acute toxicity. They are biologically degraded only very slowly.

Tetraalkylphosphonium salts are oxidized by bases (OH$^-$, RO$^-$) with elimination of alkane to give tertiary phosphine oxides only upon prolonged reaction at elevated temperature [47]. Phosphonium salts are thus more stable than the corresponding ammonium compounds under the conditions of phase-transfer catalysis.

Table 5 lists the physical properties of some phosphonium salts.

## 4.2. Production

**From Phosphines by Formation of Quaternary Compounds.** Phosphines, $R_n PH_{3-n}$, like the corresponding nitrogen compounds, form quaternary salts with acids.

$$R_n PH_{3-n} + HX \longrightarrow [R_n PH_{4-n}]\ X^-$$

R = alkyl, aryl; X = Cl, Br, I

Owing to their nucleophilicity, phosphines react with alkyl and aryl halides to give the corresponding phosphonium salts; the rate of reaction decreases in the order $R'I > R'Br > R'Cl$.

$$R_3P + R'X \longrightarrow [R_3PR']^+ X^-$$

R = alkyl, aryl, H
R'X = alkyl or benzyl halide

(Methoxymethyl)triphenylphosphonium chloride is produced by a one-pot variation of this reaction in which triphenylphosphine (TPP) is reacted with acetyl chloride and dimethoxymethane [48]. Chloromethyl methyl ether, formed as an intermediate, alkylates the TPP in a subsequent reaction.

$$Ph_3P + Cl-\overset{O}{\underset{\|}{C}}-CH_3 + CH_3OCH_2OCH_3 \longrightarrow$$
$$CH_3C(O)OCH_3 + [Ph_3PCH_2OCH_3]^+ Cl^-$$

In the formation of quaternary compounds from aryl bromides or chlorides, the so-called complex salt method is particularly suitable [49], whereby the molar ratio of phosphine to metal salt is 2:1. Nickel halides, and chlorides of cobalt, zinc, and copper(II) can be used.

$$(C_6H_5)_3P + C_6H_5Br \xrightarrow[200°C]{NiBr_2} [(C_6H_5)_4P]^+ Br^-$$

**Addition of Phosphines to Carbonyl Compounds.** A very effective method of preparing phosphonium salts is the reaction of phosphines with carbonyl compounds in the presence of at least one equivalent of acid. The carbonyl group is inserted into one or more P–H bonds, forming α-hydroxyalkylphosphonium salts.

$$PH_3 + H_2C=O + HCl \longrightarrow [P(CH_2OH)_4]^+ Cl^-$$

$$R_3P + \overset{O}{\underset{H}{\diagup}}\text{-}R' \xrightarrow{+HX} R-\overset{R}{\underset{R}{\overset{|}{P^+}}}\overset{OH}{\underset{H}{\overset{|}{-}}}R' \quad X^-$$

HX = mineral acid, carboxylic acid

**Addition of 1,3-Dienes to Halophosphines (McCormack Reaction).** 1,3-Dienes react with, e.g., dichlorophosphines, to give chlorophospholenium chlorides in a reaction analogous to the Diels–Alder reaction. The formation of the adduct takes place very slowly under normal conditions, but can be accelerated by the use of, e.g., $POCl_3$ as a solvent. Solvolysis of the product yields phospholene oxides (see Chap. 5).

$$RPCl_2 + \underset{R^2}{\overset{H_2C}{\diagup}}\underset{CH_2}{\overset{R^1}{\diagdown}} \longrightarrow \underset{R^2}{\overset{R^1}{\diagdown}}\overset{Cl}{\underset{R}{P^+\diagdown}} Cl^-$$

R = alkyl, aryl; $R^1$, $R^2$ = H, alkyl

**By Anion Exchange.** In cases where the presence of halide ions is undesirable (e.g., where they may cause corrosion), strongly basic ion exchangers are used to replace them by hydroxyl ions, which can themselves be easily exchanged for other anions. In this way, the readily accessible tetraalkylphosphonium halides can be converted to phosphonium acetates, which are used as latent hardeners and accelerators in the production of epoxy resins [50].

## 4.3. Uses

Few other classes of compounds can match the phosphonium salts for the remarkable variety of their technical applications, though the quantities involved are not always large. These include the use of phosphonium salts derived from triphenylphosphine as Wittig reagent precursors [51], and the use of tetraalkylphosphonium salts as reaction accelerators (phase-transfer catalysts [52] or promotors). The latter are used in several processes, including the production of acetic acid and acetic anhydride by the carbonylation of methanol [53], in the production of stabilizers for polypropylene, and the manufacture of isocyanurate esters [54] and polysiloxanes [55]. Owing to their high thermal stability, they are also suitable for chlorine–fluorine exchange in chloronitro aromatic compounds [56]. The quality of dithiophosphoric acids produced from $P_4S_{10}$ and alcohols is significantly improved by the presence of catalytic amounts of quaternary phosphonium compounds [57].

The biocidal properties of tetraalkylphosphonium salts are exploited in cooling water additives (Belacide, formerly Belclene 350, Ciba-Geigy [58]), in antifouling paints [59], and as additives in drilling oils.

2,4-Dichlorobenzyltributylphosphoniumchloride (Phosfleur (Perifleur Products), Chlorphonium [*115-78-6*]) is used as a growth regulator for plants of the genus chrysanthemum [60].

Technochemie has specialized in the production of quaternary compounds from triarylphosphines. Albright & Wilson produce tetrakis(hydroxymethyl)phosphonium salts as intermediates for the production of flameproofing agents for cotton textiles (Proban process [61]). Also, all phosphine producers produce phosphonium salts (American Cyanamid, Hoechst, Hokko, M & T, and Nippon).

## 5. Phosphine Oxides and Sulfides

The terms phosphine oxide and phosphine sulfide denote four-coordinate phosphorus compounds with only organic groups or hydrogen atoms present in addition to the P=O or P=S double bond.

**Table 6.** Physical properties of phosphine oxides

| Formula | CAS no. | $M_r$ | Density g/cm³ | ³¹P NMR, ppm | bp (p), °C (Pa) | mp, °C | Flash point, °C |
|---|---|---|---|---|---|---|---|
| (CH₃)₃PO | [676-96-0] | 92.08 | | +36.2 | 214 (1.03×10⁵) | 139.5 | |
| (C₈H₁₇)₃PO | [78-50-2] | 386.64 | 0.8 (50 °C) | +42.0 | 360 (1.03×10⁵) | 50 | 240 |
| sec-C₄H₉PO(C₈H₁₇)₂ | [114415-78-0] | 330.53 | 0.88 | +51.1 | 165 (70) | <−10 | >200 |
| ⌬P(CH₃)=O (cyclic) | [930-38-1] | 116.10 | 1.094 (50°C) | +71.5ᵃ | 80 (100) | 36 | 113 |
| ⌬P(CH₃)=O isomer mixture | [31563-86-7] | 116.10 | 1.12 | +71.5, +73.8ᵃ | 150 (3300) | <0 | 125 |

ᵃ In alkaline medium: 60.2/62.6 ppm.

## 5.1. Properties

Primary phosphine oxides, RH₂P=O, are only stable under normal conditions if the R group stabilizes the molecule by its steric bulk, e.g., 2,4,6-tri(*tert*-butyl)phenylphosphine. Even under mild conditions, disproportionation products (primary phosphine and phosphonous acid) are formed during oxidation of primary phosphines.

Primary phosphine sulfides are also unstable and tend to form cyclic polyphosphines at elevated temperature, with liberation of H₂S.

Secondary phosphine oxides (phosphinous acids) R₂HP=O are generally crystalline, colorless, odorless substances that are stable at room temperature. However, they disproportionate at high temperature to form secondary phosphines and phosphinic acids.

$$2\ (CH_3)_2\overset{O}{\underset{\|}{P}}H \xrightarrow{T>100\,°C} (CH_3)_2PH + (CH_3)_2\overset{O}{\underset{\|}{P}}OH$$

Tertiary phosphine oxides are, with few exceptions, crystalline substances. They are odorless, resistant to hydrolysis, and of low acute toxicity, but are not easily biologically degradable. The physical properties of some phosphine oxides are given in Table 6.

## 5.2. Production

**Oxidation of Secondary and Tertiary Phosphines.** Secondary and tertiary phosphines react smoothly with equimolar amounts of H₂O₂ in aqueous solution or with elemental sulfur in an aprotic organic solvent under mild conditions to form the corresponding secondary and tertiary phosphine oxides or sulfides. These methods are by far the most important for the production of these classes of substances.

$$R_nH_{3-n}P \xrightarrow{+\ H_2O_2/S_8} R_nH_{3-n}P=X$$

$$X = O, S;\ n = 2, 3$$

In the reaction of secondary phosphines with $H_2O_2$, the reaction conditions—especially the reaction temperature—must be chosen so that formation of phosphinic acids is minimized.

Water-soluble tertiary phosphines can be oxidized with water, with liberation of hydrogen.

$$H_3C\text{-}P(CH_2CH_2CH_2OH)(CH_2CH_2CH_2OH)(CH_3) \xrightarrow[-H_2]{H_2O/H^+} H_3C\text{-}P(=O)(CH_2CH_2CH_2OH)(CH_2CH_2CH_2OH)(CH_3)$$

$$P(CH_2OH)_3 \xrightarrow[-H_2]{H_2O/OH^-} (HOCH_2)_3P=O$$

The use of oxygen or air instead of $H_2O_2$ can be recommended only for the oxidation of secondary phosphines. When tertiary phosphines react with oxygen, the P–C bond is broken, leading to the formation of secondary phosphine oxides, phosphonates, and phosphinates, the tertiary phosphine oxides being obtained only as side products.

**By Grignard Reaction.** Phosphoryl chloride, phosphinic acid dichlorides, and phosphonic acid chlorides react with Grignard reagents with substitution of the halogen to form tertiary phosphine oxides.

$$POCl_3 \xrightarrow[-3MgXCl]{+3RMgX} R_3P=O$$

$$R^1POCl_2 \xrightarrow[-2MgXCl]{+2RMgX} R^1R_2P=O$$

$$R^1R^2POCl \xrightarrow[-MgXCl]{+RMgX} R^1R^2RP=O$$

The industrial production of trioctylphosphine oxide is carried out by this process, among others.

The reaction of the corresponding thio compounds with Grignard reagents often gives diphosphine disulfides as side products.

Secondary phosphites (dialkyl phosphonates) and tertiary phosphites can also be reacted with Grignard reagents to give tertiary phosphine oxides.

$$RO\text{-}P(=O)(H)(OR) \xrightarrow[-R^1H/-2MgX(OR)]{+3R^1MgX} R^1\text{-}P(OMgX)(R^1) \xrightarrow[-MgX_2]{+R^2X} R^1\text{-}P(=O)(R^1)\text{-}R^2$$

**Solvolysis of P–Cl Compounds.** Monohalophosphines react with water or alcohols to give secondary phosphine oxides, whereas monohalophosphonium salts (dihalophosphoranes) form tertiaryphosphine oxides under these conditions.

$$R_2PCl \xrightarrow[-RCl]{+ROH} R_2\overset{O}{\overset{\|}{P}}-H$$

$$\underset{H_3C}{\overset{H_3C}{\diagdown}}\overset{Cl}{\underset{\diagup}{P^+}}Cl^- \xrightarrow[-HCl,-RCl]{+ROH} \underset{H_3C}{\overset{H_3C}{\diagdown}}\overset{O}{\underset{\diagup}{P}} \qquad (4)$$

**By the Michaelis–Arbusov Reaction.** Esters of phosphinous acids react with alkyl halides, generally smoothly and with high yields, to form the metathesis products, with rearrangement of the organophosphorus product to give the tertiary phosphine oxide. Acylphosphine oxides may be obtained by using acyl halides.

$$\underset{R^2}{\overset{R^1}{\diagdown}}P-OR^3 \xrightarrow[-R^3Cl]{+RCl} \underset{R^2}{\overset{R^1}{\diagdown}}\overset{O}{\underset{\|}{P}}-R$$

R = alkyl, benzyl, benzoyl

**By Reaction with Carbonyl Compounds.** Carbonyl compounds react readily with secondary phosphines and phosphine oxides with substitution at the P–H bond to give the corresponding tertiary phosphines or phosphine oxides with an α-hydroxyl group.

$$\underset{CH_3}{\overset{CH_3}{\diagdown}}\overset{O}{\underset{\|}{P}}-H \xrightarrow{+H\overset{O}{\overset{\|}{C}}COOH} \underset{CH_3}{\overset{CH_3}{\diagdown}}\overset{O}{\underset{\|}{P}}-\underset{}{\overset{OH}{\underset{|}{C}}}-COOH$$

**By Reaction of Phosphinate Esters with Activated Alkenes.** Esters of phosphinous acids react smoothly with reactive alkenes in the presence of proton donors (not carboxylic acids) to give tertiary phosphine oxides. Phosphonous acid esters are converted to phosphinic acid esters.

$$\underset{H_3C}{\overset{H_3C}{\diagdown}}P-O-\underset{CH_3}{\overset{CH_3}{\diagdown}} \xrightarrow[-iso-BuOCH_3]{+\text{ acrylonitrile}\atop +\text{ methanol}} \underset{H_3C}{\overset{H_3C}{\diagdown}}\overset{O}{\underset{\|}{P}}-\text{CN}$$

## 5.3. Uses

Tertiary phosphine oxides and sulfides are used almost entirely as solvent extraction media [62]. Trioctylphosphine oxide (TOPO) is of particular importance (Cyanex 921, American Cyanamid; Hostarex PX 324, Hoechst; Hokko), being used for the extraction of metal ions [63]–[65] or carboxylic acids, alcohols, and phenol from aqueous solutions [66], [67]. Products such as Cyanex 923 (American Cyanamid, a mixture of tertiary phosphine oxides containing hexyl and octyl groups), and Hostarex PX 320 (Hoechst, *sec*-butyldioctylphosphine oxide) were also developed for this application

[68], [69]. Cyanex 471 X (American Cyanamid, triisobutylphosphine sulfide) is used for the extraction of noble metals (silver, platinum, and palladium) and mercury from strongly acidic solutions [70]. The use of tertiary phosphine oxides as catalysts is still restricted to a few reactions, though these are of economic importance. Phospholene oxides [71], which can be obtained, for example, by Equation (4), catalyze the formation of carbodiimides from isocyanates [71], and of polyamides and polyimides [72]. A further application is the production of lactones from acid chlorides [73]. Tertiary phosphine oxides can also be used for stabilizing peroxycarboxylic acids in detergents [74].

Bifunctional phosphine oxides, such as *sec*-butyl-bis(3-hydroxypropyl)phosphine oxide or 1-hydroxy-1-dimethylphosphinoacetic acid are reactive flameproofing agents [75] and herbicides [76], respectively.

Mono- and bis-acylphosphine oxides are used as initiators for the hardening of photopolymerizable materials by UV radiation [77]. They are especially suitable for hardening white paints [78] and dental materials [79].

# 6. Phosphonous Acid Derivatives

Organic derivatives of phosphonous acid are formally derived from the hypothetical phosphonous acid $H-P(OH)_2$. Structures containing three-coordinate phosphorus are only stable if both hydroxyl groups are blocked, for example, by esterification (**1**) or amide formation (**2**). Monoesters (**3**), however, contain four-coordinate phosphous and are regarded as derivatives of phosphinic acid $H_2P(=O)OH$. Phosphonous acid dihalides (dichlorophosphines) $R-PX_2$ are treated in Chapter 3.

$$R-P\begin{matrix}OR'\\OR'\end{matrix} \qquad R-P\begin{matrix}NR'_2\\NR'_2\end{matrix} \qquad R-\overset{\overset{O}{\|}}{P}\begin{matrix}OR'\\H\end{matrix}$$

$$\quad\;\; 1 \qquad\qquad\qquad 2 \qquad\qquad\quad 3$$

## 6.1. Production

Diesters of phosphonous acid are formed by reacting dihalophosphines with alcohols in the presence of bases, or by reacting phosphorous acid diester chlorides with Grignard reagents.

$$R-PCl_2 + 2R'OH \xrightarrow{base} R-P(OR')_2$$

$$Cl-P(OR')_2 + RMgX \longrightarrow R-P(OR')_2 + MgClX$$

Phosphonous acid diamides can be obtained from dihalophosphines and amines, or from phosphorous acid diamide halides and Grignard reagents.

## 6.2. Properties and Uses

Phosphonous acid derivatives are generally readily oxidized to phosphonic acid derivatives. Although the phosphorus–carbon bond is very stable, hydrolytic cleavage of the bond between the phosphorus atom and the hetero atoms takes place even under mild conditions. The tetraester of a diphosphonous acid **4** is used under the names Sandostab P-EPQ (Sandoz) and Irgafos P-EPQ (Ciba-Geigy) for the thermal stabilization of plastics [80], [81].

**4** [38613-77-3]

# 7. Phosphinic Acids and their Derivatives

Organophosphinic acids are the organic derivatives of phosphinic acid (hypophosphorous acid), $H_2P(=O)OH$, in which one or both of the hydrogen atoms on the phosphorus atom are replaced by organic groups. The free acids are of industrial importance, as are the esters, halides, and thio derivatives.

## 7.1. Properties

The phosphorus–carbon bonds in diorganophosphinic acids (with the exception of α-hydroxyalkyl, alkoxy, or trihalomethyl compounds) are very stable to hydrolysis under both acid and alkaline conditions. Long-chain substituted phosphinic acids and their dithio analogues form stable complexes with divalent cations, and are used in flotation and metal extraction. A number of phosphinic acid derivatives are intermediates in the production of flameproofing and crop protection agents.

## 7.2. Production

**By Hydrolysis of Halophosphines.** Organophosphinic acids are obtained by the hydrolysis of dihalophosphines under nonoxidizing conditions.

$$R-PCl_2 \xrightarrow[-2HCl]{+2H_2O} RP(=O)(H)(OH)$$

**By Oxidation.** Secondary phosphines, phosphine oxides, or halophosphines are oxidized to phosphinic acid derivatives by oxygen, sulfur, halogens, or other oxidizing agents. The most important industrial processes are the oxidation and sulfurization of secondary phosphines [82].

$$R_2PH \xrightarrow{O_2, H_2O_2, etc.} R_2P(=O)OH$$
$$R_2PH \xrightarrow{S_8} R_2P(=S)SH$$

**By the Michaelis–Arbusov Reaction** (see Section p. 3843). Alkyl diorganophosphinates are obtained by reacting dialkyl esters of phosphonous acid with alkyl halides.

$$R^1-P(OR^2)_2 + R^3X \longrightarrow R^1R^3P(=O)(OR^2) + R^2X$$

**By Addition to Double Bonds.** Organophosphinates RHP(=O)OH and their esters can add to compounds containing C=C, C=O, or C=N double bonds in the presence of catalysts [83]–[85]. Phosphinic acid (hypophosphorous acid) $H_3PO_2$ [6303-21-5] undergoes double addition to yield symmetrically substituted diorganophosphinates.

Dihalophosphines react with α,β-unsaturated carboxylic acids to form the dichlorides of carboxyalkylalkylphosphinic acids, which can be converted to cyclic anhydrides [86]–[88].

$$R-\underset{Cl}{\underset{|}{P}}-Cl + CH_2=CH-COOH \longrightarrow R-\underset{Cl}{\underset{|}{\overset{O}{\overset{||}{P}}}}-CH_2CH_2-\overset{O}{\overset{||}{C}}-Cl$$

$$\downarrow Ac_2O$$

$$R-\overset{O}{\overset{||}{\underset{O}{\underset{|}{P}}}}\hspace{-2pt}\begin{array}{c}\\O\end{array}$$

**From Tertiary Phosphine Oxides.** Treatment of tertiary phosphine oxides with alkali-metal hydroxides at elevated temperature usually causes the most electronegative substituent to be cleaved as an alkane or arene, forming alkali-metal salts of diorganophosphinic acids [89].

$$R_2R'P=O + NaOH \xrightarrow{200-300\,°C} R_2P(=O)ONa + R'H$$

## 7.3. Phosphinic Acids and Phosphinate Esters

Methyl phosphinic acid [*4206-94-4*], CH$_3$PH(=O)OH, is used in the production of the total herbicide Gluphosinate (Phosphinothricin) [*53369-07-6*] (Basta, Hoechst), which is used in the form of its ammonium salt [*77182-82-2*] [90], [91]. The synthesis takes place via the intermediate compound isobutyl methylphosphinate [*25296-66-6*] or the isoamyl ester [*87025-51-2*].

$$H_3C-\underset{OH}{\underset{|}{\overset{O}{\overset{||}{P}}}}-CH_2CH_2-\underset{NH_2}{\underset{|}{CH}}-\overset{O}{\overset{||}{C}}-OH$$
[*53369-07-6*]

Phenylphosphinic acid [*1779-48-2*], C$_6$H$_5$PH(=O)OH, and its sodium salt [*4297-95-4*] (Akzo, Ferro Corp.) are used to improve the stability of polyamides towards light and heat, and as antioxidants and promotors for emulsion polymerization.

Bis(hydroxymethyl)phosphinic acid [*2074-67-1*] is an intermediate in the synthesis of crop protection agents [85]. The calcium or magnesium salts (Chemische Fabrik Budenheim) are used as binders in basic fire-resistant materials [92]. The longer chain compounds bis(2,4,4-trimethylpentyl)phosphinic acid [*83411-71-6*] (Cyanex 272, American Cyanamid) and bis(2-ethylhexyl)phosphinic acid [*298-07-7*] (P-229, People's Republic of China) are used as extractants in the hydrometallurgical separation of cobalt and nickel [93].

## 7.4. Thiophosphinates

Thio derivatives of phosphinic acids are mainly used in ore and metal processing. The sodium salt of bis(2-methylpropyl)dithiophosphinic acid [*13360-80-0*] (Aerophine 3418 A) is supplied by American Cyanamid as a flo-tation agent for sulfidic ores. Bis(2,4,4-trimethylpentyl)dithiophosphinic acid [*107667-02-7*] (Cyanex 301, American Cyanamid) has better solubility in organic solvents and is suitable for extracting zinc and other heavy metals [93].

## 7.5. Other Phosphinic Acid Derivatives

The cyclic phosphinic anhydride 2-methyl- 2,5-dioxo-1,2-oxaphospholane (or 2-methyl-1,2-oxaphospholane-5-one 2-oxide) [*15171-48-9*] is of major importance in the production of washable fireproofed textiles, and are incorporated into polyester fibers by copolymerization (Trevira CS, Hoechst).

# 8. Phosphites and Hydrogenphosphonates

Phosphites are organic esters of the hypothetical phosphorous acid $P(OH)_3$ in which all three OH groups are esterified. In the mono- and diesters, the phosphorus is four-coordinate. They are therefore formal derivatives of phosphonic acid $(HO)_2P(O)H$ (phosphonates). However, both the properties and the uses of the phosphites and the hydrogenphosphonates are so closely related that treatment together is appropriate.

Hydrogenphosphonates and phosphites are mainly used as raw materials for insecticides as flameproofing agents, and as stabilizers and antioxidants for plastics.

## 8.1. Properties

Hydrogenphosphonates and phosphites are colorless liquids or solids. Owing to their structure, phosphites and phosphonic acid diesters are neutral compounds, while the monoesters have strongly acidic properties. All these compounds are readily hydrolyzable and are susceptible to oxidation. These properties make them suitable as stabilizers for plastics, in which they act by reducing peroxides and by capturing reactive radicals.

## 8.2. Production

Phosphites and hydrogenphosphonates are usually produced by reacting alcohols or phenols with phosphorus trichloride.

$$PCl_3 + 3\,ArOH \longrightarrow (ArO)_3P + 3\,HCl$$

Whereas reaction of $PCl_3$ with 3 mol phenol yields the desired triaryl phosphite with liberation of hydrogen chloride, alcohols react to form dialkyl phosphonates and alkyl chloride.

$$PCl_3 + 3\,ROH \longrightarrow \underset{RO}{\overset{RO}{>}}P\underset{H}{\overset{O}{<}} + RCl$$

To produce the trialkyl esters, it is necessary to add an equimolar amount of a substance that will neutralize acid. As a rule, ammonia or a volatile amine is used in an inert solvent in which the ammonium chloride formed is sparingly soluble and can be removed by filtration.

$$PCl_3 + 3\,ROH + 3\,NR'_3 \longrightarrow \underset{RO}{\overset{RO}{>}}P-OR + 3\,HNR'_3Cl$$

The esterification of $H_3PO_3$ with alcohols to give dialkyl phosphonates is of less importance [96]. The addition of ethylene oxide to $PCl_3$ in the presence of catalytic quantities of chloroethanol, iron chloride, or aluminum chloride produces tris(2-chloroethyl)phosphite.

$$PCl_3 + 3\,\text{(ethylene oxide)} \longrightarrow (ClCH_2CH_2O)_3P$$

Triphenyl phosphite can be reacted with alcohols to give a mixture of phenyl dialkyl and alkyl diphenyl phosphites.

$$(ArO)_3P + ROH \longrightarrow (ArO)_2POR + ArOH$$
$$(ArO)_3P + 2\,ROH \longrightarrow ArOP(OR)_2 + 2\,ArOH$$

## 8.3. Trialkyl Phosphites

Some physical properties of the most important trialkyl phosphites are listed in Table 7.

The short-chain compounds are mainly used as starting materials in the production of dialkyl esters of phosphonic acid (Michaelis – Arbusov Reaction, see Section 9.2).

$$\underset{RO}{\overset{RO}{>}}P-OR + R'X \longrightarrow R'-\underset{OR}{\overset{O}{\underset{\|}{P}}}-OR + RX$$

**Table 7.** Physical properties of trialkyl phosphites (RO)$_3$P

| R | CAS no. | $M_r$ | Density, g/cm$^3$ | bp (p), °C (Pa) | $n_D^{20}$ | Flash point,[a] °C |
|---|---|---|---|---|---|---|
| CH$_3$ | [121-45-9] | 124.08 | 1.053 | 112 (1.03 × 10$^5$) | 1.4080 | 27 |
| C$_2$H$_5$ | [122-52-1] | 166.16 | 0.963 | 156 (1.03 × 10$^5$) | 1.4127 | 54 |
| CH$_2$CH$_2$Cl | [140-08-9] | 269.51 | 1.353 | 115 (400) | 1.4876 | 190 |
| CH(CH$_3$)$_2$ | [116-17-6] | 208.24 | 0.914 | 63 (1460) | 1.4110 | 74 |
| CH$_2$(CH$_2$)$_2$CH$_3$ | [102-85-2] | 250.32 | 0.925 | 122 (930) | 1.4322 | 121 |
| CH$_2$CH(CH$_2$)$_3$CH$_3$ $\mid$ C$_2$H$_5$ | [301-13-3] | 418.64 | 0.902 | 163 (40) | 1.445 | 185 |
| iso-C$_8$H$_{17}$ | [25103-12-2] | 418.64 | 0.891 | 161 (40) | 1.450 | 146 |
| n-C$_{12}$H$_{25}$ | [3076-63-9] | 586.97 | 0.879 |  | 1.459 | 222 |

[a] Cleveland open cup.

In a further important application, they are converted into insecticidal vinyl esters of phosphoric acid [97]. Manufacturers of trialkyl phosphites include Albright & Wilson, Bayer, and Ciba-Geigy.

The long-chain compounds are mainly used as antioxidants for plastics [98], [99]. Spirocyclic compounds are also used in this application. These are obtained from pentaerythritol [115-77-5] and long-chain alcohols. Manufacturers: Albright & Wilson, Akzo, and Borg-Warner.

R = iso-C$_{10}$H$_{21}$ [26544-27-4]
C$_{18}$H$_{37}$ [3806-34-6]

[26741-53-7]

## 8.4. Triaryl Phosphites and Alkyl Aryl Phosphites

The physical properties of the more important of these compounds are given in Table 8. These compounds are used as plastics additives.

Apart from tris(nonylphenyl) phosphite (TNPP), a product derived from a mixture of nonylphenol and dinonylphenol is used.

Manufacturers (trade names): Akzo (Phosclere), Borg-Warner (Weston, Ultranox), Ciba-Geigy (Irgafos, Irgastab), Dover Chemical Corp. (Doverphos), Hoechst (Hostanox), M&T (Stavinor), Rhône-Poulenc (Garbefix), Uniroyal (Naugard).

**Table 8.** Physical properties of triarylphosphites $(R^1O)_3P$ and alkyl aryl phosphites $(R^1O)_2(R^2O)P$

| $R^1$ | $R^2$ | CAS no. | $M_r$ | Density, g/cm³ | $n_D^{20}$ | Flash point[a], °C |
|---|---|---|---|---|---|---|
| C₆H₅ | | [101-02-0] | 310.29 | 1.180 | 1.590 | 146 |
| ⟨C₆H₄⟩—C₉H₁₉ | | [26523-78-4] | 689.02 | 0.990 | 1.528 | 207 |
| (t-Bu substituted phenyl) | | [31570-04-4] | 646.94 | 1.03 | | 245 |
| C₆H₅ | iso-C₈H₁₇ | [26401-27-4] | 346.41 | 1.045 | 1.522 | 182 |
| C₆H₅ | iso-C₁₀H₂₁ | [26544-23-0] | 374.46 | 1.025 | 1.517 | 154 |
| iso-C₁₀H₂₁ | C₆H₅ | [25550-98-5] | 438.63 | 0.94 | 1.480 | 160 |

[a] Pensky–Martens closed cup.

**Table 9.** Physical properties of dialkyl and diaryl phosphonates $(RO)_2P(=O)H$

| R | CAS no. | $M_r$ | Density, g/cm³ | bp (p), °C (Pa) | $n_D^{20}$ | Flash point, °C |
|---|---|---|---|---|---|---|
| CH₃ | [868-85-9] | 110.05 | 1.201 | 170 (1.03 × 10⁵) | 1.4036 | 64[a] |
| C₂H₅ | [762-04-9] | 138.10 | 1.072 | 50 (270) | 1.4080 | 74[b] |
| CH₃(CH₂)₂CH₃ | [1809-19-4] | 194.21 | 0.995 | 118 (930) | 1.4238 | 115[a] |
| CH₂CH(CH₂)₃CH₃ \| C₂H₅ | [3658-48-8] | 306.43 | 0.973 | 163 (400) | 1.4423 | 168[c] |
| C₁₂H₂₅ | [21302-09-0] | 418.64 | 0.90 | | 1.452 | 145[a] |
| C₆H₅ | [4712-55-4] | 234.19 | 1.223 | 135 (150) | 1.558 | 143[a] |

[a] Pensky–Martens closed cup. [b] Tag closed cup. [c] Cleveland open cup.

## 8.5. Dialkyl and Diaryl Phosphonates

Table 9 gives a survey of some industrially important dialkyl and diaryl phosphonates.

The main areas of application are in the production of phosphonic acid derivatives, insecticides, and plastics additives. Technical-grade diphenyl phosphite is produced by hydrolysis of triphenyl phosphite with water, and usually contains considerable quantities of free phenol. A low-phenol product is obtained by reacting triphenyl phosphite with phosphonic acid.

## 8.6. Alkyl Phosphonates

Alkyl phosphonates are not of great industrial importance, an exception being monoethyl phosphonate, which is marketed as a fungicide in the form of the aluminum salt $Al[C_2H_5OPH(=O)O]_3$ [39148-24-8] (Fosetyl-Aluminum, Aliette, Rhône-Poulenc) [100], [101].

# 9. Phosphonic Acids and their Derivatives

Of all the various types of organophosphorus compounds, phosphonic acids and their derivatives are outstanding due to their structural variety and great economic importance. Phosphonic acid derivatives are used as crop protection agents in water treatment, in metal processing, and as flameproofing agents.

## 9.1. Properties

The dibasic phosphonic acids are mostly weaker acids than phosphoric acid. The P–C bond in phosphonic acid derivatives is generally very stable towards oxidation or hydrolysis, so that many reactions can be carried out on the organic part of the molecule. Phosphonic acids are often only slowly biodegradable, but are usually rapidly destroyed by a series of photolytic and biological degradation steps. However, compounds with electronegative substituents on the $\alpha$-carbon atom are considerably less stable.

Several di- and polyphosphonic acids exhibit complexing (sequestering) properties towards polyvalent cations, and, when added in substoichiometric amounts, prevent the precipitation of low-solubility salts such as alkaline earth sulfates or carbonates (threshold effect).

Esters of phosphonic acid are usually more stable to hydrolysis than the corresponding phosphates, but are converted into free phosphonic acids on heating under acid conditions.

## 9.2. Production

**By Oxidation.** The oxidation of low-valency organophosphorus compounds is important mainly in the case of the thiophosphonic acid chlorides. This method is used in the production of methylphosphonic acid dichloride [676-97-1] by reaction of methyldichlorophosphine with sulfuryl chloride or chlorosulfuric acid [102]. Elemental sulfur reacts with $MePCl_2$ in the presence of catalytic quantities of tetraalkylphosphonium salts [103] to form methylthiophosphonic acid dichloride [676-98-2].

$$\text{H}_3\text{C}-\underset{\text{Cl}}{\overset{\text{Cl}}{\text{P}}} \begin{array}{c} \xrightarrow{\text{HOSO}_2\text{Cl}} \\ \\ \xrightarrow{\text{S}_8,\ \text{cat.}} \end{array} \begin{array}{c} \text{H}_3\text{C}-\underset{\text{Cl}}{\overset{\text{O}\ \ \text{Cl}}{\overset{\|}{\text{P}}}} \\ \\ \text{H}_3\text{C}-\underset{\text{Cl}}{\overset{\text{S}\ \ \text{Cl}}{\overset{\|}{\text{P}}}} \end{array}$$

**From Organometallic Compounds.** Ethylthiophosphonic acid dichloride [*993-43-1*], $C_2H_5P(=S)Cl_2$, is prepared by the reaction of thiophosphoryl chloride with $(C_2H_5)_3Al_2Cl_3$ or $(C_2H_5)_3Al$ (Ethyl process) [104], [105].

**By the Michaelis–Arbusov Reaction.** One of the most important reactions in organophosphorus chemistry is the Michaelis–Arbusov reaction, in which compounds containing three-coordinate phosphorus and at least one alkoxy or alkylthio group react with an alkylating agent R'X, forming a phosphorus–carbon bond in compounds with coordination number 4 [106], [107].

$$\text{\textbackslash P-OR} \xrightarrow{\text{R'X}} \left[ -\overset{\text{R'}}{\underset{|}{\text{P}}}-\text{OR} \right]^+ X^- \longrightarrow -\overset{\text{O}}{\underset{|}{\overset{\|}{\text{P}}}}-\text{R'} + \text{RX}$$

The intermediate phosphonium compounds can only be isolated in exceptional cases. They react further by loss of alkyl halide RX. If R=R', isomerization occurs with only catalytic quantities of alkylating agent. The method offers a great variety of possibilities, owing to the large range of possible groups R and R'.

Industrially important starting materials for the Michaelis–Arbusov reaction are trialkyl phosphites $P(OR)_3$, which are converted to dialkyl esters of alkanephosphonic acid and to esters of phosphonocarboxylic acids, from which the free phosphonic acids are obtainable by hydrolysis.

$$(RO)_3P \xrightarrow{\text{cat.}} R-\overset{O}{\overset{\|}{P}}(OR)_2$$

$$(C_2H_5O)_3P + ClCH_2COOC_2H_5 \longrightarrow$$

$$(C_2H_5O)_2\overset{O}{\overset{\|}{P}}CH_2COOC_2H_5 + C_2H_5Cl$$

If α-halocarbonyl compounds are used as the alkylating agents, vinyl esters of phosphoric acid are obtained (Perkow reaction, see Chap. 10).

**By Addition to C=C Double Bonds.** Dialkyl phosphonates undergo base-catalysed addition to activated C=C double bonds. The reaction is mainly used for the production of phosphonocarboxylate esters, and can be carried out as a continuous process [108]. In the presence of a radical initiator, nonactivated alkenes can also be used.

$$\underset{RO}{\overset{RO}{>}}P\underset{H}{\overset{O}{<}} + CH_2=CH-COOR$$

$$\xrightarrow{RO^-} (RO)_2\overset{O}{\overset{\|}{P}}-CH_2CH_2-COOR$$

$$\underset{RO}{\overset{RO}{>}}P\underset{H}{\overset{O}{<}} + CH_2=CH-R' \xrightarrow{\text{Initiator}} R-CH_2CH_2-\overset{O}{\overset{\|}{P}}(OR)_2$$

**By the Quasi-Mannich Reaction.** Phosphonic (phosphorous) acid $H_3PO_3$ or its dialkyl esters react with mixtures of formaldehyde with ammonia or amines under acidic conditions to give oligo(methylenephosphonates). When oligo-(ethylene)amines are used, the industrially important poly(methylenephosphonic acids) are formed.

$$NH_3 + CH_2O + H_3PO_3 \xrightarrow{H^+} N(CH_2PO_3H_2)_3$$

**By Acylation of Phosphonic Acid.** The reaction of phosphonic acid with acetyl chloride or acetic anhydride yields 1-hydroxyethane-1,1-diphosphonic acid [2809-21-4] after hydrolysis.

$$2H_3PO_3 + \begin{matrix} H_3C-C(=O)-O \\ H_3C-C(=O)-O \end{matrix} \xrightarrow[2) H_2O]{1) \Delta} H_3C\underset{PO_3H_2}{\overset{PO_3H_2}{-}}OH + CH_3COOH$$

Alternative starting materials are $PCl_3$ and acetic acid in the presence of water [109].

The reaction can also be carried out with other carboxylic acid derivatives [110]–[113]. For example, with acetonitrile, 1-aminoethane-1,1-diphosphonic acid [15049-85-1] is obtained.

## 9.3. Phosphonic Acids and Phosphonocarboxylic Acids

Methylphosphonic acid [993-13-5] (Ciba-Geigy) is used in the production of lubricant additives and for treating textiles [114], [115]. Octylphosphonic acid [4724-48-5] (Hoechst) is used as a selective collector for the flotation of cassiterite (tin ore). Vinylphosphonic acid [1746-03-8] (Hoechst) [116], [117] or its polymers are used for the surface treatment of aluminum in the manufacture of printing plates [118]. Phenylphosphonic acid [1571-33-1] (Akzo) is used as a catalyst in the production of resins and for manufacturing stabilizers for plastics.

Salts of phosphonoformic acid and phosphonoacetic acid (**5**) have virostatic properties. Trisodium phosphonoformate (**6**) (Foscarnet) is used in the treatment of herpes (Triapten, formerly VEB Germed) or cytomegalic infections in humans (Foscavir, Astra).

$$\underset{\text{5 [4408-78-0]}}{\overset{HO}{\underset{HO}{\diagup}}\overset{O}{\overset{\|}{P}}-CH_2-\overset{O}{\overset{\|}{C}}-OH} \qquad \underset{\text{6 [34156-56-4]}}{\overset{NaO}{\underset{NaO}{\diagup}}\overset{O}{\overset{\|}{P}}\overset{O}{\diagdown}\overset{}{ONa}} \cdot 6H_2O$$

2-Phosphonobutane-1,2,4-tricarboxylic acid (**7**) [*37971-36-1*] (NH$_4$ salt [*70233-62-4*], Bayhibit AM, Bayer) is used in industrial water treatment on account of its complexing properties.

$$\underset{\text{7 [37971-36-1]}}{\overset{HO}{\underset{HO}{\diagup}}\overset{O}{\overset{\|}{P}}-\overset{CH_2-COOH}{\underset{\underset{CH_2-COOH}{|}}{\overset{|}{C}}}-COOH}$$

Geminal diphosphonic acids are used in the same application, and in washing and cleaning agents, peroxide stabilizers, and many other areas. Compounds of this type include 1-hydroxyethane-1,1-diphosphonic acid [*2809-21-4*] (Briquest ADPA, Albright & Wilson; Sequion 10 H 60, Bozetto; Turpinal SL, Henkel; Jaypol 210, M & J Polymers; Dequest 2010/2016, Monsanto; Mykon P60, Warwick) and 1-aminoethane-1,1-diphosphonic acid [*15049-85-1*] (BK Ladenburg), as well as the large group of poly(methylenephosphonic acids) produced from ammonia or ethylene amines. Some important products are nitrilotris(methylenephosphonic acid) [*6419-19-8*], ethylenediaminetetrakis(methylenephosphonic acid) [*1429-50-1*] and diethylenetriaminepentakis(methylenephosphonic acid) [*15827-60-8*]. Producers (trade names) are: Albright & Wilson (Briquest), Bozetto (Sequion), M & J Polymers (Jaypol), Monsanto (Dequest).

The total herbicide *N*-carboxymethylaminomethanephosphonic acid (*N*-phosphonomethylglycine) [*1071-83-6*], Glyphosate (Roundup, Monsanto) is outstanding for its effectiveness and ease of biodegradability. It is used in the form of its isopropylammonium salt [*38641-94-0*]. The sodium salt of 2-chloroethylphosphonic acid [*16672-87-0*] Ethepon (Ethrel, Union Carbide) causes accelerated ripening of fruit by releasing the ripening hormone ethylene in the plant.

$$\underset{\text{[1071-83-6]}}{\overset{HO}{\underset{HO}{\diagup}}\overset{O}{\overset{\|}{P}}-CH_2-NH-CH_2-COOH} \qquad \underset{\text{[16672-87-0]}}{ClCH_2CH_2-\overset{O}{\underset{OH}{\overset{\|}{P}}}\overset{}{\diagdown}OH}$$

## 9.4. Esters of Phosphonic Acid

Monoesters of phosphonic acid are used in industry only to a limited extent. Salts of short-chain alkyl alkylphosphonates have a flameproofing effect [119], [120]. The 2-ethylhexyl ester of 2-ethylhexylphosphonic acid (**8**) (PC-88A, Daihachi; Ionquest 801, Albright & Wilson) is used as an extractant for lanthanides and for the separation of cobalt and nickel [121].

**Table 10.** Physical properties of dialkyl alkyl phosphonates RP(=O)(OR)$_2$

| R | CAS no. | $M_r$ | Density, g/cm$^3$ | bp (p), °C (Pa) | $n_D^{20}$ | Flash point, °C |
|---|---|---|---|---|---|---|
| CH$_3$ | [756-79-6] | 124.08 | 1.174 | 62 (1330) | 1.4137 | 92[a] |
| C$_2$H$_5$ | [78-38-6] | 166.16 | 1.025 | 82 (1433) | 1.4148 | 90[b] |
| CH$_2$CH$_2$CH$_2$CH$_3$ | [78-46-6] | 250.32 | 0.948 | 127 (330) | 1.4310 | 125[c] |
| CH$_2$CH(CH$_2$)$_3$CH$_3$<br>\|<br>C$_2$H$_5$ | [126-63-6] | 418.64 | 0.908 | 160 (33) | 1.4481 | 158[a] |

[a] Pensky–Martens closed cup. [b] Tag closed cup. [c] Cleveland open cup.

$$\text{CH}_3(\text{CH}_2)_3\text{CHCH}_2-\overset{\overset{\text{O}}{\|}}{\underset{\text{C}_2\text{H}_5}{\text{P}}}\overset{\text{OCH}_2\text{CH}(\text{CH}_2)_3\text{CH}_3}{\underset{\text{OH}}{\overset{\|}{\underset{\text{C}_2\text{H}_5}{}}}}$$

**8** [14802-03-0]

$$\underset{\text{C}_2\text{H}_5\text{O}}{\overset{\text{C}_2\text{H}_5\text{O}}{\overset{\|}{\text{P}}}}-\text{CH}_2-\overset{\text{O}}{\overset{\|}{\text{C}}}-\text{OC}_2\text{H}_5 \qquad \underset{\text{CH}_3\text{O}}{\overset{\text{CH}_3\text{O}}{\overset{\|}{\text{P}}}}-\text{CH}_2\text{CH}_2-\overset{\text{O}}{\overset{\|}{\text{C}}}-\text{NH}-\text{CH}_2\text{OH}$$

**9** [867-13-0]          **10** [20120-33-6]

Dialkyl alkylphosphonates are used as flameproofing agents, metal extractants, plasticizers, lubricant additives, and chemical intermediates. The more important compounds are listed in Table 10 with their physical properties. Producers of the methyl compound DMMP include Akzo, Albright & Wilson, Bayer, Ciba-Geigy, and Courtaulds. Other compounds are produced by Albright & Wilson and Daihachi.

Diethyl ethoxycarbonylmethylphosphonate (**9**) (triethyl phosphonoacetate; PEE, Hoechst) is used as a stabilizer for polyesters [122]. Esters of phenylmethylphosphonic acid with bulky substituents (Irganox 1222 [976-56-7] and 1093 [3135-18-0], Ciba-Geigy) are sold as antioxidants and high-temperature stabilizers for polyamides, polyesters, polyolefins and polyurethane foams.

$$\text{HO}-\!\!\!\bigg\langle\!\!\!\bigcirc\!\!\!\bigg\rangle\!\!\!-\text{CH}_2\text{P}(\text{O})(\text{OR})_2$$

R = C$_2$H$_5$; [976-56-7]
R = C$_{18}$H$_{37}$; [3135-18-0]

A wide range of other phosphonate esters, often containing chlorine, are used as flameproofing agents. Producers (trade names) are: Akzo (Fyrol, Victastab), Albright & Wilson (Amgard, Antiblaze), Sandoz (Sandoflam 5087). Dimethyl 3-hydroxymethyl-amino-3-oxopropanephosphonate (**10**) is of particular importance in the production of washable flameproofed cellulose textiles (Pyrovatex CP, Ciba-Geigy) [123]–[125].

Table 11. Physical properties of phosphonic and thiophosphonic acid halides

| Formula | CAS no. | $M_r$ | Density, g/cm$^3$ | bp (p), °C (Pa) | $n_D^{20}$ | Flash point, °C |
|---|---|---|---|---|---|---|
| CH$_3$P(O)Cl$_2$ | [676-97-1] | 132.91 | 1.39[a] | 163 (1.03 × 10$^5$) | 1.462[a] | > 110 |
| C$_6$H$_5$P(O)Cl$_2$ | [824-72-6] | 194.99 | 1.394 | 258 (1.03 × 10$^5$) | 1.5600 | 204[b] |
| CH$_3$P(S)Cl$_2$ | [676-98-2] | 148.98 | 1.434 | 154 (1.03 × 10$^5$) | 1.548 | 76[b] |
| C$_2$H$_5$P(S)Cl$_2$ | [993-43-1] | 163.01 | 1.35 | 172 (1.03 × 10$^5$) | 1.541 | 74[b] |
| C$_6$H$_5$P(S)Cl$_2$ | [3497-00-5] | 211.04 | 1.376 | 205 (1.73 × 10$^4$) | 1.6240 | > 110 |

[a] At 38 °C.
[b] Cleveland closed cup.

## 9.5. Other Derivatives of Phosphonic Acid

Halides of phosphonic and thiophosphonic acids are mainly used in the synthesis of crop protection agents. Ehylthiophosphonic acid dichloride is used in the production of the insecticide Fonofos [944-22-9] (Dyfonate, ICI) and Trichloronat [327-98-0] (Agrisil, Phytosol, Bayer), and benzenethiophosphonic acid dichloride [126] in the production of O-2,4-dichlorophenyl O-ethyl phenylphosphonothioate (S-Seven) [3792-59-4] and EPN [2104-64-5] (both Nissan Chemical Industries).

The acid halides are produced by Akzo, Ethyl, Ferro, and Nissan Chemical Industries. Physical properties are listed in Table 11.

# 10. Esters of Phosphoric Acid

The esters of phosphoric acid constitute an important group of organophosphorus compounds with a broad application as cleaning agents and emulsifiers, textile improvers, plasticizers and flameproofing agents for plastics, anticorrosion agents, and extractants in hydrometallurgy. Phosphoric acid ester chlorides are starting materials in the production of crop protection agents. Many vinyl esters of phosphoric acid have insecticidal properties.

## 10.1. Properties

Phosphate esters are colorless liquids or crystalline or waxy solids, depending on the substituents. The triesters are neutral compounds, while the mono- and diesters are strongly acidic. The esters of phosphoric acid hydrolyze in the presence of water. The rate of hydrolysis varies widely and depends on the degree of esterification and the nature of the substituents.

## 10.2. Production

The important raw materials for the production of esters of phosphoric acid are phosphoryl chloride, phosphorus pentoxide, polyphosphoric acid, and trialkyl phosphites.

Phosphorus oxychloride reacts with phenols to form triaryl phosphates. With aliphatic alcohols, the hydrogen chloride formed must be removed by vacuum or by purging with an inert gas. Alternatively, bases may be added.

$$POCl_3 + 3\,ROH \longrightarrow (RO)_3PO + 3\,HCl$$

If a substoichiometric amount of alcohol is used, ester chlorides of phosphoric acid are formed. However, diester chlorides of phosphoric acid are better obtained by reacting dialkyl phosphonates with chlorine or sulfuryl chloride.

$$(RO)_2P(O)H + Cl_2 \longrightarrow (RO)_2P(O)Cl + HCl$$

$$(RO)_2P(O)H + SO_2Cl_2 \longrightarrow (RO)_2P(O)Cl + HCl + SO_2$$

The addition of ethylene oxide or propylene oxide to $POCl_3$ at 40–100 °C affords tris(2-chloroethyl) or tris(2-chloropropyl) phosphate. To ensure complete reaction, catalysts must be added ($AlCl_3$, $TiCl_4$, $PCl_3$).

$$POCl_3 + 3\,\text{(ethylene oxide)} \longrightarrow (ClCH_2CH_2O)_3PO$$

A mixture of mono- and diesters is formed when alcohols or phenols are reacted with phosphorus pentoxide, while the reaction with polyphosphoric acid yields monoesters of phosphoric acid and free phosphoric acid.

$$P_4O_{10} + 6\,ROH \longrightarrow 2\,(RO)_2PO(OH) + 2\,ROPO(OH)_2$$

$$H_6P_4O_{13} + 3\,ROH \longrightarrow 3\,ROPO(OH)_2 + H_3PO_4$$

An important method for the production of insecticidal vinyl phosphates is the reaction between trialkylphosphites and α-halocarbonyl compounds (Perkow Reaction) [127].

$$P(OEt)_3 + X-CH_2C(O)R \longrightarrow (EtO)_2P(O)-O-C(R)=CH_2 + EtX$$

**Table 12.** Physical properties of trialkyl phosphates (RO)$_3$PO

| R | CAS no. | $M_r$ | Density, g/cm$^3$ | bp ($p$), °C (Pa) | $n_D^{20}$ | Flash point, °C |
|---|---|---|---|---|---|---|
| CH$_3$ | [512-56-1] | 140.08 | 1.197 | 197 (1.03 × 10$^5$) | 1.3960 | none |
| C$_2$H$_5$ | [78-40-0] | 182.16 | 1.064 | 215 (1.03 × 10$^5$) | 1.4039 | 116 |
| CH$_2$CH$_2$Cl | [115-96-8] | 285.51 | 1.414 | 210–220 (2700) | 1.4720 | 252[a] |
| CH(CH$_3$)CH$_2$Cl | [13674-84-5] | 327.57 | 1.294 | [b] | 1.4625 | 218[a] |
| CH(CH$_2$Cl)$_2$ | [13674-87-8] | 430.91 | 1.513 | [b] | 1.5019 | 251[a] |
| CH$_2$(CH$_2$)$_2$CH$_3$ | [126-73-8] | 266.32 | 0.976 | 148–153 (1330) | 1.4249 | 165[a] |
| CH$_2$CH$_2$OC$_4$H$_9$ | [78-51-3] | 398.48 | 1.006 | 215–228 (532) | 1.4380 | 224[a] |
| CH$_2$CH(CH$_2$)$_3$CH$_3$<br>\|<br>C$_2$H$_5$ | [78-42-2] | 434.64 | 0.924 | 196–200 (133) | 1.4440 | 193[a] |

[a] Cleveland open cup. [b] Decomposes.

## 10.3. Trialkyl Phosphates

Table 12 gives a survey of the important trialkyl phosphates.

Triethyl phosphate is used as a catalyst in the production of acetic anhydride by the ketene process, as a desensitizing agent for peroxides, and as a solvent and plasticizer for cellulose acetate. Tributyl phosphate is used for solvent ex-traction in hydrometallurgy. Tris(2-ethylhexyl) phosphate is used as a solvent in the production of hydrogen peroxide and is also used as a plasticizer with flameproofing properties. The important producers of trialkyl phosphates are Akzo, Albright & Wilson, Bayer, Daihachi Chemical Industry, and FMC.

Tris(chloroalkyl) phosphates are used as fire retardants for polyurethane foams, polyesters, and other plastics [128]. In addition to the compounds listed in Table 12, a large number of other halogen-containing, sometimes oligomeric trialkyl phosphates with mixed ester groups are marketed as flameproofing agents. Producers (trade name) are: Albright & Wilson (Amgard), Akzo (Fyrol), Bayer (Disflamoll), Courtaulds, Great Lakes Chemical Corp. (Firemaster), Hoechst (Genomoll), Olin (Thermolin), and Sandoz (Sandoflam).

## 10.4. Triaryl and Alkyl Aryl Phosphates

Some important aromatic triesters of phosphoric acid are listed in Table 13.

These compounds are mainly used as flameproofing and plasticizing agents for plastics, as lubricants, and as low-flammability hydraulic fluids. Mixtures of isomers are mainly used, based on technical-grade alkyl phenols.

Some producers (trade names) are: Akzo (Phosflex), Albright & Wilson (Pliabrac), Bayer (Disflamoll), Ciba-Geigy (Reofos), FMC (Kronitex), Monsanto (Santicizer).

**Table 13.** Physical properties of triaryl phosphates $(R^1O)_3PO$ and alkyl aryl phosphates $(R^1O)(R^2O)_2PO$

| $R^1$ | $R^2$ | CAS no. | $M_r$ | Density, g/cm³ | bp (p), °C (Pa) | $n_D^{20}$ | Flash point, °C |
|---|---|---|---|---|---|---|---|
| $C_6H_5$ | | [115-86-6] | 326.29 | | 244 (1330) | | 223 |
| $C_6H_4CH_3$ | | [1330-78-5] | 368.37 | 1.17 | 249 (533) | 1.555 | 252 |
| $C_6H_3(CH_3)_2$ | | [25155-23-1] | 410.45 | 1.147 | 248–265 (533) | 1.553 | 263 |
| $C_6H_4CH(CH_3)_2$ | | [68937-41-7] | 452.53 | 1.16 | 220–270 (533) | 1.552 | 254 |
| $CH_2CH(CH_2)_3CH_3$<br>\|<br>$C_2H_5$ | $C_6H_5$ | [1241-94-7] | 362.41 | 1.09 | 239 (1330) decomp. | 1.508 | 224 |
| iso-$C_{10}H_{21}$ | $C_6H_5$ | [29761-21-5] | 390.46 | 1.07 | 245 (1330) decomp. | 1.507 | 241 |
| $C_6H_4CH_3$ | $C_6H_5$ | [26444-49-5] | 340.32 | 1.20 | 235–255 (1330) | 1.561 | 230 |
| $C_6H_4C(CH_3)_3$ | $C_6H_5$ | [56803-37-3] | 385.42 | 1.18 | 258 (1330) | 1.555 | 263 |

## 10.5. Mono- and Dialkyl Phosphates

Only a small number of pure compounds in this class are of industrial importance. Bis(2-ethylhexyl) phosphate [298-07-7] DEHPA, is used in hydrometallurgy as an extraction solvent for a large number of metals. It is produced by chlorinating bis(2-ethylhexyl) phosphonate [3658-48-8] to give the phosphate diester chloride, followed by hydrolysis [129], or by saponification of tris(2-ethylhexyl) phosphate [78-42-2]. It is produced by Albright & Wilson, Bayer, Daihachi, and Hoechst. Products with a high monoester content, obtained from ethoxylated alcohols, are used as defoamers in washing powders (Sokalan S, BASF).

Esters of thiophosphoric acid are used in plant protection and as flotation agents in ore preparation, lubricant additives, for solvent extraction of metals, and in rubber production.

Mixtures of mono- and diesters of phosphoric acid are of major industrial importance. Their properties can be varied over a wide range by choice of the alcohol or phenol. Ethoxylated alcohols are most commonly used. The wide range of applications includes emulsifiers in plant protection, cleaning agents, cosmetics, and in the paper and textile industries. Mixtures of mono- and diesters are also used as acid hardeners for resins. Manufacturers (trade names) include Akzo (Dapral, Victawet), Albright & Wilson (Albrite, Briphos, Duraphos), GAF (Gafac), Henkel (Disponil), Hoechst (Hostaphat, Leomin, Knapsack phosphate esters), Hüls (Marlophor), and Rhône-Poulenc (Celanol, Soprophor).

Mixtures of monoesters of phosphoric acid with free phosphoric acid are also marketed. These are mainly used as components of detergents (Hoechst).

# 11. Esters of Thiophosphoric Acid

The esters of thiophosphoric acid are formally derived by the replacement by sulfur of one or more of the oxygen atoms bonded to phosphorus in esters of phosphoric acid. Thiophosphoric acids form mono-, di-, and triesters, the location of the organic substituent being indicated as *O*- or *S*-. Of the large range of possible compounds, the derivatives of mono- and dithiophosphoric acid have attained the greatest industrial importance.

## 11.1. Properties

Thiophosphate esters generally have an unpleasant smell and in the pure state are colorless liquids or solids. The triesters are neutral, while the mono- and diesters are strongly acidic. Technical-grade *O,O*-dialkyl dithiophosphates usually have a greenish or yellowish color due to traces of heavy metals. The short-chain compounds in particular cannot be stored indefinitely, and are usually immediately converted into other products. The thermal and hydrolytic stability increases with increasing alkyl chain length and degree of esterification and upon salt formation, so that diaryl dithiophosphates are more easily hydrolyzed than alkyl esters of comparable chain length [130].

Many esters of thiophosphoric acid have high biological activity as acetylcholine esterase inhibitors and are therefore useful insecticides. However, the toxicity to warm-blooded animals is very high. The compounds are decomposed to less toxic products on contact with the soil, as they are readily hydrolyzed.

## 11.2. Production

Important raw materials for the production of esters of thiophosphoric acid include phosphorus pentasulfide, thiophosphoryl chloride, and dialkyl hydrogenphosphonates.

Phosphorus pentasulfide reacts with alcohols or phenols with loss of $H_2S$ to form the industrially important *O,O*-diesters of dithiophosphoric acid. Addition of catalytic nitrogen, phosphorus, or sulfur compounds (e.g., phosphonium halides) improves the yield of the reaction and the product quality [131]–[133].

$$P_4S_{10} + 8\,ROH \longrightarrow 4\, \begin{array}{c} RO \\ \phantom{x} \\ RO \end{array}\!\!\!P\!\!\!\begin{array}{c} S \\ \phantom{x} \\ SH \end{array} + 2\,H_2S$$

In practice, a suspension of phosphorus pentasulfide in the end product is reacted with excess alcohol at 60–80 °C. Due to their lower reactivity, phenols must be reacted at 80–120 °C. The hydrogen sulfide produced is burnt, and the sulfur dioxide formed is removed by scrubbing with alkali.

*O,O*-Dialkyl esters of dithiophosphoric acid are converted to *O,O*-diester chlorides of thiophosphoric acid by reaction with chlorine. Depending on the conditions, the byproducts are hydrogen chloride and disulfur dichloride [10025-67-9] or sulfur.

$$2 \; (RO)_2P(O)SH + 3\,Cl_2 \longrightarrow 2 \; (RO)_2P(O)Cl + S_2Cl_2 + 2\,HCl$$

$$(RO)_2P(O)SH + Cl_2 \longrightarrow (RO)_2P(O)Cl + \tfrac{1}{n}S_n + HCl$$

Another route to this class of compounds is the reaction of thiophosphoryl chloride with alcohols in the presence of acid acceptors.

$$PSCl_3 + 2\,ROH \longrightarrow (RO)_2P(O)Cl + 2\,HCl$$

Salts of thiophosphoric acid are produced by reacting dialkyl hydrogen phosphonates with sulfur in the presence of a base (e.g., ammonia). The *O,O*-diesters of thiophosphoric acid are liberated from these compounds by reaction with phosphoric acid.

$$(RO)_2P(O)H + S_8 + NH_3 \longrightarrow (RO)_2P(S)O^-\,NH_4^+$$

## 11.3. Monothiophosphates

*O,O*-Diesters of thiophosphoric acid are used as extraction solvents for heavy metals, especially zinc, or as lubricant additives. The *O,O*-diester of thiophosphoric anhydride (**11**) (Sandoflam 5060, Sandoz) is used as a halogen-free flameproofing agent for low-density polyethylene films.

**11** [4090-51-1]

The alkylation of *O,O*-dialkyl thiophosphates usually gives mixtures of *O,O,O*- and *O,O,S*-triesters of thiophosphoric acid, some of which are used in crop protection.

The *O,O*-diester chlorides of thiophosphoric acid are very important in the synthesis of crop protection agents, and are usually reacted with phenols or their salts to give insecticidal or fungicidal *O,O,O*-triesters of thiophosphoric acid (Parathion [56-38-2], Methylparathion [298-00-0]; Pyrazophos [13457-18-6]. The most important intermediates are *O,O*-dimethoxythiophosphoryl chloride [2524-03-0], $(CH_3O)_2PSCl$, ($M_r$ 160.56, *bp* 70–72 °C (270 Pa), $\varrho$ 1.305 g/cm$^3$, $n_D^{20}$ 1.4807, flash point 101 °C, colorless liquid (Cheminova), and *O,O*-diethoxythiophosphoryl chloride [2524-04-1],

$(C_2H_5O)_2PSCl$, ($M_r$ 188.61, bp 71–72 °C (93 Pa), $\varrho$ 1.200 g/cm$^3$, $n_D^{20}$ 1.4688, flash point 106 °C, colorless liquid (Albright & Wilson, Cheminova).

## 11.4. Dithiophosphates

The *O,O*-diesters of dithiophosphoric acid and aqueous solutions of sodium or ammonium dithiophosphates are used as flotation agents for sulfidic copper and cadmium ores. A new application of increasing importance is the extraction of cadmium from wet phosphoric acid by formation of insoluble complex salts [134]–[136]. The *O,O*-dialkyl dithiophosphates of short to medium chain length ($C_2$–$C_8$) are particularly effective in this application. The amount added is ca. 0.3%, depending on the origin of the crude phosphoric acid and the temperature. Preliminary reduction of the acid with iron improves the efficiency of separation of the cadmium [137].

The zinc salts of long-chain esters are used as lubricant additives to improve high-pressure lubrication properties, and useful life of the lubricant [138]. They are produced by reacting zinc oxide with the alcohol-containing *O,O*-diesters of dithiophosphoric acid. The crude product is recovered by pressure filtration, and the water liberated in the reaction and the excess alcohol are then removed by vacuum distillation.

$$\begin{matrix} RO \\ RO \end{matrix} \!\!\!\! P \!\!\!\! \begin{matrix} S \\ SH \end{matrix} + ZnO \longrightarrow \left[ \begin{matrix} RO \\ RO \end{matrix} \!\!\!\! P \!\!\!\! \begin{matrix} S \\ S^- \end{matrix} \right]_2 Zn^{2+} + H_2O$$

These products protect the metal surface by forming a sulfide and thiophosphate layer, by removing oxidation products from the lubricant, and by corrosion inhibition.

Water-insoluble *O,O*-dialkyl esters of dithiophosphoric acid can be used as extractants for the hydrometallurgical removal of heavy metals such as zinc from dilute acidic solutions, and are therefore of interest for the treatment of wastewater and polluted groundwater.

An important reaction of *O,O*-dialkyl dithiophosphates is the addition of an activated C=C double bond to form *O,O,S*-triesters of dithiophosphoric acid. Thus, *O,O*-dimethyl dithiophosphate [756-80-9] and diethyl maleate react to produce the insecticide Malathion [121-75-5].

$$(CH_3O)_2\overset{S}{\overset{\|}{P}}SH + \begin{matrix} H \\ H \end{matrix} \!\!\! \diagdown \!\!\! \diagup \!\!\! \begin{matrix} COOC_2H_5 \\ COOC_2H_5 \end{matrix}$$

$$\longrightarrow (CH_3O)_2\overset{S}{\overset{\|}{P}}SCH\overset{O}{\overset{\|}{C}}OC_2H_5 \\ \phantom{\longrightarrow (CH_3O)_2PSC}CH_2\overset{}{C}OC_2H_5 \\ \phantom{\longrightarrow (CH_3O)_2PSCHC}\overset{\|}{O}$$

The addition to C=O double bonds is the key reaction in the synthesis of the insecticide Terbufos [13071-79-9] from *O,O*-diethyldithiophosphoric acid [298-06-6], formaldehyde, and 2-methyl-2-propanethiol [75-66-1].

**Table 14.** Physical properties, O.O-dialkyl dithiophosphates $(RO)_2P(S)SH$

| R | CAS no. | $M_r$ | Density, g/cm$^3$ | bp (p), °C (Pa) | $n_D^{20}$ | Flash point, °C |
|---|---|---|---|---|---|---|
| CH$_3$ | [756-80-9] | 158.19 | 1.28 | 48 (270) | 1.5350 | 92 |
| C$_2$H$_5$ | [298-06-6] | 186.23 | 1.17 | 85 (650) | 1.5070 | 82[a] |
| CH(CH$_3$)$_2$ | [107-56-2] | 214.29 | 1.10 | 90 (400) | 1.4933 | 112[a] |
| CH$_2$CH(CH$_3$)$_2$ | [2253-52-3] | 242.34 | 1.05 | 107 (300) | 1.4908 | 136[a] |
| CH(CH$_3$)C$_2$H$_5$ | [107-55-1] | 242.34 | 1.07 | 100 (200) | 1.4928 | 79[b] |
| CH$_2$CH(CH$_2$)$_3$CH$_3$ \| C$_2$H$_5$ | [5810-88-8] | 354.56 | 0.98 | | 1.4855 | 180[a] |

[a] Cleveland open cup. [b] Pensky–Martens closed cup.

$$H_5C_2O\diagup\!\!\!\!\!\!\!P\!\!\!\!\!\diagdown^S_{SH} \;\;+ CH_2O + HS\text{-}CH(CH_3)_2$$

$$\longrightarrow\;\; H_5C_2O\diagup\!\!\!\!\!\!\!P\!\!\!\!\!\diagdown^S_{SCH_2S\text{-}CH(CH_3)_2}$$

$O,O$-Dialkyl dithiophosphates are alkylated at the sulfur atom by alkyl halides. This reaction is used to produce numerous insecticidal $O,O,S$-triesters of dithiophosphoric acid.

$$RO\diagup\!\!\!\!\!\!\!P\!\!\!\!\!\diagdown^S_{SM} + RX \longrightarrow RO\diagup\!\!\!\!\!\!\!P\!\!\!\!\!\diagdown^S_{SR} + MX$$

**Alkyl Dithiophosphates.** The physical properties of the most important $O,O$-dialkyldithiophosphoric acids are given in Table 14.

Most long-chain compounds are produced from mixtures of alcohols. Manufacturers include Albright & Wilson, American Cyanamid, Cheminova, and Hoechst. Aqueous solutions of the salts are marketed as flotation agents by American Cyanamid (Aerofloat) and Hoechst (Hostaflot). Lubricant additives based on the zinc salt are supplied by Chevron and Lubrizol.

**Aryl Dithiophosphates.** An important compound of this type is dicresyldithiophosphoric acid [27157-94-4] $(CH_3C_6H_4O)_2P(S)SH$ ($M_r$ 279.41, brown liquid), used as a flotation agent for sulfidic ores. The commercial product is not usually a pure substance, being the reaction product of a mixture of phenol, cresols, and xylenols with phosphorus pentasulfide. Both the free acid and an aqueous solution of the ammonium salt are commercial products (Aerofloat 25, 31, 242, American Cyanamid; Phosokresol B, C, E, Hoechst).

# 12. Economic Aspects

Measured by the tonnage of inorganic phosphorus-containing base materials, e.g., $PCl_3$ and $POCl_3$, used in their production, the organophosphorus products discussed here are of minor importance. Based on value, the comparison is more favorable, as they are all value-added products, the value of some specialized products being very high. However, for reasons of industrial secrecy, there is very little information available from the manufacturers concerning production quantities and values, so that some of the figures given here are only estimates.

Of the phosphines, triphenylphosphine is the largest product from the point of view of both tonnage and value, being the starting material for Wittig reagents and catalyst ligands. The output in 1990 in Europe was ca. 2000 t. Products of comparable economic importance include the fire-retardant precursors tetrakis(hydroxymethyl)phosphonium chloride and sulfate.

The organophosphorus product with by far the largest sales value is still phosphonomethyl glycine (Glyphosate, Monsanto), at ca. $10^9$ DM/a. Halogen-free phosphonic and phosphinic acid derivatives are of increasing importance in the production of nonflammable textiles. Approximately 1000 t/a of these products are produced.

The market for plastics additives (flame retardants, plasticizers and antioxidants) expressed as $PCl_3$ in the USA for 1990 is estimated at 30 000 t phosphorus trichloride, corresponding to 6800 t/a phosphorus. Of the total $PCl_3$ produced, 10 % (15 000 t/a) is converted to phosphorous acid for the production of industrial water-treatment additives. In Europe, the quantity of halogen-containing flameproofing agents based on esters of phosphoric acid was ca. 9000 t/a in 1990. Halogen-free products are estimated at 1500 t/a, and other organophosphorus plastics additives amounted to ca. 5000 t/a. In Japan in 1987, ca. 3600 t of phosphite esters, > 10 000 t halogen-free phosphate esters, and ca. 1400 t halogen-containing flameproofing agents were used. The total Japanese market for organophosphorus products was ca. 30 000 t, with a value of $180 \times 10^9$ Yen.

Apart from the esters of thiophosphoric acid used in plant protection, the most important compounds of this class are the zinc dithiophosphates used in lubricants, for which over 50 % of phosphorus pentasulfide production (ca. 65 000 t) was used in the United States in 1990. The quantity of phosphorus pentasulfide used in the production of flotation agents was < 2000 t/a. Production in developing countries with phosphate ore deposits is increasing. For example, in 1989 Chile consumed ca. 2400 t dithiophosphates, of which ca. 50 % was from its own production.

## 13. Toxicology

**Phosphines.** Tributylphosphine, a liquid with a garliclike odor, has an $LD_{50}$ of 750 mg/kg in rats following oral administration, whereas the $LD_{50}$ is slightly more than 2000 mg/kg in rabbits upon dermal application. The substance is irritating to the skin and mucous membranes. Inhalation causes nausea and, at higher concentrations, damage to the lungs and kidneys [139]. In rats, 4-h $LC_{50}$ values of 0.17 and 12.5 mg/L have been reported for phenylphosphine and triphenylphosphine, respectively. Symptomatically, mild irritation of the respiratory tract was prominent [140]. For triphenylphosphine, the $LD_{50}$ is 800–1600 mg/kg in rats following oral administration, and in rabbits > 5000 mg/kg following dermal application. Triisooctylphosphine is orally less toxic than triphenylphosphine ($LD_{50}$ 21 400 mg/kg) in acute doses, but it is resorbed better dermally ($LD_{50}$ 3970 mg/kg) and is also more irritating to the skin [141]. The TLV-TWA value for phenylphosphine is 0.23 mg/m$^3$ [142].

**Halophosphines.** For dichloroethylphosphine, the lethal dose in mice following a 10-min inhalation is 1990 mg/m$^3$. For dichloro(2-fluoroethoxy)phosphine, it is 100 mg/m$^3$ in guinea pigs [143].

**Phosphonium Compounds.** Dodecyltriphenylphosphonium bromide possesses antimycotic properties and has been used in invert soaps. In higher concentrations, it is irritating to the skin [144].

Tetrakis(hydroxymethyl) phosphonium salts (chloride, sulfate, acetate/phosphate) are used to produce crease-resistant and flame-retardant finishes on cotton textiles and cellulosic fabrics. Tetrakis(hydroxymethyl)phosphonium chloride and tetrakis(hydroxymethyl)phosphonium sulfate were tested for carcinogenicity by oral administration in rats and mice and did not cause a dose-related increase in the incidence of tumors. The mixed acetate/phosphate salt exhibited a weak promoting activity on mouse skin. Results of genotoxicity studies with these compounds were not consistent [176].

**Phosphorous–Nitrogen Compounds.** $N,N$-Bis(2-chloroethyl)tetrahydro-2$H$-1,3,2-oxazaphosphorin-2-amine 2-oxide monohydrate (cyclophosphamide) is used as a cytostatic agent in tumor therapy. In doses comparable to those given in clinical treatment, the compound causes an increased incidence of tumors in mice and rats and it is also genotoxic. Only in the organism is it metabolized to the active compound.

1,1',1''-Phosphinylidenetris(2-methylaziridine) (METEPA) is used as a cross-linking agent in dye, textile, and polymer chemistry. The acute oral toxicity in rats ($LD_{50}$) is 136 mg/kg (male animals) and 213 mg/kg (female animals). TEPA (1,1',1''-phosphinylidenetrisaziridine), which was formerly used as a flameproofing agent in the textile industry, found only limited application in the treatment of cancer, since the thio analogue was equally effective and had a lower toxicity. The $LD_{50}$ in rats is 37 mg/kg following oral administration and 87 mg/kg upon dermal application. Thio-TEPA (1,1′,1″-phosphinothioylidenetrisaziridine) is used as an antineoplastic agent. The $LD_{50}$ in mice is 46 mg/kg following oral administration. The substance is genotoxic and leads to a higher incidence of malignant tumors in mice following intraperitoneal application and in rats following intravenous administration [145]. Several cases of leukaemia have been reported in humans following treatment with Thio-TEPA [145], [177]. In contrast, hexamethylphosphoric triamide, which is used as solvent, catalyst, and stabilizer for polystyrene, is only slightly toxic in acute doses ($LD_{50}$, oral: male rats, 2650 mg/kg; female rats, 3360 mg/kg). However, upon chronic inhalation, the substance caused a dose-dependent increase in the incidence of nasal tumors in rats. Moreover, pathological changes occurred in the entire respiratory tract [146] – [148]. In the United States the substance is classified as a suspected human carcinogen [142].

**Phosphate Esters.** *Trimethyl phosphate* is metabolized to dimethyl phosphate in the body; the latter is excreted as a conjugate. The substance is only moderately toxic in acute doses ($LD_{50}$ acute, oral, rat 2000 mg/kg). Hyperexcitability and tremor are prominent symptoms, and at higher doses death occurs due to cessation of respiration. Following repeated administration, flaccid and partly spastic paralysis of the extremities have been observed in rabbits [149]. In dogs, too, trimethyl phosphate is neurotoxic. The peripheral nerve fibers were found to be damaged in histological studies [150]. Increased incidences of fibroma in male rats and increased incidences of endometrial tumors in female mice were observed after two years of oral administration of trimethyl phosphate to rats and mice. Female rats and male mice showed no increased incidence of tumors [151]. Trimethyl phosphate has been found to be a mutagen in most test systems; however, generally high doses or concentrations were required [152] – [156], [178]. A sterilizing effect was found in mice, rats, and rabbits for trimethyl phosphate, whereby the target cells were the spermatozoa in the epididymis [157]. *Dimethyl phosphate*, a metabolite of trimethyl phosphate, shows no antifertility effects in male animals, in contrast to trimethyl phosphate. It has been found to be nongenotoxic in various test systems [168].

*Triethyl phosphate*, which is metabolized to diethyl phosphate and is excreted in the urine as such together with *S*-ethyl cysteine, has an $LD_{50}$ (oral, rat) of 1310 – 1600 mg/kg. Poisoning is characterized by narcosis, excitability followed by CNS depression, loss of coordination, paresis of the hind extremities, disturbances in breathing, hypotonia, reduced cardiac activity, and reduced muscle tone. Triethyl phosphate is slightly irritating to the mucous membranes. Skin irritation and sensitizing effects have not been observed. The inhibition of choline esterases is low in comparison to other phosphate esters. Following repeated administration, triethyl phosphate causes an increase in the weight of the liver and adrenal glands with associated hepatocellular hypertrophies as the histological correlation. Triethyl phosphate is mutagenic in the Ames test and in drosophila (recessive lethal test, eye mosaic assay). However, cell transformation studies, in vivo cytogenic studies, and dominant lethal tests have given no indication of genotoxicity. The alkylating action is much weaker than that of trimethyl phosphate [158], [178].

*Tributyl phosphate* blocks cholinesterase only weakly [159]. The major metabolites are dibutyl hydrogenphosphate, butyl dihydrogenphosphate, and butyl bis(3-hydroxybutyl) phosphate [160]. The oral $LD_{50}$ in rats is between 1390 and 3350 mg/kg. Upon subchronic oral administration of tributyl phosphate, damage to the liver, kidneys and urinary bladder were observed [179]. Tributylphosphate was not neurotoxic in rats after acute or subchronic oral exposure in rats [180]. Tributylphosphate proved to be nongenotoxic in several test systems. It is not teratogenic in rats or rabbits and did not impair fertility in rats [179], [181]. Irritation of eyes and respiratory tract were observed after inhalation exposure in humans [179]. The TLV-TWA value is 2.2 mg/m$^3$ [142].

*2-Ethylhexyl diphenyl phosphate* has only a very low acute toxicity ($LD_{50}$, oral, rat and rabbit, > 24 000 mg/kg). No indications of carcinogenic effects were found in rats after two years of administration of the substance in the feed (up to 1%) [161]. 2-Ethylhexyl diphenyl phosphate was not genotoxic in the Ames test, the HGPRT test, and the in vivo chromosome aberration test in rats. No teratogenic effects were observed in rats, and development of the offspring was only impaired at maternally toxic doses [182].

*Tris(2-chloroethyl) Phosphate.* For tris(2-chloroethyl) phosphate, which is used as a flameproofing agent, $LD_{50}$ values of 430 – 1410 mg/kg (and > 2000 mg/kg in one study) have been reported for rats following oral administration. The substance is nonirritating to very mildly irritating to the skin and mucous membranes. Subchronic administration of tris(2-chloroethyl) phosphate caused increased liver weight in rats and mice. Additionally, rats showed increased kidney weights and degenerative changes in the hippocampus region of the brain. Results of mutagenicity studies on tris(2-chloroethyl) phosphate are not consistent, but the majority of tests were negative. Results of carcinogenicity studies suggest that tris(2-chloroethyl) phosphate possesses a weak tumorigenic activity, possibly via an epigenetic mechnism. When pregnant rats were treated with tris(2-chloroethyl) phosphate, toxic effects were found in the mother, but no fetotoxic or teratogenic effects were observed. Impairment of fertility was observed in a two-generation study in mice [162], [183].

*Tris (2,3-dibromopropyl) phosphate,* which was mainly used as a flameproofing agent in synthetic fibers for children's nightdresses, possesses only a low acute toxicity ($LD_{50}$, oral, rat: 5240 mg/kg) [163]. Renal inflammation and atrophy of the testicles were observed following subchronic dermal application in rabbits [164]. The compound was found to be genotoxic in several test systems [163], [165]–[167]. It is carcinogenic in mice and rats following chronic administration in the feed [163].

**Aromatic Phosphate Esters.** *Cresyl Phosphates.* Among the aromatic phosphate esters, tricresyl phosphates with one or more *o*-cresyl groups are of toxicological importance. They give a completely different clinical picture following poisoning in human beings and some animal species than that resulting from alkyl phosphate poisoning. Motor paralysis, initially of the feet, tibia and thighs, and later of the hands and arms, occurs one to three weeks after intake. The initial flaccid paralysis progresses to a spastic condition. In the past, tricresyl phosphates have caused numerous mass poisoning incidents, for example, as softeners for PVC and in contaminated edible oils and machine-gun oil. The largest incidence of poisoning occurred in 1929/1930 in the United States following consumption of ginger spirit which was adulterated with tricresyl phosphate ("ginger paralysis") [169]. In the organism, tricresyl phosphate is metabolized to the active agent by hydroxylation of the $\alpha$-C atoms in the *ortho* position, followed by ring closure. The active agent irreversibly inhibits a nonspecific esterase in the CNS (neurotoxic esterase, NTE). Other compounds producing this type of neurotoxicity, which is known as organophosphate-induced delayed polyneuropathy (OPIDP) or neuropathy (OPIDN), are related phosphates, phosphonates, and phosphoramidates. Primarily, the axons are damaged. The demyelinisation of the affected nerves, which is later visible on histological examination, should be regarded as a secondary effect [170], [184]. Tri-*o*-cresyl phosphate causes a dose-dependent decrease in sperm motility and sperm population in the epididymis in male rats [171], [172]. The TLV-TWA value has been set at 0.1 mg/m$^3$.

*Triphenyl phosphate* itself is not neurotoxic. The effects of this type that are occasionally observed are attributed to impurities [184], [185]. The TLV-TWA value for triphenyl phosphate is 3 mg/m$^3$ [142].

**Phosphites.** Due to its tendency to hydrolyse, *trimethyl phosphite* is irritating to the skin, eyes, and respiratory tract. Studies on the acute oral toxicity gave LD$_{50}$ values of 1600–2890 mg/kg in rats. Following single administration of high doses (in the LD$_{50}$ range), flaccid paralysis of the extremities was observed. Local irritating properties of the substance are prominent when the substance is inhaled repeatedly. After repeated oral administration to rats, changes of the lungs, fatty degeneration of the liver, and reduced sperm production were observed. The results of mutagenicity tests are not consistent. Upon oral administration to pregnant rats, trimethyl phosphite causes an increase in fetal resorption rate as well as anomalies in the offspring [168]. The TLA-TWA value has been set at 10 mg/m$^3$ [142]. *Diethyl phosphite* (oral LD$_{50}$ in rats 3500–5560 mg/kg) and *triethyl phosphite* (oral LD$_{50}$ in rats 1840–4000 mg/kg) are mildly irritating to the skin in rabbits and induced skin sensitization in guinea pigs. These compounds were not mutagenic in the Ames test and micronucleus test. Lung damage was observed after repeated oral administration of diethyl phosphite [186], [187]. The following LD$_{50}$ values have been reported for other aliphatic phosphites following oral administration to rats: dimethyl phosphite, 3800 mg/kg; diethyl phosphite, 5200 mg/kg; tri-*n*-butyl phosphite, 3200 mg/kg; triisooctyl phosphite, 19 200 mg/kg [173].

The minimum lethal dose in rats following subcutaneous administration is ca. 2400 mg/kg for triphenyl phosphite, 3400 mg/kg for tri-*p*-cresyl phosphite, 3400–5700 mg/kg for tri-*m*-cresyl phosphite, and 11 300 mg/kg for tri-*o*-cresyl phosphite. Poisoning was characterized initially by mild, and then severe tremors, normally of the larger muscle groups. Hyperexcitability, spasticity, ataxia, and flaccid paresis of the extremities appear several days after administration. Tri-*o*-cresyl phosphite and triphenyl phosphite exhibited the highest activity. Upon oral administration to chickens, tri-*o*-cresyl phosphate causes neurotoxic effects similar to tri-*o*-cresyl phosphate. Probably, it is oxidized to the latter in the body [174]. In the case of triphenyl phosphite, which causes ataxia in rats, chickens, and cats, the pattern of injury caused differs from that described for the tricresyl phosphates [175].

## 14. References

[1] G. M. Kosolapoff, L. Maier (eds.): *Organic Phosphorus Compounds*, Wiley-Interscience, New York 1972 – 1976.
[2] *Houben-Weyl*, 4th ed., **XII/1, XII/2**.
[3] *Houben-Weyl*, 4th ed., **E 1, E 2**.
[4] J. Emsley, D. Hall: *The Chemistry of Phosphorus*, Harper & Row, London 1976.
[5] D. E. C. Corbridge: *Phosphorus*, 4th ed., Elsevier, Amsterdam 1990.
[6] J. I. G. Cadogan: *Organophosphorus Reagents in Organic Synthesis*, Academic Press, London 1979.
[7] M. Grayson, E. J. Griffith: *Topics in Phosphorus Chemistry*, vol. **1 – 11**, Interscience Publishers, New York 1964 – 1983.
[8] *Organophosphorus Chemistry*, vol. **1 – 20**, The Royal Society of Chemistry, London 1970 – 1989.
[9] R. S. Edmundson: *Dictionary of Organophosphorus Compounds*, Chapman and Hall, London 1987.
[10] G. M. Kosolapoff, L. Maier (eds.): *Organic Phosphorus Compounds*, vol. **1**, Wiley-Interscience, New York 1972.
[11] Hoechst, DE 2 703 802, 1977 (G. Elsner, G. Heymer, H.-W. Stephan).
[12] M. C. Hoff, J. Hill, *J. Org. Chem.* **24** (1959) 356.
[13] S. H. Li, C. A. Larsen, N. I. Buchan, G. B. Stringfellow, *Conf. Ser. Inst. Phys.* **96** (1989) no. 3, 153.
[14] Deutsche Advance Produktion, DE 1 265 746, 1965 (H. J. Lorenz, A. R. Zintl, V. Franzen).
[15] M&T Chemicals, DE 2 638 720, 1975 (V. M. Chopdekar).
[16] BASF, DE 2 050 095, 1970 (H. Müller, A. Stü-binger).
[17] M. M. Rauhut, H. A. Currier, A. M. Semsel, V. P. Wystrach, *J. Org. Chem.* **26** (1961) 5138.
[18] Hoechst, US 4 324 919, 1980 (G. Elsner, H. Vollmer, E. Reutel).
[19] Hoechst, EP 0 258 693, 1987 (N. Weferling, G. Elsner, H.-W. Stephan, F.-K. Frorath).
[20] American Cyanamid Co, US 2 822 376, 1957; CA 1 151 212, 1981 (A. J. Robertson, J. C. Oppelt).
[21] W. J. Vullo, *Ind. Eng. Chem. Prod. Res. Dev.* **5** (1966) 346.
[22] W. Wolfsberger, *Chem. Ztg.* **115** (1991) 7.
[23] H. Pommer, *Angew. Chem. Int. Ed. Engl.* **16** (1977) 423.
[24] Rhône-Poulenc, FR 2 314 910, 1975 (E. G. Kuntz).
[25] Ruhrchemie, EP 0 175 919, 1985 (L. Bexten, B. Cornils, D. Kupies).
[26] Ciba-Geigy US 4 138 256, 1979 (C. Chylewski et al.).
[27] Shell Int. Res., NL 6 604 094, 1965 (R. C. Morris, J. L. van Winkle, R. F. Mason).
[28] Shell Oil, US 3 644 563, 1972 (R. S. Bauer et al.).
[29] Mobay Co., GB 1 244 416, 1971 (R. L. Sandridge).
[30] Smith-Kline Beckman, CH 634 581, 1978 (D. T. Hill, I. Lantos, B. M. Sutton).

[31] Stauffer Chemical, DE 2 046 314, 1969 (E. H. Uhing).
[32] Stauffer Chemical, EP 50 841, 1981 (A. E. Skrzec).
[33] Mobil Oil, DE 2 456 095, 1975 (A. T. Jurewicz, W. W. Kaeding).
[34] Toyama Chemical Industry, JP 44 003 354, 1969 (Y. Toyama, M. Nakabayashi, T. Uehara); *Chem. Abstr.*
[35] Hoechst, DE 2 629 299, 1976 (K. Gehrmann, A. Ohorodnik, K. H. Steil, S. Schaefer).
[36] Stauffer Chemical, US 4 409 152, 1981 (F. H. Lawrence).
[37] Nissan Chemical Industries, JP 50 0 120 054, 1975 (M. Aramaki et al.).
[38] Sandoz DE 2 152 481, 1972 (K. Hofer, G. Tscheulin).
[39] Koppers, US 3 078 304, 1963 (H. Niebergall).
[40] Hoechst, EP 93 418, 1983 (H.-J. Kleiner).
[41] Hoechst, EP 93 419, 1983 (H.-J. Kleiner).
[42] Hoechst, EP 93 420, 1983 (H.-J. Kleiner).
[43] Hoechst, EP 110 305, 1983 (H.-J. Kleiner).
[44] Cassella, DE 2 233 941, 1972 (C. Heid, D. Hoffmann, J. Polster).
[45] K. Weissermel, H.-J. Kleiner, M. Finke, U.-H. Felcht, *Angew Chem.* **93** (1981) 256–266.
[46] P. Beck in G. M. Kosolapoff, L. Maier (eds.): *Organic Phosphorus Compounds*, vol. **2**, Wiley-Interscience, New York 1972, pp. 189–508. E. V. Dehmlow, *Angew. Chem. Int. Ed. Engl.* **13** (1974) 170.
[47] M. Grayson, P. T. Keough, *J. Am. Chem. Soc.* **82** (1960) 3919–3924. D. Landini, A. Maia, A. Rampoldini, *J. Org. Chem.* **51** (1986) 3187–3191.
[48] Schering, EP 0 377 850, 1989 (H. Häuser).
[49] S. Affandi, et al., *Synth. React. Inorg. Met. Org. Chem.* **17** (1987) 307–317.
[50] Dow Chem., EP 0 019 852, 1980 (G. A. Doorikian, J. L. Bertram).
[51] I. Gosney, A. G. Rowley in [6]
[52] H. H. Freedman, *Pure Appl. Chem.* **58** (1986) 857–868.
[53] Hoechst, EP 0 240 703, 1987 (H. Erpenbach, K. Gehrmann, W. Lork, P. Prinz).
[54] Akzo, EP 0 078 567, 1982 (J. Brand).
[55] Dow Corning, US 4 433 096, 1984 (G. N. Bokerman, N. R Langley).
[56] Boots, US 4 287 374, 1980 (R. A. North).
[57] Hoechst, EP 0 036 485, 1981 (W. Krause, J. Große, W. Klose).
[58] Ciba-Geigy, US 4 874 526, 1989 (R. Grade, B. M. Thomas).
[59] Ciba-Geigy, EP 0 105 843, 1983 (W. Wehner, R. Grade).
[60] Mobil, US 3 268 323, 1966 (L. E. Goyette).
[61] D. J. Daigle, A. W. Frank, *Text. Res. J.* **52** (1982) 751–755.
[62] W. A. Rickelton, R. J. Boyle, *Sep. Sci. Technol.* **23** (1988) 1227; P. R. Danesi, *Solvent Extr. Ion Exch.* **3** (1985) 435.
[63] F. J. Hurst, D. J. Crouse, US 3 711 591.
[64] R. Marr et al., EP 9 106 118 A1, 1983.
[65] *Phosphorus Potassium* **111** (1981) 31 (part I); **112** (1981) 31 (part II).
[66] R. R. Grinstead, DE 2 423 272, 1974/1983.
[67] CTP, EP 0 038 317, 1981 (W. Kanzler, J. Schedler).
[68] American Cyanamid, EP 0 132 700, 1984 (A. J. Robertson, W. A. Rickelton).
[69] Hoechst, EP 0 259 583, 1987 (N. Weferling et al.).
[70] Mitsui Cyanamid KK, JP 19 3 846, 1985.
[71] ICI, DE OS 2 614 323, 1976 (A. Ibbotson).
[72] The Upjohn, DE-OS 2 801 701, 1978.

[73] BASF, DE 3 927 146, 1989 (W. Franzischka et al.).
[74] Degussa, DE 3 740 899, 1987 (M. Dankowski, P. Nagler, T. Lieser).
[75] FMC Corp., EP 0 095 453, 1983 (D. P. Braksmeyer).
[76] Hoechst, EP 0 196 026, 1986 (H. J. Löher, K. Bauer, H. Bieringer).
[77] Ciba-Geigy, EP 0 413 657, 1990 (W. Rutsch, K. Dietliker, R. G. Hall).
[78] BASF, US 4 710 523, 1987 (P. Lechtgen, I. Buethe, F. Jacobi, W. Trimborn).
[79] Espe, EP 0 184 095, 1985 (K. Ellrich, C. Herzig).
[80] Sandoz, DE 2 152 481, 1972 (K. Hofer, G. Tscheulin).
[81] Hoechst, EP 374 761, 1989 (M. Böhshar, H.-J. Kleiner, K. Waldmann, G. Pfahler).
[82] Cyanamid Canada, GB 2 068 381, 1981 (A. J. Robertson, T. Ozog).
[83] R. Engel in A. S. Kende (ed.): *Organic Reactions,* vol. **36,** J. Wiley & Sons, New York 1988, pp. 175–248.
[84] K. Weissermel, H.-J. Kleiner, M. Finke, U.-H. Felcht, *Angew. Chem.* **93** (1981) 256–266.
[85] Ciba-Geigy, DE 2 805 074, 1978 (L. Maier).
[86] V. K. Khairullin, T. I. Sobčuk, A. N. Pudovik, *Zh. Obshch. Khim.* **37** (1967) 710.
[87] V. K. Khairullin, R. M. Kondrat'eva, A. N. Pudovik, *Zh. Oshch. Khim.* **38** (1968) 288.
[88] Hoechst, DE 2 529 731, 1975 (E. Lohmar, K. Gehrmann, A. Ohorodnik, P. Stutzke).
[89] Hoechst, EP 159 656, 1984 (H.-J. Kleiner).
[90] Hoechst, CH 620 812, 1976 (W. Rupp et al.).
[91] P. Langelueddeke et al., *Meded. Fac. Landbouwwet. Rijksuniv. Gent* **47** (1982) 95.
[92] Chemische Fabrik Budenheim Rudolf A. Oetker, DE 3 616 168, 1986 (K. Sommer, K. Dorn, G. Scheuer, K. Götzmann).
[93] W. A. Rickelton, R. J. Boyle, *Sep. Sci. Technol.* **23** (1988) 1227–1250.
[94] Hoechst, DE 2 346 787, 1975 (H.-J. Kleiner, M. Finke, U. Bollert, W. Herwig).
[95] Hoechst, DE 2 454 189, 1976 (U. Bollert, A. Ohorodnik, E. Lohmar).
[96] Hoechst, DE-OS 1 668 031, 1967 (K. Schimmelschmidt, H.-J. Kleiner).
[97] C. Fest, K.-J. Schmidt: *The Chemistry of Organophosphorus Pesticides,* 2nd ed., Springer-Verlag, Berlin 1982, pp. 68–80.
[98] R. Gächter, H. Müller (eds.): *Plastics Additives Handbook,* Hanser Verlag, München 1989.
[99] T. J. Henman: *World Index of Polyolefin Stabilizers,* Kogan Page, London 1982.
[100] Philagro, DE-OS 2 751 035, 1977 (J. Abblard, M. Decines, R. Viricel).
[101] Philagro, DE-OS 2 911 516, 1979 (A. Bernard, A. Disdier, M. Royer).
[102] Hoechst, EP 49 343, 1981 (J. Große).
[103] Hoechst, DE 3 131 249, 1981 (W. Krause).
[104] Ethyl Corporation, DE 1 793 376, 1972 (J. B. Hinkamp, V. F. Hnizda).
[105] Ethyl Corporation, US 3 879 454, 1975 (J. B. Hinkamp, V. F. Hnizda).
[106] B. A. Arbusov, *Pure Appl. Chem.* **9** (1964) 307.
[107] A. K. Bhattacharya, G. Thyagarajan, *Chem. Rev.* **81** (1981) 415.
[108] Bayer, EP 358 022, 1989 (R. Kleinstück et al.).
[109] B. Blaser, K.-H. Worms, H.-G. Germscheid, K. Wollmann, *Z. Anorg. Allg. Chem.* **381** (1971) 247.
[110] W. Plöger, N. Schindler, K. Wollmann, K.-H. Worms, *Z. Anorg. Allg. Chem.* **389** (1972) 119.
[111] Hoechst, DE 3 047 107, 1980 (W. Klose, T. Auel).
[112] Henkel, EP 332 068, 1989 (H. Blum, S. Hemmann).
[113] H. Blum, *Z. Naturforsch. B: Anorg. Chem. Org. Chem.* **43 B** (1988) 75.
[114] Ciba-Geigy, DE 2 903 928, 1979 (H. Nachbur, P. Rohringer).
[115] Ciba-Geigy, EP 157 731, 1985 (H. O. Wirth, K. Müller).

[116] Hoechst, EP 77 460, 1980 (H.-J. Kleiner, W. Dursch).
[117] Hoechst, EP 281 122, 1987 (G. Roscher, H.-J. Kleiner, G. Ihl, H. Leipe).
[118] Kalle, US 3 468 725, 1969 (F. Uhlig).
[119] General Electric, EP 321 002, 1989 (C. A. A. Claessen).
[120] Ciba-Geigy, EP 310 559, 1989 (C. D. Weis, P. Sutter).
[121] Daihachi Chemical Industry, US 4 196 076, 1980 (A. Fujimoto, I. Miura, K. Noguchi).
[122] Akzo, DE 2 708 790, 1977 (J. Kowallik, A. Brandner).
[123] Ciba-Geigy, DE 2 013 665, 1970 (H. Nachbur, J. Kern, A. Maeder).
[124] Ciba-Geigy, DE 2 147 481, 1973 (P. Hofmann, H. Nachbur, A. Maeder).
[125] R. Aenishänslin et al., *Text. Res. J.* **39** (1969) 375.
[126] Nissan Chemical Industries, DE 2 527 650, 1974 (Y. Ura, H. Takamatsu, C. Funabashi).
[127] R. G. Harvey, E. R. De Sombre in M. Grayson, E. J. Griffith (eds.): *Topics in Phosphorus Chemistry*, vol. **1**. Wiley-Interscience, New York 1964, p. 57.
[128] G. Bohlmann: "Flame Retardant Chemicals," *Process Economics Programm,* Supplement A, SRI International, Menlo Park, Ca.
[129] Albright & Wilson, Mobil Oil, EP 33 999, 1980 (R. P. Napier, T. N. Williams Jr., B. E. Johnston, J. M. Horn).
[130] G. Côte, D. Bauer, *Anal. Chem.* **56** (1984) 2153.
[131] Produits Chimiques Ugine Kuhlmann, FR 7 220 487, 1972 (M. Démarcq).
[132] Hoechst, EP-A 36 485, 1981 (W. Krause, J. Große, W. Klose).
[133] ICI America, EP-A 285 073, 1988 (S. B. Mirviss).
[134] Hoechst, EP 85 344, 1983 (R. Gradl, G. Schimmel, W. Krause, G. Heymer).
[135] E. Jdid, J. Bessière, P. Blazy, *Ind. Miner. Tech.* 1984, 389.
[136] H. v. Plessen, G. Schimmel, *Chem. Ing. Tech.* **59** (1987) 772.
[137] Hoechst, US 4 713 229, 1985 (G. Schimmel, R. Gradl).
[138] G. E. Russell: *Process Economics Program,* Report No. 113, Stanford Research Institute, Menlo Park, Ca., 1977, p. 1, 75.
[139] W. B. Deichmann, H. W. Gerarde: *Toxicology of Drugs and Chemicals,* Academic Press, New York, 1969, pp. 600.
[140] R. S. Waritz, R. M. Brown, *Am. Ind. Hyg. Assoc. J.* **36** (1975) 452–458.
[141] *Patty's,* 3rd ed., **2 A,** 2259.
[142] American Conference of Governmental Industrial Hygienists: *TLV's — Threshold Limit Values and Biological Indices for 1993–1994,* Cincinatti, Ohio, 1993.
[143] *Ullmann,* 4th ed., **18,** 394.
[144] J. Kimmig, D. Jerchel, *Klin. Wochenschr.* **28** (1950) 429–431.
[145] IARC: *IARC Monographs on the Evaluation of the Carcinogenic Risk of Chemicals to Man,* vol. **9**, Lyon 1975.
[146] IARC: *IARC Monographs on the Evaluation of the Carcinogenic Risk of Chemicals to Man,* vol. 15, Lyon 1977.
[147] K. P. Lee, H. J. Trochimowicz, *Toxicol. Appl. Pharmacol.* **62** (1982) 90–103.
[148] K. P. Lee, H. J. Trochimowicz, *JNCI J. Natl. Cancer Inst.* **68** (1982) no. 1, 157–172.
[149] W. M. B. Deichmann, S. Witherup, *J. Pharmacol. Exp. Ther.* **88** (1946) 338–342.
[150] U. Schaeppi, G. Krinke, W. Kobel, *Neurobehav. Toxicol. Teratol.* **6** (1984) 39–50.
[151] NCI: "Bioassay of Trimethylphosphate for Possible Carcinogenicity," *DHEW Publ. (NIH) (U.S.),* NCI-CG-TR-81, 1978.
[152] T. H. Connor, *Mutat. Res.* **65** (1979) 121–131.
[153] J. Van't Hof, L. A. Schairer, *Mutat. Res.* **99** (1982) 303–315.

[154] J. F. Sina et al., *Mutat. Res.* **113** (1983) 357–391.
[155] N. Degraeve, M. C. Chollet, J. Moutschen, *J. Toxicol. Clin. Exp.* **6** (1986) 5–11.
[156] U. Graf et al., *Mutat. Res.* **222** (1989) 359–373.
[157] R. D. Harbison, C. Dwivedi, M. A. Evans, *Toxicol. Appl. Pharmacol.* **35** (1976) 481–490.
[158] Beratergremium für umweltrelevante Altstoffe (BUA) der GDCh: *Triethylphosphat (Phosphorsäure-tri-ethylester)*, VCH Verlagsgesellschaft, Weinheim 1989.
[159] J. C. Sabine, F. N. Hayes, *AMA Arch. Ind. Hyg. Occup. Med.* **6** (1952) 174–177.
[160] T. Suzuki, K. Sasaki, M. Takeda, M. Uchiyama, *J. Agric. Food Chem.* **32** (1984) 603–610.
[161] J. F. Treon, F. R. Dutra, F. P. Cleveland, *AMA Arch. Ind. Hyg. Occup. Med.* **8** (1953) 170–184.
[162] Beratergremium für umweltrelevante Altstoffe (BUA) der GDCh: *Tris(2-chlorethyl)phosphat*, VCH Verlagsgesellschaft, Weinheim 1987.
[163] IARC: *IARC Monographs on the Evaluation of the Carcinogenic Risk of Chemicals to Man*, vol. **20**, Lyon 1979.
[164] R. E. Osterberg, G. W. Bierbower, R. M. Hehir, *J. Toxicol. Environ. Health* **3** (1977) 979–987.
[165] A. Blum, B. N. Ames, *Science* **195** (1977) 17–23.
[166] M. J. Prival, E. C. McCoy, B. Gutter, H. S. Rosenkranz, *Science* **195** (1977) 76–78.
[167] E. Zeiger, D. A. Pagano, A. A. Nomeir, *Environ. Mutagen.* **4** (1982) 271–277.
[168] D. Henschler: *Gesundheitsschädliche Arbeitsstoffe. Toxikologisch-arbeitsmedizinische Begründung von MAK-Werten*, VCH Verlagsgesellschaft, Weinheim 1989.
[169] W. Forth, D. Henschler, W. Rummel: *Allgemeine und spezielle Pharmakologie und Toxikologie*, Bibliographisches Institut & F. A. Brockhaus AG, Mannheim 1987.
[170] M. K. Johnson, *CRC Crit. Rev. Toxicol.* **4** (1975) 289–316.
[171] S. G. Somkuti et al., *Toxicol. Appl. Pharmacol.* **89** (1987) 49–63.
[172] S. G. Somkuti et al., *Toxicol. Appl. Pharmacol.* **89** (1987) 64–72.
[173] L. Levin, K. L. Gabriel, *Am. Ind. Hyg. Assoc. J.* **34** (1973) 286–291.
[174] M. I. Smith, R. D. Lillie, E. Elvove, E. F. Stohlmann, *J. Pharmacol. Exp. Ther.* **49** (1933) 78–99.
[175] S. S. Padilla, T. B. Grizzle, D. Lyerly, *Toxicol. Appl. Pharmacol.* **87** (1987) 249–256.
[176] IARC: *IARC Monographs on the Evaluation of the Carcinogenic Risk of Chemicals to Man*, vol. **48**, Lyon 1990.
[177] IARC: *IARC Monographs on the Evaluation of the Carcinogenic Risk of Chemicals to Man*, vol. **50**, Lyon 1990.
[178] E. W. Vogel, M. J. M. Nivard, *Mutagenesis* **8** (1993) 57–81.
[179] Berufsgenossenschaft der chemischen Industrie: *Toxicological evaluations. No. 170: Tributylphosphat*, Springer Verlag, Berlin 1993.
[180] C. E. Healy, P. C. Beyrouty, B. R. Broxup, *Am. Ind. Hyg. Assoc. J.* **56** (1995) 349–355.
[181] T. Noda, T. Yamano, M. Shimizu, S. Morita, *Fd. Chem. Toxic.* **32** (1994) 1031–1036.
[182] Berufsgenossenschaft der chemischen Industrie: *Toxicological Evaluations, no. 194: Diphenyl-2-ethylhexylphosphat*, Springer Verlag, Berlin 1995.
[183] Berufsgenossenschaft der chemischen Industrie: *Toxicological Evaluations, no. 33: Tris(2-chlor)ethyl phosphat*, Springer Verlag, Berlin 1995.
[184] M. Lotti, *CRC Crit. Rev.* **4** (1992) 465–487.
[185] A. Nakamura, *Environ. Health Criter.* **111** (1991) 1–80.
[186] Berufsgenossenschaft der chemischen Industrie: *Toxicological Evaluations, no. 192: Triethylphosphit*, Springer Verlag, Berlin 1995.
[187] Berufsgenossenschaft der chemischen Industrie: *Toxicological Evaluations, no. 193: Diethylphosphit*, Springer Verlag, Berlin 1995.